Hormones, Genes, and Cancer

HORMONES, GENES, AND CANCER

Brian E. Henderson, M.D.
Bruce Ponder, Ph.D., FRCP, FRS
Ronald K. Ross, M.D.

2003

OXFORD
UNIVERSITY PRESS

Oxford   New York
Auckland   Bangkok   Buenos Aires   Cape Town   Chennai
Dar es Salaam   Delhi   Hong Kong   Istanbul   Karachi   Kolkata
Kuala Lumpur   Madrid   Melbourne   Mexico City   Mumbai
Nairobi   São Paulo   Shanghai   Taipei   Tokyo

Copyright © 2003 by Oxford University Press, Inc.

Published by Oxford University Press, Inc.
198 Madison Avenue, New York, New York, 10016
http://www.oup-usa.org

Oxford is a registered trademark of Oxford University Press

All rights reserved. No part of this publication may be reproduced,
stored in a retrieval system, or transmitted, in any form or by any means,
electronic, mechanical, photocopying, recording, or otherwise,
without the prior permission of Oxford University Press.

Library of Congress Cataloging-in-Publication Data
Hormones, genes, and cancer /
[edited] by Brian E. Henderson, Bruce Ponder, Ronald K. Ross.
p. cm.   Includes bibliographical references and index.
ISBN 0-19-513576-8
1. Cancer—Endocrine aspects. 2. Hormones—Carcinogenicity
3. Generative organs—Cancer—Molecular aspects.
4. Generative organs—Cancer—Chemoprevention.
I. Henderson, Brian E.
II. Ponder, B. A. J. (Bruce A. J.), 1944–
III. Ross, Ronald K. (Ronald Keith)
[DNLM: 1. Neoplasms—etiology. 2. Genes—physiology.
3. Hormones—physiology.
QZ   202 H812 2003]   RC268.2 .H676 2003
616.99′4071—dc21   2002025834

1 2 3 4 5 6 7 8 9

Printed in the United States of America
on acid-free paper

We gratefully acknowledge the contributors to this book for the high quality of their efforts. The field of hormonal carcinogenesis contains a prodigious amount of scientific effort covering a wide variety of scientific disciplines. We have tried to highlight those areas of most current interest to clinicians, geneticists, and epidemiologists dealing with human subjects. We have had to limit reference to the considerable body of animal experimentation that was so important to the development of the field of hormonal carcinogenesis.

On behalf of all the contributors to this book, we express our gratitude to the many cancer patients who have participated in epidemiological and clinical studies that have contributed so much to the recent and rapid development of the diagnosis, detection, and treatment of the hormone related cancers.

Finally, a personal message of thanks to Judith, Maggie, and Karen, for years of encouragement and support.

Preface

The genetic basis of familial breast and other hormone-related cancers was firmly established with the identification of mutations in *BRCA1*, *BRCA2*, and *TP53*. However, it has recently become clear that such genes account for only a small part of genetic susceptibility to the common forms of cancer. The information from the Human Genome Project together with rapid advances in technology for molecular genetic analysis now makes it possible to search for the genes that underlie predisposition to the majority of cancers which do not cluster in families. The advent of polymerase chain reaction and, subsequently, high-throughput genotyping and microarray technology has opened a new and exciting era in molecular medicine.

The role of hormones in cancer etiology, treatment, and prevention has been of enormous interest since the pioneering work of Huggins and Furth and others nearly 60 years ago. With the changing landscape created by the extraordinary advances in laboratory technology in the past decade, there are new opportunities to study more definitively and in greater detail the genetic pathways in hormonal carcinogenesis.

We felt it was timely to review how far we have come in fully understanding these relationships and to discuss where this important field is heading in the next decade. In this book, we have engaged experts in this field to review the state of the art of research in the molecular genetics, etiology, progression, and prevention of hormone-related cancers. We emphasize breast cancer and prostate cancer not only because these are numerically the most important hormone-related cancer sites but also because these are by far the best studied to date in terms of understanding the molecular genetic contributions to their etiology and progression. We also review briefly the current status of research on endometrial, ovarian, and testicular cancers.

Los Angeles, California	B.E.H.
Cambridge, United Kingdom	B.P.
Los Angeles, California	R.K.R.

Contents

Contributors   **xi**

1. Introduction and Background   **1**
   Brian E. Henderson and Roberta McKean-Cowdin

2. Biosynthesis, Transport, and Metabolism of Steroid Hormones   **12**
   Frank Z. Stanczyk and Philip Bretsky

3. The Nuclear Receptor Superfamily   **38**
   John M. Rosenfeld, Hung-ying Kao, and Ronald M. Evans

4. Genomic Approaches to the Genetics of Hormone-Responsive Cancer   **99**
   Joel N. Hirschhorn, Celeste Leigh Pearce, and David Altshuler

5. Breast Cancer: Epidemiology and Molecular Endocrinology   **120**
   Heather Spencer Feigelson

6. The Impact of Exogenous Hormone Use on Breast Cancer Risk   **139**
   Gillian K. Reeves, Emily Banks, and Timothy J.A. Key

7. An Overview of Genetic Predisposition and the Search for Predisposing Genes for Breast Cancer   **157**
   Bruce Ponder

8. Estrogen Biosynthesis Genes: P-450 Aromatase   **169**
   Colin D. Clyne and Evan R. Simpson

9. Estrogen Metabolism Genes: *HSD17B1* and *HSD17B2*   **181**
   Hellevi Peltoketo, Veli Isomaa, Debashis Ghosh, and Pirkko Vihko

10. Breast Cancer: Intervention in *BRCA1* and *BRCA2* Families   **199**
    Frances V. Elmslie and Rosalind A. Eeles

11. Implications of Hormones and Hormonal Risk Factors on Screening Strategies   **210**
    Giske Ursin

12. Hormonal Chemoprevention with Tamoxifen and Selective Estrogen Receptor Modulators   **218**
    David J. Bentrem and V. Craig Jordan

13. Other Hormonal Prevention Strategies 248
    Darcy V. Spicer and Malcolm C. Pike

14. Progression from Hormone-Dependent to Hormone-Independent Breast Cancer 255
    Isabell A. Schmitt, Milana Dolezal, and Michael F. Press

15. Prostate Cancer: Epidemiology and Molecular Endocrinology 273
    Ronald K. Ross, Nick M. Makridakis, and Juergen K.V. Reichardt

16. Androgen Receptor Signaling in Prostate Cancer 288
    Wayne D. Tilley, Grant Buchanan, and Gerhard A. Coetzee

17. Hereditary Prostate Cancer: The Search for Major Genes and the Role of Genes Involved in Androgen Action 316
    Bao-Li Chang, Aubrey R. Turner, William B. Isaacs, and Jianfeng Xu

18. Androgen-Independent Prostate Cancer Progression: Mechanistic Insights 331
    George V. Thomas and Charles L. Sawyers

19. Hormonal Therapies for Prostate Cancer 343
    David I. Quinn and Derek Raghavan

20. Endometrial Cancer: Epidemiology and Molecular Endocrinology 371
    Linda S. Cook, Jennifer A. Doherty, Noel S. Weiss, and Chu Chen

21. Ovarian Cancer: Epidemiology and Molecular Endocrinology 398
    Alice S. Whittemore and Valerie McGuire

22. Testicular Cancer: Epidemiology and Molecular Endocrinology 413
    Anthony J. Swerdlow

    Index 437

Contributors

David Altshuler, M.D., Ph.D.
Department of Genetics and Medicine, Harvard
  Medical School/Massachusetts General Hospital
Director, Medical and Population Genetics,
  Whitehead/MIT Center for Genome Research
Boston, MA

Emily Banks, M.B., B.S., B.Med.Sci., Ph.D.
Imperial Cancer Research Fund
Cancer Epidemiology Unit
University of Oxford
Radcliffe Infirmary
Oxford, United Kingdom

David J. Bentrem, M.D.
Department of Surgery and Molecular
  Pharmacology
Northwestern University Medical School
Robert H. Luire Comprehensive Cancer Center
Chicago, IL

Philip Bretsky, M.P.H.
Department of Preventive Medicine
University of Southern California Keck School
  of Medicine
Los Angeles, CA

Grant Buchanan, Ph.D.
Department of Medicine
University of Adelaide and Hanson Institute
Adelaide, Australia

Bao-Li Chang, M.D.
Center for Human Genomics
Wake Forest School of Medicine
Winston-Salem, NC

Chu Chen, Ph.D., D.A.B.C.C.
Fred Hutchinson Cancer Research Center
Seattle, WA

Colin D. Clyne, Ph.D.
Prince Henry's Institute of Medical Research
Melbourne, Australia

Gerhard A. Coetzee, Ph.D.
Department of Urology and Preventive Medicine
University of Southern California Keck School of
  Medicine
Norris Comprehensive Cancer Center
Los Angeles, CA

Linda S. Cook, Ph.D.
Department of Community Health Sciences
University of Calgary
Alberta Cancer Board
Calgary, Canada

Jennifer A. Doherty, M.S.
Division of Public Health Sciences
Fred Hutchinson Cancer Research Center
Seattle, WA

Milana Dolezal, M.D.
Department of Internal Medicine
University of Southern California
Los Angeles, CA

Rosalind A. Eeles, M.A., M.R.C.P., F.R.C.R., Ph.D.
Institute of Cancer Research and Royal Marsden
  NHS Trust
Sutton, United Kingdom

Frances V. Elmslie, M.D., M.R.C.P.
Department of Clinical Genetics
Guy's Hospital
London, United Kingdom

Ronald M. Evans, Ph.D
Gene Expression Laboratory
The Salk Institute/Howard Hughes Medical Institute
La Jolla, CA

Heather Spencer Feigelson, Ph.D.
Epidemiology and Surveillance Research
American Cancer Society
Atlanta, GA

Debashis Ghosh, Ph.D.
The Hauptman-Woodward Medical Research
    Institute
Buffalo, NY

Brian E. Henderson, M.D.
Department of Preventive Medicine
University of Southern California Keck School of
    Medicine
Los Angeles, CA

Joel N. Hirschhorn, M.D., Ph.D.
Department of Genetics
Harvard Medical School and Children's Hospital
Whitehead/MIT Center for Genome Research
Boston, MA

William B. Isaacs, M.D.
Department of Urology
Johns Hopkins Hospital
Brady Urological Institute
Baltimore, MD

Veli Isomaa, Ph.D.
Research Center for Molecular Endocrinology
University of Oulu
Oulu, Finland

V. Craig Jordan, Ph.D., D.Sc.
Director, Lynn Sage Breast Cancer Research
    Program
Robert H. Lurie Comprehensive Cancer Center
Northwestern University Medical School
Chicago, IL

Hung-ying Kao, Ph.D.
Department of Biochemistry
School of Medicine Case Western Reserve
    University
Cleveland, OH

Timothy J. A. Key, B.V.M.&S., M.S., D.Phil.
Imperial Cancer Research Fund
Cancer Epidemiology Unit
University of Oxford
Radcliffe Infantry
Oxford, United Kingdom

Nick M. Makridakis, Ph.D.
Department of Research Biochemistry and
    Molecular Biology
University of Southern California
Institute for Genetic Medicine
Los Angeles, CA

Valerie McGuire, Ph.D.
Division of Epidemiology
Stanford University School of Medicine
Stanford, CA

Roberta McKean-Cowdin, Ph.D.
University of Southern California Keck School of
    Medicine
Department of Preventive Medicine
Los Angeles, CA

Celeste Leigh Pearce
Norris Comprehensive Cancer Center
University of Southern California Keck School of
    Medicine
Los Angeles, CA

Hellevi Peltoketo, Ph.D.
Department of Biochemistry
University of Oulu
Oulu, Finland

Malcolm C. Pike
Preventive Medicine
University of Southern California Keck School of
    Medicine
Norris Comprehensive Center
Los Angeles, CA

Bruce Ponder, Ph.D., FRCP, FRS
Department of Oncology
University of Cambridge
Cambridge, United Kingdom

Michael F. Press, M.D., Ph.D.
University of Southern California
Norris Comprehensive Cancer Center
Los Angeles, CA

CONTRIBUTORS

David I. Quinn, M.B.B.S., Ph.D., F.R.A.C.P.
Department of Medical Oncology
University of Southern California
Norris Comprehensive Cancer Center
Los Angeles, CA

Derek Raghavan, M.B.B.S., Ph.D., F.A.C.P., F.R.A.C.P.
Department of Medical Oncology
University of Southern California
Norris Comprehensive Cancer Center
Los Angeles, CA

Gillian K. Reeves, B.S., M.S., Ph.D.
Imperial Cancer Research Fund
Cancer Epidemiology Unit
Radcliffe Infirmary
Oxford, United Kingdom

Juergen K. V. Reichardt, Ph.D.
Department of Biochemistry and Molecular Biology
University of Southern California Keck School of Medicine
Institute for Genetic Medicine
Los Angeles, CA

John M. Rosenfeld, Ph.D.
Gene Expression Laboratory
The Salk Institute
La Jolla, CA

Ronald K. Ross, M.D.
Department of Preventive Medicine
University of Southern California
Norris Comprehensive Cancer Center
Los Angeles, CA

Charles L. Sawyers, M.D.
Department of Medicine
University of California
Los Angeles, CA

Isabell A. Schmitt, M.D.
Department of Pathology
University of Southern California
Los Angeles, CA

Evan R. Simpson, Ph.D.
Prince Henry's Institute of Medical Research
Clayton, Australia

Darcy V. Spicer, M.D.
Department of Medicine
University of Southern California Keck School of Medicine
Norris Comprehensive Center
Los Angeles, CA

Frank Z. Stanczyk, Ph.D.
Departments of Obstetrics and Gynecology and of Preventive Medicine
University of Southern California
Los Angeles, CA

Anthony J. Swerdlow, Ph.D., D.M.
Department of Epidemiology
Institute of Cancer Research
Sutton, United Kingdom

George V. Thomas, M.D.
Department of Pathology and Laboratory Medicine
University of California
Los Angeles, CA

Wayne D. Tilley, Ph.D.
Dame Roma Mitchell Cancer Research Laboratories
Department of Medicine
University of Adelaide and Hanson Institute,
Adelaide, Australia

Aubrey R. Turner, M.D.
Center for Human Genomics
Wake Forest School of Medicine
Winston-Salem, NC

Giske Ursin, M.D., Ph.D.
Department of Preventive Medicine
University of Southern California
Norris Comprehensive Cancer Center
Los Angeles, CA

Pirkko Vihko, M.D., Ph.D.
Research Center for Molecular Endocrinology
University of Oulu
Oulu, Finland

Noel S. Weiss, M.D., Dr.P.H.
University of Washington
Fred Hutchinson Cancer Research Center
Seattle, WA

Alice S. Whittemore, Ph.D.
Department of Health Research and Policy
Stanford University School of Medicine
Stanford, CA

Jianfeng Xu, M.D.
Center for Human Genomics
Wake Forest School of Medicine
Winston-Salem, NC

Hormones, Genes, and Cancer

Introduction and Background

BRIAN E. HENDERSON
ROBERTA MCKEAN-COWDIN

The concept that hormones are an important part of the carcinogenic process, at least in some organs, was initiated over 100 years ago by the observations of Beatson[1] on the relationship between cancer of the breast and ovarian function. Subsequent experimental studies in rodents by Bittner[2,3] demonstrated that hormones could increase the risk of mammary cancer. In parallel, Rous and Kidd[4] and later Berenblum[5] popularized the multistage concept of carcinogenesis. Elegant studies of the carcinogenic process in skin led to the generalization that carcinogenesis was, at a basic level, a two-stage process, involving initiation and promotion. By the 1950s, it was generally believed that there were three classes of *initiators*: viruses, chemicals, and certain physical agents, such as radiation. Hormones were considered to be *promoters*, capable of accelerating the carcinogenic process, but only in the aftermath of exposure to a carcinogenic initiator. The three classes of initiator shared the property of potential direct DNA damage, i.e., genotoxicity. In contrast, hormones were generally not directly genotoxic, their proliferative effects could be reversed, and their role in carcinogenesis aborted, by their removal as a source of stimulation. Whereas carcinogen-induced tumors are autonomous at the outset, those related to excessive hormonal stimulation pass through a prolonged stage of growth and progress to autonomy only late, if at all, in the carcinogenic process. Thus, although hormones were considered to be important in accelerating an existing cancer, they were not generally believed to be primary carcinogens.

INTERNATIONAL VARIATION IN RATES

The remarkable international variation in cancer rates, for many of the common cancers, such as breast and prostate, seemed to many to support the traditional view that endogenous hormones could not be the sole or primary cause. The low rates of breast, prostate, and several other cancers in Asian populations and their increase toward Western rates upon migration to the United States or Europe suggested that chemical factors or other environmental agents were the major causes of these cancers. While cigarette smoking was one such obvious chemical carcinogen, which would help to explain the international variation in lung and other smoking-related cancer sites, the most obvious cause of the other cancers seemed to be diet, other lifestyle factors, or unidentified environmental agents. In 1964, an expert committee of the World Health Organization stated that "the categories of cancer that are thus influenced, directly or indirectly, by extrinsic factors . . . collectively account for more than three-quarters of human cancers."[6]

In their systematic review of cancer causation, Doll and Peto[7] placed the majority of the unexplained excess of cancer observed in migrating populations on dietary factors acting directly or indirectly through their potential impact on lifestyle factors (e.g., reproduction, exercise). During the past 20 years, there has been a concerted effort by epidemiologists and experimentalists to verify the role of dietary factors in the etiology of cancer. Much of this effort has been directed toward proving the detrimental effects of dietary fat and the potential protective effect of a wide range of dietary antioxidants.[8] Unfortunately, it now seems likely that dietary factors are directly related to only a relatively small number of cancers, primarily, and not surprisingly, those of the digestive tract (esophagus, stomach, and large bowel). At the same time, it appears increasingly likely that the majority, if not all, of the hormone-related cancers have little direct relationship to any particular dietary item.

HORMONAL CARCINOGENESIS

It was Furth[9] who first tentatively suggested that hormones might be directly carcinogenic not by a genotoxic mechanism per se but by influencing the rate of cell division and thereby increasing the potential for spontaneous mutations. Drawing partially on the work of others, he suggested that while DNA molecules replicate, some copying "mistakes" might go unrepaired. In fact, the chromosomal instability at mitosis, could "produce cells carrying new karyotypes," which "are potential ancestors of novel clones liable to become malignant tumors."[9] Furth[9] went on to describe five lines of evidence to support the hypothesis of carcinogenesis without an extrinsic, genotoxic carcinogen. This evidence drew heavily on his own experience with thyroid carcinogenesis in the rat.

During the early 1970s, MacMahon and colleagues[10] published a series of papers suggesting that estrogens, generally, and estradiol, specifically, could be involved in human breast cancer carcinogenesis. At the same time, there was widespread scientific interest in the publication of data from Spiegelman's laboratory[11] demonstrating a microscopically, and immunologically, identifiable murine mammary tumor virus or type B virus in human breast milk. We undertook our first epidemiological study of breast cancer in young women to address the possibility that a transmissible agent causing breast cancer might also exist in human breast milk. We were unable to substantiate this hypothesis as there was no evidence of excess risk associated with breast-feeding, and the excess familial risk of breast cancer was seen in both the paternal and the maternal family trees.[12] However, we were very impressed with the evidence supporting a role for endogenous estrogen and the key importance of age at menarche as an expression of this susceptibility. Over the subsequent 25 years, we as well as others[13,14] have continued to utilize epidemiological and serological studies to accumulate evidence that endogenous estrogens played a pivotal role in breast cancer.[15] At the same time, it became increasingly clear that endogenous hormones were likely to be important in the etiology of other hormone-related cancers, including those of the testes, ovary, endometrium, prostate, and thyroid. In the mid-1970s, there was a sudden increase in the incidence of endometrial cancer; and in a series of epidemiological studies,[16] it became clear that this "epidemic" was caused by the introduction and subsequent widespread use of estrogen-replacement therapy. Perhaps more definitively than any laboratory experiment, this epidemic demonstrated the critical role of estrogen in endometrial cancer causation and the relationship between hormone-induced increased cell proliferation and the evolution of cancer.[15] It was also obvious that there was no other extrinsic carcinogen necessary or involved in this epidemic. In other words, the traditional two-stage initiator–promoter paradigm of carcinogenesis did not apply in this model of hormonal carcinogenesis.

In 1982, we published our first paper[15] attempting to synthesize the experimental and human data into a coherent model of hormonal carcinogenesis (Fig. 1.1). There was, however, a paucity of convincing evidence on the precise relationship between specific hormones and risk of cancer in specific target organs. In general, measurement of circulating sex steroid hormones, such as estrogen, is complicated by individual, day-to-day, and laboratory variations.

INTRODUCTION AND BACKGROUND

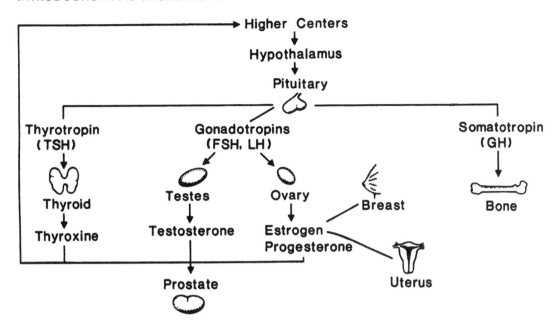

Figure 1.1 The endogenous hormones (sex steroids or pituitary peptides) that are responsible for the growth and development of the relevant end organ, e.g., breast, thyroid, prostate are the same hormones that, by causing cell proliferation in the end organ, predispose that end organ to the development of a malignant phenotype.

Seldom were results from one laboratory readily comparable to those from another, and inconsistency of findings relating hormones to cancer risk became the norm. However, in recent years, there has been a series of prospective studies supporting a direct association between circulating estrogen levels and breast cancer risk. A meta-analysis of breast cancer and serum estrogen levels was published by Thomas et al.[17] There was an average 16% higher level of the most bioactive estrogen, estradiol, in patients with breast cancer than in unaffected women ($p = 0.0003$).

An important concept related to human studies of hormonal carcinogenesis has evolved from observations such as those of Thomas et al.[17] that mean differences in circulating sex steroid levels associated with increased cancer risk are typically not large. Pike[18] made the critical observation that the age-specific incidence curve of breast cancer, as well as endometrial and ovarian cancers, fit the log incidence/log age model of Cook et al.[19] once an effect of menopause was figured into the model. Based on such a model, small differences in serum hormone levels, since they are present virtually constantly, over a lifetime can be shown mathematically to predict substantial changes in cancer risk. Thus, a 20% difference in a particular trophic hormone could translate into a two- to threefold lifetime increase in risk of a particular cancer, the actual value depending on the constant associated with the age-specific incidence of that cancer. This observation is of critical importance as we evaluate the magnitude of the effect of sequence variants in genes associated with steroid hormone biosynthesis or metabolism.

A second, and arguably relevant, concept related to human studies of hormonal carcinogenesis has gradually emerged from studies of the epidemiology of prostate cancer.[20] A series of prospective studies of circulating testosterone (T) and dihydrotestosterone (DHT) did not show the same consistent relationship of higher serum androgen levels and prostate cancer risk.[21] In their meta-analysis, Eaton et al.[21] concluded the following:

> This quantitative review reveals no convincing evidence that serum levels of endogenous sex hormones, their precursor compounds and metabolites and related binding protein, differ between men who subsequently go on to develop prostate cancer and those who do not.

Even more striking has been the observation that African Americans, who have the highest risk of prostate cancer, do not have the highest levels of circulating T or DHT.[22,23] A seminal observation by Ross et al.[20] demonstrated that Japanese men, who have a low risk of prostate cancer, have lower circulating levels of 3α,17β-androstanediol glucuronide, which derives from the intraprostatic metabolism of DHT, the major androgen in the prostate that binds to the androgen receptor. Thus, the most relevant measure of the bioactive steroid (e.g., estradiol, DHT) or polypeptide (e.g., follicle-stimulating hormone, luteinizing hormone) hormone is at the cellular level, where ligand binding to a specific receptor occurs; and this may or may not be well represented by the measurement of such hormones in serum or urine.

GENETIC BASIS OF HORMONAL CARCINOGENESIS

The identification and genetic characterization of families at high risk of breast cancer is the most striking example of the potential importance of inherited traits in breast cancer causation. The localization and sequencing of *BRCA1* and *BRCA2* has provided two important genes contributing to familial breast and ovarian cancer.[24,25] Somewhat surprisingly, mutations in these genes, which appear to be classic tumor-suppressor genes, seem to contribute little, if at all, to the causative pathway of sporadic breast cancer. Likewise, mutations in two other tumor-suppressor genes, *TP53* and *AT*, contribute to breast cancer risk in certain families but, again, not generally to the risk of sporadic breast cancer.[26]

The focus of much current breast cancer research is on susceptibility and progression, until the hormone-independent stage, by endogenous hormone stimulation (Fig. 1.2). Under the majority of circumstances, we assume that a breast epithelial cell does not contain a germline mutation in a tumor-suppressor gene such as *BRCA1*. In response to circulating steroids, (e.g., ovarian estradiol and progesterone) there is an accumulation of cell divisions in breast epithelium over many years. Each cell division carries a certain risk of a DNA copying error that is not corrected,

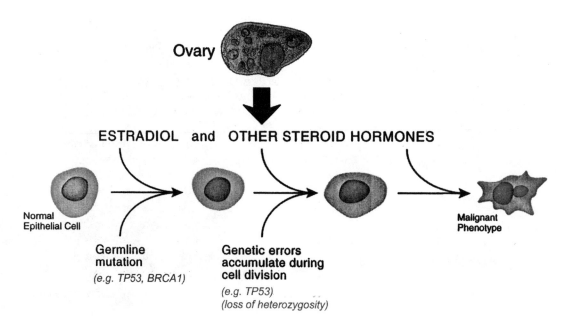

Figure 1.2 Estradiol and, to a lesser degree, other steroid hormones (e.g., progesterone) drive breast cell proliferation, which facilitates mutation, enhances fixation of mutations, or facilitates expression of genetic errors by loss of heterozygosity by defects in DNA repair. Germline mutations in relevant tumor-suppressor genes accelerate the transformation to the malignant phenotype.

INTRODUCTION AND BACKGROUND

and some of these are eventually relevant to the constellation of such mutations needed to produce the malignant phenotype. Ovarian steroid hormones drive the process of cell division directly and are, thus, the primary carcinogens. The amount of ovarian steroid hormones produced during each menstrual cycle is under strong genetic control, and the relevant genes are those in the relevant sex steroid biosynthesis and metabolism pathway (Table 1.1). We assume that there are common ($\geq 1\%$) sequence variations in these genes, which can produce meaningful differences in total ovarian steriod "exposure" over a woman's lifetime. Of course, the same, or novel, sequence variants in these genes can be associated with the progression of hormone-related cancers, as has been well documented for variants in the androgen receptor gene.[20] The details of the endocrine pathways and the relevant candidate genes will be discussed by several of the chapters in this book, so further details are not provided here. A list of some candidate endocrine genes is provided in Table 1.1, and common variants in a broader set of genes are under investigation (see Chapter 4). The important feature of this paradigm is that the candidates are predictable because they encode proteins that are part of known physiological and cellular pathways. More candidate genes will certainly emerge as we begin to understand more about the relevant signaling and transport pathways in steroid-responsive cells. One important example of such nonendocrine factors is the growth hormone family, such as insulin-like growth factor-1 and related compounds, which are discussed in Chapter 5.

With the sequencing of the human genome and the development of high-throughput DNA sequencing and genotyping technologies, there has been a rapid expansion in such studies of complex traits and multigenic diseases. The future holds considerable promise for defining genetic pathways and developing targeted diagnostic and therapeutic approaches to the hormone-related cancers.

CURRENT TRENDS IN HORMONE-RELATED CANCERS

While we propose that genetic variation influences cancer risk in hormone-responsive tissue by programming endogenous hormone production, transport, and response, other lifestyle, diagnostic, and treatment factors appear to explain short-term trends in incidence and mortality. For example, trends in hormone-related cancers, such as breast, ovarian, prostate, and uterine cancers, show that incidence and mortality rates are influenced by several factors, including changing patterns of hormone-replacement therapy, oral contraceptive use, disease-screening practices, reproductive characteristics, and other lifestyle factors that vary over time and by racial/ethnic group. During the past two decades, efforts to diagnose early stage cancer through screening have resulted in artificial increases in breast and prostate cancer incidence. The impact of changing reproductive factors and screening efforts on the incidence and mortality of each of these hormone-associated cancers is discussed below and illustrated with data from the United States Surveillance Epidemiology and End Results (SEER) program[27] and the National Center for Health Statistics.[28]

Prostate Cancer

The expansion of prostate cancer screening efforts in the United States during the mid 1980s,

Table 1.1 Examples of Candidate Genes in Five Hormone-Related Cancers

Cancer Site	Proliferative Hormone	Relevant Genes
Breast	Estrogen	*CYP17, CYP19, HSD17B1, ER*
	Progesterone	*CYP17, PR, COMT*
Endometrium	Estrogen	*CYP17, HSD17B1, HSD17B2, ER, PR*
Ovary	FSH, progesterone, inhibin B	*FSH, FSHR, HSD17B1, PR, INHB*
Prostate	Testosterone (dihydrotestosterone)	*CYP17, HSD17B3, HSD17B4, HSD17B5, SRD5A2, AR*
Testis	LH, estrogen?	*LH, LHR, CYP17*

CYP17, cytochrome P-450 17,20 lyase; *CYP19*, aromatase; *HSD17B1*, 17β-hydroxysteroid dehydrogenase type I; *ER*, estrogen receptor; *PR*, progesterone receptor; *FSH, FSHR*, follicle-stimulating hormone and receptor; *INHB*, inhibin B; *5RD5A2*, 5-α-reductase type II; *LH, LHR*, luteinizing hormone and receptor.

with emphasis on the prostate-specific antigen (PSA) assay, resulted in large increases in prostate cancer incidence through 1992 for white males and through 1993 for African-American males. Prostate cancer incidence increased 183% for white males and 136% for African-American males from 1973 to 1992, with the largest increases occurring from approximately 1988 through 1992. Rates from 1992 to 1998 show a reversal of this trend as incidence began to decrease in both white (-25.4%) and African-American (-9.9%) men (Table 1.2). The trend for prostate cancer in African-American and white men is similar, with the exception that it is delayed roughly 2 years in African Americans (Fig. 1.3), and the rate is approximately 1.5 times higher in African-American than white men.

While the value of the PSA assay in identifying cases of asymptomatic prostate cancer and preventing mortality through the detection of early-stage disease has been unclear, it now appears that prostate cancer mortality is beginning to decline. Following the introduction of the PSA assay in the United States, mortality rates associated with prostate cancer increased by 2.9% per year between 1987 and 1991, stabilized from 1991 to 1994, and then declined 4.7% per year from 1995 to 1998.[29] While it remains too early to be certain, this initial decline in prostate cancer mortality suggests that screening efforts may be effective at decreasing deaths.[30] Although it is generally accepted that detection of cancer at an early stage results in better survival than detection at a late stage, it is not clear what percent of screened men with early-stage prostate cancer would actually progress to symptomatic disease. A study of untreated stage T1a prostate cancer reported that only 15% of the men enrolled progressed to clinical disease.[31] Any long-term benefits of PSA screening on prostate cancer mortality are not yet apparent. In the next decades, prostate cancer mortality can be expected to fall to the extent that PSA screening effectively identifies the population at high risk of developing symptomatic disease and assuming that the majority of such men are effectively treated.

Breast Cancer

Breast cancer incidence and mortality rates are strongly influenced by changing reproductive patterns, diagnostic screening practices, and treatment options. The introduction of mammography across the United States in the 1980s was followed by a rapid rise in breast cancer incidence (Fig. 1.4). Incidence rates increased among African-American and white women of all ages from 1973 through 1991 (37.4% and 33.1%, respectively). From 1991 through 1998, breast cancer incidence continued to increase at a slower rate among African-American (5.7%) and white (6.5%) women aged 50 and older, while decreasing in women under the age of 50 (African American, -10.8%; white, -3.0%). The initial rise in cases in the 1980s was limited to early-stage breast cancer and in situ disease; however, more cases of stage II lymph node-positive breast cancer among white women 50–64 years old have been reported in recent years (1994–1998).[29,32] The reason for the change in incidence may be explained by the impact of screening; however, age-specific birth-cohort effects also may have contributed to the secular patterns in incidence rates.[33,34]

Breast cancer mortality rates, which increased during the 1980s and early 1990s, are now de-

Table 1.2 Incidence in the United States of Select Hormone-Related Cancers, Comparing Percentage Change for 1973–1991 to trends in 1991–1998

	African-American Incidence		White Incidence	
Cancer Site	1973/74–1991/92 (presented as 7-year interval)	1991/92–1997/98	1973/74–1991/92 (presented as 7-year interval)	1991/92–1997/98
Breast (women only)	12.6	1.3	9.9	4.5
Uterine	2.2	2.0	−11.5	2.4
Ovary	0.6	−1.0	2.9	−8.0
Prostate	39.8	−9.9	50.0	−25.4
Testis	°	°	19.5	6.6

°Numbers too small to estimate.

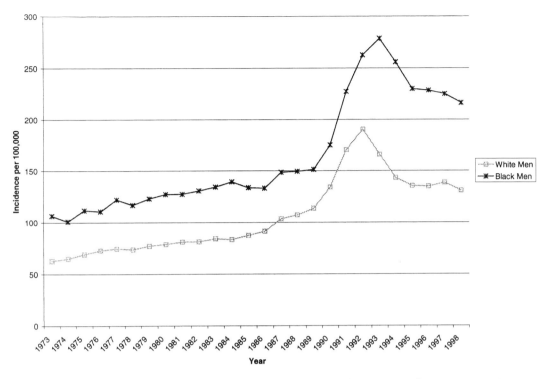

Figure 1.3 Prostate cancer incidence among African-American and white men, aged 0–85, SEER 1973–1998. (Adapted from Ries et al.[27] with permission.)

clining across age groups for white women and among African-American women under 65 years of age. The decline in mortality was first observed in white women under 55 years of age in the early 1980s, but remarkably, declining mortality rates are now seen even among women 85 years and older.[35] The reason for the declining mortality rates is most likely due to improvements in treatment, including chemotherapy and tamoxifen, although the impact of screening cannot be ruled out. Findings from the 1998 tamoxifen trial provide strong support that breast cancer is preventable among high-risk women, the risk of invasive and noninvasive breast cancers each being reduced by approximately 50%.[36] Ongoing clinical trials of tamoxifen, raloxifene, and aromatase inhibitors are providing additional breast cancer prevention and treatment options.[37]

Ovarian Cancer

Incidence rates of ovarian cancer, which are strongly influenced by both oral contraceptive use and reproductive experiences, declined in the United States from 1991 through 1998 in both African-American and white women: approximately −1% in African Americans and −8% in whites (Table 1.2). The decrease was similar for women above and below 50 years of age. Mortality rates also declined in this period (−7.1%, African Americans; −5.6%, whites) (Table 1.3). While declining rates of ovarian cancer are encouraging, the total percent change in incidence over the entire SEER tracking period (1973–1998) was relatively small (−1%, whites; 0.5%, African Americans) compared to the dramatic changes seen for breast and prostate cancers (Fig. 1.4). Declining rates of ovarian cancer are likely to be a consequence of extended oral contraceptive use, which interrupts ovulation and endogenous cyclic hormonal changes. Several explanations have been proposed to explain the association between ovarian cancer rates and ovulatory cycles, including incessant ovulation, increased exposure to gonadotropins, and increased exposure to high levels of estrogen that occur within the ovary during a normal men-

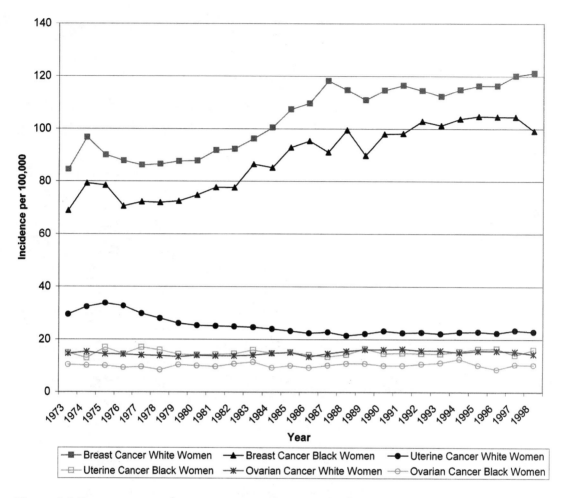

Figure 1.4 Breast, ovarian, and uterine cancer incidence among African-American and white women, ages 0–85, SEER 1973–1993. (Adapted from Ries et al.[27] with permission.)

Table 1.3 Mortality in the United States from Select Hormone-Related Cancers, Comparing Percentage Change for 1979–1991 to Trends in 1991–1998

Cancer Site	African-American Mortality		White Mortality	
	1979/80–1991/92 (presented as 7-year interval)	1991/92–1997/98	1979/80–1991/92 (presented as 7-year interval)	1991/92–1997/98
Breast (women only)	11.8	−2.5	0.6	−13.9
Uterine	−6.3	−5.5	−11.3	−2.5
Ovary	4.8	−7.1	1.2	−5.6
Prostate	17.7	−10.0	10.8	−16.2
Testis	°	°	−25.8	−16.7

°Numbers too small to estimate.

strual cycle. The incessant ovulation hypothesis supposes that each cycle of ovulation, which includes repair of ovarian surface epithelium, increases the risk of ovarian cancer.[38] Declining ovarian cancer rates would then be explained by a reduction in causative factors associated with pregnancies and long-term oral contraceptive use.

Uterine Cancer

Trends in uterine cancer incidence and mortality are influenced by factors that alter circulating estrogen and progesterone levels, including hormone replacement therapy and oral contraceptives. Risk of uterine cancer will increase if endometrial tissue is exposed to estrogen that is unopposed by progestogen.[15,39,40] Thus, estrogen-replacement therapy (ERT), in which estrogen is given to postmenopausal women without any progestogen, increases uterine cancer risk. In the United States, a doubling of estrogen prescriptions (ERT) to treat menopausal symptoms beginning in 1966 resulted in the epidemic of endometrial cancer that ended in the late 1970s,[41] when progestogens were increasingly added to ERT to reduce or possibly eliminate this risk. Rates peaked in 1975 among white women of all ages and African-American women under the age of 50 years; rates peaked among African-American women over 50 years old 2 years later in 1977. As a result of the epidemic, ERT for menopausal symptoms among women who had not had a hysterectomy was largely discontinued. Following these changes, incidence rates of uterine cancer decreased 26% from 1976 to 1987, when they began to stabilize. Incidence was slightly higher in 1998 than 1991 for both African-American and white women (2.0%, African Americans; 2.4%, whites); however, rates of uterine cancer in the 1990s were generally stable (Fig. 1.4).

In contrast to ERT, oral contraceptives decrease the risk of endometrial cancer. Oral contraceptives, which deliver an estrogen and a high-dose progestogen for 21 days of a 28-day cycle, diminish risk by decreasing the duration of exposure of the endometrium to estrogen unopposed by progestogen from the 2 weeks of a normal menstrual cycle to the 1 week when the oral contraceptive is not taken.[42–44] Endogenous estrogen levels are low during the 7 days when the oral contraceptive is not used.

Testicular Cancer

During the 1970s and 1980s, the incidence of cancer of the testis increased among white male populations of most European countries, the United States, Australia, and New Zealand.[45] An increase in the rate was first noted among men born after 1920, but this increase was consistently sustained for men born after 1950.[45] By the early 1990s, the rate of increase seemed to be abetting. A substantial body of experimental and epidemiological evidence has indicated that prenatal events are important risk factors for testicular cancer;[46–49] in utero exposure to endogenous and exogenous estrogen in particular has been implicated.[46,48,50–52]

CONCLUSIONS

The development of a hormonal carcinogenesis model began over 100 years ago with the observation that ovarian function was related to breast cancer risk. Over time, it became apparent that hormones act as carcinogens by increasing cellular proliferation and thereby increasing the chance of random DNA copy errors. Through epidemiologic and serological studies, we observed that hormones are key determinants of cancer in several hormone-sensitive tissues, including the prostate, breast, endometrium, ovary, and possibly thyroid and testes. While findings from studies including serum hormone measurements are not always comparable, several independent prospective studies have reported evidence supporting an association between circulating estrogen levels and breast cancer risk. Further, it is believed that relatively small individual differences in serum hormone levels (20%) can multiply lifetime risk for a particular cancer by two- to threefold. For cancers other than breast and uterine, the most relevant measure of the bioactive steroid or polypeptide may be at the cellular level, such as intraprostatic DHT and prostate cancer, which may not be well represented by circulating hormone levels.

The identification of *BRCA1* and *BRCA2* provided evidence of genes that directly affect the risk of hormone-related cancers, and their discovery initiated the rapidly expanding field of genetic susceptibility. Hormonal carcinogenesis is now being characterized in terms of specific sequence changes in germline and somatic cells.

Cancer incidence in hormone-sensitive tissues has been relatively stable over time, although incidence and mortality trends for several hormone-related cancers (prostate, breast, uterine) have fluctuated, driven largely by changes in disease detection methods and use of exogenous hormones, including oral contraceptives and hormone-replacement therapy. Changes in cancer rates are often interpreted in terms of changing diet and environmental pollutants; however, trends in hormone-related cancers are consistent with our knowledge of lifetime exposure to hormones (endogenous and exogenous) and improving methods of cancer screening and treatment. The definition of the genetic basis of hormonal carcinogenesis should lead to continued decreases in these cancers over the next few decades.

REFERENCES

1. Beatson GT. On the treatment of inoperable cases of carcinoma of the mamma: suggestions for a new method of treatment with illustrative cases. Lancet 2:104, 1896.
2. Bittner JJ. The causes and control of mammary cancer in mice. Harvey Lect 42:221, 1946–1947.
3. Bittner JJ. The genesis of breast cancer in mice. Texas Rep Biol Med 10:160, 1952.
4. Rous P, Kidd JG. Conditional neoplasms and subthreshold neoplastic states. J Exp Med 73:365, 1941.
5. Berenblum I. Cocarcinogenesis. Br Med Bull 4:343, 1947.
6. World Health Organization. Prevention of Cancer. Technical Report Series 267. Geneva: WHO, 1964.
7. Doll R, Peto R. The causes of cancer: quantitative estimates of avoidable risks of cancer in the United States today. J Natl Cancer Inst 66:1191–1308, 1981.
8. Henderson BE, Ross RK, Pike MC. Toward the primary prevention of cancer. Science 254:1131–1138, 1991.
9. Furth J. Hormones as etiological agents in neoplasia. In: Becker FF (ed). Cancer. A Comprehensive Treatise, vol 1, New York: Plenum, 1975, pp 75–120.
10. MacMahon B, Cole P, Brown J. Etiology of human breast cancer: a review. J Natl Cancer Inst 50:21–42, 1973.
11. Schlom J, Spiegelman S. Evidence for viral involvement in murine and human mammary adenocarcinoma. Am J Clin Pathol 60:44–56, 1973.
12. Henderson BE, Powell D, Rosario I, Keys C, Hanisch R, Young M, Casagrande J, Gerkins V, Pike MC. An epidemiologic study of breast cancer. J Natl Cancer Inst 53:609–614, 1974.
13. Henderson BE, Gerkins V, Rosario I, Casagrande J, Pike MC. Elevated serum levels of estrogen and prolactin in daughters of patients with breast cancer. N Engl J Med 293:790–795, 1975.
14. Cole P, Cramer D, Yen S, Paffenbarger R, MacMahon B, Brown J. Estrogen profiles of premenopausal women with breast cancer. Cancer Res 38:745–748, 1978.
15. Henderson BE, Ross RK, Pike MC, Casagrande JT. Endogenous hormones as a major factor in human cancer. Cancer Res 42:3232–3239, 1982.
16. Mack TM, Pike MC, Henderson BE, Pfeffer RI, Gerkins VR, Arthur M, Brown SE. Estrogens and endometrial cancer in a retirement community. N Engl J Med 294:1262–1267, 1976.
17. Thomas HV, Reeves GK, Key TJ. Endogenous estrogen and postmenopausal breast cancer: a quantitative review. Cancer Causes Control 8:922–928, 1997.
18. Pike MC. Age-related factors in cancers of the breast, ovary, and endometrium. J Chronic Dis 40(Suppl 2):59S–69S, 1987.
19. Cook PJ, Doll R, Fellingham SA. A mathematical model for the age distribution of cancer in man. Int J Cancer 4:93–112, 1969.
20. Ross RK, Coetzee GA, Pearce CL, Reichardt JK, Bretsky P, Kolonel LN, Henderson BE, Lander E, Altshuler D, Daley G. Androgen metabolism and prostate cancer: establishing a model of genetic susceptibility. Eur Urol 35:355–361, 1999.
21. Eaton NE, Reeves GK, Appleby PN, Key TJ. Endogenous sex hormones and prostate cancer: a quantitative review of prospective studies. Br J Cancer 80:930–934, 1999.
22. Platz EA, Rimm EB, Willett WC, Kantoff PW, Giovannucci E. Racial variation in prostate cancer incidence and in hormonal system markers among male health professionals. J Natl Cancer Inst 92:2009–2017, 2000.
23. Wu AH, Whittemore AS, Kolonel LN, John EM, Gallagher RP, West DW, Hankin J, Teh CZ, Dreon DM, Paffenbarger RS Jr. Serum androgens and sex hormone–binding globulins in relation to lifestyle factors in older African-American, white, and Asian men in the United States and Canada. Cancer Epidemiol Biomarkers Prev 4:735–741, 1995.
24. Miki Y, Swensen J, Shattuck-Eidens D, et al. A strong candidate for the breast and ovarian cancer susceptibility gene *BRCA1*. Science 266:66–71, 1994.

25. Wooster R, Bignell G, Lancaster J, Swift S, Seal S, Mangion J, Collins N, Gregory S, Gumbs C, Micklem G. Identification of the breast cancer susceptibility gene *BRCA2*. Nature 378:789–792, 1995.
26. Malkin D, Li FP, Strong LC, et al. Germ line *p53* mutations in a familial syndrome of breast cancer, sarcomas, and other neoplasms. Science 250: 1233–1238, 1990.
27. Ries L, Kosary C, Hankey B. In: Ries L, Kosary C, Hankey B (eds). SEER Cancer Statistics Review, 1973–1998. Bethesda, MD: National Cancer Institute, 2002. Surveillance, Epidemiology, and End Results (SEER) Program Public-Use Data (1973–1998), National Cancer Institute, DCCPS, Surveillance Research Program, Cancer Statistics Branch, released April 2001.
28. CDC/NCHS. Age-adjusted death rates for 72 selected causes, by race and sex using year 2000 standard population: United States, 1979–98. National Vital Statistics System, Mortality, 2002. Atlanta, GA; Hyattsville, MD: CDC/NCHS, 2002. http://www.cdc.gov/nchs/datawh/datawh.htm
29. Howe HL, Wingo PA, Thun MJ, Ries LA, Rosenberg HM, Feigal EG, Edwards BK. Annual report to the nation on the status of cancer (1973 through 1998), featuring cancers with recent increasing trends. J Natl Cancer Inst 93:824–842, 2001.
30. Shibata A, Whittemore AS. Re: Prostate cancer incidence and mortality in the United States and the United Kingdom. J Natl Cancer Inst 93:1109–1110, 2001.
31. Cheng L, Neumann RM, Blute ML, Zincke H, Bostwick DG. Long-term follow-up of untreated stage T1a prostate cancer. J Natl Cancer Inst 90:1105–1107, 1998.
32. Beahrs O, Henson D, Hutter R, Myers M. Manual for Staging of Cancer, 3rd ed. Philadelphia: American Joint Committee on Cancer, 1988.
33. Chu KC, Tarone RE, Kessler LG, Ries LA, Hankey BF, Miller BA, Edwards BK. Recent trends in U.S. breast cancer incidence, survival, and mortality rates. J Natl Cancer Inst 88:1571–1579, 1996.
34. Feigelson HS, Henderson BE, Pike MC. Re: Recent trends in U.S. breast cancer incidence, survival, and mortality rates. J Natl Cancer Inst 89: 1810–1812, 1997.
35. CDC/NCHS. Unpublished Table NEWSTAN 79-98L. National Vital Statistics System, Mortality, 2002. Atlanta, GA: CDC/NCHS, 2002. http://www.cdc.gov/nchs/datawh.htm
36. Fisher B, Costantino JP, Wickerham DL, et al. Tamoxifen for prevention of breast cancer: report of the National Surgical Adjuvant Breast and Bowel Project P-1 Study. J Natl Cancer Inst 90:1371–1388, 1998.
37. National Cancer Institute. Cancer facts. http://cis.nci.gov/fact/7_16.htm, 2002.
38. Henderson BE, Ross RK, Pike MC. Hormonal chemoprevention of cancer in women. Science 259:633–638, 1993.
39. Henderson BE, Ross R, Bernstein L. Estrogens as a cause of human cancer: the Richard and Hinda Rosenthal Foundation Award lecture. Cancer Res 48:246–253, 1988.
40. Key TJ, Pike MC. The dose–effect relationship between "unopposed" oestrogens and endometrial mitotic rate: its central role in explaining and predicting endometrial cancer risk. Br J Cancer 57:205–212, 1988.
41. Weiss NS, Szekely DR, Austin DF. Increasing incidence of endometrial cancer in the United States. N Engl J Med 294:1259–1262, 1976.
42. Henderson BE, Casagrande JT, Pike MC, Mack T, Rosario I, Duke A. The epidemiology of endometrial cancer in young women. Br J Cancer 47:749–756, 1983.
43. Combination oral contraceptive use and the risk of endometrial cancer. The Cancer and Steroid Hormone Study of the Centers for Disease Control and the National Institute of Child Health and Human Development. JAMA 257:796–800, 1987.
44. Weiss NS, Sayvetz TA. Incidence of endometrial cancer in relation to the use of oral contraceptives. N Engl J Med 302:551–554, 1980.
45. Bergstrom R. Increase in testicular cancer incidence in six European countries: a birth cohort phenomenon. J Natl Cancer Inst 88:727–733, 1996.
46. Depue RH. Estrogen exposure during gestation and risk of testicular cancer. J Natl Cancer Inst 71:1151–1155, 1983.
47. Depue RH. Maternal and gestational factors affecting the risk of cryptorchidism and inguinal hernia. Int J Epidemiol 13:311–318, 1984.
48. Henderson BE, Benton B, Jing J, Yu MC, Pike MC. Risk factors for cancer of the testis in young men. Int J Cancer 23:598–602, 1979.
49. Morrison AS. Cryptorchidism, hernia, and cancer of the testis. Int J Cancer 23:598–602, 1979.
50. Petridou E. Baldness and other correlates of sex hormones in relation to testicular cancer. J Natl Cancer Inst 56:731–733, 1976.
51. Leary FJ. Males exposed in utero to diethylstilbestrol. Int J Cancer 71:982–985, 1997.
52. Depue RH, Bernstein L, Ross RK, Judd HL, Henderson BE. Hyperemesis gravidarum in relation to estradiol levels, pregnancy outcome, and other maternal factors: a seroepidemiologic study. Am J Obstet Gynecol 156:1137–1141, 1987.

2

Biosynthesis, Transport, and Metabolism of Steroid Hormones

FRANK Z. STANCZYK
PHILIP BRETSKY

In the human body, a balance exists between production and clearance of steroid hormones. Production of steroid hormones occurs de novo by biosynthetic pathways in specific endocrine glands, i.e., the adrenals and ovaries in women and the adrenals and testes in men. In addition, steroid hormones can be produced in peripheral (nonendocrine gland) tissues from circulating precursors that originate from the endocrine glands. Important sites of peripheral steroid hormone formation include the liver, kidney, breast, prostate, and sexual and nonsexual skin. After steroid hormones are secreted by the endocrine glands, they enter the systemic circulation, where they are mostly bound to proteins. The low-affinity bound and non-protein-bound (free) steroids, sometimes referred to as *bioavailable steroids*, are available for binding to steroid hormone receptors (progestogen, androgen, estrogen, glucocorticoid, mineralocorticoid); and if active, they exert a biological effect. Alternatively, they undergo metabolism and are excreted primarily through the kidney and, to a lesser extent, the intestine, appearing in urine and feces, respectively.

The objectives of the present chapter are (1) to review the steroidogenic pathways of biosynthesis and metabolism, (2) to discuss the genes encoding the enzymes involved in steroid hormone biosynthesis and metabolism, and (3) to review how steroid hormones interact with proteins in the blood.

BIOSYNTHESIS OF STEROID HORMONES

Conversion of Cholesterol to Pregnenolone

The first and rate-limiting step in the biosynthesis of steroid hormones in the adrenals and gonads is the conversion of cholesterol to pregnenolone (Figs. 2.1, 2.2). This reaction occurs in the mitochondrion and is catalyzed by the cholesterol side-chain cleavage cytochrome P-450 (P450scc) enzyme (also referred to as cholesterol desmolase or cholesterol lyase) in conjunction with auxiliary electron-transferring proteins, located in the inner mitochondrial membrane. This electron-transport system consists of three protein components: reduced nicotinamide adenine dinucleotide phosphate (NADPH)-dependent reductase (ferredoxin reductase), ferredoxin, and cytochrome P-450. Electrons are

BIOSYNTHESIS, TRANSPORT, AND METABOLISM OF STEROID HORMONES

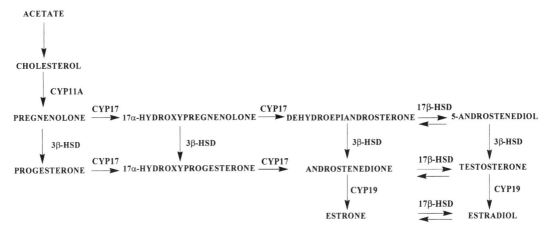

Figure 2.1 Biosynthesis of steroid hormones in the gonads.

transported from NADPH via each protein component to the terminal cytochrome P-450. The side-chain cleavage reaction can be divided into 20-hydroxylation, 22-hydroxylation, and cleavage of the bond between carbons 20 and 22. Evidence from in vivo and in vitro studies indicates that two factors control the rate at which the precursor, cholesterol, is converted to pregnenolone: the delivery of cholesterol from intracellular sources to the enzyme system and the level of P450scc. The rate-limiting step in the conversion of cholesterol to pregnenolone appears to be the transport of cholesterol from extracellular sources to the inner mitochondrial membrane and the subsequent loading of the precursor into the active site of P450scc. Intramitochondrial cholesterol movement appears to be due to coordinated activation of the recently cloned steroidogenic acute regulatory (StAR) protein and of the mitochondrial peripheral-type benzodiazepine receptor. Both the delivery of cholesterol to the enzyme and the

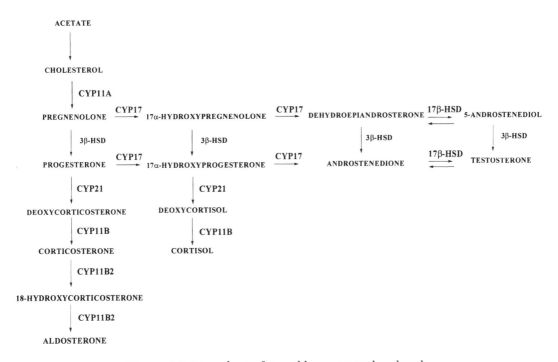

Figure 2.2 Biosynthesis of steroid hormones in the adrenals.

level of P450scc and its associated protein, ferredoxin, are primarily under the control of tropic hormones using cyclic adenosine monophosphate (cAMP) or calcium as the intracellular messenger. Tropic hormone regulation of steroid synthesis can be either *acute*, i.e., occurring within minutes and resulting in rapid steroid biosynthesis, or *chronic*, i.e., occurring over a prolonged time and resulting in continuous steroid formation.

StAR Protein

The gene encoding the StAR protein is located at chromosome 8p.11.2[1] and organized into seven exons and six introns.[2,3] Gene expression of the StAR protein is regulated by gonadotropins and corticotropin in the gonads and adrenal, respectively, via a cAMP second message.[4,5]

A mechanism for induction of *StAR* gene transcription in luteinized granulosa cells has been proposed.[4] Stimulation of *StAR* gene transcription in these cells occurs by luteinizing hormone (LH) via cAMP. The transcriptional process requires ongoing protein synthesis and takes hours. Apparently, the orphan nuclear receptor steroidogenic factor 1 (SF-1) is essential for *StAR* gene expression, perhaps in conjunction with a factor that transactivates at the proximal element.

A model has been proposed showing that the StAR protein acts in the transfer of cholesterol from the outer to the inner mitochondrial membrane.[6,7] In response to tropic hormone stimulation, a 37 kDa precursor to the StAR protein is rapidly synthesized in the cytosol and transferred to the mitochondria. The mechanism of this transport is unknown. However, there is speculation that during the transport of the 37 kDa protein "contact sites" are established between the inner and outer membranes and a processing protease makes a 32 kDa form of the precursor protein. Further processing of the latter protein gives rise to the mature 30 kDa protein, i.e., the StAR protein.

Peripheral-Type Benzodiazepine Receptors

Peripheral-type benzodiazepine receptors (PBRs) have been shown to play an important role in the regulation of cholesterol transport.[8] These receptors are highly abundant in steroidogenic cells and are found primarily on the outer mitochondrial membrane. They have been implicated in the acute stimulation of adrenal and gonadal steroidogenesis, presumably by mediating the entry and distribution of cholesterol in the mitochondria by a mechanism that has not been elucidated. Studies show that the diazepam-binding inhibitor (endozepine), which is an endogenous PBR ligand, stimulates cholesterol transport and promotes cholesterol loading to P450scc in vitro. Its presence appears to be essential for human chorionic gonadotropin–induced steroidogenesis by Leydig cells.[8] Disruption of the PBR gene in steroidogenic cells causes a dramatic reduction in both the ability of the cells to transport cholesterol in the mitochondria and steroid production, despite high levels of the StAR protein. Thus, it appears that the coordinated activation of the StAR protein and PBRs is essential for cholesterol transport and subsequent steroidogenesis.

Transfer of Cholesterol to the Mitochondrion

Steroid-synthesizing cells have at least three metabolically active pools of cholesterol: *(1)* a small, metabolically active pool of free cholesterol; *(2)* a large storage pool in which cholesterol is stored as cholesterol esters of free fatty acids; and *(3)* a fixed pool of membrane cholesterol, which is not available for steroid synthesis.[9] The free cholesterol pool is formed from endogenous conversion of acetate to cholesterol via a complex series of reactions and by hydrolysis of cholesterol esters, including those that are part of the storage pool and those that are constituents of low-density lipoprotein (LDL) and high-density lipoprotein (HDL). After cellular uptake of these lipoproteins, LDL is processed through lysozymes, whereas HDL enters the cytosolic pool directly. Dietary cholesterol incorporated into LDL is most commonly used for steroid hormone biosynthesis. Although free cholesterol represents only a small fraction of the total cholesterol pool, this fraction can be increased by cAMP stimulation of cytosolic cholesterol esterase and suppression of acyl-coenzyme A (CoA) cholesterol acyltransferase.[10,11]

It has been proposed that during processing of the StAR protein, cholesterol is transferred from the outer to the inner mitochondrial membrane. A hydrophobic core may be formed by a protein complex between the outer and inner mitochondrial membrane, through which cholesterol can pass.[8]

Ferredoxin Reductase, Ferredoxin, and P450scc

As mentioned earlier, the electron-transport system which acts in conjunction with the side-chain cleavage enzyme consists of ferredoxin reductase, ferredoxin, and the terminal cytochrome P-450. The characteristics of this system have been reviewed by Chung et al.[12] All three components are found in the adrenal, gonads, placenta, and some parts of the brain. In addition, ferredoxin and its reductase are found in the kidney and liver.

Ferredoxin is often named for the tissue from which it is purified, e.g., the name adrenodoxin is used when the source is the adrenal. Multiple copies of the *ferredoxin* gene have been found. However, the active gene is located on chromosome 11q[13] and composed of four exons and three introns, spanning more than 20 kb of DNA.[14] The gene has multiple polyadenylation signals at the 3' end, resulting in transcripts of different lengths.[15,16] The ferredoxin promoter is relatively simple, consisting of a TATA box and two Sp1-binding sites, which is sufficient to direct cAMP-dependent transcription in a steroidogenic cell-specific fashion.[12]

P450scc is encoded by the *CYP11A1* gene. The human *CYP11A1* gene is present on chromosome 15 as a single-copy gene, which consists of nine exons and eight introns, spanning 20 kb.[17,18] The 5'-flanking region of the *CYP11A1* gene is very complex and consists of many binding sites.[12] Its TATA box directs cAMP-dependent transcription in a steroidogenic cell-specific fashion. The orphan nuclear receptor SF-1 activates *CYP11A1* gene expression.[19]

Regulation of Cholesterol to Pregnenolone Conversion

P450scc plays a key role in carrying out the conversion of cholesterol to pregnenolone and is subject to many levels of regulation. In the adrenal, gonads, and placenta, P450scc is regulated by various tropic hormones, utilizing cAMP or calcium as intracellular messengers.[12] Steroid hormone biosynthesis in the adrenals is stimulated by the pituitary hormone corticotropin, which binds to its receptor at the cell surface and activates adenyl cyclase to increase intracellular cAMP. The latter compound serves as an intracellular messenger to transduce both the acute and the long-term effects of corticotropin.

The acute effect of corticotropin increases the output of steroids within minutes of its administration.[12] This effect is due to increased synthesis of StAR protein, which accelerates the transport of cholesterol into the inner mitochondrial membrane, where P450scc converts cholesterol to pregnenolone. The accelerated transport of substrate for P450scc action results in enhanced adrenal steroid hormone secretion. This rapid steroidogenic reaction is the method that cells use to respond to stress.

In contrast to the acute effect of corticotropin, its long-term action requires many hours to achieve and is due to stimulation of *CYP11A1* gene expression.[12] This relatively slow steroidogenic effect is important for maintenance of homeostasis in the body.

Another stimulator of *CYP11A1* gene expression is the potent vasoconstrictor angiotensin II, which is formed from the inactive hormone angiotensin I in a variety of tissues.[12] Angiotensin II regulates mineralocorticoid biosynthesis in the zona glomerulosa of the adrenal cortex. Its mechanism of action involves binding to its membrane receptor and activation of protein kinase C, which in turn increases intracellular calcium levels and subsequent signal transduction and gene stimulation.

In the ovary, *CYP11A1* expression is induced by follicle-stimulating hormone (FSH) and LH, resulting in biosynthesis of estrogens, progesterone, and androgens during the menstrual cycle.[12] The major intracellular mediator for *CYP11A1* stimulation by gonadotropins is cAMP.

Ferredoxin gene expression is also stimulated in the adrenal and ovary, using cAMP as intracellular messenger, in a manner similar to that described for *CYP11A1*.[12] However, in contrast to the *CYP11A1* and *ferredoxin* genes, the mechanism of action of the *ferredoxin reductase* gene does not involve a change in cAMP or calcium levels.[12] Instead, its mRNA levels are diminished, suggesting possible posttranscriptional regulation of the *ferredoxin reductase* gene.

Conversion of Pregnenolone to Progesterone

Following the conversion of cholesterol to pregnenolone by the mitochondrial side-chain cleavage system, the adrenals and gonads can transform pregnenolone to either progesterone or 17α-hydroxypregnenolone. The formation of progesterone from pregnenolone is catalyzed by the enzyme 3β-hydroxysteroid oxidoreductase [commonly referred to as 3β-hydroxysteroid dehydrogenase (3β-HSD)] in combination with $\Delta^{5,4}$-isomerase, in the presence of the cofactor (NAD^+). In addition, 3β-HSD catalyzes the formation of other Δ^4-3-ketosteroids from corresponding Δ^5-3β-hydroxysteroids, which leads to the formation of androgens, estrogens, mineralocorticoids, and glucocorticoids. Thus, the reactions involving 3β-HSD represent an obligatory step in the formation of highly biologically active steroids.

3β-HSD is expressed in the adrenal, gonads, and placenta, as well as in other tissues such as the liver and kidney, where its function has not been elucidated.[20-22] In cells, it is membrane-bound in the mitochondria and endoplasmic reticulum.[20-22] A family of closely related genes encode for 3β-HSD, and the various 3β-HSD isoforms are expressed in a tissue-specific manner involving separate regulatory mechanisms.

Presently, only two functional *3β-HSD* genes (type 1 and type 2) together with three pseudogenes have been characterized in the human genome.[23-26] Both genes are located in the chromosome 1p13.1 region, are 7.84 and 7.88 kb in length, and consist of four exons and three introns.[25,27] The type 1 and type 2 3β-HSD proteins are 93.5% homologous in amino acid sequence.[23-26] Type 1 gene expression occurs primarily in the placenta, mammary gland, and skin, whereas the type 2 isoform is expressed almost exclusively in the adrenals and gonads.

Formation of 17α-Hydroxypregnenolone, 17α-Hydroxyprogesterone, Dehydroepiandrosterone, and Androstenedione

After formation of pregnenolone and progesterone, each of these compounds can undergo hydroxylation at carbon 17 and subsequent cleavage of the side chain (carbons 20, 21) at carbon 17, forming dehydroepiandrosterone (DHEA) and androstenedione, respectively. 17α-Hydroxylation occurs through the action of 17α-hydroxylase, whereas $C_{17,20}$-lyase catalyzes the side-chain cleavage. It is now well documented that both enzymatic activities are catalyzed by a single protein, P450c17 encoded by the *CYP17* gene.[28] Since both 17α-hydroxylation and cleavage of the bond between carbons 17 and 20 are obligatory reactions in the biosynthesis of androgens and estrogens, the P450c17 protein is of fundamental importance in reproductive biology.

Cloning of cDNA from human adrenals and gonads, and subsequent nuclease protection experiments, established that there was only one species of human P450c17 mRNA, which is identical in both the adrenals and gonads.[29] Subsequent gene cloning confirmed that in humans there is only one *CYP17* gene,[30,31] which lies on chromosome 10q24.3.[32-34] Additional functional genetic evidence supporting the single P450c17 concept was provided by studies showing ablation of both 17α-hydroxylase and $C_{17,20}$-lyase activities due to mutations of this gene.[35-37]

The two activities of the *CYP17* gene appear to be regulated independently.[28] In testicular Leydig cells and ovarian theca cells, $C_{17,20}$-lyase activity is very high so that the C_{21} steroid precursors are converted to C_{19} products, with little residual 17-hydroxylated C_{21} product. In contrast, the zona fasciculata of the adrenal produces large amounts of 17-hydroxylated C_{21} steroids and very little androgen. Increasing the molar ratio of electrons to P450c17 increases the ratio of $C_{17,20}$-lyase activity to 17α-hydroxylase activity.[28] In addition, P450c17 must be phosphorylated on serine and threonine residues by a cAMP-dependent protein kinase to acquire $C_{17,20}$-lyase activity. On the basis of these findings, it has been proposed that the ratio of $C_{17,20}$-lyase to 17α-hydroxylase activity is regulated by the availability of electrons flowing to the enzymes[28]. Electron flow can be increased by an increase in molar concentrations of the flavoprotein that carries electrons to P450c17. Alternatively the $C_{17,20}$-lyase/17α-hydroxylase ratio may be regulated by alteration of serine/threonine phosphorylation.

Formation of 5-Androstene-3β,17β-Diol and Testosterone

5-Androstene-3β,17β-diol and testosterone are formed from DHEA and androstenedione, respectively, through the action of 17β-hydroxysteroid oxidoreductase, which is commonly referred to as 17β-hydroxysteroid dehydrogenase (17β-HSD); the reactions are reversible. Five isoforms of the enzyme, encoded by the *17β-HSD* gene, have been described in humans, each having cell-specific expression, substrate specificity, regulatory mechanisms, and reductase or oxidative catalytic activities. They are designated types 1–5 in the chronological order of their isolation. Chromosomal locations, enzymatic activities, and cellular distributions of the human 17β-HSD isoenzymes vary (see review by Andersson and Moghrabi[38]).

The 17β-HSD type 1 isoenzyme, originally referred to as placental estradiol-17β-dehydrogenase, was the first 17β-HSD to be purified to homogeneity and cloned.[23,39,40] The gene encoding 17β-HSD type 1 is located at chromosome 17q21, which is in close proximity to the *BRCA1* locus. The enzyme is a soluble protein with substrate preference for estrogens.[41] Its affinity for C_{18} steroids is approximately 100 times higher than for C_{19} steroids.[41] Although 17β-HSD type 1 is localized predominantly in the ovary (granulosa cells)[42] and placenta (syncytiotrophoblasts),[43] it is also expressed in some malignant epithelial cells of the breast[44] and endometrium.[45] The enzyme utilizes NADPH as cofactor,[40] and its catalytic preference is reduction.[46,47]

17β-HSD type 2, which is encoded by the gene located at chromosome 16q24,[48,49] is a microsomal enzyme[50] that preferentially utilizes NAD^+ as cofactor and catalyzes the oxidation of steroids with a hydroxyl group at carbon 17 (e.g., testosterone, estradiol) or carbon 20 (e.g., 20α-dihydroprogesterone).[50] The enzyme is distributed among many extraglandular tissues, e.g., endometrium, placenta, and liver; however, it is primarily expressed in the endometrium.[48] The level and specific activity of 17β-HSD type 2 are increased during the luteal phase of the menstrual cycle in a manner that parallels circulating progesterone levels during this period.[48]

The 17β-HSD type 3 isoenzyme is predominantly expressed in the testis.[51] It is encoded by the gene located at chromosome 9q22 and localized in the microsomal fraction of testicular homogenates.[38] The preferred cofactor for this isozyme is NADPH, which preferentially catalyzes the reduction of androstenedione to testosterone at carbon 17.[52]

Deficiency of 17β-HSD type 3 causes a form of male pseudohermaphroditism referred to as 17β-HSD deficiency,[53] in which there is a deficiency in the biosynthesis of testosterone from androstenedione. The deficiency is confined to individuals with a 46 XY karyotype; these individuals have testes, wolffian duct–derived male internal genitalia (with the exception of a prostate), female external genitalia, and gynecomastia.[54,55]

Considerably less is known about the more recently characterized isozymes, 17β-HSD type 4[56] and 17β-HSD type 5.[57] The type 4 isozyme appears to be distributed in many different tissues and localized in peroxisomes. It catalyzes the oxidation of C_{18} steroids, utilizing NAD^+ as cofactor. In contrast, the type 5 isozyme is located in cytosol liver and skeletal muscle and catalyzes the reduction of C_{19} and C_{21} steroids, utilizing NADPH as cofactor. The isozyme is encoded by a gene located at chromosome 10p14,15.

Studies show that 17β-HSDs are important in the regulation of estrogen action and preservation of tissue levels of progesterone. In the human endometrium, this is accomplished by oxidation of estradiol to estrone and of 20α-dihydroprogesterone to progesterone, respectively. The NAD^+-dependent oxidation of estradiol to estrone and conversion of 20α-dihydroprogesterone to progesterone are much greater in human endometrium during the luteal phase of the ovarian cycle, when progesterone production is high, compared to the follicular phase, which is characterized by low progesterone production.[58] Addition of progesterone 20α-dihydroprogesterone, 5α-androstene-3β,17β-diol, testosterone, or estradiol to human endometrial explants causes an increase in the oxidative activity of both the 17β-HSD and 20α-HSD enzymes.[59,60] Because of the tissue-specific expression and substrate specificity of 17β-HSDs, cells in peripheral (nonendocrine gland) tissues are provided with the necessary mechanisms to control levels of intracellular androgens and/or estrogens. This hormonal control has been termed "intracrinology" by Labrie and coworkers.[61] Through intracrine activity, androgens and/or estrogens that are produced locally exert their action inside the same cells where they are synthesized. Therefore, due to their pivotal role in the formation of active steroid hormones, 17β-HSDs allow each peripheral cell to regulate its own development, growth, and function.

Formation of Estrogens

Biosynthesis of estrone and estradiol occurs from androstenedione and testosterone through the action of a complex microsomal aromatase cytochrome P-450 (P450arom), encoded by the *CYP19* gene. The reaction involving each of the androgens to the corresponding estrogens is known as *aromatization* since it involves conversion of the A ring of the androgens to an aromatic ring. Details of most of the aromatization mechanisms have been elucidated.[62] Briefly, the following steps are involved, beginning with androstenedione as precursor. First, the methyl group at carbon 19 becomes hydroxylated, giving rise to 19-hydroxyandrostenedione. This is followed by a second hydroxylation step at the same carbon, yielding 19,19-dihydroxyandrostenedione, which may reversibly dehydrate to form 19-ketoandrostenedione. Finally, either or both of these intermediates may then undergo simultaneous breaking of the bond between carbons 10 and 19, as well as that between the 1β and 2β hydrogen bonds at carbons 1 and 2, respectively. The end product is estrone, which has an aromatic A ring without a methyl group at carbon 10. The enzymatic steps involved in the conversion of androstenedione to estrone are similar when testosterone is the precursor yielding estradiol. Estrone and estradiol are interconvertible through the action of specific 17β-HSDs.

CYP19 has been cloned and characterized. It is unusual compared to genes encoding other P-450 enzymes because of its tissue-specific expression, which appears to be regulated by tissue-specific promoters.[63] This conclusion is based on the presence of specific 5' termini in the transcripts encoding P450arom in tissues such as the ovary, placenta, and breast, as well as in adipose tissue. Thus, transcripts that are specific for proximal promoter II are found in the ovary, whereas transcripts specific for distal promoter 1.1 are found uniquely in the placenta. However, adipose tissue contains two species of transcripts containing 1.3- and 1.4-specific sequences.

Tissue-specific regulation of human *CYP19* expression is consistent with tissue-specific biosynthesis of predominant estrogens. For example, the ovary synthesizes primarily estradiol, whereas the predominant estrogens in the placenta and adipose tissue are estriol and estrone, respectively. This tissue specificity may reflect the type of substrate presented to the aromatase enzyme and/or unknown factors.

Formation of Corticosteroids

On the basis of their biological activity, corticosteroids can be divided into two types, mineralocorticoids and glucocorticoids. The mineralocorticoid biosynthetic pathway, leading to the formation of aldosterone, involves three sequential hydroxylation reactions. First, progesterone is converted to two compounds with mineralocorticoid activity, through the action of 21-hydroxylase and 11β-hydroxylase: 11-deoxycorticosterone (DOC) and corticosterone, respectively. The latter compound has approximately 10 times more mineralocorticoid activity than DOC. Subsequently, corticosterone is transformed to 18-hydroxycorticosterone via 18-hydroxylase. This product then undergoes oxidation of the hydroxyl group at carbon 18 through the action of the enzyme aldehyde synthetase (18-oxidase), to form aldosterone. This mineralocorticoid is about 10 times more potent than corticosterone.

In contrast to the mineralocorticoid pathway, the glucocorticoid pathway, leading to formation of cortisol, begins with 17-hydroxyprogesterone. The latter compound is converted to 11-deoxycortisol, which is then transformed to cortisol through the action of 21-hydroxylase and 11β-hydroxylase, respectively. These reactions are analogous to the formation of corticosterone from progesterone. Cortisol has high glucocorticoid activity, in contrast to 11-deoxycortisol, which lacks significant amounts of this activity.

21-Hydroxylase

The enzyme 21-hydroxylase is located in the endoplasmic reticulum and encoded by the *CYP21* gene, which is located on chromosome 6p21.3.[64,65] The *CYP21* gene is formed adjacent to and alternating with the *C4A* and *C4B* genes, which encode the fourth component of serum complement.[64,65] Their nucleotide sequences are 98% identical in exons and 96% identical in introns.[66,67] The preferred substrate for this enzyme is 17-hydroxyprogesterone.[68]

The 21-hydroxylase enzyme is 36% identical in amino acid sequence to the 17-hydroxylase/C$_{17-20}$-lyase enzyme.[29,69] In addition, their genes have a similar exon–intron organization, which suggests that both genes evolved from a common ancestor.[30]

11β-Hydroxylase

Humans have two distinct 11-hydroxylase isoenzymes, CYP11B1 and CYP11B2, both of which are encoded by two genes[70] on chromosome 8q21–q22[71] located about 40 kb apart.[72,73] Each gene consists of nine exons, and the location of introns in each gene is identical to that of the *CYP11A* gene, which encodes the cholesterol side-chain cleavage enzyme P450scc. The predicted sequences of the CYP11B isoenzymes are each about 36% identical to P450scc. Transcription of CYP11B1 is regulated primarily by corticotropin, whereas angiotensin II regulates CYP11B2 transcription.

The CYP11B isoenzymes are mitochondrial cytochrome P-450 enzymes. CYP11B1 is expressed at high levels and CYP11B2 at low levels in normal adrenal glands.[70] The latter isoenzyme is dramatically increased in aldosterone-secreting tumors. In vitro studies show that both isoenzymes can convert DOC and 11-deoxycortisol to corticosterone and cortisol, respectively.

18-Hydroxylase and 18-Oxidase

The CYP11B2 isoenzyme also has 18-hydroxylase and 18-oxidase activities, which convert corticosterone to aldosterone.[74–76] In contrast, the CYP11B1 isoenzyme does not form detectable amounts of aldosterone from either corticosterone or 18-hydroxycorticosterone.

TRANSPORT OF STEROID HORMONES

Certain steroid-binding proteins in the blood play an important role in the availability of endogenous and exogenously administered steroid hormones for target and metabolism cells.[77,78] Following their secretion into the circulation, certain steroids bind with high affinity ($K_a = 1 \times 10^8$ to 1×10^9) but low capacity to sex hormone–binding globulin (SHBG) or to corticosteroid-binding globulin (CBG). In addition, all steroids bind with low affinity ($K_a = 1 \times 10^4$ to 1×10^6) and high capacity to albumin. Roughly 95% of all circulating steroids are protein-bound; the remainder is unbound ("free"). Although steroid hormones can also bind to other circulating proteins, the most relevant steroid-binding proteins are SHBG, CBG, and albumin.

Sex Hormone–Binding Globulin

Properties

The glycoprotein SHBG is produced and secreted by hepatocytes.[79] It has a molecular weight of approximately 90,000 daltons, and its primary structure is known from both direct sequencing[80] and the cloning of its cDNA.[81–83] The gene encodes a single polypeptide chain, which is then glycosylated. The active binding unit is a homodimer, which contains three oligosaccharide chains that vary in content.[80,84] Each mole of the homodimer binds a single mole of steroid. SHBG has a half-life of approximately 6 days and binds with high affinity to three major circulating steroids.[78] It binds with highest affinity to dihydrotestosterone (DHT), followed by testosterone and then estradiol (K_a of 5.5×10^9, 1.6×10^9, and 6.8×10^8, respectively).[77]

Physiological Role

Traditionally, it has been thought that both SHBG and albumin are primary determinants of free steroid and that only free steroid, e.g., testosterone, is available for tissue uptake, i.e., for metabolism and target cells. This view was questioned when, in a rat model, it was demonstrated that both albumin-bound and free steroid, i.e., the non-SHBG-bound steroid fraction, are bioavailable.[85] Although this theory is consistent with the fact that albumin binds to steroids with low affinity, there is evidence that contradicts it.[86] Nevertheless, over the last 10 years or so, there has been a growing contention that the non-SHBG-bound fraction of biologically active steroids is important for metabolism and biological action in target cells. This view makes SHBG the primary regulator of the amount of steroid available for tissue uptake.

Regulation

Sex hormone–binding globulin is regulated in a complex fashion, which is poorly understood. Originally, it was thought that the major regulators of SHBG biosynthesis were estrogens and androgens. This hypothesis was based on the observation that normal women have significantly higher serum SHBG levels than normal men. However, subsequently, a number of studies showed that estrogens and androgens do not always account for the significant changes in circulating SHBG levels observed in clinical situations. For example, although patients with polycystic ovarian syndrome generally have elevated androgens and decreased SHBG levels, some have normal SHBG levels.[87,88] Similarly, the effect of estrogens on SHBG levels is inconsistent. During the latter half of pregnancy, SHBG levels rise dramatically as estrogens increase. However, in infancy, SHBG levels also rise in spite of low circulating estrogens; and in puberty, levels fall as estrogens increase in girls.

The concentration of circulating SHBG is the net result of its entry and exit from the plasma compartment. In most instances, it is difficult to determine whether changes in circulating SHBG levels are due to changes in its secretion or clearance. Although a number of factors affect circulating SHBG levels, estrogens and thyroid hormone are by far the most important. In addition, there is evidence that insulin plays an important role in SHBG regulation.[89]

Corticosteriod-Binding Globulin

Like SHBG, CBG is a glycoprotein, but it is a much smaller molecule than SHBG; it has a molecular weight of 52,000 daltons. Cloning of the cDNA of CBG demonstrated that it consists of a single polypeptide.[78] The polypeptide undergoes glycosylation, yielding five carbohydrate chains, which represent 18% of the total content of the CBG molecule. This globulin has one binding site and binds with high specificity but low capacity to corticosteroids and progesterone. The half-life of CBG has been reported to be 5 days.[78]

CBG arises from synthesis and secretion by hepatocytes where both the protein and its mRNA are present in high concentrations. Unlike SHBG, the circulating level of CBG is altered by a limited number of hormonal influences. The most important of these influences are estrogens, which result in significant increases in the concentration of CBG in the circulation.

METABOLISM OF STEROIDS

Steroid hormones undergo extensive metabolism by biochemical reactions in a variety of peripheral tissues, which include the liver, kidney, genital and nongenital skin, prostate, as well as adipose tissue. However, the liver is the primary site of steroid metabolism. Extensive steroid metabolism is due to the fact that steroids contain functional groups (e.g., double bond, ketone group, hydroxyl group) that are vulnerable to reduction or oxidation. Reduction of double bonds and ketone groups gives rise to hydrogens and hydroxyl groups that are in either the α or β orientation. Consequently, multiple isomers of a metabolite can be formed, as with progesterone metabolites. In addition to steroid metabolism involving oxidation/reduction reactions, estrogens are especially vulnerable to hydroxylation reactions, which may occur on most of the carbons of the estrogen molecule. A third important source of steroid metabolites results from conjugation reactions. For a steroid to be eliminated from the body, it must be transformed biochemically from its lipophilic form to a water-soluble form. This is achieved by forming a sulfate or glucuronide derivative of the steroid, a process referred to as *conjugation*. The conjugation reaction occurs at a hydroxyl group of the steroid molecule.

Reductases and Dehydrogenases

Although reductases and dehydrogenases are found in a variety of body tissues, their concentration predominates in certain tissues, e.g., the liver. Already discussed were 3β-HSD and 17β-HSD. The present section will discuss the 5α-reductases and 3α-HSDs. Little is known about the 5β-reductases.

5α-Reductases

Molecular cloning studies have revealed the existence of two genes that encode isoenzymes of 5α-reductases. The isoenzymes are designated types 1 and 2 and have distinct biochemical properties and tissue distributions. They are encoded by the *SRD5A1* and *SRD5A2* genes, respectively. The type 1 isoenzyme is found primarily in the liver and skin. It is optimally active at an alkaline pH. In contrast, the type 2 isoenzyme predominates in urogenital tissues and is optically active at an acidic pH. The *SRD5A2* gene is located on chromosome 2 (2p23)[90] and spans over 40 kb of genomic DNA, with five exons and four introns.[91] The type 2 isoenzyme plays a major role in prostate development and disease and is highly sensitive to inhibition by finasteride. The therapeutic benefit of finasteride in the treatment of benign prostatic hyperplasia is due to its inhibitory effect on the type 2 isoenzyme, which predominates in the prostate.

3α-Hydroxysteroid Dehydrogenases

Mammalian 3α-HSDs are expressed by genes that are members of the aldoketoreductase (AKR) 1C superfamily.[92] These enzymes work together with the 5α- and 5β-reductases to form 3α,5α- and 3α,5β-tetrahydrosteroids. In steroid target tissues, 3α-HSDs function as molecular switches and regulate steroid hormone action.

Four highly related members of the *AKR1C* gene family (*AKR1C1*, *AKR1C2*, *AKR1C3*, *AKR1C4*) have been characterized. These genes have high nucleotide sequence homology but differ with respect to the tissue in which they are expressed and the reactions that they catalyze.[93] The *AKR1C1* gene was previously referred to as *20α-HSD* and *DDH1* (dihydrodiol dehydrogenase).[94,95] The enzyme that it expresses has 20α-HSD activity. The *AKR1C2* gene expresses an enzyme that was originally shown to have high-affinity binding for the bile salts using gel-filtration chromatography of human liver cytosol.[96] Subsequently, the enzyme was purified and its gene cloned and identified as 3α-HSD type 3.[95] The gene is expressed in multiple tissues, including the liver and hormone-responsive tissues such as the prostate and breast.[97] The AKR1C2 enzyme has high affinity for DHT and may be the predominant 3α-HSD that reduces DHT in the prostate. Although the AKR1C3 enzyme was originally identified as 3α-HSD type 2, it has low affinity for DHT and is currently considered to be 17β-HSD type 5.[98] Finally, the AKR1C4 enzyme, originally identified as 3α-HSD type 1, has the highest affinity for DHT but has been identified only in the liver.[99]

Hydroxylases and Methyltransferases

Although progesterone, androgens, estrogens, and corticosteroids can undergo metabolism by hydroxylation, this reaction is predominant with estrogens in the body. The enzymes involved in the hydroxylation of estradiol and estrone are members of the cytochrome P-450 family and act in conjunction with NADPH. Although most of the oxidative metabolism of estradiol and estrone occurs in the liver, some estrogen-metabolizing isoforms of P-450 are selectively expressed in certain extrahepatic tissues of estrogen metabolism, e.g., mammary gland, uterus, and brain. The function of NADPH-dependent hydroxylation of estradiol and estrone by multiple cytochrome P-450 enzymes in target tissues or cells is largely unknown. However, certain hydroxylated estrogen metabolites may possess important biological functions that are not directly associated with the parent hormone.[100] Furthermore, some of the biological effects exerted by these estrogens may be mediated by specific intracellular receptors that are different from the classical estrogen receptors.[100]

The 2- and 4-hydroxylated estrogens undergo methylation, which is catalyzed by the enzyme catechol-*O*-methyltransferase (COMT). This enzyme also catalyzes the methylation of catecholamines and other catechols. Its gene has been mapped to 22q11.1–q11.2.[101] It is present in large amounts in the liver and kidney and is also found in significant amounts in the endometrium and mammary glands.

Conjugation and Deconjugation of Steroids

There are two major mechanisms by which steroid hormones are conjugated; they involve the formation of sulfated and glucuronidated steroids. The sulfation reaction occurs in three steps: first, adenosine triphosphate (ATP) is sulfated by sulfate ions (SO_4^{2-}) in the presence of the enzyme ATP sulfurylase, yielding adenosine-5′-phosphate (APS); second, APS then reacts with ATP through the action of APS kinase to give the active sulfate phosphoadenosine-5-phosphosulfate (PAPS); third, PAPS reacts with the hydroxyl group of a steroid, in the presence of the enzyme sulfuryl transferase. In humans, there are three members of the phenol sulfotransferase gene family: *SULT1A1*, *SULT1A2*, and *SULT1A3*.[102–104] These genes are highly homologous and colocalized on the short arm of chromosome 16. Sulfuryltransferase activity is found in the soluble fractions of cells of the liver, adrenal cortex (zona fasciculata and zona reticularis but not zona glomerulosa), and testis.

Glucuronidation is a reaction that involves the transfer of the glucuronide moiety of uridine diphosphoglucuronic acid (UDPGA) to the steroid in the presence of the enzyme glucuronyl transferase. This enzyme is encoded by the *UGT1A* gene. One member of this family of genes (*UGT1A1*) has been shown to be involved in estradiol metabolism.[105]

Very little is known about the genes responsible for the hydrolysis of steroid sulfates and glucuronides. There are two members of the sulfatase gene family, *ARSC1* and *ARSC2*, and the β-glucuronidase enzyme is encoded by the *GUSB* gene.

Metabolism of Progesterone

As mentioned earlier, progesterone undergoes extensive reduction of its double bond and ketone groups (at carbons 3 and 20). The reduced products include 5α- and 5β-pregnanedione, four different pregnanolone isomers (3α-hydroxy-5α-pregnan-20-one, 3β-hydroxy-5α-pregnan-20-one, 3α-hydroxy-5β-pregnan-20-one, 3β-hydroxy-5β-pregnan-20-one), and eight different pregnanediol isomers (5α-pregnane-3α,20α-diol, 5α-pregnane-3α,20β-diol, 5α-pregnane-3β,20α-diol, 5α-pregnane-3β,20β-diol, 5β-pregnane-3α,20α-diol, 5β-pregnane-3α,20β-diol, 5β-pregnane-3β,20α-diol, 5β-pregnane-3β,20β-diol). The enzymes involved in the formation of these products include 5α-reductase, 5β-reductase, 3α-HSD, 3β-HSD, 20α-HSD, and 20β-HSD.

Progesterone can also undergo hydroxylation, and the hydroxylated molecule may undergo reduction, as described earlier. The reduced, hydroxylated, and/or reduced and hydroxylated progesterone molecule then undergoes conjugation. The principal metabolite of progesterone in urine is 5β-pregnane-3α,20α-diol glucuronide (pregnanediol glucuronide). Urinary excretion of this metabolite correlates highly with serum progesterone levels.[106]

Metabolism of Androgens

Dihydrotestosterone Formation

Peripheral tissues have the capacity to transform testosterone to the more potent androgen DHT, through the action of the enzyme 5α-reductase. In peripheral tissues, DHT can also be formed by another pathway (Fig. 2.3), beginning with androstenedione, which is interconvertible with testosterone through the action of the enzyme 17β-HSD. Androstenedione can undergo 5α reduction in the same manner as testosterone, yielding 5α-androstane-3,17-dione, which can then be converted to DHT via 17β-HSD activity in a reaction that is reversible. In men the preferred pathway of DHT formation is from testosterone, whereas in women the formation of DHT via 5α-androstane-3,17-dione is the more important pathway.[107]

Dihydrotestosterone Metabolism

Dihydrotestosterone is metabolized rapidly in peripheral tissues. Different enzymes can use DHT as a substrate (Fig. 2.3). It can undergo reduction of the 3-ketone group to form either 5α-androstane-3α,17β-diol (also referred to as 3α-androstanediol or 3α-diol) via the enzyme 3α-HSD or 5α-androstane-3β,17β-diol (3β-androstanediol or 3β-diol) via the enzyme 3β-HSD. These reactions are reversible. In vivo, the back-conversion of 3α-diol to DHT is greater than 50%.[108] The former metabolite (3α-diol) is considered to have some androgenic activity.[109]

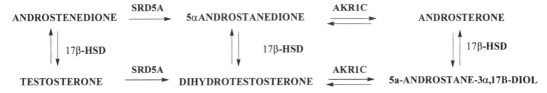

Figure 2.3 Androgen formation in peripheral tissues.

For this reason, it has been studied more extensively than its 3β epimer.

In addition to its conversion to DHT, 5α-androstane-3,17-dione can be transformed reversibly to androsterone through the action of 3α-HSD. Furthermore, androsterone can be transformed reversibly to 3α-diol via the enzyme 17β-HSD. In both men and women, serum levels of 3α-diol and androsterone are very low.

In addition to the transformations of DHT just described, DHT can be conjugated to form DHT glucuronide and DHT sulfate. Serum levels of the latter metabolite in men are relatively high (3–5 ng/ml), whereas serum DHT glucuronide levels are relatively low (Stanczyk et al., unpublished data).

All of the compounds with a hydroxyl group shown in Figure 2.3 are conjugated in peripheral tissues. The main conjugates formed from the androgens depicted in Figure 2.3 are shown in Table 2.1. Since 3α-diol has two hydroxyl groups, it is possible to form a glucuronide or sulfate group at either of the hydroxylated carbons. The predominant glucuronide form of 3α-diol is 3α-diol 17β-glucuronide.[110] This conjugate is usually referred to as 3α-androstanediol glucuronide or simply 3α-diol G.

Table 2.1 Conjugated Androgens Formed from Testosterone and Its 5α-Reduced Metabolites

Precursor	Conjugate
Testosterone	Testosterone glucuronide
	Testosterone sulfate
Dihydrotestosterone (DHT)	DHT glucuronide
	DHT sulfate
3α-Androstanediol (3α-diol)	3α-Diol-3α-glucuronide
	3α-Diol-17β-glucuronide
	3α-Diol-3α-sulfate
	3α-Diol-17β-sulfate
Androsterone	Androsterone glucuronide
	Androsterone sulfate

Circulating Markers of Dihydrotestosterone

With increasing evidence that DHT is formed peripherally, it was expected that measurements of this androgen in blood would provide special insight into disorders of peripheral androgen formation and action; however, the results have been disappointing. This is evident in peripheral disorders of DHT formation in genetic males, such as 5α-reductase deficiency or androgen resistance. In these disorders, plasma DHT levels can be normal or even increased, instead of being low.[111] This may be explained by the relatively low circulating DHT levels and very high affinity of SHBG for DHT.

The next androgen that was considered to be a potential marker of DHT formation was 3α-diol. As stated earlier, this compound not only is a direct metabolite of DHT but also has androgenic activity. However, this compound also did not prove to be a useful marker of DHT formation, probably because circulating 3α-diol levels are very low (<150 pg/ml), and it is not easy to separate the various peripheral androgenic disorders with this measurement.

Serum levels of 3α-diol G reflect 5α-reductase activity and peripheral androgen action.[111] Furthermore, 3α-diol G appears to reflect specifically 5α-reductase type 2 activity. Our data show that the decrease in serum DHT levels following finasteride treatment in men correlates highly with decreased serum 3α-diol G levels.[112]

Metabolism of Estrogens

Estradiol and estrone are metabolized primarily by hydroxylation at different carbons of the steroid nucleus. Hydroxylation occurs at carbons 1, 2, 4, 6, 7, 11, 14, 15, 16, and 18, forming a variety of metabolites (Table 2.2). In some in-

Table 2.2 Hydroxylated and Oxygenated Metabolites formed From Estradiol and Estrone

Position on Carbon	Estrogen Metabolite	Position on Carbon	Estrogen Metabolite
C-1	1-Hydroxyestrone	C-11	11β-Hydroxyestrone
			11-Ketoestrone
C-2	2-Hydroxyestrone		11β-Hydroxyestradiol
	2-Hydroxyestradiol		11-Ketoestradiol
	2-Hydroxyestriol		Δ(11)-Dehydro-17α-estradiol
			Δ(9,11)-Dehydroestrone
C-4	4-Hydroxyestrone		
	4-Hydroxyestradiol	C-14	14α-Hydroxyestrone
	4-Hydroxyestriol		14α-Hydroxyestradiol
C-6	6α-Hydroxyestrone	C-15	15α-Hydroxyestrone
	6β-Hydroxyestrone		15β-Hydroxyestrone
	6-Ketoestrone		15α-Hydroxyestradiol
	6α-Hydroxyestradiol		15α-Hydroxyestriol (estetrol)
	6β-Hydroxyestradiol		
	6-Ketoestradiol	C-16	16α-Hydroxyestrone
	6α-Hydroxyestriol		16β-Hydroxyestrone
	6-Ketoestriol		16-Ketoestrone
			16α-Hydroxyestradiol (estriol)
C-7	7α-Hydroxyestrone		16-Epiestriol
	7β-Hydroxyestrone		16-Ketoestradiol
	7α-Hydroxyestradiol		16,17-Epiestriol
	7β-Hydroxyestradiol		
	7-Ketoestradiol	C-17	17α-Estradiol
	7α-Hydroxyestriol		17-Epiestriol
		C-18	18-Hydroxyestrone

stances, the hydroxyl group is also oxidized to form a ketone group. Hydroxylation of carbon 2 or 4 of estradiol or estrone gives rise to catechol estrogens, which contain a hydroxyl group adjacent to the hydroxyl group at carbon 3 (Fig. 2.4).

Quantitatively, the two most important pathways of metabolism involve 2-hydroxylation and 16α-hydroxylation (Fig. 2.5). Major enzymes for hepatic estrogen 2-hydroxylation are found in the cytochrome P-450 1A2 and 3A fami-

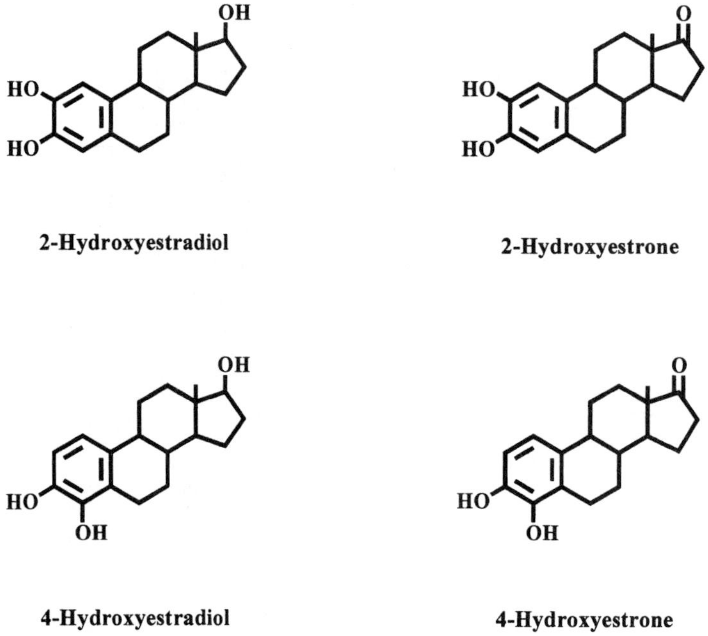

2-Hydroxyestradiol

2-Hydroxyestrone

4-Hydroxyestradiol

4-Hydroxyestrone

Figure 2.4 Catechol estrogens.

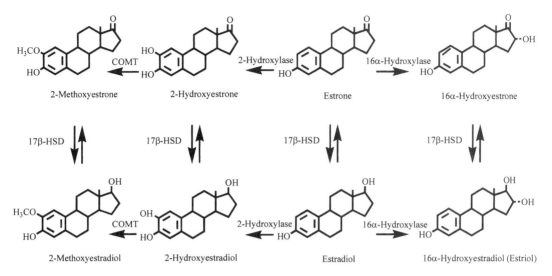

Figure 2.5 Major pathways of estrogen metabolism in humans.

lies.[113–117] Cytochrome P-450 3A is also found in several extrahepatic tissues[118] and may contribute substantially to 2-hydroxylation in these tissues. P-450 3A4 is one of the most abundant P-450s in human liver and has high 2-hydroxylase activity. The P-450 1A2 and 3A4 enzymes are encoded by the *CYP1A2* and *CYP3A4* genes, respectively.[119,120] The *CYP3A4* gene is located on chromosome 7. The *CYP1A2* gene spans approximately 7.8 kb of genomic DNA and contains seven exons. It is separated from the *CYP1A1* gene by a 23 kb segment containing no other open reading frames.[121] These two genes are in opposite orientation and share a 5′-flanking region, within which are included xenobiotic response elements. Compared to *CYP1A1*, *CYP1A2* has no substantial conservation of structure in exons 2, 4, 5, and 6 when examining both nucleotides and the total number of bases.[119]

Although estradiol and estrone are extensively hydroxylated at carbon 2 in hepatic and extrahepatic tissues, the concentrations of 2-hydroxyestradiol and 2-hydroxyestrone are very low in the systemic circulation.[122–125] This is consistent with the extremely high metabolic clearance rates of 2-hydroxyestradiol and 2-hydroxyestrone that have been reported (12,200 and 39,125 l/day, respectively).[125–127] It is believed that the metabolic clearance rates of 2-hydroxylated estradiol and estrone are so high because of the rapid metabolism of these estrogens to the corresponding 2-methoxy metabolites. It appears that there is extensive methylation of these estrogens within the circulation by red blood cells, which contain COMT.[128]

Estrone and, to a lesser extent, estradiol undergo extensive 16α-hydroxylation to form estriol. Cytochrome P-450 3A4 has strong catalytic activity for 16α-hydroxylation.[115] Once formed, estriol is cleared rapidly in nonpregnant women, so its blood concentration is low but its urinary concentration is relatively high. In contrast to the serum concentration of estriol in nonpregnancy, serum estriol levels in pregnancy are very high, especially in the third trimester. This is due to increased production of estriol originating from the fetoplacental unit. The fetal liver has very high 16α-hydroxylase activity, which converts dehydroepiandrosterone sulfate (DHEAS) to 16α-hydroxy-DHEAS. The latter compound undergoes placental transformation to 16α-hydroxyandrostenedione and subsequently to estriol through the action of the enzymes 3β-HSD-$\Delta^{5,4}$-isomerase and aromatase, respectively. Estriol is then secreted predominantly into the maternal compartment. Because approximately 90% of the precursors for estriol formation originate in the fetus, serum unconjugated estriol levels have been used to monitor fetal well-being.

16α-Hydroxylated estrogens have some unique properties. 16α-Hydroxyestrone and

16α-hydroxyestradiol are able to activate the classical estrogen receptor and have potent hormonal activity.[129] In contrast, 2-hydroxylated estrogens bind more weakly to the estrogen receptor and have lower estrogenic activity.[130,131] It has also been shown that 16α-hydroxyestrone forms a covalent reaction with the estrogen receptor.[132] A two-step reaction is involved in the formation of stable adducts of 16α-hydroxyestrone with amino-containing macromolecules (Fig. 2.6).[133] First, a Schiff base is formed between the estrogen and amino group in a reversible reaction. The Schiff base is highly unstable and rapidly converted to a stable 16-keto-17β-amino estrogen adduct via chemical rearrangement. A preliminary study suggests that 16α-hydroxyestrone may activate classical estrogen receptor–mediated oncogene expression and growth stimulation for a prolonged time.[134]

Association of 16α-Hydroxylated Estrogens with Breast Cancer Risk

On the basis of their in vivo studies, Fishman and colleagues[135] suggested that increased formation of 16α-hydroxylated estrogen metabolites might be associated with increased risk of developing breast cancer. Their initial studies showed that 2- and 16-hydroxylation of estradiol was minimally affected by age and did not differ between premenopausal and postmenopausal women.[129] However, when these enzymatic activities were compared between breast cancer patients ($n = 33$) and matched controls ($n = 10$), 16-hydroxylation was associated with increased risk of breast cancer, whereas the competing 2-hydroxylation pathway was either neutral or associated with decreased risk.[135] The investigators suggested that the breast cancer patients had an increased extent of 16α-hydroxylation prior to the onset of the disease, unless the increase was a consequence of the cancer itself. In a subsequent study, using a murine mammary tumor model, Bradlow and associates[136] reported a close correlation between the extent of tumor incidence and 16α-hydroxylation, but not 2-hydroxylation, of estradiol. Furthermore, they found that the presence of exogenous mouse mammary virus increases 16α-hydroxylation of estradiol and suggested that this provides a potential link between the hormonal and viral elements of tumorigenesis in the mouse.

Telang and colleagues[137,138] also carried out in vitro studies to support their view that 16α-hydroxylation of estrogens is associated with increased breast cancer risk. They found that the extent of 16α-hydroxylation of estradiol is significantly elevated in explant cultures of human terminal duct lobular units. In addition, 16α-hydroxylation is genotoxic and increases cell proliferation in a mouse mammary epithelial cell line.[137,138]

The above studies utilized a radiometric procedure for measuring the enzymatic activity of the oxidative metabolism of estradiol in vivo. This method involves administration of a precursor labeled with tritium at a biochemically reactive carbon site. Upon oxidation or hydroxylation at that site, the tritium is transferred to body water. The extent of a reaction is calculated from the area under the body water/specific activity curve, using an integrated equation that corrects for the rapid turnover of body water in small animals such as mice.[136] Concerns have been raised about the validity of this assay in in vivo

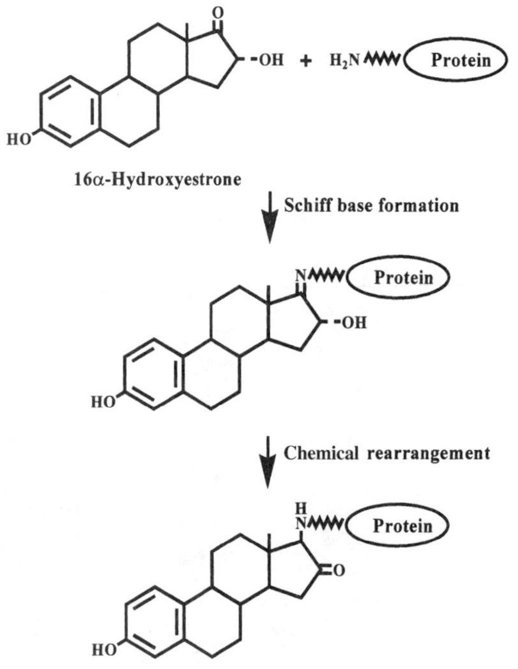

Figure 2.6 Proposed mechanism for the formation of the stable adduct of 16α-hydroxyestrone with protein.

studies that determine the release of tritium from [2-^{3}H]- or [16-^{3}H]-labeled estrogens to quantify the extent of 2- and 16α-hydroxylation, respectively.[100] A substantial amount of nonenzymatic release of tritium from [2-^{3}H]-estradiol in the radiometric assay for 2-hydroxylation of estradiol has been reported.[139–141] Also, since enolization is an essential step for the 16α-hydroxylation of estrone,[142] the tritium at the carbon 16 position may be labile.[100] These findings are supported by data showing that substantial differences exist in the rate of hepatic microsomal 2-hydroxylation of estradiol when measured by radiometric assay compared to gas chromatography–mass spectrometry (GC-MS), which quantifies the isolated estrogen directly.[143]

The 2/16α-Hydroxylated Estrogen–Breast Cancer Risk Hypothesis

On the basis of the studies described above, which showed that increased formation of 16α-hydroxylated metabolites relative to 2-hydroxylated estrogen metabolites may be associated with an elevated risk of breast cancer, it was hypothesized that a low urinary 2-hydroxyestrone to 16α-hydroxyestrone ratio should be inversely associated with breast cancer risk.[144] Development of a competitive-type enzyme immunoassay (EIA) method for quantifying these metabolites in urine[145] allowed the hypothesis to be tested rapidly and relatively inexpensively in a large number of samples. This assay was used in the study by Kabat and associates,[144] who measured the metabolites in spot urine from breast cancer cases ($n = 42$) and controls ($n = 64$), including both premenopausal and postmenopausal women. Although the 2-hydroxyestrone to 16α-hydroxyestrone ratio was not associated with breast cancer overall, the ratio in the postmenopausal group was significantly lower in the cases ($n = 23$) compared to the controls ($n = 28$). In a prospective study by Meilahn and coworkers,[146] in which the 2- and 16α-hydroxylated estrogens were quantified by EIA in spot urine from premenopausal and postmenopausal women, no difference in this ratio was observed between premenopausal cases ($n = 60$) and controls ($n = 184$). A lower (15%) average 2- to 16α-hydroxyestrone ratio was found in postmenopausal breast cancer cases ($n = 42$) compared to matched controls ($n = 139$). However, compared to postmenopausal women in the lowest tertile category of the 2- to 16α-hydroxyestrone ratio, women in the highest tertile had an odds ratio of 0.71 for breast cancer. The confidence interval was wide, and no significant difference was found. In a study by Ho and associates,[147] it was shown that urinary 2-/16α-hydroxyestrone ratios, obtained by EIA measurements, were significantly lower in women with breast cancer ($n = 65$) compared to controls ($n = 36$).

As part of a large methodological study to evaluate the reliability of sex steroid hormone assay reproducibility, the EIA method developed by Klug and coworkers[145] for quantifying 2- and 16α-hydroxyestrone was found to be problematic when these estrogens were analyzed in urine from postmenopausal women.[148] The specific concern was the lack of reproducibility in measuring the analytes at concentrations near or at the limit of assay detection, i.e., 2 ng/ml. In response to this concern, a sandwich-type enzyme-linked immunosorbent assay (ELISA) was developed with a sensitivity of 0.625 ng/ml.[149]

The validity of the assay was tested by Falk and associates[150] using five replicate urine specimens from each of five premenopausal and five postmenopausal women. 2-Hydroxyestrone and 16α-hydroxyestrone were analyzed in one of the replicate urine samples from each subject daily for 5 days. Each sample was analyzed in triplicate, and the results were averaged. The results showed that the assay coefficients of variation ranged from 10% to 20%, and intraclass correlation coefficients ranged from 85% to 95% in both groups. In addition, one urine aliquot from each of the women was used for analysis of 2-hydroxyestrone and 16α-hydroxyestrone by GC-MS. The results from this analysis were compared to corresponding values obtained by ELISA. For the 2-hydroxyestrone measurements in premenopausal women, ELISA values were significantly lower (mean 20.2 vs. 30.8 ng/ml) with the GC-MS method. In contrast, slightly lower levels of the same analyte were found in postmenopausal urine with the GC-MS method (5.2 vs. 6.1 ng/ml). Similar findings were reported for 16α-hydroxyestrone levels in premenopausal samples (mean 12.6 vs. 21.5 ng/ml). However, in the postmenopausal group, this analyte was approximately twofold higher when measured by GC-MS compared to ELISA (mean 6.72 vs. 3.22 ng/ml). Comparison of the ratios of 2-hydroxyestrone to 16α-hydroxyestrone obtained by the two different methods showed that they were similar (1.53 vs. 1.44) in the premenopausal group, but in the postmenopausal group the ratio obtained by GC-MS was approximately twofold greater with ELISA (2.11 vs. 1.02). Although discrepancies were observed in the absolute values of the two estrogen metabolites in both the premenopausal and postmenopausal samples, the values obtained by ELISA correlated well with those determined by GC-MS. Even though the ELISA method for quantifying 2-hydroxyestrone and 16α-hydroxyestrone was found to be generally reproducible and to correlate well with the GC-MS method, overall there was a large discrepancy in the absolute values of the estrogens between the two methods. Furthermore, the number of samples analyzed in the method comparison was small ($n = 10$).

Other epidemiological studies addressing the hypothesis that the ratio of 2-hydroxylation to 16α-hydroxylation of estrone is important in breast cancer development have not been supportive. Adlercreutz and coworkers[151] determined the urinary estrogen profile, consisting of 13 estrogens, by GC-MS twice in 1 year in 10 premenopausal Finnish women with breast cancer and two control groups, one consuming an omnivorous diet ($n = 12$) and the other a lactovegetarian diet ($n = 11$). They found that the ratio of 2-hydroxyestrone to 16α-hydroxyestrone was highest in the breast cancer group and lowest in the vegetarians. Thus, they could not confirm the observation of higher 16α-hydroxylation of estrogens in premenopausal breast cancer patients. Instead, they found that the relative amounts of 2-hydroxylated estrogens in the breast cancer group tended to be higher than in the controls. This finding is supported by data from Lemon and associates,[152] who carried out a retrospective meta-analysis of all epidemiological investigations since 1966 in which urinary excretion measurements of estrone, estradiol, and estriol were obtained from healthy women who were at risk for nonfamilial breast cancer as determined by the authors of each study. Their meta-analysis used data reported in 13 studies in which more than 2800 subjects aged 15 to 59 years were included and similar analytical methods (modified Brown colorimetric method) were used in most instances. A previously described and validated method[153] was used to estimate catechol estrogen excretion mathematically from urinary excretion of estrone, estradiol, and estriol based on the obligate reciprocal relationship between 2-/4-hydroxylase activity and 16α-hydroxylase activity. The meta-analysis showed that catechol estrogen excretion rose exponentially with increasing breast cancer risk and was 78%–97% higher in high-risk women of all ages and menstrual cycle phases.

Two population-based case-control studies that utilized either the EIA or ELISA method for quantifying urinary 2-hydroxyestrone and 16α-hydroxyestradiol also do not support the hypothesis that the ratio of these two metabolites is an important risk factor for breast cancer.[154,155] The results from the study using ELISA for the measurements show that the mean urinary 2-hydroxyestrone and 16α-hydroxyestrone levels were 13.8% and 12.1% higher, respectively, in cases ($n = 66$) compared to controls ($n = 76$). Contrary to the hypothesis, the ratio of 2- to 16α-hydroxyestrone was higher in cases.

One of the factors that has been attributed to the well-recognized lower risk of breast cancer in Asian women compared to non-Asian women is a difference in estrogen metabolism. Thus, studies have been conducted to quantify the levels and relative ratios of urinary 2-hydroxylated and 16α-hydroxylated estrogens in Asian women. Adlercreutz and coworkers[156] used GC-MS to measure these estrogens and compare the levels between 13 Asian premenopausal women (mostly Vietnamese) living in Hawaii and 12 Finnish premenopausal women. They found that compared to the Asian women the Finnish women had a higher extent of 2-hydroxylation and a similar extent of 16α-hydroxylation. The ratio of 2- to 16α-hydroxylation was four- to fivefold higher among the Finnish women than among the Asian women. In a similar study by Ursin and coworkers,[157] the ELISA method was used to quantify hydroxylated estrogens in Chinese ($n = 67$) and African-American or Caucasian-American ($n = 58$) postmenopausal women. No statistical differences in the ratio were found between the Chinese and American women.

On the basis of the above studies, there is insufficient evidence to support the 2-/16α-hydroxyestrone–breast cancer risk hypothesis. Further studies with more reliable assay methods are warranted.

4-Hydroxylation

Although 2-hydroxylation of estradiol and estrone is the predominant pathway of catechol estrogen formation in the liver, small amounts of 4-hydroxylated estradiol and estrone are also formed.[113,158–161] 4-Hydroxylation may be an important pathway for catechol formation in extrahepatic tissues, e.g., breast and uterus.[162,163]

4-Hydroxyestradiol has several important properties. It is similar to estradiol with respect to its availability to bind to and activate the estradiol receptor.[130,164–166] 4-Hydroxyestradiol stimulates uterine growth when injected into animals,[164,166,167] but its uterotropic potency is slightly weaker than that of estradiol.[167] Also, it inhibits COMT-catalyzed O-methylation of catecholamines.[168–170] In addition, 4-hydroxyestradiol undergoes metabolic redox cycling to generate free radicals such as superoxide and chemically reactive estrogen semiquinone/quinone intermediates.[171–173] These metabolic intermediates may damage DNA and other cellular constituents,[174–17] induce cell transformation,[178] and initiate tumorigenesis.[179–181] Furthermore, 4-hydroxyestradiol is a strong carcinogen toward the hamster kidney, under conditions in which 2-hydroxyestradiol is not carcinogenic.[182,183] The potent carcinogenicity of 4-hydroxyestradiol may be due to its potential genotoxicity and its potent growth-stimulating effect.

Both 2-hydroxyestradiol and 2-hydroxyestrone can undergo metabolic redox cycling to generate free radicals such as superoxides and the chemically reactive estrogen semiquinone/quinone intermediates,[171–173] similar to that observed with 4-hydroxyestradiol. However, unlike 4-hydroxyestradiol, 2-hydroxyestradiol and 2-hydroxyestrone have little or no tumorigenic activity toward the male Syrian hamster kidney.[182,183] This may be due to a more rapid rate of metabolism of the 2-hydroxylated estrogens by COMT-catalyzed O-methylation,[184,185] more rapid clearance,[124,186] and lower estrogenicity in target tissues.[164–166,179,187–191] Furthermore, 2-hydroxyestradiol is a potent inhibitor of tumor cell proliferation[192–197] and angiogenesis.[195,197]

CONCLUSIONS

Although pathways of steroid hormone biosynthesis in the adrenals, ovaries, and testes have been known for a number of years, recently only have we begun to understand how genes control the enzymes associated with these pathways. Even less is known about how genes control the enzymatic steps involved in steroid metabolism. This is due to the fact that there are so many metabolites formed, as evident in the discussion on estrogens. Large interindividual differences exist in the metabolism of steroid hormones, which may be reflected in interindividual differences in estrogen action. The formation of so many metabolites raises the question of why it is necessary for the human body to form all of these metabolites. The usual answer to this question is that it is nature's way of detoxifying potent biologically active hormones. However, the metabolites might have important but unrecognized biological effects that are necessary for the action of some hormones. A clearer picture of the molecular mechanisms involved in hormone biosynthesis and metabolism will help us to understand better the role of hormones in health and disease.

REFERENCES

1. Sugawara T, Holt JA, Driscoll D, et al. Human steroidogenic acute regulatory protein: functional activity in COS-1 cells, tissue-specific expression, and mapping of the gene to 8p11.2 and a pseudogene to chromosome 13. Proc Natl Acad Sci USA 92:4778–4782, 1995.
2. Sugawara T, Lin D, Holt JD, et al. Structure of the human steroidogenic acute regulatory protein (StAR) gene: StAR stimulates mitochondrial cholesterol 27-hydroxylase activity. Biochemistry 34:12506–12512, 1995.
3. Clark BJ, Soo SC, Caron KM, Ikeda Y, Parker KL, Stocco DM. Hormonal and developmental regulation of the steroidogenic acute regulatory (StAR) protein. Mol Endocrinol 9:1346–1355, 1995.
4. Sugawara T, Kiriakidou M, McAllister JM, Holt JA, Arakane F, Strauss JF III. Regulation of ex-

pression of the steroidogenic acute regulatory protein *(StAR)* gene: a central role for steroidogenic factor 1. Steroids 62:5–9, 1997.
5. Kim Y-C, Ariyoshi N, Artemenko I, Elliott ME, Bhattacharyya KK, Jefcoate CR. Control of cholesterol access to cytochrome *P450scc* in rat adrenal cells mediated by regulation of the steroidogenic acute regulatory protein. Steroids 62:10–20, 1997.
6. Stocco DM. The steroidogenic acute regulatory (StAR) protein two years later. Endocrine 6:99–109, 1997.
7. Stocco DM, Clark BJ. The role of the steroidogenic acute regulatory protein in steroidogenesis. Steroids 62:29–36, 1997.
8. Papadopoulos V, Amri H, Boujrad N, et al. Peripheral benzodiazepine receptor in cholesterol transport and steroidogenesis. Steroids 62:21–28, 1997.
9. Brown MS, Kovanen PT, Goldstein JL. Receptor-mediated uptake of lipoprotein-cholesterol and its utilization for steroid synthesis in the adrenal cortex. Recent Prog Horm Res 35:215–257, 1979.
10. Jamal Z, Suffolk RA, Boyd GS, Suckling KE. Metabolism of cholesteryl ester in monolayers of bovine adrenal cortical cells. Effect of an inhibitor of acyl-CoA: cholesterol acyltransferase. Biochim Biophys Acta 834:230–237, 1985.
11. Yeaman SJ. Hormone-sensitive lipase—a multipurpose enzyme in lipid metabolism. Biochim Biophys Acta 1052:128–152, 1990.
12. Chung B-c, Guo I-C, Chou S-J. Transcriptional regulation of the *CYP11A1* and ferredoxin genes. Steroids 62:37–42, 1997.
13. Morel Y, Picado-Leonard J, Wu D-A, et al. Assignment of the functional gene for human adrenodoxin to chromosome 11q13 → gter and of adrenodoxin pseudogenes to chromosome 20cen → q13.1. Am J Hum Genet 43:52–59, 1988.
14. Chang C-Y, Wu D-A, Lai C-C, Miller WL, Chung B-c. Cloning and structure of the human adrenodoxin gene. DNA 7:609–615, 1988.
15. Okamura T, Kagimoto M, Simpson ER, Waterman MR. Multiple species of bovine adrenodoxin mRNA: occurrence of two different mitochondrial precursor sequences associated with the same mature sequence. J Biol Chem 262:10335–10338, 1987.
16. Picado-Leonard J, Voutilainen R, Kao L-c, Chung B-c, Strauss JF III, Miller WL. Human adrenodoxin: cloning of three cDNAs and cycloheximide enhancement in JEG-3 cells. J Biol Chem 263:3240–3244, 1988.
17. Chung B-c, Matteson KJ, Voutilainen R, Mohandas TK, Miller WL. Human cholesterol side-chain cleavage enzyme. P450scc: cDNA cloning, assignment of the gene to chromosome 15 and expression in the placenta. Proc Natl Acad Sci USA 83:8962–8966, 1986.
18. Morohashi K, Sogawa K, Omura T, Fujii-Kuriyama Y. Gene structure of human cytochrome *P-450* (SCC), cholesterol desmolase. J Biochem 101:879–887, 1987.
19. Hu MC, Hsu NC, Pai CI, Wang CK, Chung B-c. Functions of the upstream and proximal steroidogenic factor 1 (SF-1)–binding sites in the CYP11A1 promoter in basal transcription and hormonal response. Mol Endocrinol 15:812–818, 2001.
20. Thomas JL, Myers RP, Strickler RC. Human placental 3β-hydroxy-5-ene-steroid dehydrogenase and steroid 5 → 4-ene isomerase: purification from mitochondria and kinetic profiles, biophysical characterization of the purified mitochondrial and microsomal enzymes. J Steroid Biochem 33:209–217, 1989.
21. Cherradi N, Defaye G, Chambaz E. Characterization of the 3β-hydroxysteroid dehydrogenase activity associated with bovine adrenocortical mitochondria. Endocrinology 134:1358–1364, 1994.
22. Sauer LA, Chapman JC, Dauchy RT. Topology of 3β-hydroxy-5-ene-steroid dehydrogenase/Δ5-Δ4-isomerase in adrenal cortex mitochondria and microsomes. Endocrinology 134:751–759, 1994.
23. Luu-The V, Labrie C, Zhao HF, et al. Characterization of cDNAs for human estradiol 17β-dehydrogenase and assignment of the gene to chromosome 17: evidence for two mRNA species with distinct 5′-termini in human placenta. Mol Endocrinol 3:1301–1309, 1989.
24. Rhéaume E, Lachance Y, Zhao H-F, et al. Structure and expression of a new complementary DNA encoding the almost exclusive 3β-hydroxysteroid dehydrogenase/Δ^5-Δ^4-isomerase in human adrenals and gonads. Mol Endocrinol 5:1147–1157, 1991.
25. Lorence MC, Corbin CJ, Kamimura N, Mahendroo MS, Mason JI. Structural analysis of the gene encoding human 3beta-hydroxysteroid dehydrogenase/delta5-4-isomerase. Mol Endocrinol 4:1850–1855, 1990.
26. Lachance Y, Luu-The V, Verreault H, et al. Structure of the human type II 3beta-hydroxysteroid dehydrogenase/delta5-delta4-isomerase (3beta-HSD) gene: adrenal and gonadal specificity. DNA Cell Biol 10:701–711, 1991.
27. Berube D, Luu-The V, Lachance Y, Gagne R, Labrie F. Assignment of the human 3beta-hydroxysteroid dehydrogenase gene *(HSDB3)* to the p13 band of chromosome 1. Cytogenet Cell Genet 52:199–200, 1989.
28. Miller WL, Auchus RJ, Geller DH. The regulation of 17,20 lyase activity. Steroids 62:133–142, 1997.
29. Chung B-c, Picado-Leonard J, Haniu M, et al. Cytochrome *P450c17* (steroid 17α-hydroxylase/17,20 lyase): cloning of human adrenal and testis cDNAs indicates the same gene is ex-

pressed in both tissues. Proc Natl Acad Sci USA 84:407–411, 1987.
30. Picado-Leonard J, Miller WL. Cloning and sequence of the human gene encoding for P450c17 (steroid 17α-hydroxylase/17,20 lyase): similarity to the gene for P450c21. DNA 6:439–448, 1987.
31. Kagimoto M, Winter JS, Kagimoto K, Simpson ER, Waterman MR. Structural characterization of normal and mutant human steroid 17-alpha-hydroxylase genes: molecular basis of one example of combined 17-alpha-hydroxylase/17,20 lyase deficiency. Mol Endocrinol 2:564–570, 1988.
32. Matteson KJ, Picado-Leonard J, Chung B, Mohandas TK, Miller WL. Assignment of the gene for adrenal P450c17 (17α-hydroxylase/17,20 lyase) to human chromosome 10. J Clin Endocrinol Metab 63:789–791, 1986.
33. Sparkes RS, Klisak I, Miller WL. Regional mapping of genes encoding human steroidogenic enzymes: P450scc to 15q23–q24, adrenodoxin to 11q22; adrenodoxin reductase to 17q24–q25; and P450c17 to 10q24–q25. DNA Cell Biol 10:359–365, 1991.
34. Fan YS, Sasi R, Lee C, Winter JSD, Waterman MR, Lin CC. Localization of the human CYP17 gene (cytochrome P450 17a to 10q24.3) by fluorescence in situ hybridization and simultaneous chromosome banding. Genomics 14:1110–1111, 1992.
35. Fardella CE, Hum DW, Homoki J, Miller WL. Point mutation Arg[440] to His in cytochrome P450c17 causes severe 17α-hydroxylase deficiency. J Clin Endocrinol Metab 79:160–164, 1994.
36. Yanase T. 17α-Hydroxylase/17,20 lyase defects. J Steroid Biochem 53:153–157, 1995.
37. LaFlamme N, Leblanc J, Mailloux J, Faure N, Labrie F, Simard J. Mutation R96W in cytochrome P450c17 gene causes combined 17α-hydroxyl/17,20 lyase deficiency in two French Canadian patients. J Clin Endocrinol Metab 81:264–268, 1996.
38. Andersson S, Moghrabi N. Physiology and molecular genetics of 17β-hydroxysteroid dehydrogenases. Steroids 62:143–147, 1997.
39. Peltoketo H, Isomaa V, Maentausta O, Vihko R. Complete amino acid sequence of human placental 17β-hydroxysteroid dehydrogenase deduced from cDNA. FEBS Lett 239:73–77, 1988.
40. Gast MJ, Sims HF, Murdock GL, Gast PM, Strauss AW. Isolation and sequencing of a complementary deoxyribonucleic acid clone encoding human placental 17β-estradiol dehydrogenase: identification of the putative cofactor binding site. Am J Obstet Gynecol 161:1726–1731, 1989.
41. Jarabak J, Sack GHJ. A soluble 17β-hydroxysteroid dehydrogenase from human placenta: the binding of pyridine nucleotides and steroids. Biochemistry 8:2203–2212, 1969.
42. Luu-The V, Labrie C, Simard J, et al. Structure of two in tandem human 17β-hydroxysteroid dehydrogenase genes. Mol Endocrinol 4:268–275, 1990.
43. Martel C, Rhéaume E, Takahashi M, et al. Distribution of 17β-hydroxysteroid dehydrogenase gene expression and activity in rat and human tissues. J Steroid Biochem Mol Biol 41:597–603, 1992.
44. Poutanen M, Isomaa V, Lehto V-P, Vihko R. Immunological analysis of 17β-hydroxysteroid dehydrogenase in benign and malignant human breast tissue. Int J Cancer 50:386–390, 1992.
45. Mäentausta O, Boman K, Isomaa V, Stendahl U, Bäckström T, Vihko R. Immunohistochemical study of the human 17β-hydroxysteroid dehydrogenase and steroid receptors in endometrial adenocarcinoma. Cancer 70:1551–1555, 1992.
46. Dumont M, Luu-The V, De Launoit Y, Labrie F. Expression of human 17β-hydroxysteroid dehydrogenase in mammalian cells. J Steroid Biochem Mol Biol 41:605–608, 1992.
47. Lin SX, Yang F, Jin JZ, et al. Subunit identity of the dimeric 17β-hydroxysteroid dehydrogenase from human placenta. J Biol Chem 267:16182–16187, 1992.
48. Casey ML, MacDonald PC, Andersson S. 17β-Hydroxysteroid dehydrogenase type 2: chromosomal assignment and progestin regulation of gene expression in human endometrium. J Clin Invest 94:2135–2141, 1994.
49. Durocher F, Morissette J, Labrie Y, Labrie F, Simard J. Mapping of the HSD17B2 gene encoding type II 17β-hydroxysteroid dehydrogenase close to D16S422 on chromosome 16q24.1–q24.2. Genomics 25:724–726, 1995.
50. Wu L, Einstein M, Geissler WM, Chan HK, Elliston KO, Andersson S. Expression cloning and characterization of human 17β-hydroxysteroid dehydrogenase type 2, a microsomal enzyme possessing 20α-hydroxysteroid dehydrogenase activity. J Biol Chem 169:12964–12969, 1993.
51. Geissler WM, Davis DL, Wu L, et al. Male pseudohermaphroditism caused by mutations of testicular 17β-hydroxysteroid dehydrogenase 3. Nature Genet 7:34–39, 1994.
52. Inano H, Tamaoki B. Testicular 17β-hydroxysteroid dehydrogenase: molecular properties and reaction mechanism. Steroids 48:1–26, 1986.
53. Andersson S, Russell DW, Wilson JD. 17β-Hydroxysteroid dehydrogenase 3 deficiency. Trends Endocrinol Metab 7:121–126, 1996.
54. Saez JM, de Peretti E, Morera AM, David M, Bertrand J. Familial male pseudohermaphroditism and gynecomastia due to a testicular 17-ketosteroid reductase defect. Studies in vivo. J Clin Endocrinal Metab 32:604–610, 1971.
55. Saez JM, Morera AM, de Peretti E, Bertrand J. Further in vivo studies in male pseudohermaphroditism with gynecomastia due to a testicular 17-ketosteroid reductase defect (com-

pared to a case of testicular feminization). J Clin Endocrinol Metab 34:598–600, 1972.
56. Adamski J, Normand T, Leenders F, et al. Molecular cloning of a novel widely expressed human 80 kDa 17β-hydroxysteroid dehydrogenase IV. Biochem J 311:437–443, 1995.
57. Zang Y, Dufort I, Soucy P, Labrie F, Luu-The V. Cloning and expression of human type V 17β-hydroxysteroid dehydrogenase [abstract]. In: Program, 77th annual meeting of the Endocrine Society 1995, p 622. Abstract nr P3-614.
58. Tseng L, Gurpide E. Estradiol and 20α-dihydroprogesterone dehydrogenase activities in human endometrium during the menstrual cycle. Endocrinology 94:419–423, 1974.
59. Tseng L, Gurpide E. Induction of human endometrial estradiol dehydrogenase by progestins. Endocrinology 97:825–833, 1975.
60. Tseng L, Gurpide E. Stimulation of various 17β- and 20α-hydroxysteroid dehydrogenase activities by progestins in human endometrium. Endocrinology 104:1745–1748, 1979.
61. Labrie F, Luu-The V, Lin S-X, et al. The key role of 17β-hydroxysteroid dehydrogenases in sex steroid biology. Steroids 62:148–158, 1997.
62. Osawa Y. Mechanism of aromatization. In: Scow RO (ed.) Endocrinology. Amsterdam: Excerpta Medica, 1973, p 814.
63. Simpson ER, Mahendroo MS, Means GD, et al. Aromatase cytochrome $P450$, the enzyme responsible for estrogen biosynthesis. Endocr Rev 15:342–355, 1994.
64. Carroll MC, Campbell RD, Porter RR. Mapping of steroid 21-hydroxylase genes adjacent to complement component C4 genes in HLA, the major histocompatibility complex in man. Proc Natl Acad Sci USA 82:521–525, 1985.
65. White PC, Grossberger D, Onufer BJ, et al. Two genes encoding steroid 21-hydroxylase are located near the genes encoding the fourth component of complement in man. Proc Natl Acad Sci USA 82:1089–1093, 1985.
66. Higashi Y, Yoshioka H, Yamane M, et al. Complete nucleotide sequence of two steroid 21-hydroxylase genes tandemly arranged in human chromosome: a pseudogene and a genuine gene. Proc Natl Acad Sci USA 83:2841–2845, 1986.
67. White PC, New MI, Dupont B. Structure of human steroid 21-hydroxylase genes. Proc Natl Acad Sci USA 83:5111–5115, 1986.
68. Tusie-Luna MT, Traktman P, White PC. Determination of functional effects of mutations in the steroid 21-hydroxylase gene *(CYP21)* using recombinant vaccinia virus. J Biol Chem 265:20916–20922, 1990.
69. Bradshaw KD, Waterman MR, Couch RT, et al. Characterization of complementary deoxyribonucleic acid for human adrenocortical 17alpha-hydroxylase: a probe for analysis of 17alpha-hydroxylase deficiency. Mol Endocrinol 1:348–354, 1987.
70. Mornet E, Dupont J, Vitek A, White PC. Characterization of two genes encoding human steroid 11beta-hydroxylase (P-450(11)beta). J Biol Chem 264:20961–20967, 1989.
71. Chua SC, Szabo P, Vitek A, et al. Cloning of cDNA encoding steroid 11beta-hydroxylase (P450c11). Proc Natl Acad Sci USA 84:7193–7197, 1987.
72. Lifton RP, Dluhy RG, Powers M, et al. Hereditary hypertension caused by chimaeric gene duplications and ectopic expression of aldosterone synthase. Nat Genet 2:66, 1992.
73. Pascoe L, Curnow KM, Slutsker L, et al. Glucocorticoid-suppressible hyperaldosteronism results from hybrid genes created by unequal crossovers between CYP11B1 and CYP11B2. Proc Natl Acad Sci USA 89:8327–8331, 1992.
74. Kawamoto T, Mitsuuchi Y, Ohnishi T, et al. Cloning and expression of a cDNA for human cytochrome P-450$_{aldo}$ as related to primary aldosteronism. Biochem Biophys Res Commun 173:309–316, 1990.
75. Ogishima T, Shibata H, Shimada H, et al. Aldosterone synthase cytochrome *P-450* expressed in the adrenals of patients with primary aldosteronism. J Biol Chem 266:10731–10734, 1991.
76. Curnow KM, Tusie-Luna MT, Pascoe L, et al. The product of the *CYP11B2* gene is required for aldosterone biosynthesis in the human adrenal cortex. Mol Endocrinol 5:1513–1522, 1991.
77. Westphal U. Steroid–Protein Interactions II. Berlin: Springer-Verlag, 1986.
78. Rosner W. Plasma steroid-binding proteins. Endocrinol Metab Clin North Am 20:697–720, 1991.
79. Khan MS, Knowles BB, Aden DP, et al. Secretion of testosterone–estradiol-binding globulin by a human hepatoma-derived cell line. J Clin Endocrinol Metab 53:448–449, 1981.
80. Walsh KA, Titani K, Takio K, Kumar S, Hayes R, Petra PH. Amino acid sequence of the sex steroid binding protein of human blood plasma. Biochemistry 25:7584–7590, 1986.
81. Gershagen S, Fernlund P, Lundwall A. A cDNA coding for human sex hormone binding globulin. Homology to vitamin K-dependent protein S. FEBS Lett 220:129–135, 1987.
82. Hammond GL, Underhill DA, Smith CL, et al. The cDNA-reduced primary structure of human sex hormone–binding globulin and location of its steroid-binding domain. FEBS Lett 215:100–104, 1987.
83. Que BG, Petra PH. Characterization of a cDNA coding for sex steroid–binding protein of human plasma. FEBS Lett 219:405–409, 1987.
84. Avvakumov GV, Zhuk NI, Strel'chyonok OA. Subcellular distribution and selectivity of the protein-binding component of the recognition system for sex-hormone-binding protein–estradiol complex in human decidual endometrium. Biochim Biophys Acta 881:489, 1986.

85. Pardridge W. Plasma protein-mediated transport of steroid and thyroid hormones. Am J Physiol 252:E156–E164, 1995.
86. Hobbs C, Hannan CE, Plymate SR. Effect of sex hormone binding globulin and albumin on steroid uptake into rat brain. J Steroid Biochem Mol Biol 42:629–635, 1992.
87. Plymate SR, Fariss B, Bassett M, Matej L. Obesity and its role in polycystic ovary syndrome. J Clin Endocrinol Metab 45:1246–1248, 1981.
88. Dunaif A, Graf M. Insulin administration alters gonadal steroid metabolism independent of changes in gonadotropin secretion in insulin-resistant women with polycystic ovary syndrome. J Clin Invest 83:23–29, 1989.
89. Plymate SR. Regulation of serum sex-hormone-binding globulin in PCOS. Androgen Excess Disord Women 48:497–505, 1997.
90. Thigpen AE, Davis DI, Milatovich A, et al. Molecular genetics of steroid 5α-reductase 2 deficiency. J Clin Invest 90:799–809, 1992.
91. Russell DW, Berman DM, Bryan JT, et al. The molecular genetics of steroid 5α-reductases. Recent Prog Horm Res 49:275–284, 1994.
92. Penning TM. Human 3alpha-hydroxysteroid dehydrogenase isoforms (AKR1C1–AKR1C4) of the aldo-keto reductase superfamily: functional plasticity and tissue distribution reveals roles in the inactivation and formation of male and female sex hormones. Biochem J 351:67–77, 2000.
93. Khanna M. Substrate specificity, gene structure, and tissue-specific distribution of multiple human 3 alpha-hydroxysteroid dehydrogenases. J Biol Chem 270:20162–20168, 1995.
94. Hara A. Relationship of human liver dihydrodiol dehydrogenases to hepatic bile acid–binding protein and an oxidoreductase of human colon cells. Biochem J 313:373–376, 1996.
95. Dufort I. Molecular cloning of human type 3 3α-hydroxysteroid dehydrogenase that differs from 20α-hydroxysteroid dehydrogenase by seven amino acids. Biochem Biophys Res Commun 228:474–479, 1996.
96. Stolz A, Sugiyama Y, Kuhlenkamp J, Kaplowitz N. Identification and purification of a 36 kDa bile acid binder in human hepatic cytosol. FEBS Lett 177:31–35, 1984.
97. Shiraishi H. Sequence of the cDNA of a human dihydrodiol dehydrogenase isoform (AKR1C2) and tissue distribution of its mRNA. Biochem J 334:399–405, 1998.
98. Dufort I. Characteristics of a highly labile human type 5 17beta-hydroxysteroid dehydrogenase. Endocrinology 140:568–574, 1999.
99. Dufort I, Labrie F, Luu-The V. Human types 1 and 3 alpha-hydroxysteroid dehydrogenases: differential lability and tissue distribution. J Clin Endocrinol Metab 86:841–886, 2001.
100. Zhu BT, Conney AH. Functional role of estrogen metabolism in target cells: review and perspectives. Carcinogenesis 19:1–27, 1998.
101. Grossman MH, Emanuel BS, Budaft ML. Chromosomal mapping of the human catechol-O-methyltransferase gene to 22q11.1–q11.2. Genomics 12:822–825, 1992.
102. Falany JL, Falany CN. Expression of cytosolic sulfotransferases in normal mammary epithelial cells and breast cancer cell lines. Cancer Res 56:1151–1155, 1996.
103. Harris RM, Waring RH, Kirk CJ, Hughes PJ. Sulfation of estrogenic alkylphenols and 17β-estradiol by human platelet sulfotransferases. J Biol Chem 275:159–166, 2000.
104. Seth P, Lunetta KL, Bell DW, et al. Phenol sulfotransferases: hormonal regulation, polymorphism, and age of onset of breast cancer. Cancer Res 60:2859–2863, 2000.
105. Guillemette C, De Vivo I, Hankinson SE, et al. Association of genetic polymorphisms in UGT1A1 with breast cancer and plasma hormone levels. Cancer Epidemiol Biomarkers Prev 10:711–714, 2001.
106. Stanczyk FZ, Miyakawa I, Goebelsmann U. Direct radioimmunoassay of urinary estrogen and pregnanediol glucuronides during the menstrual cycle. Am J Obstet Gynecol 137:443–450, 1980.
107. Stanczyk FZ, Matteri RK, Kaufman FR, Gentzschein E, Lobo RA. Androstanedione is an important precursor of dihydrotestosterone in the genital skin of women and is metabolized via 5α-androstanedione. J Steroid Biochem Mol Biol 37:129–132, 1990.
108. Horton R. Testicular steroid transport, metabolism and effects. In: Becker KL (ed). Principles and Practice of Endocrinology and Metabolism, 2nd ed. Philadelphia: JB Lippincott, 1995, pp 1042–1047.
109. Rosness PA, Eik-Nes KB. Biosynthesis of androgens. In: Martini L, Motta M (eds). Androgens and Anti-Androgens. New York: Raven Press, 1977, pp 1–9.
110. Rittmaster RS, Thompson DL, Listwak S, Loriaux DL. Androstanediol glucuronide isomers in normal men and women and in men infused with labeled dihydrotestosterone. J Clin Endocrinol Metab 66:212–216, 1988.
111. Horton R, Lobo R. Peripheral androgens and the role of androstanediol glucuronide. In: Horton R, Lobo RA (eds). Clinics in Endocrinology Metabolism, vol 15. Philadelphia: WB Saunders, 1986, pp 293–306.
112. Stanczyk FZ, Skinner EC, Mertes S, Spahn MF, Lobo RA, Ross RK. Alterations in circulating levels of androgens and PSA during treatment with finasteride in men at high risk for prostate cancer. In: Li JJ, Li SA, Gustafsson J-A, Nandi S, Sekely LI (eds). Hormonal Carcinogenesis II. New York: Springer-Verlag, 1996, pp 404–407.
113. Kerlan V, Dreano Y, Bercovici JP, Beaune PH, Floch HH, Berthou F. Nature of cytochrome P450 involved in the 2-/4-hydroxylations of

estradiol in human liver microsomes. Biochem Pharmacol 44:1745–1756, 1992.
114. Ball SE, Forrester LM, Wolf CR, Back DJ. Differences in the cytochrome *P-450* isoenzymes involved in the 2-hydroxylation of oestradiol and 17α-ethinyloestradiol. Biochem J 267:221–226, 1990.
115. Shou M, Korzekwa KR, Brooks EN, Krausz KW, Gonzalez FJ, Gelboin HV. Role of human hepatic cytochrome *P-450* 1A2 and 3A4 in the metabolic activation of estrone. Carcinogenesis 18:207–214, 1997.
116. Guengerich FP. Oxidation of 17α-ethynylestradiol by human liver cytochrome *P-450*. Mol Pharmacol 33:500–508, 1988.
117. Aoyama T, Korzekwa K, Gillette J, Gelboin HV, Gonzalez FJ. Estradiol metabolism by complementary deoxyribonucleic acid–expressed human cytochrome *P-450s*. Endocrinology 126:3101–3106, 1990.
118. de Waziers I, Cugnenc PH, Yang CS, Leroux JP, Beaune PH. Cytochrome *P-450* isoenzymes, epoxide hydrolase and glutathione transferases in rat and human hepatic and extrahepatic tissues. J Pharmacol Exp Ther 253:387–394, 1990.
119. Ikeya K, Jaiswal AK, Owens RA, Jones JE, Nebert DW, Kimura S. Human *CYP1A2*: sequence, gene structure, comparison with the mouse and rat orthologous gene, and differences in liver 1A2 mRNA expression. Mol Endocrinol 3:1399–1408, 1989.
120. Wrighton SA, Vandenbranden M. Isolation and characterization of human fetal liver cytochrome *P450HLp2*: a third member of the *P450III* gene family. Arch Biochem Biophys 268:144–151, 1989.
121. Corchero J, Pimprale S, Kimura S, Gonzalez FJ. Organization of the *CYP1A* cluster on human chromosome 15: implications for gene regulation. Pharmacogenetics 11:1–6, 2001.
122. Ball P, Emons G, Haupt O, Hoppen H-O, Knuppen R. Radioimmunoassay of 2-hydroxyestrone. Steroids 31:249–258, 1978.
123. Emons G, Ball P, Knuppen R. Radioimmunoassays of catecholestrogens. In: Merriam GR, Lipsett MB (eds). Catecholestrogens. New York: Raven Press, 1983, pp 71–81.
124. Emons G, Merriam R, Pfeiffer D, Loriaux DL, Ball P, Knuppen R. Metabolism of exogenous 4- and 2-hydroxyestradiol in human male. J Steroid Biochem 28:499–504, 1987.
125. Kono S, Brandon DD, Merriam GR, Loriaux DL, Lipsett MB. Low plasma levels of 2-hydroxyestrone are consistent with its rapid metabolic clearance. Steroids 36:463–472, 1980.
126. Longcope C, Femino A, Flood C, Williams KIH. Metabolic clearance rate and conversion ratios of ^{3}H-2-hydroxyestrone in normal men. J Clin Endocrinol Metab 54:347–380, 1982.
127. Kono S, Merriam GR, Brandon D, Loriaux DL, Lipsett MB, Fujino T. Radioimmunoassay and metabolic clearance rate of catecholestrogens, 2-hydroxyestrone and 2-hydroxyestradiol in man. J Steroid Biochem 19:627–633, 1983.
128. Bates GW, Edman CD, Porter JC, MacDonald PC. Metabolism of catechol estrogen by human erythrocytes. J Clin Endocrinol Metab 45:1120–1123, 1977.
129. Fishman J, Martucci CP. Biological properties of 16α-hydroxyestrone: implications in estrogen physiology and pathophysiology. J Clin Endocrinol Metab 51:611–615, 1980.
130. van Aswegen CH, Purdy RH, Wittliff JL. Binding of 2-hydroxyestradiol and 4-hydroxyestrdiol to estrogen receptor human breast cancers. J Steroid Biochem 32:485–492, 1989.
131. Fishman J, Osborne MP, Telang NT. The role of estrogen in mammary carcinogenesis. Ann NY Acad Sci 768:91–100, 1995.
132. Swaneck GE, Fishman J. Covalent binding of the endogenous estrogen 16α-hydroxyestrone to estradiol receptor in human breast cancer cells: characterization and intranuclear localization. Proc Natl Acad Sci USA 85:7831–7835, 1988.
133. Miyairi S, Ichikawa T, Nambara T. Structure of the adduct of 16α-hydroxyestrone with a primary amine: evidence for the Heyns rearrangement of steroidal D-ring α-hydroxyamines. Steroids 56:361–366, 1991.
134. Hsu C-J, Kirkman BR, Fishman J. Differential expression of oncogenes c-*fos*, c-*myc* and neu/Her-2 induced by estradiol and 16α-hydroxyestrone in human cancer cell line [abstract]. In: Program, 73rd annual Endocrine Society meeting, Washington DC, 1991. Abstract nr 586.
135. Schneider J, Kinne D, Fracchia A, Pierce V, Bradlow HL, Fishman J. Abnormal oxidative metabolism of estradiol in women with breast cancer. Proc Natl Acad Sci USA 79:3047–3051, 1982.
136. Bradlow HL, Hershcopf RJ, Martucci CP, Fishman J. Estradiol 16alpha-hydroxylation in the mouse correlates with mammary tumor incidence and presence of murine mammary tumor virus: a possible model for the hormonal etiology of breast cancer in humans. Proc Natl Acad Sci USA 82:6295–6299, 1985.
137. Telang NT, Axelrod DM, Wong GY, Bradlow HL, Osborne MP. Biotransformation of estradiol by explant culture of human mammary tissue. Steroids 56:37–43, 1991.
138. Telang NT, Suto A, Wong GY, Osborne MP, Bradlow HL. Induction of estrogen metabolite 16α-hydroxyestrone of genotoxic damage and aberrant proliferation in mouse mammary epithelial cells. J Natl Cancer Inst 84:634–638, 1992.
139. Hersey RM, Gunsalus P, Lloyd T, Weisz J. Catechol estrogen formation by brain tissue: a comparison of the release of tritium from [2-^{3}H] 2-hydroxyestradiol formation from [6,7-H]

140. Hersey RM, Williams KIH, Weisz J. Catechol estrogen formation by brain tissue: characterization of a direct product isolation assay for estrogen-2- and 4-hydroxyestradiol formation by rabbit hypothalami in vitro. Endocrinology 109:1902–1911, 1981.
140. Hersey RM, Williams KIH, Weisz J. Catechol estrogen formation by brain tissue: characterization of a direct product isolation assay for estrogen-2- and 4-hydroxyestradiol formation by rabbit hypothalamus. Endocrinology 109:1912–1920, 1981.
141. Jellinck PH, Hahn EF, Norton BI, Fishman J. Catechol estrogen formation and metabolism in brain tissue: comparison of tritium release from different portions of ring A of the steroids. Endocrinology 115:1850–1856, 1985.
142. Fishman J. Stereochemistry of enolization of 17-keto steroids. J Org Chem 31:520–523, 1966.
143. Sepkovic DW, Bradlow HL, Michnovicz J, Murtezani S, Levy I, Osborne MP. Catechol estrogen production in rat microsomes after treatment with indole-3-carbinol, ascorbigen or β-naphthaflavone: a comparison of stable isotope dilution gas chromatography–mass spectrometry and radiometric methods. Steroids 59:318–323, 1994.
144. Kabat GC, Chang CJ, Sparano JA, Sepkovic DW, Hu XP, Khalil A, et al. Urinary estrogen metabolites and breast cancer: a case-control study. Cancer Epidemiol Biomarkers Prev 6:505–509, 1997.
145. Klug TL, Bradlow HL, Sepkovic DW. Monoclonal antibody–based enzyme immunoassay for simultaneous quantitation of 2- and 16α-hydroxyestrone in urine. Steroids 59:648–655, 1994.
146. Meilahn EN, De Stavola B, Allen DS, et al. Do urinary estrogen metabolites predict cancer? The Guernsey III cohort follow-up. Br J Cancer 78:1250–1255, 1998.
147. Ho GH, Luo XW, Ji CY, Foo SC, Ng EH. Urinary 2/16 α-hydroxyestrone ratio: correlation with serum insulin-like growth factor binding protein-3 and a potential biomarker of breast cancer risk. Am Acad Med Singapore 27:294–299, 1998.
148. Ziegler RG, Rossi SC, Fears TR, et al. Quantifying estrogen metabolism: an evaluation of the reproducibility and validity of enzyme immunoassays for 2-hydroxyestrone and 16α-hydroxyestrone in urine. Environ Health Perspect 105:607–614, 1997.
149. Bradlow HL, Sepkovic DW, Klug T, Osborne MP. Application of an improved ELISA assay to the analysis of urinary estrogen metabolites. Steroids 63:406–413, 1998.
150. Falk RT, Rossi SC, Fears TR, et al. A new ELISA kit for measuring urinary 2-hydroxyestrone, 16α-hydroxyestrone, and their ratio: reproducibility, validity, and assay performance after freeze–thaw cycling and preservation by boric acid. Cancer Epidemiol Biomarkers Prev 9:81–87, 2000.
151. Adlercreutz H, Fotsis T, Höckerstedt K, et al. Diet and urinary estrogen profile in premenopausal omnivorous and vegetarian women and in premenopausal women with breast cancer. J Steroid Biochem 34:527–530, 1989.
152. Lemon HM, Heidel JW, Rodriguez-Sierra JFR. Increased catechol estrogen metabolism as a risk factor of nonfamilial breast cancer. Cancer 69:457–465, 1992.
153. Lemon HM, Heidel J, Rodriguez-Sierra JF. A method for estimation of catechol estrogen metabolism from excretion of non-catechol estrogens. Cancer 68:444–450, 1991.
154. Ursin G, London S, Stanczyk FZ, et al. A pilot study of urinary estrogen metabolites (16α-OHE$_1$ and 2-OHE$_1$) in postmenopausal women with and without breast cancer. Environ Health Perspect 105:601–605, 1997.
155. Ursin G, London S, Stanczyk FZ, Gentzschein E, Paganini-Hill A, Ross RK, Pike MC. Urinary 2-hydroxyestrone/16α-hydroxyestrone ratio and risk of breast cancer in postmenopausal women. J Natl Cancer Inst 91:1067–1072, 1999.
156. Adlercreutz H, Gorbach SL, Goldin BR, Woods MN, Dwyer JT, Hämäläinen E. Estrogen metabolism and excretion in Oriental and Caucasian women. J Natl Cancer Inst 86:1076–1082, 1994.
157. Ursin G, Wilson M, Henderson BE, et al. Do urinary estrogen metabolites reflect the differences in breast cancer risk between Singapore Chinese and United States African-American and white women? Cancer Res 61:3326–3329, 2001.
158. Suchar LA, Chang RL, Rosen RT, Lech J, Conney AH. High performance liquid chromatography separation of hydroxylated estradiol metabolites: formation of estradiol metabolites by liver microsomes from male and female rats. J Pharmacol Exp Ther 272:197–206, 1995.
159. Zhu BT, Roy D, Liehr JG. The carcinogenic activity of ethinyl estrogens is determined by both their hormonal characteristics and their conversion to catechol metabolites. Endocrinology 132:577–583, 1993.
160. Hammond DK, Zhu BT, Wang MY, Ricci MJ, Liehr JG. Cytochrome *P450* metabolism of estradiol in hamster liver and kidney. Toxicol Appl Pharmacol 145:54–60, 1997.
161. Dannan GA, Porubek DJ, Nelson SD, Waxman DJ, Guengerich FP. 17β-Estradiol 2- and 4-hydroxylation catalyzed by rat hepatic cytochrome *P-450*: roles of individual forms, inductive effects, developmental patterns, and alterations by gonadectomy and hormone replacement. Endocrinology 118:1952–1960, 1986.
162. Liehr JG, Ricci MJ, Jefcoate CR, Hannigan EV, Hokanson JA, Zhu BT. 4-Hydroxylation of estradiol by human uterine myometrium and myoma microsomes: implications for the mechanism of uterine tumorigenesis. Proc Natl Acad Sci USA 92:9220–9224, 1995.
163. Hayes CL, Spink DC, Spink BC, Cao JQ,

Walker NJ, Sutter TR. 17β-Estradiol hydroxylation catalyzed by human cytochrome P450IB1. Proc Natl Acad Sci USA 93:9776–9781, 1996.
164. Ball P, Knuppen R. Catecholoestrogens (2- and 4-hydroxyoestrogens): chemistry, biogenesis, metabolism, occurrence and physiological significance. Acta Endocrinol (Copenh) 232:1–127, 1980.
165. MacLusky NJ, Barnea ER, Clark CR, Naftolin F. Catechol estrogens and estrogen receptors. In: Merram GR, Lipsett MB (eds). Catechol Estrogens. New York: Raven Press, 1983, pp 151–165.
166. Martucci C, Fishman J. Uterine estrogen receptor binding of catecholestrogens and of estetrol (1,3,5(10) estratriene-3,15α,16α,17β tetrol). Steroids 27:325–333, 1976.
167. Franks S, MacLusky NJ, Naftolin F. Comparative pharmacology of oestrogens and catecholestrogens: actions on the immature uterus in vivo and in vitro. J Endocrinol 94:91–98, 1982.
168. Ball P, Knuppen R, Haupt M, Breuer H. Interactions between estrogens and catechol amines. III. Studies on the methylation of catechol estrogens, catechol amines and other catechols by the catechol O-methyltransferase of human liver. J Clin Endocrinol Metab 34:736–746, 1972.
169. Breuer H, Köster G. Interaction between estrogens and neurotransmitters at the hypophysial–hypothalamic level. J Steroid Biochem 5:961–967, 1974.
170. Ghraf R, Hiemke CH. Interaction of catechol estrogens with catecholamine synthesis and metabolism. In: Merriam GR, Lipsett MB (eds). Catechol Estrogens. New York: Raven Press, 1983, pp 177–187.
171. Liehr JG, Ulubelen AA, Strobel HW. Cytochrome P-450-mediated redox cycling of estrogen. J Biol Chem 261:16865–16870, 1986.
172. Liehr JG, Roy D. Free radical generation by redox cycling of estrogens. Free Radic Biol Med 8:415–423, 1990.
173. Liehr JG. Genotoxic effects of estrogens. Mutat Res 238:269–276, 1990.
174. Nutter LM, Ngo EO, Abul-Hajj YJ. Characterization of DNA damage induced by 3,4-estrone-o-quinone in human cells. J Biol Chem 266:16380–16386, 1991.
175. Han X, Liehr JG. DNA single-strand breaks in kidneys of Syrian hamsters treated with steroidal estrogens: hormone-induced free radical damage preceding renal malignancy. Carcinogenesis 15:997–1000, 1994.
176. Han X, Liehr JG. 8-Hydroxylation of guanine bases in kidney and liver DNA of hamsters treated with estradiol: role of free radicals in estrogen-induced carcinogenesis. Cancer Res 54:5515–5517, 1994.
177. Cavalieri EL, Stack DE, Devanesan PD, et al. Molecular origin of cancer: catechol estrogen-3,4-quinones as endogenous tumor initiators. Proc Natl Acad Sci USA 94:10937–10942, 1997.
178. Hayashi N, Hasegawa K, Komine A, et al. Estrogen-induced cell transformation and DNA-adduct formation in cultured Syrian hamster embryo cells. Mol Carcinogen 16:149–156, 1996.
179. Yager JD, Liehr JG. Molecular mechanisms of estrogen carcinogenesis. Annu Rev Pharmacol Toxicol 36:203–232, 1996.
180. Liehr JG. Mechanism of metabolic activation and inactivation of catechol estrogens: a basis of genotoxicity. Polycyclic Aromatic Compounds 6:229–239, 1994.
181. Cavalieri EL. Minisymposium on endogenous carcinogens: the catechol estrogen pathway. Polycyclic Aromatic Compounds 6:223–228, 1994.
182. Li JJ, Li SA. Estrogen carcinogenesis in Syrian hamster tissues: role of metabolism. Fed Proc 46:1858–1863, 1987.
183. Liehr JG, Fang WF, Sirbasku DA, Ulubelen AA. Carcinogenicity of catechol estrogens in Syrian hamsters. J Steroid Biochem 24:353–356, 1986.
184. Li SA, Purdy RH, Li JJ. Variations in catechol-O-methyltransferase activity in rodent tissues: possible role in estrogen carcinogenicity. Carcinogenesis 10:63–67, 1989.
185. Roy D, Weisz J, Liehr JG. The O-methylation of 4-hydroxyestradiol is inhibited by 2-hydroxyestradiol: implications for estrogen-induced carcinogenesis. Carcinogenesis 11:450–462, 1990.
186. Lipsett MB, Merriam GR, Kono S, Brandon DD, Pfeiffer DG, Loriaux DL. Metabolic clearance of catechol estrogens. In: Merriam GR, Lipsett MB (eds). Catechol Estrogens. New York: Raven Press, 1983, pp. 105–114.
187. Fishman J. Biological action of catecholestrogens. J Endocrinol 85:59P-65P, 1981.
188. Schutze N, Vollmen G, Tiemann I, Geiger M, Knuppen R. Catecholstrogens are MCF-7 cell estrogen agonists. J Steroid Biochem Mol Biol 46:781–789, 1993.
189. Schutze N, Vollmer G, Knuppen R. Catecholestrogens are agonists of estrogen receptor-dependent gene expression in MCF-7 cells. J Steroid Biochem Mol Biol 48:453–461, 1994.
190. Schneider J, Huh MM, Bradlow HL, Fishman J. Antiestrogen action of 2-hydroxyestrone on MCF-7 human breast cancer cells. J Biol Chem 259:4840–4845, 1984.
191. Vandewalle B, Lefebvre J. Opposite effects of estrogen and catecholestrogen on hormone-sensitive breast cancer cell growth and differentiation. Mol Cell Endocrinol 61:239–246, 1989.
192. Seegers JC, Aveling M-L, van Aswegen CH, Cross M, Koch F, Joubert WS. The cytotoxic effects of estradiol-17β, catecholestradiols and methoxyestradiols on dividing MCF-7 and Hela cells. J Steroid Biochem 32:797–809, 1989.
193. Lottering M-L, Haag M, Seegers JC. Effects of 17β-estradiol metabolites on cell cycle events in MCF-7 cells. Cancer Res 52:5926–5932, 1992.

194. D'Amato RJ, Lin CM, Flynn E, Folkman J, Hamel E. 2-Methoxyestradiol, an endogenous mammalian metabolite, inhibits tubulin polymerization by interacting at the colchicine site. Proc Natl Acad Sci USA 91:3964–3968, 1994.
195. Fotsis T, Zang Y, Pepper MS, et al. The endogenous oestrogen metabolite 2-methoxyoestradiol inhibits angiogenesis and suppresses tumor growth. Nature 368:237–239, 1994.
196. Hamel E, Lin CM, Flynn E, D'Amato RJ. Interaction of 2-methoxyestradiol an endogenous mammalian metabolite, with unpolymerized tubulin and tubulin polymers. Biochemistry 35:1304–1310, 1996.
197. Klauber N, Parangi S, Flynn E, Hamel E, D'Amato RJ. Inhibition of angiogenesis and breast cancer in mice by the microtubule inhibitors 2-methoxyestradiol and Taxol. Cancer Res 57:81–86, 1997.

3

The Nuclear Receptor Superfamily

JOHN M. ROSENFELD
HUNG-YING KAO
RONALD M. EVANS

Lipophilic hormones, including steroids, thyroids, and retinoids, regulate complex processes in cellular differentiation, metabolic homeostasis, and animal development. Their ability to diffuse into target cells and bind with high affinity and specificity to their cognate intracellular receptors allows these molecules to coordinate physiological responses by regulating the expression of gene networks. The proteins directly involved in transducing these hormonal signals are known broadly as *nuclear receptors* (NRs). Upon binding to their specific ligand, NRs modulate the transcription of target genes through interactions with discrete DNA-binding sites known as *hormone response elements*, *cofactor molecules*, and *general transcriptional machinery*. As ligand-activated transcription factors, hormone NRs are indispensable for proper growth and development and often perform essential roles in specifying cell lineages, inducing differentiation, and controlling cellular growth and function. The multitude of mechanisms by which this class of transcription factors orchestrates complex transcriptional control currently makes this protein family one of the most intensely studied areas of molecular biology.

The molecular cloning of a cDNA encoding the glucocorticoid receptor (GR) represented a critical advance in modern molecular endocrinology.[1,2] Perhaps more than any other discovery, the molecular characterization of this receptor allowed the subsequent cloning of the majority of the remaining receptors by virtue of homology within the highly conserved DNA-binding domain (DBD). Since the isolation of the GR cDNA in 1985, more than 60 distinct members of the hormone NR superfamily have been identified.[3,4] This number defines the NRs as the largest single family of transcription factors. Indeed, sequencing of the entire genome of the nematode *Caenorhabditis elegans* indicates that more than 1.5% of the coding sequence is comprised of putative NR encoding genes.[5] The *Drosophila* genome sequence predicts a much smaller number of NR encoding genes (27 in total), and human sequencing efforts predict 49 unique receptor genes.[6,6a] As we shall see in later sections, many of these receptors have no known ligands.

Steroid hormone-related diseases were recognized long before the identification of their cognate intracellular receptors, including generalized resistance to thyroid hormone and vitamin A and vitamin D deficiencies.[7] These diseases often involve insensitivity of receptors to endogenous steroid and vitamin hormones as a result of genetic mutation of the receptor or other proteins that function upstream in the regulation of hormone production or function. Furthermore, modulation of receptor function in certain physiological or pathophysiological states is often clinically beneficial. For these reasons, pharmacological administration of natural or synthetic ligands for several receptors is effective in a variety of clinical syndromes, including diseases associated with reproduction and fertility, immune function, obesity, diabetes, heart disease, and cancer. Identification of natural, endogenous ligands and design of synthetic ligands for nuclear receptors is therefore an important area of research not only for understanding the physiology of NRs, but also for pharmacological intervention in a number of medical syndromes.

In this chapter, we shall explore the NR superfamily in regard to evolution, structure–function relationships, cofactor requirement, chemical/structural models of receptor function, and physiology. Due to the large number of members in this superfamily, several methods for classifying receptors have been developed, including consideration of dimerization status, DNA-binding specificity, and the nature of the chemical ligands to which the receptors bind. In addition to classical steroid and thyroid hormone receptors, we will present recent findings regarding the function of many of the more newly discovered *orphan* receptors (i.e., those that do not bind to classical steroid- or thyroid-derived compounds). We will conclude our discussion of the superfamily by describing molecular approaches to elucidating the functions of orphan receptors, which is currently one of the most challenging and exciting aspects of NR biology.

STRUCTURE AND FUNCTION OF NUCLEAR RECEPTORS

Evolution of the Nuclear Receptor Superfamily

As mentioned above, the NR superfamily is the largest family of transcription factors. This extensive diversity is the result of individual gene duplications and mutations selected for by functional pressure throughout evolution. Because mammalian NRs are dispersed throughout multiple chromosomes of the genome, it is likely that early metazoans experienced genomic duplications that facilitated this distribution.[8] Despite this diversity, however, NRs share conserved structural elements, which has allowed the identification and cloning of novel receptors by virtue of sequence similarity. The most highly conserved portion of nuclear receptors is the DBD, which often displays >90% amino acid sequence identity between *orthologous* (functionally identical genes across species) and *paralogous* (related genes within a species) family members. While the zinc finger DNA-binding motif is present within transcription factors of virtually all eukaryotes, the NR DBD, containing two Cys_2:Cys_2 zinc fingers, is apparently restricted to metazoans, with orthologous or related genes present in nematodes, arthropods, fish, birds, amphibians, and mammals.[5] This finding is supported by polymerase chain reaction–directed searches for more ancient NR DBDs, utilizing degenerate primers directed to the highly conserved P and D boxes within the zinc fingers.[9] The sequence of the more weakly conserved ligand-binding domain (LBD) of these receptors provides a sequence determinant that is utilized to define subtypes of receptor classes. Sequence identity within the LBD is often in the range of 80%–90% for related subtypes such as retinoic acid receptors (RARs) α, β, and γ within the same species and 50%–70% across species. Interestingly, LBDs that exhibit as low as 15% amino acid identity retain nearly superimposable structural features when analyzed by X-ray crystallography.[10] It is therefore likely that the overall structural fold of this domain persists throughout the superfamily, despite low sequence similarity. A consensus signature motif of residues has been derived based on comparative sequence analysis of multiple-receptor LBDs.[11] Amino acid substitutions within this signature motif, and elsewhere within the ligand-binding pocket, can alter ligand-binding specificity while retaining the overall structural features of the LBD, suggesting that differences in ligand structures are reflected by the relatedness of the receptor. Examination of an evolutionary tree of NRs reveals the diversity of the superfamily by the number of paralogous members within receptor classes, with each receptor gene often encoding multiple splice variants in a tissue-specific or developmentally regulated manner (Fig. 3.1). Due to the large number of receptor genes, many of which have been isolated and named independently by different methods, a systematic nomenclature for designating novel receptors based on sequence similarity has been adopted.[4]

Detailed comparative sequence analysis of receptors across species predicts that the DNA-binding function of these proteins evolved prior to ligand-binding capacity. All NRs from organisms that evolved prior to chordates, arthropods, and nematodes are more similar to orphan receptors that lack known ligands than to steroid or thyroid family members, suggesting that the prototypes of NRs were orphans.[9,13,14] Steroid hormone receptors are apparently restricted to vertebrates, with the exception of the insect ecdysone receptor (EcR), representing the prototypical steroid receptor. The insect EcR, which most closely resembles the farnesoid X receptor (FXR) in primary sequence, is required for virtually every aspect of insect larval metamorphosis.[15] The primary function of EcR in this process is to induce expression of numerous orphan NRs (several of which have vertebrate homologs of unknown function) and establish a temporal regulatory hierarchy of NR activity.[16] Interestingly, EcR itself requires heterodimerization with ultraspiracle (USP), a *Drosophila* retinoid X receptor (RXR) homolog, to facilitate DNA binding and transactivation.[17,18] The observation that USP can heterodimerize with EcR, mammalian receptors, and other insect orphan receptors provides compelling functional evidence that heterodimerization is an evolutionarily conserved mechanism of gene regulation, as has been observed for members of the basic helix-loop-helix leucine zipper transcription factors.[19] Furthermore, a *Drosophila* relative of the mammalian nuclear corepressor silencing mediator of retinoic acid and thyroid hormone receptors (SMRT) is also involved in modulating EcR:USP activity, suggesting that cofactor association is also a conserved mechanism of gene regulation.[20] Unlike the vertebrate RXRs, however, USP does not bind to the RXR ligand 9-*cis* RA.[21] In this respect, USP is functionally more similar to orphan receptors, supporting the notion that the ligand-binding function is restricted to vertebrates. In fact, several *Drosophila* orphan receptors lack an LBD entirely (i.e., the gap genes *Kni, Knrl, Egon*), representing true functional intermediates between zinc finger transcription factors and ligand-dependent NRs. While it is possible that the more ancient orphan receptors that encode an LBD respond to specific ligands, it is probable that many of them regulate gene expression in a ligand-independent manner. These mechanisms could include constitutive activity, regulated heterodimerization of subunits, binding site exclusion and competition, or posttranslational modifications.

Nuclear Receptors Contain Modular Functional Domains

Nuclear receptors typically contain five to six modular domains that are functionally distinct (denoted A–F) (Fig. 3.2). The amino-terminal region (or A/B domain) of the receptors is highly variable in sequence and size and contains a ligand-independent activation function (AF-1). The C domain harbors a highly conserved DBD that targets NRs to their cognate chromosomal binding sites. For some receptors, this region also contains nuclear localization sequence and dimerization activity.[22] The hinge domain (D domain) is the most variable region among receptors and represents a flexible linker between the two core structures, the DBD and the LBD. Recent structure data have indicated an important role for this domain in the stabilization of the LBD.[23] The LBD (E domain) binds lipophilic chemical ligands with high affinity, ensuring both specificity and selectivity of the physiological response. In addition, this domain confers dimerization, cofactor association, heat shock protein (hsp) binding, and nuclear localization properties for most class I receptors (see below, Hormone Response Elements and the DNA-Binding Domain). Perhaps most importantly, the LBD contains the ligand-dependent activation function (AF-2), which has been implicated in transcriptional activation through coactivator recruitment. Several receptors, including the RARs, also contain a carboxyl-terminal F domain, which is highly divergent among receptors and of as yet undetermined function.

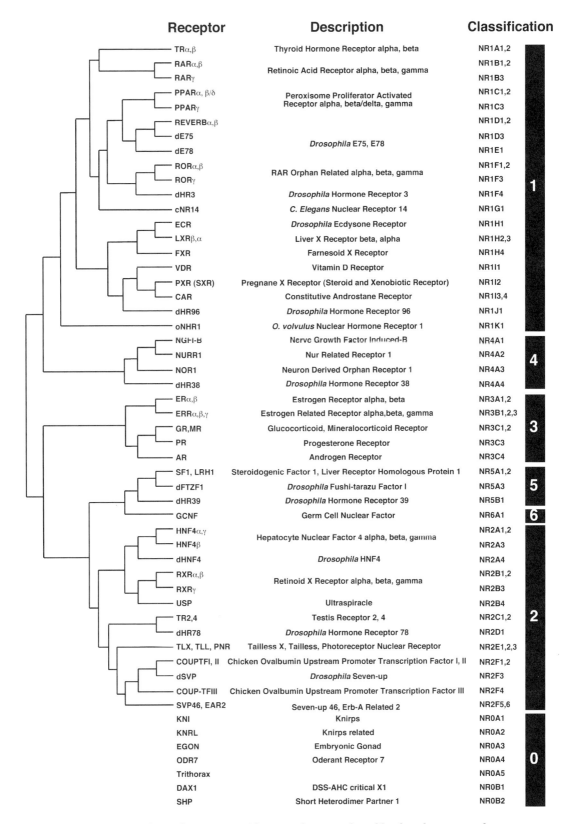

Figure 3.1 Descriptions for nuclear receptor abbreviations, as well as the gene classification scheme devised by the Nuclear Receptors Nomenclature Committee are included. Orphan receptors of the NR0 class are not included in the cladogram, due to lack of DNA-binding domain or ligand-binding domain encoding sequences. Tree is not drawn to scale. NR, nuclear receptor; DSS-AHC critical X1, dosage sensitive sex reversal adrenal hypoplasia congenita critical X chromosome region 1. (Adapted from Figures 1 and 2 of Anonymous.[4])

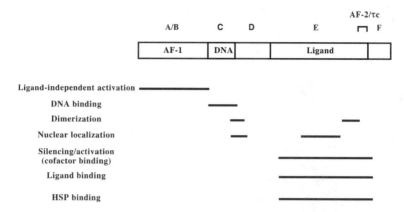

Figure 3.2 Schematic representation of nuclear receptor (NR) functional domains. A typical NR contains five to six functional domains, denoted as A to F. The A/B domain possesses ligand-independent transactivation activity (AF-1). The C domain is responsible for DNA binding and plays an auxiliary role in dimerization. The D domain contains a linker sequence between domains C and E and may play a role in stabilization of the ligand-binding domain (LBD). The E domain (LBD) harbors multiple activities, including dimerization, ligand binding, and silencing/activation (and therefore cofactor binding). In the case of class I receptors, the E domain also contains a nuclear localization sequence and heat shock protein (HSP)-binding activity. The function of the F domain is currently unknown.

Hormone Response Elements and the DNA-Binding Domain

The concept that hormone NRs function as sequence-specific DNA-binding transcription factors was substantiated when a hormone response element (HRE) for the GR was first identified by mutational analysis of the hormone-responsive long terminal repeat (LTR) of the mouse mammary tumor virus (MMTV) (Fig. 3.3).[24] Since the establishment of this concept, the DBD has been an area of intense investigation to determine how specific recognition of its HRE target are conferred. The DBD is comprised of two highly conserved zinc fingers, and sequence homology within this region differentiates NRs from other zinc finger DNA-binding proteins (Fig. 3.3B).[25]

Several approaches have been employed to identify and characterize HREs, including (1) DNAse I protection (footprinting), (2) electrophoresis gel mobility assay (EMSA), (3) transcription activity–based transient transfection assays, and (4) in vitro binding site selection assays. Characteristically, NRs bind to HREs as either a monomer or a dimer. The HREs are comprised of either a direct repeat (DR), an inverted repeat (IR), or an everted repeat (ER) of a consensus half-site sequence. Molecular and structural studies of NR DBDs bound to HREs in-

Figure 3.3 The nuclear receptor DNA-binding domain and the hormone response element (HRE). **A:** The HREs are composed of either direct repeats (DRs), inverted repeats (IRs), or everted repeats (ERs). The preferred half-site sequence for class I receptors is 5′-AGAACA-3′, while the preferred half-site sequence for class II, III, and IV is 5′-AGGTCA-3′. **B:** The nuclear receptor DNA-binding domain (DBD) is distinguished by two zinc fingers. The zinc fingers of human thyroid receptor beta (TRβ) are depicted. Residues contributing to the P box and D box are indicated as open squares. **C:** A list of individual nuclear receptor P and D box sequences along with their HREs are shown as are their DNA recognition motif. GR, glucocorticoid receptor; PR, progesterone receptor; MR, mineralocorticoid receptor; AR, androgen receptor; RAR, retinoic acid receptor; ER, estrogen receptor; VDR, vitamin D receptor; PPAR, peroxisome proliferator-activated receptor; LXR, liver X receptor; FXR, farnesoid X receptor; CAR, constitutive androstane receptor; RXR, retinoid X receptor; HNF4, hepatocyte nuclear factor 4; COUP-TF, chicken ovalbumin upstream promoter transcription factor; SF-1, steroidogenic factor 1.

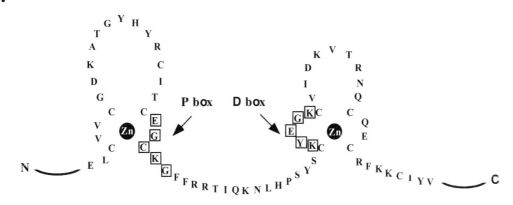

dicate that half-site recognition is specified by a region known as the *P box* (Fig. 3.3B,C). The P box typically consists of five amino acids situated at the end of the amino-terminal (proximal) zinc finger and makes base-specific contacts within the major groove of the HRE.[26] The D box, located at the amino-terminal end of the C-terminal (distal) finger, appears to contribute to DBD HRE recognition by detecting the spacing between the two half-sites.[27,28] The D box has also been shown to be essential for cooperative binding of the GR LBD to the glucocorticoid response element (GRE) sequence, supporting the hypothesis that it may participate in receptor dimerization.[29]

Receptor Classification by Mode of Action

Nuclear receptors can be grouped into four classes based on their DNA-binding and dimerization properties (Fig. 3.4).[25] Generally speaking, class I receptors include GRs, mineralcorticoid receptors (MRs), progesterone receptors (PRs), and androgen receptors (ARs), which function as homodimers and bind to IRs separated by 3 base pairs (i.e., IR-3). An exception to this rule is the estrogen receptors (ERs) which can form either heterodimers or homodimers on IR-3.[30–32] Class II receptors require heterodimerization with RXR family members and bind to DRs, varying from DR-1 to DR-5. The spacing, in general, specifies the identity of the dimerization partner. Class III receptors, including hepatocyte nuclear factor 4 (HNF-4), chicken ovalbumin upstream promoter-transcription factor (COUP-TF), and germ cell nuclear factor (GCNF), bind to DRs primarily as homodimers.[33–35] Class IV receptors, such as steroidogenic factor 1 (SF-1), dosage sensitive sex reversal adrenal hypoplasia congenita X chromosome region 1 (DAX-1), retinoid-related orphan receptor α (RORα), the ER-related receptors (ERRs), and tailless related receptor X (TLX), bind to extended half-site sequences as monomers.[16,36–38] Specificity and selectivity for this receptor class may involve sequences flanking the half-site. Many of the receptors in classes III and IV lack known ligands and are therefore classified as orphan nuclear receptors. Their mode of DNA binding, dimerization status, and mode of activation are less well characterized than those of classes I and II.

Class I Receptors

Class I receptors, with the exception of ER, are unique in their ability upon hormone binding to bind imperfect IR-3 elements exclusively as homodimers. The optimal response elements for GR, PR, and AR are strikingly similar and contain the consensus sequence AGAACA(n3)TGTTCT.[39,40] Introduction of PR or AR into cells stimulates transactivation of the GRE of the LTR of the MMTV in a ligand-dependent manner.[41–44] These observations illustrate a specificity problem, i.e., how glucocorticoid, progesterone, or androgen responsiveness is defined in a given IR-3-containing sequence. One plausible mechanism to solve steroid-specific gene activation is by differential expression of the receptors. For example, expression of glucocorticoid-regulated genes in hepatoma cells can be upregulated by expression of PR, which is normally not ex-

Figure 3.4 Members of the nuclear receptor superfamily can be grouped into four classes based on their DNA-binding and dimerization properties.[51] Class I receptors include steroid receptors that bind DNA as homodimers on the inverted repeat 3 (IR3) response element. The half-site sequence for the estrogen receptor (ER) is more similar to non-class I receptors. Receptors ERα and ERβ can also form heterodimers. Class II receptors form heterodimers with retinoid X receptors (RXRs) and bind DNA with direct repeats (DRs). Class III receptors can bind DRs as homodimers, heterodimers with RXR, or monomers. Class IV receptors bind to extended half-sites as monomers. (Adapted from Figure 2 of Mangelsdorf and Evans.[51]) GR, glucocorticoid receptor; MR, mineralocorticoid receptor; PR, progesterone receptor; AR, androgen receptor; TR, thyroid hormone receptor; RAR, retinoic acid receptor; VDR, vitamin D receptor; PPAR, peroxisome proliferator-activated receptor; LXR, liver X receptor; FXR, farnesoid X receptor; CAR, constitutive androstane receptor; PXR, pregnane X receptor; COUPTF, chicken ovalbumin upstream promoter transcription factor; HNF4, hepatocyte nuclear factor 4; GCNF, germ cell nuclear factor; NGFI-B, nerve growth factor inducible B; TR2, 4, testis receptor 2, 4; TLX, tailless X related; ERR, estrogen related receptor; SF-1, steroidogenic factor 1; ROR, retinoid related orphan receptor.

Class I Receptors
(Homodimers)

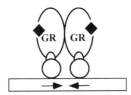

GR	Glucocorticoid
MR	Mineralocorticoid
PR	Progesterone
AR	Androgen
ER	Estrogen

Class III Receptors
(Mixed Dimer Properties)

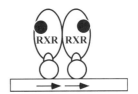

RXR	9-cis Retinoic Acid
COUPTF	?
HNF4	?
GCNF	?
NGFI-B	?
TR2, 4	?

Class II Receptors
(Heterodimers)

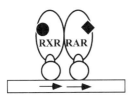

TR	Thyroid Hormone
RAR	All-trans Retinoic Acid
VDR	1,25-dihydroxy Vitamin D_3
PPAR	Eicosanoids, Fatty Acids
EcR	Ecdysone
LXR	Oxysterols
FXR	Bile Acids
CAR	Androstanes
PXR	Steroids, Xenobiotics

Class IV Receptors
(Monomers)

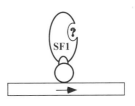

TLX	?
ERR	?
SF-1	?
ROR	?
Rev-erb	?

pressed in these cells.[45] Other mechanisms accounting for selectivity, such as selective steroid transport and selective inactivation, have also been proposed.[46,47] Additional specificity can be achieved through modulation of NR activity by additional promoter-specific transcription factors (see below, The Chromatin Link). Unlike the GR, AR, MR, and PR, the ER is structurally more related to class II receptors than to class I receptors. It recognizes IR-3s with the half-site sequence AGGTCA, which is identical to the half-site used by NRs of classes II–IV.

Class II–IV Receptors

The consensus HREs for non-class I receptors are DRs of the core AGGTCA half-site. The vitamin D receptor (VDR) and thyroid hormone receptors (TRs) heterodimerize with RXRs and bind DR-3 or DR-4 elements, respectively,[48] whereas PPAR:RXRs recognize DR-1.[49,50] Natural response elements for RAR appear to be more flexible and include DR-1, DR-2, and DR-5. Together, these have expanded the mode of DNA binding for NRs to the so-called 1-to-5 rule.[51] Although the 1-to-5 rule is predictive for most response elements, subtle differences in the sequence of the half-site and the 5' extension of these response elements are also important and can contribute to the specificity and activity of the heterodimers.[26,52] In addition to DR motifs, there are a variety of complex HREs regulated by NRs of this class. For example, RXR:TR, RXR:RAR, and RXR:RXR can recognize and activate transcription from a synthetic IR (IR-0).[53,54] Furthermore, the RAR:RXR heterodimer binds and activates an ER-8 sequence on the γF-crystallin promoter.[55] Similar to Class I receptors, Class II NRs can also recognize similar response elements, conferring multiple ligand responsiveness to certain response elements. This cross signalling at the level of the response element has been demonstrated for the pregnane X receptor (PXR), constitutive androstane receptor (CAR), and VDR on an ER-6/DR-4 element within the cytochrome P450 3A4 (CYP3A4) promoter, as well as for PXR, CAR, and FXR on an ER-6 within the adenosine triphosphate (ATP) binding cassette C2 (ABCC2) promoter.[56,56a,56b,57] It is likely that recognition of these and other composite elements is also affected by DNA and chromatin structure surrounding these more flexible response regions.

The DNA-binding properties of class III and class IV receptors are even more complex. It has been shown that HNF4 and COUP-TFI bind DR-1 in the human cytochrome CYP2D6 promoter whereas HNF4 and COUP-TFII synergistically activate transcription of a DR-1 element in the CYP7A1 promoter.[34] Besides their unique heterodimerization capacity on DR-3, DR-4, and DR-5 elements, the promiscuous RXRs can also form homodimers on IR-0 or DR-1 elements.[51,58] Rev-erb and ROR bind DNA as monomers to the sequence AATGT-AGGTCA and coexpression of Rev-erb suppresses the transcriptional activation by ROR.[59] Receptors such as nerve growth factor induced-B (NGFI-B) and its family members Nur related receptor-1 (Nurr-1) and neuron derived orphan receptor 1 (NOR-1) can bind to half-sites as monomers, homodimers, as well as heterodimers among family members or with RXRs.[60–63] Orphan receptors testis receptor 2 (TR2) and TR4, two closely related proteins, bind DNA from DR-1 to DR-5 with high affinity.[64,65] However, it appears that besides binding DNA as homodimers, TR2 and TR4 form heterodimers in standard protein–protein interactions as well as gel shift assays.[66,67] Intriguingly, heterodimers seemed dominant on a direct repeat spaced by five nucleotides (DR-5), albeit in transient transfections. It is therefore difficult to classify the DNA binding characteristics of these NRs.

Atypical Nuclear Receptors

Several structurally divergent members of the NR superfamily have been isolated. The human gene DAX-1, which lacks a conventional zinc finger DBD, encodes one of these atypical NRs. DAX-1 contains a region consisting of four repeats of alanine and glycine-rich sequences that likely binds DNA. It is responsible for dosage-sensitive sex- and X-linked adrenal hypoplasia, an inherited disorder of adrenal gland development.[68] The LBD of DAX-1 is similar to typical members of the NR superfamily.

The heterodimeric partner of DAX-1, SHP (short heterodimer partner), is structurally similar,[69] containing a putative LBD but lacking a conventional DBD. Also, SHP has been proposed to function as a negative regulator of NR signaling pathways because it interacts with and inhibits the transcriptional activity of various nonsteroid receptors, including TRs, RARs, and agonist-bound ERα and ERβ. This inhibition is caused in part by *SHP*-associated repression activity and by competition with coactivators for receptor binding.[70–72]

The AF-1 Domain

The amino-terminal AF-1 domain of some NRs harbors a ligand-independent activation function that has been demonstrated to be cell-specific (Table 3.1). This region differs in length and sequence among the various receptors. Some AF-1 domains are also targets of phosphorylation, which can have profound effects on receptor function. The mechanisms by which NR AF-1s stimulate transcription may include direct contact with the TFIID-containing basal transcriptional machinery, indirect contact with the basal machinery through intermediary factors, or phophorylation-dependent mechanisms of transcriptional regulation. Descriptions and examples of these mechanisms are considered below.

Using a cell-free transcription system and transient transfection assays, the AF-1 domain of GR was shown to possess ligand-independent transactivation activity when separated from the rest of the receptor molecule.[90–92] Many factors, including components of the basal transcription apparatus transcription factor IID (TFIID), and TATA binding protein (TBP), have been postulated to mediate the GR AF-1 function based on their physical interaction with the AF-1 do-

Table 3.1 Ligand-Independent Transactivation by Nuclear Receptors

Receptor	Phosphorylation	Kinase	Cofactor	Reference
Steroid receptors				
ERα	Ser^{104}, Ser^{106}	Cyclin A–CDK2		73
	Ser^{167}	pp90rsk1	?	74
	Ser^{118}	MAPK	?	75, 76
	Ser^{122}	PKCδ	?	77
			TAFII30	78, 79
			P160	80
			P68	81
				82
ERβ	Ser^{124}	MAPK	SRC-1	161
PR			SRC-1	83
			SRA	82
AR			P160	84
			TFIIF	85
			SRA	82
GR			TFIID, TBP	90, 93, 94
			DRIP150	95
			SRA	82
MR		PKA		86
Nonsteroid receptors				
RARα	Ser^{77}	CDK7	TFIIH	96
PPARγ			CBP/p300	162
			PGC-1	87
PPARα	Ser^{12}, Ser^{21}	MAPK		88
HNF4			CBP	89, 161
			PC4, TBP, ADA2	164
			TAFII31, TAFII80	164
			TFIIB, TFIIH-p62	164
SF-1	Ser^{203}	MAPK	GRIP1	161

Ligand-independent activity is mediated by associated cofactors or by interacting with basal transcription factors. While this activity has been mapped to the AF-1 domain for some receptors, it may require the ligand-binding domain for others. Phosphorylation status also plays a role in the AF-1 function for some receptors. ER, estrogen receptor; PR, progesterone receptor; AR, androgen receptor; GR, glucocorticoid receptor; MR, mineralocorticoid receptor; RAR, retinoic acid receptor; PPAR, peroxisome proliferator-activated receptor; HNF4, hepatocyte nuclear factor 4; SF-1, steroidogenic factor 1; Ser, serine; CDK, cyclin dependent kinase; pp90rsk1, protein phosphatase 90 ribosomal S6 kinase1; MAPK, mitogen activated protein kinase; PKC/PKA, protein kinase C/A; TAF, transcription associated factor; SRC-1, steroid receptor coactivator 1; SRA, steroid receptor RNA activator; TF, transcription factor; TBP, TATA binding protein; DRIP, vitamin D receptor interacting protein; CBP, cyclic adenosine monophosphate (cAMP) response element binding protein (CREB) binding protein; PC4, positive cofactor 4; ADA, alteration deficiency activator; GRIP, glucocorticoid receptor interacting protein.

main.[93,94] The GR AF-1 domain interacts with DRIP150, a subunit of the DRIP (vitamin D receptor–interacting protein) complex, in a ligand-independent manner (see below, Nuclear Receptor Coactivators).[95]

In addition to direct contacts with the basal machinery, phosphorylation of the AF-1 domain often modifies receptor function. For example, RARα is phosphorylated at Ser[77] by cyclin-dependent kinase 7 (cdk7) both in vitro and in vivo. This phosphorylation is required for transactivation activity as mutation from serine to alanine dramatically reduces this activity.[96] The phosphorylation and transactivation activity of RARα is enhanced upon overexpression of cdk7 and the associated general transcription factor TFIIH. Together, these results support a model in which ligand-independent AF-1 activity of RARα is mediated through its association with cdk7 and associated TFIIH. Other mechanisms of transcriptional activation by the AF-1 domain involving transcriptional co-activators are described below (see Coactivators Involved in Transcriptional Activation by the AF-1 Domain).

The Ligand-Binding Domain Contains the Ligand-Dependent AF-2 Function

The LBD of NRs encodes multiple functions, including activation and silencing (co-factor binding), dimerization, hsp binding, and nuclear localization. Class I receptors distinguish themselves from other members of the NR family by their cellular localization. Unlike other NRs, which primarily localize in the nucleus, most unliganded class I receptors are associated with a large, multisubunit complex of chaperones, including Hsp90 and the immunophilin Hsp56.[97] Hormone binding triggers a multistep process that involves a change in receptor conformation and dissociation of several hsps, receptor dimerization, nuclear translocation, DNA binding to cognate chromosomal sites, and transcriptional regulation. An autonomous activation domain (AF-2/τc) has been characterized in the carboxyl-terminal region of the E domain of several NRs corresponding to a conserved amphipathic α helix that is essential for AF-2 functions, including coactivator binding and ligand-dependent transcriptional activation. Interestingly, a subset of orphan receptors are missing this core motif, raising the possibility that these receptors may not require ligand binding for their function.

Atomic Structure of the Nuclear Receptor Ligand-Binding Domain

Receptor–hormone interactions were initially explored using genetic and biochemical techniques to assess the consequences of radiolabeled ligand binding to classical hormone receptor LBDs. The identification of whole new families of orphan nuclear receptors, however, accelerated the quest to obtain X-ray crystallographic and nuclear magnetic resonance spectroscopic structural models of both apo and holo LBDs. Similarly, the identification of ligands that antagonized the transcriptional activity of receptors, while binding with similar affinity as receptor agonists, necessitated a molecular explanation for ligand-dependent transcriptional activation. In general, the amino acid sequence of the LBD is poorly conserved between receptor families (roughly 15%–60% amino acid identity). Despite the large degree of sequence divergence between families of NR LBDs, sequence comparison does reveal conserved primary sequence as well as secondary structural elements within the carboxyl-terminal half of the receptors.[11] Based on LBD sequence alignments and crystal structures of multiple receptor LBDs, the canonical LBD consists of roughly 12 α helices interrupted by three short β strands, which organize into antiparallel sheets in the crystal structure (Fig. 3.5A).[98] A conserved sequence signature motif overlaps helices 3 and 4 in the primary structure, and contacts within these crystals suggest that residues in this region participate in both ligand stabilization and co-factor interaction. Indeed, comparison of the crystal structures of the ER, PR, RAR, RXR, and peroxisome proliferator–activated receptor (PPAR) LBDs reveals a remarkable conservation of overall structure, despite significant sequence divergence.[99] Structural interfaces include a receptor dimerization interface surrounding helix 11 (which also participates in ligand contacts) and a canonical NR fold formed by helices 3, 4, and 12. The hydrophobic cleft formed at this interface participates in both ligand stabilization

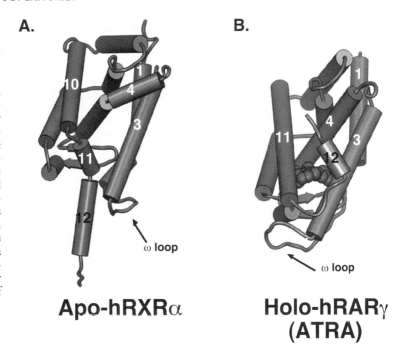

Figure 3.5 Nuclear receptor transposition of activation function 2 (AF-2) is induced by ligand occupancy. Comparison of the apo-human retinoid X receptor alpha (apo-hRXRα) (A) and holo-human retinoic acid receptor gamma (holo-hRARγ) (B) crystal structures. Schematic diagrams of the RXRα ligand-binding domain (LBD) in the apo configuration and RARγ LBD bound to all-*trans*-retinoic acid (ATRA) as monomers are depicted with α helices as tubes, β sheets as arrows, and the ligand as a space-filled model.[98,101] Select helices are numbered for reference, and the position of the ω loop is indicated.

and coactivator interaction, which is supported by mutational analysis.[100] Mutagenesis experiments map the ligand-dependant activation function (AF-2) of NRs to residues within helix 12, including a highly conserved glutamate that is required for coactivator association. Multiple hydrophobic residues throughout the LBD contribute to the structure of the ligand-binding pocket, the volume of which is tailored for occupancy by the hydrophobic ligand in the holo configuration. Ligand specificity, therefore, is largely dictated by the identity of select residues within the hydrophobic ligand-binding pocket, as well as by specific contacts with residues of helix 12.

Comparison of the apo-RXRα and holo-RARγ bound to all-*trans* retinoic acid (ATRA) LBD structures demonstrates the dramatic conformational shift associated with ligand binding, otherwise known as NR *transformation* (compare Fig. 3.5A,B).[101] In apo-RXRα, the AF-2 domain (helix 12) adopts an extended, solvent-exposed conformation, apparently allowing ligand access to the ligand-binding pocket core. Additionally, an extensive ω loop between helices 2 and 3 is folded away from the LBD. Upon ligand entry into the binding pocket, a structural rearrangement in helices 3, 10, and 11 results in a dramatic rotation that flips the ω loop (in the holo-RARγ structure) nearer to the LBD core and repositions helix 12 (and hence AF-2) in close proximity to helices 3, 4, and 11. In this mouse trap model of ligand binding, the role of AF-2 is to form a lid over the ligand-binding pocket, as well as to contribute stabilizing interactions with the ligand itself. This overall rearrangement in the position of helix 12 is thought to render the complex more resistant to protease digestion, which is the hallmark of NR transformation. The functional consequence of repositioning helix 12 involves cofactor release and exchange, as we shall see in the following section.

NUCLEAR RECEPTOR COFACTORS

A large body of evidence suggests that common transcriptional intermediary cofactors are required for transcriptional activity by NRs.[102–104] Using a series of biochemical approaches and yeast molecular genetics, receptor-associated cofactors have been isolated in both the absence (corepressors) and the presence of ligand (coactivators).[105–108] Homologs of both coactivators and corepressors have also been isolated from invertebrates.[20,109] It remains to be seen how conserved the mechanisms of transcriptional regulation by NRs are throughout evolution.

Nuclear Receptor Coactivators

Transcriptional activation by NRs is a complex process as NR ligand-dependent activation is thought to require the assembly of a multicomponent complex containing an array of coactivator proteins, many of which possess intrinsic enzymatic activity, as well as a variety of interaction motifs. Sequence analyses of receptor-interacting protein 140 (RIP140) and p160 family proteins identified multiple copies of a conserved motif, LXXLL (where L is leucine and X can be any amino acid), termed the *NR box*. This motif is essential to mediate and specify binding of coactivators to liganded NR LBDs.[110] Mutational studies indicate that mutation of leucine residues to alanine within the motif dramatically reduces the affinity of coactivators for NRs.[111] The amino acids flanking LXXLL also play a role in modulating affinity and specificity in NR selection and binding. This conclusion is based on observations that some LXXLL-containing proteins do not bind to NRs and that different receptors display preferences for different LXXLL-containing α helices. The functional significance of this motif is best illustrated by its presence in other classes of coactivator proteins, such as cyclic adenosine monophosphate (cAMP) response element (CRE) binding protein (CREB) binding protein (CBP)/p300, TR-associated protein (TRAP)/DRIPs, and E3 ubiquitin-protein ligase (or E6-AP).

A list of members of the expanding NR coactivator family is shown in Table 3.2. In considering this protein family, we will discuss a subset of those that have been well characterized in regard to NR interaction and activation. Originally, CBP was characterized as a coactivator required for transcriptional activation by CREB.[141,142] Also, CBP participates in transactivation by a variety of other transcription factors, including p53, nuclear factor κB (NF-κB), activator protein 1 (AP-1), and NRs.[143,144] It directly interacts with RXR, TR, and ER through its amino-terminal domain, which contains an NR box (LXXLL) required for NR interaction.[133,134] p300 is a closely related protein that shares many functional properties of CBP, including potentiation of transactivation by diverse transcription factors.[114] Interestingly, p300 is essential for RA-dependent activation but is not required for CREB stimulation.[145] Reduction at the RNA level of p300 but not CBP inhibits RA-induced differentiation of embryonic carcinoma F9 cells.[146] Similarly, RA-dependent upregulation of the cell-cycle inhibitors p21^{Cip} and p27^{Kip} have distinct requirements for p300 and CBP, respectively. These results indicate that CBP and p300 are not functionally redundant but play subtly distinct physiological roles.

The p300/CBP-associated factor (PCAF) was initially isolated as a factor associated with CBP/p300.[132] This interaction can be disrupted in the presence of the viral oncoprotein E1A, which affects normal cellular growth and differentiation by E1A recruitment of cellular CBP. Also, PCAF interacts with RXR:RAR heterodimers as well as the class I receptors ER, GR, and AR and can enhance retinoid-dependent promoter activity.[147] Microinjection of anti-PCAF antibodies into living cells blocks ligand-dependent NR activation, and this effect can be reversed by introducing a PCAF expression vector.[148]

The p160 family of coactivators consists of three members, designated steroid receptor coactivator 1–3 (SRC-1–3).[149,150] SRC-2 represents the molecules known variously as activating cofactor of thyroid and retinoid receptors (ACTR), receptor-associated coactivator 3 (RAC3), amplified in breast cancer 1 (AIB1), thyroid receptor activator molecule 1 (TRAM1), and CBP cointegrator protein (p/CIP). SRC-3 is also known as the glucocorticoid receptor interacting protein (GRIP1) and transcription intermediary factor 2 (TIF2). In general, these proteins interact with NRs in an AF-2- and ligand-dependent manner. Several NRs appear to activate transcription constitutively, independent of apparent ligand availability. The class IV receptors ERR1, ERR2 and ERR3, for example, seem to activate transcription by recruiting p160 family coactivators through the LBD.[151,152] Class I and class II receptors, however, bind p160 family proteins in a ligand-dependent manner.

Targeted disruption of the *SRC-1* gene in mouse revealed that the *SRC*-null mutant is viable and fertile but displays partial hormone re-

Table 3.2 Nuclear Receptors Coactivators

Coactivator	Receptor Interaction	Functional Properties	References
SRC-1	PR, ER, RAR RXR, TR, GR	HAT, interacts with CBP/p300, autonomous activation domain, KO exhibits generalized hormone insensitivity	111–116
SRC-2/ACTR/RAC3/ AIB1/P/CIP/TRAM-1	RAR, TR, RXR, ER	HAT, interacts with CBP/p300 and PCAF, autonomous activation domain, overexpressed in breast tumors and breast cancer cell lines, KO mice are viable, but show dwarfism and reproductive defects	117–122
SRC-3/GRIP1/TIF2	GR, TR, ER, AR	Interacts with CBP	123, 124
TIF1	ER, RAR, RXR	Potentiates RXR/RAR AF2 in yeast, inhibits RXR function when overexpressed in mammalian cells	125
Trip-1/Sug-1	TR, RAR, RXR	Sug1 is a component of yeast 26S proteasome, contains ATPase domain	126, 127
TRIP230	TR	Selectively coactivates TR in specific cell types	128
ARA-70	AR	Coactivates AR in prostate cells	129, 130
TRAP/DRIP/ARC	TR, VDR	Purified as TR and VDR ligand-dependent associated complex, coactivates VDR on chromatin templates, the 220 Kd subunit PBP/TRAP220/RB18A/DRIP230/TRIP2/CRSP200 binds PPARγ, RAR, RXR, TRβ, and ER, TRAP220 KO displays heart failure and neuronal defects in TR function, overexpressed in breast tumors and breast cancer cell lines	107, 108, 131 156
PCAF	PR, TR, RAR, GR, AR, ER	HAT, interacts with CBP and ACTR	132, 147
CBP/p300	RAR, RXR, ER, PPAR, TR	HAT, contains autonomous activation domains	114, 133, 134
E6-AP	AR, ER, PR, GR	Coactivates steroid receptors, E3 ubiquitin protein ligase	135, 136
BRG-1	GR, ER	Required for chromatin remodeling by GR	137
PGC-1	PPARγ	Expressed in brown adipose tissue and skeletal muscle, cold inducible	87
NCoA-62	VDR, RAR, RXR	Coactivates VDR strongly, GR, RAR, and ER weakly	138
ARIP3	AR	Expressed in testis	139
NSD1	RAR, TR, RX, ER	Interacts independent of ligands	140

SRC-1, steroid receptor coactivator 1; ACTR, activating cofactor of thyroid and retinoid receptors; RAC, receptor associated cofactor; AIB, amplified in breast cancer; P/CIP, p300/cyclic adenosine monophosphate (cAMP) response element binding protein (CREB) binding protein cointegrator protein; TRAM, thyroid receptor activator molecule; GRIP, glucocorticoid receptor interacting protein; TIF, transcription intermediary factor; TRIP, thyroid receptor interacting protein; SUG, suppressor of Gal4; ARA, androgen receptor activator; TRAP, thyroid receptor associated protein; DRIP, vitamin D receptor interacting protein; ARC, activator recruited cofactor; CBP, cyclic adenosine monophosphate (cAMP) response element binding protein (CREB) binding protein; PCAF, p300-CBP associated factor; E6-AP, E6 associated protein; BRG, brahma related gene; PGC, peroxisome proliferator-activated receptor coactivator; NcoA, nuclear receptor coactivator; ARIP, androgen receptor interacting protein; NSD, nuclear receptor binding SET domain protein; SET, suppressor of variegation 3-9, E(z) and Trithorax; PR, progesterone receptor; ER, estrogen receptor; RAR, retinoic acid receptor; RXR, retinoid X receptor; TR, thyroid receptor; GR, glucocorticoid receptor; AR, androgen receptor; VDR, vitamin D receptor; PPAR, peroxisome proliferator-activated receptor; HAT, histone acetyltransferase; NR, nuclear receptor; KO, knock-out; AF-2, activation function 2; ATPase, adenosine triphosphate hydrolase; PBP, PPAR binding protein, CRSP, coactivator required for Sp1.

sistance in target organs, including mammary gland, prostate, and testis, in response to progesterone and estrogen.[116] Amplification or overexpression of AIB1/ACTR/RAC3 was observed in approximately 70% of breast cancer tumors examined.[117] Interestingly, genetic disruption of SRC-2 in mice resulted in a pleiotropic phenotype displaying dwarfism, delayed puberty, reduced female reproductive function, and blunted mammary gland development.[153] These results suggest that p160 proteins may have distinct physiological functions. While p160 proteins are generally defined by a high degree of amino acid similarity within their functional domains, the spacing of these domains as well as variable sequence regions most likely provide specificity determinants that make these proteins functionally distinct.

Using biochemical purification schemes, multicomponent complexes have been isolated based on their ligand-dependent association with TR or VDR.[107,108] These two complexes, TRAP and DRIP share many of the same components and enhance VDR transactivation activity on a chromatin template.[131,154] The components of DRIP/TRAP overlap considerably with another novel cofactor complex, activator-recruited cofactor (ARC), which was identified based on its association with several other activators.[131] Similar to SRC-2, overexpression of PPAR binding protein (PBP), a component of the TRAP/DRIP complex, has also been detected in 24%–30% of breast tumors.[155] Genetic disruption of TRAP220/PBP revealed that null mice die during an early gestational stage with heart failure, neuronal development with extensive apotosis, and a prominent defect in TR function.[156]

Taken together, these observations suggest that NRs interact with multiple distinct coactivator proteins simultaneously to orchestrate transcriptional activation. These co-activators are not unique to NRs; they also play a role in diverse signaling pathways. The nature of the specificity of these interactions as well as functional redundance among these proteins are the focuses of current research.

Structure of the Coactivator:Receptor Complex

Crystallographic consideration of coactivator recruitment may help to understand how individual receptor LBDs utilize and select among the complex array of available coactivator proteins. In addition to holo-LBD crystal structures, cocrystal structures of receptor LBDs complexed to NR box–containing peptide fragments of coactivators have been solved for ERα:GRIP-1, TRβ:GRIP-1, and PPARγ:SRC-1.[111,115,157] One interesting feature of both the apo- and holo-PPARγ (bound to the thiazolidinedione BRL49653, also known as rosiglitazone) structures is the extensive volume of the ligand-binding pocket for this receptor. The physical observation of this large binding pocket confirms structure–activity studies demonstrating that a variety of structurally diverse compounds can bind to members of this receptor family with low affinity and further supports a role for PPARs as general lipid sensors.[158,159] Co-crystals of liganded PPARγ LBD bound to a peptide containing the LXXLL receptor interaction domain of the coactivator SRC-1 illustrate the coactivator–receptor interaction (Fig. 3.6A). E471, a highly conserved residue among receptor AF-2 domains, participates in a polar interaction with K301 (helix 3) and the SRC-1 peptide, locking the coactivator within the hydrophobic groove formed by helices 3, 4, and 12 as a charged clamp. Furthermore, the leucine side chains of the co-activator peptide are orientated toward the LBD, buried within the hydrophobic groove. This docking of the coactivator presumably defines a combinatorial surface for interaction with other cofactors, ultimately resulting in a positive transcriptional response. A possible explanation for the presence of multiple LXXLL motifs within coactivator proteins was suggested by an observation of the PPARγ:SRC-1 co-crystal. In this study, PPARγ crystallized as a homodimer, and interestingly, a second NR box from the SRC-1 peptide occupied a similar position in the dimer partner. This suggested that a single coactivator polypeptide may span the molecule to contact both coactivator surfaces in a receptor homo- or heterodimer and influence the ligand responsiveness of both dimer partners. Future structural and biochemical studies will be required to confirm the validity and functional significance of this interesting observation.

Structural Basis of Chemical Nuclear Receptor Antagonism

Thus, ligand binding results in a conformational rearrangement in the positioning of the AF-2 domain of the receptor, which subsequently provides an interaction surface for coactivator association. Intensive definition of structure–activity relationships among NR LBDs and their cognate ligands, as well as significant efforts in synthetic organic chemistry, have produced a num-

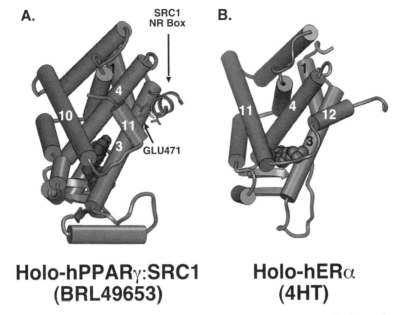

Figure 3.6 Activation function 2 (AF-2) forms a favorable interaction with the nuclear receptor (NR) box only in the presence of receptor agonists. Comparison of the holo human peroxisome proliferator-activated receptor gamma (hPPARγ) human steroid receptor coactivator 1 (hSRC1) cocrystal and antagonist-bound human estrogen receptor alpha (hERα) crystal structures. Schematic diagrams of the hPPARγ ligand-binding domain (LBD) complexed with the synthetic agonist BRL49653 (Rosiglitazone) and the LXXLL motif of hSRC-1 peptide[115] and the hERα LBD bound to the antagonist 4-hydroxytamoxifen[157] as monomers are depicted with α helices as tubes, β sheets as arrows, and ligands as space-filled models. Select helices are numbered for reference, and side chains are illustrated for the leucine residues within the SRC1 NR box and the conserved AF-2 glutamic acid 471 of PPARγ. In this rendering, the PPAR helix 11 corresponds to the ER helix 12, which encodes the AF-2 function.

ber of compounds that are high-affinity ligands for NRs but do not elicit positive transcriptional activity. Understanding how these chemical antagonists function at the molecular level is critical to understanding the mechanistic aspects of NR function and physiology. Structural analyses of the ERα LBD bound to natural and synthetic agonists, as well as the estrogen antagonists raloxifene and 4-hydroxytamoxifen (known collectively as selective estrogen receptor modulators) provide a model for co-activator recruitment and antagonist action. Specifically, these crystal structures suggest that antagonist binding displaces the AF-2 from its normal holo position (Fig. 3.6B). This distortion of the LBD fold by bulky ligand substitutions either prevents dissociation of co-repressors or disrupts the coactivator interface sufficiently to prevent LXXLL docking. These structures can be used to molecularly model receptor–agonist, –antagonist, or –partial agonist interactions, allowing directed searches for compounds that influence receptor function through coactivator and corepressor interactions.

Coactivators Involved in Transcriptional Activation by the AF-1 Domain

An emerging theme for the regulation of AF-1 activity is the involvement of receptor phosphorylation (Table 3.1). The AF-1 domains of AR, PR, ERα, ERβ, and SF-1 recruit p160 coactivator proteins to potentiate their activation function.[73] Phosphorylation of the receptor AF-1 domain appears to enhance affinity for p160 family proteins, thereby potentiating the activation function.[161,162] Other mechanisms for AF-1 function include the recruitment of helicases and other coactivators such as CBP/p300.[163,89,81,164]

In addition to protein moieties, a steroid receptor RNA activator (SRA) was identified based on its ability to enhance the activity of the AF-1 domain of PR in a yeast two-hybrid assay.[82]

It specifically potentiates the transactivation activity of steroid receptors. Introduction of stop codons within the transcript or treatment with the protein synthesis inhibitor cycloheximide does not block its functions, suggesting that SRA acts as an RNA transcript. Additionally, biochemical fractionation reveals that SRA is a component of distinct ribonucleoprotein complexes, one of which appears to contain SRC-1. Although how SRA activates transcription by steroid receptors is unclear, it represents a novel mechanism by which NRs regulate transcription.

Nuclear Receptor Corepressors

The notion that certain NRs can actively repress basal transcription was derived from the study of the oncogene v-erbA, encoded by avian erythroblastosis virus (AEV).[165,166] This study demonstrated that, unlike its cellular counterpart TRα, v-erbA encodes a constitutive transcriptional repressor due to loss of ligand-binding activity. It was later shown that this silencing effect by receptors is mediated by cofactors. Four mammalian NR co-repressors have been isolated, including the small unique NR corepressor (SUN-CoR), Alien, SMRT, and NR corepressor (N-CoR), based on their ability to bind unliganded receptor and their autonomous repression activity (Table 3.3).[105,106,167,168]

Initially, SUN-CoR was isolated by yeast two-hybrid screens for its ability to interact with Rev-ErbA.[167] It potentiates transcriptional repression by TR and Rev-ErbA and represses transcription when fused to a heterologous DBD. Interestingly, SUN-CoR also interacts with N-CoR. Alien interacts with TR, DAX-1, and COUP-TFI but not with RXR, RARα, or GR.[168] Addition of thyroid hormone leads to dissociation of Alien from TR. Alien is a highly conserved protein showing 90% identity between the human and Drosophila homologs. Drosophila Alien shows similar activities in that it interacts in a hormone-sensitive manner with TR and harbors an autonomous silencing function. The mechanism underlying its repression activity, however, has not been demostrated.

The nuclear protein SMRT and its isoform thyroid receptor associated cofactor 2 (TRAC2) were isolated by a yeast two-hybrid assay using RXR as bait.[105,169–171] It contains four autonomous repression domains and two receptor interaction domains,[170] and can interact with a panel of receptors, including unliganded RAR, TR, PPARα, PPARγ, Rev-erbA, and COUP-TFI.[105,172–176] In general, unliganded class I receptors do not associate with corepressors since they localize in the cytoplasm. However, certain antagonist-bound steroid receptors, including ER and PR, bind SMRT.[177–179] The NR corepressor N-CoR was identified by a combination of biochemical and yeast two-hybrid assays.[106] Studies of gene disruption of N-CoR in mice suggest a role in erythrocyte and thymocyte development as well as in

Table 3.3 Nuclear Receptor Corepressors

Corepressor	Receptor Interaction	Functional Properties	References
SMRT	COUP-TF, ER, PPAR, RAR, Rev-erbA, TR	Associates with both class I and class II HDACs Binds to antagonist-bound ER and PR	105, 167–173 177, 196, 198 196, 198
N-CoR	COUP-TF, DAX-1, RAR, Rev-erbA, TR	Originally identified as RXR:TR-interacting factor Highly similar to SMRT Associates with both class I and class II HDACs Required for CNS functions, erythrocyte and thymocyte development	106, 170, 172 175–178, 196, 198 204, 207
SUN-COR	Rev-erbA, TR	Interacts with N-CoR and SMRT in vitro and with endogenous N-CoR in cells.	166
Alien	COUP-TF, DAX-1, TR, VDR	Highly conserved between fly and human	165
SMRTER	EcR	Interacts with human TRβ and RARα Recruits deacetylases	20

SMRT, silencing mediator of retinoic acid and thryoid hormone receptors; N-Cor, nuclear receptor corepressor; SUN-COR, small unique nuclear receptor corepressor; SMRTER, SMRT related ecdysone receptor associated factor; TR, thyroid hormone receptor; RAR, retinoic acid receptor; ER, estrogen receptor; PPAR, peroxisome proliferator-activated receptor; COUP-TF, chicken ovalbumin upstream promoter transcription factor; DAX-1, dosage sensitive sex reversal adrenal hypoplasia congenita X region 1; VDR, vitamin D receptor; EcR, ecdysone receptor; HDAC, histone deacetylase; PR, progesterone receptor; CNS, central nervous system.

central nervous system (CNS) functions.[180] Despite their structural similarity and overlapping function, SMRT and N-CoR differ in several ways, including their differential affinity for individual receptors. For example, DAX-1 binds to N-CoR but not SMRT.[181] Like CBP and p300, SMRT and N-CoR interact with diverse classes of transcription factors and function as platform proteins.[182–187]

A SMRT/N-CoR-related protein, SMRTER (SMRT-related EcR-interacting factor), was isolated from *Drosophila*.[20] It is a large protein that shares limited overall sequence similarity with SMRT and N-CoR but retains common functionality of SMRT/N-CoR. It interacts with fly NRs and possesses autonomous repression activity. Remarkably, SMRTER confers repression activity upon the EcR through an evolutionarily conserved repression pathway (see below, The Chromatin Link).

Detailed mapping of the receptor interaction domains of SMRT/N-CoR has identified two short peptide motifs that contain a hydrophobic core ($\Phi XX\Phi\Phi$) similar to that found in coactivators. Both motifs are both necessary and sufficient for mediating corepressor binding to unliganded receptors and sense the presence of ligand by dissociating from the receptor.[188–190] These two motifs, termed interaction domain-1 and 2 (17 and 19 amino acids, respectively), are conserved in both position and sequence between N-CoR and SMRT. Secondary structure prediction suggests that they are likely to adopt an amphipathic α-helical conformation. These findings further extend the analogy with the helical coactivator LXXLL motifs. Furthermore, the receptor-interacting surface of corepressors is likely to be larger than that for coactivators. Mutations in the amino acids that directly participate in coactivator binding also disrupt corepressor association. These results suggest a direct mechanistic link between activation and repression via competition for a common or at least partially overlapping binding site for corepressors and coactivators.

The Chromatin Link

Eukaryotic DNA is highly organized into a higher-order structure termed *chromatin*, with repeats of a protein–DNA complex that consists of a histone octamer (nucleosome) wrapped by 146 base pairs of DNA. Acetylation of histones H3 and H4 is thought to generate an open chromatin conformation, which allows access of the transcriptional apparatus to the regulatory or promoter regions of genes. As such, hyperacetylated chromatin is often associated with transcriptionally active genes, while transcriptionally silent chromatin loci are usually hypoacetylated. The levels of histone acetylation are determined by the interplay between two opposite enzymatic activities, histone acetyltransferases (HATs) and histone deacetylases (HDACs). The identification of HATs and HDACs represents a milestone in the understanding of the mechanism of transcriptional regulation by NRs. To date, NR coactivators CBP/p300, PCAF, p160, SRC-1, and ACTR have been demonstrated to possess intrinsic HAT activity.[118,132,191–193] The HAT activity of coactivators appears to be crucial for coactivator function as mutations that abolish HAT activity dramatically reduce transcriptional activation. Several groups have concluded that CBP/p300, p160 family proteins, and PCAF are part of a coactivator complex that is essential for NR activation. Indeed, RAR transactivation activity requires PCAF HAT activity but not CBP HAT activity. Conversely, CREB requires CBP HAT activity but not PCAF HAT activity.[148] Similarly, MyoD-mediated p21 expression and cell cycle arrest are dependent on PCAF HAT but not p300 HAT activity.[194] These observations suggest that these HAT activities may provide overlapping but distinct and highly specific activities and that different HATs are responsible for activation mediated by different transcriptional regulators. Using the chromatin immunoprecipitation assay, which allows evaluation of the acetylated state of NR target genes, in vivo evidence of acetylated chromatin associated with NR target genes has been reported.[195] Interestingly, acetylation of activating cofactor of thyroid and retinoid receptors (ACTR) by p300 regulates its ability to associate with NRs.

Genetic analysis of mutants defective in transcriptional regulation from yeast and the identification of the first mammalian histone deacetylase, HDAC1, have had a great impact on our understanding of transcriptional repression by NRs.[196] Two classes of structurally distinct histone deacetylases have been isolated from both mammalian cells and yeast.[197–203] The structure and primary sequence of mammalian class I HDACs (HDAC1, -2, -3, and -8) are similar to those of the yeast Rpd3 protein. Class II HDACs, including HDAC4, -5, -6, and -7, are larger and homologous to yeast Hda1 protein. The yeast Sin3 and Rpd3 (HDAC1 homolog) proteins negatively regulate the expression of a variety of genes. Genetic complementation experiments indicate that Rpd3 and Sin3 are involved in the same genetic repression pathway.[204–207] Thus, the HDAC repression pathway is a generally conserved mechanism of transcriptional control from yeast to vertebrates. Several independent studies have concluded that certain class II NRs actively repress expression of target genes through the recruitment of a histone deacetylase complex containing SMRT/N-CoR, mSin3A, and both classes of histone deacetylases.[202,203,208–211] Further evidence for the evolutionary conservation of the repression pathway includes the functional homology of the *Drosophila* corepressor SMRTER, as discussed previously.[20] Taken together, these studies support a model wherein transcriptional regulation by class II NR is mediated through ligand-dependent dissociation of HDAC-containing corepressor complexes and association of coactivator complexes containing an array of HAT activities (Fig. 3.7).

Atypical Cofactors

Although originally identified as a ligand-dependent coactivator,[212] RIP140 has unusual properties that suggest that it may modulate receptor function in a distinct manner from other co-activators and corepressors. For example, RIP140 (when overexpressed) strongly suppresses the transactivation activities of several NRs in the presence of their cognate ligands, including ER, RAR X, liver X receptor alpha (LXRα), and PPARγ. Unlike the p160 family proteins, which contain three LXXLL motifs located in the central region, RIP140 has nine LXXLLs distributed throughout the whole molecule. This unique feature may be responsible for its ability to interact with other unliganded receptors, such as LXR, the PPARs, TR2, and TR4.[213] Intriguingly, this ligand-independent interaction did not occur with other receptors, such as TR, RAR, RXR, and ER. The carboxyl terminus of RIP140, where no LXXLL is found, can also interact strongly with liganded RXR–RAR heterodimers. When tethered to DNA, RIP140 encodes intrinsic *trans*-repressive activity. In summary, RIP140 represents a distinct class of cofactor and may play a more active role in modulating receptor activities than the passive competitor for members of the p160 family.

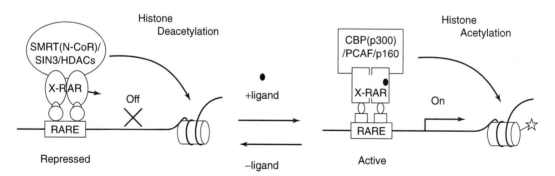

Figure 3.7 A model for transcriptional regulation by class II nuclear receptors. Transcriptional activity of class II nuclear receptors is mediated by associated cofactors. Unliganded receptors repress basal transcription through association with corepressor complexes that contain silencing mediator of retinoic acid and thyroid receptors (SMRT), mSin3, and both classes of histone deacetylases (HDACs). Upon ligand binding, receptors dissociate with corepressor complexes and recruit coactivator complexes that contain an array of histone acetyltransferases. For simplicity, basal transcription machinery is not shown in this diagram. N-Cor, nuclear receptor corepressor; RAR, retinoic acid receptor; RARE, retinoic acid response element; CBP, cyclic adenosine monophosphate (cAMP) response element binding protein (CREB) binding protein; PCAF, p300-CBP associated factor.

OTHER ASPECTS OF TRANSCRIPTIONAL CONTROL BY NUCLEAR RECEPTORS

Nuclear Receptor Ligands

Consideration of the diverse stereospecific structures of the known ligands for NRs reflects the diversity and size of the superfamily (Figs. 3.8–3.10). As discussed previously, the ligand-binding function of NRs evidently evolved in very small steps, yielding a large family of proteins capable of discriminating minor differences in chemical structures. The majority of known ligands are lipophilic, polycyclic planar compounds containing a variable number of ring structures, side chains, and alcohol, carboxylate, or ketone functional groups. Many of the receptors (including the class I receptors, VDR, LXR, EcR, FXR, PXR, and CAR) recognize unique metabolites of the mevalonate pathway that share a three- or four-ring cholesterol-like structure. The natural sterol moieties are recognized with high affinity by the classical steroid receptors, with dissociation constants in the low to high nanomolar range, making these receptors exquisitely sensitive to changes in steroid levels. The liver, kidney, adrenal cortex, and other steroidogenic tissues are the main sites of synthesis of these compounds, which are delivered to target organs bound to a variety of serum carrier proteins.[213A] The TRs, RARs, and PPARs recognize distinct hydrophobic compounds, derived from metabolic pathways distal to or distinct from the mevalonate pathway, including thyroxin, retinol, and eicosanoid metabolites, respectively. For many of the classical steroid hormones, the physiologically active receptor ligands are distinct metabolites of the parental steroid, providing a mechanism for regulation and specificity of ligand production by metabolism. For example, dihydrotestosterone has dramatically higher affinity for AR than testosterone, and VDR is activated by 1,25-dihydrotachysterol and not vitamin D.[214] Ligands for a subset of xenobiotic orphan receptors may also include metabolic products of select P-450 steroid hydroxylases that resemble steroid metabolites (i.e., PXR/SXR, CAR). Certain steroid receptors are somewhat promiscuous with respect to ligand specificity and are capable of binding to identical ligands. Examples of this cross-specificity include mineralocorticoid receptor high-affinity binding to cortisol and the ability of both GRs and PRs to bind the synthetic glucocorticoid antagonist RU486 with relatively high affinity.[215,216] For some receptors, such as the retinoid receptors and PPARs, fatty acyl chain length and degree of saturation can also influence binding affinity.[158,217] Furthermore, orthologs within a particular family can selectively recognize distinct ligands, as is the case for PPARs, or exhibit overlapping binding specificity, as is the case for the retinoid receptors ER and TR.[21,218]

For many receptors, synthetic agonists and antagonists have been developed as therapeutic tools to modulate NR function. For example, detailed structure–function analyses of retinoid receptor ligands has yielded numerous compounds that exhibit overlapping specificity with RAR and RXR subtypes, as well as distinct isoform specific activities.[219] These classes of receptors are particularly amenable to such chemical probing, due to the ability of RAR and RXR paralogs to discriminate between isomeric forms of retinoic acid. In this manner, chemical probes that can antagonize the activity of receptor subtypes among this diverse family have been developed, and these pharmacological tools can be utilized to explore receptor-specific functions as well as potential therapeutic utility.[220] As mentioned previously, many of the orphan nuclear receptors lack a known physiological ligand and may function as either constitutive activators, repressors, or metabolic sensors of dietary compounds.

Heterodimerization and Allosterism

The allosteric interaction between RXR and its heterodimeric partner is specified by the particular heterodimeric partner's DNA-binding properties and ligand specificity. Together, they determine the transcriptional potency of the heterodimers.[221,222] The uniqueness of class II receptors is manifested by their ability to heterodimerize with the common partner RXR and the responsiveness of the heterodimers to RXR-specific ligands. Two classes of RXR-dependent heterodimers can be distinguished based on

Receptor	Natural	Agonist	Antagonist
ER	β-estradiol	Diethylstilbestrol	Hydroxytamoxifen
PR	Progesterone	Norethrindrone	RU486
AR	Dihydrotestosterone	R1881	Cyproterone Acetate
GR	Cortisol	Dexamethasone	RU486
MR	Aldosterone	Fluorocortisol	Spironolactone
VDR	1,25-dihydroxy-cholcalciferol	Dihydrotachysterol	
TR	3,4,3'-L-triiodothyronine	3,5-dimethyl-3-isopropylthyronine	

their ability to respond to RXR ligands.[221] For example, when RXR heterodimerizes with VDR, TR, or RAR on DR-3, DR-4, or DR-5 response elements, respectively, the heterodimers do not respond to the RXR agonists (LG268 or LG69) and are nonpermissive. The inability to respond to RXR agonists is mediated by two distinct mechanisms. First, the ligand-binding capacity of RXR is dramatically decreased.[60,223] Second, the partner recruits corepressors in the unliganded state (in the case of TR).[223] The nonpermissive heterodimers bind DRs such that RXR occupies the 5′ half-site and the partner occupies the 3′ half-site.[28,224,225] In contrast, PPAR:RXR and LXR:RXR function as permissive heterodimers that bind DR-1 and DR-4, respectively, and can be activated by 9-*cis* RA.[60,226] In the case of PPAR:RXR, the heterodimer binds DR-1 in a reverse orientation such that PPAR binds the 5′ half-site and RXR binds the 3′ half-site.

A somewhat unexpected observation came from analysis of RXR:RAR heterodimers that direct distinct transcriptional responses on differentially spaced half-sites. On DR-5 (which is found in the *RARβ2* gene promoter), RXR:RAR heterodimers repress transcription in the absence of hormone and activate transcription upon RAR-specific agonist binding. However, the RXR:RAR heterodimer represses transcription in both the absence and the presence of hormone when bound to DR-1 [found in the promoter of the *CRBPII* (cytosolic retinol-binding protein II) gene]. This allosteric control is due to the inability of RXR to bind ligand, the constitutive association of co-repressor with the heterodimer when bound to DR-1, and the requirement of RAR binding to the 5′ half-site of DR-1.[227]

Studies on the effects of synthetic retinoids for RXR and RAR suggested that agonists and antagonists exert differential allosteric regulation on the transcriptional activity of the heterodimers. Based on their structure–function relationship, it is conceivable that structurally distinct ligands can differentially modulate transcriptional activity (or cofactor association) of the heterodimers since both repression and activation activities are mapped to overlapping regions within the LBD. A unique form of allosteric control has been identified in RXR:RAR signaling.[220,228] LG100754 is an RXR-specific antagonist that occupies the LBD but cannot activate RXR homodimers. Unexpectedly, this ligand acts as an agonist of RXR:RAR and PPAR:RXR. This evidence suggests that ligand binding of one subunit (RXR) results in a linked conformational change in the second, noncovalently bound subunit (RAR) of the heterodimer. Consequently, this conformational change results in ligand-dependent dissociation of co-repressors and association of coactivators and generates a transcriptionally active heterodimer.

Crosstalk Between Nuclear Receptors and Other Transcription Factors

Nuclear receptors can also influence the activities of other transcription factors in either a positive or a negative manner (Fig. 3.11). In some contexts, this crosstalk does not require the DBD. The best-characterized examples of NR crosstalk are between GR and AP-1 and between GR and NF-κB.[252–255]

The GR induces differentiation and represses the production of proinflammatory cytokines, while AP-1 promotes proliferation and inflammatory responses.[256–258] The latter is a sequence-specific transcription factor composed of either homo- or heterodimers among members within the Jun family (c-Jun, JunB, and JunD) or among proteins of the Jun and Fos (c-Fos, FosB, Fra1, and Fra2) families.[259,260] It receives extracellular signals through protein kinase signaling cas-

---

Figure 3.8 Chemical structures of steroid and thyroid hormone receptor ligands. Natural hormones and examples of high-affinity synthetic agonists and antagonists are presented for each of the steroid and thyroid classes of receptor. *Agonist activity* is defined as the ability to induce receptor-dependent transcription of natural or synthetic response element reporters in cotransfection experiments and, in most cases, is supported by in vitro ligand-receptor binding studies. ER, estrogen receptor; PR, progesterone receptor; AR, androgen receptor; GR, glucocorticoid receptor; MR, mineralocorticoid receptor; VDR, vitamin D receptor; TR, thyroid hormone receptor.

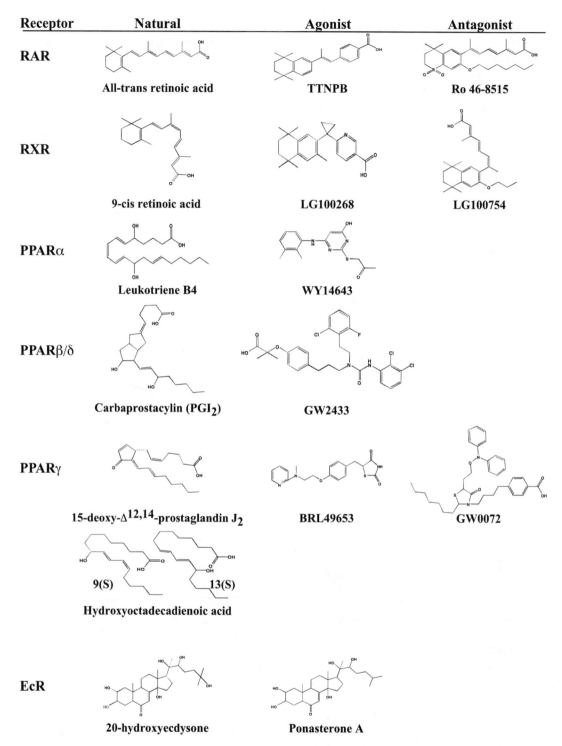

Figure 3.9 Chemical structures of nonsteroid receptor ligands. Proposed natural ligands and select high-affinity synthetic agonists and antagonists are presented, where available. RAR, retinoic acid receptor; RXR, retinoid X receptor; PPAR, peroxisome proliferator-activated receptor; EcR, ecdysone receptor.

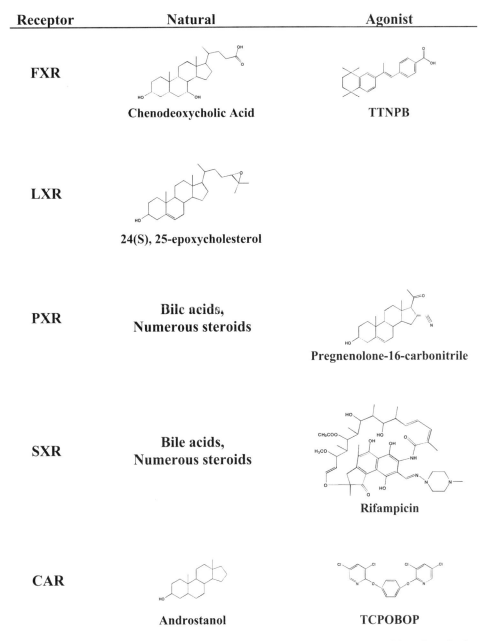

Figure 3.10 Chemical structures of select orphan receptor ligands. Proposed natural ligands and select high-affinity synthetic agonists are presented, where available. FXR, farnesoid X receptor; LXR, liver X receptor; PXR, pregnane X receptor; SXR, steroid and xenobiotic receptor; CAR, constitutive androstane receptor.

cades, such as mitogen-activated protein kinase (MAPK).[258] While inhibition of AP-1 activity was initially described for GR, it has now been shown that overexpression of certain receptors, such as PPARα, PPARγ, and TR, also antagonizes AP-1 activity. The GR may also directly interact with DNA-bound AP-1 through its DBD and the bZip domain of AP-1. This interaction results in a "repressive" conformation such that the overall activity of AP-1 is compromised. A special type of transrepression occurs in the so-called composite response elements, where the binding sites of GR and AP-1 partially overlap. Interestingly, GR acts synergistically when AP-1

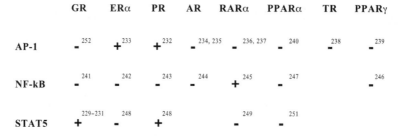

Figure 3.11 Cross-talk between nuclear receptors and activator protein 1 (AP-1), nuclear factor kappa B (NF-kB), and signal transducer and activator of transcription 5 (STAT5) documenting how nuclear receptors may influence the activity of other transcription factors. The effects of nuclear receptor cross-talk are indicated as +, indicating synergistic activation, and −, indicating antagonism. Numbers represent primary reference citations. GR, glucocorticoid receptor; ER, estrogen receptor; PR, progesterone receptor; AR, androgen receptor; RAR, retinoic acid receptor; PPAR, peroxisome proliferator-activated receptor; TR, thyroid hormone receptor.

is composed of a Jun dimer but represses activity when AP-1 contains a Jun:Fos dimer.

Other nonexclusive mechanisms for this transcriptional crosstalk have also been proposed. Both NRs and AP-1 requires the common coactivator CBP for transcriptional activation. It has been postulated that the antagonism between NRs and AP-1 derives from competition for CBP because overexpression of CBP relieves GR-dependent AP-1 repression.[134] Yet another mechanism for NR involvement in AP-1 activity involves the Jun amino-terminal kinase (JNK) signaling pathway, which plays a pivotal role in regulating AP-1 activity.[261] Phosphorylation of c-Jun at serine-73 and serine-63 by JNK is required for maximum AP-1 transcriptional activity. This modification results in a higher affinity of c-Jun for CBP. Hormone-activated NRs can inhibit the JNK signal-transduction pathway by upregulation of the expression of MAPK phosphatases, which negatively modulate the activity of JNK. Reciprocally, activation of JNK and overexpression of AP-1 apparently inhibit GR-mediated transactivation on promoters containing HREs. Two lines of evidence suggest that NR-mediated transactivation and AP-1 repression may be two separable functions. One is that GR mutations within the DBD that impair dimerization and therefore both DNA-binding and transcriptional activity are still able to repress AP-1 activity.[262,263] The other is that transactivation-defective TRα also possesses the ability to repress AP-1 activity.[264] These results indicate that the mechanism of transrepression between GR and AP-1 is complex and can occur at multiple regulatory pathways.

The NF-κB/Rel family of transcription factors participates in the activation of immune and inflammatory responses by upregulation of the expression of several inflammatory cytokines and chemokines. These include interleukin-2 (IL-2), IL-6, IL-8, and enzymes such as the inducible form of nitric oxide synthase.[265,266] In quiescent cells, inactive NF-κB is sequestered in the cytoplasm through association of inhibitory proteins, termed IκBs. The IκB proteins regulate the DNA binding and subcellular localization of NF-κB by masking a nuclear localization sequence located at the C terminus of NF-κB. Activation of NF-κB is achieved through proteolytic degradation of IκB, which is triggered by phosphorylation of IκB by IκB kinases, IKK-1 and IKK-2. The level of IκB is controlled, at least in part, by GR.[267,268] Induction of IκB by overexpression of GR can then prevent NF-κB translocation into the nucleus. However, IκB-independent repression of NF-κB by NRs has been proposed.[269] Two observations analogous to AP-1 transrepression activity have also been reported. First, a direct interaction between the receptor and NF-κB subunits has been shown and may contribute to the negative crosstalk.[270,271] Second, like AP-1, NF-κB activation requires CBP/p300 and may compete for its binding. In general, the mechanisms of transrepression by GR to AP-1 and NF-κB are similar. As a result, determination of the exact mechanism of NR crosstalk has proven difficult due to a multiplicity of common interactions between transcription factors. Because of these interactions with common cofactors, traditional methodologies of exploring transcriptional mechanism by transient transfection and over-expression are unsatisfactory in themselves and must be accompanied by careful analysis of endogenous mRNA and protein levels of the relevant factors in each system utilized. As we shall see in the next section, NRs control so many aspects of physiology and cellular function that all cell processes and conditions must be considered, such as cell cycle, differentiation state, and nutritional availability (or serum conditions).

NUCLEAR RECEPTOR PHYSIOLOGY—CONTROL OF GENE NETWORKS

Significant amounts of data concerning the molecular mechanisms of receptor function have been obtained from in vitro methods, yeast two-hybrid assays, or mammalian tissue culture interaction studies. While impressive advances have occurred in the field of NR transcription, the physiology that manifests itself by virtue of expression of these receptors is more difficult to characterize. The frequent occurrence of the typical HRE within genomic promoter regulatory loci raises significant regulatory specificity issues in lieu of our current progress in unambiguous direct target gene identification. Many of these target genes contain multiple receptor response elements or otherwise respond to multiple receptor ligands through the same response element. The complexity of hormone responsiveness of the MMTV LTR remains an instructive illustration of the conundrum of receptor specificity, where nearly overlapping or otherwise identical binding sites respond to multiple distinct hormone response pathways.[272] Because many NRs are bound to DNA in the unliganded state (i.e., receptor classes II, III, and IV), however, transcriptional activation through ligand binding is not the only mechanism for control of gene expression by NRs. Indeed, several of these receptors can have both direct and indirect transcriptional repressive functions when bound to their cognate response elements. In this manner, the ability of steroid hormones and non-hormonal lipid-soluble metabolites to activate NRs involves not only activation of gene expression via binding to positive response elements but also repression of otherwise basal or activated transcription via both DNA-dependent and independent mechanisms. This ability to both directly and indirectly alter cellular activities in response to ligands is the hallmark of NR physiology.[273]

As a class, NRs are involved in multiple aspects of biology, including distinct roles in embryogenesis, organogenesis, development, cellular proliferation, differentiation, and metabolism. This extensive range of functional roles is achieved partly by expression of a large number of unique NRs as well as by developmentally regulated and tissue-specific expression of each receptor and its splice isoforms. This complexity is further amplified, considering tissue-specific expression of cofactors, posttranslational modification, and regulation of receptor function. As primary effectors of initiating transcriptional regulation of whole gene networks, numerous studies have confirmed that gene mutation and/or alteration of NR activity can result in disease states specific to the context of NR expression within that cell, tissue, or organ. Furthermore, these alterations in normal physiological function can have profound influences on growth and development. The following description of NR physiology includes receptor expression patterns, transgenic mouse gene knockout phenotypes, putative target genes, and human molecular genetics of receptor genes. In most cases, we have restricted our presentation of target genes to include only those genes where discrete response elements have been defined and shown to respond to NR ligands in a transcriptionally positive, receptor-dependant manner. An emerging theme throughout the following exploration of NR physiology is the frequent involvement of NRs in aberrant physiology or pathophysiological syndromes, including oncogenesis.

Sex Steroid Hormone Receptors

Estrogen, Progesterone, and Androgen Receptors

The steroid and corticosteroid hormone receptors are responsible for controlling a number of reproductive and homeostatic systems, respectively. The sex steroid receptors ER, PR, and AR direct transcription of genes involved in sexual development, differentiation, and specification (Tables 3.4, 3.5). As a result of their growth-promoting activities in these processes, these receptors are pharmacologically targeted in a number of neoplastic reproductive tissues. Receptors ERα and ERβ are predominantly expressed in reproductive tissues, as well as in vasculature, cardiac muscle, and bone. Their activity is subject to estradiol availability, which is regulated by hormonal cascades generated from the ovary and other steroidogenic tissues. Confirmation of the role of ERs in the regulation of the reproductive tract was achieved by targeted deletion of ERα and ERβ alleles in transgenic mice, which develop normally but are infertile or suffer ovarian dysfunction, respectively.[296,297] In addition to reproductive maintenance, specialized roles for ERs include regulation of bone density and involvement in vascular function. Apparently, ERα also plays a physiological role in spermatogenesis since both male and female mice are infertile. In addition to infertility, ERα knockout mice have diminished bone density, as expected from the causal relationship between the loss of estrogen production following menopause and the occurrence of osteoporosis. Analysis of gene expression in ERα knockouts also demonstrates loss of induction of suspected target genes (PR, *lactoferrin, prolactin*) and further reveals a role for ERs in the negative regulation of gonadotropin gene expression in the hypothalamic–pituitary axis.[286,289] The reproductive phenotype of ERβ knockout mice is considerably less severe than that of ERα-deficient animals. Although ERα and ERβ share overlapping expression patterns and functional roles, they are not completely redundant since ERβ cannot rescue the reproductive phenotype of the ERα knockout. Compound knockouts display phenotypes similar to individual receptor knockouts, with one exception: although the reproductive tract of compound mutants develops normally, ovarian tissues transdifferentiate to form cells and structures characteristic of testes, suggesting that loss of both receptors may result in sex reversal.[285] Conversely, these receptors are also apparently not required for survival since compound knockouts are viable. Because ERα and ERβ are capable of heterodimerization, it is possible that the spectrum of phenotypes observed in these animals reflects partially penetrant phenotypes due to loss of one functional subunit in the heterodimer.[30] Future studies may resolve this isoform specificity, as well as address the ER activity in regulating CNS function and behav-

Table 3.4 Summary of Class I Steroid Hormone Receptor Proposed Functions and Representative Target Genes

Receptor	Binding Site	Principal Functions	Target Genes	References
ERα,β	IR-3 (AGGTCA)	Reproductive regulation	Myc, *lactoferrin*, oxytocin, PR, cathepsin, EB1, vitellogenin, PRL, ovalbumin, pS2	282, 289
PR	IR-3 (AG^A/$_T$ACA)	Reproductive regulation	MMTV LTR myc, jun, FAS, ovalbumin	287, 291
AR	IR-3 (AGAACA)	Sexual determination	MMTV LTR, probasin, C(3), Slp, PSA, p21, KLK2, AKT	284, 294
GR	IR-3 (AGAACA)	Stress response, immune regulation, metabolism, CNS development	MMTV LTR, MTIIA, tyrosine and alanine aminotransferase, tyrosine oxidase, PEPCK, GH, Na/K-ATPaseβ1, POMC (−) PRL (−)	275, 279
MR	IR-3 (AGAACA)	Salt balance	Na/K-ATPase	292, 293

Receptor response elements are presented as orientation, spacing, and half-site composition of repeats. Proposed functions and select transcriptional target genes, where available, are presented. References are select reviews that cover the material presented. We apologize to our many colleagues whose work is presented. Space limitations require us to cite relevant reviews covering the summarized material only. ER, estrogen receptor; PR, progesterone receptor; AR, androgen receptor; GR, glucocorticoid receptor; MR, mineralocorticoid receptor; IR, inverted repeat; CNS, central nervous system; PRL, prolactin; MMTV, mouse mammary tumor virus; LTR, long terminal repeat; FAS, fatty acid synthase; PSA, prostate specific antigen; MTIIA, metallothionein IIA; PEPCK, phosphoenolpyruvate carboxykinase; GH, growth hormone; Na/K-ATPase, sodium potassium dependent adenosine triphosphatase; POMC, proopiomelanocortin; (−), negative transcriptional activity.

Table 3.5 Summary of Class I Steroid Hormone Receptor Knockout Phenotypes, Expression Patterns and Human Mutations

Receptor	Knockout Phenotype	Expression Pattern	Human Mutations	References
ERα	Viable, male/female sterility follicular development impaired, ovarian and uterine defects, impaired spermatogenesis, reduced bone density, absence of mammary glands, agressive behavior	UT, OV, PG, MG, CNS, H, B	Nonsense mutation, E_2-insensitive, multiple mutations in breast tumors	274, 280, 281
ERβ	Viable, fertile, impaired ovulation α/β: viable, infertile, ovary transdifferentiation to seminiferous tubules	Similar to ERα		285, 286
PR	viable, female sterility ovulation, leutinization, uterine and mammary gland defects thymic involution impaired	UT, OV, PG, MG, THY		276, 278
AR		TS, PG, CNS, SC	Multiple mutations conferring androgen resistance, gene amplification in prostate carcinoma	288, 295
GR	Post-natal lethality organ maturation impaired (lung, adrenal gland) elevated plasma corticosteroids, corticotropin	Ubiquitous	Point mutations, glucocorticoid resistance in leukemic cell lines	277, 283
MR	Postnatal lethality, pseudohypoaldosteronism, elevated renin, angiotensin, aldosterone	PIT, H, K, CO, CNS		290

Phenotypes of homozygous null mutations of the indicated mouse receptor genes are summarized, as are expression patterns and known human mutations for individual receptor genes, where available. Compound mutant phenotypes are indicated for the estrogen receptor. ER, estrogen receptor; PR, progesterone receptor; AR, androgen receptor; GR, glucocorticoid receptor; MR, mineralocorticoid receptor; B, bone; CNS, central nervous system; CO, colon; H, heart; K, kidney; MG, mammary gland; PIT, pituitary gland; PG, prostate gland; SC, spinal cord; TS, testes; THY, thymus; UT, uterus; E2, β-estradiol.

ior.[298] (A detailed discussion of estrogen receptor transactivation mechanisms can be found in Chapter 14.)

The concerted action of estrogen and progesterone signaling is required to maintain the reproductive tract in females. Ovarian biosynthesis of these hormones is stimulated by CNS-regulated pulses of pituitary-derived leutinizing hormone (LH) and follicle-stimulating hormone (FSH). The NRs for these steroids influence production and metabolism of their respective hormones through a complex regulatory circuit involving ovarian and adrenal steroid and peptide hormone biosynthesis. The PR gene is a direct transcriptional target of ER, and the transcriptional activity of PR is often associated with antagonism of ER target genes through an unusual mechanism of transcriptional repression.[299] Two receptor mRNAs, encoding PRa and PRb, are transcribed from alternate promoters within the PR gene, and these receptor isoforms have partially distinct transcriptional properties.[300,301] While many aspects of PR target gene specificity and selectivity are still poorly understood, endocrinological examination of estrogen and progesterone signaling suggest that ER is important for the proliferative phase of reproductive maintenance, while PR directs the differentiating aspects of reproductive cellular function.

Given the interdependence of estrogen and progesterone signaling in regulating reproductive function and maintenance, it is perhaps not surprising that similar (albeit less severe) reproductive phenotypes are observed in PR knockout mice.[276] A role for PR in thymic involution, which occurs following pregnancy and provides protection for the fetus against maternal immunity, has been established by analysis of loss of PR activity in thymic epithethelium of knockout animals.[302] Other NRs are also involved in regulating the maturation and apoptosis of immune cells, suggesting that steroids and corticosteroids are instrumental in regulating several aspects of thymic function.

The AR is a major participant in male sex determination and differentiation in response to dihydrotestosterone binding. As mentioned previously, dihydrotesterone is the active AR metabolite, derived by metabolism of testosterone produced in males by the testicular Leydig cells. Testosterone is also a precursor steroid for estrogen production by peripheral tissue metabolism (Fig. 3.12).[303,304] Multiple mutations within the human X-linked AR gene have been characterized, the majority of which result in androgen-resistance or androgen-insensitivity syndrome. This insensitivity gives rise to a spectrum of sex-determination phenotypes, often resulting in complete feminine testicularization.[305] Transcriptional targets of androgen action include proteins involved in cell cycle control, sper-

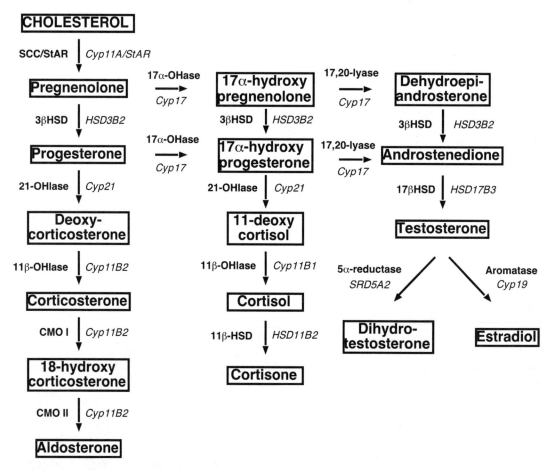

Figure 3.12 Steroid hormone biosynthesis from cholesterol. Cholesterol serves as the precursor to the production of progesterone, aldosterone, cortisone, estrogen and testosterone. Enzymatic activities are indicated in bold, and gene designations for the proposed activities are indicated in italics. (Pathway information adapted from Figures 3 and 4 from Biason-Lauber.[304]) SCC, cholesterol side chain cleavage enzyme; StAR, steroidogenic acute regulatory protein; 3βHSD, 3-beta-hydroxysteroid dehydrogenase; 21-OHlase, 21-hydroxylase; 11β-OHlase, 11n-beta-hydroxylase; CMO, corticosterone methyl oxidase; 17α-OHlase, 17-alpha-hydroxylase; 17βHSD, 17-beta-hydroxysteroid dehydrogenase.

matogenesis, and sex determination. The response elements for ER, PR, and AR are nearly identical, and it is likely that androgens and antiandrogens influence the transcriptional activity of genes in these other pathways as well. (A detailed discussion of androgen receptor transactivation mechanisms can be found in Chapter 16.)

Sex Steroid Hormone Receptors and Cancer

Because of their involvement in the regulation of growth control and differentiation during development of the reproductive system, ER, PR, and AR activities are targets of therapeutic intervention in cancers associated with these tissues. Both ER and PR are expressed in breast and uterine cancers, and activation of AR has been implicated in prostatic cancer.[306,307] Oncogenes such as c-*myc*, c-*fos*, and c-*jun* are both direct and indirect transcriptional targets for these receptors in these tissues; and it is likely that these and other cell cycle regulators are influenced by steroid action.[308–310] Clinical use of antagonists and partial agonists of these receptors, such as 4-hydroxytamoxifen and RU486, results in reduction of tumor progression in patients whose tumors are characterized as hormone-responsive. Hormone responsiveness of these tumors often declines following treatment, probably due to second-site mutations that alter the ligand-binding capacity of the receptor.[7] Amplification of receptor and co-activator genes can be detected in malignant cells, supporting a direct role for NRs in tumor progression and oncogenicity.[117] While the molecular mechanisms by which these receptors promote aberrant proliferation are still under evaluation, it is clear that antagonism of their activity can positively influence growth arrest and has a beneficial impact on prognosis of these cancers.[311]

Corticoid, Vitamin D, and Thyroid Hormone Receptors

Glucocorticoid and Mineralocorticoid Receptors

Glucocorticoids are involved in a wide array of physiological responses, including regulation of glucose, lipid, and protein metabolism, stress response, immune function, and mood (Tables 3.4, 3.5). Given the wide array of responses that these hormones (produced primarily by the adrenal gland) elicit, it is not surprising that the GR is expressed widely and that its function has been extensively explored. As discussed above (see Crosstalk Between Nuclear Receptors and Other Transcription Factors), analysis of bona fide target genes for GR has been complicated by the multiple mechanisms of transcriptional control that this receptor exhibits. For gluconeogenic enzymes, the GR generally activates transcription through positively acting response elements. However, many of the actions of glucocorticoids are inhibitory, and this is partially achieved through binding to negative response elements within the promoters for genes such as prolactin and proopiomelanocortin. Most of the antiinflammatory effects of glucocorticoids involve another mechanism of repression, which does not seem to require the DBD of this receptor and probably involves protein–protein interactions of GR with subunits of the AP-1 and NF-κB transcription factors described earlier.[283,312] Elegant studies evaluating the functional consequences of GR inactivation in mice demonstrate that GR function is essential for viability but that its DBD is indeed dispensable for most of these essential functions.[313,314] This was demonstrated by comparing the phenotypes of animals that had homozygous null GR alleles with those that had wild-type alleles replaced by a mutant allele that prevented homodimerization and impaired DNA binding. The only phenotype common to the two mutant lines of mice was lack of GC-dependent induction of gluconeogenic enzymes, suggesting that many of the physiological functions of GRs are mediated by DNA-independent mechanisms of transcriptional regulation. As discussed previously, NRs probably acquired DNA-binding activity prior to ligand-binding activity during evolution. In this case, it seems that under certain physiological states, regulation of transcription by the LBD (i.e., cross-talk) supercedes the functionality of the DBD. As with several other NRs, two alternatively spliced mRNAs are generated from the GR gene, designated GRα and GRβ, which differ in length at the 3′ end of the transcript. While the in vivo significance of expression of these isoforms has not been thoroughly explored, preliminary studies suggest that these isoforms have differential transcriptional activity and that truncated GRβ, which lacks an AF-2 region, functions as a dominant negative subunit.[315]

Mineralocorticoid receptors are primarily involved in the regulation of salt balance. In addition to being activated by the mineralocorticoid aldosterone, MR binds to and is activated by glucocorticoids. Since MRs are expressed in many tissues that also express GRs, it is probable that MR also activates transcription through GREs.[316,317] However, in tissues where MR function is critical, such as in the gut and kidney, glucocorticoids are metabolized via the action of 11β-hydroxysteroid dehydrogenase, ensuring MR–aldosterone sensitivity.[318] Currently, little is known about bona fide MR target genes, other than that loss of the receptor through gene knockout results in severely impaired regulation of salt balance.[319] In any event, it is clear that the CNS-related activities of these two receptors ensure appropriate regulation of peptide and steroid hormone production and response from the hypothalamic–pituitary–adrenal axis.

Vitamin D_3 and Thyroid Receptors

The vitamin D_3 receptor is involved in bone mineralization and calcium deposition in response to levels of 1,24-dihydrotachysterol (produced by the liver and kidney), as illustrated by kindred studies of mutations within the human VDR gene (Tables 3.6, 3.7).[320] These mutations fall into two classes: point mutations within the DBD that interfere with vitamin D–dependent transcription and mutations that affect ligand binding, rendering individuals insensitive to the actions of vitamin D_3. Both mutations result in clinical phenotypes of general vitamin D_3 resistance, including hypocalcemia and rickets.[359] As a dimerization partner with RXR, VDR is also a candidate for hormone therapy in certain cancers. For example, activation of VDR in breast, lung, prostate, and colon tumors can have antiproliferative effects, which may influence RXR activity in these cells.[360] Additionally, VDR can activate the cyclin-dependent kinase inhibitor p21, providing another mechanism for influence on cell cycle activities.[361]

Thyroid hormone receptors (TRα and TRβ) exhibit multiple modes of DNA binding, including both homodimerization and RXR heterodimerization on a number of diverse response elements, where they can both activate and silence transcription, depending on the promoter context. The TRs are activated by triiodothyronine (T3), and physiological roles for these receptors include regulation of thyroid and pituitary tropic hormone production, affecting a variety of CNS and metabolic functions. Dominant human mutations within the *TRβ* gene (but not *TRα*) have been characterized, and individuals carrying these mutations display clinical symptoms of generalized resistance to thyroid hormone.[362] The mutations interfere with ligand binding but not dimerization or DNA binding and likely represent dominant negative mutations.

Thyroid Receptors and Cancer

An oncogenic form of *TRα* (*v-erbA*) is required for transformation by the avian erythroleukemia retrovirus (AEV).[363] *v-erbA* is expressed as a gag fusion at its amino terminus, followed by the receptor sequence, which has lost nine amino acids at its C terminus. As a result of this mutation, *v-erbA* has lost the ability to bind and respond to T3 but retains the ability to bind DNA in a sequence-specific manner. The disease manifested by this virus (which also expresses *v-erbB*, an oncogenic, constitutively active tyrosine kinase) is characterized by the presence of immature erythroblasts, committed cells with self-renewal capability that are blocked in differentiation. *v-erbA* contributes to the leukemogenic phenotype by transcriptional repression of erythroid-specific genes, acting as a dominant silencer by interfering with the differentiating properties of splenic–erythrocytic TR and RAR, and can also influence AP-1 activity.[364] Importantly, the *v-ErbA* td359 mutant, which fails to block differentiation, also fails to repress transcription and contains a mutation that diminishes the affinity of *v-ErbA* for the corepressor SMRT.[105] Other dominant negative mutations within both *TRα* and *TRβ* have been identified in human hepatocellular carcinoma cells.[365] Clearly, the activity of TRs impacts multiple growth-related processes

Table 3.6 Summary of Class II Nonsteroid Hormone Receptor Proposed Functions and Representative Target Genes

Receptor	Binding Site	Principal Functions	Target Genes	References
VDR	DR-3	Bone mineralization, ossification, calcium and phosphate metabolism	p21/WAF1, VD24-hydroxylase, osteocalcin, osteopontin	323, 324, 336
TRα, β (v-erbA)	DR-4, ER-6 ER-8, IR-0	Thyroid metabolism	carbonic anhydrase, GH, malic enzyme, MHC-A, MBP TRH (−), TSHα,β (−)	325, 327
RARα, β, γ	DR-2, DR-5 ER-8, IR-0 °DR-1 (−)	Vitamin A metabolism, embryonic development CNS development and regulation, myeloid differentiation	CRBPI, CRABPII°, RAR-α,β,γ, HOX1.6, ADH3, p21, CD38	321, 326, 329
<u>**RXR**α,β,γ</u>	DR-1, IR-0	Heterodimerization partner for class II NRs, vitamin A metabolism, embryonic development and differentiation	CRBPII, CRABPII, MHCI, (dependant upon dimer partner)	51, 327, 330
PPARα	DR-1	Peroxisome proliferation, fatty acid β-oxidation	AOX, CYP4A, LXRα, PEPCK2, BIF, FAT	332, 335, 335a
PPARδ	DR-1	Uterine implantation, cell cycle control		
PPARγ	DR-1	Adipogenesis, cellular differentiation, lipid and glucose metabolism	aP2, CD36, LXRα FATP, PEPCK1, LPL, adipsin, leptin (−)	
LXRα, β	DR-4	Cholesterol catabolism	CYP7A, ABCA1, ABC8, CETP SREBP-1, ApoE	328, 333, 333a
FXR	IR-0, IR-1, ER-6	Bile acid metabolism	IBABP, SHP, NTCP, OATP, ABCC2	331, 335a
PXR/SXR	DR-3, ER-4 ER-6	Steroid, xenobiotic metabolism	CYP3A4, 2A/2B	348, 348a
CAR	DR-3, DR-4, DR-5, ER-6	Steroid, xenobiotic metabolism	CYP2A/2B, 3A	322, 334

Response elements for receptors that heterodimerize with the retinoid X receptor (RXR) are presented, as are proposed functions and select transcriptional target genes. Response elements are designated direct repeat (DR), inverted repeat (IR), or everted repeat (ER), with spacing between half-sites as indicated. Boldface indicates permissivity for RXR ligands, and underlining indicates ability to homodimerize. References are select reviews that cover the material presented. We apologize to our many colleagues whose work is presented. Space limitations require us to cite relevant reviews covering the summarized material only. VDR, vitamin D receptor; TR, thyroid hormone receptor; RAR, retinoic acid receptor; RXR, retinoid X receptor; PPAR, peroxisome proliferator-activated receptor; LXR, liver X receptor; FXR, farnesoid X receptor; PXR/SXR, pregnane X/steroid and xenobiotic receptor; CAR, constitutive androstane receptor; CNS, central nervous system; VD24-hydroxylase, vitamin D 24-hydroxylase; GH, growth hormone; MHC-A, myosin heavy chain A; MBP, myelin basic protein; TRH, thyrotropic releasing hormone; TSH, thyroid stimulating hormone; CRBPI, cellular retinol binding protein I; CRABPII, cellular retinoic acid binding protein II; RAR, retinoic acid receptor; HOX, homeobox 1.6; ADH3, alcohol dehydrogenase 3; MHC, major histocompatibility complex; AOX, acyl CoA oxidase; CYP, cytochrome p450; PEPCK, phosphoenolpyruvate carboxykinase; BIF, bifunctional enzyme; FAT, fatty acid translocator; FATP, fatty acid transport protein; LPL, lipoprotein; ABC, adenosine triphosphate binding casette; CETP, cholesterol ester transfer protein; SREBP1, sterol regulatory element binding protein 1; ApoE, apolipoprotein E; IBABP, intestinal bile acid binding protein; SHP, short heterodimer partner, OATP, organic anion transport protein; NTCP, sodium taurocholate carrier protein. Negative (−) represents negative transcriptional activity from this promoter depending upon response element.

and plays crucial roles as both an activator and a silencer of transcription.

Retinoid Receptors

Retinoic acid, the active metabolite of dietary vitamin A, exerts profound effects systemically and in a variety of target tissues when administered in pharmacological doses in rodents and humans.[366] The morphogen-like activity of this hormone is primarily transduced intracellularly through the activity of two families of retinoid nuclear receptors, the RARs and RXRs. Both families consist of three orthologous genes each, α, β, γ, several of which also express differentially spliced isoforms. With gradedly overlapping patterns of expression throughout the body, nearly all tissues express at least one class of re-

Table 3.7 Summary of Class II Nonsteroid Hormone Receptor Knockout Phenotypes, Expression Patterns, and Human Mutations

Receptor	Knockout Phenotype	Expression Pattern	Human Mutations	References
VDR	Postnatal lethality, rickets symptomology	B, K, CNS	Point mutations, vitamin D_3 resistance	324, 347
TRα	Viable, reduced body temperature, arrythmia	Ubiquitous		
TRβ	Viable, deaf compound α/β: retarded growth and bone maturation, defects in fertility and thyroid-pituitary axis	CNS, PIT	Point mutations, resistance to thyroid hormone	339, 341
RARα	Decreased viability, homeotic transformations, male sterility, congenital defects	Ubiquitous	Multiple translocations and protein fusions associated with acute promyelocytic leukemia	338, 346, 356
RARβ	Viable, long-term memory defects	Neural tube, interdigital mesenchyme, neuroepithelia LU, I, genital, inner ear epithelia, CNS		340, 344, 349, 350
RARγ	Decreased viability, homeotic transformations, male sterility, congental defects	Neural tube, bone & facial cartilage, epidermis		
RXRα	Embryonic lethal, E10.5–17.5, myocardial thinning, retinal defects	L, K, SP, visceral tissues, epidermis		342, 343, 350
RXRβ	Viable, male infertility, abnormal spermatid maturation	Ubiquitous		
RXRγ	Viable, long-term memory defects	M, CNS, PIT		
PPARα	Viable, disturbed peroxisomal proliferation, fatty acid metabolism when challenged by fasting/high-fat diet	L, K, H, BAT, I		345, 352, 353, 355, 357
PPARβ/δ	Viable, reduction in adipose tissue semi-penetrant lethality in some genetic backgrounds	Ubiquitous		
PPARγ	Embryonic lethality at E10 placental defects, cardiomyopathy, later postnatal lethality, lipodystrophy and multiple hemorrhages	AD, MP, LI, PL	Point mutations in LBD in colon adenocarcinomas, insulin resistance, translocation in thyroid carcinoma	
LXRα	Viable, impaired hepatic function when challenged by high-cholesterol diet	L, K, I, SP, AD		351
FXR	Viable, hypercholesterolemia, elevated fecal and urinary bile acids	L, I, K, AG		354, 354a
PXR/SXR	Viable, impaired xenobiotic response	L, I, K		348, 358
CAR	Viable, impaired xenobiotic response	L, I, K, H, M		334, 334a

Phenotypes of homozygous null mutations of the indicated mouse receptor genes are summarized, as are expression patterns and known human mutations for individual receptor genes, where available. VDR, vitamin D receptor; TR, thyroid hormone receptor; RAR, retinoic acid receptor; RXR, retinoid X receptor; PPAR, peroxisome proliferator-activated receptor; LXR, liver X receptor; FXR, farnesoid X receptor; PXR/SXR, pregnane X/steroid and xenobiotic receptor; CAR, constitutive androstane receptor; B, bone; K, kidney; CNS, central nervous system; I, intestine; PIT, pituitary gland; LU, lung; SP, spleen; M, muscle; L, liver; H, heart; BAT, brown adipose tissue; AD, adipose; MP, macrophage; LI, large intestine; PL, placenta; AG, adrenal gland.

ceptor of each family, ensuring global RA responsiveness.[367] Circulating RA levels are tightly regulated, as is the transport and metabolism of this vitamin.[368] As described below, the isomers of RA that activate the RAR and RXR receptors (all-*trans* RA and 9-*cis* RA, respectively), transduce both direct and cross-coupling signals in a variety of visceral, neural, and epithelial tissues that specify programs of development, differentiation, and cognitive function (Tables 3.6, 3.7). The spectrum of phenotypes observed in fetal vitamin A deficiency syndrome are mirrored in various compound transgenic knockout mice, confirming the requirement of these receptors for regulating RA signaling.

Retinoic Acid Receptors

All members of the RAR family are capable of binding to both all-*trans* RA and 9-*cis* RA, the two isomers produced by the metabolism of the active derivative of dietary vitamin A, retinol.[217] The abundance and tissue distribution of these compounds are regulated partially by the RARs themselves as RXR heterodimers and by RXR homodimers.[52] These various receptor dimers participate in complex tissue-specific transcriptional regulation of the cellular retinol and RA binding proteins (CRBPs and CRABPs) and metabolic enzymes that catabolize retinoids, ensuring appropriate distribution of and protection from the teratogenic effects of retinol esters in development. Developmental studies suggest that a morphogen gradient of RA activity throughout the developing embryo is critical for proper limb formation, axial patterning, neural crest cell migration, and specification of cognitive, sensory, and motor neurons.[340] Also, RARs are involved in specialized functions in the differentiation of neural and visceral epithelia and the epidermis. In addition to these developmental and diffentiative roles, many molecular studies demonstrate that RAR activity can modulate the function of its dimeric partner, RXR, primarily due to interference by virtue of high-affinity dimerization between RAR and RXR. Since RAR:RXR dimers are more stable than RXR homodimers, overexpression of RAR can effectively compete for the intracellular pool of RXR molecules and reduce the activity of RXR-mediated activation from RXR-specific response elements.[58] This suggests that physiological levels of expression of RAR and RXR subunits in each cell dictate the responsiveness of that particular cell to both all-*trans* RA and 9-*cis* RA. Indeed, evaluation of the expression patterns for members of each of these families suggests complex, partially overlapping but non-redundant functions for each of the particular RAR subtypes (Table 3.7). The plethora of genetic data on single or multiple knockouts for each of the many RARα, -β, and -γ subtypes and isoforms also demonstrate both essential overlapping and distinct functions for each family member. Although deletion of individual receptors is not associated with lethality, careful analysis of compound mutants suggests a profound role for these receptors in the specification and organogenesis of multiple systems.[369–371]

Retinoic Acid Receptors and Cancer

In addition to analysis of RAR function by gene knockout technology in mice, several genetic lesions within RAR loci are found in human patients who exhibit acute promyelocytic leukemia (APL). This disease, which represents about 5% of acute myeloblastic leukemias in adults, is characterized by abnormal accumulation of myeloid precursors blocked at the promyelocytic stage of maturation, resulting in severe leukocytosis and hemorrhage.[349] Five types of chromosomal translocation have been described in APL, which produce chimeric proteins containing either promyelocytic leukemia (PML), promyelocytic leukemia zinc finger (PLZF), nucleophosmin (NPM), nuclear meiotic apparatus (NuMA), or signal transducer and activator of transcription 5β (STAT5β) fused to RARα (Fig. 3.13). These fusion partners are structurally unrelated, although all are nuclear proteins that contain a homodimerization domain. The two best-studied chimeric receptors, PML-RARα and PLZF-RARα, retain identical sequences from the RAR gene, including the DBD and LBD, while acquiring additional sequences from PML and PLZF, respectively, at their amino termini. Both chimeric receptors are leukemogenic at physiological concentrations of RA but behave differently when exposed to pharmacological does of RA. Because of its altered stoichiometric interaction with corepressors, PML-RAR responds

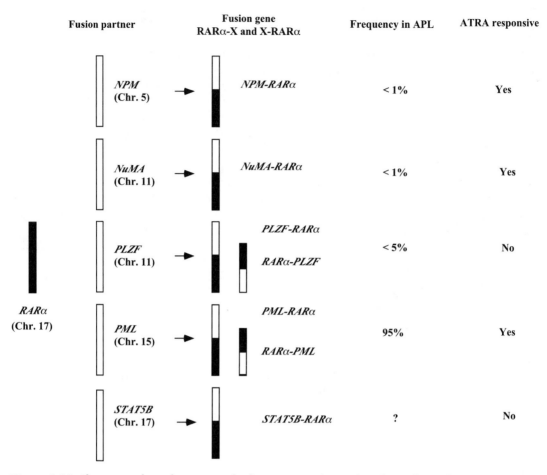

Figure 3.13 Chromosomal translocation involved in acute promyelocytic leukemia (APL). Five types of chromosomal translocations have been described in APL. These translocations involve retinoic acid receptor alpha (RARα) in chromosome 17 fused to different partners, including promyelocytic leukemia (PML), promyelocytic leukemia zinc finger protein (PLZF), nucleophosmin (NPM), nuclear meiotic apparatus (NuMA), and signal transducer and activator of transcription 5B (STAT5b). Interestingly, while patients with PML-RARα, NPM-RARα, or NuMA-RARα respond to pharmacological concentrations of retinoic acid, patients with PLZF-RARα or STAT5B do not respond to retinoic acid at all. (Adapted from Figure 1 of Lin, et al.[348]) Chr., chromosome; ATRA, all-trans retinoic acid.

only to pharmacological does of RA.[372] The lack of RA responsiveness of PLZF-RARα is due to an additional interaction surface between corepressor complexes and the amino terminus of PLZF, which is insensitive to ligand binding of the chimeric receptor. As a consequence, PLZF-RARα functions as a dominant negative form of RAR and constitutively represses RAR target genes. The mechanism underlying the leukemogenesis of PML-RARα is not yet known, but the possibility that PML-RARα interferes with the normal function of PML, such as growth inhibition and apotosis promotion, may partly account for its oncogenicity.

Retinoid X Receptors

As common heterodimerization partners for members of the class II NRs, RXR activity is involved in nearly all nonsteroid NR-dependent signaling pathways (Table 3.6, Fig. 3.4). Its involvement in RAR, VDR, TR, and orphan receptor physiology is well documented: RXR heterodimerization is required for the activity of at least 11 disinct receptor pathways.[51,373] Despite expression of at least one RXR isoform in every tissue, null mutation of RXRα is embryonic lethal at embryonic day (E) 10.5 to E17.5 due to failure of placentogenesis and cardiac func-

THE NUCLEAR RECEPTOR SUPERFAMILY

tion (Table 3.7).[374,375] The RXR genes, like the RAR family, are expressed widely, in a partially overlapping but subtly distinct pattern in a number of tissues. Both RARα and RXRβ are ubiquitously expressed, while RXRα and RXRγ expression is restricted to distinct organ systems.[21,376] Also similar to RARs, RXR single knockout animals exhibit distinct phenotypes, demonstrating that the RXRs are not functionally redundant but have evolved sufficiently following gene duplication to perform distinct cellular functions. Examples of this specialization include roles for RXRβ and -γ in long-term memory and spermatogenesis, respectively.[377,378] Some of the phenotypes observed in RXR-deficient animals are linked to its involvement in RAR signaling and retinoid metabolism. Evaluation of knockout phenotypes, therefore, must consider potential inactivation of partner function through RXR loss of function. In fact, in at least one case (PPARγ), the RXRα phenotype has been nearly phenocopied by knockout of a dimer partner, illustrating the apparent auxiliary function that RXR exhibits as a master regulator of nonsteroidal receptor function.

As homodimers, RXRs regulate vitamin A metabolism and availability through activation of CRBP and CRABP expression from DR-1 elements.[58] As discussed above, this regulation is critically dependent on the cellular RXR/RAR ratio as well as the ATRA/9-*cis* ratio since RXR:RAR can function negatively through DR-1. Since most cells express higher levels of RXR, this allows the possibility of 9-*cis*-dependent RXR involvement in dimer partner signaling pathways. As shown in Table 3.6 and Figure 3.4, a subset of the RXR-requiring NR, are permissive for RXR signaling.[51,221] This permissivity, or ability to activate transcription in response to 9-*cis* RA, has important pharmacological implications for diseases or pathophysiological states associated with RARs, PPARs, LXRs, FXRs, and other receptors, as we shall see below.

Orphan Nuclear Receptors

Steroid and thyroid receptors were originally identified as purified proteins that were capable of binding to radiolabeled hormone ligands. With the advent of molecular cloning, cDNAs encoding novel families of receptors were isolated by virtue of homology similarity within the DBD or by interaction screens utilizing RXRs. Many of these receptors have no known ligand and, therefore, represent orphan NRs that participate in as yet unidentified signaling pathways. For a few of these receptors, this reverse endocrinology approach has revealed previously unknown intracrine or paracrine signaling pathways that are modulated by the activities of the receptors (Tables 3.6, 3.7). For many others, both the putative ligands that activate them and their targets of action remain unknown and represent challenges for future studies to unravel the physiology or pathophysiological circuits that they regulate.

Lipid Receptors: Peroxisome Proliferator–Activated Receptors

The PPARs were first identified as an activity that mediated the dramatic proliferation of rat hepatic peroxisomes in response to fibrates, industrial plasticizers, and herbicides.[379] Two physiologically interesting results arise from treatment with these xenobiotic compounds in rodents. The peroxisomal response results in reduction of serum triglyceride levels by virtue of β and ω oxidation of long chain fatty acids within the peroxisomal and mitochondrial compartments, and prolonged treatment of rodents with these compounds directly correlated with hepatocarcinogenicity.[380–383] Subsequent cloning efforts produced three PPAR cDNAs, designated α, β/δ (NUC-1, FAAR), and γ, which exhibit similar DNA-binding specificities but distinct expression patterns and ligand-binding profiles.[158,384,385] Activity of PPARα promotes fatty acid catabolism by activating expression of genes required for fatty acid uptake and long chain fatty acid β and ω oxidation, presumably in response to dietary, circulating, or intracellular concentrations of fatty acids. Although potent synthetic ligands for PPARα activation have been produced, the identity of endogenous ligands remains obscure: PPARα can be activated by a number of diverse long chain polyunsaturated fatty acids, such as linoleate, arachidonate, and the prostaglandin metabolite leukotriene B_4, at high concentrations.[386,387] Targeted disruption of this receptor in the mouse suggests that its function is not essential for survival as null homozygotes are phenotypically normal. However, these animals are insensitive to the actions of peroxisome proliferators and, when challenged with fasting or high-fat diets, experience perturbations in lipid and glucose homeostasis.[388,389] These observations suggest that PPARα not only functions as a xenobiotic receptor but also may play a significant role in lipid homeostasis in response to unknown physiological ligands.

While PPARα contributes to systemic fatty acid disposal through oxidation, PPARγ regulates and promotes lipid storage by its activities in adipocytes and macrophages. The γ isoform was initially thought to be an adipose-specific NR, whose activation by the prostaglandin metabolite 15-deoxy-$\Delta^{12,14}$-prostaglandin J$_2$ (or synthetic thiazolidinedione) resulted in differentiation of multiple cell types to an adipocyte phenotype.[390–395] The adipogenic process minimally requires insulin and glucocorticoid signaling, as well as the activities of multiple basic helix-loop-helix-leucine zipper transcription factors of the CAAT/enhancer binding protein and sterol regulatory element binding protein class.[396] In addition to its established role in adipogenesis, however, PPARγ performs unique roles in macrophage function, intestinal mucosal proliferation/differentiation, and placentogenesis.[239,397–399] Activation of PPARγ in these tissues may not induce the same program of target genes observed in adipocyte differentiation, suggesting the existence of cell-specific, differential gene targets. One possible explanation for this cell specificity is the identification of tissue specific coactivators as candidates for this activity have been identified.[87] Analysis of PPARγ function by gene knockout in mice reveals part of the essential function of PPARγ: disruption of both PPARγ alleles results in embryonic lethality at E10 due to lack of placentogenesis and myocardial thinning, similar to RXRα-deficient embryos.[399] Furthermore, rescue of this early lethality by tetraploid rescue reveals the requirement of PPARγ for adipogenesis since these animals fail to form fat pads. While these studies demonstrate the requirement for PPARγ in adipogenesis (i.e., absence of fat pads), it remains unclear what other systems or tissues are affected by loss of PPARγ function.

PEROXISOME PROLIFERATOR–ACTIVATED RECEPTORS AND OBESITY: INSULIN SIGNALING

The observations that the thiazolidinedione class of Type 2 diabetic non-insulin-dependent diabetes mellitus insulin sensitizers act as direct PPARγ ligands implicates PPARγ as a NR involved in glucose as well as lipid homeostasis.[400] Although adipose tissue and obesity contribute to insulin resistance, the molecular details concerning this relationship remain unclear.[401] Several possibilities for the involvement of PPARγ activity in insulin and glucose resistance have been suggested, such as regulation of adipocyte size, glucose transporters, and circulating leptin and free fatty acid levels.[402–405] Since PPARγ activity may directly or indirectly contribute to regulation of several of these parameters, it is likely that this receptor will continue to be targeted in the pharmacological intervention of Type 2 diabetes and obesity.[406] Additionally, several studies have implicated PPARγ as a regulator of macrophage foam cell generation, a process directly correlated to the incidence of atherosclerosis, linking PPARγ activity to yet another disease state associated with lipid dysregulation.

PEROXISOME PROLIFERATOR–ACTIVATED RECEPTORS AND CANCER

The ability of PPARγ to promote and establish differentiated phenotypes in at least two cell types (i.e., adipose and macrophage) suggests that PPARγ activators may also be useful to block proliferation in malignant cells that express this receptor. In fact, preliminary studies demonstrate that in breast cancer and liposarcoma cell lines PPARγ and/or RXR ligands are capable of limiting cell growth by promoting differentiation.[407,408] This differentiation therapy is similar in concept to RA treatment in APL, where activation of the NR interferes with or supercedes proliferative growth signals by unknown mechanisms. The ability of PPARγ to promote differentiation, however, is likely tissue- and context-dependent. For example, activation of PPARγ in the colon of the adenomatous polyposis coli (APC)-deficient Min mouse (a model of familial adenomatous polyposis) results in an increase in colon polyp number.[397,409]

However, studies in colon cancer cell lines found that thiazolidinedione treatment correlated with cell growth arrest.[409a] Further analysis of PPARγ transcriptional activity and function in these tissues is therefore required to qualify this receptor as a potential target for treatment of these cancers.

Relatively little is known about the physiology of the more widely expressed PPARβ/δ, which can be activated by prostaglandin I_2 (PGI_2) or carbaprostacylin, a synthetic nonmetabolizable derivative of PGI_2. Evaluation of mice deficient in cyclooxygenase-2 (COX-2), an inducible cyclooxygenase required for prostaglandin synthesis, suggests that PPARβ/δ activity may be required for uterine implantation during pregnancy.[410] Interestingly, COX-2 activity is also involved in colon carcinogenesis in that suppression of COX-2 enzyme activity in intestinal mucosa reduces polyp numbers in the Min mouse.[411] A link between the effects of nonsteroidal anti-inflammatory drugs on colon carcinogenesis may involve regulation of PPARβ/δ expression by APC and β-catenin.[412] PPARβ/δ is also active in wound healing, and can be induced by cAMP signaling.[412a]

Liver X and Farnesoid X Receptors

Another class of orphan receptors that bind nonsteroidal ligands, the liver X and farnesoid X receptors (LXRs and FXR), are involved in the regulation of cholesterol catabolism and bile acid synthesis, respectively (Tables 3.6, 3.7). Isoforms LXRα and -β are activated by stereospecific cholesterol metabolites, such as 22(R)-hydroxycholesterol and 24(S),25-epoxycholesterol in the liver and kidney.[413] Once activated, these receptors (in combination with another orphan receptor, liver receptor homologue 1 [LRH-1]) induce transcription of the rate-limiting enzyme of bile acid synthesis, cytochrome P-450 cholesterol 7α-hydroxylase (CYP7A). This feed-forward pathway of cholesterol clearance is apparently kept in check by the actions of another receptor, FXR, which is activated by conjugated bile acid metabolites such as chenodeoxycholate.[414–416] When bile acids accumulate (subsequent to LXR activation), FXR is activated and stimulates transcription of intestinal bile acid binding protein, which effectively buffers the cell from toxic bile compounds. In addition, FXR activates expression of yet a fourth orphan receptor, SHP-1, which then dimerizes with the constitutively active LRH-1 and attenuates LRH/LXR-mediated transcription of CYP7A, preventing further synthesis of bile acids from cholesterol.[417] Despite its crucial role in gating bile acid biosynthesis and cholesterol clearance, LXRα is apparently not essential for survival as deletion of mouse LXRα yields phenotypically normal mice.[418] However, when challenged with a high-cholesterol diet, these animals accumulate cholesterol in the liver due to failure to induce CYP7A activity, in spite of the presence of functional LXRβ.

Oxysterols apparently also upregulate expression of proteins involved in sterol/lipid transport. For example, certain ABC transporters involved in lipid transport and cholesterol ester transfer proteins are activated through LXR signaling.[419–421] An additional level of regulation may include transcriptional induction of LXRs by PPAR family members activated by certain fatty acids.[422] The involvement of these receptors in the tightly regulated process of enterohepatic cholesterol metabolism suggests that pharmacological modulation of their activity could dramatically affect lipoprotein metabolism in disease states such as hypertriglyceridemia, hypercholesterolemia, and atherosclerosis.

PXR/SXR, CAR

The PPAR, LXR and FXR receptors are apparently activated by metabolites of fatty acids, oxysterols, and bile acids, respectively, suggesting that they function as homeostatic sensors within these metabolic pathways. For PPARα, activity is influenced by xenobiotic compounds as well as putative endogenous fatty acids, suggesting that this receptor may be involved in P-450 induction in response to dietary or environmental toxins. Another class of orphans that exhibits liver-specific expression and is involved in induction of various P-450 enzymes includes PXR/SXR and CAR receptors (Tables 3.6, 3.7). As transcriptional regulators of endo and xenobiotic metabolism, all of these receptors collectively represent potential pharmaceutical targets with respect to improving co-consumed drug action, as well as treatment of various hepatic disorders.[422a]

The broad specificity of ligands that can activate PXR has implicated this receptor as a sensor for hepatic steroids and xenobiotics. The ligands that can activate PXR include C18, C19, and C21 steroid agonists and antagonists, estranes, androstanes, pregnanes, phytoestrogens, the antibiotic rifampicin and many other pharmaceuticals.[57] Once activated, PXR (and its human homologue SXR) directly induce expression of the CYP3A family of steroid hydroxylases.[334,423] Although PXR and SXR are highly conserved within both the DBD and the LBD, they exhibit distinct ligand-binding specificity. While PXR also responds positively to both steroid agonists and antagonists of the C21 class, a high-affinity ligand, pregnenolone-16-carbonitrile (a glucocorticoid antagonist), has low affinity for the SXR LBD. Also, PXR has low affinity for rifampicin, a high-affinity agonist for SXR. Despite the divergence of ligand-binding activity, both receptors activate similar target genes in their respective hosts, perhaps indicating differential toxicity between mouse and human in response to xenochemicals and steroid metabolites. The promiscuous binding activity of these receptors also supports the concept of PXR and SXR as broad-specificity steroid and xenobiotic sensors.

Another receptor, CAR, is also involved in induction of a class of xenobiotic and steroid metabolizing enzymes of the CYP2A class. It represents an unusual orphan receptor, being regulated on one level by chemically induced cytoplasmic to nuclear translocation. Its activity can also be affected positively and negatively by distinct chemical ligands that occupy its ligand-binding pocket.[322] The naturally occuring compound androstenol is an antagonist of constitutive CAR transactivation, while other synthetic chemicals positively affect its activity. CAR is also responsible for the transcriptional induction of CYP2A and CYP2B family members by phenobarbital.[424] PXR and CAR can share response elements on some promoters, including CYP2A/B and CYP3A promoters. This overlapping responsiveness to distinct chemical signals provides a metabolic safety net regarding xenobiotic metabolism.[334,424a,b]

Other Orphan Receptors

Preliminary analysis of the activities and expression patterns of members of the large family of orphan receptors suggests that many are involved in the regulation of liver-specific metabolism (i.e., HNF-4, PPARα, LXR, FXR, LRH-1, SHP-1, SXR, CAR), while others seem to play roles in steroidogenesis, sexual differentiation, and germ cell development (i.e., TR2, TR4, SF-1, GCNF, NGFI-B, DAX-1) or neural function and specification (ERR, ROR, TLX, photoreceptor nuclear receptor [PNR]) (Tables 3.8, 3.9). It is likely that these receptors integrate and regulate gene expression by virtue of regulation of the activities of some of the previously mentioned NRs. Several orphans (COUPs, Rev-ErbA) seem to function exclusively as negative regulators of transcription, silencing in *trans* or bound to DNA as homodimers, RXR heterodimers, and even monomers.[3] At least one receptor (CAR) is a constitutive activator, whose activity is repressed in a ligand-dependent manner.[322] Several structurally divergent orphan receptors have been isolated. For example, DAX-1 lacks a conventional zinc-finger DBD but contains a region consisting of four repeats of alanine and glycine-rich sequences that is involved in DNA binding. Deletion or mutation of DAX-1 is responsible for dosage-sensitive sex- and X-linked adrenal hypoplasia, an inherited disorder of adrenal gland development.[68] In contrast to the DBD, the LBD of DAX-1 is similar to typical members of the NR superfamily. The heterodimeric partner of DAX-1, SHP is structurally similar to DAX-1, containing a putative LBD but lacking a conventional DBD.[69] It has also been proposed to function as a negative regulator of NR signaling pathways because it interacts with and inhibits transcriptional activity of various other receptors, including TRs, RARs, and agonist-bound ERα and ERβ. This inhibition is due to in part to SHP-associated repression activity and to competition with coactivators for receptor binding.[70–72,456] SHP also plays an important role in CYP7A and bile acid metabolism, as mentioned in the LXR and FXR section.

In many cases, these atypical orphan receptors seem to function as adapters, bridging NR

Table 3.8 Summary of Class III and Class IV Orphan Receptor Proposed Functions and Representative Target Genes

Receptor	Binding Site	Proposed Function	Target Genes	References
Class III: Orphan (Homodimeric) Nuclear Receptors				
NGFI-B/	DR-5,	Steroidogenesis, stress response,	Steroid 21-hydroxylase (P450c21),	424, 426
Nurr1/NOR1	Dimer/ monomer	T-cell apoptosis	tyrosine hydroxylase	
RevErbα, β	Dimer/ monomer			427, 429
COUPα,β,γ	DR-1, RXR dimerization	Neural cell migration Gene silencing	Ovalbumin, NGFI-A, ApoCII (±)	73, 436, 440
HNF-4α, γ	DR-1	Glucose/FA metabolism, liver differentiation	ApoCII, ApoCIII, transthyretin, α1 antitrypsin, HIV LTR	3, 428
TR2, TR4	Multiple elements, can heterodimerize			65
GCNF	Dimer/ monomer	Reproductive regulation, steroidogenesis		443
Class IV: Orphan (Monomeric) Nuclear Receptors				
SF-1/FTZ-F1/LRH-1		Reproductive regulation, steroidogenesis, sterol metabolism	MIS, DAX-1, steroid 21-hydroxylase (P450c21), CYP17, CYP7A, CYP2A	434
ERR1, 2, 3		Placentogenesis	ERRs, MCAD, pS2, lactoferrin, TR, osteopontin	151, 439, 450
RORα,β,γ		CNS function	Gamma crystallin	431, 433
TLX		CNS function, nervous system patterning	Pax2	454
DAX-1/ SHP-1	Distinct DBD	Reproductive regulation dominant negative activity	Determined by dimer partner	449

Modes of DNA binding are presented, including binding site motif and ability to homo- and heterodimerize. Boldface indicates permissivity of retinoid X receptor (RXR) ligands (for those that dimerize with RXR), and underlining indicates ability to homodimerize. Proposed functions are presented, as are select transcriptional target genes, where available. References are select reviews that cover the material presented. We apologize to our many colleagues whose work is presented. Space limitations require us to cite relevant reviews covering the summarized material only. NGFI-B, nerve growth factor induced receptor B; Nurr1, Nur related receptor 1; NOR, neuron derived orphan receptor; COUP-TF, chicken ovalbumin upstream promoter transcription factor; HNF4, hepatocyte nuclear factor 4; TR2/TR4, testis receptor 2, 4; GCNF, germ cell nuclear factor; SF-1, steroidogenic factor 1; FTZ-F1, fushi tarazu factor 1; LRH1, liver receptor homologous protein 1; ERR, estrogen related receptor; ROR, retinoic acid related orphan receptor; TLX, tailless related receptor X; DAX-1, dosage sensitive sex reversal adrenal hypoplasia X chromosome region 1; SHP, small heterodimer partner; RXR, retinoid X receptor; DR, direct repeat; DBD, DNA binding domain; POMC, proopiomelanocortin; apo, apolipoprotein; HIV, human immunodeficiency virus; LTR, long terminal repeat; MIS, mullerian inhibiting substance; MCAD, medium chain acyl-coA-dehydrogenase. Positive and negative symbols (±) indicates positive and negative activity from this promoter, depending upon dimerization state and response element.

and common cofactor interactions, a functional and evolutionary step between the receptor and the coactivators/corepressors. Future studies will be aimed at elucidating functions for these orphan receptors, in the hope of identifying new physiological circuits and mechanisms of receptor regulation.

Orphans and Cancer

One example of a true orphan receptor that is involved in oncogenesis is a chromosomal translocation that involves the NR NOR-1. The Ewing sarcoma-NOR-1 (EWS) gene fusion generated by the t(9;22) chromosomal translocation

Table 3.9 Summary of Class III and Class IV Orphan Receptor Knockout Phenotypes, Expression Patterns, and Human Mutations

Receptor	Knockout Phenotype	Expressed Pattern	Human Mutations	References
NGFI-B/ Nurr1/ NOR1	Viable, elevated Nurr1 Nurr^1, postnatal lethality, neural agenesis	AG, THY, PHN	Chromosomal translocation in EMC, fusion with EWS	432, 435, 441
COUP-TF	TFI, Perinatal lethality, TFII, in utero lethality, neural cell migration, heart, endothelia	CNS, H, L, LU, K, vascular endothelia, overlapping expression for TFI, II		455
HNF-4α	Embryonic lethality at E8.5	L, I, K		453
TR2,4		TS		65, 430
GCNF		OV, TS		443
SF-1/FTZ-F1	Postnatal lethality, agenesis of gonads and adrenals	AG, CNS, G		446, 448
ERR1, 2, 3	ERR2, Embryonic lethality, placental defect	ERR1, ubiquitous ERR2/3, restricted		437, 451, 451a
RORα,β,γ	RORα, Staggerer mouse, cerebellar ataxia, dendritic atrophy, circadian function RORγ, lymphogenesis defects	CNS, LY, BO		442, 452
TLX	Viable, agressive, defective limbic system, blindness	CNS, retina		438, 454
PNR		Retina, PC	Mutation in enhanced S cone syndrome, visual impairment	447
DAX-1	Viable, males sterile	AG, CNS, G	Loss of function mutations, X-linked hypogonadism	444, 445
SHP-1	Viable, bile acid accumulation, loss of protection against dietary bile acids	L, K, I, P, LU, B, CNS	Mild obesity in Japanese	445a, 445b, 445c

Phenotypes of homozygous null mutations of the indicated mouse receptor genes are summarized where available, as are expression patterns for individual receptor genes. NGFI-B, nerve growth factor induced receptor B; Nurr1, Nur related receptor 1; NOR, neuron derived orphan receptor; COUP-TF, chicken ovalbumin upstream promoter transcription factor; HNF4, hepatocyte nuclear factor 4; TR2/TR4, testis receptor 2, 4; GCNF, germ cell nuclear factor; SF-1, steroidogenic factor 1; FTZ-F1, fushi tarazu factor 1; LRH1, liver receptor homolgous protein 1; ERR, estrogen related receptor; ROR, retinoic acid related orphan receptor; TLX, tailless related receptor X; DAX-1, dosage sensitive sex reversal adrenal hypoplasia X chromosome region 1; SHP, small heterodimer partner; AG, adrenal gland, THY, thymus; PHN, paraventricular hypothalamic neuron; CNS, central nervous system; H, heart; L, liver; LU, lung; K, kidney; I, intestine; TS, testis; OV, ovary; G, gonads; LY, lymph; B, bone; EMC, extraskeletal myxoid chondrosarcoma; EWS, ewing sarcoma.

found in extraskeletal myxoid chondrosarcomas encodes a fusion protein containing the amino-terminal domain of the EWS protein fused to the whole coding sequence of the orphan NR NOR-1 (also called translocated in extraskeletal chondrosarcoma, or TEC).[457] Extraskeletal myxoid chondrosarcomas are soft tissue tumors of chondroblastic origin, which occur primarily in muscle. The EWS protein possesses a conserved RNA-recognition motif and can bind RNA in vitro. However, the resulting fusion protein loses this motif, which is replaced by the entire TEC/NOR-1 protein. The mechanism underlying the pathogenesis is poorly understood, but the potent transcriptional activity of the fusion protein may account for its oncogenic potential in chrondrosarcomas by activating transcription of target genes involved in cell proliferation. Further analysis of orphan receptor function may implicate other orphans in similar pathophysiological states, such as Nurr1 involvement in neural agenesis and dopaminergic signaling.[457a]

ORPHAN NUCLEAR RECEPTORS: REVERSE ENDOCRINOLOGY AND PHARMACOLOGY

As stated above, several NRs that were once considered orphans are now known to play important physiological and pathophysiological roles in response to xenobiotics, dietary and cellular metabolites, presumably as paracrine or intacrine mediators. Indeed, whole families of receptors have emerged that exhibit neural or steroidogenic expression patterns and likely govern various aspects of sexual development, hormonal regulation, or behavior. However, of the more than 60 NRs that have been identified, most have unknown functions in regard to both ligand and target gene specificity (cf. Fig. 3.4, Tables 3.8, 3.9). Many of these receptors have apparent homologs in fish, birds, amphibians, insects, and worms. While homology cloning of receptors by conventional molecular biology techniques has produced a plethora of related receptors of novel or unknown function, the emergence of functional genomics renders this fairly recent practice somewhat extinct. In nematodes, genome sequencing has identified 200 putative NRs. While the number of receptors employed by fly, mouse and man is significantly smaller, the remaining orphan receptors represent endocrinology waiting to be discovered. Future challenges in NR research are to tease out the physiology of these receptors with regard to ligand requirement, identification and specificity, cofactor requirement, target gene identification, and receptor cross-coupling. Current approaches used to address these functions are addressed below.

Ligand-Binding Assays

Many orphan receptors that have a putative LBD can be assessed for ligand-binding activity in the absence of the DBD, by fusing the LBD to well-characterized heterologous DBDs, such as that of Gal4. For example, cotransfection of tissue culture cells with a Gal4 LBD expression plasmid in combination with a reporter plasmid containing Gal4 DNA-binding sites can indicate ligand binding by production of the reporter molecule, such as β-galactosidase, luciferase, or green fluorescent protein. This and other bioactivity-guided ligand-binding assays allow the screening of complex chemical libraries or endogenously synthesized compounds for ligand-binding activity in mammalian cells (for review, see Kliewer et al.[331] and Blumberg and Evans[373]). Compounds identified in this manner can be further tested by direct radiolabeled ligand-binding competition assays and additional transfection experiments with full-length receptors, to confirm functional interactions. More sophisticated reporter systems may also predict agonist and antagonist activities for the receptor LBD, by scoring for coactivator or corepressor association. As discussed previously, structural determinations and molecular modeling can be used to predict candidate ligand structures in more rational structure–activity ligand screens.[458,459] Once ligand-binding activity is qualified, further studies in cell culture and animals with the candidate ligand may assess the physiological or phenotypic impact on the cell, tissue, or animal tested and can be further explored for safety and efficacy of modulation of receptor function in physiological or pathophysiological states.

Proximity Assays

Elaborations on the Gal4-based screening systems have produced several in vitro proximity assays for ligand-binding activity, which can include assessment of coactivator association in response to ligand treatment. One such technique, the scintillation proximity assay, involves a more sensitive evaluation of ligand–receptor interaction by immobilizing recombinant LBD on beads coated with scintillation compound. Binding of radioactively labeled ligands can then be scored by the scintillation activity of the receptor-bound beads.[460] Additional assays, such as the coactivator-dependent receptor ligand assay (CARLA) and fluorescence resonance energy transfer (FRET), rely on the ability of the ligand to induce physical interaction between recombinant LBD protein and the coactivator NR boxes.[159,414,460] The readout for these interaction assays is provided by glutathione-S-transferase pull-down of the coactivator (labeled coactivator, CARLA) or by FRET between fluorophor-labeled LBD and coactivator peptides. When performed in multiwell plates, high-throughput screens for compounds that influence receptor–ligand or receptor–NR box interactions are possible. This type of molecular interaction screen can be adapted to analyze corepressor dissociation as well, in effect allowing screening for modulators of multiple receptor functions. In this manner, screens for differential or preferential cofactor recruitment can also contribute to the development of selective receptor behaviors.

Interaction Assays

Other types of interaction assay, including FRET, yeast, and mammalian two-hybrid and other molecular interaction assays, will continue to explore the involvement of corepressors and co-activators in orphan receptor biology. For each new receptor, there exists the possibility to identify receptor and tissue-specific cofactors that interact with AF-1, DBD, LBD, and AF-2. Heterodimerization partners can also be identified using these interaction assays. The multicomponent nature of the NR repressed and activated complexes may be systematically explored via protein–protein interactions assays, as well as by other biochemical means that can reveal order of addition and stoichiometric relationships of these receptor complexes.[461,462]

Target Genes: Differential Display and Expression Analysis

One of the largest barriers to NR analysis is the identification of bona fide target genes. Treatment of cultured cells and animals with nuclear receptor ligands can have effects on transcription that do not involve the receptor under investigation. Conversely, hyperstimulation of an individual receptor can have consequences upon the activity of other NRs or genomic acting factors indirectly, by sequestering common cofactors. These and other mechanisms of transrepression or otherwise indirect transcriptional effects necessitate isolation of the individual specific nuclear receptor binding motifs within the regulatory regions of candidate target genes.

Equally important is in vivo demonstration of receptor involvement by loss of induction upon genetic loss of the receptor in knockout models, supported by induction that is insensitive to cyclohexamide treatment of cells.

Many of the current approaches for identifying receptor targets rely on the construction of subtracted cDNA libraries from primary cultured cells or tissues that are tailored to the receptor in question for differential display gene analysis. Generally, this requires cDNA synthesis from mRNA of samples treated in the presence or absence of ligand (necessitating prior ligand identification for orphan receptors), cDNA subtraction, and polymerase chain reaction (PCR) amplification with random or specific oligonucleotide primers.[463] If ligands are unavailable, constitutively active receptors can also be introduced into cells or animals to promote differential expression. Amplicons obtained in this manner are then subcloned, sequenced, and qualified by Northern analysis, *in situ* hybridization, or real time quantitative PCR to determine the validity of the ligand response. Once cloned and verified, database searches may facilitate analysis of target function and reproducibility and suggest possible mechanisms or pathways for receptor function. Approaches of this type, known variously as differential display, representational difference analysis (RDA), and suppression subtractive hybridization (SSH), have proven useful for identifying targets for several receptors or receptor-related activities,[464–466] as well as differentially expressed proteins in various stages of cancer progression.[467]

Another approach for target gene identification that is becoming more popular with the advent of expressed sequence tag (EST) databases and genome sequencing is global gene expression analysis on DNA microchip arrays, or oligonucleotide gene chips. In this approach, mRNA is isolated from cells or tissues in the presence or absence of ligand (or other treatment: including dominant negative or positive receptors), the populations fluorescently labeled, and hybridized to silica-based chips containing ordered oligonucleotides or cDNAs from EST or other gene libraries. Comparison of hybridization signals produces a pattern of genes whose expression is up or downregulated in a ligand or treatment dependent manner. Similar to differential display, microarray analysis has been adapted to explore expression of cancer cells, yeast responses to external stimuli, as well as developmental regulation of simple organisms.[468] This system is essentially limited only by the number of clones analyzed and comparisons performed, resulting in the generation of large relational databases of NR-mediated expression patterns.

Assessment of Function Through Gene Targeting in Mice

Traditionally, transgenic mice have been useful for analyzing loss-of-function phenotypes by targeted gene knockout.[469] Similar gene transfer or knock-in techniques have been employed to analyze gain of function (receptor overexpression), loss of function (overexpression of dominant negative proteins), as well as mutant NR function. As such, knockouts of orphan receptors have provided informative phenotypes for a number of receptors (Table 3.9). When receptor knockout results in an embryonic or perinatal lethal phenotype, however, assessment of receptor function in later developmental or adult stages is difficult, if not impossible. Use of conditional knockout or inducible alleles can facilitate this analysis, by examining phenotypes in developmental or tissue-specific knockouts, utilizing *cre* recombinase-based recombination and P-lox alleles or inducible gene expression systems.[470–472] In this manner, receptors that are essential for embryonic or organ development can be analyzed for loss of function following the critical developmental period during which their function is required. This type of approach has proven useful for examining the functions of orphan nuclear receptors such as ERR2 and HNF-4α, knockouts of which exhibit early lethality.[3,473] Additionally, these genetic approaches can be modified to assess the consequences of expressing constitutively active or dominant negative alleles of receptors in a tissue-targeted and controlled manner, allowing careful anal-ysis of receptor physiology and cross-coupling activities.

The ability to manipulate the mouse genome in combination with the methodologies described above will allow analysis of newly discovered orphan receptors and their involvement in animal physiology and pathophysiology. Synthetic agonist and antagonist development may provide tools to modulate the activity of receptors implicated in disease states and provide further relevant functional information regarding orphan NR biology.

ACKNOWLEDGMENTS

The authors wish to thank Dr. Michael Downes and Dr. Marc Tini for critical reading of the manuscript.

REFERENCES

1. Hollenberg SM, Weinberger C, Ong ES, et al. Primary structure and expression of a functional human glucocorticoid receptor cDNA. Nature 318:635–641, 1985.
2. Miesfeld R, Rusconi S, Godowski PJ, et al. Genetic complementation of a glucocorticoid receptor deficiency by expression of cloned receptor cDNA. Cell 46:389–399, 1986.
3. Enmark E and Gustafsson JA. Orphan nuclear receptors—the first eight years. Mol Endocrinol 10:1293–1307, 1996.
4. Anonymous. A unified nomenclature system for the nuclear receptor superfamily. Cell 97:161–163, 1999.
5. Clarke ND, Berg JM. Zinc fingers in Caenorhabditis elegans: finding families and probing pathways. Science 282:2018–2022, 1998.
6. Adams MD, Celniker SE, Holt RA, et al. The genome sequence of Drosophila melanogaster. Science 287:2185–2195, 2000.
6a. Robinson-Rechavi M, Carpentier AS, Duffraisse M, et al. How many nuclear hormone receptors are there in the human genome? Trends Genet 17:554–556, 2001.
7. Tenbaum S and Baniahmad A. Nuclear receptors: structure, function and involvement in disease. Int J Biochem Cell Biol 29:1325–1341, 1997.
8. Ohno S. Evolution by Gene Duplication. New York: Springer-Verlag, 1970.
9. Escriva H, Safi R, Hanni C, et al. Ligand binding was acquired during evolution of nuclear receptors. Proc Natl Acad Sci USA 94:6803–6808, 1997.
10. Tanenbaum DM, Wang Y, Williams SP, et al. Crystallographic comparison of the estrogen and progesterone receptor's ligand binding domains. Proc Natl Acad Sci USA 95:5998–6003, 1998.
11. Wurtz JM, Bourguet W, Renaud JP, et al. A canonical structure for the ligand-binding domain of nuclear receptors. Nat Struct Biol 3:206, 1996.
13. Laudet V, Hänni C, Coll J, et al. Evolution of the nuclear receptor gene superfamily. EMBO J 11:1003–1013, 1992.
14. Zilliacus J, Carlstedt-Duke J, Gustafsson JA, et al. Evolution of distinct DNA-binding specificities within the nuclear receptor family of transcription factors. Proc Natl Acad Sci USA 91:4175–4179, 1994.

15. Thummel CS. From embryogenesis to metamorphosis: the regulation and function of Drosophila nuclear receptor superfamily members. Cell 83:871–877, 1995.
16. Yu RT, McKeown M, Evans RM, et al. Relationship between Drosophila gap gene tailless and a vertebrate nuclear receptor Tlx. Nature 370:375–379, 1994.
17. Oro AE, McKeown M, Evans RM. Relationship between the product of the Drosophila ultraspiracle locus and the vertebrate retinoid X receptor. Nature 347:298–301, 1990.
18. Yao TP, Segraves WA, Oro AE, et al. Drosophila ultraspiracle modulates ecdysone receptor function via heterodimer formation. Cell 71:63–72, 1992.
19. Gallant P, Shiio Y, Cheng PF, et al. Myc and Max Homologs in Drosophila. Science 274:1523–1527, 1996.
20. Tsai CC, Kao HY, Yao TP, et al. SMRTER, a Drosophila nuclear receptor coregulator, reveals that EcR-mediated repression is critical for development. Mol Cells 4:175–186, 1999.
21. Mangelsdorf DJ, Borgmeyer U, Heyman RA, et al. Characterization of three RXR genes that mediate the action of 9-cis retinoic acid. Genes Dev 6:329–344, 1992.
22. Beato M. Gene regulation by steroid hormones. Cell 56:335–344, 1989.
23. Pissios P, Tzameli I, Kushner P, et al. Dynamic stabilization of nuclear receptor ligand binding domains by hormone or corepressor binding. Mol Cells 6:245–253, 2000.
24. von der Ahe D, Janich S, Scheidereit C, et al. Glucocorticoid and progesterone receptors bind to the same sites in two hormonally regulated promoters. Nature 313:706–709, 1985.
25. Mangelsdorf DJ, Thummel C, Beato M, et al. The nuclear receptor superfamily: the second decade. Cell 83:835–839, 1995.
26. Zilliacus J, Wright AP, Carlstedt-Duke J, et al. Structural determinants of DNA-binding specificity by steroid receptors. Mol Endocrinol 9:389–400, 1995.
27. Umesono K, Evans RM. Determinants of target gene specificity for steroid/thyroid hormone receptors. Cell 57:1139–1146, 1989.
28. Perlmann T, Rangarajan PN, Umesono K, et al. Determinants for selective RAR and TR recognition of direct repeat HREs. Genes Dev 7:1411–1422, 1993.
29. Dahlman-Wright K, Grandien K, Nilsson S, et al. Protein-protein interactions between the DNA-binding domains of nuclear receptors: influence on DNA-binding. J Steroid Biochem Mol Biol 45:239–250, 1993.
30. Cowley SM, Hoare S, Mosselman S, et al. Estrogen receptors alpha and beta form heterodimers on DNA. J Biol Chem 272:19858–19862, 1997.
31. Pettersson K, Grandien K, Kuiper GG, et al. Mouse estrogen receptor beta forms estrogen response element-binding heterodimers with estrogen receptor alpha. Mol Endocrinol 11:1486–1496, 1997.
32. Tremblay GB, Tremblay A, Labrie F, et al. Dominant activity of activation function 1 (AF-1) and differential stoichiometric requirements for AF-1 and -2 in the estrogen receptor alpha-beta heterodimeric complex. Mol Cell Biol 19:1919–1927, 1999.
33. Chen F, Cooney AJ, Wang Y, et al. Cloning of a novel orphan receptor (GCNF) expressed during germ cell development. Mol Endocrinol 8:1434–1444, 1994.
34. Cairns W, Smith CAD, McLaren AW, et al. Characterization of the human cytochrome P4502D6 promoter. A potential role for antagonistic interactions between members of the nuclear receptor family. J Biol Chem 271:25269–25276, 1996.
35. Stroup D and Chiang JY. HNF4 and COUP-TFII interact to modulate transcription of the cholesterol 7alpha-hydroxylase gene (CYP7A1). J Lipid Res 41:1–11, 2000.
36. Wilson TE, Paulsen RE, Padgett KA, et al. Participation of non-zinc finger residues in DNA binding by two nuclear orphan receptors. Science 256:107–110, 1992.
37. Gearing KL, Göttlicher M, Widmark E, et al. Fatty acid activation of the peroxisome proliferator activated receptor, a member of the nuclear receptor gene superfamily. J Nutr 124(Suppl 8):1284S–1288S, 1994.
38. Giguère V, Tini M, Flock G, et al. Isoform-specific amino-terminal domains dictate DNA-binding properties of ROR alpha, a novel family of orphan hormone nuclear receptors. Genes Dev 8:538–553, 1994.
39. Roche PJ, Hoare SA, Parker MG. A consensus DNA-binding site for the androgen receptor. Mol Endocrinol 6:2229–2235, 1992.
40. Lieberman BA, Bona BJ, Edwards DP, et al. The constitution of a progesterone response element. Mol Endocrinol 7:515–527, 1993.
41. Darbre P, Page M, King RJ. Androgen regulation by the long terminal repeat of mouse mammary tumor virus. Mol Cell Biol 6:2847–2854, 1986.
42. Cato AC, Weinmann J. Mineralocorticoid regulation of transcription of transfected mouse mammary tumor virus DNA in cultured kidney cells. J Cell Biol 106:2119–2125, 1988.
43. Ham J, Thomson A, Needham M, et al. Characterization of response elements for androgens, glucocorticoids and progestins in mouse mammary tumour virus. Nucleic Acids Res 16:5263–5276, 1988.
44. Beato M, Chalepakis G, Schauer M, et al. DNA regulatory elements for steroid hormones. J Steroid Biochem 32:737–747, 1989.
45. Strahle U, Boshart M, Klock G, et al. Glucocorticoid- and progesterone-specific effects are

determined by differential expression of the respective hormone receptors. Nature 339:629–632, 1989.
46. Funder JW. Mineralocorticoids, glucocorticoids, receptors and response elements. Science 259:1132–1133, 1993.
47. Kralli A, Bohen SP, Yamamoto KR. LEM1, an ATP-binding-cassette transporter, selectively modulates the biological potency of steroid hormones. Proc Natl Acad Sci USA 92:4701–4705, 1995.
48. Umesono K, Murakami KK, Thompson CC, et al. Direct repeats as selective response elements for the thyroid hormone, retinoic acid, and vitamin D3 receptors. Cell 65:1255–1266, 1991.
49. Durand B, Saunders M, Leroy P, et al. All-*trans* and 9-*cis* retinoic acid induction of CRABPII transcription is mediated by RAR-RXR heterodimers bound to DR1 and DR2 repeated motifs. Cell 71:73–85, 1992.
50. Kliewer SA, Umesono K, Mangelsdorf DJ, et al. Retinoid X receptor interacts with nuclear receptors in retinoic acid, thyroid hormone and vitamin D3 signalling. Nature 355:446–449, 1992.
51. Mangelsdorf DJ, Evans RM. The RXR heterodimers and orphan receptors. Cell 83:841–850, 1995.
52. Mangelsdorf DJ, Umesono K, Evans RM. The retinoid receptors. In: Sporn MB, Roberts AB, Goodman DS (eds). The Retinoids: Biology, Chemistry, and Medicine. New York: Raven Press, 1994, 319–349.
53. Glass CK, Holloway JM, Derary OV, et al. The thyroid hormone receptor binds with opposite transcriptional effects to a common sequence motif in thyroid hormone and estrogen response elements. Cell 54:313–323, 1988.
54. Umesono K, Giguere V, Glass CK, et al. Retinoic acid and thyroid hormone induce gene expression through a common responsive element. Nature 336:262–265, 1988.
55. Tini M, Otulakowski G, Breitman ML, et al. An everted repeat mediates retinoic acid induction of the gamma F-crystallin gene: evidence of a direct role for retinoids in lens development. Genes Dev 7:295–307, 1993.
56. Pascussi JM, Jounaidi Y, Drocourt L, et al. Evidence for the presence of a functional pregnane X receptor response element in the *CYP3A7* promoter gene. Biochem Biophys Res Commun 260:377–381, 1999.
56a. Makishima M, Lu TT, Xie W, et al. Vitamin D receptor as an intestinal bile acid sensor. Science 296:1313–1316, 2002.
56b. Kast HR, Goodwin B, Tarr PT, et al. Regulation of multidrug resistance-associated protein 2 (ABCC2) by the nuclear receptors pregnane X receptor, farnesoid X-activated receptor, and constitutive androstane receptor. J Biol Chem 277:2908–2915, 2002.
57. Blumberg B, Sabbagh WJ, Juguilon H, et al. SXR, a novel steroid and xenobiotic-sensing nuclear receptor. Genes Dev 12:3195–3205, 1998.
58. Mangelsdorf DJ, Umesono K, Kliewer SA, et al. A direct repeat in the cellular retinol-binding protein type II gene confers differential regulation by RXR and RAR. Cell 66:555–561, 1991.
59. Forman BM, Chen J, Blumberg B, et al. Crosstalk among ROR alpha 1 and the Rev-erb family of orphan nuclear receptors. Mol Endocrinol 8:1253–1261, 1994.
60. Forman BM, Umesono K, Chen J, et al. Unique response pathways are established by allosteric interactions among nuclear hormone receptors. Cell 81:541–550, 1995.
61. Perlmann T, Vennstrom B. The sound of silence. Nature 377:387–388, 1995.
62. Okabe T and Nawata H. NGFI-B/nur77 family involved in T-cell apoptosis [in Japanese]. Nippon Rinsho 54:1768–1772, 1996.
63. Maira M, Martens C, Philips A, et al. Heterodimerization between members of the Nur subfamily of orphan nuclear receptors as a novel mechanism for gene activation. Mol Cell Biol 19:7549–7557, 1999.
64. Lee CH, Chang L, Wei LN. Molecular cloning and characterization of a mouse nuclear orphan receptor expressed in embryos and testes. Mol Reprod Dev 44:305–314, 1996.
65. Lee CH, Chinpaisal C, Wei LN. A novel nuclear receptor heterodimerization pathway mediated by orphan receptors TR2 and TR4. J Biol Chem 273:25209–25215, 1998.
66. Chinpaisal C, Chang L, Hu X, et al. The orphan nuclear receptor TR2 suppresses a DR4 hormone response element of the mouse *CRABP-I* gene promoter. Biochemistry 36:14088–14095, 1997.
67. Chang C, Pan HJ. Thyroid hormone direct repeat 4 response element is a positive regulatory element for the human TR2 orphan receptor, a member of steroid receptor superfamily. Mol Cell Biochem 189:195–200, 1998.
68. Zanaria E, Muscatelli F, Bardoni B, et al. An unusual member of the nuclear hormone receptor superfamily responsible for X-linked adrenal hypoplasia congenita. Nature 372:635–641, 1994.
69. Seol W, Mahon MJ, Lee YK, et al. Two receptor interacting domains in the nuclear hormone receptor corepressor RIP13/N-CoR. Mol Endocrinol 10:1646–1655, 1996.
70. Seol W, Choi HS, Moore DD. An orphan nuclear hormone receptor that lacks a DNA binding domain and heterodimerizes with other receptors. Science 272:1336–1339, 1996.
71. Seol W, Hanstein B, Brown M, et al. Inhibition of estrogen receptor action by the orphan receptor SHP (short heterodimer partner). Mol Endocrinol 12:1551–1557, 1998.
72. Lee YK, Dell H, Dowhan DH, et al. The orphan nuclear receptor SHP inhibits hepatocyte nuclear factor 4 and retinoid X receptor transacti-

vation: two mechanisms for repression. Mol Cell Biol 20:187–195, 2000.
73. Rogatsky I, Trowbridge JM, Garabedian MJ. Potentiation of human estrogen receptor alpha transcriptional activation through phosphorylation of serines 104 and 106 by the cyclin A-CDK2 complex. J Biol Chem 274:22296–22302, 1999.
74. Joel PB, Smith J, Sturgill TW, et al. pp90rsk1 regulates estrogen receptor-mediated transcription through phosphorylation of Ser-167. Mol Cell Biol 18:1978–1984, 1998.
75. Kato S, Endoh H, Masuhiro Y, et al. Activation of the estrogen receptor through phosphorylation by mitogen-activated protein kinase. Science 270:1491–1494, 1995.
76. Bunone G, Briand PA, Miksicek RJ, et al. Activation of the unliganded estrogen receptor by EGF involves the MAP kinase pathway and direct phosphorylation. EMBO J 15:2174–2183, 1996.
77. Lahooti H, Thorsen T, Aakvaag A. Modulation of mouse estrogen receptor transcription activity by protein kinase C delta. J Mol Endocrinol 20:245–259, 1998.
78. Jacq X, Brou C, Lutz Y, et al. Human TAFII30 is present in a distinct TFIID complex and is required for transcriptional activation by the estrogen receptor. Cell 79:107–117, 1994.
79. Verrier CS, Roodi N, Yee CJ, et al. High-mobility group (HMG) protein HMG-1 and TATA-binding protein-associated factor TAF(II)30 affect estrogen receptor-mediated transcriptional activation. Mol Endocrinol 11:1009–1019, 1997.
80. Webb P, Nguyen P, Shinsako J, et al. Estrogen receptor activation function 1 works by binding p160 coactivator proteins. Mol Endocrinol 12:1605–1618, 1998.
81. Endoh H, Maruyama K, Masuhiro Y, et al. Purification and identification of p68 RNA helicase acting as a transcriptional coactivator specific for the activation function 1 of human estrogen receptor alpha. Mol Cell Biol 19:5363–5372, 1999.
82. Lanz RB, McKenna NJ, Onate SA, et al. A steroid receptor coactivator, SRA, functions as an RNA and is present in an SRC-1 complex. Cell 97:17–27, 1999.
83. Onate SA, Boonyaratanakornkit V, Spencer TE, et al. The steroid receptor coactivator-1 contains multiple receptor interacting and activation domains that cooperatively enhance the activation function 1 (AF1) and AF2 domains of steroid receptors. J Biol Chem 273:12101–12108, 1998.
84. Alen P, Claessens F, Verhoeven G, et al. The androgen receptor amino-terminal domain plays a key role in p160 coactivator-stimulated gene transcription. Mol Cell Biol 19:6085–6097, 1999.
85. McEwan IJ, Gustafsson J. Interaction of the human androgen receptor transactivation function with the general transcription factor TFIIF. Proc Natl Acad Sci USA 94:8485–8490, 1997.
86. Massaad C, Houard N, Lombes M, et al. Modulation of human mineralocorticoid receptor function by protein kinase A. Mol Endocrinol 13:57–65, 1999.
87. Puigserver P, Wu Z, Park CW, et al. A cold-inducible coactivator of nuclear receptors linked to adaptive thermogenesis. Cell 92:829–839, 1998.
88. Juge-Aubry CE, Hammar E, Siegrist-Kaiser C, et al. Regulation of the transcriptional activity of the peroxisome proliferator-activated receptor alpha by phosphorylation of a ligand-independent trans-activating domain. J Biol Chem 274:10505–10510, 1999.
89. Green VJ, Kokkotou E, Ladias JA. Critical structural elements and multitarget protein interactions of the transcriptional activator AF-1 of hepatocyte nuclear factor 4. J Biol Chem 273:29950–29957, 1998.
90. Hollenberg SM, Giguere V, Segui P, et al. Colocalization of DNA-binding and transcriptional activation functions in the human glucocorticoid receptor. Cell 49:39–46, 1987.
91. Freedman LP, Yoshinaga SK, Vanderbilt JN, et al. In vitro transcription enhancement by purified derivatives of the glucocorticoid receptor. Science 245:298–301, 1989.
92. Tsai SY, Srinivasan G, Allan GF, et al. Recombinant human glucocorticoid receptor induces transcription of hormone response genes in vitro. J Biol Chem 265:17055–17061, 1990.
93. McEwan IJ, Wright AP, Dahlman-Wright K, et al. Direct interaction of the tau 1 transactivation domain of the human glucocorticoid receptor with the basal transcriptional machinery. Mol Cell Biol 13:399–407, 1993.
94. Ford J, McEwan IJ, Wright AP, et al. Involvement of the transcription factor IID protein complex in gene activation by the N-terminal transactivation domain of the glucocorticoid receptor in vitro. Mol Endocrinol 11:1467–1475, 1997.
95. Hittelman AB, Burakov D, Iniguez-Lluhi JA, et al. Differential regulation of glucocorticoid receptor transcriptional activation via AF-1-associated proteins. EMBO J 18:5380–5388, 1999.
96. Rochette-Egly C, Adam S, Rossignol M, et al. Stimulation of RAR alpha activation function AF-1 through binding to the general transcription factor TFIIH and phosphorylation by CDK7. Cell 90:97–107, 1997.
97. Ylikomi T, Wurtz JM, Syvälä H, et al. Reappraisal of the role of heat shock proteins as regulators of steroid receptor activity. Crit Rev Biochem Mol Biol 33:437–466, 1998.
98. Bourguet W, Ruff M, Chambon P, et al. Crystal structure of the ligand-binding domain of the human nuclear receptor RXR-alpha. Nature 375:377–382, 1995.
99. Weatherman RV, Fletterick RJ, Scanlan TS. Nuclear-Receptor Ligands and Ligand-Binding Domains. Ann Rev Biochem 68:559–581, 1999.

100. Moras D, Gronemeyer H. The nuclear receptor ligand-binding domain: structure and function. Curr Opin Cell Biol 10:384–391, 1998.
101. Renaud JP, Rochel N, Ruff M, et al. Crystal structure of the RAR-gamma ligand-binding domain bound to all-*trans* retinoic acid. Nature 378:681–689, 1995.
102. Meyer ME, Gronemeyer H, Turcotte B, et al. Steroid hormone receptors compete for factors that mediate their enhancer function. Cell 57:433–442, 1989.
103. Casanova J, Helmer E, Selmi-Ruby S, et al. Functional evidence for ligand-dependent dissociation of thyroid hormone and retinoic acid receptors from an inhibitory cellular factor. Mol Cell Biol 14:5756–5765, 1994.
104. Baniahmad A, Leng X, Burris TP, et al. The tau 4 activation domain of the thyroid hormone receptor is required for release of a putative corepressor(s) necessary for transcriptional silencing. Mol Cell Biol 15:76–86, 1995.
105. Chen JD, Evans RM. A transcriptional co-repressor that interacts with nuclear hormone receptors. Nature 377:454–457, 1995.
106. Horlein AJ, Naar AM, Heinzel T, et al. Ligand-independent repression by the thyroid hormone receptor mediated by a nuclear receptor co-repressor. Nature 377:397–404, 1995.
107. Fondell JD, Ge H, Roeder RG. Ligand induction of a transcriptionally active thyroid hormone receptor coactivator complex. Proc Natl Acad Sci USA 93:8329–8333, 1996.
108. Rachez C, Suldan Z, Ward J, et al. A novel protein complex that interacts with the vitamin D3 receptor in a ligand-dependent manner and enhances VDR transactivation in a cell-free system. Genes Dev 12:1787–1800, 1998.
109. Akimaru H, Chen Y, Dai P, et al. *Drosophila* CBP is a co-activator of cubitus interruptus in hedgehog signalling. Nature 386:735–738, 1997.
110. Heery DM, Kalkhoven E, Hoare S, et al. A signature motif in transcriptional co-activators mediates binding to nuclear receptors. Nature 387:733–736, 1997.
111. Darimont BD, Wagner RL, Apriletti JW, et al. Structure and specificity of nuclear receptor-coactivator interactions. Genes Dev 12:3343–3356, 1998.
112. Onate SA, Tsai SY, Tsai MJ, et al. Sequence and characterization of a coactivator for the steroid hormone receptor superfamily. Science 270:1354–1357, 1995.
113. Smith CL, Onate SA, Tsai MJ, et al. CREB binding protein acts synergistically with steroid receptor coactivator-1 to enhance steroid receptor-dependent transcription. Proc Natl Acad Sci USA 93:8884–8888, 1996.
114. Yao TP, Ku G, Zhou N, et al. The nuclear hormone receptor coactivator SRC-1 is a specific target of p300. Proc Natl Acad Sci USA 93:10626–10631, 1996.
115. Nolte RT, Wisely GB, Westin S, et al. Ligand binding and co-activator assembly of the peroxisome proliferator–activated receptor-gamma. Nature 395:137–143, 1998.
116. Xu J, Qiu Y, DeMayo FJ, et al. Partial hormone resistance in mice with disruption of the steroid receptor coactivator-1 (SRC-1) gene. Science 279:1922–1925, 1998.
117. Anzick SL, Kononen J, Walker RL, et al. AIB1, a steroid receptor coactivator amplified in breast and ovarian cancer. Science 277:965–968, 1997.
118. Chen H, Lin RJ, Schiltz RL, et al. Nuclear receptor coactivator ACTR is a novel histone acetyltransferase and forms a multimeric activation complex with P/CAF and CBP/p300. Cell 90:569–580, 1997.
119. Li H, Gomes PJ, Chen JD. RAC3, a steroid/nuclear receptor-associated coactivator that is related to SRC-1 and TIF2. Proc Natl Acad Sci USA 94:8479–8484, 1997.
120. Takeshita A, Cardona GR, Koibuchi N, et al. TRAM-1, A novel 160-kDa thyroid hormone receptor activator molecule, exhibits distinct properties from steroid receptor coactivator-1. J Biol Chem 272:27629–27634, 1997.
121. Torchia J, Rose DW, Inostroza J, et al. The transcriptional co-activator p/CIP binds CBP and mediates nuclear-receptor function. Nature 387:677–684, 1997.
122. Suen CS, Berrodin TJ, Mastroeni R, et al. A transcriptional coactivator, steroid receptor coactivator-3, selectively augments steroid receptor transcriptional activity. J Biol Chem 273:27645–27653, 1998.
123. Hong H, Kohli K, Trivedi A, et al. GRIP1, a novel mouse protein that serves as a transcriptional coactivator in yeast for the hormone binding domains of steroid receptors. Proc Natl Acad Sci USA 93:4948–4952, 1996.
124. Voegel JJ, Heine MJ, Zechel C, et al. TIF2, a 160 kDa transcriptional mediator for the ligand-dependent activation function AF-2 of nuclear receptors. EMBO J 15:3667–3675, 1996.
125. Le Douarin B, Zechel C, Garnier JM, et al. The N-terminal part of TIF1, a putative mediator of the ligand-dependent activation function (AF-2) of nuclear receptors, is fused to B-raf in the oncogenic protein T18. EMBO J 14:2020–2033, 1995.
126. Lee JW, Ryan F, Swaffield JC, et al. Interaction of thyroid-hormone receptor with a conserved transcriptional mediator. Nature 374:91–94, 1995.
127. Rubin DM, Coux O, Wefes I, et al. Identification of the gal4 suppressor Sug1 as a subunit of the yeast 26S proteasome. Nature 379:655–657, 1996.
128. Chang KH, Chen Y, Chen TT, et al. A thyroid hormone receptor coactivator negatively regulated by the retinoblastoma protein. Proc Natl Acad Sci USA 94:9040–9045, 1997.

129. Yeh S and Chang C. Cloning and characterization of a specific coactivator, ARA70, for the androgen receptor in human prostate cells. Proc Natl Acad Sci USA 93:5517–5521, 1996.
130. Miyamoto H, Yeh S, Wilding G, et al. Promotion of agonist activity of antiandrogens by the androgen receptor coactivator, ARA70, in human prostate cancer DU145 cells. Proc Natl Acad Sci USA 95:7379–7384, 1998.
131. Naar AM, Beaurang PA, Zhou S, et al. Composite co-activator ARC mediates chromatin-directed transcriptional activation. Nature 398: 828–832, 1999.
132. Yang XJ, Ogryzko VV, Nishikawa J, et al. A p300/CBP-associated factor that competes with the adenoviral oncoprotein E1A. Nature 382:319–324, 1996.
133. Chakravarti D, LaMorte VJ, Nelson MC, et al. Role of CBP/P300 in nuclear receptor signalling. Nature 383:99–103, 1996.
134. Kamei Y, Xu L, Heinzel T, et al. A CBP integrator complex mediates transcriptional activation and AP-1 inhibition by nuclear receptors. Cell 85:403–414, 1996.
135. Huibregtse JM, Scheffner M, Beaudenon S, et al. A family of proteins structurally and functionally related to the E6-AP ubiquitin-protein ligase. Proc Natl Acad Sci USA 92:5249, 1995.
136. Imhof MO, McDonnell DP. Yeast RSP5 and its human homolog hRPF1 potentiate hormone-dependent activation of transcription by human progesterone and glucocorticoid receptors. Mol Cell Biol 16:2594–2605, 1996.
137. Fryer CJ, Archer TK. Chromatin remodelling by the glucocorticoid receptor requires the BRG1 complex. Nature 393:88–91, 1998.
138. Baudino TA, Kraichely DM, Jefcoat SC, Jr., et al. Isolation and characterization of a novel coactivator protein, NCoA-62, involved in vitamin D–mediated transcription. J Biol Chem 273: 16434–16441, 1998.
139. Moilanen AM, Karvonen U, Poukka H, et al. A testis-specific androgen receptor coregulator that belongs to a novel family of nuclear proteins. J Biol Chem 274:3700–3704, 1999.
140. Huang N, vom Baur E, Garnier JM, et al. Two distinct nuclear receptor interaction domains in NSD1, a novel SET protein that exhibits characteristics of both corepressors and coactivators. EMBO J 17:3398–3412, 1998.
141. Chrivia JC, Kwok RPS, Lamb N, et al. Phosphorylated CREB binds specifically to the nuclear protein CBP. Nature 365:855–859, 1993.
142. Kwok RP, Lundblad JR, Chrivia JC, et al. Nuclear protein CBP is a coactivator for the transcription factor CREB. Nature 370:223–226, 1994.
143. Goldman PS, Tran VK, Goodman RH. The multifunctional role of the co-activator CBP in transcriptional regulation. Recent Prog Horm Res 52:103–119, 1997.
144. Giordano A, Avantaggiati ML. p300 and CBP: partners for life and death. J Cell Physiol 181: 218–230, 1999.
145. Yao TP, Oh SP, Fuchs M, et al. Gene dosage-dependent embryonic development and proliferation defects in mice lacking the transcriptional integrator p300. Cell 93:361–372, 1998.
146. Kawasaki H, Eckner R, Yao TP, et al. Distinct roles of the co-activators p300 and CBP in retinoic-acid-induced F9-cell differentiation. Nature 393:284–289, 1998.
147. Blanco JC, Minucci S, Lu J, et al. The histone acetylase PCAF is a nuclear receptor coactivator. Genes Dev 12:1638–1651, 1998.
148. Korzus E, Torchia J, Rose DW, et al. Transcription factor-specific requirements for coactivators and their acetyltransferase functions. Science 279:703–707, 1998.
149. McKenna NJ, Lanz RB, O'Malley BW. Nuclear receptor coregulators: cellular and molecular biology. Endocr Rev 20:321–344, 1999.
150. McKenna NJ, Xu J, Nawaz Z, et al. Nuclear receptor coactivators: multiple enzymes, multiple complexes, multiple functions. J Steroid Biochem Mol Biol 69:3–12, 1999.
151. Hong H, Yang L, Stallcup MR. Hormone-independent transcriptional activation and coactivator binding by novel orphan nuclear receptor ERR3. J Biol Chem 274:22618–22626, 1999.
152. Xie W, Hong H, Yang NN, et al. Constitutive activation of transcription and binding of coactivator by estrogen-related receptors 1 and 2. Mol Endocrinol 13:2151–2162, 1999.
153. Xu J, Liao L, Ning G, et al. The steroid receptor coactivator SRC-3 (p/CIP/RAC3/AIB1/ACTR/TRAM-1) is required for normal growth, puberty, female reproductive function, and mammary gland development. Proc Natl Acad Sci USA 97:6379–6384, 2000.
154. Rachez C, Lemon BD, Suldan Z, et al. Ligand-dependent transcription activation by nuclear receptors requires the DRIP complex. Nature 398:824–828, 1999.
155. Zhu Y, Qi C, Jain S, et al. Amplification and overexpression of peroxisome proliferator-activated receptor binding protein (PBP/PPARBP) gene in breast cancer. Proc Natl Acad Sci USA 96:10848–10853, 1999.
156. Ito M, Yuan CX, Okano HJ, et al. Involvement of the TRAP220 component of the TRAP/SMCC coactivator complex in embryonic development and thyroid hormone action. Mol Cell 5:683–693, 2000.
157. Shiau AK, Barstad D, Loria PM, et al. The structural basis of estrogen receptor/coactivator recognition and the antagonism of this interaction by tamoxifen. Cell 95:927–937, 1998.
158. Forman BM, Chen J, Evans RM. Hypolipidemic drugs, polyunsaturated fatty acids, and eicosanoids are ligands for peroxisome proliferator-

activated receptors alpha and delta. Proc Natl Acad Sci USA 94:4312–4317, 1997.
159. Krey G, Braissant O, L'Horset F, et al. Fatty acids, eicosanoids, and hypolipidemic agents identified as ligands of peroxisome proliferator-activated receptors by coactivator-dependent receptor ligand assay. Mol Endocrinol 11:779–791, 1997.
160. Brzozowski AM, Pike AC, Dauter Z, et al. Molecular basis of agonism and antagonism in the oestrogen receptor. Nature 389:753–758, 1997.
161. Hammer GD, Krylova I, Zhang Y, et al. Phosphorylation of the nuclear receptor SF-1 modulates cofactor recruitment: integration of hormone signaling in reproduction and stress. Mol Cell 3:521–526, 1999.
162. Tremblay A, Tremblay GB, Labrie F, et al. Ligand-independent recruitment of SRC-1 to estrogen receptor beta through phosphorylation of activation function AF-1. Mol Cell 3:513–519, 1999.
163. Yoshida E, Aratani S, Itou H, et al. Functional association between CBP and HNF4 in transactivation. Biochem Biophys Res Commun 241:664–669, 1997.
164. Gelman L, Zhou G, Fajas L, et al. p300 interacts with the N- and C-terminal part of PPARgamma2 in a ligand-independent and -dependent manner, respectively. J Biol Chem 274:7681–7688, 1999.
165. Damm K. ErbA: tumor suppressor turned oncogene? Faseb J. 7:904–909, 1993.
166. Damm K, Evans RM. Identification of a domain required for oncogenic activity and transcriptional suppression by v-erbA and thyroid-hormone receptor alpha. Proc Natl Acad Sci USA 90:10668–10672, 1993.
167. Zamir I, Dawson J, Lavinsky RM, et al. Cloning and characterization of a corepressor and potential component of the nuclear hormone receptor repression complex. Proc Natl Acad Sci USA 94:14400–14405, 1997.
168. Dressel U, Thormeyer D, Altincicek B, et al. Alien, a highly conserved protein with characteristics of a corepressor for members of the nuclear hormone receptor superfamily. Mol Cell Biol 19:3383–3394, 1999.
169. Sande S, Privalsky M. Identification of TRACs, a family of co-factors that associate with, and modulate the activity of, nuclear hormone receptors. Mol Endocrinol 10:813–825, 1996.
170. Ordentlich P, Downes M, Xie W, et al. Unique forms of human and mouse nuclear receptor corepressor SMRT. Proc Natl Acad Sci USA 96:2639–2644, 1999.
171. Park EJ, Schroen DJ, Yang M, et al. SMRTe, a silencing mediator for retinoid and thyroid hormone receptors-extended isoform that is more related to the nuclear receptor corepressor. Proc Natl Acad Sci USA 96:3519–3524, 1999.
172. Downes M, Burke LJ, Bailey PJ, et al. Two receptor interaction domains in the corepressor, N-CoR/RIP13, are required for an efficient interaction with Rev-erbA alpha and RVR: physical association is dependent on the E region of the orphan receptors. Nucleic Acids Res 24:4379–4386, 1996.
173. Shibata H, Nawaz Z, Tsai SY, et al. Gene silencing by chicken ovalbumin upstream promoter-transcription factor I (COUP-TFI) is mediated by transcriptional corepressors, nuclear receptor-corepressor (N-CoR) and silencing mediator for retinoic acid receptor and thyroid hormone receptor (SMRT). Mol Endocrinol 11:714–724, 1997.
174. Yan ZH, Karam WG, Staudinger JL, et al. Regulation of peroxisome proliferator-activated receptor alpha-induced transactivation by the nuclear orphan receptor TAK1/TR4. J Biol Chem 273:10948–10957, 1998.
175. Robinson CE, Wu X, Nawaz Z, et al. A corepressor and chicken ovalbumin upstream promoter transcriptional factor proteins modulate peroxisome proliferator-activated receptor-gamma2/retinoid X receptor alpha-activated transcription from the murine lipoprotein lipase promoter. Endocrinology 140:1586–1593, 1999.
176. Gurnell M, Wentworth JM, Agostini M, et al. A dominant-negative peroxisome proliferator-activated receptor gamma (PPARgamma) mutant is a constitutive repressor and inhibits PPARgamma-mediated adipogenesis. J Biol Chem 275:5754–5759, 2000.
177. Jackson TA, Richer JK, Bain DL, et al. The partial agonist activity of antagonist-occupied steroid receptors is controlled by a novel hinge domain-binding coactivator L7/SPA and the corepressors N-CoR or SMRT. Mol Endocrinol 11:693–705, 1997.
178. Lavinsky RM, Jepsen K, Heinzel T, et al. Diverse signaling pathways modulate nuclear receptor recruitment of N-CoR and SMRT complexes. Proc Natl Acad Sci USA 95:2920–2925, 1998.
179. Wagner BL, Norris JD, Knotts TA, et al. The nuclear corepressors NCoR and SMRT are key regulators of both ligand- and 8–bromo-cyclic AMP-dependent transcriptional activity of the human progesterone receptor. Mol Cell Biol 18:1369–1378, 1998.
180. Jepsen K, Hermanson O, Onami TM, et al. Combinatorial roles of the nuclear receptor corepressor in transcription and development. Cell 102:753–763, 2000.
181. Crawford PA, Dorn C, Sadovsky Y, et al. Nuclear receptor DAX-1 recruits nuclear receptor corepressor N-CoR to steroidogenic factor 1. Mol Cell Biol 18:2949–2956, 1998.
182. Dhordain P, Lin RJ, Quief S, et al. The LAZ3(BCL-6) oncoprotein recruits a SMRT/mSIN3A/histone deacetylase containing com-

plex to mediate transcriptional repression. Nucleic Acids Res 26:4645–4651, 1998.
183. Huynh KD, Bardwell VJ. The BCL-6 POZ domain and other POZ domains interact with the co-repressors N-CoR and SMRT. Oncogene 17:2473–2484, 1998.
184. Kao HY, Ordentlich P, Koyano-Nakagawa N, et al. A histone deacetylase corepressor complex regulates the Notch signal transduction pathway. Genes Dev 12:2269–2277, 1998.
185. Muto A, Hoshino H, Madisen L, et al. Identification of Bach2 as a B-cell-specific partner for small maf proteins that negatively regulate the immunoglobulin heavy chain gene 3′ enhancer. EMBO J 17:5734–5743, 1998.
186. Asahara H, Dutta S, Kao HY, et al. Pbx-Hox heterodimers recruit coactivator-corepressor complexes in an isoform-specific manner. Mol Cell Biol 19:8219–8225, 1999.
187. Bailey P, Downes M, Lau P, et al. The nuclear receptor corepressor N-CoR regulates differentiation: N-CoR directly interacts with MyoD. Mol Endocrinol 13:1155–1168, 1999.
188. Hu X and Lazar MA. The CoRNR motif controls the recruitment of corepressors by nuclear hormone receptors. Nature 402:93–96, 1999.
189. Nagy L, Kao H-Y, Love JD, et al. Mechanism of corepressor binding and release from nuclear hormone receptors. Genes Dev 13:3209–3216, 1999.
190. Perissi V, Staszewski LM, McInerney EM, et al. Molecular determinants of nuclear receptor-corepressor interaction. Genes Dev 13:1999.
191. Bannister AJ, Kouzarides T. The CBP co-activator is a histone acetyltransferase. Nature 384:641–643, 1996.
192. Ogryzko VV, Schiltz RL, Russanova V, et al. The transcriptional coactivators p300 and CBP are histone acetyltransferases. Cell 87:953–959, 1996.
193. Spencer TE, Jenster G, Burcin MM, et al. Steroid receptor coactivator-1 is a histone acetyltransferase. Nature 389:194–198, 1997.
194. Puri PL, Avantaggiati ML, Balsano C, et al. p300 is required for MyoD-dependent cell cycle arrest and muscle-specific gene transcription. EMBO J 16:369–383, 1997.
195. Chen H, Lin RJ, Xie W, et al. Regulation of hormone-induced histone hyperacetylation and gene activation via acetylation of an acetylase. Cell 98:675–686, 1999.
196. Taunton J, Hassig CA, Schreiber SL. A mammalian histone deacetylase related to the yeast transcriptional regulator Rpd3p. Science 272:408–411, 1996.
197. Yang WM, Yao YL, Sun JM, et al. Isolation and characterization of cDNAs corresponding to an additional member of the human histone deacetylase gene family. J Biol Chem 272:28001–28007, 1997.
198. Fischle W, Emiliani S, Hendzel MJ, et al. A new family of human histone deacetylases related to *Saccharomyces* cerevisiae HDA1p. J Biol Chem 274:11713–11720, 1999.
199. Grozinger CM, Hassig CA, Schreiber SL. Three proteins define a class of human histone deacetylases related to yeast Hda1p. Proc Natl Acad Sci USA 96:4868–4873, 1999.
200. Verdel A, Khochbin S. Identification of a new family of higher eukaryotic histone deacetylases. Coordinate expression of differentiation-dependent chromatin modifiers. J Biol Chem 274:2440–2445, 1999.
201. Hu E, Chen Z, Fredrickson T, et al. Cloning and characterization of a novel human class I histone deacetylase that functions as a transcription repressor. J Biol Chem 275:15254–15264, 2000.
202. Huang EY, Zhang J, Miska EA, et al. Nuclear receptor corepressors partner with class II histone deacetylases in a Sin3-independent repression pathway. Genes Dev 14:45–54, 2000.
203. Kao HY, Downes M, Ordentlich P, et al. Isolation of a novel histone deacetylase reveals that class I and class II deacetylases promote SMRT-mediated repression. Genes Dev 14:55–66, 2000.
204. Vidal M, Gaber RF. RPD3 encodes a second factor required to achieve maximum positive and negative transcriptional states in *Saccharomyces cerevisiae*. Mol Cell Biol 11:6317–6327, 1991.
205. Vidal M, Strich R, Esposito RE, et al. RPD1 (SIN3/UME4) is required for maximal activation and repression of diverse yeast genes. Mol Cell Biol 11:6306–6316, 1991.
206. Bowdish KS, Mitchell AP. Bipartite structure of an early meiotic upstream activation sequence from Saccharomyces cerevisiae. Mol Cell Biol 13:2172–2181, 1993.
207. Vannier D, Balderes D, Shore D. Evidence that the transcriptional regulators SIN3 and RPD3, and a novel gene (SDS3) with similar functions, are involved in transcriptional silencing in S. cerevisiae. Genetics 144:1343–1353, 1996.
208. Alland L, Muhle R, Hou H, Jr., et al. Role for N-CoR and histone deacetylase in Sin3-mediated transcriptional repression. Nature 387:49–55, 1997.
209. Heinzel T, Lavinsky RM, Mullen TM, et al. A complex containing N-CoR, mSin3 and histone deacetylase mediates transcriptional repression. Nature 387:43–48, 1997.
210. Li H, Leo C, Schroen DJ, et al. Characterization of receptor interaction and transcriptional repression by the corepressor SMRT. Mol Endocrinol 11:2025–2037, 1997.
211. Nagy L, Kao HY, Chakravarti D, et al. Nuclear receptor repression mediated by a complex containing SMRT, mSin3A, and histone deacetylase. Cell 89:373–380, 1997.
212. Cavailles V, Dauvois S, L'Horset F, et al. Nuclear factor RIP140 modulates transcriptional

213. activation by the estrogen receptor. EMBO J 14:3741–3751, 1995.
213. Lee CH, Chinpaisal C, Wei LN. Cloning and characterization of mouse RIP140, a corepressor for nuclear orphan receptor TR2. Mol Cell Biol 18:6745–6755, 1998.
213a. Englebienne P. The serum steroid transport proteins: biochemistry and clinical significance. Mol Aspects Med 7:313–396, 1984.
214. Haussler MR, Haussler CA, Jurutka PW, et al. The vitamin D hormone and its nuclear receptor: molecular actions and disease states. J Endocrinol 154(Suppl):S57–S73, 1997.
215. Beaumont K, Fanestil DD. Characterization of rat brain aldosterone receptors reveals high affinity for corticosterone. Endocrinology 113: 2043–2051, 1983.
216. Benhamou B, Garcia T, Lerouge T, et al. A single amino acid that determines the sensitivity of progesterone receptors to RU486. Science 255:206–209, 1992.
217. Heyman RA, Mangelsdorf DJ, Dyck JA, et al. 9-cis retinoic acid is a high affinity ligand for the retinoid X receptor. Cell 68:397–406, 1992.
218. Mangelsdorf DJ, Kliewer SA, Kakizuka A, et al. Retinoid receptors. Recent Prog Horm Res 48:99–121, 1993.
219. Gehin M, Vivat V, Wurtz JM, et al. Structural basis for engineering of retinoic acid receptor isotype-selective agonists and antagonists. Chem Biol 6:519–529, 1999.
220. Lala DS, Mukherjee R, Schulman IG, et al. Activation of specific RXR heterodimers by an antagonist of RXR homodimers. Nature 383: 450–453, 1996.
221. Leblanc BP, Stunnenberg HG. 9-cis retinoic acid signaling: changing partners causes some excitement. Genes Dev 9:1811–1816, 1995.
222. Lefstin JA, Yamamoto KR. Allosteric effects of DNA on transcriptional regulators. Nature 392:885–888, 1998.
223. Kurokawa R, Soderstrom M, Horlein A, et al. Polarity-specific activities of retinoic acid receptors determined by a co-repressor. Nature 377: 451–454, 1995.
224. Kurokawa R, Yu VC, Naar A, et al. Differential orientations of the DNA-binding domain and carboxy-terminal dimerization interface regulate binding site selection by nuclear receptor heterodimers. Genes Dev 7:1423–1435, 1993.
225. Zechel C, Shen XQ, Chen JY, et al. The dimerization interfaces formed between the DNA binding domains of RXR, RAR and TR determine the binding specificity and polarity of the full-length receptors to direct repeats. EMBO J 13:1425–1433, 1994.
226. Willy PJ, Umesono K, Ong ES, et al. LXR, a nuclear receptor that defines a distinct retinoid response pathway. Genes Dev 9:1033–1045, 1995.
227. Kurokawa R, DiRenzo J, Boehm M, et al. Regulation of retinoid signalling by receptor polarity and allosteric control of ligand binding. Nature 371:528–531, 1994.
228. Schulman IG, Li C, Schwabe JW, et al. The phantom ligand effect: allosteric control of transcription by the retinoid X receptor. Genes Dev 11:299–308, 1997.
229. Lechner J, Welte T, Doppler W. Mechanism of interaction between the glucocorticoid receptor and Stat5: role of DNA-binding. Immunobiology 198:112–123, 1997.
230. Cella N, Groner B and Hynes NE. Characterization of Stat5a and Stat5b homodimers and heterodimers and their association with the glucocortiocoid receptor in mammary cells. Mol Cell Biol 18:1783–1792, 1998.
231. Wyszomierski SL, Yeh J, Rosen JM. Glucocorticoid receptor/signal transducer and activator of transcription 5 (STAT5) interactions enhance STAT5 activation by prolonging STAT5 DNA binding and tyrosine phosphorylation. Mol Endocrinol 13:330–343, 1999.
232. Bamberger AM, Bamberger CM, Gellersen B, et al. Modulation of AP-1 activity by the human progesterone receptor in endometrial adenocarcinoma cells. Proc Natl Acad Sci USA 93:6169–6174, 1996.
233. Webb P, Nguyen P, Valentine C, et al. The estrogen receptor enhances AP-1 activity by two distinct mechanisms with different requirements for receptor transactivation functions. Mol Endocrinol 13:1672–1685, 1999.
234. Kallio PJ, Poukka H, Moilanen A, et al. Androgen receptor-mediated transcriptional regulation in the absence of direct interaction with a specific DNA element. Mol Endocrinol 9:1017–1028, 1995.
235. Aarnisalo P, Santti H, Poukka H, et al. Transcription activating and repressing functions of the androgen receptor are differentially influenced by mutations in the deoxyribonucleic acid-binding domain. Endocrinology 140:3097–3105, 1999.
236. Schule R, Rangarajan P, Yang N, et al. Retinoic acid is a negative regulator of AP-1-responsive genes. Proc Natl Acad Sci USA 88:6092–6096, 1991.
237. Yang-Yen HF, Zhang XK, Graupner G, et al. Antagonism between retinoic acid receptors and AP-1: implications for tumor promotion and inflammation. New Biol 3:1206–1219, 1991.
238. Zhang XK, Wills KN, Husmann M, et al. Novel pathway for thyroid hormone receptor action through interaction with jun and fos oncogene activities. Mol Cell Biol 11:6016–6025, 1991.
239. Ricote M, Li AC, Willson TM, et al. The peroxisome proliferator-activated receptor-gamma is a negative regulator of macrophage activation. Nature 391:79–82, 1998.
240. Delerive P, De Bosscher K, Besnard S, et al. Peroxisome proliferator–activated receptor alpha negatively regulates the vascular inflamma-

tory gene response by negative cross-talk with transcription factors NF-kappaB and AP-1. J Biol Chem 274:32048–32054, 1999.
241. Scheinman RI, Gualberto A, Jewell CM, et al. Characterization of mechanisms involved in transrepression of NF-kappa B by activated glucocorticoid receptors. Mol Cell Biol 15:943–953, 1995.
242. Stein B, Yang MX. Repression of the interleukin-6 promoter by estrogen receptor is mediated by NF-kappa B and C/EBP beta. Mol Cell Biol 15:4971–4979, 1995.
243. Kalkhoven E, Wissink S, van der Saag PT, et al. Negative interaction between the RelA(p65) subunit of NF-kappaB and the progesterone receptor. J Biol Chem 271:6217–6224, 1996.
244. Palvimo JJ, Reinikainen P, Ikonen T, et al. Mutual transcriptional interference between RelA and androgen receptor. J Biol Chem 271:24151–24156, 1996.
245. Segars JH, Nagata T, Bours V, et al. Retinoic acid induction of major histocompatibility complex class I genes in NTera-2 embryonal carcinoma cells involves induction of NF-kappa B (p50–p65) and retinoic acid receptor beta–retinoid X receptor beta heterodimers. Mol Cell Biol 13:6157–6169, 1993.
246. Chinetti G, Griglio S, Antonucci M, et al. Activation of proliferator–activated receptors alpha and gamma induces apoptosis of human monocyte-derived macrophages. J Biol Chem 273:25573–25580, 1998.
247. Poynter ME, Daynes RA. Peroxisome proliferator-activated receptor alpha activation modulates cellular redox status, represses nuclear factor-kappaB signaling, and reduces inflammatory cytokine production in aging. J Biol Chem 273:32833–32841, 1998.
248. Stoecklin E, Wissler M, Schaetzle D, et al. Interactions in the transcriptional regulation exerted by Stat5 and by members of the steroid hormone receptor family. J Steroid Biochem Mol Biol 69:195–204, 1999.
249. Widschwendter M, Widschwendter A, Welte T, et al. Retinoic acid modulates prolactin receptor expression and prolactin-induced STAT-5 activation in breast cancer cells in vitro. Br J Cancer 79:204–210, 1999.
250. Ricote M, Huang J, Fajas L, et al. Expression of the peroxisome proliferator-activated receptor gamma (PPARgamma) in human atherosclerosis and regulation in macrophages by colony stimulating factors and oxidized low density lipoprotein. Proc Natl Acad Sci USA 95:7614–7619, 1998.
251. Zhou YC, Waxman DJ. Cross-talk between janus kinase-signal transducer and activator of transcription (JAK-STAT) and peroxisome proliferator-activated receptor-alpha (PPARalpha) signaling pathways. Growth hormone inhibition of PPARalpha transcriptional activity mediated by STAT5b. J Biol Chem 274:2672–2681, 1999.
252. Cato AC, Wade E. Molecular mechanisms of anti-inflammatory action of glucocorticoids. Bioessays 18:371–378, 1996.
253. Gottlicher M, Heck S, Herrlich P. Transcriptional cross-talk, the second mode of steroid hormone receptor action. J Mol Med 76:480–489, 1998.
254. Resche-Rigon M, Gronemeyer H. Therapeutic potential of selective modulators of nuclear receptor action. Curr Opin Chem Biol 2:501–507, 1998.
255. McKay LI, Cidlowski JA. Molecular control of immune/inflammatory responses: interactions between nuclear factor-kappa B and steroid receptor-signaling pathways. Endocr Rev 20:435–459, 1999.
256. Yamamoto KR. Steroid receptor regulated transcription of specific genes and gene networks. Annu Rev Genet 19:209–252, 1985.
257. Didonato JA, Saatcioglu F and Karin M. Molecular mechanisms of immunosuppression and anti-inflammatory activities by glucocorticoids. Am J Respir Crit Care Med 154:S11–S15, 1996.
258. Karin M, Liu ZG, Zandi E. AP-1 function and regulation. Curr Opin Cell Biol 9:240–246, 1997.
259. Vogt PK, Bos TJ. jun: oncogene and transcription factor. Adv Cancer Res 55:1–35, 1990.
260. Angel P, Karin M. The role of Jun, Fos and the AP-1 complex in cell-proliferation and transformation. Biochim. Biophys Acta 1072:129–157, 1991.
261. Caelles C, Gonzalez-Sancho JM, Munoz A. Nuclear hormone receptor antagonism with AP-1 by inhibition of the JNK pathway. Genes Dev 11:3351–3364, 1997.
262. Heck S, Kullmann M, Gast A, et al. A distinct modulating domain in glucocorticoid receptor monomers in the repression of activity of the transcription factor AP-1. EMBO J 13:4087–4095, 1994.
263. Helmberg A, Auphan N, Caelles C, et al. Glucocorticoid-induced apoptosis of human leukemic cells is caused by the repressive function of the glucocorticoid receptor. EMBO J 14:452–460, 1995.
264. Saatcioglu F, Lopez G, West BL, et al. Mutations in the conserved C-terminal sequence in thyroid hormone receptor dissociate hormone-dependent activation from interference with AP-1 activity. Mol Cell Biol 17:4687–4695, 1997.
265. Verma IM, Stevenson JK, Schwarz EM, et al. Rel/NF-kappaB/I kappaB family: intimate tales of association and dissociation. Genes Dev 9:2723–2735, 1995.
266. Baeuerle PA, Baltimore D. NF-kappaB: ten years after. Cell 87:13–20, 1996.
267. Auphan N, DiDonato JA, Rosette C, et al. Immunosuppression by glucocorticoids: inhibition of NF-kappaB activity through induction of I kappa B synthesis. Science 270:286–290, 1995.

268. Scheinman RI, Cogswell PC, Lofquist AK, et al. Role of transcriptional activation of IkappaB alpha in mediation of immunosuppression by glucocorticoids. Science 270:283–286, 1995.
269. Heck S, Bender K, Kullmann M, et al. I kappaB alpha-independent downregulation of NF-kappaB activity by glucocorticoid receptor. EMBO J 16:4698–4707, 1997.
270. Wissink S, van Heerde EC, Schmitz ML, et al. Distinct domains of the RelA NF-kappaB subunit are required for negative cross-talk and direct interaction with the glucocorticoid receptor. J Biol Chem 272:22278–22284, 1997.
271. McKay LI, Cidlowski JA. Cross-talk between nuclear factor-kappa B and the steroid hormone receptors: mechanisms of mutual antagonism. Mol Endocrinol 12:45–56, 1998.
272. Atchison Ml. Enhancers: mechanisms of action and cell specificity. Annu Rev Cell Biol 4:127–153, 1988.
273. Lin RJ, Kao HY, Ordentlich P, et al. The transcriptional basis of steroid physiology. Cold Spring Harb Symp Quant Biol 63:577–585, 1998.
274. Korach KS. Insights from the study of animals lacking functional estrogen receptor. Science 266:1524–1527, 1994.
275. Bamberger CM, Schulte HM, Chrousos GP. Molecular determinants of glucocorticoid receptor function and tissue sensitivity to glucocorticoids. Endocr Rev 17:245–261, 1996.
276. Lydon JP, DeMayo FJ, Conneely OM, et al. Reproductive phenotpes of the progesterone receptor null mutant mouse. J Steroid Biochem Mol Biol 56:67–77, 1996.
277. DeRijk R, Sternberg EM. Corticosteroid resistance and disease. Ann Med 29:79–82, 1997.
278. Humphreys RC, Lydon JP, O'Malley BW, et al. Use of PRKO mice to study the role of progesterone in mammary gland development. J Mammary Gland Biol Neoplasia 2:343–354, 1997.
279. McEwan IJ, Wright AP, Gustafsson JA. Mechanism of gene expression by the glucocorticoid receptor: role of protein-protein interactions. Bioessays 19:153–160, 1997.
280. Rissman EF, Early AH, Taylor JA, et al. Estrogen receptors are essential for female sexual receptivity. Endocrinology 138:507–510, 1997.
281. Rissman EF, Wersinger SR, Taylor JA, et al. Estrogen receptor function as revealed by knockout studies: neuroendocrine and behavioral aspects. Horm Behav 31:232–243, 1997.
282. Giguère V, Tremblay A, Tremblay GB. Estrogen receptor beta: re-evaluation of estrogen and antiestrogen signaling. Steroids 63:335–339, 1998.
283. Tronche F, Kellendonk C, Reichardt HM, et al. Genetic dissection of glucocorticoid receptor function in mice. Curr Opin Genet Dev 8:532–538, 1998.
284. Brinkmann AO, Blok LJ, de Ruiter PE, et al. Mechanisms of androgen receptor activation and function. J Steroid Biochem Mol Biol 69:307–313, 1999.
285. Couse JF, Hewitt SC, Bunch DO, et al. Postnatal sex reversal of the ovaries in mice lacking estrogen receptors α and β. Science 286:2328–2331, 1999.
286. Couse JF, Korach KS. Estrogen receptor null mice: what have we learned and where will they lead us? Endocr Rev 20:358–417, 1999.
287. Giangrande PH, McDonnell DP. The A and B isoforms of the human progesterone receptor: two functionally different transcription factors encoded by a single gene. Recent Prog Horm Res 54:291–313, 1999.
288. Gottlieb B, Pinsky L, Beitel LK, et al. Androgen insensitivity. Am J Med Genet 89:210–217, 1999.
289. Warner M, Nilsson S, Gustafsson JA. The estrogen receptor family. Curr Opin Obstet Gynecol 11:249–254, 1999.
290. Berger S, Bleich M, Schmid W, et al. Mineralocorticoid receptor knockout mice: lessons on Na+ metabolism. Kidney Int 57:1295–1298, 2000.
291. Conneely OM, Lydon JP, De Mayo F, et al. Reproductive functions of the progesterone receptor. J Soc Gynecol Investig 7:S25–32, 2000.
292. Funder JW. Aldosterone and mineralocorticoid receptors: orphan questions. Kidney Int 57:1358–1363, 2000.
293. Kolla V, Litwack G. Transcriptional regulation of the human Na/K ATPase via the human mineralocorticoid receptor. Mol Cell Biochem 204:35–40, 2000.
294. Prins GS. Molecular biology of the androgen receptor. Mayo Clin Proc 75(Suppl):S32–S35, 2000.
295. Tindall DJ. Androgen receptors in prostate and skeletal muscle. Mayo Clin Proc 75(Suppl):S26–S30, 2000.
296. Lubahn DB, Moyer JS, Golding TS, et al. Alteration of reproductive function but not prenatal sexual development after insertional disruption of the mouse estrogen receptor gene. Proc Natl Acad Sci USA 90:11162–11166, 1993.
297. Krege JH, Hodgin JB, Couse JF, et al. Generation and reproductive phenotypes of mice lacking estrogen receptor beta. Proc Natl Acad Sci USA 95:15677–15682, 1998.
298. Ogawa S, Chan J, Chester AE, et al. Survival of reproductive behaviors in estrogen receptor b gene-deficient (bERKO) male and female mice. Proc Natl Acad Sci USA. 96:12887–12892, 1999.
299. Kraus WL, Weis KE, Katzenellenbogen BS. Inhibitory cross-talk between steroid hormone receptors: differential targeting of estrogen receptor in the repression of its transcriptional activity by agonist- and antagonist-occupied progestin receptors. Mol Cell Biol 15:1847–1857, 1995.
300. Kastner P, Krust A, Turcotte B, et al. Two distinct estrogen-regulated promoters generate transcripts encoding the two functionally differ-

ent human progesterone receptor forms A and B. EMBO J 9:1603–1614, 1990.
301. Vegeto E, Shahbaz MM, Wen DX, et al. Human progesterone receptor A form is a cell- and promoter-specific repressor of human progesterone receptor B function. Mol Endocrinol 7:1244–1255, 1993.
302. Tibbetts TA, DeMayo F, Rich S, et al. Progesterone receptors in the thymus are required for thymic involution during pregnancy and for normal fertility. Proc Natl Acad Sci USA 96:12021–12026, 1999.
303. Hamilton DW, Ofner P. Androgen action and target-organ androgen metabolism. Basic Reproductive Medicine, Reproductive Function in Men 2:142–174, 1982.
304. Biason-Lauber A. Molecular medicine of steroid hormone biosynthesis. Mol Aspects Med 19:155–220, 1998.
305. McPhaul MJ, Marcelli M, Zoppi MS, et al. Genetic basis of endocrine disease 4:the spectrum of mutations in the androgen receptor gene that cause androgen resistance. J Clin Endocrinol Metab 76:17–23, 1993.
306. Craft N, Sawyers CL. Mechanistic concepts in androgen-dependence of prostate cancer. Cancer Metastasis Rev 17:421–427, 1998.
307. Osborne CK. Steroid hormone receptors in breast cancer management. Breast Cancer Res Treat 51:227–238, 1998.
308. Dubik D, Dembinski TC, Shiu RP. Stimulation of c-*myc* oncogene expression associated with estrogen-induced proliferation of human breast cancer cells. Cancer Res 47:6517–6521, 1987.
309. Bhattacharyya N, Ramsammy R, Eatman E, et al. Protooncogene, growth factor, growth factor receptor, and estrogen and progesterone receptor gene expression in the immature rat uterus after treatment with estrogen and tamoxifen. J Submicrosc Cytol Pathol 26:147–162, 1994.
310. Duan R, Porter W, Samudio I, et al. Transcriptional activation of c-*fos* protooncogene by 17beta-estradiol: mechanism of aryl hydrocarbon receptor–mediated inhibition. Mol Endocrinol 13:1511–1521, 1999.
311. Mitlak BH, Cohen FJ. Selective estrogen receptor modulators: a look ahead. Drugs 57:653–663, 1999.
312. Karin M. New twists in gene regulation by glucocorticoid receptor: is DNA binding dispensable? Cell 93:487–490, 1998.
313. Cole TJ, Blendy JA, Monaghan AP, et al. Targeted disruption of the glucocorticoid receptor gene blocks adrenergic chromaffin cell development and severely retards lung maturation. Genes Dev 9:1608–1621, 1995.
314. Reichardt HM, Kaestner KH, Tuckermann J, et al. DNA binding of the glucocorticoid receptor is not essential for survival. Cell 93:531–541, 1998.
315. Oakley RH, Sar M, Cidlowski JA. The human glucocorticoid receptor beta isoform. Expression, biochemical properties, and putative function. J Biol Chem 271:9550–9559, 1996.
316. Evans RM. Molecular characterization of the glucocorticoid receptor. Recent Progr Horm Res 45:1–27, 1989.
317. Pearce D. A mechanistic basis for distinct mineralocorticoid and glucocorticoid receptor transcriptional specificities. Steroids 59:153–159, 1994.
318. Funder JW, Pearce PT, Smith R, et al. Mineralocorticoid action: target tissue specificity is enzyme, not receptor, mediated. Science 242:583–585, 1988.
319. Berger S, Bleich M, Schmid W, et al. Mineralocorticoid receptor knockout mice: pathophysiology of $Na+$ metabolism. Proc Natl Acad Sci USA 95:9424–9429, 1998.
320. Hughes MR, Malloy PJ, Kieback DG, et al. Point mutations in the human vitamin D receptor gene associated with hypocalcemic rickets. Science 242:1702–1705, 1988.
321. Chambon P. A decade of molecular biology of retinoic acid receptors. Faseb J 10:940–954, 1996.
322. Forman BM, Tzameli I, Choi HS, et al. Androstane metabolites bind to and deactivate the nuclear receptor CAR-beta. Nature 395:612–615, 1998.
323. Issa LL, Leong GM, Eisman JA. Molecular mechanism of vitamin D receptor action. Inflamm Res 47:451–475, 1998.
324. Jones G, Strugnell SA, DeLuca HF. Current understanding of the molecular actions of vitamin D. Physiol Rev 78:1193–1231, 1998.
325. Koenig RJ. Thyroid hormone receptor coactivators and corepressors. Thyroid 8:703–713, 1998.
326. Nagpal S, Chandraratna RA. Vitamin A and regulation of gene expression. Curr Opin Clin Nutr Metab Care 1:341–346, 1998.
327. Niles RM. Control of retinoid nuclear receptor function and expression. Subcell Biochem 30:3–28, 1998.
328. Peet DJ, Janowski BA, Mangelsdorf DJ. The LXRs: a new class of oxysterol receptors. Curr Opin Genet Dev 8:571–575, 1998.
329. Hurst RE, Waliszewski P, Waliszewska M, et al. Complexity, retinoid-responsive gene networks, and bladder carcinogenesis. Adv Exp Med Biol 462:449–467, 1999.
330. Johnson A and Chandraratna RA. Novel retinoids with receptor selectivity and functional selectivity. Br J Dermatol 140(Suppl 54):12–17, 1999.
331. Kliewer SA, Lehmann JM, Willson TM. Orphan nuclear receptors: shifting endocrinology into reverse. Science 284:757–760, 1999.
332. Michalik L, Wahli W. Peroxisome proliferator-activated receptors: three isotypes for a multitude of functions. Curr Opin Biotechnol 10:564–570, 1999.
333. Russell DW. Nuclear orphan receptors control cholesterol catabolism. Cell 97:539–542, 1999.

333a. Lu TT, Repa JJ, Mangelsdorf DJ. Orphan nuclear receptors as eLiXiRs and FiXeRs of sterol metabolism. J Biol Chem 276:37735–37738, 2001.
334. Waxman DJ. *P450* gene induction by structurally diverse xenochemicals: central role of nuclear receptors CAR, PXR, and PPAR. Arch Biochem Biophys 369:11–23, 1999.
334a. Wei P, Zhang J, Egan-Hafley M, et al. The nuclear receptor CAR mediates specific xenobiotic induction of drug metabolism. Nature 407:920–923, 2000.
335. Corton JC, Anderson SP, Stauber A. Central role of peroxisome proliferator-activated receptors in the actions of peroxisome proliferators. Annu Rev Pharmacol Toxicol 40:491–518, 2000.
335a. Chawla A, Repa JJ, Evans RM, et al. Nuclear receptors and lipid physiology: opening the X-files. Science 294:1866–1870, 2001.
336. Kato S. The function of vitamin D receptor in vitamin D action. J Biochem (Tokyo). 127:717–722, 2000.
337. Wolffe AP, Collingwood TN, Li Q, et al. Thyroid hormone receptor, v-ErbA, and chromatin. Vitam Horm 58:449–492, 2000.
338. Chatterjee VK, Beck-Peccoz P. Hormone-nuclear receptor interactions in health and disease. Thyroid hormone resistance. Baillieres Clin Endocrinol Metab 8:267–283, 1994.
339. Jameson JL. Mechanisms by which thyroid hormone receptor mutations cause clinical syndromes of resistance to thyroid hormone. Thyroid 4:485–492, 1994.
340. Kastner P, Mark M, Chambon P. Nonsteroid nuclear receptors: what are genetic studies telling us about their role in real life? Cell 83:859–869, 1995.
341. Kopp P, Kitajima K, Jameson JL. Syndrome of resistance to thyroid hormone: insights into thyroid hormone action. Proc Soc Exp Biol Med 211:49–61, 1996.
342. Sucov HM, Lou J, Gruber PJ, et al. The molecular genetics of retinoic acid receptors: cardiovascular and limb development. Biochem Soc Symp 62:143–156, 1996.
343. Rowe A. Retinoid X receptors. Int J Biochem Cell Biol. 29:275–278, 1997.
344. Schulman IG and Evans RM. Retinoid receptors in development and disease. Leukemia. 11(Suppl 3):376–377, 1997.
345. Fruchart JC, Duriez P, Staels B. Peroxisome proliferator-activated receptor-alpha activators regulate genes governing lipoprotein metabolism, vascular inflammation and atherosclerosis. Curr Opin Lipidol 10:245–257, 1999.
346. Gothe S, Wang Z, Ng L, et al. Mice devoid of all known thyroid hormone receptors are viable but exhibit disorders of the pituitary-thyroid axis, growth, and bone maturation. Genes Dev 13:1329–1341, 1999.
347. Kato S, Takeyama K, Kitanaka S, et al. In vivo function of VDR in gene expression-VDR knock-out mice. J Steroid Biochem Mol Biol 69:247–251, 1999.
348. Kliewer SA, Lehmann JM, Milburn MV, et al. The PPARs and PXRs: nuclear xenobiotic receptors that define novel hormone signaling pathways. Recent Prog Horm Res 54:345–367, 1999.
348a. Xie W, Evans RM. Orphan nuclear receptors: the exotics of xenobiotics. J Biol Chem 276:37739–37742, 2001.
349. Lin RJ, Egan DA, Evans RM. Molecular genetics of acute promyelocytic leukemia. Trends Genet 15:179–184, 1999.
350. Mark M, Ghyselinck NB, Wendling O, et al. A genetic dissection of the retinoid signalling pathway in the mouse. Proc Nutr Soc 58:609–613, 1999.
351. Repa JJ, Mangelsdorf DJ. Nuclear receptor regulation of cholesterol and bile acid metabolism. Curr Opin Biotechnol 10:557–563, 1999.
352. Rocchi S, Auwerx J. Peroxisome proliferator-activated receptor-gamma: a versatile metabolic regulator. Ann Med 31:342–351, 1999.
353. Tontonoz P, Nagy L. Regulation of macrophage gene expression by peroxisome-proliferator-activated receptor gamma: implications for cardiovascular disease. Curr Opin Lipidol 10:485–490, 1999.
354. Chawla A, Saez E, Evans RM. Don't know much bile-ology. Cell 103:1–4, 2000.
354a. Sinal CJ, Tohkin M, Miyata M, et al. Targeted disruption of the nuclear receptor FXR/BAR impairs bile acid and lipid homeostasis. Cell 102:731–744, 2000.
355. Escher P, Wahli W. Peroxisome proliferator-activated receptors: insight into multiple cellular functions. Mutat Res 448:121–138, 2000.
356. Forrest D, Vennstrom B. Functions of thyroid hormone receptors in mice. Thyroid 10:41–52, 2000.
357. Kadowaki T. Insights into insulin resistance and type 2 diabetes from knockout mouse models. J Clin Invest 106:459–465, 2000.
358. Xie W, Barwick JL, Downes M, et al. Humanized xenobiotic response in mice expressing nuclear receptor SXR. Nature 406:435–439, 2000.
359. Whitfield GK, Selznick SH, Haussler CA, et al. Vitamin D receptors from patients with resistance to 1,25-dihydroxyvitamin D3: point mutations confer reduced transactivation in response to ligand and impaired interaction with the retinoid X receptor heterodimeric partner. Mol Endocrinol 10:1617–1631, 1996.
360. Carlberg C, Mathiasen IS, Saurat JH, et al. The 1,25-dihydroxyvitamin D3 (VD) analogues MC903, EB1089 and KH1060 activate the VD receptor: homodimers show higher ligand sensitivity than heterodimers with retinoid X receptors. J Steroid Biochem Mol Biol 51:137–142, 1994.

361. Liu M, Lee MH, Cohen M, et al. Transcriptional activation of the Cdk inhibitor p21 by vitamin D3 leads to the induced differentiation of the myelomonocytic cell line U937. Genes Dev 10:142–153, 1996.
362. Collingwood TN, Adams M, Tone Y, et al. Spectrum of transcriptional, dimerization, and dominant negative properties of twenty different mutant thyroid hormone beta-receptors in thyroid hormone resistance syndrome. Mol Endocrinol 8:1262–1277, 1994.
363. Weinberger C, Thompson CC, Ong ES, et al. The c-erb-A gene encodes a thyroid hormone receptor. Nature 324:641–646, 1986.
364. Sharif M, Privalsky ML. V-erbA and c-erbA proteins enhance transcriptional activation by c-jun. Oncogene 7:953–960, 1992.
365. Lin KH, Zhu XG, Shieh HY, et al. Identification of naturally occurring dominant negative mutants of thyroid hormone alpha 1 and beta 1 receptors in a human hepatocellular carcinoma cell line. Endocrinology 137:4073–4081, 1996.
366. Gudas LJ. Retinoids and vertebrate development. J Biol Chem 269:15399–15402, 1994.
367. Sucov HM, Evans RM. Retinoic acid and retinoic acid receptors in development. Mol Neurobiol 10:169–184, 1995.
368. Evans RM. The molecular basis of signaling by vitamin A and its metabolites. Harvey Lect 90:105–117, 1994.
369. Lohnes D, Mark M, Mendelsohn C, et al. Function of the retinoic acid receptors (RARs) during development (I). Craniofacial and skeletal abnormalities in RAR double mutants. Development 120:2723–2748, 1994.
370. Mendelsohn C, Lohnes D, Décimo D, et al. Function of the retinoic acid receptors (RARs) during development (II). Multiple abnormalities at various stages of organogenesis in RAR double mutants. Development 120:2749–2771, 1994.
371. Sucov HM, Izpisúa-Belmonte JC, Gañan Y, et al. Mouse embryos lacking RXR alpha are resistant to retinoic-acid-induced limb defects. Development 121:3997–4003, 1995.
372. Lin RJ, Evans RM. Acquisition of oncogenic potential by RAR chimeras in acute promyelocytic leukemia through formation of homodimers. Mol Cell 5:821–830, 2000.
373. Blumberg B, Evans RM. Orphan nuclear receptors—new ligands and new possibilities. Genes Dev 12:3149–3155, 1998.
374. Kastner P, Grondona JM, Mark M, et al. Genetic analysis of RXR alpha developmental function: convergence of RXR and RAR signaling pathways in heart and eye morphogenesis. Cell 78:987–1003, 1994.
375. Sucov HM, Dyson E, Gumeringer CL, et al. RXR alpha mutant mice establish a genetic basis for vitamin A signaling in heart morphogenesis. Genes Dev 8:1007–1018, 1994.
376. Krezel W, Dupe V, Mark M, et al. RXR gamma null mice are apparently normal and compound RXR alpha$^{+/-}$/RXR beta$^{-/-}$/RXR gamma$^{-/-}$ mutant mice are viable. Proc Natl Acad Sci USA 93:9010–9014, 1996.
377. Kastner P, Mark M, Leid M, et al. Abnormal spermatogenesis in RXR beta mutant mice. Genes Dev 10:80–92, 1996.
378. Chiang MY, Misner D, Kempermann G, et al. An essential role for retinoid receptors RARbeta and RXRgamma in long-term potentiation and depression. Neuron 21:1353–1361, 1998.
379. Issemann I, Green S. Activation of a member of the steroid hormone receptor superfamily by peroxisome proliferators. Nature 347:645–650, 1990.
380. Green S, Tugwood JD, Issemann I. The molecular mechanism of peroxisome proliferator action: a model for species differences and mechanistic risk assessment. Toxicol Lett 64–65 Spec No: 131–139, 1992.
381. Green S. Peroxisome proliferators: a model for receptor mediated carcinogenesis. Cancer Surv 14:221–232, 1992.
382. Lemberger T, Desvergne B, Wahli W. Peroxisome proliferator-activated receptors: a nuclear receptor signaling pathway in lipid physiology. Ann Rev Cell Dev Biol 12:335–363, 1996.
383. Seedorf U. Peroxisomes in lipid metabolism. J Cell Biochem Suppl 30–31:158–167, 1998.
384. Kliewer SA, Forman BM, Blumberg B, et al. Differential expression and activation of a family of murine peroxisome proliferator-activated receptors. Proc Natl Acad Sci USA 91:7355–7359, 1994.
385. Willson TM, Brown PJ, Sternbach DD, et al. The PPARs: from orphan receptors to drug discovery. J Med Chem 43:527–550, 2000.
386. Göttlicher M, Widmark E, Li Q, et al. Fatty acids activate a chimera of the clofibric acid-activated receptor and the glucocorticoid receptor. Proc Natl Acad Sci USA 89:4653–4657, 1992.
387. Devchand PR, Keller H, Peters JM, et al. The PPARalpha-leukotriene B4 pathway to inflammation control. Nature 384:39–43, 1996.
388. Peters JM, Hennuyer N, Staels B, et al. Alterations in lipoprotein metabolism in peroxisome proliferator-activated receptor alpha-deficient mice. J Biol Chem 272:27307–27312, 1997.
389. Aoyama T, Peters JM, Iritani N, et al. Altered constitutive expression of fatty acid-metabolizing enzymes in mice lacking the peroxisome proliferator-activated receptor alpha (PPARalpha). J Biol Chem 273:5678–5684, 1998.
390. Tontonoz P, Graves RA, Budavari AI, et al. Adipocyte-specific transcription factor ARF6 is a heterodimeric complex of two nuclear hormone receptors, PPAR gamma and RXR alpha. Nucleic Acids Res 22:5628–5634, 1994.
391. Tontonoz P, Hu E, Graves RA, et al. mPPAR gamma 2: tissue-specific regulator of an adipocyte enhancer. Genes Dev 8:1224–1234, 1994.

392. Forman BM, Tontonoz P, Chen J, et al. 15-Deoxy-delta 12, 14-prostaglandin J2 is a ligand for the adipocyte determination factor PPAR gamma. Cell 83:803–812, 1995.
393. Hu E, Tontonoz P, Spiegelman BM. Transdifferentiation of myoblasts by the adipogenic transcription factors PPAR gamma and C/EBP alpha. Proc Natl Acad Sci USA 92:9856–9860, 1995.
394. Kliewer SA, Lenhard JM, Willson TM, et al. A prostaglandin J2 metabolite binds peroxisome proliferator-activated receptor gamma and promotes adipocyte differentiation. Cell 83:813–819, 1995.
395. Wu Z, Xie Y, Bucher NL, et al. Conditional ectopic expression of C/EBP beta in NIH-3T3 cells induces PPAR gamma and stimulates adipogenesis. Genes Dev 9:2350–2363, 1995.
396. Spiegelman BM, Hu E, Kim JB, et al. PPAR gamma and the control of adipogenesis. Biochimie 79:111–112, 1997.
397. Saez E, Tontonoz P, Nelson MC, et al. Activators of the nuclear receptor PPARgamma enhance colon polyp formation. Nat Med 4:1058–1061, 1998.
398. Tontonoz P, Nagy L, Alvarez JG, et al. PPARgamma promotes monocyte/macrophage differentiation and uptake of oxidized LDL. Cell 93:241–252, 1998.
399. Barak Y, Nelson MC, Ong ES, et al. PPARg is required for placental, cardiac, and adipose tissue development. Molec Cell 585–594:1999.
400. Lehmann JM, Moore LB, Smith-Oliver TA, et al. An antidiabetic thiazolidinedione is a high affinity ligand for peroxisome proliferator-activated receptor gamma (PPAR gamma). J Biol Chem 270:12953–12956, 1995.
401. Spiegelman BM. PPAR-gamma: adipogenic regulator and thiazolidinedione receptor. Diabetes 47:507–514, 1998.
402. De Vos P, Lefebvre AM, Miller SG, et al. Thiazolidinediones repress ob gene expression in rodents via activation of peroxisome proliferator-activated receptor gamma. J Clin Invest 98:1004–1009, 1996.
403. Long SD, Pekala PH. Regulation of GLUT4 gene expression by arachidonic acid. Evidence for multiple pathways, one of which requires oxidation to prostaglandin E2. J Biol Chem 271:1138–1144, 1996.
404. Qian H, Hausman GJ, Compton MM, et al. Leptin regulation of peroxisome proliferator-activated receptor-gamma, tumor necrosis factor, and uncoupling protein-2 expression in adipose tissues. Biochem Biophys Res Commun 246:660–667, 1998.
405. Kubota N, Terauchi Y, Miki H, et al. PPARg mediates high-fat diet-induced adipocyte hypertrophy and insulin resistance. Mol Cell 4:597–609, 1999.
406. Komers R, Vrána A. Thiazolidinediones—tools for the research of metabolic syndrome X. Physiol Res 47:215–225, 1998.
407. Tontonoz P, Singer S, Forman BM, et al. Terminal differentiation of human liposarcoma cells induced by ligands for peroxisome proliferator-activated receptor gamma and the retinoid X receptor. Proc Natl Acad Sci USA 94:237–241, 1997.
408. Mueller E, Sarraf P, Tontonoz P, et al. Terminal differentiation of human breast cancer through PPAR gamma. Mol Cells 1:465–470, 1998.
409. Lefebvre A, Chen I, Desreumaux P, et al. Activation of the peroxisome proliferator-activated receptor g promotes the development of colon tumors in C57BL/6J-APC Min/+ mice. Nature Med 4:1053–1057, 1998.
409a. Sarraf P, Mueller E, Jones D, et al. Differentiation and reversal of malignant changes in colon caner through PPARg. Nature Med 4:1046–1052, 1998.
410. Lim H, Gupta RA, Ma WG, et al. Cyclo-oxygenase-2-derived prostacyclin mediates embryo implantation in the mouse via PPARdelta. Genes Dev 13:1561–1574, 1999.
411. Oshima M, Dinchuk JE, Kargman SL, et al. Suppression of intestinal polyposis in Apc delta716 knockout mice by inhibition of cyclooxygenase 2 (COX-2). Cell 87:803–809, 1996.
412. He TC, Chan TA, Vogelstein B, et al. PPARdelta is an APC-regulated target of nonsteroidal anti-inflammatory drugs. Cell 99:335–345, 1999.
412a. Tan NS, Michalik L, Noy N, et al. Critical roles of PPAR beta/delta in keratinocyte response to inflammation. Genes Dev 15:3263–3277, 2001.
413. Janowski BA, Willy PJ, Devi TR, et al. An oxysterol signalling pathway mediated by the nuclear receptor LXR alpha. Nature 383:728–731, 1996.
414. Makishima M, Okamoto AY, Repa JJ, et al. Identification of a nuclear receptor for bile acids. Science 284:1362–1365, 1999.
415. Parks DJ, Blanchard SG, Bledsoe RK, et al. Bile acids: natural ligands for an orphan nuclear receptor. Science 284:1365–1368, 1999.
416. Wang H, Chen J, Hollister K, et al. Endogenous bile acids are ligands for the nuclear receptor FXR/BAR. Mol Cell 3:543–553, 1999.
417. Lu TT, Makishima M, Repa JJ, et al. Molecular basis for feedback regulation of bile acid synthesis by nuclear receptors. Mol Cell 6:507–515, 2000.
418. Peet DJ, Turley SD, Ma W, et al. Cholesterol and bile acid metabolism are impaired in mice lacking the nuclear oxysterol receptor LXR alpha. Cell 93:693–704, 1998.
419. Costet P, Luo Y, Wang N, et al. Sterol-dependent transactivation of the ABC1 promoter by the liver X receptor/retinoid X receptor. J Biol Chem 275:28240–28245, 2000.
420. Luo Y, Tall AR. Sterol upregulation of human CETP expression in vitro and in transgenic mice

by an LXR element. J Clin Invest 105:513–520, 2000.
421. Venkateswaran A, Repa JJ, Lobaccaro JM, et al. Human white/murine ABC8 mRNA levels are highly induced in lipid-loaded macrophages. A transcriptional role for specific oxysterols. J Biol Chem 275:14700–14707, 2000.
422. Tobin KA, Steineger HH, Alberti S, et al. Cross-talk between fatty acid and cholesterol metabolism mediated by liver X receptor-alpha. Mol Endocrinol 14:741–752, 2000.
422a. Willson TM, Jones SA, Moore JT, et al. Chemical genomics: functional analysis of orphan nuclear receptors in the regulation of bile acid metabolism. Med Res Rev 21:513–522, 2001.
423. Kliewer SA, Moore JT, Wade L, et al. An orphan nuclear receptor activated by pregnanes defines a novel steroid signaling pathway. Cell 92:73–82, 1998.
424. Honkakoski P, Zelko I, Sueyoshi T, et al. The nuclear orphan receptor CAR-retinoid X receptor heterodimer activates the phenobarbital-responsive enhancer module of the CYP2B gene. Mol Cell Biol 18:5652–5658, 1998.
424a. Xie W, Barwick JL, Simon CM, et al. Reciprocal activation of xenobiotic response genes by nuclear receptors SXR/PXR and CAR. Genes Dev 14:3014–3023, 2000
424b. Moore LB, Parks DJ, Jones SA, et al. Orphan nuclear receptors constitutive androstane receptor and pregnane X receptor share xenobiotic and steroid ligands. J Biol Chem 275: 15122–15127, 2000.
425. Maruyama K, Tsukada T, Ohkura N, et al. The NGFI-B subfamily of the nuclear receptor superfamily (review). Int J Oncol 12:1237–1243, 1998.
426. Fernandez PM, Brunel F, Jimenez MA, et al. Nuclear receptors Nor1 and NGFI-B/Nur77 play similar, albeit distinct, roles in the hypothalamo-pituitary-adrenal axis. Endocrinology 141:2392–2400, 2000.
427. Chawla A, Lazar MA. Induction of Rev-ErbA alpha, an orphan receptor encoded on the opposite strand of the alpha-thyroid hormone receptor gene, during adipocyte differentiation. J Biol Chem 268:16265–16269, 1993.
428. Sladek FM. Orphan receptor HNF-4 and liver-specific gene expression. Receptor 3:223–232, 1993.
429. Dumas B, Harding HP, Choi HS, et al. A new orphan member of the nuclear hormone receptor superfamily closely related to Rev-Erb. Mol Endocrinol 8:996–1005, 1994.
430. Hirose T, Fujimoto W, Tamaai T, et al. TAK1: molecular cloning and characterization of a new member of the nuclear receptor superfamily. Mol Endocrinol 8:1667–1680, 1994.
431. Carlberg C, Wiesenberg I. The orphan receptor family RZR/ROR, melatonin and 5-lipoxygenase: an unexpected relationship. J Pineal Res 18:171–178, 1995.

432. Crawford PA, Sadovsky Y, Woodson K, et al. Adrenocortical function and regulation of the steroid 21-hydroxylase gene in NGFI-B-deficient mice. Mol Cell Biol 15:4331–4316, 1995.
433. Tini M, Fraser RA, Giguere V. Functional interactions between retinoic acid receptor-related orphan nuclear receptor (ROR alpha) and the retinoic acid receptors in the regulation of the gamma F-crystallin promoter. J Biol Chem 270:20156–20161, 1995.
434. Ikeda Y. SF-1: a key regulator of development and function in the mammalian reproductive system. Acta Paediatr Jpn 38:412–419, 1996.
435. Murphy EP, Dobson AD, Keller C, et al. Differential regulation of transcription by the NURR1/NUR77 subfamily of nuclear transcription factors. Gene Expr 5:169–179, 1996.
436. Qiu Y, Krishnan V, Pereira FA, et al. Chicken ovalbumin upstream promoter-transcription factors and their regulation. J Steroid Biochem Mol Biol 56:81–85, 1996.
437. Luo J, Sladek R, Bader JA, et al. Placental abnormalities in mouse embryos lacking the orphan nuclear receptor ERR-beta. Nature 388:778–782, 1997.
438. Monaghan AP, Bock D, Gass P, et al. Defective limbic system in mice lacking the tailless gene. Nature 390:515–517, 1997.
439. Sladek R, Bader JA and Giguere V. The orphan nuclear receptor estrogen-related receptor alpha is a transcriptional regulator of the human medium-chain acyl coenzyme A dehydrogenase gene. Mol Cell Biol 17:5400–5409, 1997.
440. Tsai SY, Tsai MJ. Chick ovalbumin upstream promoter-transcription factors (COUP-TFs): coming of age. Endocr Rev 18:229–240, 1997.
441. Zetterstrom RH, Solomin L, Jansson L, et al. Dopamine neuron agenesis in Nurr1-deficient mice. Science 276:248–250, 1997.
442. Dussault I, Fawcett D, Matthyssen A, et al. Orphan nuclear receptor ROR alpha-deficient mice display the cerebellar defects of staggerer. Mech Devel 70:147–153, 1998.
443. Greschik H, Schüle R. Germ cell nuclear factor: an orphan receptor with unexpected properties. J Mol Med 76:800–810, 1998.
444. Bassett JH, O'Halloran DJ, Williams GR, et al. Novel DAX1 mutations in X-linked adrenal hypoplasia congenita and hypogonadotrophic hypogonadism. Clin Endocrinol (Oxf) 50:69–75, 1999.
445. Goodfellow PN, Camerino G. DAX-1, an "antitestis" gene. Cell Mol Life Sci 55:857–863, 1999.
445a. Wang L, Lee YK, Bundman D, et al. Redundant pathways for negative feedback regulation of bile acid production. Dev Cell 2:721–731, 2002.
445b. Kerr TA, Saeki S, Schneider M, et al. Loss of nuclear receptor SHP impairs but does not eliminate negative feedback regulation of bile acid synthesis. Dev Cell 2:713–720, 2002.

445c. Nishigori H, Tomura H, Tonooka N, et al. Mutations in the small heterodimer partner gene are associated with mild obesity in Japanese subjects. Proc Natl Acad Sci USA 98:575–580, 2001.
446. Hammer GD, Ingraham HA. Steroidogenic factor-1: its role in endocrine organ development and differentiation. Front Neuroendocrinol 20:199–223, 1999.
447. Kobayashi M, Takezawa S, Hara K, et al. Identification of a photoreceptor cell-specific nuclear receptor. Proc Natl Acad Sci USA 96:4814–4819, 1999.
448. Luo X, Ikeda Y, Lala D, et al. Steroidogenic factor 1 (SF-1) is essential for endocrine development and function. J Steroid Biochem Mol Biol 69:13–18, 1999.
449. Parker KL, Schedl A, Schimmer BP. Gene interactions in gonadal development. Annu Rev Physiol 61:417–433, 1999.
450. Vanacker JM, Bonnelye E, Chopin-Delannoy S, et al. Transcriptional activities of the orphan nuclear receptor ERR alpha (estrogen receptor-related receptor-alpha). Mol Endocrinol 13:764–773, 1999.
451. Heard DJ, Norby PL, Holloway J, et al. Human ERRgamma, a third member of the estrogen receptor-related receptor (ERR) subfamily of orphan nuclear receptors: tissue-specific isoforms are expressed during development and in the adult. Mol Endocrinol 14:382–392, 2000.
451a. Giguere V. To ERR in the estrogen pathway. Trends Endocrinol Metab 13:220–225, 2002.
452. Kurebayashi S, Ueda E, Sakaue M, et al. Retinoid-related orphan receptor gamma (RORgamma) is essential for lymphoid organogenesis and controls apoptosis during thymopoiesis. Proc Natl Acad Sci USA 97:10132–10137, 2000.
453. Li J, Ning G, Duncan SA. Mammalian hepatocyte differentiation requires the transcription factor HNF-4alpha. Genes Dev 14:464–474, 2000.
454. Yu RT, Chiang MY, Tanabe T, et al. The orphan nuclear receptor Tlx regulates Pax2 and is essential for vision. Proc Natl Acad Sci USA 97:2621–2625, 2000.
455. Zhou C, Tsai SY, Tsai M. From apoptosis to angiogenesis: new insights into the roles of nuclear orphan receptors, chicken ovalbumin upstream promoter-transcription factors, during development. Biochim Biophys Acta 1470:M63–68, 2000.
456. Johansson L, Bavner A, Thomsen JS, et al. The orphan nuclear receptor SHP utilizes conserved LXXLL-related motifs for interactions with ligand-activated estrogen receptors. Mol Cell Biol 20:1124–1133, 2000.
457. Sjogren H, Meis-Kindblom J, Kindblom LG, et al. Fusion of the EWS-related gene TAF2N to TEC in extraskeletal myxoid chondrosarcoma. Cancer Res 59:5064–5067, 1999.
457a. Backman C, Perlmann T, Wallen A, et al. A selective group of dopaminergic neurons express Nurr1 in the adult mouse brain. Brain Res 851:125–132, 1999.
458. Brown PJ, Smith-Oliver TA, Charifson PS, et al. Identification of peroxisome proliferator-activated receptor ligands from a biased chemical library. Chem Biol 4:909–918, 1997.
459. Berger J, Leibowitz MD, Doebber TW, et al. Novel peroxisome proliferator-activated receptor (PPAR) gamma and PPARdelta ligands produce distinct biological effects. J Biol Chem 274:6718–6725, 1999.
460. Nichols JS, Parks DJ, Consler TG, et al. Development of a scintillation proximity assay for peroxisome proliferator-activated receptor gamma ligand binding domain. Anal Biochem 257:112–119, 1998.
461. Ito M, Yuan CX, Malik S, et al. Identity between TRAP and SMCC complexes indicates novel pathways for the function of nuclear receptors and diverse mammalian activators. Mol Cell 3:361–370, 1999.
462. Guenther MG, Lane WS, Fischle W, et al. A core SMRT corepressor complex containing HDAC3 and TBL1, a WD40-repeat protein linked to deafness. Genes Dev 14:1048–1057, 2000.
463. Diatchenko L, Lau YF, Campbell AP, et al. Suppression subtractive hybridization: a method for generating differentially regulated or tissue-specific cDNA probes and libraries. Proc Natl Acad Sci USA 93:6025–6030, 1996.
464. Hu E and Spiegelman BM. Identification of novel genes involved in adipose differentiation by differential display. Methods Mol Biol 85:195–204, 1997.
465. Motojima K. Peroxisome proliferator-activated receptor (PPAR)-dependent and -independent transcriptional modulation of several nonperoxisomal genes by peroxisome proliferators. Biochimie 79:101–106, 1997.
466. Kuang WW, Thompson DA, Hoch RV, et al. Differential screening and suppression subtractive hybridization identified genes differentially expressed in an estrogen receptor-positive breast carcinoma cell line. Nucleic Acids Res 26:1116–1123, 1998.
467. Nelson PS, Gan L, Ferguson C, et al. Molecular cloning and characterization of prostase, an androgen-regulated serine protease with prostate-restricted expression. Proc Natl Acad Sci USA 96:3114–3119, 1999.
468. White KP, Rifkin SA, Hurban P, et al. Microarray analysis of *Drosophila* development during metamorphosis. Science 286:2179–2184, 1999.
469. Capecchi MR. The new mouse genetics: altering the genome by gene targeting. Trends Genet 5:70–76, 1989.
470. No D, Yao TP, Evans RM. Ecdysone-inducible gene expression in mammalian cells and transgenic mice. Proc Natl Acad Sci USA 93:3346–3351, 1996.

471. Barlow C, Schroeder M, Lekstrom-Himes J, et al. Targeted expression of Cre recombinase to adipose tissue of transgenic mice directs adipose-specific excision of loxP-flanked gene segments. Nucleic Acids Res 25:2543–2545, 1997.
472. Chen J, Kubalak SW, Chien KR. Ventricular muscle-restricted targeting of the RXRalpha gene reveals a non-cell-autonomous requirement in cardiac chamber morphogenesis. Development 125:1943–1949, 1998.
473. Chen WS, Manova K, Weinstein DC, et al. Disruption of the *HNF-4* gene, expressed in visceral endoderm, leads to cell death in embryonic ectoderm and impaired gastrulation of mouse embryos. Genes Dev 8:2466–2477, 1994.

4

Genomic Approaches to the Genetics of Hormone-Responsive Cancer

JOEL N. HIRSCHHORN
CELESTE LEIGH PEARCE
DAVID ALTSHULER

Over the last 20 years, there has been remarkable progress in our understanding of the molecular basis of cellular function and disease, particularly within the area of cancer biology.[1–3] Despite this rich understanding of fundamental biological mechanisms, however, we remain surprisingly ignorant of the specific molecular defects responsible for risk and progression of disease. To the extent that we can uncover why one person is predisposed to cancer and another less or why one tumor displays a more malignant form and another less, we would transform our understanding of etiological mechanisms, improve targeting of prevention and treatment, and direct the search for novel therapies. The goal of this chapter is to review methods for identifying the genetic basis of hormone-responsive cancers and to describe how recent developments in genetics and genomics promise to accelerate the search for the root causes of disease.

Human disease typically results from a combination of inherited differences in gene sequence (genetics), external influences (environment), and random fluctuation. (We leave to the reader and philosophers the question of where free will and behavior belong in this schema.) In some circumstances, such as the role of smoking in lung cancer,[4] a single environmental exposure can make a major contribution to disease risk. In other cases, such as early-onset breast cancer, exposure to a mutation in a single gene can be both necessary and sufficient to cause disease.[5,6] In the population, however, most cancers are attributable neither to a single identifiable environmental factor nor to mutation of a single gene. Rather, cancer typically results from a combination of factors: inborn and somatically acquired mutations in gene sequence, one or more environmental factors, and certainly an element of bad luck. Such disorders are referred to as *complex traits.* This complexity has thus far befuddled researchers: where the web of causality is broad, any single strand plays only a minor role and eludes most efforts at discovery.

FAMILIAL CLUSTERING OF HORMONE-RESPONSIVE CANCERS

Our primary topic of discussion is the role of inherited variation in gene sequence and its impact on cancer risk in the general population.

We begin by examining the evidence that genetics (inherited variation in genome sequence) actually plays a role in the risk of cancer. The strongest data involve familial aggregation, showing that rates of disease are higher in the families of cancer patients than in the population at large. Of course, families are characterized by shared environment as well as shared genes. These factors are typically disentangled by comparing rates of concordance in monozygotic (MZ) and dizygotic (DZ) twin pairs.[7] Monozygotic twins are 100% identical in genome sequence, whereas DZ twins are identical (on average) across only 50% of their DNA; the method relies on the assumption that both types of twins share environmental influences to a similar degree. To the greatest degree possible, comparison of MZ to DZ twins separates the effects of genes and the environment.

Multiple large studies[8] indicate that the risk of the most common hormone-responsive cancers (breast, prostate, and ovary) is significantly influenced by inherited differences in genome sequence. Specifically, the MZ twin of a patient with cancer is two to three times more likely to get the same cancer than a DZ twin. These studies suggest that 27%–57% of the variation in population risk is due to gene effects and the remainder to non-genetic (environmental and random) factors.[9,10] While these studies show that inherited factors play a causal role, they also demonstrate that genes are far from the whole story: in MZ twins, the rate of concordance for prostate cancer is on the order of 25%, showing that even genetically identical people can have very different outcomes with regard to cancer.[9,10] Moreover, in most cases, the genetic contribution is not attributable to a single gene (which would produce a recognizable mendelian pattern of inheritance) but rather must be divided among a larger number of more modest genetic contributions.

IDENTIFYING THE GENES UNDERLYING COMPLEX, POLYGENIC DISORDERS

To date, successful efforts to identify genes for human diseases have focused on disorders caused by highly penetrant mutations,[11] including many genes that cause familial cancer.[5,6,12–15] The approach relies on the following simple idea: within each family, different individuals who have the same genetic disease are likely to share the same mutated copy of the gene responsible. That is, affected individuals will display "allele sharing" in excess of that predicted by random segregation of alleles. Given the low rate of recombination in a single meiosis, a sparse map of genetic markers (one every 10 million bases or so) can be used to comprehensively search the entire genome for regions displaying excess allele sharing.

Based on this foundation, the following highly successful approach to positional cloning was developed and codified: (1) type a sparse set of markers in multiplex families (those with multiple affected individuals), (2) localize the disease gene to a chromosomal segment showing excess allele sharing, (3) test additional markers to map individual recombinants and more precisely delimit the location of the disease gene, (4) identify candidate genes in this "critical region," and (5) screen these genes for mutations (examining affected family members) to identify the etiological variants. Typically, proof of causality (for a gene and for the mutations it carries) relies on the discovery of multiple independent mutations (in the different families examined), each of which changes the sequence of the encoded protein, cosegregates with disease, and has a functional consequence that can be recognized based on primary sequence data (e.g., deletions, splice site mutations, stop codons or alteration of highly conserved residues) or tests of protein function.

For common diseases that are complex in causality, however, there has been very limited success from this approach.[16] We suggest three main reasons for the limited progress to date. First, where disease is caused by the combined (or heterogeneous) action of many genes, affected relatives will not necessarily inherit the same mutant copy of the same gene. Thus, cosegregation in families offers limited power to localize disease genes in polygenic disorders. Moreover, the power of linkage studies is particularly limited for very common alleles as linkage relies on unilineal inheritance of each causal allele. To the extent that disease-causing mutations might be common in the population, they can be inherited from both the paternal and maternal lineages, further diluting any excess allele sharing attributable to the allele.[17] Second, even in the case where linkage analysis successfully localizes disease gene(s) to a chromosomal region, locating the culprit within the implicated interval is intrinsically more difficult for complex, polygenic traits. In a simple trait, the near-perfect correlation of genotype and phenotype means that individual recombination events in a family can be used to precisely bound the location of a disease-causing mutation relative to sites of recombination. In polygenic traits, in contrast, the relationship between genotype and phenotype is probabilistic, rather than deterministic: some individuals inherit a causal mutation but remain healthy and, conversely, many cases of disease will exist absent any given mutation of interest. Thus, single recombination events provide little or no information about the location of disease genes since the presence or absence of the causal mutation is not absolutely predictive of disease. Third, the identification and recognition of causal mutations may be problematic in the case of complex traits. Monogenic traits often involve complete disruptions of gene function and are thus often caused by gross (and therefore easily recognized) alterations in gene sequence: stop codons, deletions, and nonconservative amino acid substitutions. In such situations, the sequencing of a modest number of cases can convincingly demonstrate the presence of an unusual number of evidently deleterious mutations. It is far from clear, however, that the sequence changes contributing to complex traits will be so easily recognized from the primary sequence. If many of the mutations underlying complex traits turn out to be modest amino acid alterations or noncoding regulatory mutations, it will be much more difficult to recognize them from sequence data alone. For these reasons, sequence information from a small number of cases is unlikely to demonstrate the "smoking gun" in the case of a complex trait.

Given these considerable challenges, how is it that the genes contributing to complex traits will be found? At a minimum, variants responsible for an increase in risk should exist at a higher frequency in cases than in healthy controls, and this frequency deviation can be measured in association studies of sufficient size (and given appropriate controls). This will require, first, identification of putative disease-causing variants and, second, comparison of their frequency between individuals with and without the disease. Until recently, such studies were severely limited in breadth (the number of genes and variants that could be discovered) and depth (the number of individuals in whom these variants could be tested for association to disease). The limited scope of these studies was particularly problematic because we do not know which genes need to be examined for mutations nor the characteristics of disease-causing alleles (common or rare, in coding or noncoding regions).

CORRELATING GENETIC VARIATION AND DISEASE

These limitations have been significantly diminished through the sequencing of the human genome (and that of other organisms), the development of high-throughput tools for collecting genomic information, and sophisticated methods for data analysis. These developments make conceivable the comprehensive characterization of common genetic variation at each locus, the systematic testing of these variants for association to disease in large clinical populations, and the extension of these studies from a small number of candidate genes to a genome-wide scale.

The ultimate implementation of this approach will be unbiased and genomewide in scale, examining all genetic variations (both common and rare) across the entire set of human genes. For the foreseeable future, however, such studies present hurdles, both practical and analytic, that have yet to be surmounted. In the near term, two shortcuts may provide successes while the technology for such unbiased, comprehensive studies is developed.

Candidate Genes

First, given the depth of current knowledge of disease mechanisms as well as clues provided from family-based linkage studies and other genomic methods, we can prioritize subsets of genes for immediate study. For example, classical epidemiology has identified hormonal pathways that modulate cancer risk, such as the clear role of parity and contraceptive use in the risk of ovarian cancer (see Chapters 5, 6, 15, 20, 21, and 22 for reviews). These data demonstrate that manipulation of hormonal pathways can influence ovarian cancer risk. It is also clear that individuals vary in their menstrual patterns and exposure to endogenous sex steroid hormones. Thus, one important hypothesis is that genetic variation in hormonal pathways might explain some of the interindividual risk of ovarian and breast cancers.

Other candidate genes have been suggested from a range of biological investigations. As discussed in detail elsewhere (Chapters 10 and 17), genes such as *BRCA1*, *BRCA2*, and *HPC1* have been identified based on the study of rare familial cancer syndromes. The pathways of tissue-specific carcinogenesis identified by these studies represent a powerful clue to the mechanisms underlying common forms of disease. Of course, traditional laboratory investigation of cancer biology has leapt forward in cell culture and animal models, providing a detailed framework of cell signaling, cell cycle regulation, and DNA repair. Although beyond the scope of this chapter, the insights into cancer biology are both detailed and broad and provide a rich framework to guide the selection of candidate genes for subsequent association studies.

Common Variants

The second shortcut is to study the effect of common polymorphisms on disease risk. At a minimum, common polymorphisms are much simpler to study than rare variants: their existence can be determined and their frequency accurately assessed in a modest number of individuals. Perhaps more importantly, population genetic research has demonstrated that the vast majority ($\approx$90%) of all human genetic variation (as measured by heterozygosity) is attributable to alleles present in greater than 1% of the population. That is, not only are common variants the easiest to study (which would be of minor interest if such variants represented only a small fraction of all genetic variations) but they also account for the overwhelming majority of all human genetic diversity. Of course, this overall pattern of human genetic variation (which is dominated by variation that is evolutionarily neutral) cannot be directly extrapolated to define the characteristics of mutations affecting the risk of common disease. Only the discovery of a large number of disease-causing mutations can resolve the considerable debate about the role of common mutations in common diseases. However, it is helpful at least to clearly articulate the so-called common variant–common disease hypothesis:[18–20] given that most evolutionarily neutral genetic variation is due to alleles that are common in the population, diseases that were selectively neutral or advantageous during human evolution (i.e., not selectively disadvantageous[21]) will be largely attributable to the segregation of common mutations. This hypothesis has suggested an initial approach to determining the genetic basis of common disease: identify high-priority candidate genes and/or genomic regions and test common variation within these genes or regions for a role in disease using association studies.

In the sections that follow, we explore both genomic strategies for identifying candidate genes and regions and methods for testing common genetic variation in such genes or regions for association to disease.

GENOMIC APPROACHES TO IDENTIFYING CANDIDATE GENES

Genome Sequence Information

Human Genome Sequence

The complete human genome sequence will reveal and make finite the universe of genes that might possibly underlie inherited variation in any trait of interest. The first draft sequences of the human genome were presented early in 2001,[22,23] and both drafts were similar in their completeness.[24] (The publicly funded project continues to collect data, which are released immediately upon discovery.) At the end of 2001, the public human genome sequence captured more than 97.8% of the euchromatic portion of human chromosomes, with 63% in finished form (for current statistics, see http://www.ncbi.nlm.nih.gov/genome/seq/HsHome.shtml). In addition, draft genome sequences of the laboratory mouse have been produced by both a private company and a public consortium (http://www.ensembl.org/Mus_musculus/), and much progress is being made in generating sequences of at least four other vertebrate genomes (rat, zebrafish, fugu, and tetraodon).

At present, it is estimated that the human genome contains some 30,000–40,000 genes,[22,23] although this number remains a topic of ongoing debate and study.[25,26] Gene number can be estimated by several methods:[22,23] homology to expressed sequence tags and mRNAs, cross-species conservation, and computational recognition of genic motifs. Gene annotation methods can be validated, furthermore, according to two definitions of *gene*. The first is simply to prove that the annotated gene is expressed as mRNA in human tissue, e.g., using expressed sequence tag sequencing,[27] high-density arrays,[28] or polymerase chain reaction–based methods. Such evidence does not address whether the expressed sequences are functional but shows that they are expressed. The evidence in favor of a gene is enhanced when the segment is not only expressed as mRNA but also conserved in sequence across species.

The classical and most stringent test is to demonstrate directly that manipulation of a putative gene results in a functional consequence. Since there is no way to know which phenotypes to query, however, and since many genes may have redundant functions or are necessary only under certain conditions, functional tests can be interpreted only in the affirmative. Increasingly high-throughput methods for manipulating gene function may make more practical the widespread testing of predicted genes for function under a range of conditions in experimental models (see section 2.5, below).

Whatever the gene number, the number of unique functional proteins is certainly even larger, due to alternative splicing and the creation of multi-peptide macromolecular complexes. Moreover, defining the gene set is simply the first step toward annotating these genes with functional descriptions. Thus, even an initial description of the complete human gene set is a task that will continue well into the coming decade.

Given our current knowledge of biological pathways, however, the human genome sequence can already be used to generate hypotheses based on biochemical function. That is, if an investigator believes that genes involved in DNA repair or steroid hormone biosynthesis are likely involved in cancer biology, it is now a simple matter to identify rapidly and with great completeness all such genes based on annotation or sequence similarity within and across species. Since such hypotheses are by nature speculative, they are of greatest value when combined with information on chromosomal segregation, mRNA and protein expression, and functional tests (Fig. 4.1).

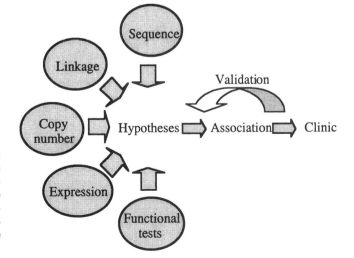

Figure 4.1 Five genomic approaches that can be used to identify candidate genes for association studies: direct examination of genome sequence, linkage analysis in families, detection of variation in gene copy number, global analysis of gene expression, and large-scale manipulation of gene function.

Comparative Genomics

One increasingly powerful approach to expanding our understanding of gene function is the comparison of genomes from different species. To identify functional sequences, a genome sequence can be examined for motifs that are strongly conserved across species boundaries. The underlying assumption is that sequences involved in gene function will be less divergent than randomly selected sequences, due to the action of purifying selection against deleterious mutations. With the sequencing of the mouse and three fish genomes to multifold coverage, such cross-species comparisons can now be applied to the human genome on a global scale. Regions that are highly conserved between related species represent either coding regions or functionally relevant non-coding (presumably regulatory) motifs. It appears that approximately 50% of regions showing high levels of mouse–human sequence conservation fall outside of currently annotated coding regions, and it is likely that many of these represent novel regulatory sequences (K. Lindblad-Toh, M. Zody, and E. Lander, personal communication).[29] More importantly, there is growing evidence that conserved noncoding sequences often represent bona fide regulatory domains. Loots, Rubin and colleagues[30] studied one such conserved sequence in transgenic mice, proving that this previously uncharacterized segment of DNA (identified purely by genome sequence comparison) coordinately regulates the expression of interleukins 4, 5, and 13.

In the second form of comparative genomics, knowledge of pathways in model systems is combined with sequence homology to generate functional hypotheses about human gene function. One of the remarkable findings of the last decade is the extent to which biochemical functions of genes can be conserved across species. For example, the DNA mismatch repair system is conserved from yeast to humans, and genes identified by analysis of mismatch repair in yeast have been shown to be mutated in patients with hereditary nonpolyposis colon cancer.[31] Systematic programs of mutational and functional analysis[32,33] ensure that the breadth and depth of knowledge of biochemical pathways should expand dramatically in the future. Given the conservation of biological mechanisms, this information will likely be directly relevant to the study of human cancer.

In summary, the sequence of the human genome, especially when compared to that of other organisms and annotated with functional information, offers an increasingly comprehensive database for generating hypotheses about gene function and then designing experiments to test those hypotheses. These hypotheses in turn identify candidate genes that can be tested for a role in determining cancer risk.

Linkage Analysis in Families

Linkage analysis examines patterns of segregation in *multiplex* families (those in which more than one individual has the phenotype of interest). The fundamental goal of linkage analysis is to find chromosomal regions in which alleles are coinherited by the affected individuals more frequently than predicted by chance; such regions are more likely to contain genetic variants that cause disease. The most important advantage of linkage analysis is that it is genomewide in scale and, therefore, capable of discovering genes not already expected to play a role in disease. In addition, linkage analysis is robust to allelic (but not locus) heterogeneity. That is, if a single gene is mutated to cause a substantial proportion of cases but in each family the particular mutation (allele) is unique, linkage will be undiminished in its power. Linkage studies have great power where alleles are rare in the population and strong in effect and have, thus, revolutionized the study of single-gene ("mendelian") disorders. Linkage analysis has limitations, however, which include its diminished power for genes of modest effect[20] and for alleles that are common in the population. These characteristics have biased our knowledge of genetic mechanisms to those operative in families with rare mutations of high impact, which may or may not be representative of the mechanisms operating in the general population. Finally, as a practical matter, the regions of allele sharing identified by linkage analysis tend to be large in size (spanning tens of millions of base pairs); thus, even in the best of circumstances, linkage circumscribes (but can never specify) the location of the culprit in disease.

Because linkage analysis is unbiased and genomewide, it represents a critical starting point for any study of disease genetics. This approach has been used successfully to identify *BRCA1* and *BRCA2*, implicated in breast and ovarian cancers.[5,6] Although linkage studies have implicated several chromosomal regions in prostate cancer,[34–39] only two genes have been identified to date. *ELAC2* (*HPC2*) was identified in 2001,[38] with both rare and common mutations that showed association to disease, although the consistency of this association remains unclear.[40–42] Mutations in *RNASEL* were proposed (via positional cloning) to be responsible for a subset of familial cases of prostate cancer.[43] While these are exciting developments, it is not yet clear whether these genes or the pathways they identify play a significant role in the inherited basis of the common forms of these diseases.

In summary, linkage studies combined with a genome sequence (to identify the specific genes in each genomic region) serve as a practical, genomewide approach to identifying regions that are more likely to contain genes contributing to disease. Genes in these regions become *positional candidates* (those whose residence in an implicated chromosomal region highlights their potential role in disease).[44]

Variation in Gene Copy Number

Tumors are often characterized by variation in chromosomal copy number, with loci harboring dominant oncogenes amplified in tumors and regions with tumor-suppressor genes deleted. Initially, chromosomal karyotype was used to search for gross alterations in ploidy, leading to a limited number of spectacular discoveries such as the Philadelphia chromosome in chronic myelogenous leukemia. Subsequently, genome-wide methods were developed to search for alterations in copy number. Two methods that have been used successfully are analysis of loss of heterozygosity (LOH) and comparative genomic hybridization (CGH).

When a cell is heterozygous for a loss-of-function mutation in a tumor-suppressor gene, loss of the wild-type allele (due to either a segmental deletion or a gene-conversion event) can cause a transformed phenotype. This phenomenon is termed *loss of heterozygosity* and can be detected (if sufficiently large) by genotyping either microsatellites or single nucleotide polymorphisms (SNPs).[45] The approach has been limited in the past by the low resolution of microsatellite polymorphism maps, but high-density SNP maps increase the precision with which such regions are defined. As regions displaying LOH are discovered in tumors, it will be possible to test the genes they harbor as candidates in genetic association studies.

Another approach to identifying regions of either gene amplification or deletion in tumors across the entire genome is CGH.[46] In this technique, DNA from patients is labeled and hybridized back to the human genome; the amount of label at each location indicates the copy number of sequences in that region. The original description of CGH was labor-intensive and did not allow for high-resolution identification of regions with increased or decreased copy number. Newer methods use hybridization to high-density spotted arrays of clones (rather than metaphase chromosomal spreads), increasing the resolution to which gene amplifications and deletions could be detected and making this method more promising for the identification of candidate genes to be studied.[47]

In summary, tumors are characterized by amplification and deletion of regions harboring oncogenes and tumor-suppressor genes. It is reasonable to suspect that genes undergoing such changes in copy number may also influence cancer risk through inherited differences in DNA sequence variation. By searching for variation in copy number over a large number of tumors and on an increasingly fine scale, it should become possible to identify chromosomal regions that are likely to contain genes relevant for genetic and functional study.

Global Analysis of Gene Expression

We previously considered how inheritance of chromosomal regions in affected individuals or in tumor tissue can be used to build hypotheses about gene function. Instead of studying variation in gene sequence, a complementary approach is to analyze variation in gene expression as mRNA and protein. Because altering gene transcription is a ubiquitous mechanism for cell signaling and regulation, many cellular behaviors are reflected in altered patterns of gene expression. Investigating this phenomenon has traditionally been limited, however, by two fundamental constraints: only a small subset of genes was known (and thus could be studied) and, even among known genes, it was necessary to assay them one at a time. With increasingly complete databases of gene sequences, however, it has become possible to design expression probes for essentially all human genes. When combined with high-throughput methods, such as high-density arrays to measure transcript levels, these sequence resources can transform expression monitoring into an unbiased, genomewide approach.

The development of high-density array technology has dramatically expanded the breadth of methods for studying variation in gene expression. Specifically, nucleic acids of known sequence are deposited in specified locations on a solid support, and labeled experimental samples are hybridized to the array. After washing to eliminate nonspecific binding, the concentration of each mRNA can be determined based on the location and intensity of the label. The power and generality of this approach has been dramatically enhanced by the development of methods (photolithographic and robotic) that place tens to hundreds of thousands of such probes on individual glass slides.

Already, high-density arrays are finding diverse application in biomedical research.[48] For example, expression profiling has been used to identify genes whose expression is characteristic of cells in a given developmental or clinical state. Using this approach, there are now multiple examples where tumor diagnosis and classification can be made based solely on gene expression profiles, including leukemia,[49] breast cancer,[50] and B-cell lymphomas.[51] More importantly, gene expression profiles can be used to identify not only cells characteristic of known states but also previously unrecognized subtypes of cancer. This has made it possible to identify subgroups of patients with greater homogeneity of outcomes and etiological mechanisms.[51–56]

The clinical nosology of cancer has traditionally been based on site of origin and appearance under the light microscope and may be only weakly correlated with the diversity of underlying biological mechanisms. If so, then mechanistic heterogeneity within a single diagnostic category may be responsible for much of the complexity of "complex" diseases. Under this hypothesis, genomic tools that reveal biologically relevant subgroups of tumors may play an important role in revealing the underlying diagnostic categorization of disease, resulting in more homogeneous groups with simpler genetic architecture. Thus, a new taxonomy of cancer, based on molecular signatures rather than visual inspection, may prove to be a critical step in the path toward identifying the genes responsible for cancer.

Gene expression monitoring is useful not only for developing markers and classifying disease but also for generating functional hypotheses. For example, microarrays offer an efficient approach to capturing the complete list of transcriptional changes accompanying the transition from carcinoma in situ to a more invasive form or after the experimental manipulation of a hormonal pathway or expression of an oncogene. In this way, an initial hypothesis about a functional pathway of interest can be expanded to include other genes that are transcriptionally coregulated or downstream targets.

In a few cases, it has been possible to confirm functional hypotheses first suggested by expression analysis. For example, the gene encoding RhoC (*ARHC*) was identified as an expression correlate of tumor metastasis in a melanoma model; critically, blockade of RhoC diminished metastasis and activation enhanced metastasis in this model.[57] As yet, few such functional hypotheses have been validated, due to the relative mismatch between our ability to rapidly generate such hypotheses using expression microarrays and our more limited ability to test them in the laboratory. However, methods for systematically testing gene function on a global scale are rapidly evolving and are discussed briefly in the next section.

Large-Scale Manipulation of Gene Function

To date, most genomic studies in mammals have been observational: this is because the tools for measuring genetic information in mammals are currently much more robust and scalable than are the tools for manipulating genetic information. This mismatch is creating a major bottleneck in biomedical research as hypothesis generation races ahead of hypothesis testing. Technologies for altering gene function are rapidly improving, however, due to both increasing efficiency of existing methods and the development of novel approaches. Two approaches to testing hypotheses on a genomic scale, discussed briefly below, are large-scale mutagenesis and exogenous addition of modulators of gene or protein activity.

Genomewide mutagenesis has long been a mainstay of genetic analysis in single-cell and invertebrate model systems. It is only recently, however, that widespread mutagenesis has become possible in two vertebrate systems: zebrafish[58] and the laboratory mouse.[59] Chemical agents, radiation, and/or insertion of exogenous DNA sequences are used to disrupt gene function randomly; and the progeny are then examined for the phenotype of interest. Animals with the phenotype of interest are then studied to identify the mutations responsible (using traditional positional cloning approaches). This approach is unbiased with regard to pathway and genomewide in scope, although it will reveal only genes which, when mutated, result in a monogenic form of the disease. Mutagenesis will offer increasing power as genome sequences, dense variation maps, and other such tools become generally available to hasten the positional cloning process. The approach is limited by the generation time of the animals, the cost of breeding and housing large numbers of mutated progeny, and the labor of screening each mutated animal for phenotypic variation.

An alternative approach is to use nongenetic methods to disrupt gene function in a targeted manner. Two such approaches have recently improved in their power and scope. First, exogenously applied nucleic acid probes can be used to target the disruption of genes, based on either antisense RNA approaches[60] or the more recently described RNA interference (RNAi).[61] The latter appears particularly promising because of its great utility in invertebrates, increasingly well-understood mechanism of action,[62–64] and the likelihood that it represents an endogenous and well-conserved mechanism for gene regulation in vivo.[65–68] Second, chemical screening can be used to identify small molecules that interact with and modulate proteins of interest. In the so-called chemical genetics paradigm,[69] small molecule screening replaces mutagenesis as a general paradigm for widespread (and conditional) disruption of gene activity. As increasingly diverse chemical libraries are paired with miniaturized high-throughput screening assays, chemical genomics may become a powerful tool for broad-based manipulation of gene function.

In summary, these three approaches (mutagenesis, RNAi, and small molecule screening) offer the potential for widespread manipulation of gene function in cell culture and model system organisms. The phenotypes induced by these manipulations indicate the biological role of the genes under study, thereby leading to candidate genes that can be studied for genetic variation and association to disease.

COMMON GENETIC VARIATION AND ITS ASSOCIATION WITH DISEASE

The preceding section has illustrated the diversity of genomic methods for identifying large numbers of candidate genes for any disease of interest. Below, we describe novel approaches to comprehensively characterize the common genetic variation at each candidate locus and to test it for association with disease.

Association Studies: From Genotype to Phenotype

Genetic association studies are simple in conception: one or more putative causal variants are identified and their frequency is compared in populations with and without the trait of interest. In fact, all human genetic approaches conclude with an association study, for both simple and complex traits, whether guided by linkage analysis or a hypothesis about a candidate gene; the association of genotype with phenotype indicts a gene as causing disease.[70] Traditionally, however, the term *genetic association study* has referred to a population-based design in which common variants in one or a small number of genes are examined for association to a polygenic disease. The approach has been constrained by our limited knowledge of the genome and its variation, as well as the rudimentary tools available for experimental analysis. Given the limited number of possibilities that could practically be tested, association was at best a scattershot approach to finding disease-causing alleles.

The strength and attraction of association lies in its simplicity and statistical power. As association studies represent a direct test of each allele, they offer the greatest sensitivity to detect the relationship between a variant and disease.[20] This statistical power will likely be necessary as available data suggest that common, multifactorial diseases may well be largely attributable to the contribution of many variants of individually modest effect. Examples of associations include variation in *APOE* (encoding apolipoprotein E) with Alzheimer's disease,[71] *F5* (encoding factor V) with deep venous thrombosis,[72] and *PPARG* [encoding peroxisome proliferator–activated receptor γ (PPARγ)] with Type 2 diabetes.[17] In each of these cases, it was possible to demonstrate association using only hundreds or thousands of patients, whereas many more would have been needed in a linkage study. For the most extreme example, the common Pro12Ala variant in PPARγ, a p value of $\approx 10^{-3}$ was obtained in at least four separate studies of a few thousand patients,[17,73–75] whereas it was calculated that over 3 million sibpairs would be needed to obtain a LOD score of 3.0 in a linkage analysis.[17] Given the trivial fraction of genes and of common variants that have thus far been tested by association (certainly less than 0.01% of all common human genetic variation), these successful examples must represent the tip of an iceberg of significant magnitude.

If statistical power and anecdote tip the balance in favor of association studies with common variants, then incompleteness and irreproducibility offer a substantial counterbalance. First, while available methods may soon allow comprehensive surveys of common variation, such studies are incomplete if they do not query all the rare sequence variants that might contribute to disease. As described below, however, the study of common sequence variation may offer a surprisingly comprehensive approach to assessing human genetic diversity, at least for traits that were neutral or advantageous throughout evolution. Second, association studies have been (and will remain) irreproducible unless we can describe and overcome all possible confounders that can cause false associations. In the following two sections, we discuss each of these topics in detail. First, we discuss the recently described features of human sequence variation that should allow systematic association studies even without the discovery of every variant in the gene or region of interest. Second, we discuss the possible confounders of association studies and the current approaches that can be utilized to avoid the confusion they create.

Human Genome Sequence Variation

Patterns of Sequence Variation

When any two copies of the human genome are compared, their sequence similarity is striking. When randomly sampled from a multiethnic population, any two copies differ at a rate of only 8×10^{-4} per base pair, or 1:1250 bases examined.[23,76] (The rate of sequence diversity across two randomly chosen chromosomes is termed *heterozygosity*.) In the coding regions of genes, heterozygosity is even lower, 1:2000 bases compared across two chromosomes;[18,23,77–79] the lower rate of heterozygosity in coding regions is thought to be due to purifying selection against alterations in amino acid sequence.[18,78] Thus, amino acid altering variants make up less than 1% of all sequence diversity, although they represent a somewhat higher proportion of functionally important changes.

The striking lack of diversity in human populations is unusual: e.g., our closest primate relatives (chimpanzee, gorilla, and orangutan) display 3–10 times higher levels of sequence variation than do humans.[80,81] The low rate of sequence diversity likely reflects the small size of the ancestral founder population from which current humans are derived.

Most of this sequence variation (approximately 90%) is attributable to single-base changes, where one nucleotide is swapped for an alternate at a single position. Less frequent classes of variants include insertions or deletions of one or a few nucleotides, variations in the number of tandem repeated sequences, or more complex changes involving multiple bases. A single position where both alternate nucleotides are present in the population at a frequency of greater than 1 in 100 is an SNP.

Remarkably, approximately 90% of all human heterozygosity is attributable to variants with a population frequency greater than 1%. The contribution of common alleles can be directly demonstrated in two types of survey of human variation. In one approach, segments of genes are resequenced in a large number of individuals; these studies show that most of the total variation in a population is due to common, rather than rare, variants.[18,78] The second experimental approach directly assesses heterozygosity by discovering variants as random heterozygous sites (i.e., comparing two randomly chosen chromosome pairs) and then by typing these SNPs in an independent sample to determine their allele frequency. Tens of thousands of SNPs have been processed through such a protocol, with ≈90% of variants displaying population frequencies greater than 1%).[76,82,83,83a]

These studies demonstrate that most human sequence variation is explained by common variants: a collection of all human genetic variation with frequency greater than 1% would capture perhaps 90% of all human heterozygosity. As discussed above, the common disease–common variant hypothesis states that variation explaining susceptibility to common diseases (which are not deleterious from an evolutionary standpoint) should therefore reflect the predominantly common nature of neutral human sequence variation. That is, since most human variation is explained by common variation, it is likely that common variants will explain a sizable fraction of the genetic risk of common disease (unless those diseases were disadvantageous during the evolution of the ancestral human population).

Identification of common human sequence variants is a critical first step toward testing the role of variants in disease pathogenesis. Recent efforts have sought to generate just such a catalogue of common human genetic diversity. The vast majority of variants have been identified by automated, high-throughput approaches with objectively applied computational algorithms for sequence alignment and polymorphism discovery.[82,84–87] The two largest SNP discovery efforts were performed by the SNP Consortium (TSC) and BAC overlap projects, together identifying over 2.2 million candidate SNPs in the human genome (L. Stein, personal communication).[76] Systematic testing has shown that fewer than 5% of these candidate SNPs represent false-positives and that ≈90% have frequencies greater than 1% when tested in large, multiethic population samples.[76,83,83a] As expected, based on their method of ascertainment (discovery in two or a small number of chromosomes), these SNPs display a broad range of allele frequencies, with approximately 50% of the polymorphic SNPs having frequencies greater than 20% (in any given population).

Based on estimates of human heterozygosity and the allele frequency distribution of SNPs, it has been estimated[88] that the human genome contains some 10–11 million SNPs in total, of which 2.2 million are currently represented in the public domain. (A similar number of additional SNPs were reported to be identified by the private company Celera and can be accessed for a fee.) Combining public and for-profit SNP databases, it is likely that as many as 30%–40% of all SNPs are already known, with a considerably higher percentage of those with the highest population frequencies.[88] Furthermore, there are nascent efforts to resequence every human exon in a diverse population panel, which might soon result in a comprehensive catalogue of common coding region variation.

Even if every common human variant was discovered, it will likely not be necessary to type every variant in each disease study. This is because neighboring variants are often correlated with each other (a phenomenon called linkage disequilibrium, or LD). Taking LD into account, a subset of SNPS may be adequate to capture the information carried by nearly all common variants, including those yet to be discovered. The extent and consequences of LD between neighboring common variants are discussed in the next section.

Linkage Disequilibrium and Human Haplotype Structure

Linkage disequilibrium (also termed allelic association) is the observation that closely linked variants are often correlated in the population: individuals who carry a given SNP allele at one site are much more likely than expected by chance to carry a particular allele at another, nearby locus. This correlation reflects the fact that each chromosome in the current population traces its ancestry back to a small number of common ancestors. Because there has been limited time, and thus opportunity for mutation and recombination, since these common ancestors, ancestral segments (termed *haplotypes*) are often inherited en bloc by many individuals in the population. A dense but generic set of polymorphic markers allows identification of these ancestral haplotypes, allowing each haplotype to be tested (along with all of the variants that it carries) for correlation to disease. The advantage of this indirect approach is that it does not require the prior discovery of each functional variant, only the haplotypes on which they reside. However, the critical questions are in the details: how strong are the correlations among nearby SNPs, and to what extent can a region be parsed into ancestral haplotype blocks?

Until recently, there has been only theoretical speculation and anecdotal information about human haplotype structure, with available data stressing the variability across the genome and across populations. In an influential publication, Kruglyak[89] used coalescent simulations to predict that (under certain assumptions about human population history and recombination) "useful" LD would typically extend over only a few kilobases in the human genome. Kruglyak[89] noted that deviations from his assumption, e.g., due to population expansion or contraction or to hot spots of recombination, would alter significantly these predictions; but at that time there were insufficient data to further constrain the simulations.

Empirical studies of LD, furthermore, were limited (due to the available technology) in scope, typically examining only a sparse collection of variants, a single genomic region, or a single population sample.[90–106] Thus, drawing general conclusions has been problematic. Advances in technology have facilitated large-scale surveys of human haplotype structure, examining dense maps of polymorphisms, broad samples of the human genome, and multiple population samples.[83a,107–110] These studies have revealed a strikingly simple pattern of human genetic variation, with long-range haplotypes that should dramatically increase the power and efficiency of genetic association studies.

The simplest way to examine LD is to consider the correlation between randomly selected pairs of markers. When examined in samples of European and Asian origin, pairwise LD is strong over distances of tens of thousands of base pairs.[83a,107–109] When samples from West Africa were examined for a large number of loci, pairwise LD extended over considerably shorter distances.[83a,107] While these trends are clear, there is tremendous scatter among regions.

The broad scatter in pairwise LD becomes more understandable when local patterns of LD are examined. The data typically reveal "blocks" of contiguous markers showing near-complete LD, ranging in size from 5 to >150 kb.[83a,104,106,108,110] These blocks of strong LD are interspersed with sites at which allelic associations precipitously break down, indicating sites of historical recombination among haplotypes. Moreover, these haplotype blocks are larger in European and Asian samples [on average, 20[83a]] than in samples with more recent African ancestry [on average, 11 kb[83a]]. These two features of human haplotype structure, that there are regions of variable length over which allelic associations are strong interspersed with sites of historical recombination, underlie the tremendous scatter in pairwise measures of LD. The difference between population samples is likely due to population contractions in the history of the current human population.[107,111]

Finally, within each block of strong LD, very limited haplotype diversity is observed, typically only four to five high-frequency haplotypes that explain >90% of all samples in the population.[83a,108,110] Haplotype diversity is somewhat higher in African samples (five to six common haplotypes) than in European and Asian samples (four to five common haplotypes).[83a] Given the large size and low haplotype diversity of these blocks, it is possible to select a limited number of haplotype tag SNPs[109,110] that capture all of the haplotype diversity of a region in only a few, well-chosen markers. For example, if a haplotype block is 30 kb in length, it will typically contain 100 or so common variants and numerous rare changes; but if all of these SNPs cluster into only four or five common haplotypes, it is possible to perform an initial screen of considerable completeness simply by typing the three or four SNPs that between them tag each of the common ancestral haplotypes encountered in the block. By discovering and exploiting these patterns, it should be possible to achieve a 10- to 30-fold reduction in the number of SNPs required to undertake studies of common variants across the genome.

In summary, each region of the genome can be parsed into haplotype blocks (regions over which there has been little or no historical recombination) and sites of historical recombination. Within each block, there are only a few common haplotypes that capture 90% or more of all common genetic variation. By empirically determining these patterns, it is possible to select haplotype tag SNPs that efficiently capture most of the genetic diversity of any given genomic region. These patterns hold across all examined human populations, although the size of haplotype blocks is smaller and the diversity of haplotypes greater in samples from Africa compared to Europe or Asia.

Given the size and diversity of human haplotypes, it appears that in European and Asian populations, something like 300,000 SNPs, carefully selected after empirical determination of haplotype patterns across the genome, will ultimately allow well powered association studies that span the human genome. In samples from Africa, a somewhat larger number of SNPs may be needed because of the greater diversity of haplotypes. (Note that it will take the detailed genotyping of 1–2 million SNPs across the genome to find the 300,000–500,000 SNPs that tag all of the haplotypes in the population.) For the study of a single gene, this translates into 5–10 carefully chosen and empirically validated haplotype tag SNPs that will be needed to survey genetic variation in a disease study. These data indicate a comprehensive and efficient approach to studies of common human sequence diversity, without the bias involved in selecting only a single class of variants, e.g., those that alter an amino acid, prior to study.

Single Nucleotide Polymorphism Genotyping Technology

Utilizing the resources developed around common variants (the catalogue of variants and their relationship to each other within haplotypes) requires methods to rapidly, accurately, and cheaply determine the genotypes of these variants in large population samples. If 5–10 SNPs need to be examined in a population of several thousand individuals to comprehensively survey a gene, then approximately 20,000–50,000 genotypes will be required to test common variation in a single gene. To test a pathway of 100 or more candidate genes, several million genotypes will be required. In addition, to study the entire human genome, literally billions of genotypes will likely be required.

Fortunately, significant technological advances have been achieved over the last decade, making possible the comprehensive study of sets of genes. As recently as a year or two ago, it took months of dedicated effort to genotype one variant in 1000–2000 individuals at a cost of $\approx$$5.00 per genotype. Now, the same amount of genotyping can easily be done in a single day at a cost of $0.50 or less per genotype. Many SNP genotyping technologies exist and have been comprehensively reviewed.[112] Each method has different advantages and disadvantages, but the most useful methods display the following characteristics: low cost (significantly less than $1.00 per genotype), automated workflow, robust assay design, relatively high throughput (at least 5000 genotypes per day per person, including time for data processing and analysis), low DNA usage (5 ng or less per genotype), and high accuracy (at least 99%). These methods, combined with the common variant resources described above, have finally made possible large-scale comprehensive association testing of high-priority candidate genes.

Regardless of these technological considerations, scientists performing association studies must overcome the evident and widespread irreproducibility for their results to be embraced by the broader scientific community. As many of the likely confounders have been identified and can be studied, a straightforward path to their elimination can be proposed. These confounders and their implications for effective study design are examined in the next section.

Sources of Inconsistent Association Results

Inadequate Statistical Power

Complex genetic diseases are caused by multiple genetic loci that contribute to disease risk.

Although in some rare cases a single mutation may confer a large relative risk (e.g., *BRCA1*), most cancer cases appear not to be explained by such highly penetrant mutations. Specifically, given the modest results obtained with linkage studies (from all but specially selected familial cases), it appears that there are few loci that confer more than a modest effect on individual risk. Accordingly, studies to find these variants will need to involve large samples. Of course, the exact sample sizes required cannot be calculated without knowing the relative risks that will typically be encountered. However, the few examples that are already known suggest that effects of less than twofold may be typical, meaning that sample sizes in the thousands will be needed to achieve significance.

Under the assumption of modest effects, nearly all studies to date have been drastically underpowered. Consequently, most of the studies that manage to achieve statistical significance have done so because they considerably overestimated the genetic effect of the allele being studied. (This phenomenon is analogous to the well-known "winner's curse" in auction theory, in which the winning bidder, by virtue of having the highest estimate of the value of the item, usually ends up overpaying.) This overestimation of effect size was observed for the association between the *PPARG* Pro12Ala polymorphism and Type 2 diabetes;[17] a similar phenomenon was noted in recent reviews of association studies (J.N. Hirschhorn, personal communication).[113]

Several steps can be taken to counter the problem of inadequate power. First, and most importantly, large sample sizes can and must become the standard in the field. Second, all available published data can be combined using meta-analysis to further increase power. (The limitations of meta-analysis are significant, however: inconsistencies in phenotypic definition, genotype assignment, and publication bias.) Third, positive studies should be assumed to have overestimated the strength of the association, especially when it is the first positive report or there are numerous negative reports for the same association. Accordingly, sample sizes for replication studies need to be sharply adjusted upward (compared to sample size estimates obtained using the risk model estimated by the original report).

Population Stratification

Population stratification can occur when the population under study is derived from more than one ethnic group. If disease risk happens to be higher for one of the ethnic groups and if the frequencies of particular alleles vary as well, spurious association may arise.[114] There are documented examples of false-positive associations due to this phenomenon, including that of Type 2 diabetes and the Gm locus in Pima Indians[115] and that between variation in the *DRD2* gene and alcoholism.[116] The degree to which ethnic confounding (*population stratification*) generates false-positives is uncertain, however, and remains a matter of heated debate.[117]

Luckily, there are now multiple methods to avoid or, at least, detect the confounding effect of population stratification. First, a variety of family-based methods are available that are immune to confounding by population stratification (e.g., transmission disequilibrium testing).[118–121] However, the most powerful of these methods (utilizing affected offspring and their parents) is less practical for late-onset diseases such as cancer, where parents are often unavailable at the time of diagnosis. Family-based methods using siblings[122,123] and more complex pedigrees are available, although these are less efficient than case-control methods and recruitment of family members is often difficult. Second, several methods are now available to directly assess the degree of stratification present in a sample of unrelated individuals. These are based on the genotyping of random markers (un-

Table 4.1 Causes of Inconsistent Association Results

Inadequate statistical power
Population stratification
False-positives due to statistical fluctuation (multiple-hypothesis testing)
True differences between populations
    Variable linkage disequilibrium between markers and causal variants
    Unmeasured modifiers (gene–gene and gene–environment interactions)

linked to any disease locus) in the study sample.[124–126] Using these markers, it is possible to construct an empirical null distribution of association between cases and controls. If the observed distribution of alleles matches the theoretical expectation for unlinked markers, then no stratification is present and the study can go forward with little or no fear of false-positives from ethnic confounding. If stratification is detected, there are two options. First, the observed deviation from the theoretical null distribution can be used to adjust p values upward, correcting for the degree of stratification. This approach is simple, but it requires an increased sample size to attain the lower p values needed to achieve significance. Alternatively, cases and controls can be rematched based on the information gathered from typing the random markers, balancing the ethnic background of the case and control samples.[127] This approach probably requires a larger initial pool of controls from which to match the cases but avoids sacrificing statistical power by eliminating stratification from the outset.

Statistical Thresholds for Interpretation: Priors Are Low

Fundamentally, however, the largest obstacle to robust association studies is the large size of the human genome and, thus, the low likelihood of stumbling across an actual causal allele. This is often described as the "multiple hypothesis testing" problem: each study may involve multiple genes, multiple SNPs at each gene, and multiple phenotypes examined, increasing the likelihood of encountering a false-positive result due to statistical fluctuation. We believe that this formulation is misleading, however, in that it highlights the intent of the investigator, rather than the relevant quantity: the prior likelihood that the variant causes disease and the adjustment of this estimate based on the observed data. Below, we describe a simple Bayesian approach to the problem, which helps to clarify the statistical thresholds that will likely be needed to obtain robust association results and emphasizes the need for caution in interpreting weakly positive association studies.

It is currently estimated that the human genome contains approximately 30,000 genes. Absent any other knowledge of biology, each would have an equal a priori likelihood of containing variation that confers risk of disease. If common variation at each gene is typically contained in one of ≈5–10 ancestral segments, then the prior probability of a true positive will be the number of true positives in the genome divided by the 150,000–300,000 ancestral segments that need to be examined to find them. Assuming optimistically that there are between 10 and 100 alleles in the genome that are causally associated with disease susceptibility and could be discovered with the sample sizes employed, the prior probability of association by typing any given variant is at best between 1/1500 and 1/30,000. (One can arrive at similar numbers using different assumptions. For example, there are approximately 60,000–100,000 missense polymorphisms in the human population; if 10–100 of them are detectably associated with disease, then the a priori probability that a missense SNP will be associated with disease varies from 1/600 to 1/10,000.) Note that this optimistically assumes that there are only 100 or fewer mutations across which the population risk of disease is spread and that the studies to be undertaken are powered to discover them. More likely, if there are truly 100 or more alleles causing disease, the impact of many of these alleles will be so small that only studies of unprecedented size could have power to find them.

Under these assumptions, Bayes' theorem shows that a nominal p value of 0.05 corresponds to a posterior probability of true association of between 0.07% and 1.3%. Thus, even under a very optimistic scenario, a single nominal p value of 0.05 for a randomly selected candidate gene will result in a vanishingly small chance of a true association. Of course, optimistic investigators will protest that their candidate genes are not selected randomly, but rather with knowledge of gene function. However, an honest appraisal of the difficulty of extrapolating from pathway biology to population variation shows that the problem remains. For any disease, there are without a doubt hundreds of candidate genes, and thus thousands of candidate variants, that could be (and will soon be) examined for their role in disease. In addition, a substantial proportion of causal alleles will be found in the vast majority of human genes not yet recognized as candidates for the disease in question. Thus, even if one optimistically assumes that 50% of all mutations will be in already identified candidate genes and that there are 300 such candidates (over which the 50% will be spread), the likelihood of a p value of 0.05 reflecting a true association rises only to between 1 in 3 and 1 in 30. Moreover, these likelihoods must be further diminished by the fact that most studies are underpowered, so the probability of achieving a p value of 0.05 even in the presence of a true association is far less than 1. In other words, even under the most optimistic scenario, a single p value of 0.05 will represent a false-positive association at least two out of three times and more likely 99 times of 100. These estimates fit well with recent meta-analyses of the association literature, which suggest that perhaps 5%–20% of the many hundreds of published genetic associations are in fact reproducible on repeated testing (J.N. Hirschhorn et al., personal communication).[113]

In light of this model, we estimate that the p value threshold required to infer a high likelihood of a true association is less than 10^{-4}. If one discounts the ability to pick candidate genes (or once the candidate genes have been exhausted), then the required p values are closer to 10^{-6}. Finally, if the false-positive rate is inflated for other reasons (undisclosed multiple phenotypes having been tested or confounding by population stratification), the thresholds for significance will need to be lowered still further. These calculations suggest that, in the absence of widely accepted thresholds of significance, single positive reports of association should be viewed skeptically, unless the p values are quite low (on the order of 10^{-4}) or the prior probability of association is believed to be remarkably high. If the latter is the case, the investigator should explicitly and quantitatively argue why a lower threshold is appropriate. To achieve p values this low for alleles of modest effect, very large sample sizes and/or combined analysis of multiple samples will be required.

These difficulties are increased exponentially when gene–gene or gene–environment interactions are considered. The prior probability of any given gene–gene or gene–environment pair explaining a large degree of risk is vanishingly small. Thus, in the absence of a very strong prior hypothesis (e.g., a known biochemical or physical interaction between proteins) or an extremely impressive p value, associations of this nature are more likely to represent false-positives arising from aggressive dredging of data, rather than a biologically meaningful relationship to disease.

A simple but straightforward approach that lessens (but does not eliminate) the multiple hypothesis testing inherent in pairwise tests is to first identify alleles (or environmental factors) that are definitively associated with disease and to then search only for genetic or environmental modifiers of these associations. This approach avoids testing most of the numerous hypotheses required by an exhaustive test of gene–gene and gene–environment combinations. Conversely, if an exhaustive pairwise search is performed, it should be interpreted as a hypothesis-generating exercise rather than as a meaningful test of any specific hypothesis. However, hypothesis generation has to be viewed with caution since common sense and past experience tells that the community of scientists, physicians, and journalists is much more likely to take a published report as a hypothesis confirmation than a hypothesis test.

True Differences Between Populations

It is clear from the preceding section that low prior probabilities, multiple hypothesis testing, lack of power, and the "winner's curse" are more than adequate to explain a great deal of variability in association studies. Nevertheless, one of the most frequently cited reasons for inconsistencies in association studies is methodological or population-specific differences between studies. It is important to highlight that there are few or no examples where a polymorphism that is present in multiple populations has been definitively proven to increase risk of disease in one population and not in another. In theory, however, it remains possible that differences between populations could contribute to the variability in association study results. Three such differences are discussed below.

First, phenotypic differences between association studies are often cited to explain differences in results. Certainly, reliable and precise phenotypic measurements are critical for genetic studies. As an example, repeat measurements of blood pressure over time greatly reduce the measurement error and increase the likelihood of identifying underlying genetic causes of hypertension.[128] However, it is not clear to what degree phenotypic variability has thus far led to inconsistency in association studies.

A second potential source of differences between studies is a difference in patterns of LD between the populations. If the marker being studied is not the causal variant but rather is correlated with the causal variant, the correlation could be strong in one population but not another. This difference would lead to a greater likelihood of detecting association in the population with stronger LD. The extent to which this phenomenon has actually caused false associations is not clear, but patterns of LD certainly can vary between major ethnic groups (and certainly allele frequencies, upon which the strength of correlation depends, do vary between major ethnic groups).[83a,111] Now that it is possible to empirically measure the extent of LD around an associated allele and to test haplotypes (rather than individual alleles), this potential unmeasured confounder can be controlled.

A last potential difference between populations is unmeasured environmental or genetic modifiers. If the relevant environmental modifier is known and has been measured, it can be taken into account. However, there are certainly many environmental modifiers of disease risk that are unknown and that may differ significantly between populations. Similarly, if gene–gene interactions are important in pathogenesis and if the relevant allele frequencies differ substantially between populations, then associations might vary due to these unmeasured genetic modifiers.

The possibility of unmeasured modifiers is often used to justify a claimed association in light of failure by others to replicate. In fact, this is the exact opposite of the correct Bayesian analysis. In the presence of gene–gene or gene–environment interactions, the likelihood of finding an association is decreased (because its effect is diffused by unmeasured confounders and, thus, power is diminished). If the prior probability of detection is lower, then the posterior probability of a false-positive is higher. Thus, invoking gene–gene or gene–environment interactions diminishes, not increases, the likelihood that failure to replicate was due to the original report representing a true, rather than a false-positive, association. Until these unmeasured modifiers can be measured, great care must be taken in interpreting irreproducible associations to disease in light of proposed but undocumented gene–gene or gene–environment interactions.

CONCLUSIONS

The recent revolution in genomics and human genetics has paved the way for new approaches to determining the genetic basis of common diseases. Genomewide collections of candidate genes can now be generated using a variety of approaches. Moreover, the recognition that common sequence variants explain the vast bulk of total human genetic diversity offers the possibility that they may also explain much of the inherited variation in susceptibility to common diseases. Using newly emerging resources (the human genome sequence, extensive catalogues of common human genetic variants, an understanding of the patterns of correlation between neighboring variants, and SNP genotyping technology), it is now possible to comprehensively test this hypothesis for sets of candidate genes and eventually the entire human genome. If we can rigorously interpret the resulting associations, these tools offer a powerful addition to the armamentarium of human genetic research and promise the elucidation of the genetic basis of human cancer in the years to come.

REFERENCES

1. Hoeijmakers JH. Genome maintenance mechanisms for preventing cancer. Nature 411: 366–374, 2001.
2. Taipale J, Beachy PA. The Hedgehog and Wnt signalling pathways in cancer. Nature 411:349–354, 2001.
3. Evan GI, Vousden KH. Proliferation, cell cycle and apoptosis in cancer. Nature 411:342–348, 2001.
4. IARC. Tobacco Smoking. Monographs on the Evaluation of the Carcinogenic Risk of Chemicals to Humans. Lyon: IARC, 1986.
5. Wooster R, Bignell G, Lancaster J, et al. Identification of the breast cancer susceptibility gene *BRCA2*. Nature 378:789–792, 1995.

6. Miki Y, Swensen J, Shattuck-Eidens D, et al. A strong candidate for the breast and ovarian cancer susceptibility gene *BRCA1*. Science 266:66–71, 1994.
7. Martin N, Boomsma D, Machin G. A twin-pronged attack on complex traits. Nat Genet 17:387–392, 1997.
8. Risch N. The genetic epidemiology of cancer: interpreting family and twin studies and their implications for molecular genetic approaches. Cancer Epidemiol Biomarkers Prev 10:733–741, 2001.
9. Lichtenstein P, Holm NV, Verkasalo PK, et al. Environmental and heritable factors in the causation of cancer—analyses of cohorts of twins from Sweden, Denmark, and Finland. N Engl J Med 343:78–85, 2000.
10. Page WF, Braun MM, Partin AW, et al. Heredity and prostate cancer: a study of World War II veteran twins. Prostate 33:240–245, 1997.
11. Collins FS. Of needles and haystacks: finding human disease genes by positional cloning. Clin Res 39:615–623, 1991.
12. Kinzler KW, Nilbert MC, Su LK, et al. Identification of *FAP* locus genes from chromosome 5q21. Science 253:661–665, 1991.
13. Gagel RF. *ret* protooncogene mutations and endocrine neoplasia—a story intertwined with neural crest differentiation. Endocrinology 137:1509–1511, 1996.
14. Chandrasekharappa SC, Guru SC, Manickam P, et al. Positional cloning of the gene for multiple endocrine neoplasia-type 1. Science 276:404–407, 1997.
15. Malkin D, Li FP, Strong LC, et al. Germ line *p53* mutations in a familial syndrome of breast cancer, sarcomas, and other neoplasms. Science 250:1233–1238, 1990.
16. Altmuller J, Palmer LJ, Fischer G, et al. Genomewide scans of complex human diseases: true linkage is hard to find. Am J Hum Genet 69:936–950, 2001.
17. Altshuler D, Hirschhorn JN, Klannemark M, et al. The common PPARγ Pro12Ala polymorphism is associated with decreased risk of type 2 diabetes. Nat Genet 26:76–80, 2000.
18. Cargill M, Altshuler D, Ireland J, et al. Characterization of single-nucleotide polymorphisms in coding regions of human genes. Nat Genet 22:231–238, 1999.
19. Lander ES. The new genomics: global views of biology. Science 274:536–539, 1996.
20. Risch N, Merikangas K. The future of genetic studies of complex human diseases. Science 273:1516–1517, 1996.
21. Pritchard JK. Are rare variants responsible for susceptibility to complex diseases? Am J Hum Genet 69:124–137, 2001.
22. Lander ES, Linton LM, Birren B, et al. Initial sequencing and analysis of the human genome. Nature 409:860–921, 2001.
23. Venter JC, Adams MD, Myers EW, et al. The sequence of the human genome. Science 291:1304–1351, 2001.
24. Aach J, Bulyk ML, Church GM, et al., Computational comparison of two draft sequences of the human genome. Nature 409:856–859, 2001.
25. Hogenesch JB, Ching KA, Batalov S, et al. A comparison of the *Celera* and *Ensembl* predicted gene sets reveals little overlap in novel genes. Cell 106:413–415, 2001.
26. Yeh RF, Lim LP, Burge CB. Computational inference of homologous gene structures in the human genome. Genome Res 11:803–816, 2001.
27. Adams MD, Soares MB, Kerlavage AR, et al. Rapid cDNA sequencing (expressed sequence tags) from a directionally cloned human infant brain cDNA library. Nat Genet 4:373–380, 1993.
28. Shoemaker DD, Schadt EE, Armour CD, et al. Experimental annotation of the human genome using microarray technology. Nature 409:922–927, 2001.
29. Frazer KA, Ueda Y, Zhu Y, et al. Computational and biological analysis of 680 kb of DNA sequence from the human 5q31 cytokine gene cluster region. Genome Res 7:495–512, 1997.
30. Loots GG, Locksley RM, Blankespoor CM, et al. Identification of a coordinate regulator of interleukins 4, 13, and 5 by cross-species sequence comparisons. Science 288:136–140, 2000.
31. Eshleman JR, Markowitz SD. Mismatch repair defects in human carcinogenesis. Hum Mol Genet 5:1489–1494, 1996.
32. Winzeler EA, Shoemaker DD, Astromoff A, et al. Functional characterization of the *S. cerevisiae* genome by gene deletion and parallel analysis. Science 285:901–906, 1999.
33. Vidal M. A biological atlas of functional maps. Cell 104:333–339, 2001.
34. Gibbs M, Chakrabarti L, Stanford JL, et al. Analysis of chromosome 1q42.2–43 in 152 families with high risk of prostate cancer. Am J Hum Genet 64:1087–1095, 1999.
35. Berthon P, Valeri A, Cohen-Akenine A, et al. Predisposing gene for early-onset prostate cancer, localized on chromosome 1q42.2–43. Am J Hum Genet 62:1416–1424, 1998.
36. Berry R, Schroeder JJ, French AJ, et al. Evidence for a prostate cancer-susceptibility locus on chromosome 20. Am J Hum Genet 67:82–91, 2000.
37. Smith JR, Freije D, Carpten JD, et al. Major susceptibility locus for prostate cancer on chromosome 1 suggested by a genome-wide search. Science 274:1371–1374, 1996.
38. Tavtigian SV, Simard J, Teng DH, et al. A candidate prostate cancer susceptibility gene at chromosome 17p. Nat Genet 27:172–180, 2001.
39. Xu J, Meyers D, Freije D, et al. Evidence for a prostate cancer susceptibility locus on the X chromosome. Nat Genet 20:175–179, 1998.
40. Suarez BK, Gerhard DS, Lin J, et al. Polymor-

phisms in the prostate cancer susceptibility gene *HPC2/ELAC2* in multiplex families and healthy controls. Cancer Res 61:4982–4984, 2001.
41. Wang L, McDonnell SK, Elkins DA, et al. Role of *HPC2/ELAC2* in hereditary prostate cancer. Cancer Res 61:6494–6499, 2001.
42. Rokman A, Ikonen T, Mononen N, et al. *ELAC2/HPC2* involvement in hereditary and sporadic prostate cancer. Cancer Res 61:6038–6041, 2001.
43. Carpten J, Nupponen N, Isaacs S, et al. Germline mutations in the *ribonuclease L* gene in families showing linkage with *HPC1*. Nat Genet 30:181–184, 2002.
44. Collins FS. Positional cloning moves from perditional to traditional. Nat Genet 9:347–350, 1995.
45. Lindblad-Toh K, Winchester E, Daly MJ, et al. Large-scale discovery and genotyping of single-nucleotide polymorphisms in the mouse. Nat Genet 24:381–386, 2000.
46. Kallioniemi A, Kallioniemi OP, Sudar D, et al. Comparative genomic hybridization for molecular cytogenetic analysis of solid tumors. Science 258:818–821, 1992.
47. Pollack JR, Perou CM, Alizadeh AA, et al. Genome-wide analysis of DNA copy-number changes using cDNA microarrays. Nat Genet 23:41–46, 1999.
48. Brown PO, Botstein D. Exploring the new world of the genome with DNA microarrays. Nat Genet 21:33–37, 1999.
49. Golub TR, Slonim DK, Tamayo P, et al. Molecular classification of cancer: class discovery and class prediction by gene expression monitoring. Science 286:531–537, 1999.
50. Perou CM, Jeffrey SS, van de Rijn M, et al. Distinctive gene expression patterns in human mammary epithelial cells and breast cancers. Proc Natl Acad Sci USA 96:9212–9217, 1999.
51. Alizadeh AA, Eisen MB, Davis RE, et al. Distinct types of diffuse large B-cell lymphoma identified by gene expression profiling. Nature 403:503–511, 2000.
52. Pomeroy SL, Tamayo P, Gaasenbeek M, et al. Prediction of central nervous system embryonal tumour outcome based on gene expression. Nature 415:436–442, 2002.
53. Shipp MA, Ross KN, Tamayo P, et al. Diffuse large B-cell lymphoma outcome prediction by gene-expression profiling and supervised machine learning. Nat Med 8:68–74, 2002.
54. Armstrong SA, Staunton JE, Silverman LB, et al. MLL translocations specify a distinct gene expression profile that distinguishes a unique leukemia. Nat Genet 30:41–47, 2002.
55. Bhattacharjee A, Richards WG, Staunton J, et al. Classification of human lung carcinomas by mRNA expression profiling reveals distinct adenocarcinoma subclasses. Proc Natl Acad Sci USA 98:13790–13795, 2001.
56. van 't Veer LJ, Dai H, van de Vijver MJ, et al. Gene expression profiling predicts clinical outcome of breast cancer. Nature 415:530–536, 2002.
57. Clark EA, Golub TR, Lander ES, Hynes RO. Genomic analysis of metastasis reveals an essential role for RhoC. Nature 406:532–535, 2000.
58. Patton EE, Zon LI. The art and design of genetic screens: zebrafish. Nat Rev Genet 2:956–966, 2001.
59. Brown SD, Balling R. Systematic approaches to mouse mutagenesis. Curr Opin Genet Dev 11:268–273, 2001.
60. Matteucci MD, Wagner RW. In pursuit of antisense. Nature 384(Suppl 6604):20–22, 1996.
61. Fire A, Xu S, Montgomery MK, et al. Potent and specific genetic interference by double-stranded RNA in Caenorhabditis elegans. Nature 391:806–811, 1998.
62. Zamore PD, Tuschl T, Sharp PA, et al. RNAi: double-stranded RNA directs the ATP-dependent cleavage of mRNA at 21 to 23 nucleotide intervals. Cell 101:25–33, 2000.
63. Elbashir SM, Harborth J, Lendeckel W, et al. Duplexes of 21-nucleotide RNAs mediate RNA interference in cultured mammalian cells. Nature 411:494–498, 2001.
64. Elbashir SM, Lendeckel W, Tuschl T, et al. RNA interference is mediated by 21- and 22-nucleotide RNAs. Genes Dev 15:188–200, 2001.
65. Ruvkun G. Molecular biology. Glimpses of a tiny RNA world. Science 294:797–799, 2001.
66. Lagos-Quintana M, Rauhut R, Lendeckel W, et al. Identification of novel genes coding for small expressed RNAs. Science 294:853–858, 2001.
67. Lau NC, Lim le EP, Weinstein EG, et al. An abundant class of tiny RNAs with probable regulatory roles in *Caenorhabditis elegans*. Science 294:858–862, 2001.
68. Lee RC, Ambros V. An extensive class of small RNAs in *Caenorhabditis elegans*. Science 294:862–864, 2001.
69. Stockwell BR. Chemical genetics: ligand-based discovery of gene function. Nat Rev Genet. 1:116–125, 2000.
70. Altshuler D, Daly M, Kruglyak L. Guilt by association. Nat Genet 26:135–137, 2000.
71. Strittmatter WJ, Roses AD. Apolipoprotein E and Alzheimer's disease. Annu Rev Neurosci 19:53–77, 1996.
72. Dahlback B. Resistance to activated protein C caused by the factor VR506Q mutation is a common risk factor for venous thrombosis. Thromb Haemost 78:483–488, 1997.
73. Douglas JA, Erdos MR, Watanabe RM, et al. The peroxisome proliferator-activated receptor-gamma2 Pro12Ala variant: association with type 2 diabetes and trait differences. Diabetes 50:886–890, 2001.
74. Mori H, Ikegami H, Kawaguchi Y, et al. The Pro12 → Ala substitution in PPAR-gamma is associated

with resistance to development of diabetes in the general population: possible involvement in impairment of insulin secretion in individuals with type 2 diabetes. Diabetes 50:891–894, 2001.
75. Deeb SS, Fajas L, Nemoto M, et al., A Pro12Ala substitution in PPARgamma2 associated with decreased receptor activity, lower body mass index and improved insulin sensitivity. Nat Genet 20:284–287, 1998.
76. Sachidanandam R, Weissman D, Schmidt SC, et al. A map of human genome sequence variation containing 1.42 million single nucleotide polymorphisms. Nature 409:928–933, 2001.
77. Li WH, Sadler LA. Low nucleotide diversity in man. Genetics 129:513–523, 1991.
78. Halushka MK, Fan JB, Bentley K, et al. Patterns of single-nucleotide polymorphisms in candidate genes for blood-pressure homeostasis. Nat Genet 22:239–247, 1999.
79. Cambien F, Poirier O, Nicaud V, et al. Sequence Diversity in 36 Candidate Genes for Cardiovascular Disorders. Am J Hum Genet 65:183–191, 1999.
80. Kaessmann H, Heissig F, von Haeseler A, et al. DNA sequence variation in a non-coding region of low recombination on the human X chromosome. Nat Genet 22:78–81, 1999.
81. Kaessmann H, Wiebe V, Weiss G, et al. Great ape DNA sequences reveal a reduced diversity and an expansion in humans. Nat Genet 27:155–156, 2001.
82. Altshuler D, Pollara VJ, Cowles CR, et al. An SNP map of the human genome generated by reduced representation shotgun sequencing. Nature 407:513–516, 2000.
83. Marth G, Yeh R, Minton M, et al. Single-nucleotide polymorphisms in the public domain: how useful are they? Nat Genet 27:371–372, 2001.
83a. Gabriel SB, Schaffner SF, Nguyen H, et al. The structure of haplotype blocks in the human genome. Science 296:2225–2229, 2002.
84. Buetow KH, Edmonson MN, Cassidy AB. Reliable identification of large numbers of candidate SNPs from public EST data. Nat Genet 21:323–325, 1999.
85. Irizarry K, Kustanovich V, Li C, et al. Genome-wide analysis of single-nucleotide polymorphisms in human expressed sequences. Nat Genet 26:233–236, 2000.
86. Marth GT, Korf I, Yandell MD, et al. A general approach to single-nucleotide polymorphism discovery. Nat Genet 23:452–456, 1999.
87. Mullikin JC, Hunt SE, Cole CG, et al. An SNP map of human chromosome 22. Nature 407:516–520, 2000.
88. Kruglyak L, Nickerson DA. Variation is the spice of life. Nat Genet 27:234–236, 2001.
89. Kruglyak L. Prospects for whole-genome linkage disequilibrium mapping of common disease genes. Nat Genet 22:139–144, 1999.
90. Chakravarti A, Buetow KH, Antonarakis SE, et al. Nonuniform recombination within the human beta-globin gene cluster. Am J Hum Genet 36:1239–1258, 1984.
91. Tishkoff SA, Dietzsch E, Speed W, et al. Global patterns of linkage disequilibrium at the CD4 locus and modern human origins. Science 271:1380–1387, 1996.
92. Clark AG, Weiss KM, Nickerson DA, et al. Haplotype structure and population genetic inferences from nucleotide-sequence variation in human lipoprotein lipase. Am J Hum Genet 63:595–612, 1998.
93. Kidd KK, Morar B, Castiglione CM, et al. A global survey of haplotype frequencies and linkage disequilibrium at the *DRD2* locus. Hum Genet 103:211–227, 1998.
94. Tishkoff SA, Goldman A, Calafell F, et al. A global haplotype analysis of the myotonic dystrophy locus: implications for the evolution of modern humans and for the origin of myotonic dystrophy mutations. Am J Hum Genet 62:1389–1402, 1998.
95. Zerba KE, Ferrell RE, Sing CF. Genetic structure of five susceptibility gene regions for coronary artery disease: disequilibria within and among regions. Hum Genet 103:346–354, 1998.
96. Collins A, Lonjou C, Morton NE. Genetic epidemiology of single-nucleotide polymorphisms. Proc Natl Acad Sci USA 96:15173–15177, 1999.
97. Rieder MJ, Taylor SL, Clark AG, Nickerson DA. Sequence variation in the human angiotensin converting enzyme. Nat Genet 22:59–62, 1999.
98. Boehnke M. A look at linkage disequilibrium [news; comment]. Nat Genet 25:246–247, 2000.
99. Dunning AM, Durocher F, Healey CS, et al. The extent of linkage disequilibrium in four populations with distinct demographic histories. Am J Hum Genet 67:1544–1554, 2000.
100. Eaves IA, Merriman TR, Barber RA, et al. The genetically isolated populations of Finland and sardinia may not be a panacea for linkage disequilibrium mapping of common disease genes. Nat Genet 25:320–323, 2000.
101. Jorde LB. Linkage disequilibrium and the search for complex disease genes. Genome Res 10:1435–1444, 2000.
102. Martin ER, Lai EH, Gilbert JR, et al. SNPing away at complex diseases: analysis of single-nucleotide polymorphisms around APOE in Alzheimer disease. Am J Hum Genet 67:383–394, 2000.
103. Taillon-Miller P, Bauer-Sardina I, Saccone NL, et al. Juxtaposed regions of extensive and minimal linkage disequilibrium in human Xq25 and Xq28 [see comments]. Nat Genet 25:324–328, 2000.
104. Templeton AR, Clark AG, Weiss KM, et al. Recombinational and mutational hotspots within the human lipoprotein lipase gene. Am J Hum Genet 66:69–83, 2000.

105. Abecasis GR, Noguchi E, Heinzmann A, et al. Extent and distribution of linkage disequilibrium in three genomic regions. Am J Hum Genet 68:191–197, 2001.
106. Subrahmanyan L, Eberle MA, Clark AG, et al. Sequence Variation and Linkage Disequilibrium in the Human T-Cell Receptor beta (TCRB) Locus. Am J Hum Genet 69:381–395, 2001.
107. Reich DE, Cargill M, Bolk S, et al. Linkage disequilibrium in the human genome. Nature 411:199–204, 2001.
108. Patil N, Berno AJ, Hinds DA, et al. Blocks of limited haplotype diversity revealed by high-resolution scanning of human chromosome 21. Science 294:1719–1723, 2001.
109. Johnson GC, Esposito L, Barratt BJ, et al. Haplotype tagging for the identification of common disease genes. Nat Genet 29:233–237, 2001.
110. Daly MJ, Rioux JD, Schaffner SF, Hudson TJ, Lander ES. High-resolution haplotype structure in the human genome. Nat Genet 29:229–232, 2001.
111. Cavalli-Sforza LL, Menozzi P, Piazza A. The History and Geography of Human Genes. Princeton University Press, Princeton, NJ, 1994.
112. Syvanen AC. Accessing genetic variation: genotyping single nucleotide polymorphisms. Nat Rev Genet 2:930–942, 2001.
113. Ioannidis JP, Ntzani EE, Trikalinos TA, Contopoulos-Ioannidis DG. Replication validity of genetic association studies. Nat Genet 29:306–309, 2001.
114. Thomas D, Witte JS. Population Stratification: A Problem for Case-Control Studies of Candidate-Gene Associations? Cancer Epidemiol Prev Biomarkers 11:505–512, 2002.
115. Knowler WC, Williams RC, Pettitt DJ, Steinberg AG. Gm3;5,13,14 and type 2 diabetes mellitus: an association in American Indians with genetic admixture. Am J Hum Genet 43:520–526, 1988.
116. Gelernter J, O'Malley S, Risch N, et al. No association between an allele at the D2 dopamine receptor gene (*DRD2*) and alcoholism. JAMA 266:1801–1807, 1991.
117. Morton NE, Collins A. Tests and estimates of allelic association in complex inheritance. Proc Natl Acad Sci USA 95:11389–11393, 1998.
118. Allison DB. Transmission-disequilibrium tests for quantitative traits. Am J Hum Genet 60:676–690, 1997.
119. Spielman RS, Ewens WJ. The TDT and other family-based tests for linkage disequilibrium and association. Am J Hum Genet 59:983–989, 1996.
120. Ewens WJ, Spielman RS. The transmission/disequilibrium test: history, subdivision, and admixture. Am J Hum Genet 57:455–464, 1995.
121. Spielman RS, McGinnis RE, Ewens WJ. Transmission test for linkage disequilibrium: the insulin gene region and insulin-dependent diabetes mellitus (IDDM). Am J Hum Genet 52:506–516, 1993.
122. Spielman RS, Ewens WJ. A sibship test for linkage in the presence of association: the sib transmission/disequilibrium test [see comments]. Am J Hum Genet 62:450–458, 1998.
123. Horvath S, Laird NM. A discordant-sibship test for disequilibrium and linkage: no need for parental data. Am J Hum Genet 63:1886–1897, 1998.
124. Pritchard JK, Rosenberg NA. Use of unlinked genetic markers to detect population stratification in association studies. Am J Hum Genet 65:220–228, 1999.
125. Reich DE, Goldstein DB. Detecting association in a case-control study while correcting for population stratification. Genet Epidemiol 20:4–16, 2001.
126. Devlin B, Roeder K. Genomic control for association studies. Biometrics 55:997–1004, 1999.
127. Pritchard JK, Stephens M, Donnelly P. Inference of population structure using multilocus genotype data. Genetics 155:945–959, 2000.
128. Levy D, DeStefano AL, larson MG, et al. Evidence for a gene influencing blood pressure on chromosome 17. Genome scan linkage results for longitudinal blood pressure phenotypes in subjects from the Framingham Heart Study. Hypertension 36:477–483, 2000.

5

Breast Cancer: Epidemiology and Molecular Endocrinology

HEATHER SPENCER FEIGELSON

Breast cancer is the most common cancer in women worldwide.[1] In many developing countries, including Asian countries that were believed to have a low risk of breast cancer, incidence is on the rise. Figure 5.1 shows that in countries like Japan, Singapore, and Mexico, the standardized mortality rates for breast cancer are steadily increasing. In fact, rates in Singapore are beginning to approach U.S. rates. The breast cancer mortality rates in Westernized countries, like the United States and the United Kingdom, have leveled off and even started to decline (Fig. 5.2). This decline in mortality has been attributed more to improvements in detection and treatment, rather than to factors that influence breast cancer incidence.[2]

FACTORS THAT INFLUENCE BREAST CANCER RISK

Endogenous Hormones

A large and compelling body of epidemiological and experimental data implicates estrogens in the etiology of human breast cancer.[3] Animal studies have repeatedly demonstrated that estrogens can induce and promote mammary tumors in rodents and that removing the ovaries or administering an antiestrogenic drug has the opposite effect.[4] The hormone-dependent nature of breast cancer is reflected in the age-incidence curve,[5] and the most widely accepted risk factors for breast cancer can be thought of as measures of the cumulative "dose" of estrogen that breast epithelium is exposed to over time (Table 5.1).

The findings of 29 epidemiological studies of endogenous hormones and postmenopausal breast cancer have been summarized in a meta-analytic review.[6] Taken together, the six prospective studies included in the analysis show that postmenopausal women who subsequently develop breast cancer have a 15% higher mean serum estradiol concentration than unaffected women ($p = 0.0003$).

The role of hormones other than estrogens is less clear.[7-14] Whether elevated progesterone levels increase breast cancer risk is unclear[15] but recent experimental data suggest that progestins are breast mitogens and, as such, are likely to increase risk.[16] One might expect that lower levels of sex hormone–binding globulin (SHBG) would result in increased risk of breast cancer because it would imply higher estrogen bioavailability. However, some studies have reported a positive association between SHBG and risk of breast cancer,[8,13] while others have shown an inverse association.[7,10,14] Similarly, testosterone has been shown to increase risk in some,[7,8,13,14] but not all,[10] studies; and only limited data on androstenedione exist.[10,12] Further work is nec-

BREAST CANCER: EPIDEMIOLOGY AND MOLECULAR ENDOCRINOLOGY 121

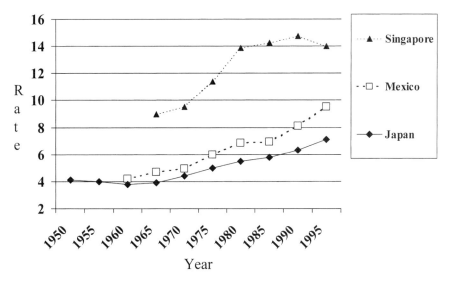

Figure 5.1 Age-standardized breast cancer mortality rates per 100,000 for Singapore (1965–1995), Mexico, and Japan (1950–1995). (From WHO Databank, http://www-depdb.iarc.fr)

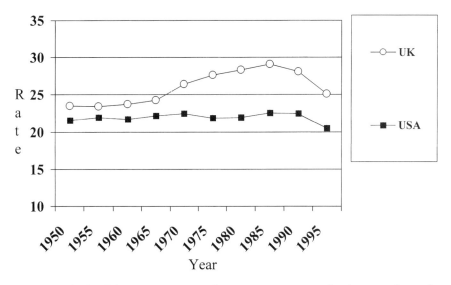

Figure 5.2 Age-standardized breast cancer mortality rates per 100,000 for the United Kingdom and the United States, 1950–1995. (From WHO Databank, http://www-depdb.iarc.fr)

Table 5.1 Factors that Influence Breast Cancer Risk

Established factors
- Genetic susceptibility and family history
- Endogenous steroid hormone levels
- Age at menarche
- Age and type of menopause
- Age at first birth and number of births
- Height, weight, and body size
- Level of physical activity
- Lactation
- Alcohol consumption
- Exogenous hormone use (hormone-replacement therapy, combination oral contraceptives)

Possible factors
- Insulin-like growth factors
- Dietary components, including fat, fiber, and soy

essary to resolve the conflicting findings of the role of SHBG in breast cancer and to confirm the reported association with testosterone and other androgens.

Age at Menarche

Endogenous estrogen levels and breast cancer risk are strongly influenced by age at menarche. In general, a decrease of about 20% in breast cancer risk results from each year that menarche is delayed. For a fixed age at menarche, women who establish regular menstrual cycles within 1 year of the first menstrual period have more than double the risk of breast cancer of women with a 5-year or more delay in onset of regular cycles.[17,18] Women with early menarche (age 12 or younger) and rapid establishment of regular cycles had an almost fourfold increased risk of breast cancer compared to women with later menarche (age 13 or older) and a longer duration of irregular cycles.

Twin studies have suggested a major genetic component to the onset of menarche.[19] However, evidence also supports an important role for environmental factors, especially physical activity, in the onset of ovulation. Girls who engaged in regular ballet dancing, swimming, or running experience a considerable delay in the onset of menses.[20,21] Even moderate physical activity during adolescence can lead to anovular cycles.[22] Girls who engaged in regular, moderate physical activity (averaging at least 600 kcal of energy per week) were 2.9 times more likely than girls who engaged in lesser amounts of physical activity to be anovular.

Age at Menopause

Early menarche and late menopause increase the number of ovulatory cycles during which a woman is exposed to high levels of estrogen. It has been estimated that women who experience natural menopause (as defined by cessation of periods) before the age of 45 have only one-half the breast cancer risk of those whose menopause occurs after the age of 55.[23] Artificial menopause, by either bilateral oophorectomy or pelvic irradiation, also markedly reduces breast cancer risk. The effect appears to be slightly greater than that of natural menopause, probably because surgical removal of the ovaries causes an abrupt cessation of hormone production, whereas some hormone production continues for a few months or years after a natural menopause.

Parity

Early age at first birth (i.e., before age 20) reduces the risk of breast cancer by about 50% relative to nulliparous women. Full-term pregnancies at later ages add smaller increments of protection, and women who have a first pregnancy over the age of 30 are actually at higher risk of breast cancer than nulliparous women.[24] This paradoxical effect of a later first full-term pregnancy has been repeatedly confirmed in epidemiological studies. Furthermore, a recent full-term pregnancy also increases risk.[25] Among women giving birth during the previous 3 years, breast cancer risk is nearly three times higher than that of women of the same age, parity, and age at first birth whose most recent birth occurred at least 10 years earlier. First-trimester abortions, whether spontaneous or induced and occurring before the first full-term pregnancy,[26,27] have also been associated in some studies with increased breast cancer risk, although the interpretation of these studies has been questioned.[28]

Based on these results, it appears that two contradictory effects of pregnancy on risk of breast cancer are particularly notable during a first pregnancy: a short-term increase in risk, followed in the long term by a substantial reduction in risk.[24] This apparent paradox has a physiological explanation based on patterns of estrogen and prolactin secretion and metabolism during pregnancy. During the first trimester of pregnancy, the level of free estradiol rises rapidly. However, as a pregnancy progresses, prolactin and free estradiol levels lower and SHBG levels rise, yielding a net overall benefit with respect to the endogenous estrogen profile. Perhaps more importantly, the effect of a first pregnancy may be to cause some premalignant cells to terminally differentiate, thereby losing their malignant potential.

Anthropometric Factors

Height, weight, body mass (measured most commonly by body mass index, or BMI), fat distribution patterns, and weight change have been shown to influence breast cancer risk.

The data on breast cancer risk and height have been relatively consistent in demonstrating an increase in risk with increased adult height. A pooled analysis of seven prospective cohort studies of height and breast cancer risk reported relative risks for breast cancer, after adjusting for other risk factors, of 1.02 per 5 cm of height among premenopausal women [95% confidence interval (CI) 0.96–1.10] and 1.07 among postmenopausal women (95% CI 1.03–1.12).[29] The relationship between greater height and risk of breast cancer is hypothesized to be due to the influence of growth hormone, insulin-like growth factor-I (IGF-I),[30] or possibly in utero influences on ductal stem cells.[31]

The relationship between weight and breast cancer risk is dependent on age. In premenopausal women, high body weight does not increase the risk of breast cancer. Pooled results have shown an inverse association between risk of breast cancer in premenopausal women and body weight.[29] Compared to women weighing less than 60 kg, the risk of breast cancer was 0.58 (95% CI 0.40–0.83) among women weighing 80 kg or more. A similar relationship was observed for BMI. This protective effect of obesity among premenopausal women may result from the increased frequency of anovulatory cycles associated with high body weight.[32] However, this protective effect is reversed among postmenopausal women.

Pooled multivariate analysis of seven prospective cohort studies demonstrated that among postmenopausal women increasing body weight and increasing BMI showed a strong linear association with increased breast cancer risk.[29] Compared to the referent weight of <60 kg, postmenopausal women weighing more than 80 kg had a 25% increase in risk of breast cancer (p for trend = 0.003). Similarly, postmenopausal women with BMI >33 are at 27% higher risk compared to women with BMI <21 (p for trend = 0.001). Increasing risk of breast cancer among postmenopausal women with increasing weight is due to increasing circulating estrogen, which results from extraglandular aromatization of plasma androstenedione to estrone catalyzed by the aromatase P450 enzyme. Obesity also is associated with decreased SHBG production and increased proportions of free and albumin-bound estrogens.

Several investigators have reported that women with increased central adiposity have increased risk of breast cancer independent of adult adiposity compared to women whose fat is distributed over the hips and lower extremities.[33] This observation is likely explained by the multiple hormonal changes that accompany central adiposity, including decreased SHBG and increased androgen levels. There has also been a consistent association between adult weight gain and breast cancer risk. This observation may be attributable to the fact that adult weight gain reflects mainly an increase in body fat, rather than body weight, which includes both lean and fat mass.[33]

Physical Activity

As described above, regular physical activity delays the onset of menarche and the establishment of regular ovulatory cycles and, thus, should reduce the risk of breast cancer. Exercise may also decrease risk of breast cancer by increasing the length of menstrual cycles, decreasing the frequency of ovulatory cycles, and reducing circulating ovarian hormone levels.[21,34–36] However, evidence of such a link between exercise in adulthood and reduced risk of breast cancer has been inconsistent. As reviewed by Friedenreich et al.,[37] 23 of 35 studies conducted on the association between physical activity and breast cancer have demonstrated a reduction in risk among those women who are most active in occupational and/or recreational activities. Two of the studies found slightly increased risk, while the remaining 10 found no association. These inconsistencies are likely due, in part, to the complexities of collection information on exercise patterns.[38] These inconsistencies may also be due to the differing effects of exercise on different times of reproductive life. Physical activity may affect breast cancer risk through different physiological mechanisms in pre- and postmenopausal women. Further, perimenarchal and premenopausal periods may be more susceptible to ovulatory disruption than other periods of active ovulation. Given the small number of breast cancer risk factors that are modifiable, further research is needed to resolve these inconsistencies and arrive at a clear understanding of how regular physical activity may modify breast cancer risk.

Lactation

Lactation has been increasingly reported to provide protection against breast cancer development. If the cumulative number of ovulatory cycles is directly related to breast cancer risk, a beneficial effect of long duration of nursing would be expected since nursing results in a delay in re-establishing ovulation following a completed pregnancy. With only a small proportion of mothers having a large number of cumulative nursing months, most epidemiological studies have been unable to provide precise estimates of the effects of lactation on breast cancer risk. However, studies in non-Western populations have consistently reported an inverse relationship between lactation and breast cancer risk.[25,39–43] Further, when attention is focused on premenopausal women, studies have fairly consistently shown a 20%–30% reduction in risk among women who have ever breast-fed.[44] London[45] noted that recent changes in breast-feeding, i.e., "on-demand" feeding rather than scheduled feedings, may result in a longer anovulatory phase and, in part, may explain why early studies failed to show an association with postmenopausal breast cancer. More recent studies[44,46,47] have also demonstrated a protective effect of ever-breast-feeding, with reduced risk of breast cancer. In a large population-based case-control study, Newcomb et al.[44] found that breast-feeding for at least 2 weeks was associated with a slightly reduced risk of breast cancer [odds ratio (OR) = 0.87, 95% CI 0.78–0.96] after adjustment for age, parity, age at first birth, and other relevant risk factors.

Alcohol

One of the few modifiable risk factors that have been associated consistently with breast cancer is alcohol consumption. Women who drink on average one alcoholic beverage daily have 10%–30% higher risk of incident breast cancer than nondrinkers.[48,49] It has been hypothesized that drinking in early adult life may be particularly deleterious for breast cancer risk;[50] studies of the relationship between alcohol and premenopausal breast cancer are fewer and less consistent than studies of postmenopausal women.[51] However, recent studies suggest that there is no increased risk from alcohol consumption among premenopausal women. A pooled analysis of six prospective studies of mainly postmenopausal women did not find evidence of effect modification by menopausal status,[48] and a recent study examining alcohol and breast cancer mortality in a large prospective cohort found no increased risk among premenopausal women.[52] Further, a cohort study among women 25–42 years of age found that neither alcohol consumption in the previous year nor average lifetime alcohol consumption was associated with increased risk of incident breast cancer.[51]

Several hypotheses have been developed to explain how alcohol may increase the risk of breast cancer.[53,54] The hypothesis that alcohol increases circulating hormone levels has received the most attention. Short-term feeding experiments, cross-sectional data from postmenopausal women, and animal models have been used to demonstrate the effect of alcohol on serum hormones. A recent short-term feeding experiment among 51 healthy postmenopausal women found that consuming 15 or 30 g of alcohol per day resulted in statistically significant increases in serum estrone sulfate and dehydroepiandrosterone sulfate (DHEAS) concentrations.[55] These results are consistent with previous results from a feeding study among premenopausal women that found an increase in DHEAS concentration among women consuming 30 g of alcohol per day[56] and with a cross-sectional analysis of serum hormone levels in which estrone sulfate levels were positively correlated with self-reported alcohol consumption.[57]

A second hypothesis suggests that the characteristics of the alcohol–breast cancer relationship can be explained by IGFs. Yu[53] reasons that alcohol may increase the production of IGFs, which in turn, increase the risk of breast cancer. Evidence of the role of IGFs is based, in part, on the observation in some studies[48] that women with the greatest alcohol consumption do not show a higher risk of breast cancer. Since IGFs are produced almost entirely in the liver, IGF production could be impaired by heavy alcohol consumption.

Several studies have reported an interaction between dietary folate and alcohol.[58–61] All four consistently suggest that the increased risk of breast cancer associated with alcohol consumption may be reduced by adequate folate intake. The fact that our food supply is now being supplemented with folate as a means to reduce birth defects may have an additional benefit on breast cancer.[61]

The Insulin-Like Growth Factor Family

Evidence is accumulating that insulin-like growth factors (IGFs) may play an important role in breast carcinogenesis. The IGF family includes the polypeptide ligands IGF-I and IGF-II; two receptors, IGF-IR and IGF-IIR; six binding proteins (IGFBP-1 through IGFBP-6); and a large group of IGFBP proteases, which degrade IGFBPs to increase bioactive IGF. The actions of both IGF-I and IGF-II are mediated through IGF-IR, which is located on the cell membrane. The IGFs play an important role in regulating cell proliferation, differentiation, apoptosis, and transformation.[62] As a potent mitogen, IGF-I increases DNA synthesis and accelerates the progression of cell division by stimulating the expression of cyclin D1.[63]

Levels of circulating IGF-I change substantially over time. Expression of the *IGF-I* gene is regulated primarily by growth hormone, and IGF-I increases slowly from birth to puberty, surges at puberty, then declines with age.[64] In addition to growth hormone, estrogen and other hormones, tamoxifen, and oral contraceptives interact with members of the IGF family and influence IGF expression in breast tissue.[64] Animal studies have suggested that dietary factors may induce different patterns of IGF-I transcription, and energy restriction has also been shown to modulate circulating IGF levels. Anthropometric factors, physical activity, alcohol, and smoking have been reported to affect the level of IGF-I, although much of the evidence is weak.[64]

Epidemiological studies have begun to investigate the role of circulating IGFs and IGFBPs, as well as polymorphisms in the genes that encode them, in the etiology of breast cancer. Most studies have focused on the association with serum or plasma levels of IGF-I, IGFBP-3, or their ratio, which is a proposed proxy for biologically active IGF-I. Hankinson et al.[65] found a markedly increased risk of premenopausal breast cancer [relative risk (RR) = 7.28, 95% CI 2.40–22.0] for women with the highest levels of plasma IGF-I but no association among postmenopausal women. Risk of breast cancer was also elevated among premenopausal women with a high ratio of IGF-I to IGFBP-3 (RR = 2.46, 95% CI 0.97–6.24) compared to those with a low ratio.

Studying this pathway presents important challenges to understanding the role of IGFs in carcinogenesis. In this complex family, risk of cancer could be mediated through any of the members, the IGFs, the receptors, or the binding proteins, or through interactions involving all three with endogenous hormones or dietary components. Accurately measuring plasma or serum levels can be problematic and may introduce significant misclassification into the study. Further, a one-time measurement of IGFs may not capture the relevant time period of exposure. For example, for breast cancer the relevant time period to measure IGFs may be premenopausal or even perimenarchal, when it is difficult or even impossible to obtain a blood sample.

Dietary Fat and Fiber

Much attention has been focused on dietary differences, particularly in fat consumption, to explain both the international pattern of breast cancer occurrence and changes in rates of breast cancer following migration to high-risk countries from low-risk countries.[66,67] International breast cancer mortality rates correlate highly with per capita consumption of fat in the diet (correlation coefficient $r = 0.93$).[67] When international breast cancer incidence rates rather than mortality rates are considered, the magnitude of the correlation is still quite high ($r = 0.84$). The correlation of fat consumption with international breast cancer mortality remains highly significant even after statistical adjustment for body weight and age at menarche. However, such ecological studies suffer from many well-known limitations, including the *ecological fallacy*, which results from attempting to make inappropriate conclusions at the individual level from summary data applied to a large geographic area. Summary indices used for ecological comparisons do not usually measure the true exposure of interest. For example, per capita consumption of fat is often taken from national food disappearance data and does not measure actual consumption or take into account how much food is thrown out instead of consumed.

Many case-control studies of fat consumption and breast cancer have found only small differences between cases and controls, generally no larger than the differences in total caloric consumption. However, Howe and colleagues[68] combined 12 large case-control studies representing populations with a wide range of dietary habits and underlying rates of breast cancer to study the diet–breast cancer relationship. They found that the breast cancer risk of postmenopausal women was positively associated with both total fat intake (RR = 1.46 for 100 g/day, $p = 0.0002$) and saturated fat intake (RR = 1.57 for highest quintile of intake, $p < 0.0001$). Nonetheless, cohort studies that have examined total fat, saturated fat, or vegetable fat[69–72] have found little or no difference in breast cancer risk over a wide range of fat intakes.[73]

High-fiber diets may protect against breast cancer, perhaps because fiber reduces the intestinal reabsorption of estrogens excreted via the biliary system.[70] In one animal study, a high-fiber diet was associated with a reduced incidence of mammary cancer.[74] Assessment of fiber intake in epidemiological studies has been problematic because of a paucity of data on the fiber content of individual foods and disagreement about the most appropriate methods of biochemical analysis to determine different types of fiber.

There have been several attempts to demonstrate a reduction in serum estrogen levels following dietary interventions that reduce fat or increase fiber intake.[75] A recent meta-analysis of several studies demonstrated a 7.4% average reduction in estradiol levels of premenopausal women and a 23% reduction in postmenopausal women following an intervention of dietary fat intake.[75] This analysis could not distinguish between a direct dietary effect on hormone level vs. an indirect effect through disruption of ovulatory cycles in premenopausal women. If such interventions can be successful at reducing serum estrogens over an extended period of time, whether directly or indirectly, then a reduction in breast cancer risk should be observed. The substantial effect of fat reduction on estrogen levels reported for postmenopausal women is largely dependent on the results of a single study[76] and, thus, must be viewed with caution.

Phytoestrogens

Dietary *phytoestrogens*, plant substances that are structurally or functionally similar to estrogen, have been proposed to act as estrogen antagonists in breast, prostate, and endometrial cells. However, epidemiological studies of dietary phytoestrogens and breast cancer have been inconclusive. Some studies indicate that soy intake may reduce a woman's risk of premenopausal breast cancer;[77–79] however, at least one study found no association between dietary soy and breast cancer.[80] One study has also reported a decrease in risk for postmenopausal breast cancer.[79] Ingram et al.[81] re-examined the association between phytoestrogens and breast cancer by measuring urinary excretion rates of two classes of phytochemicals and found that an inverse relationship exists with the risk of both pre and postmenopausal breast cancer and urinary excretion. It remains untested whether the measured phytochemicals actually serve as markers of other correlated dietary components, such as fiber, which also has been postulated to reduce breast cancer risk.[82]

The results of animal experiments and dietary intervention studies using soy products are also inconsistent. Experimental studies show that phytoestrogens can exhibit both an estrogenic and an antiestrogenic effect, depending on the study conditions.[83–94] Two dietary studies did find that premenopausal women on soy-rich diets had lower levels of leutinizing hormone, follicle-stimulating hormone, and progesterone and longer menstrual cycles.[95,96] Of potential concern are two reports that found women on soy-rich diets had elevated numbers of hyperplastic epithelial cells in their breast fluid[97] and significantly increased rates of breast lobular epithelial proliferation.[98]

Xenobiotic Pesticides

Both natural and synthetic environmental estrogens mimic the estrogenic activity of steroid hormones. Dietary phytoestrogens, as discussed above, have been proposed to act as estrogen antagonists in breast cells, potentially protecting these tissues from cancer formation. In contrast, *xenobiotics*, environmental estrogens such as pesticides, have been proposed to act as estrogen agonists, possibly increasing the risk of cancer formation. Although we would expect that a weak estrogen, whether from natural or synthetic sources, would act in a similar manner when applied to the same system, the effect is difficult to predict due to variable characteristics of the estrogen-like substances and the test systems.

Current scientific evidence does not support a relationship between weak environmental estrogens and cancer risk. Concern over environmental contaminants with estrogenic potential became an issue in the 1990s. Fueling these concerns, a number of studies reported associations between environmental estrogens and increased risks of human breast cancer.[99,100] Although the activity level of most environmental pesticides is at least 1000 times less than that of the endogenous human steroid hormone 17β-estradiol, there was uncertainty over whether combinations of pesticides found in the environment could act in concert to produce stronger effects. An early study suggested that a panel of chemical pesticides could act synergistically in competitive estrogen receptor binding and estrogen-responsive yeast assays.[101] However, the findings were not supported by subsequent studies using over 10 different estrogen-responsive assays.[102,103]

If the mechanism of action of environmental estrogens on cancer risk is strictly hormonal, it is unlikely that estrogenic pesticides and dietary phytoestrogens would exclusively produce opposite effects. Evaluation of in vitro and in vivo work for both phytoestrogens and xenobiotics as a whole may lead to a more objective interpretation of the current data and assessment of what questions remain unanswered.

Hormone-Replacement Therapy

Hormone-replacement therapy (HRT) will be discussed only briefly here as a full discussion appears in chapter 6. According to a meta-analysis including over 160,000 women, current or recent use of HRT increases the risk of breast cancer in relation to increasing duration of use.[104] For women whose last use of HRT was less than 5 years before diagnosis, risk increased by 2.3% ($p = 0.0002$) for each year of use. However, women who stopped using HRT 5 or more years before diagnosis had no increased risk, regardless of duration of use. After taking these timing factors into account, no other index of timing was important, including age at first use or time between menopause and first use.

Although this combined analysis was large and detailed, it may still fail to determine the true risk of breast cancer that can be attributed to HRT since many differences between HRT users and nonusers exist. Users of HRT may have different opportunities for breast cancer diagnosis. For example, they may have more frequent mammographic and physician examinations. Women with a family history of breast cancer are more likely to be never-users, and HRT users are likely to be of higher social class and education. Laya et al.[105] have provided direct evidence that current HRT use reduces the sensitivity and specificity of mammographic screening, most likely by increasing the radiographic density of the breast. Finally, genetic determinants, like those that determine endogenous hormone levels, may play a role in determining HRT use.[106] This is described in more detail below (see Multigenic Models of Susceptibility).

The vast majority of data from this combined analysis pertained to estrogen-only replacement therapy (ERT). Combination HRT (CHRT), in which a progestin is given sequentially or continuously with estrogen during a monthly cycle, has grown rapidly in popularity in the past two decades. Although the combined analysis suggested that risk of breast cancer associated with CHRT might be greater than for ERT, there were few long-term users of CHRT available for analysis, so the risk estimates are statistically imprecise.[104]

Studies are now beginning to appear in the literature that demonstrate this increased risk associated with CHRT. Ross and colleagues[107] reported that for each 5 years of use, risk was four times greater for CHRT users than for ERT users. Specifically, the OR per 5 years of use for CHRT was 1.24 (95% CI 1.07–1.45) and for ERT it was 1.06 (95% CI 0.97–1.15). Schairer et al.[108] observed similar risks among a cohort of 46,355 women in the Breast Cancer Detection Demonstration Project. The RR increased by 0.01 per year of estrogen-only use and by 0.08 per year of CHRT use.

Oral Contraceptives

Results from a recent meta-analysis of 54 studies indicate that a modestly increased risk for breast cancer was associated with current (RR = 1.24, $p < 0.00001$) and recent (RR = 1.16, $p < 0.00001$) combination oral contraceptive (COC) use.[109] This excess risk does not persist, and there is no evidence of an increased risk 10 or more years after cessation of COC use. However, the degree of the association was modified by age at first use of COCs. For recent users, risk was greatest for those who began COCs before the age of 20 and tended to decline with increasing age at diagnosis. Total duration of COC use was not associated with increased risk of breast cancer once recency of use was taken into account. There is little information about cancer risk 10 years after cessation of COC use. Moreover, most women who stopped use 10 or more years ago had used COCs for only short periods of time. In the next decade, women who began use as teenagers will reach their late 40s and early 50s. At that time, it will be important to re-examine the effects of long-term and early use of COCs.

Family History and Inherited Susceptibility

As detailed in chapter 7, family history of breast cancer is associated with increased risk. This is especially so if the history includes a first-degree family member who was affected at an early age or had bilateral disease.

Much attention is given to "hereditary" breast cancer when, in fact, those cases that can be attributed to a single-gene mutation are very rare. Epidemiological evidence suggests that BRCA1 and BRCA2 together account for only about 5% of all breast cancers.[110] Germline p53 mutations and perhaps the ATM gene account for an even smaller proportion of breast cancers.[111–113]

Although BRCA1 and BRCA2 are often described together, there are important epidemiological distinctions between them. The BRCA1 gene has been associated with increased risk of ovarian cancer, while BRCA2 has been suggested to play a role in male breast cancer and possibly other cancers, such as pancreatic and prostate. The BRCA1 gene is associated with an early age at onset, but this is less clear for tumors associated with BRCA2. The histology of BRCA1- and BRCA2-associated tumors differs from sporadic cases and from each other.

Lifetime risk estimates for women who carry a mutation in either BRCA gene are changing with increased study. The lifetime risk estimate for breast cancer from BRCA1 from family-based studies is 80%–90% and 64% for ovarian cancer by age 70.[114] However, non-family-based estimates are much lower. In one study of Jewish women, risks for breast cancer were 56% and 16% for ovarian cancer among women with a BRCA1 or BRCA2 mutation, respectively.[115] In a study of women referred to breast clinics, only 16% of breast cancer patients with a positive family history carried a BRCA1 mutation.[116] In a hospital-based sample of ovarian cancer patients, 3% had a mutant BRCA1 gene.[117] As additional population-based estimates are reported, these risk estimates will continue to be refined. An important challenge remains to find the genetic and environmental co-factors that interact with BRCA1 and BRCA2 and lead to variations in gene penetrance and expression.

MULTIGENIC MODELS OF SUSCEPTIBILITY

Although there is evidence that hormonal secretion and metabolism can be environmentally influenced, e.g., through diet and physical activity, the control of hormonal patterns is largely genetically regulated. It has been hypothesized that a multigenic model of breast cancer predisposition can be developed that includes polymorphisms in genes involved in estrogen biosynthesis and intracellular binding.[118] This model would include functionally relevant polymorphisms that would act together and in combination with established risk factors to define a high-risk profile for breast cancer. A key assumption in this multigenic model is that variation in genes that encode critically important enzymes in estradiol biosynthesis would individually provide only modest differences in the rate of biosynthesis. Presumably, there would be limited evolutionary tolerance for major variation in hormone synthesis that could disrupt reproductive ability. However, a combination of genes, each with minor variation in expressed activity, could provide a degree of separation of risk that would be clinically useful. These small variations could result in a large cumulative effect after several decades. For example, the model of breast tissue age by Pike et al.[5] demonstrates that a 20% difference in levels of circulating estrogen can result in a more than twofold increase in lifetime breast cancer risk.

Although many candidate genes for such a model exist (See Chapter 2 for more details on the steroid pathways), the genes originally proposed included three of interest: the 17β-hydroxysteroid dehydrogenase 1 (*HSD17B1*) gene, the cytochrome P450c17α (*CYP17*) gene, and the estrogen receptor α (*ESR1*) gene (Fig. 5.3). These genes were selected not only because of their known function but also because polymorphisms had already been identified that could be functionally relevant. Data have been published on the role of *CYP17* and *HSD17B1* and breast cancer risk.[119] Huang and colleagues[120] have published findings of a similar model with the estrogen-metabolizing genes *CYP17*, *CYP1A1* (which participates in estrogen hydroxylation), and catechol-O-methyltransferase (*COMT*), which encodes the enzyme responsible for O-methylation, leading to inactivation of catechol estrogen. Several candidate genes for such a model are described in subsequent chapters, and the numerous studies of other polymorphisms are reviewed elsewhere.[121]

The CYP17 Gene

At present, data to support the role of *CYP17* are most compelling. As summarized below, it has been shown to be associated with the risk of breast cancer, serum hormone levels in pre- and postmenopausal women, estrogen metabolites measured in urine, age at menarche, and use of HRT.

The *CYP17* gene codes for the cytochrome P450c17α enzyme, which mediates both steroid 17α-hydroxylase and 17,20-lyase activities and

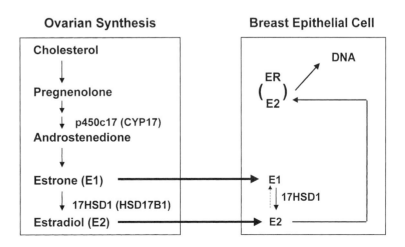

Figure 5.3 Schematic presentation of estrogen metabolism in the ovaries and breast epithelium and three candidate genes that may play a role in breast cancer etiology. The genes of interest are the cytochrome P450c17α (*CYP17*) gene, the 17β-hydroxy-steroid dehydrogenase 1 (*HSD17B1*) gene, and the estrogen receptor α (*ESR1*) gene.

functions at key branch points in human steroidogenesis.[122] The 5'-untranslated region of *CYP17* contains a single-base pair polymorphism 34 bp upstream from the initiation of translation and 27 bp downstream from the transcription start site (T27C).[123] This base pair change creates a recognition site for the MspAI restriction enzyme and has been used to designate two alleles, *A1* (the published sequence) and *A2*.

An association between risk of breast cancer and this *CYP17* polymorphism was first reported in 1997.[124] In a case-control study of incident breast cancer among Asian, African-American, and Latina women, a 2.5-fold increased risk of advanced breast cancer was observed among women who carry the *CYP17 A2* allele. This study also presented preliminary evidence suggesting that *CYP17* may be associated with age at menarche. The reduced risk of breast cancer associated with later age at menarche was largely limited to *A1/A1* women (OR = 0.47, 95% CI 0.22–0.98 for breast cancer and later age at menarche) compared to women who carried the *A2* allele (OR = 0.80, 95% CI 0.51–1.27).

These results suggested that serum hormone levels may differ by *CYP17* genotype. In a follow-up study, it was reported that *CYP17* genotype was associated with serum estradiol (E_2) and progesterone levels among young nulliparous women.[125] As shown in Figure 5.4, serum E_2 measured around day 11 of the menstrual cycle was 11% and 57% higher ($p = 0.04$), respectively, among women hetero- and homozygous for the *CYP17 A2* allele compared to *A1/A1* women. Similarly, around cycle day 22, E_2 was 7% and 28% higher ($p = 0.06$) and progesterone (shown in Fig. 5.5) was 24% and 30% higher ($p = 0.04$). These data provide direct evidence of genetic control of serum hormone levels.

The *CYP17* gene has also been associated with use of HRT.[106] Among 749 postmenopausal women randomly selected from a larger multiethnic cohort, those who carried the *CYP17 A2/A2* genotype were about half as likely as those with the *A1/A1* genotype to be current HRT

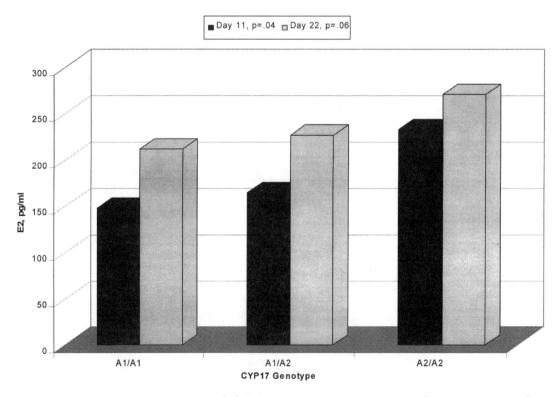

Figure 5.4 Geometric mean serum estradiol (E_2) concentrations among young nulliparous women on days 11 and 22 of the menstrual cycle by *CYP17* genotype.

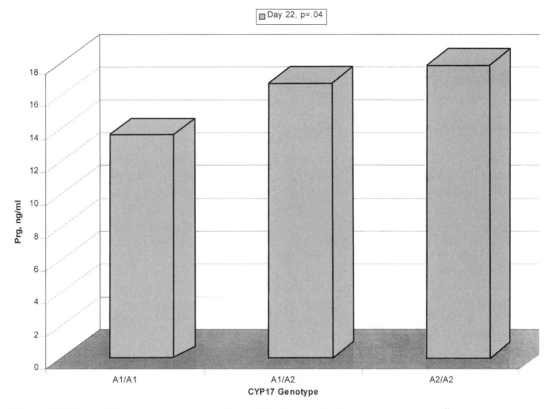

Figure 5.5 Geometric mean serum progesterone (Prg) concentrations among young nulliparous women on day 22 of the menstrual cycle by *CYP17* genotype.

users (OR = 0.52, 95% CI 0.31–0.86). Presumably, women with the A2/A2 genotype have fewer indications for HRT use due to their higher level of endogenous hormones.

Since this original study of *CYP17* was published, at least 11 other studies have reported on *CYP17* and breast cancer.[120,126–131] The results of these studies are largely negative and suggest heterogeneity by ethnicity. However, there may be several reasons for the discrepant results. The majority of the studies are small. Several of the smaller studies[120,128,131] found a modest elevation in breast cancer risk with the *CYP17* A2 allele in some subgroups, but those findings did not reach statistical significance. The two largest studies conducted to date found no association.[126,129] However, Dunning et al.[126] did not examine the possible confounding effects of HRT, which may mask the *CYP17*–breast cancer association. Haiman and colleagues[129] gave adequate consideration to potential confounding and did not find an association between *CYP17* and breast cancer. However, their results are compatible with the potential modification of breast cancer risk due to late age at menarche.

Others have shown an association between *CYP17* and breast cancer only among specific age groups of women. Kristensen et al.[130] and Miyoshi et al.[132] suggest that the effect of *CYP17* may be limited to older cases (i.e., over 55 years of age at diagnosis), while Bergman-Jungestrom et al.[127] and Spurdle et al.[133] found increased risk among premenopausal women.

More consistent data are accumulating to suggest that *CYP17* is a modifier of other breast cancer risk factors, such as age at menarche and parity.[121,134] At least three studies have shown that the protective effect of later onset of menarche was limited to women with the *A1/A1* genotype.[124,129,134] One study has shown that *CYP17* genotype was associated with estrogen metabolites measured in urine.[135] The ratio between 2-hydroxyestrone (2OHE) and 16α-hydroxyestrone (16αOHE) demonstrated a dose–response relationship where women with the *A1/A1* genotype had the highest urinary ratio of 2OHE to 16αOHE (median = 1.47) and women with the *A2/A2* had the lowest ratio (median = 1.21, $p = 0.01$). Lower 2OHE:16αOHE ratios may be associated with increased risk of breast cancer.[136,137] Thus, this observation is compatible with the hypothesis that the *CYP17 A2* allele confers a higher risk of breast cancer.

An important piece of information about *CYP17* is still missing: What is the functional relevance of this polymorphism? Or what is it marking? The T → C polymorphism in *CYP17*, converting the sequence CACT into CACC, does not influence Sp-1 binding, as had been suggested based on its similarity to other known Sp-1-binding sequences.[130] Gene function studies are needed to determine if the *A2* allele confers specifically a higher expression level of *CYP17*.[130] Such studies will need to be carefully designed and evaluated because one would not expect these polymorphisms to confer a large difference in circulating hormone levels. Standard assays may not detect the relatively small differences in activity we would predict from the epidemiological data. Emerging methods for haplotype analysis may also help to explain and resolve the inconsistencies reported in studies of *CYP17*.

Other Candidate Genes

Work is emerging on other candidate genes that may fit into this model. Carrying high-risk alleles of *CYP17* and *HSD17B1* together increased breast cancer risk among 850 incident cases and 1508 controls.[119] The *HSD17B1* gene encodes the 17HSD type 1 enzyme, which catalyzes the final step of E_2 biosynthesis, namely, the conversion of estrone to the more biologically active E_2. Type 1 is expressed in both normal and malignant breast epithelium.[138] Several polymorphisms have been identified in *HSD17B1*, including a common polymorphism in exon 6 that results in an amino acid change from serine (allele A) to glycine (allele G) at position 312.[138,139] Although current evidence indicates that this amino acid change may not affect the catalytic or immunological properties of the enzyme,[140] an early report suggested that individuals who are homozygous for serine are at marginally significantly increased risk for breast cancer.[138]

In the study, the *CYP17 A2* allele and the *HSD17B1 A* allele were designated as the high-risk alleles.[119] Subjects were then classified by number of high-risk alleles. After adjusting for age, weight, and ethnicity, carrying one or more high-risk alleles increased the risk of advanced breast cancer in a dose–response fashion. The risk among women carrying four high-risk alleles was 2.76 (95% CI 1.07–7.12) compared to those who carried none. This risk was largely limited to women who were not taking HRT and was most pronounced among those weighing 170 pounds (approximately 80 kg) or less.

A polymorphism in the *COMT* gene has been investigated in at least three studies.[141–143] Alleles of *COMT* can be designated as high-activity (the wild-type allele) or low-activity. It has been hypothesized that the low-activity alleles lead to an increased risk of breast cancer secondary to the accumulation of catechol estrogens. Published results on the association between *COMT* and breast cancer have been inconsistent. Two studies[141,143] reported opposite effects, and a third found no association and no evidence of effect modification with other risk factors.[142]

Others have examined the possible role of *CYP1A1*, which is among the major enzymes participating in estrogen hydroxylation, in breast cancer etiology.[144–148] Several polymorphisms in *CYP1A1* have been described, and two have been associated with breast cancer risk in some,[144,146,148] but not all,[145,147] studies. The strongest associations for *CYP1A1* and breast cancer are limited to women who smoke.

CONCLUDING REMARKS

The primary risk factors for breast cancer stem from prolonged endogenous and exogenous hormonal exposures. However, the genetic basis of endogenous hormone levels as an important risk factor for breast cancer has been recognized only recently. Long thought to be largely an environmentally caused cancer, it is now increasingly obvious that genetic susceptibility, through germline polymorphisms in metabolic genes, plays a critical role. Certainly the complete multigenic model of breast cancer susceptibility would include several genes. The molecular model that is emerging illustrates the importance of collaborative efforts between multiple specialties and may serve as a template for establishing models for other hormone-dependent cancers. Further study is necessary to determine which genes consistently predict breast cancer risk and carry additional supporting evidence, such as functional laboratory data, associations with serum hormone levels, or effects on known breast cancer risk factors.

REFERENCES

1. Parkin DM, Stjernsward J, Muir CS. Estimates of the worldwide frequency of twelve major cancers. Bull WHO 62:163–182, 1984.
2. Chu K, Tarone R, Kessler L, et al. Recent trends in U.S. breast cancer incidence, survival, and mortality rates. J Natl Cancer Inst 88:1571–1579, 1996.
3. Henderson BE, Ross RK, Bernstein L. Estrogens as a cause of human cancer: the Richard and Hinda Rosenthal Foundation Award lecture. Cancer Res 48:246–253, 1988.
4. Dao T. The role of ovarian hormones in mammary carcinogenesis. In: Pike M, Siiteri P, Welsch C (eds). Hormones and Breast Cancer (Banbury Report No. 8). Cold Spring Harbor, NY: Cold Spring Harbor Laboratory Press, 1981, pp 281–295.
5. Pike MC, Krailo MD, Henderson BE, Casagrande JT, Hoel DG. Hormonal risk factors, breast tissue age and age-incidence of breast cancer. Nature 330:767–770, 1983.
6. Thomas HV, Reeves GK, Key TJ. Endogenous estrogen and postmenopausal breast cancer: a quantitative review. Cancer Causes Control 8:922–928, 1997.
7. Thomas HV, Key TJ, Allen DS, et al. A prospective study of endogenous serum hormone concentrations and breast cancer risk in postmenopausal women on the island of Guernsey. Br J Cancer 76:401–405, 1997.
8. Thomas H, Key T, Allen D, et al. A prospective study of endogenous serum hormone concentrations and breast cancer risk in premenopausal women on the island of Guernsey. Br J Cancer 75:1075–1079, 1997.
9. Toniolo P, Levitz M, Seleniuch-Jacquotte A, et al. A prospective study of endogenous estrogens and breast cancer in postmenopausal women. J Natl Cancer Inst 87:190–197, 1995.
10. Lipworth L, Adami H, Trichopoulos D, Carlstrom K, Mantzoros C. Serum steroid hormone levels, sex hormone–binding globulin, and body mass index in the etiology of postmenopausal breast cancer. Epidemiology 7:96–100, 1996.
11. Hankinson S, Willett W, Michaud D, et al. Plasma prolactin levels and subsequent risk of breast cancer in postmenopausal women. J Natl Cancer Inst 91:629–634, 1999.
12. Hankinson SE, Willett WC, Manson JE, et al. Plasma sex steroid hormone levels and risk of breast cancer in postmenopausal women. J Natl Cancer Inst 90:1292–1299, 1998.
13. Dorgan J, Longcope C, Stephenson H, et al. Relation of prediagnostic serum estrogen and androgen levels to breast cancer risk. Cancer Epidemiol Biomarkers Prev 5:533–539, 1996.
14. Berrino F, Muti P, Micheli A, et al. Serum sex hormone levels after menopause and subsequent breast cancer. J Natl Cancer Inst 88:291–291, 1996.
15. Pike MC, Spicer DV, Dahmoush L, Press MF. Estrogens, progestogens, normal breast cell proliferation, and breast cancer risk. Epidemiol Rev 15:17–35, 1993.
16. Cline JM, Soderqvist G, von Schoultz E, Skoog L, von Schoultz B. Effects of hormone replacement therapy on the mammary gland of surgically postmenopausal cynomolgus macaques [see comments]. Am J Obstet Gynecol 174:93–100, 1996.
17. Henderson BE, Pike MC, Casagrande JT. Breast cancer and the oestrogen window hypothesis [letter]. Lancet 2:363–364, 1981.
18. Henderson BE, Pike MC, Ross RK. Epidemiology and risk factors. In: Bonadonna G (ed). Breast Cancer: Diagnosis and Management. New York: John Wiley and Sons, 1984, pp 1–16.
19. Kaprio J, Rimpela A, Winter T, Viken RJ, Rimpela M, Rose RJ. Common genetic influences on BMI and age at menarche. Hum Biol 67:739–753, 1995.
20. Frisch RE, Gotz-Welbergen AV, McArthur JW, et al. Delayed menarche and amenorrhea of college athletes in relation to age of onset of training. JAMA 246:1559–1563, 1981.
21. Frisch RE, Wyshak G, Vincent L. Delayed menarche and amenorrhea in ballet dancers. N Engl J Med 303:17–19, 1980.
22. Bernstein L, Ross RK, Lobo RA, Hanisch R, Krailo MD, Henderson BE. The effects of moderate physical activity on menstrual cycle patterns in adolescence: implications for breast cancer prevention. Br J Cancer 55:681–685, 1987.
23. Trichopoulos D, MacMahon B, Cole P. Menopause and breast cancer risk. J Natl Cancer Inst 48:605–613, 1972.
24. MacMahon B, Cole P, Lin TM, et al. Age at first birth and breast cancer risk. Bull WHO 43:209–221, 1970.
25. Yuan JM, Yu MC, Ross RK, et al. Risk factors for breast cancer in Chinese women in Shanghai. Cancer Res 48:1949–1953, 1983.
26. Pike MC, Henderson BE, Casagrande JT, Rosario I, Gray GE. Oral contraceptive use and early abortion as risk factors for breast cancer in young women. Br J Cancer 43:72–76, 1981.

27. Bruzzi P, Negri E, La Vecchia C, et al. Short term increase in risk of breast cancer after full term pregnancy. BMJ 297:1096–1098, 1988.
28. Wingo PA, Newsome K, Marks JS, Calle EE, Parker SL. The risk of breast cancer following spontaneous or induced abortion [published erratum appears in Cancer Causes Control 8:260, 1997]. Cancer Causes Control 8:93–108, 1997.
29. van den Brandt P, Spiegelman D, Yaun S, et al. Pooled analysis of prospective cohort studies on height, weight, and breast cancer risk. Am J Epidemiol 152:514–527, 2000.
30. Li C, Malone K, White E, Daling J. Age when maximum height is reached as a risk factor for breast cancer among young U.S. women. Epidemiology 8:559–565, 1997.
31. Trichopoulos D, Lipman R. Mammary gland mass and breast cancer risk. Epidemiology 3:523–526, 1992.
32. Willett WC, Browne ML, Bain C, et al. Relative weight and risk of breast cancer among premenopausal women. Am J Epidemiol 122:731–740, 1985.
33. Friedenreich CM. Review of anthropometric factors and breast cancer risk. Eur J Cancer Prev 10:15–32, 2001.
34. Feicht CB, Johnson TS, Martin BJ, Sparkes KE, Wagner WW Jr. Secondary amenorrhoea in athletes [letter]. Lancet 2:1145–1146, 1978.
35. Broocks A, Pirke KM, Schweiger U, et al. Cyclic ovarian function in recreational athletes. J Appl Physiol 68:2083–2086, 1990.
36. Bullen BA, Skrinar GS, Beitins IZ, von Mering G, Turnbull BA, McArthur JW. Induction of menstrual disorders by strenuous exercise in untrained women. N Engl J Med 312:1349–1353, 1985.
37. Friedenreich CM, Bryant HE, Courneya KS. Case-control study of lifetime physical activity and breast cancer risk. Am J Epidemiol 154:336–347, 2001.
38. Bernstein L, Ross RK. Re: Physical activity and breast cancer risk in a cohort of young women [letter, comment]. J Natl Cancer Inst 90:1907–1909, 1998.
39. Romieu I, Hernandez-Avila M, Lazcano E, et al. Breast cancer and lactation history in Mexican women. Am J Epidemiol 143:543–552, 1996.
40. Tao SC, Yu MC, Ross RK, et al. Risk factors for breast cancer in Chinese women of Beijing. Int J Cancer 42:495–498, 1988.
41. Yoo KY, Tajimi K, Kuroishi T, et al. Independent protective effect of lactation against breast cancer: a case-control study in Japan. Am J Epidemiol 135:726–733, 1992.
42. Rosero-Bixby L, Oberle MW, Lee NC. Reproductive history and breast cancer in a population of high fertility, Costa-Rica, 1984–85. Int J Cancer 40:747–754, 1987.
43. Land CE, Hayakawa N, Machado SG, et al. A case-control interview study of breast cancer among Japanese A-bomb survivors. II. Interactions with radiation dose. Cancer Causes Control 5:167–176, 1994.
44. Newcomb PA, Egan KM, Titus-Ernstoff L, et al. Lactation in relation to postmenopausal breast cancer. Am J Epidemiol 150:174–182, 1999.
45. London SJ. Breast-feeding and breast cancer [letter, comment]. N Engl J Med 330:1682–1684, 1994.
46. Enger SM, Ross RK, Paganini-Hill A, Bernstein L. Breastfeeding experience and breast cancer risk among postmenopausal women. Cancer Epidemiol Biomarkers Prev 7:365–369, 1998.
47. Freudenheim JL, Marshall JR, Vena JE, et al. Lactation history and breast cancer risk. Am J Epidemiol 146:932–938, 1997.
48. Smith-Warner SA, Spiegelman D, Yaun SS, et al. Alcohol and breast cancer in women: a pooled analysis of cohort studies [see comments]. JAMA 279:535–540, 1998.
49. Longnecker M. Alcoholic beverage consumption in relation to risk of breast cancer: meta-analysis and review. Cancer Causes Control 5:73–82, 1994.
50. Colditz GA, Frazier AL. Models of breast cancer show that risk is set by events of early life: prevention efforts must shift focus. Cancer Epidemiol Biomarkers Prev 4:567–571, 1995.
51. Garland M, Hunter DJ, Colditz GA, et al. Alcohol consumption in relation to breast cancer risk in a cohort of United States women 25–42 years of age. Cancer Epidemiol Biomarkers Prev 8:1017–1021, 1999.
52. Feigelson H, Calle E, Robertson A, Wingo P, Thun M. Alcohol consumption increases the risk of fatal breast cancer (United States). Cancer Causes Control 12:1, 2001.
53. Yu H. Alcohol consumption and breast cancer risk [letter, comment]. JAMA 280:1138–1139, 1998.
54. Schatzkin A, Longnecker M. Alcohol and breast cancer. Where are we now and where do we go from here? Cancer 74:1101–1110, 1994.
55. Dorgan JF, Baer DJ, Albert PS, et al. Serum hormones and the alcohol–breast cancer association in postmenopausal women. J Natl Cancer Inst 93:710–715, 2001.
56. Reichman ME, Judd JT, Longcope C, et al. Effects of alcohol consumption on plasma and urinary hormone concentrations in premenopausal women. J Natl Cancer Inst 85:722–727, 1993.
57. Hankinson SE, Willett WC, Manson JE, et al. Alcohol, height, and adiposity in relation to estrogen and prolactin levels in postmenopausal women. J Natl Cancer Inst 87:1297–1302, 1995.
58. Zhang S, Hunter DJ, Hankinson SE, et al. A prospective study of folate intake and the risk of breast cancer. JAMA 281:1632–1637, 1999.
59. Rohan TE, Jain MG, Howe GR, Miller AB. Di-

etary folate consumption and breast cancer risk. J Natl Cancer Inst 92:266–268, 2000.
60. Negri E, La Vecchia C, Franceschi S. Re: Dietary folate consumption and breast cancer risk. J Natl Cancer Inst 92:1270–1271, 2000.
61. Sellers T, Kushi L, Cerhan J, et al. Dietary folate intake, alcohol, and risk of breast cancer in a prospective study of postmenopausal women. Epidemiology 12:420–428, 2001.
62. Jones J, Clemmons D. Insulin-like growth factors and their binding proteins: biological actions. Endocr Rev 16:3–34, 1995.
63. Furlanetto R, Harwell S, Frick K. Insulin-like growth factor-I induces cyclin-D1 expression in MG63 human osteosarcoma cells in vitro. Mol Endocrinol 8:510–517, 1994.
64. Yu H, Rohan T. Role of the insulin-like growth factor family in cancer development and progression. J Natl Cancer Inst 92:1472–1489, 2000.
65. Hankinson S, Willett W, Colditz G, et al. Circulating concentrations of insulin-like growth factor-1 and risk of breast cancer. Lancet 351:1393–1396, 1998.
66. Muir C, Waterhouse J, Mack T, Powell J, Whelan S. Cancer Incidence in Five Continents, vol V. Lyon: IARC, 1987.
67. Armstrong B, Doll R. Environmental factors and cancer incidence and mortality in different countries, with special reference to dietary practices. Int J Cancer 15:617–631, 1975.
68. Howe GR, Hirohata T, Hislop TG, et al. Dietary factors and risk of breast cancer: combined analysis of 12 case-control studies [see comments]. J Natl Cancer Inst 82:561–569, 1990.
69. Howe GR, Friedenreich CM, Jain M, Miller AB. A cohort study of fat intake and risk of breast cancer [see comments]. J Natl Cancer Inst 83:336–340, 1991.
70. Hunter DJ, Willett WC. Diet, body build, and breast cancer. Annu Rev Nutr 14:393–418, 1994.
71. Mills PK, Beeson WL, Phillips RL, Fraser GE. Dietary habits and breast cancer incidence among Seventh-day Adventists. Cancer 64:582–590, 1989.
72. Willett WC, Stampfer MJ, Colditz GA, Rosner BA, Hennekens CH, Speizer FE. Dietary fat and the risk of breast cancer. N Engl J Med 316:22–28, 1987.
73. Holmes MD, Hunter DJ, Colditz GA, et al. Association of dietary intake of fat and fatty acids with risk of breast cancer. JAMA 281:914–920, 1999.
74. Cohen LA, Kendall ME, Zang E, Meschter C, Rose DP. Modulation of N-nitrosomethylurea-induced mammary tumor promotion by dietary fiber and fat [see comments]. J Natl Cancer Inst 83:496–501, 1991.
75. Wu AH, Pike MC, Stram DO. Meta-analysis: dietary fat intake, serum estrogen levels, and the risk of breast cancer [see comments]. J Natl Cancer Inst 91:529–534, 1999.
76. Heber D, Ashley JM, Leaf DA, Barnard RJ. Reduction of serum estradiol in postmenopausal women given free access to low-fat high-carbohydrate diet. Nutrition 7:137–140, 1991.
77. Lee HP, Gourley L, Duffy SW, Esteve J, Lee J, Day NE. Dietary effects on breast-cancer risk in Singapore [see comments]. Lancet 337:1197–1200, 1991.
78. Hirose K, Tajima K, Hamajima N, et al. A large-scale, hospital-based case-control study of risk factors of breast cancer according to menopausal status. Jpn J Cancer Res 86:146–154, 1995.
79. Wu AH, Ziegler RG, Horn-Ross PL, et al. Tofu and risk of breast cancer in Asian-Americans. Cancer Epidemiol Biomarkers Prev 5:901–906, 1996.
80. Yuan JM, Wang QS, Ross RK, Henderson BE, Yu MC. Diet and breast cancer in Shanghai and Tianjin, China. Br J Cancer 71:1353–1358, 1995.
81. Ingram D, Sanders K, Kolybaba M, Lopez D. Case-control study of phyto-oestrogens and breast cancer [see comments]. Lancet 350:990–994, 1997.
82. De Stefani E, Correa P, Ronco A, Mendilaharsu M, Guidobono M, Deneo-Pellegrini H. Dietary fiber and risk of breast cancer: a case-control study in Uruguay. Nutr Cancer 28:14–19, 1997.
83. Barnes S, Grubbs C, Setchell K, Carlson J. Soybeans inhibit mammary tumors in models of breast cancer. Clin Biol Res 347:329–353, 1990.
84. Constantinou A, Thomas C, Mehta RG, Runyan C, Moon R. The effect of genistein and daidzein on MNU-induced mammary tumors in rats. Proc Am Assoc Cancer Res 36:109–115, 1995.
85. Constantinou AI, Mehta RG, Vaughan A. Inhibition of N-methyl-N-nitrosourea-induced mammary tumors in rats by the soybean isoflavones. Anticancer Res 16:3293–3298, 1996.
86. Makela S, Davis VL, Tally WC, Korkman J, Salo L. Dietary estrogens act through estrogen receptor–mediated processes and show no antiestrogenicity in cultured breast cancer cells. Environ Health Perspect 572–578, 1994.
87. Martin PM, Horwitz KB, Ryan DS, McGuire WL. Phytoestrogen interaction with estrogen receptors in human breast cancer cells. Endocrinology 103:1860–1867, 1978.
88. Peterson G, Barnes S. Genistein inhibition of the growth of human breast cancer cells: independence from estrogen receptors and the multi-drug resistance gene. Biochem Biophys Res Commun 179:661–667, 1991.
89. Murrill WB, Brown NM, Zhang JX, Manzolillo PA, Barnes S, Lamartiniere CA. Prepubertal genistein exposure suppresses mammary cancer and enhances gland differentiation in rats. Carcinogenesis 17:1451–1457, 1996.
90. Lamartiniere CA, Moore JB, Brown NM, Thompson R, Hardin MJ, Barnes S. Genistein suppresses mammary cancer in rats. Carcinogenesis 16:2833–2840, 1995.

91. Fournier DB, Erdman JW Jr, Gordon GB. Soy, its components, and cancer prevention: a review of the in vitro, animal, and human data. Cancer Epidemiol Biomarkers Prev 7:1055–1065, 1998.
92. Ito A, Goto T, Okamoto T, Yamada K, Roy G. A combined effect of tamoxifen (Tam) and miso for the development of mammary tumors induced with MNU in SD rats. Proc Am Assoc Cancer Res 37:271, 1996.
93. Wang C, Kurzer MS. Phytoestrogen concentration determines effects on DNA synthesis in human breast cancer cells. Nutr Cancer 28:236–247, 1997.
94. Welshons WV, Rottinghaus GE, Nonneman DJ, Dolan-Timpe M, Ross PF. A sensitive bioassay for detection of dietary estrogens in animal feeds. J Vet Diagn Invest 2:268–273, 1990.
95. Lu LJ, Anderson KE, Grady JJ, Nagamani M. Effects of soya consumption for one month on steroid hormones in premenopausal women: implications for breast cancer risk reduction. Cancer Epidemiol Biomarkers Prev 5:63–70, 1996.
96. Cassidy A, Bingham S, Setchell KD. Biological effects of a diet of soy protein rich in isoflavones on the menstrual cycle of premenopausal women [see comments]. Am J Clin Nutr 60:333–340, 1994.
97. Petrakis NL, Barnes S, King EB, et al. Stimulatory influence of soy protein isolate on breast secretion in pre- and postmenopausal women. Cancer Epidemiol Biomarkers Prev 5:785–794, 1996.
98. McMichael-Phillips DF, Harding C, Morton M, et al. Effects of soy-protein supplementation on epithelial proliferation in the histologically normal human breast. Am J Clin Nutr 68:1431S–1435S, 1998.
99. Falck F Jr, Ricci A Jr, Wolff MS, Godbold J, Deckers P. Pesticides and polychlorinated biphenyl residues in human breast lipids and their relation to breast cancer. Arch Environ Health 47:143–146, 1992.
100. Wolff MS, Toniolo PG, Lee EW, Rivera M, Dubin N. Blood levels of organochlorine residues and risk of breast cancer [see comments]. J Natl Cancer Inst 85:648–652, 1993.
101. Arnold SF, Klotz DM, Collins BM, Vonier PM, Guillette LJ Jr, McLachlan JA. Synergistic activation of estrogen receptor with combinations of environmental chemicals [see comments, retracted by McLachlan JA, Science 277:462–463, 1997]. Science 272:1489–1492, 1996.
102. Ashby J, Lefevre PA, Odum J, Harris CA, Routledge EJ, Sumpter JP. Synergy between synthetic oestrogens [letter]? Nature 385:494, 1997.
103. Ramamoorthy K, Wang F, Chen IC, et al. Potency of combined estrogenic pesticides [letter, comment]. Science 275:405–406, 1997.
104. Collaborative Group on Hormonal Factors in Breast Cancer. Breast cancer and hormone replacement therapy: collaborative reanalysis of data from 51 epidemiological studies of 52,705 women with breast cancer and 108,411 women without breast cancer. Lancet 350:1047–1059, 1997.
105. Laya MB, Larson EB, Taplin SH, White E. Effect of estrogen replacement therapy on the specificity and sensitivity of screening mammography [see comments]. J Natl Cancer Inst 88:643–649, 1996.
106. Feigelson HS, McKean-Cowdin R, Pike MC, et al. Cytochrome P450c17alpha gene (*CYP17*) polymorphism predicts use of hormone replacement therapy. Cancer Res 59:3908–3910, 1999.
107. Ross RK, Paganini-Hill A, Wan PC, Pike MC. Effect of hormone replacement therapy on breast cancer risk: estrogen versus estrogen plus progestin. J Natl Cancer Inst 92:328–332, 2000.
108. Schairer C, Lubin J, Troisi R, Sturgeon S, Brinton L, Hoover R. Menopausal estrogen and estrogen–progestin replacement therapy and breast cancer risk. JAMA 283:485–491, 2000.
109. Collaborative Group on Hormonal Factors in Breast Cancer. Breast cancer and hormonal contraceptives: collaborative reanalysis of individual data on 53,297 women with breast cancer and 100,239 women without breast cancer from 54 epidemiological studies. Lancet 347:1713–1727, 1996.
110. Ford D, Easton DF, Peto J. Estimates of the gene frequency of *BRCA1* and its contribution to breast and ovarian cancer incidence. Am J Hum Genet 57:1457–1462, 1995.
111. Easton D, Ford D, Peto J. Inherited susceptibility to breast cancer. Cancer Surv 18:95–113, 1993.
112. Borresen AL, Andersen TI, Garber J, et al. Screening for germ line *TP53* mutations in breast cancer patients. Cancer Res 52:3234–3236, 1992.
113. Sidransky D, Tokino T, Helzlsouer K, et al. Inherited *p53* gene mutations in breast cancer. Cancer Res 52:2984–2986, 1992.
114. Ford D, Easton DF, Bishop DT, Narod SA, Goldgar DE. Risks of cancer in *BRCA1*-mutation carriers. Breast Cancer Linkage Consortium. Lancet 343:692–695, 1994.
115. Struewing JP, Hartge P, Wacholder S, et al. The risk of cancer associated with specific mutations of *BRCA1* and *BRCA2* among Ashkenazi Jews [see comments]. N Engl J Med 336:1401–1408, 1997.
116. Couch FJ, DeShano ML, Blackwood MA, et al. *BRCA1* mutations in women attending clinics that evaluate the risk of breast cancer [see comments]. N Engl J Med 336:1409–1415, 1997.
117. Stratton JF, Gayther SA, Russell P, et al. Contribution of *BRCA1* mutations to ovarian cancer. N Engl J Med 336:1125–1130, 1997.
118. Feigelson HS, Ross RK, Yu MC, Coetzee GA, Reichardt JK, Henderson BE. Genetic suscep-

tibility to cancer from exogenous and endogenous exposures. J Cell Biochem Suppl 25:15–22, 1996.
119. Feigelson H, McKean-Cowdin R, Coetzee G, Stram D, Kolonel L, Henderson B. Building a multigenic model of breast cancer susceptibility: *CYP17* and *HSD17B1* are two important candidates. Cancer Res 61:785–789, 2001.
120. Huang C-S, Chern H-D, Chang K-J, Cheng C-W, Hsu S-M, Shen C-Y. Breast cancer risk associated with genotype polymorphisms of the estrogen-metabolizing genes *CYP17*, *CYP1A1*, and *COMT*: a multigenic study on cancer susceptibility. Cancer Res 59:4870–4875, 1999.
121. Thompson P, Ambrosone C. Molecular epidemiology of genetic polymorphisms in estrogen metabolizing enzymes in human breast cancer. J Natl Cancer Inst Monogr 27:125–134, 2000.
122. Brentano ST, Picado-Leonard J, Mellon SH, Moore CC, Miller WL. Tissue-specific, cyclic adenosine 3′,5′-monophosphate–induced, and phorbol ester repressed transcription from the human P450c17 promoter in mouse cells. Mol Endocrinol 4:1972–1979, 1990.
123. Carey AH, Waterworth D, Patel K, et al. Polycystic ovaries and premature male pattern baldness are associated with one allele of the steroid metabolism gene *CYP17*. Hum Mol Genet 3:1873–1876, 1994.
124. Feigelson HS, Coetzee GA, Kolonel LN, Ross RK, Henderson BE. A polymorphism in the *CYP17* gene increases the risk of breast cancer. Cancer Res 57:1063–1065, 1997.
125. Feigelson HS, Shames LS, Pike MC, Coetzee GA, Stanczyk FZ, Henderson BE. Cytochrome P450c17alpha gene (*CYP17*) polymorphism is associated with serum estrogen and progesterone concentrations. Cancer Res 58:585–587, 1998.
126. Dunning AM, Healey CS, Pharoah PD, et al. No association between a polymorphism in the steroid metabolism gene *CYP17* and risk of breast cancer. Br J Cancer 77:2045–2047, 1998.
127. Bergman-Jungestrom M, Gentile M, Lundin A-C, Group S-EBC, Wingren S. Association between *CYP17* gene polymorphism and risk of breast cancer in young women. Int J Cancer 84:350–353, 1999.
128. Helzlsouer KJ, Huang HY, Strickland PT, et al. Association between *CYP17* polymorphisms and the development of breast cancer. Cancer Epidemiol Biomarkers Prev 7:945–949, 1998.
129. Haiman CA, Hankinson SE, Spiegelman D, et al. The relationship between a polymorphism in *CYP17* with plasma hormone levels and breast cancer. Cancer Res 59:1015–1020, 1999.
130. Kristensen VN, Haraldsen EK, Anderson KB, et al. *CYP17* and breast cancer risk: the polymorphism in the 5′ flanking area of the gene does not influence binding to Sp-1. Cancer Res 59:2825–2828, 1999.
131. Weston A, Pan CF, Bleiweiss IJ, et al. *CYP17* genotype and breast cancer risk. Cancer Epidemiol Biomarkers Prev 7:941–944, 1998.
132. Miyoshi Y, Iwao K, Ikeda N, Egawa C, Noguchi S. Genetic polymorphism in *CYP17* and breast cancer risk in japanese women. Eur J Cancer 36:2375–2379, 2000.
133. Spurdle A, Hopper J, Dite G, et al. *CYP17* promoter polymorphism and breast cancer in Australian women under age forty years. J Natl Cancer Inst 92:1674–1681, 2000.
134. Mitrunen K, Jourenkova N, Kataja V, et al. Steroid metabolism gene *CYP17* polymorphism and the development of breast cancer. Cancer Epidemiol Biomarkers Prev 9:1343–1348, 2000.
135. Jernstrom H, Vesprini D, Bradlow H, Narod S. Re: *CYP17* promoter polymorphism and breast cancer in Australian women under age forty years. J Natl Cancer Inst 93:554–555, 2001.
136. Meilahn E, DeStavola B, Allen D, et al. Do urinary oestrogen metabolites predict breast cancer? Guernsey III cohort follow-up. Br J Cancer 78:1250–1255, 1998.
137. Muti P, Bradlow H, Micheli A, et al. Estrogen metabolism and risk of breast cancer: a prospective study of the 2:16alpha-hydroxyestrone ratio in premenopausal and postmenopausal women. Epidemiology 11:635–640, 2000.
138. Mannermaa A, Peltoketo H, Winqvist R, et al. Human familial and sporadic breast cancer: analysis of the coding regions of the 17beta-hydroxysteroid dehydrogenase 2 gene (*EDH17B2*) using a single-strand conformation polymorphism assay. Hum Genet 93:319–324, 1994.
139. Normand T, Narod S, Labrie F, Simard J. Detection of polymorphisms in the estradiol 17beta-hydroxysteroid dehydrogenase 2 gene at the *EDH17B2* locus on 17q11–q21. Hum Mol Genet 2:479–483, 1993.
140. Puranen T, Poutanen M, Peltoketo H, Vihko P, Vihko R. Site-directed mutagenesis of the putative active site of human 17beta-hydroxysteroid dehydrogenase type 1. Biochem J 304:289–293, 1994.
141. Lavigne JA, Helzlsouer KJ, Huang HY, et al. An association between the allele coding for a low activity variant of catechol-O-methyltransferase and the risk for breast cancer. Cancer Res 57:5493–5497, 1997.
142. Millikan RC, Pittman GS, Tse CK, et al. Catechol-O-methyltransferase and breast cancer risk. Carcinogenesis 19:1943–1947, 1998.
143. Thompson PA, Shields PG, Freudenheim JL, et al. Genetic polymorphisms in catechol-O-methyltransferase, menopausal status, and breast cancer risk. Cancer Res 58:2107–2110, 1998.
144. Ishibe N, Hankinson SE, Colditz GA, et al. Cigarette smoking, cytochrome P450 1A1 polymorphisms, and breast cancer risk in the Nurses' Health Study. Cancer Res 58:667–671, 1998.
145. Bailey LR, Roodi N, Verrier CS, Yee CJ, Dupont

WD, Parl FF. Breast cancer and *CYP1A1*, *GSTM1*, and *GSTT1* polymorphisms: evidence of a lack of association in Caucasians and African Americans. Cancer Res 58:65–70, 1998.
146. Ambrosone CB, Freudenheim JL, Graham S, et al. Cytochrome P4501A1 and glutathione S-transferase (M1) genetic polymorphisms and postmenopausal breast cancer risk. Cancer Res 55:3483–3485, 1995.
147. Rebbeck TR, Rosvold EA, Duggan DJ, Zhang J, Buetow KH. Genetics of *CYP1A1*: coamplification of specific alleles by polymerase chain reaction and association with breast cancer. Cancer Epidemiol Biomarkers Prev 3:511–514, 1994.
148. Taioli E, Trachman J, Chen X, Toniolo P, Garte SJ. A *CYP1A1* restriction fragment length polymorphism is associated with breast cancer in African-American women. Cancer Res 55:3757–3758, 1995.

6

The Impact of Exogenous Hormone Use on Breast Cancer Risk

GILLIAN K. REEVES
EMILY BANKS
TIMOTHY J.A. KEY

The importance of endogenous hormones in the etiology of breast cancer is evident from the strong relationships between breast cancer risk and certain aspects of a woman's reproductive history, such as age at menarche, age at first birth, and age at menopause.[1] More recently, the risk of developing breast cancer among postmenopausal women has also been shown to increase substantially with increasing levels of circulating estradiol,[2] thus providing more direct evidence for the role of hormones in the development of the disease. The relationship between endogenous hormones and breast cancer risk leads naturally to the question of whether exogenous hormones, such as oral contraceptives and postmenopausal hormone therapy, might also have an effect on breast cancer risk.

Over 60 epidemiological studies worldwide have examined breast cancer risk in relation to either hormonal contraceptive use or postmenopausal hormone therapy. There have been collaborative reanalyses of data on breast cancer risk in relation to both hormonal contraceptives[3,4] and postmenopausal hormone therapy,[5] which have incorporated data from the vast majority of these original studies. In addition, a detailed review of the relationship between both these types of hormonal preparations and cancer has been compiled by the International Agency for Research on Cancer (IARC).[6] The latter review includes details of many of the individual studies and summarizes the results of the main collaborative reanalyses. Three large randomized controlled trials of postmenopausal hormone therapy have also published results with respect to breast cancer and their findings have been reviewed recently.[7] Other specific types of exogenous hormone that have been examined in relation to breast cancer risk include diethylstilbestrol, which was commonly used during pregnancy in the 1950s, and fertility-enhancing drugs, such as clomiphene citrate, which have been used since the late 1960s.[8] However, the amount of data on breast cancer risk in relation to these latter two types of hormonal preparation is substantially less than that available for hormonal contraceptives and postmenopausal hormone therapy.

The aim of this chapter is to review the evidence pertaining to the association of breast cancer risk with use of hormonal contraceptives, postmenopausal hormone therapy, and other types of exogenous hormone. In presenting the evidence on breast cancer risk in relation to hormonal contraceptives and postmenopausal hormone therapy, we have drawn heavily on the results of the collaborative reanalyses and the review of randomized trials of postmenopausal hormone therapy, and the reader is referred to the original publications for details of the individual studies. To put the findings for each class of exogenous hormones into context, each section begins with a brief outline of patterns of use and ends with a summary of the possible explanations for the findings and their implications in terms of disease incidence.

HORMONAL CONTRACEPTIVES

Background

Female sex hormones were first licensed for use as contraceptives in 1960. By far the most commonly used type of oral contraceptive has been the combined oral contraceptive, in which both an estrogen and a progestagen are given concurrently for 21 or 22 of the 28 days of the monthly cycle. Estimates suggest that over 60 million women worldwide are currently using this type of contraceptive.[9] Another class of oral contraceptives, the sequential oral contraceptives, differed from the combined type in that it contains estrogen alone for at least part of the monthly cycle. These preparations were taken off the market in the late 1970s after they were associated with an increased risk of endometrial cancer; they are not discussed further here. Other types of hormonal contraceptive contain progestagen alone; these preparations may be given orally (commonly known in this form as the minipill), via injection (depot medroxyprogesterone acetate and norethisterone oenanthate), or via subcutaneous implant. Compared to use of combined oral contraceptives, use of progestagen-only oral contraceptives has generally been low. Figures from the *Health Survey for England 1995*[10] show that, at that time, about 4% of English women aged 16–54 were currently taking progestagen-only oral contraceptives, but use of this type of contraceptive has been increasing over recent years. Use of injectable contraceptives is also becoming more common: about 12 million women worldwide are estimated to be using these contraceptives, the vast majority of which are progestagen-only preparations.[11] Given these patterns of use, it is not surprising that the majority of the available epidemiological data on breast cancer risk relates to the use of combined oral contraceptives. There is limited information about progestagen-only oral contraceptives and injectable depot-progestagens, and these types of contraceptive are considered separately.

Combined Oral Contraceptives

Almost all studies of combined oral contraceptives and breast cancer have published results on the risk of breast cancer associated with ever use of such contraceptives. Overall it appears that women who have used combined oral contraceptives are at a very slightly increased risk of breast cancer [relative risk (RR) = 1.07, 95% confidence interval (CI) 1.03–1.10], although estimates from individual studies are by no means consistent (Fig. 6.1). This is perhaps not surprising since ever use is a very crude measure of exposure and is likely to represent quite different patterns of use in different studies.

When breast cancer risk is examined in relation to various aspects of oral contraceptive use, it is clear that recency of use is the aspect most closely associated with breast cancer risk. Figure 6.2 shows the pattern of breast cancer risk in users of combined oral contraceptives compared to never-users according to time since last use. It illustrates the fact that while women are taking combined oral contraceptives they experience a small increase in the risk of breast cancer (RR = 1.24, 95% CI 1.16–1.32) but that once they cease use this excess risk declines gradually so that by about 10 years after ceasing use they are no longer at any excess risk. Once recency of use is taken into account, no other aspect of combined oral contraceptive use appears to materially alter a woman's risk of breast cancer. In particular, the magnitude of the excess RR of breast cancer among current and recent users of combined oral contraceptives is not affected by a woman's total duration of use. The slight increase in breast cancer risk associated with current or recent use of combined oral contraceptives has been found consistently across

Figure 6.1 Relative risk (RR) of breast cancer in ever-users of combined oral contraceptives compared with never-users. Each relative risk estimate and its 99% confidence interval (CI) is plotted as a black square and a line. The area of each square is proportional to the amount of statistical information available for that particular estimate. Amer Canc Soc, American Cancer Society; Canadian NBSS, Canadian National Breast Screening Study; CASH, Cancer and Steroid Hormones; FPA, Family Planning Association; RCGP, Royal College of General Practitioners; SE, standard error; WISH, Women's Interview Study of Health; WHO, World Health Organisation. (Adapted from the Collaborative Group on Hormonal Factors in Breast Cancer.[3])

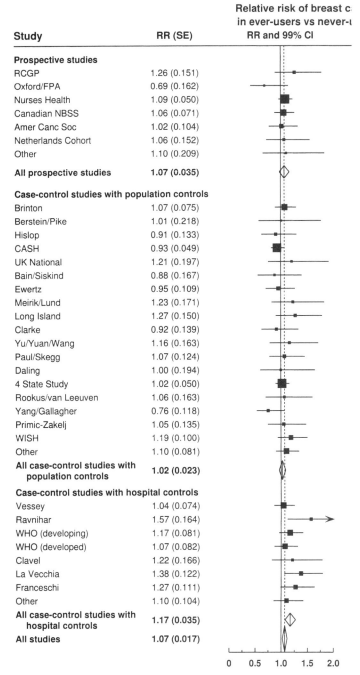

studies, even though many of the study-specific results are not individually statistically significant. This consistency is in contrast with the substantial variability in results for ever vs. never use between studies, suggesting that when the relevant index of use has been identified, the results from different studies are compatible.

Given the finding of an excess risk of breast cancer among current and recent users of oral contraceptives, it is of interest to know whether this effect is modified by certain factors such as a woman's reproductive history or her family history of breast cancer. The collaborative reanalysis found no evidence that the magnitude of the

Time since last use	RR (FSE)	RR and 99% FCI
Never	1.00 (0.014)	
Current user	1.24 (0.038)	
1-4 years	1.16 (0.032)	
5-9 years	1.07 (0.024)	
10-14 years	0.98 (0.022)	
≥15 years	1.03 (0.025)	

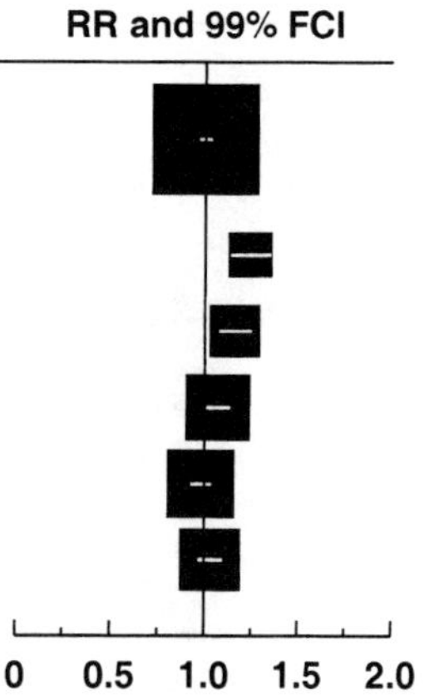

Figure 6.2 Relative risk (RR) of breast cancer according to time since last use of combined oral contraceptives. Relative risk estimates are presented in the form of floating absolute risks with corresponding floated standard errors (FSE) and floated 99% confidence intervals (FCI).[3] (Adapted from the Collaborative Group on Hormonal Factors in Breast Cancer.[3])

excess risk of breast cancer associated with recent use of combined oral contraceptives varied according to any aspect of a woman's reproductive history or a large number of other characteristics examined, including family history of breast cancer, ethnic origin, and menopausal status.

Certain types of preparation may differ in their effects on breast cancer risk, by virtue of either the specific estrogen or progestagen used or the variations in the dose in which either component is given. Information on the type or dose of estrogen or progestagen in the combined oral contraceptive that each woman had used was available for only 27 of the 54 studies in the collaboration. Based on these data, there is no clear indication that breast cancer risk varies according to the type of estrogen and progestagen used. To assess the effect of dose, the collaborative reanalysis also classified combined oral contraceptives into three broad categories of estrogen dose (<50, 50, >50 μg), which to a large extent also reflected progestagen dose. Analyses of the effects of dose within categories of time since last use of combined oral contraceptives did not, however, provide any evidence for an increase in risk with increasing dose within any category of time since last use (Fig. 6.3).

Time since last use and hormonal dose	RR (FSE)	RR and 99% FCI
NEVER-USER	1.00 (0.020)	
Last use <5 years ago		
Low dose	1.16 (0.048)	
Medium dose	1.26 (0.044)	
High dose	1.27 (0.074)	
Last use 5-9 years ago		
Low dose	0.99 (0.065)	
Medium dose	1.10 (0.045)	
High dose	1.02 (0.060)	
Last use ≥10 years ago		
Low dose	1.05 (0.070)	
Medium dose	0.98 (0.038)	
High dose	0.91 (0.038)	

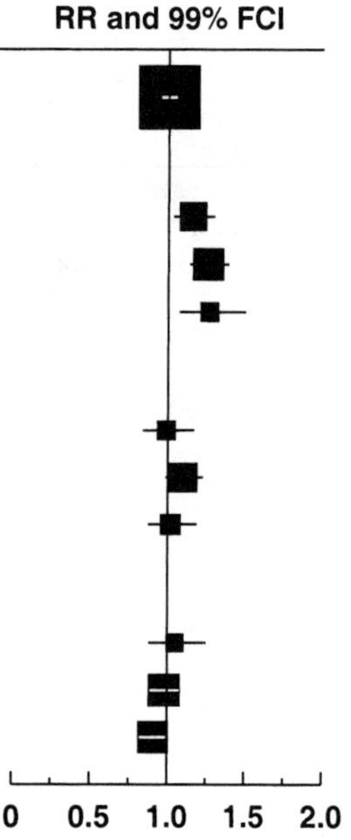

Figure 6.3 Relative risk (RR) of breast cancer according to time since last use of combined oral contraceptives and hormonal dose of the preparation mostly used. FSE, floated standard error; FCI, floated confidence interval. (Adapted from the Collaborative Group on Hormonal Factors in Breast Cancer.[3])

While it is clear that breast cancer risk is increased in current and recent users of combined oral contraceptives but not in past users, the precise pattern of risk in relation to recency of use is different for localized cancers compared to those which have spread beyond the breast. These relationships are illustrated in Figure 6.4, which is based on data from 24 of the 54 studies in the collaborative reanalysis. It can be seen from Figure 6.4 that the overall excess risk of breast cancer among current and recent users is largely due to an excess risk of localized disease. Among past users, although there is no overall excess risk of breast cancer, this masks a slight increase in the risk of localized disease and a slight deficit of advanced disease.

Since the Collaborative Group's report, the National Institute of Child Health and Human Development Women's Contraceptive and Reproductive Experiences (CARE) study[12] has presented their findings on breast cancer and oral contraceptives. This multicentre, population-based case-control study, which included over 4500 cases and a similar number of controls, found no significant increase in the risk of breast cancer among current users (RR = 1.0, 95% CI 0.8–1.3) or past users (RR = 0.9, 95% CI 0.8–1.0), when compared with never-users of oral contraceptives. However, only about 5% of women in this study were current users and the findings for this group of users do not differ significantly from those of the Collaborative Group. In accordance with the Collaborative Group, the study did not find any consistent increase in risk with increase in duration of use or with higher doses of estrogen.

Contraceptives Containing Only Progestagens

Few individual studies have examined the risk of breast cancer with respect specifically to use of contraceptives containing only progestagens, and in the collaborative reanalysis use of the two main types of such contraceptive, namely, progestagen-only oral contraceptives and injectable progestagens (depot-progestagens), was limited. Use of progestagen-only oral contraceptives was reported by only 1253 women (0.8%), mainly from the United Kingdom, Scandinavia, and New Zealand, and use of depot-progestagens was reported by only 2274 women (1.5%), mainly from New Zealand and Thailand.

When all studies with relevant data on use of progestagen-only oral contraceptives are combined (Fig. 6.5), there is a slight increase in the risk of breast cancer associated with ever use (RR = 1.12, 95% CI 1.00–1.26). When the data on depot-progestagens are similarly combined, there is little evidence of an increased risk of breast cancer in ever-users (Fig. 6.6) with an overall RR of 1.05 (95% CI 0.90–1.23).

In general, analyses in relation to specific aspects of use of either progestagen-only oral contraceptives or depot-progestagens have not revealed any relationship between duration of use and breast cancer risk. Although for both types of contraceptive there is some evidence that the risk of breast cancer is increased in women who began or ceased use recently, decreasing with time since first and last use (Figs. 6.7, 6.8), the small numbers of women involved make it difficult to distinguish between the effects of time

	(a) Cancers localised to breast		(b) Cancers spread beyond breast	
Time since last use	RR (FSE)	RR and 99% FCI	RR (FSE)	RR and 99% FCI
Never-user	1.00 (0.025)		1.00 (0.028)	
<5 years	1.21 (0.043)		1.09 (0.046)	
5–9 years	1.07 (0.036)		0.96 (0.039)	
≥10 years	1.04 (0.026)		0.93 (0.028)	

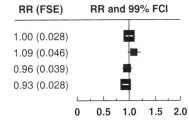

Figure 6.4 Relative risk (RR) of (a) localized disease and (b) spread disease according to time since last use of combined oral contraceptives. FSE, floated standard error; FCI, floated confidence interval. (Adapted from the Collaborative Group on Hormonal Factors in Breast Cancer.[3])

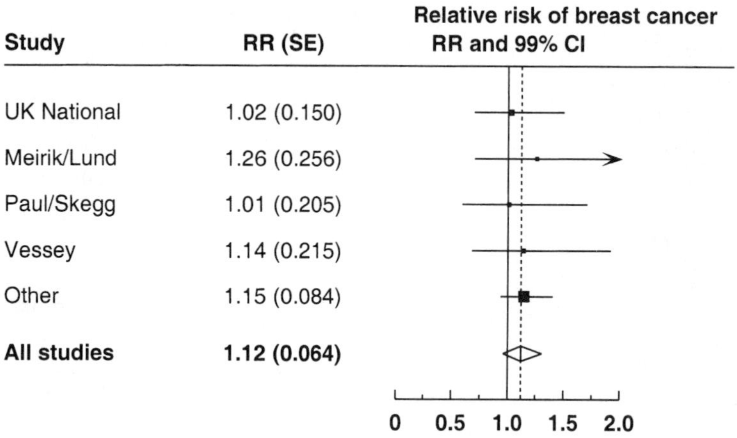

Study	RR (SE)	Relative risk of breast cancer RR and 99% CI
UK National	1.02 (0.150)	
Meirik/Lund	1.26 (0.256)	
Paul/Skegg	1.01 (0.205)	
Vessey	1.14 (0.215)	
Other	1.15 (0.084)	
All studies	**1.12 (0.064)**	

Figure 6.5 Relative risk (RR) of breast cancer in ever-users compared to never-users of progestagen-only oral contraceptives. SE, standard error; CI, confidence interval. (Adapted from the Collaborative Group on Hormonal Factors in Breast Cancer.[3])

since first and last use. Use of progestagen-only contraceptives is likely to be related to use of combined oral contraceptives. However, when analyses of breast cancer risk in relation to timing of use of progestagen-only contraceptives are restricted to women who had not used combined oral contraceptives within the last 5 years, the patterns of risk remain the same.[4]

Implications

Current evidence suggests that, for combined oral contraceptives, recency of use is the key determinant of breast cancer risk. While women are taking combined oral contraceptives there is a small but definite increase in the risk of having breast cancer diagnosed relative to never-users; but once use ceases, this increase declines, and by about 10 years after ceasing use past users are no longer at increased risk. There is considerably less information about the effects of progestagen-only oral contraceptives and depot-progestagens, but the limited available data suggest that their effects might be similar to those of combined preparations.

The explanation for these findings is not yet clear, but the fact that tumors in users of combined oral contraceptives are more likely to be localized to the breast than those in never-users raises the possibility that women who have taken combined oral contraceptives are more likely to have their tumors detected at an earlier stage. The collaborative reanalysis examined the use of screening mammograms among controls for all studies with relevant information but found no evidence that recent users were more likely to report having had a mammogram than never-users. Alternatively, the findings may be due to the biological effects of hormonal contraceptives or to a combination of factors.

Whatever the real explanation for the findings, it is clear that because breast cancer is rare among women in their 20s and 30s compared to women in their 40s, the excess number of breast

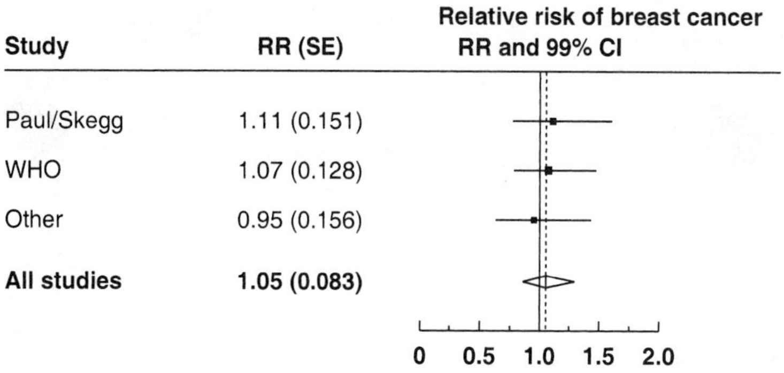

Study	RR (SE)	Relative risk of breast cancer RR and 99% CI
Paul/Skegg	1.11 (0.151)	
WHO	1.07 (0.128)	
Other	0.95 (0.156)	
All studies	**1.05 (0.083)**	

Figure 6.6 Relative risk (RR) of breast cancer in ever-users compared to never-users of depot-progestagen contraceptives. SE, standard error; CI, confidence interval; WHO, World Health Organization. (Adapted from the Collaborative Group on Hormonal Factors in Breast Cancer.[4])

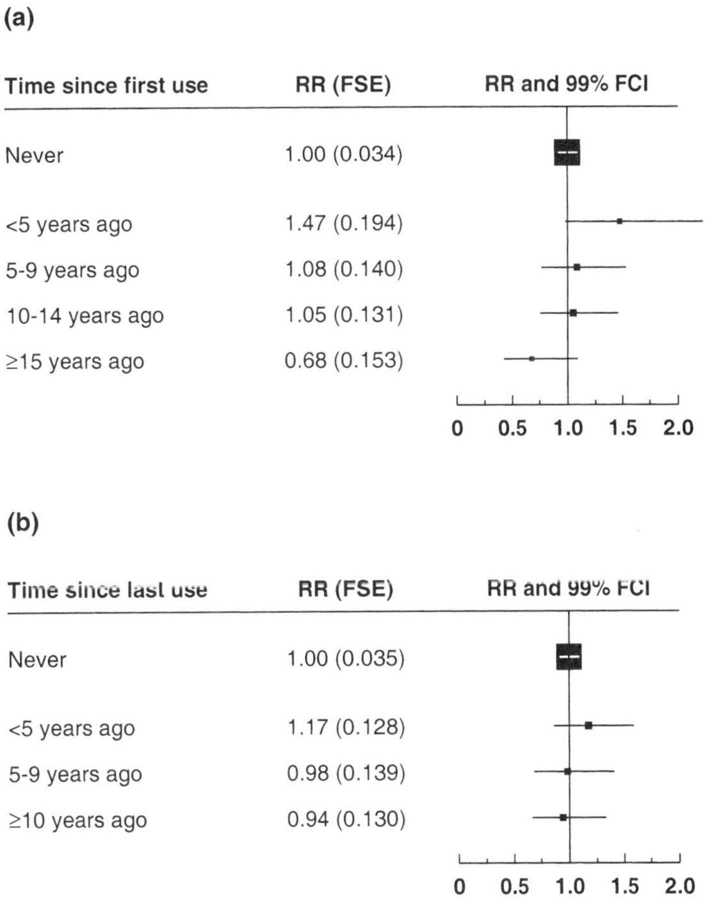

Figure 6.7 Relative risk (RR) of breast cancer in ever-users of progestagen-only oral contraceptives according to (a) time since first use and (b) time since last use. FSE, floated standard error; FCI, floated confidence interval. (Adapted from the Collaborative Group on Hormonal Factors in Breast Cancer.[4])

(a)

Time since first use	RR (FSE)	RR and 99% FCI
Never	1.00 (0.034)	
<5 years ago	1.47 (0.194)	
5-9 years ago	1.08 (0.140)	
10-14 years ago	1.05 (0.131)	
≥15 years ago	0.68 (0.153)	

(b)

Time since last use	RR (FSE)	RR and 99% FCI
Never	1.00 (0.035)	
<5 years ago	1.17 (0.128)	
5-9 years ago	0.98 (0.139)	
≥10 years ago	0.94 (0.130)	

cancers associated with recent use of combined oral contraceptives in young women (<40) is small in relation to a woman's cumulative risk of developing the disease by age 50. In addition, this excess is largely due to localized disease. Figure 6.9 illustrates the excess number of breast cancers that might be expected to be associated with use of combined oral contraceptives from age 25 to 29 based on the results of the collaborative reanalysis. It shows that by 10 years after ceasing use there is an estimated excess of five breast cancers per 10,000 users, but by 20 years after ceasing use the excess is no longer significant. Therefore, although it is expected that there would be a small, short-term increase in the number of breast cancers diagnosed in women who had used combined oral contraceptives in their 20s compared to never-users, there is no significant excess among these women in the long term. Since there is little information on women who had ceased use 20 or more years previously, estimates of cumulative risk beyond the age of 50 are unreliable.

RISK OF BREAST CANCER IN RELATION TO POSTMENOPAUSAL HORMONE THERAPY

Background

Postmenopausal hormone therapy is taken here to refer to the use of preparations containing estrogen(s), with or without progestagen(s), around the time of the menopause, primarily for the relief of menopausal symptoms. Use of such therapy has increased dramatically in the last decade such that by 1998 around one-third of women aged 50–64 in England were estimated to be using it.[13] In addition, recent estimates suggest that between 20% and 30% of women aged 45–64 in many developed countries, other

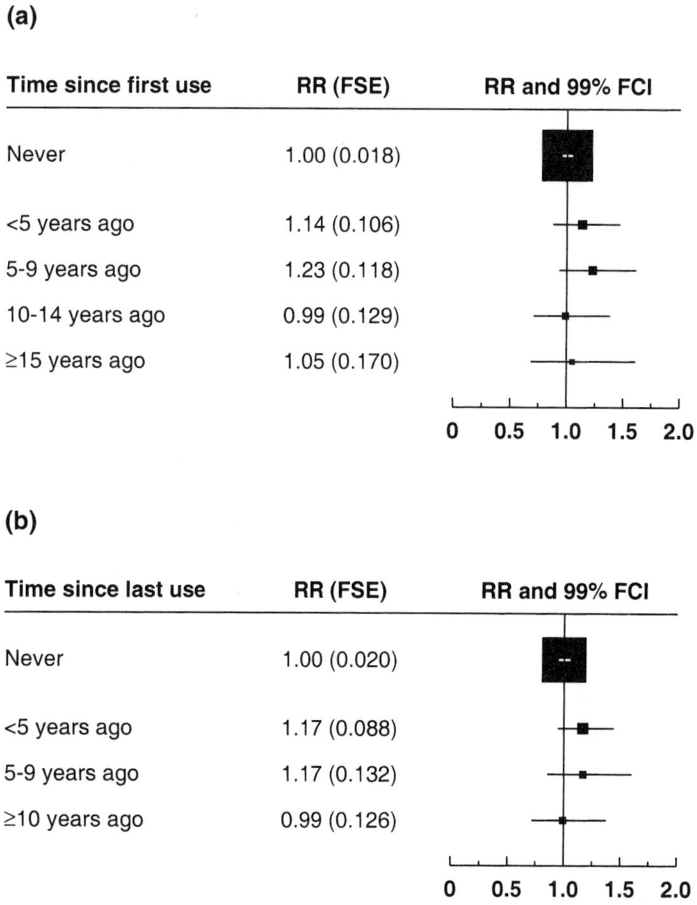

(a)

Time since first use	RR (FSE)	RR and 99% FCI
Never	1.00 (0.018)	
<5 years ago	1.14 (0.106)	
5-9 years ago	1.23 (0.118)	
10-14 years ago	0.99 (0.129)	
≥15 years ago	1.05 (0.170)	
		0 0.5 1.0 1.5 2.0

(b)

Time since last use	RR (FSE)	RR and 99% FCI
Never	1.00 (0.020)	
<5 years ago	1.17 (0.088)	
5-9 years ago	1.17 (0.132)	
≥10 years ago	0.99 (0.126)	
		0 0.5 1.0 1.5 2.0

Figure 6.8 Relative risk (RR) of breast cancer in ever-users of depot-progestagen contraceptives according to (a) time since first use and (b) time since last use. FSE, floated standard error; FCI, floated confidence interval. (Adapted from the Collaborative Group on Hormonal Factors in Breast Cancer.[4])

than the former socialist economies, are now using postmenopausal hormone therapy.[6] When it was first used in the 1930s, postmenopausal hormone therapy was administered as estrogen alone. In 1975, reports were published showing a marked elevation in the risk of endometrial cancer in women using such preparations, and it was subsequently found that adding a progestagen to estrogen attenuated this risk. In Europe, combined estrogen and progestagen quickly became the standard postmenopausal hormone therapy for women with a uterus. In the United States, while use of combined preparations gradually increased, estrogen alone continued to be prescribed to women with a uterus, often accompanied by endometrial monitoring. A greater shift toward use of combined estrogen and progestogen in the United States occurred in the mid-1990s, when randomized controlled trial evidence demonstrated the high risk of endometrial abnormalities with the use of estrogen alone.

Since use is most common around the time of the menopause and usually continues for a few years, the time of peak use is when a women is in her early 50s. Postmenopausal hormone therapy use is consistently more common in women who have had a surgical rather than a natural menopause. The relationship between hormone use and other factors varies from country to country; in general, users have a lower body mass index and a more favorable cardiovascular risk profile than nonusers.

This section summarizes the evidence for the relationship between postmenopausal hormone therapy and breast cancer. Most of the available observational data on such therapy relates to use of estrogen-only therapy. Even in studies where a small proportion of women would be expected to have used combination therapy, results do not

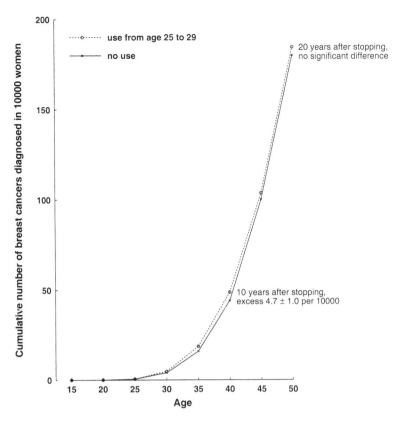

Figure 6.9 Comparison of the cumulative number of breast cancers expected among never-users of combined oral contraceptives and women who had used them between ages 25 and 29. Estimated numbers for 10,000 women based on breast cancer rates typical of Europe and North America.[4] (Adapted from the Collaborative Group on Hormonal Factors in Breast Cancer.[3])

always distinguish between the two types. The following section, therefore, describes breast cancer risk in relation to use of any postmenopausal hormone therapy, although this can largely be taken to represent estrogen-only therapy. The available evidence on the risk of breast cancer associated with specific types of postmenopausal hormone therapy is dealt with below (see Comparison of Risks According to Type of Postmenopausal Hormone Therapy Used).

Use of Any Type of Postmenopausal Hormone Therapy

Most studies of postmenopausal hormone therapy and breast cancer have found a small increase in the risk of breast cancer associated with ever use (Fig. 6.10), with an overall average risk of breast cancer in ever-users compared to never-users of 1.14 (95% CI 1.09–1.49). Further examination of breast cancer risk according to the pattern of use of postmenopausal hormone therapy shows that this excess risk is largely confined to current and recent users and among these women the risk increases with increasing duration of use (Fig. 6.11). Indeed, in current users and those who ceased use less than 5 years previously, the RR of breast cancer is estimated to increase by 2.3% (95% CI 1.1%–3.6%) for each year of use. In contrast, women who have ceased use 5 or more years previously do not appear to be at any overall increase in risk, nor does their risk increase with increasing duration of use of postmenopausal hormone therapy.

This pattern of increasing breast cancer risk with increasing duration of use of postmenopausal hormone therapy among current and recent users has been found consistently across studies and within most subgroups of women, such as those with a natural menopause

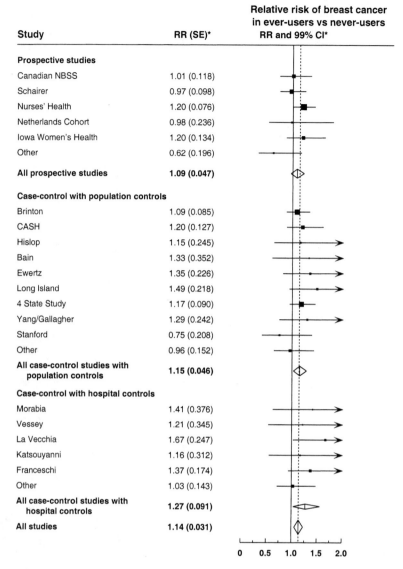

Figure 6.10 Relative risk (RR) of breast cancer in ever-users compared to never users of postmenopausal hormone therapy. Canadian NBSS, Canadian National Breast Screening Study; CASH, Cancer and Steroid Hormones; CI, confidence interval; SE, standard error. (Adapted from the Collaborative Group on Hormonal Factors in Breast Cancer.[5])

and those with a surgical menopause, women with a family history of breast cancer and those without a family history, etc. In fact, the only factor that has been shown to modify the effect of postmenopausal hormone therapy is body mass index, the effects of long durations of current or recent use being significantly more pronounced in women of low body mass index than in women of relatively high body mass index.

In 1998, a large Swedish population based case-control study published new findings on breast cancer and post-menopausal hormone therapy,[14] the main results being based on over 3300 women aged 50–74 with invasive breast cancer and over 3400 women of a similar age without breast cancer. In this study, breast cancer risk was positively associated with duration of use of any type of postmenopausal hormone therapy (excluding low, potency estrogens such as oral or vaginal estriol) with an RR of 2.43 (95%

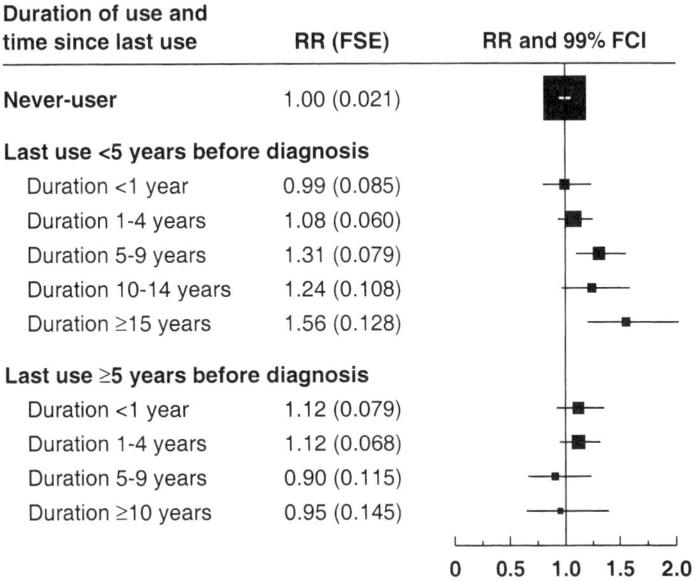

Figure 6.11 Relative risk (RR) of breast cancer by duration of use within categories of time since last use of postmenopausal hormone therapy. FSE, floated standard error; FCI, floated confidence interval. (Adapted from the Collaborative Group on Hormonal Factors in Breast Cancer.[5])

Duration of use and time since last use	RR (FSE)
Never-user	1.00 (0.021)
Last use <5 years before diagnosis	
Duration <1 year	0.99 (0.085)
Duration 1-4 years	1.08 (0.060)
Duration 5-9 years	1.31 (0.079)
Duration 10-14 years	1.24 (0.108)
Duration ≥15 years	1.56 (0.128)
Last use ≥5 years before diagnosis	
Duration <1 year	1.12 (0.079)
Duration 1-4 years	1.12 (0.068)
Duration 5-9 years	0.90 (0.115)
Duration ≥10 years	0.95 (0.145)

CI 1.79–3.3) among women who had used either type of preparation for 10 or more years compared to never-users. When the effect of duration of use was examined according to recency of use, a similar trend in risk with duration of use was observed both among women who were current users or who had ceased use less than 5 years previously (RR per year of use = 1.03, 95% CI 1.00–1.06) and in women who had ceased use 5 or more years previously (RR = 1.04, 95% CI = 0.97–1.11), although the trend of increasing risk with increasing duration of use observed among past users was not statistically significant. The Swedish study also found that use of postmenopausal hormone therapy was more strongly associated with breast cancer risk among relatively lean women than among relatively obese women. The results of the Swedish study can, therefore, be viewed as being broadly compatible with those of the collaborative reanalysis, although the magnitudes of the excess risks were somewhat greater than those found in the collaborative reanalysis.

To date, three large randomized controlled trials of postmenopausal hormone therapy have published results on breast cancer and their findings have been summarized in a recent review.[7] Two of these studies, The Heart and Estrogen/progestagen Replacement Study (HERS) and the Women's Estrogen for Stroke Trial (WEST), recruited women with previous cardiovascular disease while the Women's Health Initiative (WHI) recruited healthy women. Both HERS and WHI used a combined therapy for the treated arm of the their trials while women in the WEST trial were treated with estrogen alone. Based on a total of 205 cases of breast cancer, the overall relative risk of breast cancer associated with an average follow-up period of almost 5 years in these studies is 1.27 (1.03–1.56). This estimate is, however, based on an "intention to treat" analysis and may be an underestimate of the true effect due to non-compliance in the latter stages of the trials.

As with hormonal contraceptives, there is some evidence to suggest that the pattern of risk in relation to use of postmenopausal hormone therapy is different for localized cancers compared to cancers that have spread beyond the breast. Figure 6.12 shows the RR of both localized cancer and cancer that has spread beyond the breast, according to time since last use of postmenopausal hormone therapy, based on results from the collaborative reanalysis. It can be seen from Figure 6.12 that among current or recent users the overall excess risk of breast cancer appears to be due to an excess risk of localized disease. The information on the RR of breast cancer according to tumor spread in past users is limited, but there does not appear to be

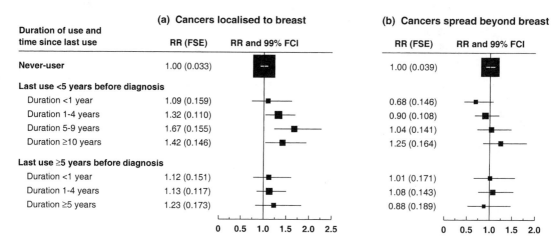

Figure 6.12 Relative risk (RR) of (a) localized disease and (b) spread disease by duration of use within categories of time since last use of postmenopausal hormone therapy. FSE, floated standard error; FCI, floated confidence interval. (Adapted from the Collaborative Group on Hormonal Factors in Breast Cancer.[5])

any increase in risk among such users either for localized or for spread disease.

Comparison of Risks According to Type of Postmenopausal Hormone Therapy Used

In the collaborative reanalysis, information on the constituents of preparations used most was available for just under one-half of the eligible women. The vast majority of these women had used estrogen-only therapy (80%) with only 12% reporting having mostly used preparations containing both estrogen and progestagen. When the effects of recency and duration of use were examined separately within these data for preparations containing estrogen alone, preparations containing progestagen (with or without an estrogen) and preparations containing estrogen in combination with some other compounds, there were no marked differences across these three categories of hormone therapy (Table 6.1). Among current and recent users of 5 or more years duration, use of preparations containing estrogen in combination with compounds other than progestagen was associated with the highest risk of breast cancer but as this group rep-

Table 6.1 Relative Risk (RR) of Breast Cancer by Time Since Last Use, Duration of Use, and Type and Dose of Postmenopausal Hormone Therapy Mainly Used

Type and Dose of Postmenopausal Hormone Therapy	Current Use or Last Use 1–4 Years Before Diagnosis		Last Use ≥5 years Before Diagnosis
	Duration <5 years	Duration ≥5 years	
	RR (SE)	RR (SE)	RR (SE)
Estrogen alone			
Total	0.99 (0.08)	1.34 (0.09)	1.12 (0.11)
Conjugated			
≤0.625 mg	0.77 (0.13)	1.64 (0.25)	1.45 (0.22)
≥1.25 mg	0.94 (0.17)	1.42 (0.16)	0.90 (0.24)
Unknown dose	1.18 (0.18)	1.18 (0.14)	0.82 (0.19)
Other estrogen	1.15 (0.17)	1.26 (0.21)	1.22 (0.21)
Estrogen and progestagen or progestagen alone	1.15 (0.19)	1.53 (0.33)	1.30 (0.46)
Estrogen and/or other	0.88 (0.26)	2.57 (0.38)	0.99 (0.32)

SE, standard error.
Source: Adapted from the Collaborative Group on Hormonal Factors in Breast Cancer.[5]

resents a somewhat heterogeneous group of compounds this finding is difficult to interpret. Among this same group of users, combined estrogen-progestagen therapy was also associated with a somewhat greater risk of breast cancer than was estrogen-only therapy, but this difference was not statistically significant. There was no indication of any important differences in risk according to type and dose of estrogen used in preparations containing estrogen alone (Table 6.1).

Information on the type of preparation used was available for almost all women in the Swedish study.[14] In this study, almost twice as many women had ever used combined therapy than had ever used estrogen-only therapy. In addition, about 20% of all women had used so-called low-potency estrogens (orally administered estriol or vaginal treatment), but exposure to these types of preparation was generally ignored in the main analyses. The risk of breast cancer was shown to increase with increasing duration of use of both estrogen-only therapy (RR per year of use = 1.03, 95% CI 0.98–1.08) and combined therapy (RR per year of use = 1.07, 95% CI 1.02–1.11); the magnitude of these trends did not differ significantly. Ever use of preparations containing only progestagens was also associated with an increase in breast cancer risk (RR = 1.59, 95% CI 1.05–2.41), but since only 100 women had used these types of preparations exclusively, this relationship could not be studied further. More detailed analyses revealed that the increase in risk with increasing duration of use of combined therapies was mainly evident for compounds containing testosterone-derived progestins and that the trend associated with use of such preparations was greater when progestins were administered continuously as opposed to cyclically. However, these subgroup analyses are necessarily based on relatively small numbers and, therefore, difficult to interpret.

Since the collaborative reanalysis, three studies have published new results on breast cancer risk according to type of postmenopausal hormone therapy used. The first, which is an extended follow-up of a large cohort within the Breast Cancer Detection Demonstration Project[15] (a U.S. breast cancer screening program), with over 2000 cases of breast cancer, has shown that the risk of breast cancer is increased both in recent users of estrogen-only therapy (RR = 1.2, 95% CI 1.0–1.4) and in recent users of combined therapy (RR = 1.4, 95% CI 1.1–1.8). However, when the trend in breast cancer risk with increasing duration of use was estimated separately for users of each type of therapy, the percentage increase in risk associated with each additional year of use was greater for users of combined therapy (RR per year of use = 1.08, 95% CI 2%–16%) than for users of estrogen-only therapy (RR per year of use = 1.01, 95% CI 1.00–1.03). The second, which was a large population-based case-control study containing information on over 1800 breast cancer cases and 1600 controls,[16] also found that the increase in risk of breast cancer with increasing duration of use was significantly greater for combined therapy (RR per 5 years of use = 1.24, 95% CI 1.07–1.45) than for estrogen-only therapy (RR per 5 years of use = 1.06, 95% CI 0.97–1.15). The third study, which was a U.S. population based case-control study of 5298 post-menopausal women with breast cancer and 5571 women without breast cancer,[17] reported that estrogen-progestin use that was both recent and of greater than 5 years duration was associated with a bigger increase in risk (RR = 1.57, 95% CI 1.15–2.14) than similar use of estrogen alone (RR = 1.39, 95% CI 1.17–1.65).

For the randomized trials, almost all the evidence on postmenopausal hormone therapy relates to use of a single combined regime, namely equine estrogen and medroxyprogesterone. Thus although these results can usefully be compared with those from observational data on estrogen only therapy and other combined preparations, there is at present little randomized data on any other type of preparation.

Implications

The fact that results of randomized controlled trials have confirmed the findings from observational studies on the association between breast cancer and postmenopausal hormone therapy means that it is now reasonable to conclude that the associations noted in observational studies are true effects of such therapy and not due to bias or confounding.

Since postmenopausal hormone therapy is generally intended to supplement falling levels of circulating ovarian hormones at the menopause, it might reasonably be expected that while women are using such therapy the beneficial effects of the menopause on breast cancer risk will be delayed. To a certain extent, this is borne out by the data in that current or recent use of postmenopausal hormone therapy increases the risk of breast cancer by about 2.3% per year of use, which is comparable to the 2.8% increase in risk estimated to apply for each year later that menopause occurs.[5] It is perhaps less clear what this analogy with the effects of delayed menopause should predict about the risk of breast cancer in past users, but the majority of the data suggest that the excess risk of breast cancer associated with long duration of use declines with increasing time after cessation of use.

The epidemiological data are also consistent in showing that the effect of postmenopausal hormone therapy is more marked in women who are not overweight compared to overweight women. It has long been known that breast cancer risk in postmenopausal women increases with increasing body mass index and that increasing body mass index causes an increase in endogenous estradiol and a decrease in sex hormone–binding globulin, resulting in a substantial increase in bioavailable estradiol.[18] A possible explanation for the modifying effect of body mass index, therefore, is that in women who are not overweight postmenopausal hormone therapy may cause a marked increase in serum estrogen levels and, therefore, in breast cancer risk, whereas in overweight women, who already have comparatively high levels of endogenous estradiol, the effect of additional increases in estrogen levels on breast cancer risk is relatively less marked.

Until relatively recently, most postmenopausal hormone therapy use was in the form of estrogen alone, and observational studies have had limited power to examine the effects of combined estrogen–progestagen preparations. However, studies that have examined the risk of breast cancer according to different types of therapy have consistently shown a greater increase in the risk of breast cancer associated with use of combined therapy than with use of estrogen-only preparations. By contrast, data from randomized controlled trials mostly pertain to a single type of combined therapy and the relatively larger risks that have been reported by such trials may reflect the use of combined rather than estrogen alone therapy.

Because the background risk of breast cancer is much higher among women in their 40s and 50s than among women in their 20s and 30s, the number of diagnoses of breast cancer associated with use of postmenopausal hormone therapy is considerably greater than that associated with use of hormonal contraceptives. Figure 6.13, based on the results of the collaborative group, illustrates the numbers of breast cancer diagnoses likely to be associated with different patterns of postmenopausal hormone therapy use. According to average incidence rates in North America and the United Kingdom, one might expect about 63 cancers to be diagnosed among 1000 never-users of postmenopausal therapy by the age of 70. By comparison, the number of cancers that would be expected by the age of 70 among 1000 women who had used postmenopausal hormone therapy for 10 years is 69. Thus, 10 years of postmenopausal hormone therapy use is associated with an estimated cumulative excess of six cancers per 1000 women. Since these estimates are based on the overall findings of the collaborative reanalysis, they can effectively be taken to represent the cumulative risk associated with estrogen-only therapy. If the higher risk of breast cancer among users of combined therapy that has been observed in recent studies is substantiated, then clearly the estimates of cumulative risk associated with this type of therapy would be greater. At the time at which the trials were set up, the combined preparation used in both the HERS and the WHI trial was the most widely used therapy in the United States. Estimates from the WHI suggest than among women in their 60s, use of such combined therapy over a 5-year period is associated with about four extra breast cancers per 1000 women.

RISK OF BREAST CANCER IN RELATION TO OTHER EXOGENOUS HORMONES

Diethylstilbestrol

Diethylstilbestrol (DES) is a potent synthetic estrogen. Between the late 1940s and the early

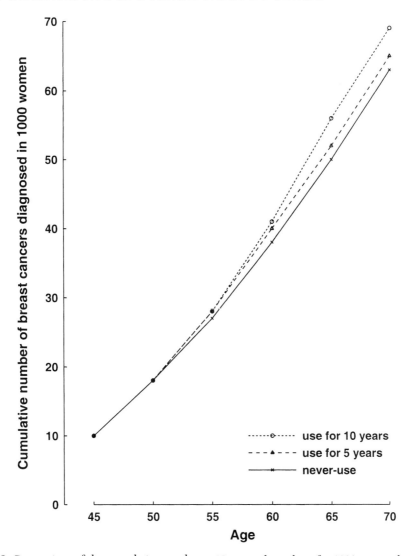

Figure 6.13 Comparison of the cumulative number of breast cancers expected among never-users of postmenopausal hormone therapy and women who had used it for durations of 5 and 10 years from age 50. Estimated numbers for 1000 women based on breast cancer rates typical of Europe and North America. (Adapted from the Collaborative Group on Hormonal Factors in Breast Cancer.[5])

1960s, DES was widely prescribed to pregnant women in an attempt to reduce the risk of miscarriages, although by the early 1950s clinical trials had already shown that it was not effective for this purpose. Up to 2.5 million women in the United States alone may have used DES during pregnancy. By the early 1970s, several epidemiological studies had shown that DES caused an increased risk of clear cell adenocarcinomas of the vagina and cervix in women who had been exposed in utero, and DES use in pregnancy was banned in the United States in 1971, although use in Europe continued until 1978.[19]

Several randomized controlled trials and cohort studies have been analyzed to investigate whether women who used DES while pregnant have an increased risk of subsequently developing breast cancer. Overall, the results suggest that breast cancer risk is increased by about 30%.[19–22] The total number of exposed cases studied is relatively small, and the details on doses are limited; therefore, it has not been possible to establish the relationships of risk with time since use of DES or dose.

The effect of DES on breast cancer risk is reasonably firmly established, but this exposure has several differences from exposures to other exogenous estrogens. For example, DES was given at high doses for a short period during pregnancy, whereas hormonal contraceptives and postmenopausal hormone therapy are given at low doses for a long period in nonpregnant women. The biological effects of DES may also differ from those of other exogenous estrogens because, in addition to its estrogenic effects, DES can have other effects, including causing sister chromatid exchange, unscheduled DNA synthesis, chromosomal aberration, disruption of the mitotic spindle, and aneuploidy.[23]

Fertility-Enhancing Drugs

Women who undergo treatment for infertility can be exposed to a variety of hormonally active drugs, including clomiphene citrate, human menopausal gonadotrophin, and gonadotrophin-releasing hormone. Studies examining the relationship between use of fertility drugs and breast cancer risk have been hampered by small numbers of women and the inability to adequately control for all important confounding factors. Bearing this in mind, their findings have been generally reassuring, showing no significant increase in the risk of breast cancer in infertile women treated with fertility drugs compared to infertile women who did not receive such treatment.[24–27] However, a recent study reported a significant, transient excess of breast cancer in the 12 months following ovulation stimulation for in vitro fertilization compared to the general female population,[25] and there remains considerable uncertainty about the effect of fertility drugs on breast cancer risk.

CONCLUSIONS

It is clear from this review that hormonal contraceptives, postmenopausal hormone therapy, and DES are associated with an increase in breast cancer risk. In contrast, results with respect to fertility-enhancing drugs, although broadly reassuring, must still be viewed as equivocal. In terms of future research, the most urgent questions concern hormonal contraceptives and postmenopausal hormone therapy since these preparations are currently used by substantial numbers of women throughout the world.

For the most part, the effects of hormonal contraceptives and postmenopausal hormone therapy on breast cancer risk appear to be remarkably consistent across groups of women identified by various characteristics and with different background risks of breast cancer. However, a large number of genes are now known to be involved in hormone metabolism and this raises the question of whether some women are more susceptible than others to the effects of exogenous hormones.[28] Existing studies of breast cancer risk in relation to relevant candidate genes are unlikely to have the power to examine such gene–environment interactions and much larger studies will be required in the future to address these issues.

As far as the effects of hormonal contraceptives on breast cancer risk are concerned, there are several areas where more data are needed. Firstly, since use of contraceptives containing only progestagens is rising rapidly, more data are needed to establish whether their effects on breast cancer are really similar to those of combined preparations. Secondly, better data are required on whether women who have used combined oral contraceptives are more likely to have their cancers detected earlier. Finally, data on women whose use ceased more than 20 years ago will be available within the next few years, and this will allow the long-term effects of oral contraceptives on breast cancer risk to be studied in more detail than has previously been possible.

It is now clear that the risk of breast cancer is increased among long-term users of postmenopausal hormone therapy. Although the vast majority of the epidemiological evidence for this association is based on women who had taken preparations containing estrogen alone, more recent studies have consistently found a greater risk of breast cancer in users of combined preparations than in women using estrogen-only preparations. Moreover, results of randomized controlled trials that have predominantly used a single type of combined therapy show a clear increase in the risk of breast cancer among users compared to a placebo group. Given the trend of increasing use of combination therapy, these observations are clearly disconcerting, particularly for women with an intact uterus, for whom estrogen-only therapy is normally contraindicated. There is, therefore, an urgent need for more data on women who have used combined preparations, to provide more accurate and detailed information on the risk of breast cancer associated with such therapy.

REFERENCES

1. Kelsey JL, Gammon MD, John EM. Reproductive factors and breast cancer. Epidemiol Rev 15:36–47, 1993.
2. The Endogenous Hormones and Breast Cancer Collaborative Group. Endogenous sex hormones and breast cancer in postmenopausal women: reanalysis of nine prospective studies. JNCI 94:606–616, 2002.
3. Collaborative Group on Hormonal Factors in Breast Cancer. Breast cancer and hormonal contraceptives: collaborative re-analysis of individual data on 53,297 women with breast cancer and 100,239 women without breast cancer from 54 epidemiological studies. Lancet 347:1713–1724, 1996.
4. Collaborative Group on Hormonal Factors in Breast Cancer. Breast cancer and hormonal contraceptives: further results. Contraception 54(Suppl):1S–106S, 1996.
5. Collaborative Group on Hormonal Factors in Breast Cancer. Breast cancer and hormone replacement therapy: collaborative re-analysis of data from 51 epidemiological studies of 52,705 women with breast cancer and 108,411 women without breast cancer. Lancet 350:1047–1059, 1997.
6. International Agency for Research on Cancer. Hormonal Contraception and Postmenopausal Hormonal Therapy. IARC Monogr Eval Carcinog Risks Hum 72, 1999.
7. Beral V, Banks E, Reeves GK. Evidence from randomised trials on the long-term effects of hormone replacement therapy. Lancet 360:942–944, 2002.
8. Cohen J, Forman R, Harlap S, Johannisson E, Lunenfeld B, de Mouzon J, Pepperell R, Tarlatzis B, Templeton A. IFFS expert group report on the Whittemore study related to the risk of ovarian cancer associated with the use of infertility agents. Hum Reprod 8:996–999, 1993.
9. Wharton C, Blackburn R. Lower dose pills. Popul Rep A7:1–31, 1989.
10. Prescott-Clarke P, Primatesta P (eds) Health Survey for England 1995, Volume I. Findings; Volume II: Survey Methodology and Documentation. London: The Stationary Office, 1997.
11. Lande RE. A new era for injectables. Popul Rep K5:1–31, 1995.
12. Marchbanks PA, McDonald JA, Wilson HG, et al. Oral contraceptives and the risk of breast cancer. N Engl J Med 346:2025–2032, 2002.
13. Million Women Study Collaborators (Writing Group: Banks E, Beral V, Reeves G). The Million Women Study: design and characteristics of the study population. Breast Cancer Res 1:73–80, 1999.
14. Magnusson C, Baron JA, Correia N, et al. Breast cancer risk following long-term estrogen and estrogen-progestin replacement therapy. Int J Cancer 81:339–344, 1999.
15. Schairer C, Lubin J, Troisi R, Sturgeon S, Brinton L, Hoover R. Menopausal estrogen and estrogen-progestin replacement therapy and breast cancer risk. JAMA 283:485–491, 2000.
16. Ross RK, Paganini-Hill A, Wan PC, Pike, MC. Effect of hormone replacement therapy on breast cancer risk: estrogen versus estrogen plus progestin. J Natl Cancer Inst 92:328–332, 2000.
17. Newcomb PA, Titus-Ernstoff L, Egan KM, Trentham-Dietz A, Baron JA, Storer BE, Willett WC Stampfer MJ. Postmenopausal estrogen and progestin use in relation to breast cancer risk. Cancer Epidemiol Biomarkers Prev 11:593–600, 2002.
18. Siiteri PK, Hammond GL, Nisker JA. Increased availability of serum estrogens in breast cancer: a new hypothesis. In: Pike, MC, Siiteri PK, Welsch CW (eds). Banbury Report 8. Hormones and Breast Cancer. New York: Cold Spring Harbor Laboratory, 1981, pp 87–106.
19. Giusti RM, Iwamoto K, Hatch EE. Diethylstilboestrol revisited: a review of the long-term health effects. Ann Internal Med 122:778–788, 1995.
20. IARC Monogr Estrogens, Progestins and Combinations. Suppl 7:273–278, 1987.
21. Calle EE, Mervis CA, Thun M, Rodriguez C, Wingo PA, Heath CW Jr. Diethylstilboestrol and risk of fatal breast cancer in a prospective cohort of US women. Am J Epidemiol 144:645–652, 1996.
22. Titus-Ernstoff L, Hatch EE, Hoover RN, et al. Long-term cancer risk in women given diethylstilbestrol (DES) during pregnancy. Br J Cancer 84:126–133, 2001.
23. Marselos M, Tomatis L. Diethylstilboestrol. I. Pharmacology, toxicology and carcinogenicity in humans. Eur J Cancer 28A:1182–1189, 1992.
24. Modan B, Ron E, Lemer-Geva L, Blumstein T, Menczer J, Rabinovici J, Oelsner G, Freedman L, Mashiach S, Lunenfeld B. Cancer incidence in a cohort of infertile women. Am J Epidemiol 147:1038–1042, 1998.
25. Venn A, Watson L, Bruinsma F, Giles G, Healy D. Risk of cancer after use of fertility drugs with in-vitro fertilization. Lancet 354:1586–1590, 1999.

26. Ricci E, Parazzini F, Negri E, Marsico S, La Vecchia C. Fertility drugs and the risk of breast cancer. Hum Reprod 14:1653–1655, 1999.
27. Potashnik G, Lerner-Geva L, Genkin L, Chetrit A, Lunenfeld E, Porath A. Fertility drugs and the risk of breast and ovarian cancers: results of a long-term follow-up study. Fertil Steril 71:853–859, 1999.
28. Dunning AM, Healey CS, Pharoah PDP, et al. A Systematic Review of Genetic Polymorphisms and Breast Cancer Risk. Cancer Epidemiol Biomarkers Prev 8:843–854, 1999.

7

An Overview of Genetic Predisposition and the Search for Predisposing Genes for Breast Cancer

BRUCE PONDER

Breast cancer, like most common cancers, tends to cluster in families.[1] There is evidence, discussed below, that most of the family clustering results from inheritance rather than shared environment. If we knew the genes that underlie the family clustering, the knowledge might be applied in several ways:

1. *Understanding of cancer development; application to treatment and prevention*. The genes can be thought of either as part of the sequence of events that must be accumulated to turn a normal cell into a cancer cell or as outside the pathway but influencing it;[2] in either case, they may suggest targets for treatment or prevention.
2. *Epidemiology*. Analytic epidemiology relies on identifying associations between exposures' and disease. If the population of individuals with cancer is heterogeneous, correlations may be weakened. Genetic classification may resolve some of this heterogeneity. Knowledge of the mechanism of action of a genetic variant may also suggest an environmental cause with which the gene could interact.
3. *Assessment of individual risk*. If individuals can be identified through genetics to be at higher (or lower) than average risk, this knowledge may be used to target interventions.

These issues are discussed further below.

In this chapter, we consider the evidence for inherited predisposition, the types of gene that might be involved, and how to search for those that are yet undiscovered.

EVIDENCE FOR INHERITED PREDISPOSITION TO BREAST CANCER

The evidence comes from familial clustering. It is clearest in families such as that shown in Figure 7.1, where several individuals are affected in successive generations. Such a pattern can only be due to a highly penetrant, dominantly inherited predisposing mutation. The gene can be identified empirically using linkage analysis, followed by positional cloning; indeed, two such genes have so far been identified, *BRCA1* and *BRCA2*.[3]

Mutations in *BRCA1* or *BRCA2* account for the majority of families with multiple cases of either breast cancer alone or breast and ovarian

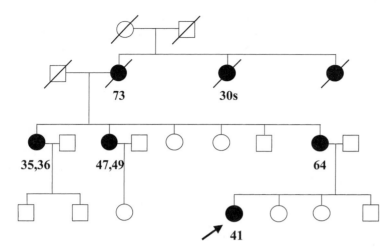

Figure 7.1 Familial breast cancer. Individuals were diagnosed with primary breast cancer at the ages shown. A familial pattern such as this is strongly suggestive of autosomal dominant predisposition by a single predisposing gene. These cases proved to be due to mutation in *BRCA1*.

cancer. It is perhaps surprising that they account for only a small part of the excess familial clustering of these cancers in the population as a whole. The overall extent of familial clustering can be measured by ascertaining a large series of breast cancer cases from the population and comparing the number of cases of breast cancer in their close relatives (in practice, mothers and sisters) with that expected from the rate in the population as a whole. In one such study of 1500 cases, there were 177 breast cancers in relatives compared with 106 expected, giving an excess of 71 (equivalent to a familial relative risk of 1.7).[4] When the 1500 breast cancer probands were systematically tested for mutations in the *BRCA1* and *BRCA2* genes, the risk to relatives in cases with a *BRCA* gene mutation was greater than that in cases where no mutation was detected; but even so, the *BRCA* mutation families accounted for only 12 of the 71 excess cases in relatives. When a correction is made for the imperfect sensitivity of *BRCA1* and *BRCA2* mutation detection, this provides an estimate that only 15%–20% of the excess familial risk is attributable to families with highly penetrant mutations in *BRCA1* and *BRCA2*. The same result was obtained in another study.[5]

The evidence that the large component of familial clustering that is not attributable to *BRCA1* and *BRCA2* also has a genetic basis, coming both from twin studies[6–8] and from mathematical modeling of the observed patterns of breast cancer within families.[9] In twin studies, the key measure is the risk to the second member of a pair of monozygotic (MZ) twins compared with the second member of a dizygotic (DZ) pair. Monozygotic twins have identical genes, whereas DZ twins share on average 50% of their genes; but both sets are expected to have shared environment to a similar degree. A greater concordance between MZ twins, which is what is observed,[6,7] therefore argues for predominantly genetic effects.

What Types of Genes Might Be Involved?

For simplicity, it will be assumed in the following discussion that most of the familial clustering of breast cancer that is not explained by mutations in *BRCA1* and *BRCA2* does indeed have a genetic basis, and we will focus on the possible nature of those genes.[2,10,11] Familial nongenetic factors may exist; ultimately, they should be identified and fitted into any risk assessment.

There are three broad possibilities for the types of genetic variation that might account for non *BRCA1/2* familial clustering. These are illustrated diagrammatically in Figure 7.2. (*1*) At one end of the spectrum are genes bearing mutations similar to those in *BRCA1* or *BRCA2*; i.e., uncommon, and conferring a moderately increased risk to the individual. These genes can be designated *BRCA3, 4 . . . BRCAX*. It is likely that any such genes will confer less extreme risks than those seen with mutations in *BRCA1* and *BRCA2*. (*2*) At the other end of the spectrum, the effects may be caused by normal genetic variation within the population. To illustrate this

SEARCH FOR PREDISPOSING GENES FOR BREAST CANCER

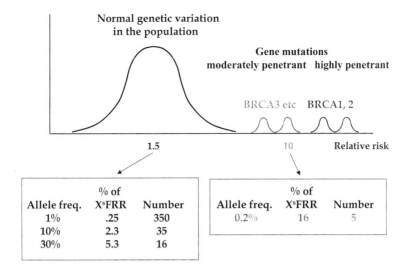

Figure 7.2 *Top:* The spectrum of genetic variants or mutations conferring risk of breast cancer. To the right, rare gene mutations conferring moderate or high levels of individual risk (relative risks of about 10-fold over individuals who do not have the mutation); to the left, the range of normal genetic variation in the population in which the combination of variants, individually of small effect, confers individual risk. *Bottom:* The number of variants/mutations with different allele frequencies, given a relative risk of either 1.5 or 10, that would be needed to account for the observed excess familial relative risk that is not accounted for by *BRCA1* and *BRCA2*. For each putative variant or mutant allele, the allele frequency, the percent of the excess familial relative risk accounted for, and hence the total number of such variants required are shown.

concept, if identical twins have identical faces but the rest of us (including siblings) have different faces, that implies that the variety of faces is the product of variation in many different genes. As variety of faces, so variety of risks of cancer. Most genes are present within the population in the form of several alleles, which differ only at one or a few positions in the sequence. If some of these alleles have slightly altered function, then possibly a combination will determine risk of breast cancer. This model predicts that multiple-case families will be rare because to be at high risk an individual must have inherited each of several different variants. It is unlikely that this would occur in several close relatives. (3) An intermediate position is to suppose that there could be several *BRCA*-like genes but possibly commoner and of much weaker effect. In this case, small family clusters would be more frequent than multiple-case families (because most people who have inherited the faulty gene still would not get breast cancer). If the genes are common, some family clusters might result from the combined effects of more than one gene.

It is possible to estimate the number and properties of the genes that would be required to account for the observed twofold familial risk. The genes are described in terms of the frequency of the predisposing allele (whether rare mutation or common variant) and its "strength" in terms of how much it increases the risk of breast cancer. Some possibilities are shown in Figure 7.2. The real situation is, however, unknown. Because the best strategy to search for the genes depends on their properties, several attempts have been made to estimate these properties by model fitting.[9,12,13]

Model Fitting

The pattern in which breast cancers occur within families will differ if the familial effect is the result of shared environment or of dominant or recessive predisposing genes. If it is genetic, the pattern will again be different if a single gene or if combinations of genes are operating within the family. The pattern that is actually observed in the families of a large unselected series of breast cancer cases is compared with the predictions made by mathematical models that assume different genetic and environmental contributions.

Two studies[12,13] have been reported in which the whole set of breast cancer cases has been tested for mutation in *BRCA1* and *BRCA2*, and individuals who test positive are removed from the series. These studies address the question of interest, which is the properties of genes that cause familial clustering not attributable to *BRCA1* and *BRCA2*. In one, based on breast cancer cases diagnosed below 40 years of age from Australia,[12] the best-fitting model was of a single common recessive predisposing gene, with an allele frequency in the population of 27% and a breast cancer risk in homozygotes of 42% by age 70. In a UK series of breast cancer cases diagnosed before age 55,[13] the best-fitting models were either a single recessive gene or a polygenic model in which the familial clustering resulted from the combined effects of several genes, each of small effect. When the same models were applied to a set of multiple-case families, the polygenic model fitted better than the recessive model as an explanation for non-*BRCA*-related familial clustering. This analysis also suggested that the multiple genes that underlie non-*BRCA* familial clustering would also act as modifiers of breast cancer risk in *BRCA*-mutation families.

It is difficult to draw firm conclusions from model fitting of this sort. Comparison of the Australian and UK breast cancer sets shows that in the Australian set the risk to siblings of a case was clearly greater than the risk to the mother, the pattern expected of recessive predisposition, whereas this effect was much less marked in the UK set. The explanation is unclear. Nongenetic confounding effects, such as differences in parity between the mothers and sisters in the series or cohort effects of changing environment with time, may be involved. In a meta-analysis of large breast cancer series, there was no significant difference in the risk to siblings or mothers of a case, suggesting that recessive effects may not play a large part in susceptibility to breast cancers in general. A polygenic model, though not proven for breast cancer until the genes are identified, would be consistent with what is known of the determinants of complex phenotypes in other areas of genetics (see, e.g., Mackay[14]).

THE SEARCH FOR PREDISPOSING GENES

Different strategies must be used to search for different types of gene.

Linkage in Multiple-Case Families

Mutations such as those in *BRCA1* and *BRCA2* will cause multiple-case families, and the genes can be sought using classical linkage analysis. The great advantage of this approach over those described below for common, weak genetic variants is that an empirical search across the entire genome is possible with several hundred evenly spaced linkage markers and a good expectation of obtaining a result. This is in contrast to the association study (see below), where reliance must be placed either on guessing and then testing individual "candidate" genes or on the use of a much larger set of genetic markers.

The linkage strategy is being used to search for *BRCA3* and similar genes, using multiple-case families in which no mutation in *BRCA1* or *BRCA2* has been identified. To date, although there have been reports of linkage, none has proved to be reproducible.[15] It may be simply that the right markers have, by chance, not yet been tried or that such a gene is not the right genetic model. More probably, if *BRCA3* is the correct model, the problem is heterogeneity; i.e., there are several genes, each accounting for a subset of families. In this situation, the statistical power of linkage analysis is greatly reduced. The remedy will be to assemble a larger family set.

Sibling Pair Analysis

A similar linkage-based approach is sibling pair analysis.[16] In this case, linkage analysis is carried out in a larger set of siblings, both of whom have breast cancer, searching for markers that are shared between the affected sibs more frequently than would be expected by chance. This strategy is designed to search for genes which are less strongly predisposing than *BRCA3* and thus less likely to give rise to families with multiple affected individuals. The analysis can be designed to search for dominant or recessive effects. Although several sibling pair studies are in progress, no positive findings have been reported. Again, the likely existence of several predisposing genes will greatly weaken the power of the analysis.

Founder Populations

The main limitations of the linkage-based approaches described above are genetic heterogeneity and the difficulty of ascertaining families of sufficient size. Populations that have arisen from a small number of founders and have remained in genetic isolation can potentially overcome these difficulties.[17] Because of the small founding population, it is likely that only one or a few predisposing genes will be present and that these will account for most of the familial cases within the population, thus reducing the effects of genetic heterogeneity. Families that now appear as separate groups are likely to trace back to a common ancestor and to carry the same copy of the predisposing gene (*identity by descent*). They thus form effectively a single extended family for the purposes of classical linkage analysis, which is carried out by searching for haplotypes (and thus chromosomal regions that share the same ancestry) present at a significantly greater frequency in breast cancer cases than controls (*identity by descent mapping*). The requirement for this strategy to succeed is that there should be a gene or genes within the population that cause a sufficient proportion of breast cancers for the haplotype sharing to be confidently recognized above the background. Examples of populations that might be suitable include Finns, Icelanders, Quebecois, and Sardinians; but no confirmed gene identifications for breast cancer have yet been made by this approach.

Association Studies

An association study is the most powerful means to search for common, weakly predisposing alleles. These alleles are envisaged to cause increases in individual risk of the order of 1.3- to 2-fold, corresponding to absolute risks of breast cancer by age 70 of around 7%–11% in predisposed individuals. Clearly, such a predisposing allele will seldom generate a family with multiple affected individuals. The analysis is therefore based on individuals in a case-control design in which the frequency of alleles in successive genes is compared between large series of cancer cases and unaffected controls. A consistent finding that a particular allele is present more often in cases than controls implies that that allele, or one closely associated with it, is involved in predisposition.

Because this approach is now being used widely to search for breast cancer genes, it will be discussed in some detail. Association studies for disease genes are generally based on the common variant–common disease hypothesis. This concept is illustrated in Figure 7.3. Briefly stated, it is as follows: spontaneous mutations constantly give rise to genetic variants. During the early history of modern populations, some of these variants became fixed in the population, as a result of factors that may include selective advantage or population bottlenecks and expansions, so that they are now present at frequencies ranging from a few percent upward. In the present population, some of these variants may have properties that are disadvantageous in that they predispose to common diseases; combinations of these variants underlie differences in disease susceptibility within the population. As already stated, the aim of the association study is to demonstrate that variants of a specific gene are present significantly more often in individuals with the disease than in controls.[18]

Whereas linkage studies in families can use markers spaced at intervals of several million base pairs, association studies will probably need markers at intervals of 10–15 kb or less. The reason is that between individuals in even quite a large family there will have been few meiotic recombinants, so the chromosomes can be compared by mapping on quite a coarse scale. Between individuals in a population, however, there may be many recombinations, leading to differences in their chromosomal maps on a much finer scale. Mapping at evenly spaced intervals of 30 kb would require 100,000 markers across the genome. Analysis on this scale is currently impossible, primarily because of cost. Association studies are therefore currently based on testing a series of candidate genes.

The study is relatively straightforward if there is a variant of the gene that is known or suspected to be directly responsible for predisposition to the disease. Usually, however, this is not

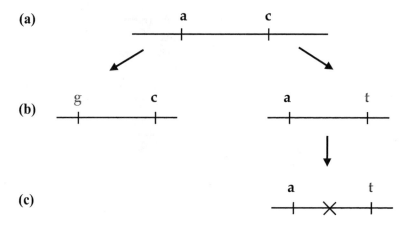

Figure 7.3 The concept of the association study. One DNA strand of a small chromosomal region is illustrated (it might be, e.g., part of the estrogen receptor gene). *(a)* An ancestral copy of the chromosome, perhaps 100,000 years ago in an African human population. The chromosome is passed down through the population, and with time, base changes occur in the DNA sequence as a result of mutation. Thus, at the stage represented in *(b)*, there are three copies of the chromosome in the population, defined by sequence polymorphism; there are 3 haplotypes, ac, gc, and at. Now suppose a further variant, X, arises on the chromosome marked by the haplotype (at). This variant predisposes to disease. It may be rapidly lost from the population; or it may increase in frequency and come to be a common genetic variant, predisposing to a common disease (e.g., breast cancer) in the modern population. If variant X is unknown, an association study may be designed to test whether X is commoner in cases of the disease than in controls; if it is, we infer that it is causally associated. However, if X is not known, the markers a and t separately or the haplotype (at) may be used instead as proxy markers for X. The assumptions and pitfalls in using these proxy markers are discussed in the text and in Figure 7.4.

the case: several variants may be known, but there is often no evidence that any of them is directly responsible for the disease. In this situation, these variants must be tested in the hope that one or more of them will provide an adequate marker for the suspected but still unidentified active variant (Fig. 7.3c). Whether or not a given variant will act as an adequate marker depends on several factors, not all of which can be predicted when the study is carried out (Fig. 7.4). Unless the variant is a perfect marker, at least some loss of statistical power will result. These uncertainties complicate the design and interpretation of these studies.

The main sources of difficulty (see Figs. 7.3 and 7.4) are as follows:

1. The marker genetic variant will reflect perfectly the active variant only if they have been perfectly coinherited through successive generations of the population. In this case, they are said to be in complete *linkage disequilibrium*. However, unless they are physically extremely close, it is likely that they will become separated by recombination at meiosis; depending on how often and when in the history of the population this has occurred, their association may be weakened even to the point where it is not detectable.

2. The association will also be weakened if the active variant has arisen independently on more than one occasion in the history of the population because then it is likely to be associated with markers on two or more different versions of the chromosome in the population.

3. Even if the marker and active variant have each occurred only once in the population, if they arose at very different times, their frequencies are likely to be very different (in general, recent variants will have a low frequency because they subtend a smaller segment of the population expansion). This can considerably weaken the power of the statistical analysis to detect association. Thus, e.g., a frequency of 0.5 vs. 0.1 for the marker and

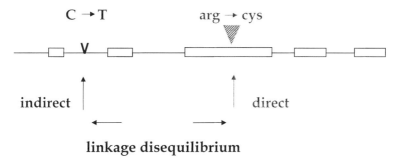

Figure 7.4 Direct and indirect (proxy) markers in association studies. The *horizontal line* represents a segment of a gene, with exons shown as *boxes*. A polymorphic DNA sequence variant that results in the coding change arg → cys is assumed to confer susceptibility to the disease (corresponding to X in Figure 7.3). If the intronic C → T polymorphism is to be used as an indirect marker, it will be maximally efficient if it precisely reflects the presence of the arg → cys variant on the same chromosome (i.e., the two variants are in complete linkage disequilibrium). Situations in which this association will be weakened or destroyed are described in the text.

active variant would require a roughly 10 times larger number of individuals to detect association between them at a given confidence level compared with the case where both are present at equal frequency.

Each of these problems goes away if the active variant is used directly in the association study. It clearly does not matter in this case, e.g., how many times the variant has arisen in the population, so long as it is the same variant. Otherwise, the problems are not so easily solved. Information about the extent of linkage disequilibrium across genes and their regulatory regions (where important variants might well lie) is still fragmentary;[19] what there is suggests that it is different in different genes. There is, in principle, no way of knowing whether presumed active variants have arisen once or many times and at what point in the history of the population. If the common variant–common disease hypothesis is wrong[20] and most polygenic susceptibility is based on a large number of recently arisen, individually rare variants, the association study design will fail.

The practical consequences of this for the current search for breast cancer genes are threefold: *(i)* candidate genes cannot easily be excluded without further information about linkage disequilibrium and certainly not by analysis of one or two variants in small case-control sets, as is the case for most studies published to date; *(ii)* the strategy for design of future studies should probably be, first, to identify the full set of common (say greater than 1% or 5% frequency) variants in the gene to be evaluated, then to construct the haplotypes defined by those variants and to define the extent of linkage disequilibrium, using haplotypes rather than single variants in the analysis of association to ensure both that the gene is covered to the maximum extent and that power losses due to differences in frequency between the marker and active variants is minimized;[21] *(iii)* until these issues are better understood and a more complete map of genetic variants and of linkage disequilibrium is available, design of association studies on a whole genome rather than candidate gene basis is premature.

Other Aspects of the Design and Interpretation of an Association Study

Phenotype

Association studies are conventionally designed using the "all or nothing" end point of cancer occurrence to define the case and control populations. Other end points may, however, have advantages. Breast cancer is almost certainly heterogeneous in its etiology. Greater power may, in principle, be obtained by defining subsets of breast cancer, e.g., by age, molecular markers (estrogen receptor-positive or -negative), or histological type. There are obvious dangers in defining these subsets post hoc. Other aspects of the cancer clinical phenotype may be used as end points, e.g., survival. Cancer represents the final result of many biological processes and, thus, might be regarded as the most distant read-out from the causative genetic variation. Intermediate phenotypes, such as mammographic patterns, age at menarche, or levels of estrogenic hormones, may be more closely correlated with a given genetic variation; and if the phenotype can be treated quantitatively as a continuous variable rather than dichotomously as with cancer, there will again be greater statistical power.

Case and Control Sets

Case and control sets should be as closely similar as possible in genetic background (i.e., population of origin). Because the frequency of most genetic variants differs from population to population, spurious associations between the occurrence of cancer and the frequency of a variant may occur if cases and controls are drawn from different genetic backgrounds. Such considerations are particularly important in the assembly of large case/control sets from major hospital centers in countries like the United States, where urban populations often are quite diverse. Testing for spurious association between variants that are known to be genetically unlinked can and should be carried out within case and control sets to check for the presence of these effects.[22]

Power Issues

The size of the case/control sets that is needed will depend on the size of effect that it is hoped to detect. Illustrative power calculations are shown in Figure 7.5. Small studies risk missing small effects but also are more prone to provide spurious positive associations as a result of chance. It is important that all positive results be replicated in independent case/control sets. A meta-analysis of association studies for breast cancer up to early 1999 showed that, while several positive results had been obtained, none could be consistently replicated by other research groups.[23]

Choice of Candidate Genes

Any gene that is known, or can be presumed, to be involved in processes that contribute to the development of breast cancer is a plausible candidate. Such genes may be thought of in three groups:[2] *(1)* those (e.g., *BRCA1*, *BRCA2*) that are part of the sequence of genetic events that leads from a normal cell to a cancer cell, *(2)* those in which variation may alter the probability that the key genetic alterations will occur (DNA repair genes, genes involved in metabolism of carcinogens), and *(3)* those in which variation may affect the probability that a clone of partially altered cells may emerge and progress to cancer (genes involved in paracrine signaling, immune response, angiogenesis).

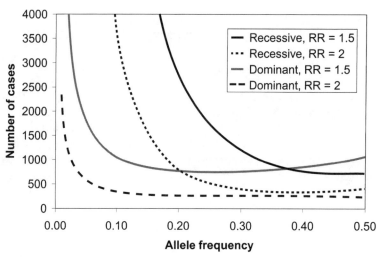

Figure 7.5 Numbers of cases and controls needed for association studies, to detect a variant of given allele frequency and relative risk (RR).

The history of the search for disease genes suggests that guessing such candidates is not often successful; indeed, the lack of clear-cut positive results to date in association studies for breast cancer reinforces this. Alternative ways to identify candidates must therefore be considered. They include (1) the use of mouse models (where loci that are associated with susceptibility or resistance can be mapped empirically and the genes form the corresponding genomic region in humans assessed as candidates;[24] (2) clues from regions of consistent somatic change in tumors, whether amplification or loss, especially if there is biased involvement of one allele; (3) a search (in humans or experimental animals, depending on feasibility) for inherited variation in specific assayable phenotypes, such as radiation response, inflammation, and angiogenesis, as a basis for a subsequent gene search based on that phenotype; and (4) the use of what in animal systems is called a "sensitized screen" approach,[25] searching by linkage methods for evidence of genetic modifiers of the phenotype in carriers of known highly penetrant predisposing genes such as BRCA1 and BRCA2. One can hope that it will eventually be possible to avoid the problems of guessing candidate genes and to use an empirical whole-genome approach, but for the reasons given earlier, this is not yet practicable.

Analysis and Presentation of Results

Conventionally, the results are expressed as the odds ratio for cancer in individuals who carry one or two copies of the less common allele, where the risk in the common homozygotes is set at 1.0. The results may be expressed as the risk attached to the variant allele or to each genotype separately. Significance can be assessed from the confidence interval around the point estimate or by calculating a p value. This raises the question of how the p values should be evaluated in a situation where there is multiple-testing. Because tests of several variants within a single gene cannot be assumed to be independent of one another, the basis for a multiple testing correction is not clear. At the odds ratios which might plausibly exist and be of interest to identify, series of the sizes of a few hundred to a thousand or so cases and controls, commonly analyzed at present, will not yield very small p values. Analysis of very large series, however, is expensive. The most satisfactory solution is probably to set criteria for evaluation of a first-pass analysis which will provide the optimum retention of true positive results for further testing and confirmation and rejection of true negatives. What these criteria should be has not been clearly enunciated. They will depend on the properties of the genetic variants that are thought to be important, and these will differ according to the application that is envisaged.

POTENTIAL APPLICATION OF GENETIC PREDISPOSITION TO BREAST CANCER PREVENTION

High-Penetrance Genes

Mutations in BRCA1 and BRCA2 define individuals who have a substantial lifetime risk of breast and ovarian cancer and a somewhat increased risk of various other cancers.[26,27] These risks are high enough for most individuals to be concerned about them, to want to know their mutation status, and to be interested in screening or prevention.[28] The main problems in developing screening or prevention strategies lie in the assembly of sufficiently large cohorts of individuals at risk to provide adequately powered studies to test efficacy and in the design of the studies because the intervention may be well separated in time from the end point of cancer incidence. Even if intermediate markers can be found to give early indications of response, intermediate markers of toxicity (the nature of which may be difficult to predict) are less easily envisaged.

Low-Penetrance Genes

The value of identifying these genes has been the subject of argument. Bell[29] views the understanding of the polygenic basis of susceptibility to common disease as "a new horizon in medicine." Sceptics, such as Holzman and Marteau,[30] urge us to "look beyond the hype." They argue that the effect of such genetic variants is weak in comparison to lifestyle and environmental factors and unlikely to be of real consequence to the individual.

The question is different depending on whether it is seen from the point of view of the individual or of the population. From the perspective of the individual, the difference between the views of Bell[29] and Holzmann and Marteau[30] may lie in whether the effects of common genetic variation are considered as a whole or one gene at a time. Taken one gene at a time, Holzmann and Marteau[30] are right: the effects are small and unlikely to engage the individual. A dominantly acting variant allele with an odds ratio for breast cancer of 1.3 and a frequency of 10% would increase the risk of breast cancer by age 70 in the individual who had it from 5.7% (the risk in the population as a whole) to about 8%, probably not sufficient to be of much interest. However, if one considers the possible effect of all genetic variants together, the picture changes. An analogy would be genes as a hand of cards dealt at birth. Some cards (2, 3, 4 . . .) are "low-risk" and some (J, Q, K, A) are "high-risk;" the rest lie in between. Most individuals will receive a mixed hand and be at near average risk; but some will be dealt mostly high cards or low cards and lie at the extremes. The question is, then, how wide is the distribution of risk in the population (see Fig. 7.2). If there is not much difference between those with a hand of 2s and 3s and those with kings and aces, then from the point of view of individual risk estimation, the genes are not of much interest. In fact, modeling based on the population-based studies described earlier suggests that the distribution may be quite wide. There may be as much as a 40-fold difference in risk between the top 20% and bottom 20% of the population, assuming the polygenic dominant model and multiplicative interaction, slightly less assuming additive genetic effects.[31]

If correct, this model has implications for the individual and for the population. It predicts that most breast cancer incidence occurs within a minority of the population who are at high risk. For example, according to the model (and it is only a model), 1 in 10 women will have a 12% or greater risk of breast cancer by age 70, and this 10% of the population will account for almost half (46%) of all breast cancer cases. Similar conclusions, that most breast cancer occurs in a minority of highly predisposed women, come from a different route, the twin study by Peto and Mack.[6] In contrast, the Gail model, which is based on clinical risk factors only, has nowhere near the power to discriminate individual risks or to identify small high-risk groups within the population.[32] The public health implications of discovering the genetic variants that contribute to risk are therefore considerable. How practicable this is will depend on the numbers and frequency of the variant alleles that are to be discovered. If a few tens of variants account for most of the genetic risk, it should be feasible; but if there are large numbers of genes involved, and particularly if there are large numbers of individually rare predisposing alleles of each gene,[20] it will be extremely difficult. In the current enthusiasm for association studies and low-penetrance genetic effects, it should perhaps be remembered that, even in experimental systems, the molecular variants that underlie quantitative genetic variation have proved elusive (see, e.g., Flint and Mott[33]).

Apart from their contribution to individual risk, single genetic variants may have public health importance through definition of a different property, the population-attributable fraction (PAF). This corresponds to the fraction of total disease incidence that would be prevented if that particular variant, or its effects, were eliminated. (The concept is somewhat confusing because the total of all PAFs for factors that contribute to a disease will exceed 100%; for breast cancer, e.g., possession of two X chromosomes has a PAF of about 99%, but there are clearly other significant risk factors as well.) What is important is that a common variant, even with a low odds ratio for individual risk, may have a substantial PAF. For example, a dominant allele with frequency 0.5 and odds ratio of 1.3 has a PAF of 16%. Such an allele may not provide much "enrichment" in the sense of concentrating a sizeable proportion of cancer incidence in a minority of the population who can be targeted for intervention. However, it may still be important if the nature of the allele suggests an intervention based on the mechanism of predisposition, which could be used for prevention in those who are predisposed. Indeed, the importance may be greater than just that suggested by the size of the PAF. The example of cholesterol in cardiovascular disease suggests that the benefits of such an intervention may be present even in individuals who lack the high-risk allele: if this is so, the value of having identified the predisposing genetic variant is that it indicates a mechanism of increased risk.

By What Criteria Should a Genetic Variant Be Assessed in an Association Study?

Conventionally, the results of association studies are expressed as the odds ratio for the variant in relation to the disease, but the foregoing discussion suggests that this may not be the only relevant measure. A very high odds ratio will be relevant, certainly; but unless the allele is rare, it will usually be detected without difficulty. In terms of genetic risk profiling, a relevant measure is the fraction of the excess familial risk explained. This will be determined in part by odds ratio and in part by allele frequency (see examples in Fig. 7.2). For commoner alleles, the PAF may be of interest. It may be helpful to design and report studies in terms of their power to detect each of these properties at a certain magnitude of effect and confidence level.

SUMMARY

Highly predisposing mutations in known genes such as *BRCA1*, *BRCA2*, and possibly *chk2* and *ATM*, account for only some 20% of the familial clustering of breast cancer, most of which probably has a genetic basis. The nature of the genes that may account for the remaining 80% is still unknown. It is likely (but not certain) that the spectrum of normal genetic variation in the population, rather than rare genetic mutations, plays a substantial part in determining risk. Model fitting suggests that an individual profile of such variation could account for large differences in susceptibility and that much of breast cancer incidence occurs in a predisposed minority of the population.

The current strategy most widely used to identify the putative predisposing genetic variants is the association study, a case-control design in which the frequency of the variant is compared between a series of breast cancer cases and controls. So far, these studies have had rather little success. Several issues in the design and conduct of the studies remain to be addressed.

ACKNOWLEDGMENTS
B.P. is a Gibb Fellow of Cancer Research UK.

REFERENCES

1. Pharoah P, Day NE, Duffy S, Easton DF, Ponder BAJ. Family history and the risk of breast cancer: a systematic review of meta-analysis. Cancer 71:800–809, 1997.
2. Ponder BAJ. Cancer genetics. Nature 411:336–341, 2001.
3. Walesh PL, King M-C. *BRCA1* and *BRCA2* and the genetics of breast and ovarian cancer. Hum Molec Genet 10:705–713, 2001.
4. Anglian Breast Cancer Study Group. Prevalence and penetrance of *BRCA1* and *BRCA2* in a population based series of breast cancer cases. Br J Cancer 83:1301–1308, 2000.
5. Peto J, Collins N, Barfoot R, et al. Prevalence of *BRCA1* and *BRCA2* gene mutations in patients with early onset breast cancer. J Natl Cancer Inst 91:943–949, 1999.
6. Peto J, Mack TM. High constant incidence in twins and other relatives of women with breast cancer. Nat Genet 26:411–414, 2000.
7. Lichtenstein P, Hulm NV, Verkasalao PK. Environmental and heritable factors in the causation of cancer. N Engl J Med 343:78–85, 2000.
8. Risch N. The genetic epidemiology of cancer: incorporating family and twin studies and their implications for molecular genetic approaches. Cancer Epidemiol Biomarkers Prev 10:733–741, 2001.
9. Antoniou AC, Pharoah P, McMullan G, et al. Evidence for further breast cancer susceptibility genes in addition to *BRCA1* and *BRCA2* in a population based study. Genet Epidemiol 21:1–18, 2001.
10. Hopper JL. More breast cancer genes? Breast Cancer Res 3:154–157, 2001.
11. Easton DF. How many more breast cancer predisposition genes are there? Breast Cancer Res 1:14–17, 1999.
12. Cui J, Antoniou A, Dita GS, et al. After *BRCA1* and *BRCA2*—what next? Am J Hum Genet 68:420–431, 2001.
13. Antoniou C, Pharoah P, McMullan G, et al. A comprehensive model for familial breast cancer incorporating *BRCA1*, *BRCA2* and other genes. Br J Cancer 86:76–83, 2002.
14. Mackay TFC. Quantitative trait loci in *Drosophila*. Nat Rev Genet 2:11–20, 2001.
15. Thompson D, Szabo C, Mangion J, et al. Evaluation of linkage of breast cancer to the putative *BRCA3* locus on chromosome 13q21 in 128 multiple case families from the Breast Cancer Linkage Consortium. Proc Natl Acad Sci USA 99:827–831, 2002.
16. Risch N, Merikangas K. The future of genetic studies of complex human diseases. Science 273:1516–1517, 1996.

17. Wright AF, Carothers AD, Piratsu M. Population choice in mapping genes for complex diseases. Nat Genet 23:397–403, 1999.
18. Cardon LR, Bell JI. Association study designs for complex diseases. Nat Rev Genet 2:91–99, 2001.
19. Goldstein DB. Islands of linkage disequilibrium. Nat Genet 29:109–112, 2001.
20. Wright AF, Hastie ND. Complex genetic diseases: controversy over the Croesus code. Genome Biol 2:2007.1–2007.8, 2001.
21. Johnson GCL, Esposito L, Barratt BJ, et al. Haplotype tagging for the identification of cancer disease genes. Nat Genet 29:233–237, 2001.
22. Goode EL, Dunning AM, Healey CS, et al. Assessment of population stratification in a large population-based cohort. IGES-126. Genet Epidemiol 21:139–183, 2001.
23. Dunning AM, Healey CS, Pharoah PDP, et al. A systematic review of genetic polymorphisms and breast cancer risk. Cancer Epidemiol Biomarkers Prev 8:843–854, 1999.
24. Balmain A, Nagasi H. Cancer resistance genes in mice: models for the study of tumour modifiers. Trends Genet 14:139–144, 1998.
25. Martin A, Nadeau JH. Sensitised polygenic trait analysis. Trends Genet 17:727–731, 2001.
26. Ford D, Easton DF, Bishop DT, Narod SA, Goldgar DE. Risks of cancer in *BRCA1* mutation carriers. Breast Cancer Linkage Consortium. Lancet 343:692–695, 1994.
27. Breast Cancer Linkage Consortium. Cancer risks in *BRCA2* mutation carriers. J Natl Cancer Inst 91:1310–1316, 1999.
28. Ponder BAJ. Genetic testing for cancer risk. Science 278:1050–1054, 1997.
29. Bell J. The new genetics in clinical practice. BMJ 316:618–620, 1998.
30. Holzman NA, Marteau TM. Will genetics revolutionise medicine? N Engl J Med 343:141–144, 2000.
31. Pharoah PDP, Antoniou A, Bobrow M, Zimmern RL, Easton DF, Ponder BAJ. Polygenic susceptibility to breast cancer and implications for prevention. Nat Genet 31:33–36, 2002.
32. Rockhill B, Spiegelman D, Byrne C, Hunter DJ, Colditz GA. Validation of the Gail et al model of breast cancer risk prediction and implications for chemoprevention. J Natl Cancer Inst 93:358–366, 2001.
33. Flint J, Mott R. Finding the molecular basis of quantitative traits: successes and pitfalls. Nat Rev Genet 2:437–445, 2001.

8

Estrogen Biosynthesis Genes: P-450 Aromatase

COLIN D. CLYNE
EVAN R. SIMPSON

Recent models of estrogen insufficiency have challenged traditional concepts of estrogen action.[1–3] Studies of humans with natural mutations in the genes that encode the estrogen receptor[4] or cytochrome P-450 aromatase[5] and of estrogen receptor and aromatase knockout mouse models[6–11] have revealed hitherto unexpected roles for estrogen in both males and females. For example, the critical role of estrogen in the spermatogenic process[12] suggests that this hormone might more appropriately be termed an androgen in this context.[13] Additionally, in postmenopausal women (as in men), estrogen no longer functions as a circulating endocrine hormone; rather, it is produced in a variety of extragonadal sites (e.g., adipose tissue or bone), where it acts locally as a paracrine, autocrine, or intracrine factor.[2,3,14] This has great significance for our understanding of the biology of estrogen-dependent diseases such as breast cancer.

SYNTHESIS OF ESTROGENS

Estrogens are synthesized from C_{19} precursors produced by the ovary (in premenopausal women), the adrenal cortex (in postmenopausal women and men), or the testis (in men). The first step in the pathway, common to the synthesis of all steroid hormones, is cleavage of the cholesterol side chain to yield pregnenolone (Fig. 8–1).[15] This reaction is catalyzed by the cholesterol side chain cleavage enzyme (P450scc) and requires 22-hydroxylase, 20α-hydroxylase, and 20:22 lyase activities,[16] all of which occur at a single active site.[17] The C_{21} steroids pregnenolone and progesterone are then converted to the C_{19} androgens dehydroepiandrosterone and androstenedione by 17α-hydroxylase (P450c17). These are two-step reactions that require 17α-hydroxylase and c17,20-lyase activities, both of which are characteristic of P450c17.[18] The final step in estrogen synthesis is catalyzed by P450 aromatase (P450arom), which converts androstenedione and testosterone to estrone and 17β-estradiol, respectively. In premenopausal women, the major source of circulating estrogens is the ovary, which expresses aromatase in the granulosa cell compartment. In postmenopausal women (and men), however, the major site of aromatization is adipose tissue,[19,20] where estrogen synthesis is dependent on the availability of circulating C_{19} precursors derived from the ovary and/or adrenal cortex.

AROMATASE AND THE *CYP19* GENE

Aromatase is a member of the cytochrome P-450 superfamily, which comprises more than

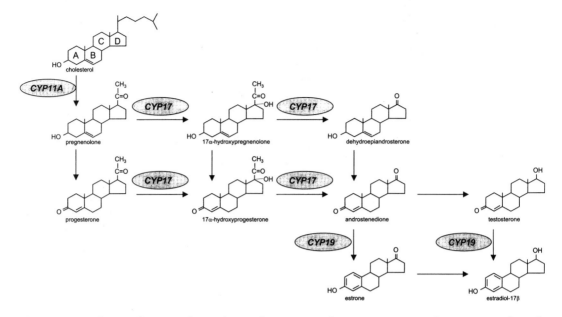

Figure 8.1 Pathways of estrogen biosynthesis. The activities of *CYP11A*, *CYP17* and *CYP19* are indicated.

480 heme-containing proteins in some 74 families. The aromatase reaction is unlike most other typical P-450 reactions and involves a unique three-step mechanism. The first two oxidative steps are followed by aromatization of the A ring to produce the phenolic A ring characteristic of estrogens.[21,22] Reduced nicotinamide adenine dinucleotide phosphate (NADPH) supplies reducing equivalents for this process via NADPH reductase. In humans, aromatase is expressed in a number of different tissues, including the ovary (granulosa and luteal cells), testes (Sertoli cells, Leydig cells, and spermatocytes), placenta, bone (osteoblasts, chondrocytes, and lining cells), brain,[23] and to a lesser extent skin and fetal liver.

Aromatase is encoded by the *CYP19* gene, which maps to chromosome 15q21.2 in humans.[24,25] *CYP19* was cloned in the late 1980s by three independent groups[26–28] and is a relatively large gene, spanning 123 kb.[25] The coding region spans nine exons beginning with exon II (Fig. 8.2). Upstream of exon II are a number of alternative untranslated first exons, which are spliced onto the 5′ region of the transcript in a tissue-specific manner.[29–31] Each of these alternative exons I is associated with its own unique

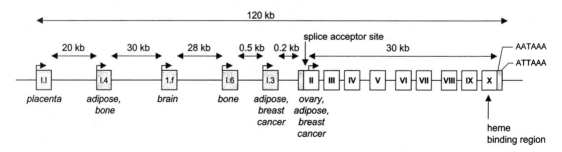

Figure 8.2 Structure of the human *CYP19* gene. The coding region comprises nine exons (II–X). Untranslated exons are shaded. The major placental promoter, I.1, is the most distally located (approximately 89 kb), and the ovarian promoter II is the most proximal. Even though each tissue expresses a unique first exon 5′-untranslated region, by splicing into a highly promiscuous splice acceptor site (AG/A\GACT) of exon II, the coding region and the translated protein product are identical in all tissues.

promoter region. For example, *CYP19* transcripts in the placenta contain untranslated exon I.I at the 5′ end. Expression in placenta is driven by a strong distal promoter, promoter I.I[30] upstream of exon I.I, which lies almost 90 kb upstream of exon II. In contrast, transcripts in ovary and testes contain 5′ sequences derived from immediately upstream of the transcription-initiation site. This is because a proximal promoter, promoter II, drives expression in the gonads.[32] Adipose tissue transcripts contain yet another distal untranslated exon located 20 kb downstream of exon I.I, namely, exon I.4.[33] Again, exon I.4 is associated with a unique promoter region, promoter I.4. Adipose tissue transcripts also contain sequences derived from untraslated exons II and I.3. As discussed below, the ratio of promoter I.4- to promoter II/I.3-derived transcripts changes in certain pathological states, including breast cancer. Several other untranslated exons have been characterized, including brain- (exon 1f) and bone- (exon I.6) specific sequences.[34,35] Although *CYP19* transcripts in different tissues therefore have unique 5′ termini, the coding region and, thus, the protein expressed are identical in each case. However, the use of multiple alternative promoters, each regulated by unique cohorts of hormones and transcription factors, permits a remarkable degree of flexibility and complexity in the regulation of aromatase expression.

TISSUE-SPECIFIC EXPRESSION OF AROMATASE

Expression in Ovary

Aromatase is expressed in the preovulatory follicles and corpora lutea of premenopausal women. Expression in granulosa cells is regulated by follicle-stimulating hormone (FSH), which increases intracellular cyclic adenosine monophosphate (cAMP) through activation of adenylyl cyclase.[36] Two distinct elements within promoter II mediate the FSH-induced cAMP signal (Fig. 8.3). The first element, located at position $-211/-202$ (TGCACGTCA), resembles the consensus palindromic cAMP response element (CRE), differing only by the inclusion of

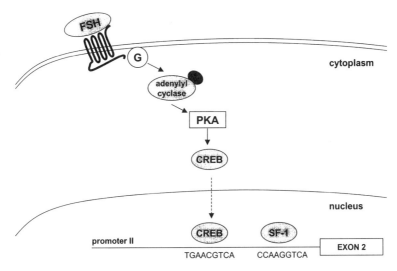

Figure 8.3 Regulation of aromatase expression in ovary. In ovarian granulosa cells, expression is regulated primarily by promoter II. Follicle-stimulating hormone (FSH) binds to its receptor and activates adenylyl cyclase, which results in formation of cyclic adenosine monophosphate (cAMP) and activation of protein kinase A (PKA), which then phosphorylates cAMP response element binding protein (CREB), which binds to a cAMP response element (CRE) on promoter II. Downstream of the CRE is a steroidogenic factor 1 (SF-1) binding site. Binding of both of these factors to their response elements is presumed to result in recruitment of the cohort of coactivators, which leads to activation of promoter II-specific expression.

an extra cytosine (underlined). Electrophoretic mobility shift analysis (EMSA) revealed that a number of proteins derived from bovine luteal cell nuclear extracts bind to this element, although one in particular, CRE-binding protein (CREB), bound only in the presence of cAMP.[37] Mutation of this CRE abolished the ability of promoter II luciferase reporter genes to respond to cAMP.[37] Thus, FSH-induced aromatase expression in ovary occurs via binding of CREB, activated by phosphorylation in response to cAMP.

A second promoter element, required for basal expression from promoter II, also contributes to cAMP inducibility. Located at position −130, this sequence (CAAGGTCA) resembles a recognition site for monomeric members of the nuclear receptor family and binds the orphan receptor steroidogenic factor-1 (SF-1/Ad4BP, NR5A1,[38,39] which is a key regulator of many other steroid hydroxylases and critical for the development of the adrenal and gonads.[40] Mutation of the promoter II SF-1 binding site markedly reduced both basal and cAMP-induced activity of promoter II reporter genes, indicating the importance of SF-1 for *CYP19* transcription from promoter II in the ovary.[39]

Expression in Adipose Tissue

Aromatase is expressed in the undifferentiated stromal mesenchymal cells of adipose tissue (preadipocytes) and not in the mature lipid-laden adipocytes.[41] Under normal circumstances, adipose stromal cells contain *CYP19* transcripts derived primarily from promoter I.4. This TATA-less promoter is stimulated by class 1 cytokines [e.g., oncostatin M and interleukin-6 (IL-6)] and tumor necrosis factor-α (TNF-α). In addition, activity of promoter I.4 has an obligate requirement for glucocorticoids.[42] The action of class 1 cytokines is transduced by the JAK-STAT pathway (Fig. 8.4). Specifically, activation of gp130-coupled receptors in adipose stromal cells leads to tyrosine phosphorylation and activation of JAK-1 and subsequent activation of signal transducer and activator of transcription 3 (STAT3) homodimers, which bind to an interferon-γ-activating sequence (GAS element) within promoter I.4.[43] The action of TNF-α is less well understood. Treatment of adipose stromal cells with TNF-α leads to rapid elevation in intracellular levels of *c-fos* and *c-jun*, which appear to bind as a heterodimer to an imperfect activating protein-1 (AP-1) site upstream of the GAS element.[44] More recent studies have shown that TNF-α induces other early-response genes of the *egr-1* family that can bind and activate promoter I.4 through a GC-rich sequence located within exon I.4.[45] This GC-rich sequence also constitutes a consensus Sp1 binding site. Since mutation of the site markedly diminishes basal transcription[42] as well as TNF-α-induced transcription,[45] it seems likely that Sp1 is involved in the regulation of promoter I.4 along with egr-1 and AP-1.

As mentioned above, the action of class 1 cytokines and TNF-α requires the presence of glucocorticoids, although glucocorticoids by themselves are ineffective. This permissive effect of glucocorticoids is mediated by a glucocorticoid response element (GRE) that binds the activated glucocorticoid receptor (GR).[42] Functional interactions between GR and STAT proteins have been reported for a number of genes,[46] and preliminary studies suggest that STAT3 and GR may interact physically, at least in yeast, to regulate promoter I.4. Since STAT proteins are known to also interact directly with Sp1,[47] one can envisage the existence of a large transcriptional complex consisting of promoter-bound STAT3, GR, Sp1, and associated coregulator proteins, such as p300, coordinating the transcriptional response to cytokines and glucocorticoids.

Although the majority of *CYP19* transcripts in adipose stromal cells contain exon I.4-specific sequences, transcripts derived from promoters II and I.3 are also expressed.[48,49] As in the ovary, promoters II and I.3 are stimulated by cAMP. Interestingly, phorbol esters markedly potentiate the effect of cAMP in adipose stromal cells, whereas in the ovary they inhibit promoter II.[50] Thus, activation of protein kinase A (PKA) and PKC signaling pathways maximally stimulates aromatase expression via promoters II and I.3. To identify endogenous factors that activate these pathways in adipose stromal cells and, thus, regulate aromatase expression, we examined a variety of hormones and neurotransmitters for effects on aromatase activity.[51,52] The most potent agent was prostaglandin E_2 (PGE$_2$),

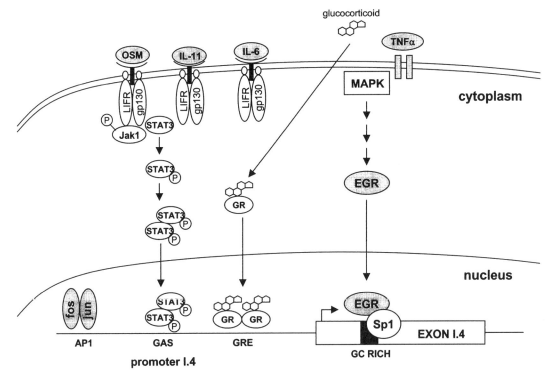

Figure 8.4 Regulation of aromatase gene expression from promoter I.4 in human adipose stromal cells. Jak1 kinase is bound to the common receptor subunit gp130 and activated following ligand binding and receptor dimerization as a consequence of phosphorylation on tyrosine residues. Signal transducer and activator of transcription 3 (STAT3) is recruited to binding sites on gp130 and phosphorylated on tyrosine residues by Jak1. These phosphotyrosine residues are recognized by SH2-homology domains on STAT3, resulting in dimerization followed by translocation to the nucleus and binding to the GAS element of promoter I.4. Following binding of glucocorticoid (GC) receptors to the glucocorticoid response element (GRE) and Sp1 to its site on untranslated exon I.4, activation of transcription of the aromatase gene from promoter I.4 is initiated. Tumor necrosis factor α (TNFα) action is presumed to be through sequential activation of mitogen-activated protein kinases (MAPKs) and early response genes, including the *egr-1* family members c-*fos* and c-*jun*. GAS, interferon-γ-activating sequence; OSM, oncostatin M; IL, interleukin; LIFR, leukemia-inhibiting factor receptor.

which binds to E-prostanoid $(EP)_1$ and EP_2 receptors in adipose stromal cells that couple to PKA and PKC, respectively. Since PGE_2 is a major secretory product of breast tumor epithelial cells, this hormone has great potential to influence aromatase expression in breast adipose tissue containing a tumor.

Expression in Breast Cancer

Estrogen levels are consistently elevated in breast tissue of cancer patients compared to normal individuals.[53,54] The presence of a breast tumor enhances the local expression of aromatase three- to fourfold in the surrounding breast adipose tissue, to levels equivalent to those within the tumor itself. Furthermore, there is a gradient of aromatase expression with the tumor as its focus.[55–57] Breast tumors produce factors that stimulate the proliferation of mesenchymal cells surrounding them, the *desmoplastic reaction*; but these factors also stimulate aromatase expression within the proliferating mesenchymal cells. Evidence from several independent laboratories suggests that aromatase expression in both breast tumors and adipose tissue surrounding breast tumors is regulated differently compared to normal adipose tissue and involves promoter switching. Thus, normal adipose tissue uses promoter I.4 to drive *CYP19* expression, whereas breast tumors and adipose tissue surrounding breast tumors use promoter II.[48,49,58]

As discussed above, promoter II is strongly induced by PGE_2, which is itself produced by breast tumor epithelium, fibroblasts, or macrophages recruited to the tumor site. Thus, the tumor environment creates a positive feedback loop whereby tumor-derived factors such as PGE_2 increase the local estrogen content, leading to further tumor growth and development (Fig. 8.5).

Activation of promoter II is therefore a key event in the overexpression of aromatase in breast tumors and adipose fibroblasts, although the molecular mechanisms by which promoter switching occur are currently unclear. One possibility is that the PGE_2-induced increase in intracellular cAMP levels leads to activation and binding of CREB to the promoter II CRE in a situation analogous to the regulation of promoter II by FSH in the ovary. Indeed, it appears that in breast cancer cell lines an additional CRE, located within promoter I.3 ($-66/-59$), contributes to cAMP-induced transcription from promoters II and I.3.[59] This CRE-like sequence overlaps a binding site for the zinc-finger transcriptional factor Snail (SnaH), which was identified by yeast one-hybrid screen using the CRE as a probe.[60] SnaH acts as a repressor of promoter I.3 activity and was expressed at higher levels in normal breast epithelial cell and stromal fibroblast cell lines than in breast cancer cell lines.[60] Thus, downregulation of SnaH in stromal cells adjacent to malignant breast epithelial cells might contribute to induction of aromatase through promoter I.3.

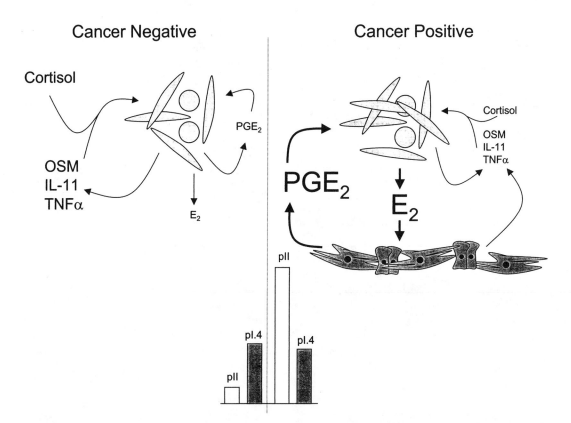

Figure 8.5 Proposed regulation of aromatase gene expression in breast adipose tissue from cancer-free individuals and from those with breast cancer. In the former case, expression is stimulated primarily by class I cytokines or tumor necrosis factor α (TNFα) produced locally in the presence of systemic glucocorticoids. As a consequence, promoter I.4-specific transcripts of aromatase predominate. In the latter case, prostaglandin E_2 (PGE_2) produced by the tumorous epithelium, tumor-derived fibroblasts, and/or macrophages recruited to the tumor site is the major factor stimulating aromatase expression, as evidenced by the predominance of promoter II and I.3-specific transcripts of aromatase. OSM, oncostatin M; E_2, estradiol; IL, interleukin.

Another study has implicated CCAAT/enhancer binding protein (C/EBP) in this promoter switching. Conditioned medium from T47D breast cancer cells caused a marked induction of C/EBPβ, but not C/EBPα, or C/EBPδ, in cultured adipose stromal cells.[61] It was suggested that the promoter switch is mediated at least in part by the tumor-induced upregulation and enhanced binding of C/EBPβ to a promoter II element at position −317/−304. This group also showed that TNF-α and IL-11 secreted by malignant breast epithelial cells inhibit adipocyte differentiation by selectively downregulating C/EBPα and peroxisome proliferator–activated receptor γ (PPARγ) and proposed that this is a mechanism for the desmoplastic reaction, namely, the proliferation of preadipocyte fibroblasts.[62]

Aromatase expression in the ovary is critically dependent on SF-1 binding to a nuclear receptor half-site within promoter II. Since adipose fibroblasts do not express SF-1, the role of this half-site in expression from promoter II in adipose stromal cells is currently under investigation. One study has implicated liver receptor homologue-1 (LRH-1, NR5A2[38]) in transcriptional regulation through this site.[63] LRH-1 and SF-1 are the two human homologues of the *Drosophila* nuclear receptor fushi tarazu F1 (Ftz-F1)[64] and share DNA-binding and transactivation properties. LRH-1 is expressed at high levels in human preadipocytes but not mature adipocytes and strongly activates expression of promoter II reporter genes in 3T3-L1 preadipocytes.[63] Several other candidate transcription factors have been proposed, based on their interaction with the nuclear receptor half-site in yeast one-hybrid screens.[65] Factors identified included EAR-2, EAR-3 (chicken ovalbumin upstream promoter-transcription factor 1, or COUP-TF1), retinoic acid receptor γ (RARγ), and p120E4F, although the major protein interacting with the site was found to be estrogen receptor–related receptor α (ERRα).[65] ERRα moderately stimulated expression of aromatase promoter reporter genes in SK-BR-3 breast cancer cells and bound directly to the promoter in EMSA.[65]

Thus, at present, the nature of the factors responsible for the enhanced aromatase expression in breast tissue observed in the presence of a tumor and the mechanisms driving the promoter switching remain under active investigation. The known pathways of promoter II regulation in adipose stromal cells are summarized in Figure 8.6.

INHIBITION OF AROMATASE

Inhibition of estrogen action is generally the first-line adjuvant therapy for patients with metastatic estrogen receptor (ER)-positive breast cancers. Recent trials have demonstrated superiority of aromatase inhibitors such as anastrozole over traditional estrogen receptor antagonists in this setting,[66] and there is therefore much interest in the development of more effective aromatase inhibitors, both as treatment and as preventive agents.[67] Estrogen receptor antagonists and aromatase enzyme inhibitors inhibit estrogen action in a global fashion, particularly in bone tissue; and this inhibition may cause osteoporosis.[68] Although this might be of less consequence in advanced postmenopausal breast cancer, the effects of estrogen deprivation in bone of young postmenopausal women could be significant. In addition, other sequelae of estrogen deprivation with anti-estrogen therapy for breast cancer such as hepatic steatosis[69–71] and cognitive impairment,[72,73] have been described. For these reasons, it is desirable to design selective aromatase modulators (SAMs) that target the aberrant overexpression of aromatase in malignant breast epithelial cells and surrounding fibroblasts, while sparing other sites of estrogen action, such as bone.[2,13] The concept of SAMs is based on the facts that (1) local aromatase activity is the main source of estrogens in breast and bone and (2) the mechanism of aromatase expression is different in breast adipose stromal cells, osteoblasts, and brain. Therefore, drugs that act at the level of transcription from the individual *CYP19* promoter(s) have the potential to modulate aromatase expression and activity in a highly tissue-selective fashion.

The first candidate SAMs are ligands for the transcription factor PPARγ. Factors that stimulate aromatase activity in preadipocytes, e.g., TNF-α and class 1 cytokines, also inhibit adipocyte differentiation in model systems such as 3T3-L1 preadipocytes.[74,75] Conversely, it

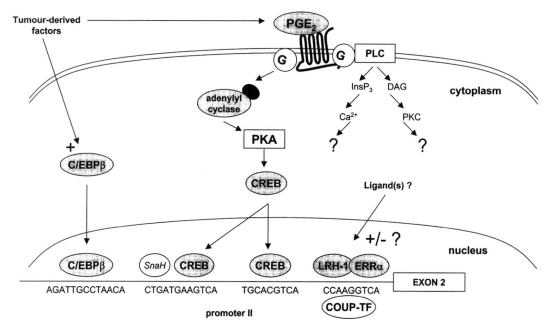

Figure 8.6 Pathways regulating aberrant overexpression of aromatase from promoter II in adipose tissue. Prostaglandin E_2 (PGE_2), derived from tumorous epithelial cells or infiltrating macrophages, activates adenylyl cyclase, leading to phosphorylation and binding of cyclic adenosine monophosphate (cAMP) response element binding protein (CREB) to two distinct cAMP response elements (CREs) within promoter II. Receptors for PGE_2 also activate phospholipase C (PLC) with resultant increases in intracellular calcium and protein kinase C (PKC) activity, which maximally induce transcription from promoter II by as yet unknown mechanisms. Tumor-derived factors also induce expression of C/EBPβ, which contributes to promoter II transcription. The proximal nuclear receptor half-site binds a number of proteins, including LRH-1 and ERRα. Activity of these monomeric nuclear receptors could be modified by phosphorylation, ligand binding, or co-regulator recruitment, possibly in response to tumor-derived factors. The net activity of promoter II is thus determined by the balance of positive (gray) and negative (white) transcription factor binding. DAG, diacylglycerol; $InsP_3$, inositol 1,4,5-triphosphate; C/EBP, CCAAT/enhancer binding protein; ERR, estrogen receptor–related receptor; LRH-1, liver receptor homologue-1; COUP-TF, chicken ovalbumin upstream promoter-transcription factor.

follows that factors that stimulate adipogenesis might inhibit aromatase activity in preadipocytes. PPARγ is a key transcription factor that controls adipose differentiation,[76] and is activated by thiazolidinediones, such as troglitazone and rosiglitazone, as well as the putative endogenous ligand 15-deoxy-Δ^{12-14}-prostaglandin J_2.[77] These PPARγ ligands (and ligands for retinoid X receptor α, the heterodimer partner of PPARγ) inhibit aromatase expression in adipose stromal cells. This has been shown at the level of aromatase activity, mRNA expression, and expression of promoter I.4 reporter constructs.[78] In addition, aromatase expression induced by cAMP and phorbol esters via promoter II is inhibited by PPARγ ligands,[79] indicating that such inhibition would be relatively specific for aromatase expression induced by tumor-derived PGE_2. Even more specific would be drugs that inhibit transcription from promoter II exclusively. It has been suggested that the genomic region surrounding the SF-1 binding site within promoter II can function as a silencer element in certain breast cancer cell lines.[80] Although the activity of promoter II is likely regulated by the balance of positive and negative transcription factors competing for binding to this site, it will be important to identify endogenous proteins that repress transcription from promoter II through this site. If this factor(s) proves to be a member of the nuclear receptor family, as seems likely given the DNA recognition sequence, then its ligands would be good candidates for specific inhibitors of aromatase expression in cancerous breast tissue.

REFERENCES

1. Simpson ER, Clyne C, Speed C, Rubin G, Bulun S. Tissue-specific estrogen biosynthesis and metabolism. Ann NY Acad Sci 949:58–67, 2001.
2. Simpson E, Rubin G, Clyne C, Robertson K, O'Donnell L, Jones M, Davis S. The role of local estrogen biosynthesis in males and females. Trends Endocrinol Metab 11:184–188, 2000.
3. Simpson E, Rubin G, Clyne C, Robertson K, O'Donnell L, Davis S, Jones M. Local estrogen biosynthesis in males and females. Endocr Relat Cancer 6:131–137, 1999.
4. Smith EP, Boyd J, Frank GR, Takahashi H, Cohen RM, Specker B, Williams TC, Lubahn DB, Korach KS. Estrogen resistance caused by a mutation in the estrogen-receptor gene in a man. N Engl J Med 331:1056–1061, 1994.
5. Morishima A, Grumbach MM, Simpson ER, Fisher C, Qin K. Aromatase deficiency in male and female siblings caused by a novel mutation and the physiological role of estrogens. J Clin Endocrinol Metab 80:3689–3698, 1995.
6. Couse JF, Hewitt SC, Bunch DO, Sar M, Walker VR, Davis BJ, Korach KS. Postnatal sex reversal of the ovaries in mice lacking estrogen receptors alpha and beta. Science 286:2328–2331, 1999.
7. Krege JH, Hodgin JB, Couse JF, Enmark E, Warner M, Mahler JF, Sar M, Korach KS, Gustafsson JA, Smithies O. Generation and reproductive phenotypes of mice lacking estrogen receptor beta. Proc Natl Acad Sci USA 95:15677–15682, 1998.
8. Lubahn DB, Moyer JS, Golding TS, Couse JF, Korach KS, Smithies O. Alteration of reproductive function but not prenatal sexual development after insertional disruption of the mouse estrogen receptor gene. Proc Natl Acad Sci USA 90: 11162–11166, 1993.
9. Fisher CR, Graves KH, Parlow AF, Simpson ER. Characterization of mice deficient in aromatase (ArKO) because of targeted disruption of the *cyp19* gene. Proc Natl Acad Sci USA 95:6965–6970, 1998.
10. Nemoto Y, Toda K, Ono M, Fujikawa-Adachi K, Saibara T, Onishi S, Enzan H, Okada T, Shizuta Y. Altered expression of fatty acid–metabolizing enzymes in aromatase-deficient mice. J Clin Invest 105:1819–1825, 2000.
11. Honda S, Harada N, Ito S, Takagi Y, Maeda S. Disruption of sexual behavior in male aromatase-deficient mice lacking exons 1 and 2 of the *cyp19* gene. Biochem Biophys Res Commun 252:445–449, 1998.
12. Robertson KM, O'Donnell L, Jones ME, Meachem SJ, Boon WC, Fisher CR, Graves KH, McLachlan RI, Simpson ER. Impairment of spermatogenesis in mice lacking a functional aromatase (*cyp19*) gene. Proc Natl Acad Sci USA 96:7986–7991, 1999.
13. Simpson E, Davis S. Why do the clinical sequelae of estrogen deficiency affect women more than men? J Clin Endocrinol Metab 83:2214, 1998.
14. Labrie F, Belanger A, Cusan L, Gomez JL, Candas B. Marked decline in serum concentrations of adrenal C19 sex steroid precursors and conjugated androgen metabolites during aging. J Clin Endocrinol Metab 82:2396–2402, 1997.
15. Simpson ER, Boyd GS. The cholesterol side-chain cleavage system of bovine adrenal cortex. Eur J Biochem 2:275–285, 1967.
16. Burstein S, Gut M. Intermediates in the conversion of cholesterol to pregnenolone: kinetics and mechanism. Steroids 28:115–131, 1976.
17. Duque C, Morisaki M, Ikekawa N, Shikita M. The enzyme activity of bovine adrenocortical cytochrome P-450 producing pregnenolone from cholesterol: kinetic and electrophoretic studies on the reactivity of hydroxycholesterol intermediates. Biochem Biophys Res Commun 82:179–187, 1978.
18. Picado-Leonard J, Miller WL. Cloning and sequence of the human gene for P450c17 (steroid 17alpha-hydroxylase/17,20lyase): similarity with the gene for P450c21. DNA 6:439–448, 1987.
19. Siiteri PK, Macdonald PC. Role of extraglandular estrogen in human endocrinology. In: Green RO, Astwoon EB (eds). Handbook of Physiology. Female Reproductive System, sect 7, vol II. Washington DC: American Physiological Society, 1973, pp 619–629.
20. Simpson ER, Zhao Y, Agarwal VR, Michael MD, Bulun SE, Hinshelwood MM, Graham-Lorence S, Sun T, Fisher CR, Qin K, Mendelson CR. Aromatase expression in health and disease. Recent Prog Horm Res 52:185–213, 1997.
21. Cole PA, Robinson CH. A peroxide model reaction for placental aromatase. J Am Chem Soc 110: 1284–1285. 1988.
22. Akhtar M, Calder MR, Corina DL, Wright JN. Mechanistic studies on C-19 demethylation in oestrogen biosynthesis. Biochem J 201:569–580, 1982.
23. Bulun SE, Simpson ER. Regulation of aromatase expression in human tissues. Breast Cancer Res Treat 30:19–29, 1994.
24. Chen SA, Besman MJ, Sparkes RS, Zollman S, Klisak I, Mohandas T, Hall PF, Shively JE. Human aromatase: cDNA cloning, Southern blot analysis, and assignment of the gene to chromosome 15. DNA 7:27–38, 1988.
25. Sebastian S, Bulun SE. A highly complex organization of the regulatory region of the human *CYP19* (aromatase) gene revealed by the Human Genome Project. J Clin Endocrinol Metab 86:4600–4602, 2001.

26. Means GD, Mahendroo MS, Corbin CJ, Mathis JM, Powell FE, Mendelson CR, Simpson ER. Structural analysis of the gene encoding human aromatase cytochrome P-450, the enzyme responsible for estrogen biosynthesis. J Biol Chem 264:19385–19391, 1989.
27. Harada N, Yamada K, Saito K, Kibe N, Dohmae S, Takagi Y. Structural characterization of the human estrogen synthetase (aromatase) gene. Biochem Biophys Res Commun 166:365–372, 1990.
28. Toda K, Terashima M, Kawamoto T, Sumimoto H, Yokoyama Y, Kuribayashi I, Mitsuuchi Y, Maeda T, Yamamoto Y, Sagara Y. Structural and functional characterization of human aromatase P-450 gene. Eur J Biochem 193:559–565, 1990.
29. Means GD, Kilgore MW, Mahendroo MS, Mendelson CR, Simpson ER. Tissue-specific promoters regulate aromatase cytochrome P450 gene expression in human ovary and fetal tissues. Mol Endocrinol 5:2005–2013, 1991.
30. Mahendroo MS, Means GD, Mendelson CR, Simpson ER. Tissue-specific expression of human P-450AROM. The promoter responsible for expression in adipose tissue is different from that utilized in placenta. J Biol Chem 266:11276–11281, 1991.
31. Toda K, Shizuta Y. Molecular cloning of a cDNA showing alternative splicing of the 5′-untranslated sequence of mRNA for human aromatase P-450. Eur J Biochem 213:383–389, 1993.
32. Jenkins C, Michael D, Mahendroo M, Simpson E. Exon-specific Northern analysis and rapid amplification of cDNA ends (RACE) reveal that the proximal promoter II (PII) is responsible for aromatase cytochrome P450 (CYP19) expression in human ovary. Mol Cell Endocrinol 97:R1–R6, 1993.
33. Mahendroo MS, Mendelson CR, Simpson ER. Tissue-specific and hormonally controlled alternative promoters regulate aromatase cytochrome P450 gene expression in human adipose tissue. J Biol Chem 268:19463–19470, 1993.
34. Honda S, Harada N, Takagi Y. Novel exon 1 of the aromatase gene specific for aromatase transcripts in human brain. Biochem Biophys Res Commun 198:1153–1160, 1994.
35. Shozu M, Zhao Y, Bulun SE, Simpson ER. Multiple splicing events involved in regulation of human aromatase expression by a novel promoter, I.6. Endocrinology 139:1610–1617, 1998.
36. Steinkampf MP, Mendelson CR, Simpson ER. Regulation by follicle-stimulating hormone of the synthesis of aromatase cytochrome P-450 in human granulosa cells. Mol Endocrinol 1:465–471, 1987.
37. Michael MD, Michael LF, Simpson ER. A CRE-like sequence that binds CREB and contributes to cAMP-dependent regulation of the proximal promoter of the human aromatase P450 (CYP19) gene. Mol Cell Endocrinol 134:147–156, 1997.
38. A unified nomenclature system for the nuclear receptor superfamily. Cell 97:161–163, 1999.
39. Michael MD, Kilgore MW, Morohashi K, Simpson ER. Ad4BP/SF-1 regulates cyclic AMP–induced transcription from the proximal promoter (PII) of the human aromatase P450 (CYP19) gene in the ovary. J Biol Chem 270:13561–13566, 1995.
40. Luo X, Ikeda Y, Parker KL. A cell-specific nuclear receptor is essential for adrenal and gonadal development and sexual differentiation. Cell 77:481–490, 1994.
41. Price T, O'Brien S, Dunaif A, Simpson ER. Comparison of aromatase cytochrome P450 mRNA levels in adipose tissue from the abdomen and buttock using competitive polymerase chain reaction amplification [abstract]. Proc Soc Gynecol Invest p179:1992.
42. Zhao Y, Mendelson CR, Simpson ER. Characterization of the sequences of the human CYP19 (aromatase) gene that mediate regulation by glucocorticoids in adipose stromal cells and fetal hepatocytes. Mol Endocrinol 9:340–349, 1995.
43. Zhao Y, Nichols JE, Bulun SE, Mendelson CR, Simpson ER. Aromatase P450 gene expression in human adipose tissue. Role of a Jak/STAT pathway in regulation of the adipose-specific promoter. J Biol Chem 270:16449–16457, 1995.
44. Zhao Y, Nichols JE, Valdez R, Mendelson CR, Simpson ER. Tumor necrosis factor-alpha stimulates aromatase gene expression in human adipose stromal cells through use of an activating protein-1 binding site upstream of promoter 1.4. Mol Endocrinol 10:1350–1357, 1996.
45. Papamakarios T, Speed CJ, Clyne CD, Simpson ER. Regulation of P450 aromatase expression by TNFα in human adipose stromal cells [abstract]. Proc Endocr Soc Aust 44:201, 2001.
46. Cella N, Groner B, Hynes NE. Characterization of Stat5a and Stat5b homodimers and heterodimers and their association with the glucocorticoid receptor in mammary cells. Mol Cell Biol 18:1783–1792, 1998.
47. Look DC, Pelletier MR, Tidwell RM, Roswit WT, Holtzman MJ. Stat1 depends on transcriptional synergy with Sp1. J Biol Chem 270:30264–30267, 1995.
48. Agarwal VR, Bulun SE, Leitch M, Rohrich R, Simpson ER. Use of alternative promoters to express the aromatase cytochrome P450 (CYP19) gene in breast adipose tissues of cancer-free and breast cancer patients. J Clin Endocrinol Metab 81:3843–3849, 1996.
49. Harada N, Utsumi T, Takagi Y. Tissue-specific expression of the human aromatase cytochrome P-450 gene by alternative use of multiple exons 1 and promoters, and switching of tissue-specific exons 1 in carcinogenesis. Proc Natl Acad Sci USA 90:11312–11316, 1993.
50. Mendelson CR, Corbin CJ, Smith ME, Smith J, Simpson ER. Growth factors suppress and phor-

bol esters potentiate the action of dibutyryl adenosine 3′,5′-monophosphate to stimulate aromatase activity of human adipose stromal cells. Endocrinology 118:968–973, 1986.
51. Mendelson CR, Smith ME, Cleland WH, Simpson ER. Regulation of aromatase activity of cultured adipose stromal cells by catecholamines and adrenocorticotropin. Mol Cell Endocrinol 37:61–72, 1984.
52. Zhao Y, Agarwal VR, Mendelson CR, Simpson ER. Estrogen biosynthesis proximal to a breast tumor is stimulated by PGE_2 via cyclic AMP, leading to activation of promoter II of the CYP19 (aromatase) gene. Endocrinology 137:5739–5742, 1996.
53. Thorsen T, Tangen M, Stoa KF. Concentration of endogenous oestradiol as related to oestradiol receptor sites in breast tumor cytosol. Eur J Cancer Clin Oncol 18:333–337, 1982.
54. van Landeghem AA, Poortman J, Nabuurs M, Thijssen JH. Endogenous concentration and subcellular distribution of estrogens in normal and malignant human breast tissue. Cancer Res 45:2900–2906, 1985.
55. Bulun SE, Simpson ER. Competitive reverse transcription-polymerase chain reaction analysis indicates that levels of aromatase cytochrome P450 transcripts in adipose tissue of buttocks, thighs, and abdomen of women increase with advancing age. J Clin Endocrinol Metab 78:428–432, 1994.
56. Harada N. Aberrant expression of aromatase in breast cancer tissues. J Steroid Biochem Mol Biol 61:175–184, 1997.
57. Zhou D, Zhou C, Chen S. Gene regulation studies of aromatase expression in breast cancer and adipose stromal cells. J Steroid Biochem Mol Biol 61:273–280, 1997.
58. Zhou C, Zhou D, Esteban J, Murai J, Siiteri PK, Wilczynski S, Chen S. Aromatase gene expression and its exon I usage in human breast tumors. Detection of aromatase messenger RNA by reverse transcription-polymerase chain reaction. J Steroid Biochem Mol Biol 59:163–171, 1996.
59. Zhou D, Chen S. Identification and characterization of a cAMP-responsive element in the region upstream from promoter 1.3 of the human aromatase gene. Arch Biochem Biophys 371:179–190, 1999.
60. Okubo T, Truong TK, Yu B, Itoh T, Zhao J, Grube B, Zhou D, Chen S. Down-regulation of promoter 1.3 activity of the human aromatase gene in breast tissue by zinc-finger protein, snail (SnaH). Cancer Res 61:1338–1346, 2001.
61. Zhou J, Gurates B, Yang S, Sebastian S, Bulun SE. Malignant breast epithelial cells stimulate aromatase expression via promoter II in human adipose fibroblasts: an epithelial–stromal interaction in breast tumors mediated by CCAAT/enhancer binding protein beta. Cancer Res 61:2328–2334, 2001.
62. Meng L, Zhou J, Sasano H, Suzuki T, Zeitoun KM, Bulun SE. Tumor necrosis factor alpha and interleukin 11 secreted by malignant breast epithelial cells inhibit adipocyte differentiation by selectively down-regulating CCAAT/enhancer binding protein alpha and peroxisome proliferator–activated receptor gamma: mechanism of desmoplastic reaction. Cancer Res 61:2250–2255, 2001.
63. Clyne CD, Speed CJ, Zhou J, Simpson ER. Liver receptor homologue-1 (LRH-1) regulates expression of aromatase in preadipocytes. J Biol Chem 277:20591–20597, 2002.
64. Lavorgna G, Ueda H, Clos J, Wu C. FTZ-F1, a steroid hormone receptor-like protein implicated in the activation of fushi tarazu. Science 252:848–851, 1991.
65. Yang C, Zhou D, Chen S. Modulation of aromatase expression in the breast tissue by ERR alpha-1 orphan receptor. Cancer Res 58:5695–5700, 1998.
66. Munster PN, Horton J. Tamoxifen vs the aromatase inhibitors: news from San Antonio, 2001. Cancer Control 8:478–479, 2001.
67. Park WC, Jordan VC. Selective estrogen receptor modulators (SERMS) and their roles in breast cancer prevention. Trends Mol Med 8:82–88, 2002.
68. Lindsey AM, Gross G, Twiss J, Waltman N, Ott C, Moore TE. Postmenopausal survivors of breast cancer at risk for osteoporosis: nutritional intake and body size. Cancer Nurs 25:50–56, 2002.
69. Murata Y, Ogawa Y, Saibara T, Nishioka A, Fujiwara Y, Fukumoto M, Inomata T, Enzan H, Onishi S, Yoshida S. Unrecognized hepatic steatosis and non-alcoholic steatohepatitis in adjuvant tamoxifen for breast cancer patients. Oncol Rep 7:1299–1304, 2000.
70. Pinto HC, Baptista A, Camilo ME, de Costa EB, Valente A, de Moura MC. Tamoxifen-associated steatohepatitis—report of three cases. J Hepatol 23:95–97, 1995.
71. Oien KA, Moffat D, Curry GW, Dickson J, Habeshaw T, Mills PR, MacSween RN. Cirrhosis with steatohepatitis after adjuvant tamoxifen. Lancet 353:36–37, 1999.
72. Paganini-Hill A, Clark LJ. Preliminary assessment of cognitive function in breast cancer patients treated with tamoxifen. Breast Cancer Res Treat 64:165–176, 2000.
73. van Dam FS, Schagen SB, Muller MJ, Boogerd W, Wall E.v.d., Droogleever Fortuyn ME, Rodenhuis S. Impairment of cognitive function in women receiving adjuvant treatment for high-risk breast cancer: high-dose versus standard-dose chemotherapy. J Natl Cancer Inst 90:210–218, 1998.
74. Petruschke T, Hauner H. Tumor necrosis factor-alpha prevents the differentiation of human adipocyte precursor cells and causes delipidation of newly developed fat cells. J Clin Endocrinol Metab 76:742–747, 1993.

75. Kawashima I, Ohsumi J, Mita-Honjo K, Shimoda-Takano K, Ishikawa H, Sakakibara S, Miyadai K, Takiguchi Y. Molecular cloning of cDNA encoding adipogenesis inhibitory factor and identity with interleukin-11. FEBS Lett 283:199–202, 1991.
76. Brun RP, Spiegelman BM. PPARgamma and the molecular control of adipogenesis. J Endocrinol 155:217–218, 1997.
77. Kliewer SA, Lenhard JM, Willson TM, Patel I, Morris DC, Lehmann JM. A prostaglandin J_2 metabolite binds peroxisome proliferator–activated receptor gamma and promotes adipocyte differentiation. Cell 83:813–819, 1995.
78. Rubin GL, Zhao Y, Kalus AM, Simpson ER. Peroxisome proliferator–activated receptor gamma ligands inhibit estrogen biosynthesis in human breast adipose tissue: possible implications for breast cancer therapy. Cancer Res 60:1604–1608, 2000.
79. Rubin GL, Duong JH, Clyne CD, Speed CJ, Gong C, Simpson ER. Ligands for the peroxisome proliferator–activated receptor γ and the retinoid X receptor inhibit aromatase cytochrome P450 (*CYP19*) expression mediated by promoter II in human breast adipose. Endocrinology, 143:2863–2871, 2002.
80. Zhou D, Chen S. Characterization of a silencer element in the human aromatase gene. Arch Biochem Biophys 353:213–220, 1998.

9

Estrogen Metabolism Genes: *HSD17B1* and *HSD17B2*

HELLEVI PELTOKETO
VELI ISOMAA
DEBASHIS GHOSH
PIRKKO VIHKO

Oophorectomy in a woman's early adult years substantially lowers her risk of breast cancer.[1–4] This is an example of the abundant epidemiological and endocrine data which indicate that estrogens are involved in sporadic breast cancer and promote growth.[5–8] Oophorectomy also reduces the risk of breast cancer among *BRCA1* mutation carriers,[9] which hints at the general role of estrogens in the proliferation of breast epithelial cells. Due to the striking effect of ovariectomy, it is assumed that the majority of breast cancers originate as estrogen-dependent lesions and later progress gradually to a hormone-independent state. Since 1896,[10] estrogen-deprivation treatment has been used for the treatment of breast cancer. On average, one-third of nonselected tumors[11] and the majority of estrogen and progesterone receptor-positive tumors[12,13] in advanced disease respond to endocrine therapy based on blocking estrogen receptors and/or deprivation of the estrogenic ligand.

In the circulation, estrogens are transported by sex hormone–binding globulin[14] and albumin to the vicinity of their target cells, after which the hormones freely diffuse through cell membranes into the cells. Serum concentrations of estrogen and steroid metabolism in the target or surrounding cells therefore determine the availability of estrogens in breast tissue. The ovary is the single primary source of estradiol (E_2) in the circulation of premenopausal women, but circulating estrone (E_1) and androgens originating from the adrenal gland are also converted to E_2 in peripheral tissues such as adipose tissue and muscles, including smooth muscle cells of the vena cava.[15–18] Breast adipose and epithelial cells also contain enzymes needed for the production in situ of E_2 from circulating precursors,[7,16] which may further enhance estrogen action in the tissue. After menopause, in particular, estrogen biosynthesis in peripheral tissues has a major role in estrogen action.[19] Finally, nutriment may contain compounds with estrogenic and/or antiestrogenic effects.[20]

An enzyme group affecting the availability of biologically active estrogens and androgens is the family of 17β-hydroxysteroid dehydrogenases (17HSDs). These catalyze the interconversions between 17β-hydroxysteroids and 17-ketosteroids (Fig. 9.1) and, because both estrogens and androgens have the highest affinity toward their receptors in the 17β-hydroxy form, regulate the biological activity of sex hormones. Re-

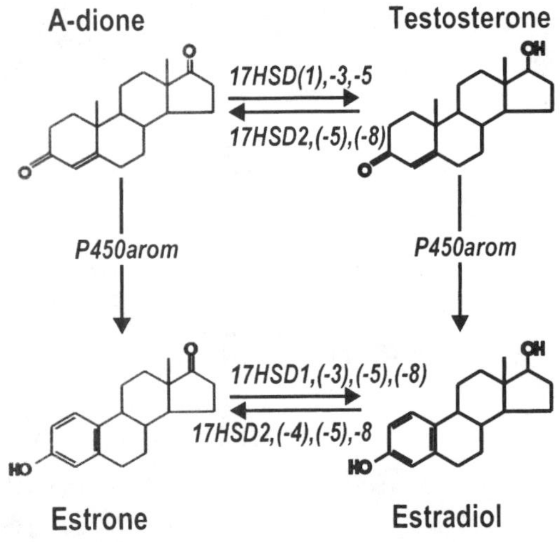

Figure 9.1 Enzymatic reactions catalyzed by P-450 aromatase and human 17β-hydroxysteroid dehydrogenases (17HSDs). Enzymatic activities that are not obviously the main function of the enzyme are shown in parenthesis. A-dione, androstenedione.

ductive 17HSD activities, i.e., 17-ketosteroid reductase (17KSR) activities, are part of the steroidogenic pathway leading to the biosynthesis of E_2 and testosterone (T) in the gonads. Certain extragonadal tissues, including breast and prostate tissues, also express reductive 17HSDs,[21–23] thus, the low-activity circulating precursors are additionally converted to their more potent forms in the target tissues of hormone action. The 17HSDs, with mainly oxidative activities, in turn tend to decrease the potency of estrogens and androgens and consequently may protect tissues from excessive hormonal action. Several tissues contain 17HSD enzymes leading to the biosynthesis of E_2 as well as 17HSD enzymes that bring about the inactivation of hormones or decrease their activity. Other enzymes, such as P-450 aromatase, estrogen sulfatase, and estrogen sulfotransferase, also modulate androgens and estrogens. Hence, the amount of biologically active hormone in a cell is dependent on the transfer of hormones and the sum of the actions of different enzyme types in the cell.

An increasing number of enzymes with 17HSD and/or 17KSR activities is being described.[24–27] Types 1, 2, and 8 of 17HSD (17HSD1, 17HSD2, 17HSD8) are primarily involved in the modulation of estrogen action in humans (Fig. 9.1).[24] In the present chapter, we concentrate on 17HSD1 and -2, whose function in peripheral estrogen action is most extensively characterized. We particularly focus on the role of 17HSDs in supplying the ligand for the estrogen receptor.

ENZYMATIC CHARACTERISTICS AND FUNCTION OF 17β-HYDROXYSTEROID DEHYDROGENASES 1 AND 2

Human 17HSD1 and -2 belong to the same dehydrogenase reductase protein family but differ from each other in many respects (Table 9.1). The main difference between the enzymes is the direction of their enzymatic activities. Human 17HSD1 mainly catalyzes reduction of E_1 to E_2,[28,29] preferring the phosphorylated form of nicotinamide-adenine dinucleotide, NADPH, as a cofactor.[30] In cultured cells, the human type 1 enzyme is also capable of reducing androstenedione (A-dione) and androstanedione to some extent, but it clearly prefers phenolic, i.e., estrogenic, substrates over androgens.[29,31] Type 2 17HSD predominantly catalyzes opposite reactions to 17HSD1, converting E_2 to E_1, T to A-dione, and dihydrotestosterone to androstanedione;[32] and it acts most efficiently in the presence of the nonphosphorylated form of the cofactor NAD^+.[33] Since intracellular concentrations of NADPH and NAD^+ are remarkably higher than those of $NADP^+$ and NADH, the preference of 17HSD1 toward reductive reactions and that of 17HSD2 toward oxidative reactions in vivo may be due to the cofactor specificities of the enzymes.[30,33]

Type 1 17HSD is more limitedly expressed than the type 2 enzyme, which may reflect their roles in estrogen metabolism. Type 1 17HSD is an essential part of the E_2 production machinery, and it is most abundantly expressed in the granulosa cells of the ovary[34] and the syncytiotrophoblasts of the placenta,[35] which secrete E_2 into the circulation. In addition, the type 1 enzyme contributes to the estrogen response by converting E_1 to E_2 locally in certain targets of estrogen action, such as breast tissue (Table 9.1 and references therein).

Table 9.1 Comparison of Human 17HSD1 and 17HSD2

	17HSD1	17HSD2	References
Tertiary structure	Homodimer	Not known	136
Size of subunit° (kDa)	35	43	32, 73, 137
Abundant expression	Ovary, placenta	Placenta, liver, gastrointestinal tract, kidney	34, 35, 38, 39, 67, 75, 101, 138
Expression also detected†	Breast, testis, blood vessels, bone	Uterus, breast, prostate	18, 23, 55, 66, 67, 70
Subcellular localization	Cytosol	Endoplasmic reticulum	139
Preferred cofactor	NADPH (NADH)	NAD^+	33
Catalytic preference in vivo‡	Reductase	Dehydrogenase	32, 56
Substrate specificity	E_1	E_2, T, DHT, (20α-P)	32, 56
Functions	Production of E_2 to the circulation and to the E_2-sythesizing and surrounding cells	Reduction of biological activity of estrogens and androgens	

°Predicted from the amino acid sequence.
†In the case of 17HSD1, only tissues in which expression of the 1.3 kb mRNA and/or 17HSD1 protein have been detected are listed because the 2.3 kb 17HSD1 mRNA is probably translated to the protein poorly.
‡Activities have been measured in cultured cells transfected with the cDNA in question. A substrate has been added to the media, and the product formed has been collected and measured at certain points of time.
DHT, 5α-dihydrotestosterone; E_1, estrone; E_2, estradiol; 20α-P, 20α-dihydroprogesterone; T, testosterone; NAD^+, oxidized nicotinamide adenine dinucleotide; NADH, reduced NAD; NADPH, reduced NAD phosphate.

Opposite to 17HSD1, 17HSD2 may be involved in the degradation and excretion of E_2 and T. The type 2 enzyme may restrict access of active sex steroids into the circulation and may protect target tissues of hormone action from excessive sex hormone influence by catalyzing androgens and estrogens into less active forms. Type 2 17HSD is expressed in a wide variety of tissues, such as breast, uterus, prostate, liver, and kidney (Table 9.1 and references therein). Typically, the type 2 enzyme is expressed in epithelial cells, such as the surface epithelial cells of the stomach, small intestine, and colon and the epithelium of the urinary bladder.[36,37] In the placenta, the type 2 enzyme may limit the access of fetal androgens to maternal tissue and the access of maternal estrogen into the fetus and, thus, act as a barrier between the fetus and mother.[38,39]

17β-HYDROXYSTEROID DEHYDROGENASES IN OVARIAN FUNCTION

During the reproductive years of women, the ovary and placenta are the most important single sources of the estrogens secreted into the circulation. Estradiol is produced in the ovarian granulosa cells of developing follicles from androgenic precursors of thecal origin.[40,41] The granulosa cells are rich in P-450 aromatase and 17HSD1, which convert A-dione to E_1 and E_1 to E_2, respectively. Expression of P-450 aromatase and 17HSD1 is under strict control and associated with the developmental phase of follicles. The maturation stage of the follicles, in turn, is related to their capacity to produce E_2.[34,42,43] Expression of 17HSD1 has also been correlated with malignant transformation of the surface epithelium of the human ovary.[44] Type 17HSD may increase the E_2 concentration in the epithelial cells and thus promote ovarian cancer progression.

In addition to the granulosa cells around the antrum, 17HSD1 is abundantly expressed in the cumulus cells,[34] which form the corona radiata encircling the ovum. Cumulus cells also express 17HSD8, which prefers the opposite activity to 17HSD1, i.e., catalysis of the reaction from E_2 to E_1.[45] These two enzymes conceivably regulate the amount of E_2 available for the ovum, but their exact function is not yet characterized. However, 17HSD8 was originally known as Ke6 protein, and its deficiency has been linked to the development of recessive polycystic kidney disease in mice.[46,47]

The third 17HSD abundantly expressed in the ovary is 17HSD7, which catalyzes the conversion of E_1 to E_2 in corpora lutea.[48] Expression of the enzyme is remarkable during rodent pregnancy, in particular.[48-50] Preliminary data, however, suggest that the human type 7 counterpart may have a function different from that of the rodent enzyme (Törn SJ, Nokelainen PA, Vihko PT, unpublished data).

17β-HYDROXYSTEROID DEHYDROGENASES IN BREAST EPITHELIAL CELLS

Both 17HSD1 and 17HSD2 are expressed in normal breast tissue,[23,51] and recent investigations suggest that the oxidative (type 2) activity is the dominant form in nontumorous breast cells.[51-53] In a study in which a total of 22 samples was analyzed, 17HSD1 protein was detected in most of the normal breast samples from women having regular menstrual cycles and in all samples from women using hormonal contraception.[23] The type 1 enzyme is expressed in epithelial cells of ducts and alveoli throughout the menstrual cycle, and it correlates with the presence of estrogen receptor. Immunohistochemical staining of 17HSD1 has also been observed in breast samples from postmenopausal women.[23]

The presence of 17HSD2 mRNA in a normal breast epithelial sample has been shown using in situ hybridization.[51] Also, 17HSD1 and 17HSD2 mRNAs have been detected in human mammary epithelial cell lines[51] and primary cultures[53] derived from women undergoing reduction mammoplasty. Oxidative 17HSD activity leading to the conversion of E_2 to E_1 is up to 50 times more predominant over reductive 17HSD activity in these cells.[51-53]

Both 17HSD1 and 17HSD2 are also expressed in malignant breast cells. About 50% of malignant breast specimens show positive immunohistochemical staining for 17HSD1.[54,55] The breast cancer cell lines analyzed[51,56] have been found to express either 17HSD1, 17HSD2, or both; and tumor-derived primary cultures have been found to possess reductive and/or oxidative 17HSD activity.[53] Concentrations and activities of the 17HSD1 and -2 enzymes vary from one sample and cell line to another, but certain primary cultures,[53] cell lines,[56] and individual breast tissue samples[57] have been found to contain the type 1 enzyme very abundantly. Expression and activity of 17HSD1 has sometimes been observed to be remarkably high, especially in primary cultures derived from stromal cells of malignant tissue[53] and in tissue samples associated with strong stromal proliferation,[57] even though linkage between the expression of 17HSD1 and that of the proliferation marker Ki has not been detected.[55]

Opposite to samples of normal breast tissue, the expression and activity of 17HSD1 is mostly dominant over oxidative 17HSD activity in malignant cells.[52,53,56] Estradiol has been shown to accumulate in breast tissue and in the malignant tissue itself, in particular,[58-60] which also implies 17HSD1 activity. Moreover, in the presence of 17HSD1, administration of E_1 results in similar growth of breast cancer cells, which E_2 causes alone, while E_1 does not have the same effect in control cells without 17HSD1.[61] Altogether, the data suggest that intracellular biosynthesis has an impact on the estrogen response. A predominance of 17HSD1 in malignant breast tissue may lead to increased estrogen-dependent proliferation and progress of cancer, while oxidative 17HSD2 may protect normal breast cells from an excessive E_2 effect.

17β-HYDROXYSTEROID DEHYDROGENASES 1 AND 2 IN OTHER TISSUES

In addition to breast tissue, both 17HSD1 and 17HSD2 are expressed in another typical target tissue of female sex steroids, the endometrium. Using immunological methods, 17HSD1 has been localized in the surface and glandular epithelial cells when progesterone was present at concentrations characteristic of an ovulatory cycle.[62] During the proliferative phase of the menstrual cycle, 17HSD1 staining was absent or minimal. However, it was maximal during the mid-secretory phase and disappeared rapidly thereafter. An antiprogestin given to women shortly after ovulation blocked the expression of 17HSD1 during the luteal phase of the cycle, suggesting progesterone receptor–mediated induction of the type 1 enzyme.[63] Using in situ hybridization, the mRNA for 17HSD2 has also been localized in epithelial cells of the endometrium; but in contrast to 17HSD1, 17HSD2 expression was highest at the end of the cycle.[38] Both 17HSD1 and 17HSD2 have been detected in human endometrium during the luteal phase of the cycle using Northern analysis, and 17HSD2 appears to be the predominant 17HSD,[56] in accordance with the preferred oxidation of 17β-hydroxysteroids in endometrial tissue.[64] Differential expression of the two 17HSD enzymes with opposite activities in the same cell types may effectively modulate intracellular E_2 concentrations during the menstrual cycle and thereby improve the control of E_2 influence during menstrual variations.

There are few data on the expression of 17HSD1 and 17HSD2 in postmenopausal endometrium and endometrial carcinoma. Antibodies against 17HSD1 have revealed the enzyme in about 50% of carcinoma specimens.[65] There was marked heterogeneity among 17HSD1-positive malignant tissue specimens; staining intensity varied greatly, showing both stained and unstained malignant epithelial cells.[65] In endometrial cancer cell lines, very little type 1 activity has been detected, while some cell lines have shown high type 2 enzyme-like activity and expression of 17HSD2 mRNA.[56] Of male sex steroid target tissues, 17HSD2 has been detected in normal and malignant prostate. Higher expression of 17HSD2 has been detected in benign prostatic hyperplasia compared with prostatic carcinoma.[66] PC3, a prostate cancer cell line, also expresses 17HSD2.[56]

The type 2 enzyme is more widely distributed in tissues of humans and other animal species than 17HSD1. Strong expression has been detected in human liver, placenta, and small intestine and lower levels in the colon, kidney, and pancreas.[56,67] Additional information on the physiological importance of the type 2 enzyme has been obtained by localizing it in mouse tissues: the enzyme is expressed in several epithelial layers of the gastrointestinal and urogenital tracts.[36] The localization and intensity of 17HSD type 2 mRNA expression were, furthermore, identical in the gastrointestinal tracts of both male and female mice.[36] During embryogenesis, its expression strongly increased in parallel with the fetal development of gastrointestinal organs.[68] All of these data suggest a role for 17HSD2 in the inactivation of sex steroids and steroid-like compounds present in the intestinal contents. Both human and mouse 17HSD2 enzymes are also highly expressed in the placenta.[38,68,69] Hence, 17HSD2 may maintain a barrier to the highly active 17β-hydroxy forms of sex steroids between mother and fetus.

HSD17B1 AND *HSD17B2*, THE GENES ENCODING 17β-HYDROXYSTEROID DEHYDROGENASES 1 AND 2

Despite the similarity of the substrates and cofactors that 17HSD1 and 17HSD2 recognize, the enzymes share low overall similarity.[32] The amino acid sequence identity between 17HSD1 and 17HSD2 peptides is as low as 23%, even though certain amino acids critical for function are conserved in both.[32] The structures of the *HSD17B1* and *HSD17B2* genes also differ remarkably: *HSD17B1* consists of six exons and five introns with a total length of 3.2 kb,[70,71] while *HSD17B2* is at least 40 kb in size and comprises five exons and four introns (Table 9.2).[72] *HSD17B2* is a single-copy gene, and *HSD17B1* in turn is located in a duplicated region, of 12.4 kb, which also contains the *HSD17BP1* gene. *HSD17B1* and *HSD17BP1* share 89% overall

Table 9.2 Comparison of Human *HSD17B1* and *HSD17B2* Genes

	HSD17B1	HSD17BP1°	HSD17B2	References
Size (kb)	3.2	3.2	>40	70–72
Number of exons	6	6	5	70–72
Size of mRNAs (kb)	1.3, 2.3		1.5 (two splicing variants of same size)	72, 73
Chromosomal localization	17q21	17q21	16q24.1–2	81, 140

°Data based on comparison with *HSD17B1*.

identity, but the function, if any, of the latter gene is unknown.

Both *HSD17B1* and *HSD17B2* are transcribed into two mRNA transcripts but by different mechanisms. The two 17HSD1 mRNAs, 1.3 and 2.3 kb in size, are a result of the presence of two transcription-start points,[70,73] whereas the two 17HSD2 mRNAs, both 1.5 kb in size, are splicing variants.[72] The two 17HSD1 mRNAs contain the same coding area, but only the amount of the 1.3 kb transcript correlates with the concentration of 17HSD1 protein.[56,74] Therefore, the presence of 2.3 kb 17HSD1 mRNA does not always indicate the presence of 17HSD1 enzyme in a tissue. It is also the 1.3 kb 17HSD1 mRNA whose transcription is subject to various forms of regulation,[74–78] which emphasizes the importance of this transcript. Alternative splicing of *HSD17B2* leads to two mRNAs with somewhat different coding areas. One of the two 17HSD2 mRNAs matches the identified type 2 cDNA and encodes an enzymatically active protein, while the protein product of the other transcript does not possess 17HSD activity and is of unknown function.[72]

The genes *HSD17B1* and *HSD17B2* are localized at chromosomal areas 17q21[75,79,80] and 16q24.1–q24.2,[67,81] respectively. Frequent loss of heterozygosity (LOH) at 16q, and at 16q24 in particular, is among the numerous chromosomal aberrations linked to sporadic breast[82–86] and prostate[87–91] carcinomas. In prostate cancer, LOH at 16q24 is generally correlated with poor prognosis, i.e., clinically aggressive behavior of the disease, metastatic disease and higher tumor grade.[89,90,92,93] However, no somatic mutations changing the amino acid sequence have been found in *HSD17B2* in prostate cancer samples,[90] suggesting that *HSD17B2* is not the gene targeted by LOH at 16q24 in prostate cancer. Deletion of *HSD17B2* with other genes at the region, however, may accelerate the progression of prostate cancer as well as that of breast cancer since the protective effect of 17HSD2 against excessive androgen and estrogen action will be missing.

Altogether, 52 independent familial breast cancer samples have been screened for mutations in *HSD17B1*, but only polymorphisms similar to those in control samples have been found.[80,94–96] The polymorphisms include two nucleotide changes which result in an altered amino acid sequence, but the amino acid variations Ala^{237} to Val and Ser^{312} to Gly do not affect the enzymatic activity of 17HSD1.[97] However, a rare nucleotide change detected in the TATA box in the promoter of *HSD17B1* may decrease transcription of the gene to some extent.[95] Amplification of *HSD17B1*, and consequently increased 17HSD1 expression and E$_2$ production, would more likely be harmful to the individual than partially reduced gene expression. So far, no sign of *HSD17B1* amplification in breast cancer samples[79] or cancer cell lines abundantly expressing 17HSD1[98] has been observed. Alterations in the structure of *HSD17B1* are thus not linked to breast or any other cancer.

REGULATION OF 17β-HYDROXYSTEROID DEHYDROGENASE 1 EXPRESSION

Besides genetic alterations, abnormal regulation of gene transcription can result in changed protein expression and biological response. An example is the mouse gene for 17HSD8, also known as the *ke6* gene, in which no nucleotide alterations have been detected but lack of the regulatory protein of which probably results in recessive polycystic kidney disease.[47] To elucidate the mechanisms resulting in the expression, or possible overexpression, of 17HSD1, regulation of *HSD17B1* expression in various cell types has been characterized.

In granulosa cells, expression of 17HSD1 and P-450 aromatase is largely regulated by similar gonadotropin-dependent systems, which are modulated by growth factors, members of the inhibin/activin family, and steroid hormones.[41,99–102] The primary inducer of 17HSD1 expression is follicle-stimulating hormone, which acts via the protein kinase A pathway. Protein kinase C–dependent inhibition, estrogens, androgens, and autocrine/paracrine growth factors present in the ovary further modulate the extent of induction.[100,102,103] Luteinizing factors subsequently cause a decline of 17HSD1 in the follicle.[103]

Retinoic acids, adenylate cyclase, protein kinase C activators, and several growth factors increase expression of 17HSD1 synergistically in JEG-3 cells, which are cancer cells of placental origin.[75,76,78,98,104,105] Similar to granulosa cells, the adenylate cyclase–dependent pathway may represent the primary mechanism behind induction of 17HSD1 in the placenta,[105] and other factors assist in maintaining the high 17HSD1 expression detected in syncytiotrophoblasts. Several paracrine, autocrine, and nutritional factors that regulate 17HSD1 expression in the placenta and/or ovaries are also present in breast tissue, as are their receptors. Regulation of 17HSD1 expression in breast cells has, however, characteristics divergent from those in granulosa and/or placental cells. First, expression of 17HSD1 in normal breast tissue is not related to the menstrual cycle.[23] In addition, cyclic adenosine monophosphate (cAMP) analog alone does not influence the type 1 concentration, and it decreases retinoic acid–stimulated 17HSD1 expression in T47D breast cancer cells.[78] Finally, neither epidermal growth factor nor phorbol myristate acetate increases type 1 expression or potentiates the influence of retinoic acids on 17HSD1 in the breast cell line, as they do in choriocarcinoma cells.[78]

Transcriptional Regulation of HSD17B1 Gene Expression

At the transcriptional level, three functional elements of HSD17B1 have been identified upstream of the cap site for the 1.3 kb 17HSD1 mRNA (Fig. 9.2). The fragment −78/+9 can initiate the transcription of HSD17B1, the region from −661 to −392 enhances the function of the promoter, while the action of the area between the enhancer and the promoter is to decrease transcripition of the HSD17B1 gene.[106–108] The identity between HSD17B1 and HSD17BP1 in these regions is as high as 98%, whereas the overall identity between the genes is lower, 89%. Variation of a few nucleotides in the regulatory elements is apparently enough to limit the transcription of HSD17BP1.[95,108]

The region −78/+9 alone induces detectable reporter gene expression in several cell lines[109] and together with the HSD17B1 enhancer or the simian virus 40 enhancer, e.g., drives reporter gene expression very efficiently.[106] The HSD17B1 promoter contains a typical TATA box-like sequence and a GC-rich area, comprising binding sites for Sp and activating protein-2 (AP-2) transcription factors.[70,71,107] The same nucleotide variation in the TATA box that has been identified in a small fraction of women with familial breast cancer as well as in control individuals also occurs in HSD17BP1 and leads to decreased transcription efficiency of the gene.[95]

The binding sites for Sp and AP-2 are located adjacent to each other, and the factors compete for binding to the region (Fig. 9.2).[107] The Sp motif is recognized by two Sp factors, Sp1 and Sp3; and mutation of the motif abolishes detectable binding of the factors and decreases promoter activity from 40% to 70% in two choriocarcinoma cell lines expressing 17HSD1. In contrast, introduction of mutations to the AP-2 element results in increased promoter activity, which is associated with increased binding of Sp1 and Sp3 to their cognate sequence. The data thus suggest that AP-2 represses the function of the HSD17B1 promoter by preventing binding to the Sp motif.[107]

The region upstream of the Sp and AP-2 motifs, located between −113 and −78, reduces gene transcription. Deletion of the fragment causes a significant increase in reporter gene expression in all cell lines tested.[106,109] A response element situated at the HSD17B1 silencer region is a binding motif for GATA factors (Fig. 9.2).[107] The transcription factors GATA-2 and, in particular, GATA-3 can bind their cognate sequence, and mutation of the GATA motif leads to decreased binding of the GATA proteins. Moreover, the GATA motif–mutated HSD17B1 fragment drives reporter gene activity as efficiently as the corresponding constructs missing the HSD17B1 silencer element. The data thus indicate that GATA-2 and GATA-3 can repress the function of the HSD17B1 gene and that the GATA motif may be an essential part of the HSD17B1 silencer.[107] The transcription factors AP-2, Sp1, and Sp3 are widely distributed; and the latter two bind to the GC-rich Sp1 motif with identical affinity.[110] Also, GATA-2 and GATA-3 are expressed in several tissues and cell types, including placental trophoblasts.[111,112] The factors activate a wide array of promoters, interact with each other as well as with other regulatory factors, and consequently mediate cell- and

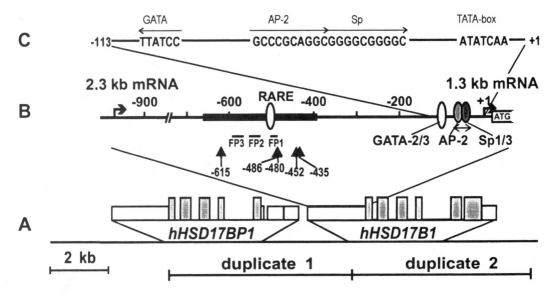

Figure 9.2 Schematic presentation of the *HSD17B1* gene and its regulatory elements. *Level A:* As a result of ancient gene duplication, the *HSD17B1* gene exists in tandem with the *HSD17BP1* gene. Gray boxes demonstrate the coding areas of exons in the *HSD17B1* gene and analogous areas in the *HSD17BP1* gene. White boxes show the noncoding areas of the exons. *Level B:* Several regulatory elements and binding motifs have been identified between the two transcription-start sites marked with bold arrows. Black box represents the *HSD17B1* enhancer, which contains a retinoic acid response element (RARE), shown as an ellipse. Regions protected in DNase I footprinting analyses (FP1–3) are marked with bars, and the positions of five nucleotides that are different in the *HSD17B1* enhancer and the *HSD17BP1* analog are shown with arrows. The positions of the binding sites for GATA, Sp, and activating protein-2 (AP-2) factors are also marked with ellipses. ATG is the translation-initiation codon. *Level C:* Detailed representation of the 5′-proximal region of the *HSD17B1* gene and the binding motifs. (From Peltoketo et al.[141] with permission.)

gene-specific effects on the promoter of the target genes.[112–121]

The *HSD17B1* enhancer remarkably increases transcription efficiency in both orientations, regardless of the promoter type it has linked or the distance between the promoter and the enhancer.[106] Reporter gene transcription is increased by the *HSD17B1* enhancer in choriocarcinoma cells, in particular; and the action of the enhancer runs in parallel with endogenous 17HSD1 expression in these cells.[106] Hence, the enhancer is likely to be responsible for the abundant expression of 17HSD1 in choriocarcinoma cells and in the placenta. The *HSD17B1* enhancer also amplifies the function of the *HSD17B1* promoter in breast cancer T47D cells; but in certain breast cancer cell lines, such as BT20, the function of the enhancer and 17HSD1 expression do not correlate very well. This suggests that additional regulatory elements of the *HSD17B1* gene may function in some breast cancer cell lines.

The *HSD17B1* enhancer contains several subelements that interact with each other (Fig. 9.2). A retinoic acid response element (RARE) located within the enhancer[106] is probably involved in the retinoic acid-induced expression of 17HSD1 which has been reported to take place.[78,122] The retinoid X receptor α (RXRα)/RARα dimer binds to the *HSD17B1* RARE, and administration of retinoic acid increases the activity of the *HSD17B1* enhancer.[78,109] Moreover, mutations in the RARE nullify the response to retinoic acids completely, which verifies direct regulation of the *HSD17B1* gene by retinoic acids.[106] The *HSD17B1* enhancer also contains three other binding sites, identified by DNase I footprinting.[108] The binding site named FP2 and a short palindromic area within it, especially, are essential for the function of the enhancer since replacing them with a nonsense sequence abolishes enhancer activity almost completely.[108] In addition, cytosine and guanidine at positions

−480 and −486, respectively, are indispensable for proper function of the *HSD17B1* enhancer. The nucleotides were replaced by guanidine and adenine in *HSD17BP1*, and as a result the *HSD17BP1* enhancer had only 10% of the activity of the *HSD17B1* enhancer.[108]

Altogether, regulation of *hHSD17B1* gene expression is a collective outcome of function of the enhancer containing several interacting subunits and a RARE, a silencer element with a GATA motif, a proximal promoter region with competing Sp and AP-2 sites, and possibly as yet unidentified regions. Cell-specific expression of *HSD17B1* is a result of the concentrations, activation/inactivation, and mutual interactions of the endocrine, paracrine, and autocrine factors regulating 17HSD1 expression and the factors binding the regulatory elements of *HSD17B1*. They may also give rise to the possible overexpression of the type 1 enzyme in the case of cancer, e.g., rather than changes in the *HSD17B1* gene.

17β-HYDROXYSTEROID DEHYDROGENASE 1 AS A TARGET OF ENDOCRINE THERAPY: STRUCTURE OF THE ENZYME

Since 17HSD1 is needed for ovarian E_2 production, controlled partial inhibition of the type 1 enzyme may lead to a lowered serum E_2 concentration and could thus be useful in the prevention or treatment of breast cancer in premenopausal women. More importantly, because 17HSD1 is expressed in a notable proportion of breast tumors, the enzyme is also a target for local inhibition of E_2 biosynthesis in breast tissue of both pre- and postmenopausal women. Structural and functional studies of 17HSD1 and 17HSD2 are aimed at the development of 17HSD1 inhibitors that are specific for 17HSD1, i.e., would have little or no affinity for the other 17HSDs, especially 17HSD2 and other oxidative 17HSDs. Knowledge of the structure of 17HSDs and other steroidogenic enzymes also assists in the design of inhibitors that do not otherwise prevent or modulate steroidogenesis and that lack agonist effects on the estrogen receptor.

Structural Studies

Detailed knowledge of the biochemistry and molecular biology of 17HSD1 has grown rapidly in the last few years, culminating in the determination of the three-dimensional structure of the human enzyme.[123] This has led to an atomic level description of the E_2 binding pocket of the enzyme and an understanding of its mechanism of action and of the molecular basis for its estrogen specificity.[28,29,124] In addition, studies on complexes of 17HSD1 with E_2 and/or $NADP^+$,[125–127] various mutant complexes,[128] and structure–function analysis through site-directed substitutions and enzyme chimeras,[28,124] have further clarified the mechanism of action and provided additional insights into the origin of estrogenic/androgenic specificities of the enzyme. Type 1 17HSD belongs to the short-chain dehydrogenase/reductase (SDR) family,[129] requiring a Tyr-X-X-X-Lys motif and a Tyr-Lys-Ser catalytic triad for activity.[123,129–131]

At 2.20 Å resolution, the structure of human 17HSD1 revealed a fold characteristic of SDRs.[123] The active site contains the Tyr-X-X-X-Lys sequence and a Ser residue. The structure also contains three α helices and a helix-turn-helix motif not observed in previously reported SDR structures. These helices, located at one end of the substrate-binding cleft away from the catalytic triad, restrict access to the active site and appear to influence substrate specificity. The side chain His^{221} plays a critical role in substrate recognition. A model of E_1-to-E_2 transition has been proposed in which residues Tyr^{155}, Lys^{159}, and Ser^{142} have functional roles in the catalysis of the oxidoreductive reaction. It has been shown by site-directed mutational studies that Tyr^{155}, Lys^{159}, and His^{221} are essential for the activity of the enzyme.[124] The structure of the active site provides a rational basis for designing more specific inhibitors. The proposed mechanism of action of the SDRs has been verified by other structural and mutational studies.[28,124,132]

The structure of the ternary complex of 17HSD1 with 3-hydroxyestra-1,2,5,7-tetraen-17-one (equilin) and $NADP^+$ has been determined.[133] Equilin is one of the major components of estrogens used in estrogen-replacement therapy (ERT), in conjunction with E_1 and 17βdihydroequilin. These conjugated estrogens are administered under the commercial name Premarin (Wyeth-Ayerst, Philadelphia, PA) as salts of their sulfate esters, which are subsequently hydrolyzed to free estrogens. The chemical structure of equilin differs from that of the substrate E_1 only by the presence of a C7=C8 double bond. Inhibition of 17HSD1 enzyme activity by 77% has been achieved with a 1 μM concentration of equilin, establishing it as a potent inhibitor of E_1-to-E_2 reduction.

Molecular Basis of Ligand Recognition

Figure 9.3a is a schematic diagram of the substrate-binding cleft. It consists of a hydrophobic pocket with two hydrophilic ends: to the right is the catalytic machinery consisting of the catalytic triad Tyr[155]–Ser[142]–Lys[159], and to the left is the 3-OH group recognizing His[221] and Glu[284]. The 17-keto oxygen accepts protons from catalytic residues Tyr[155] and Ser[142] (2.7 and 2.8 Å, respectively) at the catalytic end of the steroid-binding cleft. The 3-OH group of the ligand makes a bifurcated hydrogen bond with His[221] and Glu[282] (2.9 and 2.9 Å, respectively) at the recognition end of the cleft. Interestingly, a similar mechanism of recognition of estrogenic ligands by hydrogen bonding with their characteristic 3-OH groups is utilized by the estrogen receptor (ER) as well. In the crystal structure of the complex of E_2 with the human ERα ligand-binding domain, shown schematically in Figure 9.3b,[134] the 3-OH group is hydrogen-bonded to a glutamic acid side chain and a water molecule while a histidine side chain hydrogen-bonds to 17β-OH. In 17HSD1, a deprotonated histidine side chain, His[221], is critical to the recognition of the 3-OH group of the substrate E_1, accepting a proton from it. As E_2 is a more potent ligand for transcriptional activation of the ER than E_1, the histidine side chain is used to discriminate a hydroxyl group from a ketone group. A histidine side chain, His[524], positioned near the D ring, accepts a proton from 17β-OH, while a glutamic acid accepts a proton from 3-OH. Thus, a mechanism similar to recognition of the estrogenic 3-OH of the substrate in 17HSD1 is employed by the ER for recognition of its natural agonist E_2.

With the exception of these two hydrophilic ends, the rest of the steroid-binding cleft is almost exclusively hydrophobic. Residues Val[143], Met[147], Leu[149], Pro[150], Asn[152], and Tyr[218] make up the protein surface in the vicinity of the β face of the nearly coplanar A-B rings of the ligand. Residues Val[225] and Pro[187] are within van der Waals contact distances to the equilin α face, whereas residues Leu[262], Leu[263], and Met[279] are at the floor of the cleft. Recognition of the planar A ring and near coplanarity of A-B rings through shape complementarity is another mechanism employed by the enzyme for its estrogenic specificity. In addition to the hydrophobic environment, Phe[192] and Met[193] in the substrate-entry loop (residues 186–201) line the entry path. In the apoenzyme structure, the substrate-entry loop adopts an open conformation, providing unrestricted access to the active-site cleft.[123] However, the substrate-entry loop forms a closed conformation in the 17HSD1–equilin complex, where the polypeptide chain with residues 186–201 moves toward the catalytic cleft, restricting access to the active site, as depicted in Figure 9.3a. In this closed conformation, Phe[192] and Met[193] make van der Waals contacts (3.9 and 4.2 Å, respectively) to the ligand molecule.

The orientation of C17=O of equilin relative to the C4-hydride is more acute (52.7 degrees) than in the above scenario, owing to differences in puckering of the C-D ring system. The three-dimensional structures of the substrate, E_1, and equilin are strikingly different at the C-D ring systems because of the presence of the C7=C8 double bond in equilin. The difference in torsion angle C7—C8—C9—C11 (−179 degrees for E_1 and 121 degrees for equilin[135]) caused by the C7=C8 double bond results in 0.9 Å displacement between the C17 carbon atoms. The origin of the inhibitory property of equilin, from the structural perspective, is a manifestation of its altered C-D ring structure and location and orientation of its C17 keto group, with respect to the catalytic machinery at the catalytic end of the active site.

Ligand–Entry Loop Interactions

The open and closed conformations of the substrate-entry loop are shown in Figure 9.4, where the backbones of the apo- and holoenzyme

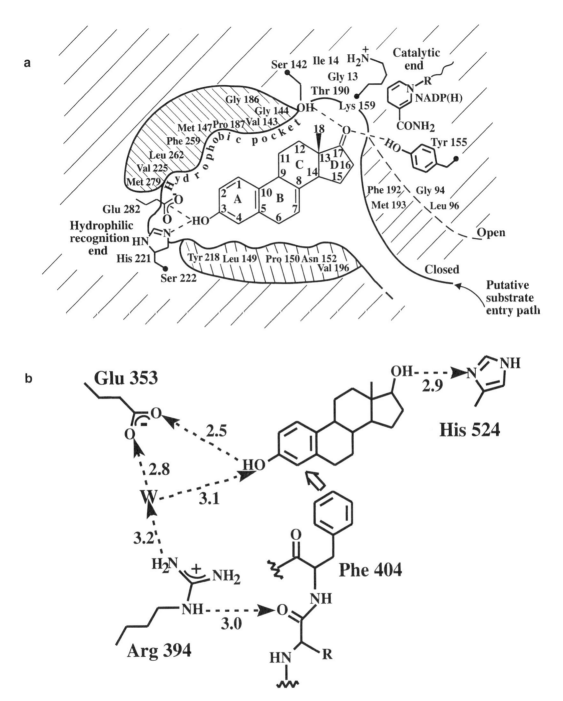

Figure 9.3 (*a*) Schematic diagram of the active site of the 17β-hydroxysteroid dehydrogenase 1 (17HSD1)–equilin–oxidized nicotinamide adenine dinucleotide phosphate (NADP$^+$) ternary complex. Residues belonging to the hydrophilic catalytic and recognition ends, as well as residues lining the hydrophobic surrounding, are shown. (From Sawicki et al.[133] with permission.) (*b*) Ligand-binding interactions in human estrogen receptor-α describing the crystal structure of the 17β-estradiol complex of the estrogen receptor ligand-binding domain. (From Figure 3A of Tanenbaum et al.[134] with permission.)

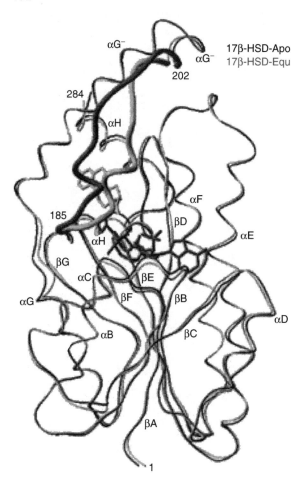

Figure 9.4 Superposition of the backbones of structures of 17β-hydroxysteroid dehydrogenase 1 (17HSD1) apoenzyme and ternary complex with equilin and oxidized nicotinamide adenine dinucleotide phosphate (NADP$^+$). Substrate-entry loops (residues 286–201) in open and closed forms are shown in thicker cross-sections.

structures are superimposed. The overall tertiary structures of the apoenzyme and the equilin complex are nearly identical, except for the substrate-entry loop (shown as a thicker cross-section of the backbone) between strand βF and helix αG″. The loop packs against both equilin and NADP$^+$ via Phe^{192} and Met^{193}. Both Phe^{192} and Met^{193} line the substrate-entry path and have van der Waals contacts with the D ring of equilin and the nicotinamide head group of NADP$^+$. This closing of the substrate-entry loop effectively traps the ligand in the steroid-binding cleft by occluding the entry path shown in Figure 9.3a.

The molecular architecture of the substrate-binding pocket of 17HSD1 has provided the basis for the rational design of ligands that would compete with substrates for the active site. Tightening of the flexible substrate-entry loop on the ligand upon its entry to the active-site[133] reveals the dynamic nature of the active site cleft and provides additional design ideas. It is likely that interaction of the enzyme atoms from this region with the ligand is partially responsible for substrate selectivity. Interactions of ligands with the residues lined along the substrate-entry path could be utilized to enhance the selectivity and specificity of ligands.

REFERENCES

1. Helmrich SP, Shapiro S, Rosenberg L, et al. Risk factors for breast cancer. Am J Epidemiol 117:35–45, 1983.
2. Lippman ME. Epidemiology of breast cancer. In: Lippman ME, Lichter AS, Danforth DN (eds). Diagnosis and Management of Breast Cancer. Philadelphia: WB Saunders, pp 1–9.
3. Brinton LA, Schairer C, Hoover RN, Fraumeni JJF. Menstrual factors and risk of breast cancer. Cancer Invest 6:245–254, 1988.

4. Schairer C, Persson I, Falkeborn M, Naessen T, Troisi R, Brinton LA. Breast cancer risk associated with gynecologic surgery and indications for such surgery. Int J Cancer 70:150–154, 1997.
5. Cullen KJ, Lippman ME. Estrogen regulation of protein synthesis and cell growth in human breast cancer. Vitam Horm 45:127–158, 1989.
6. Vihko R, Apter D. Endogenous steroids in the pathophysiology of breast cancer. Crit Rev Oncol Hematol 9:1–16, 1989.
7. Bulun SE, Price TM, Aitken J, Mahendroo MS, Simpson ER. A link between breast cancer and local estrogen biosynthesis suggested by quantification of breast adipose tissue aromatase cytochrome P450 transcripts using competitive polymerase chain reaction after reverse transcription. J Clin Endocrinol Metab 77:1622–1628, 1993.
8. Marshall E. The politics of breast cancer. Science 259:616–638, 1993.
9. Rebbeck TR, Levin AM, Eisen A, et al. Breast cancer risk after bilateral prophylactic oophorectomy in BRCA1 mutation carriers. J Natl Cancer Inst 91:1475–1479, 1999.
10. Beatson GT. On the treatment of inoperable cases of carcinoma of the mamma. Suggestion for a new method of treatment, with illustrative cases. Lancet 2:104–107, 1896.
11. Santen RJ, Manni A, Harvey H, Redmond C. Endocrine treatment of breast cancer in women. Endocr Rev 11:221–265, 1990.
12. Sedlacek SM, Horowitz KB. The role of progestins and progesterone receptors in the treatment of breast cancer. Steroids 44:467–484, 1994.
13. Osborne CK. Steroid hormone receptors in breast cancer management. Breast Cancer Res Treat 51:227–238, 1998.
14. Fortunati N. Sex hormone–binding globulin: not only a transport protein. What news is around the corner. J Endocrinol Invest 22:223–234, 1999.
15. Grodin JM, Siiteri PK, MacDonald PC. Source of estrogen production in postmenopausal women. J Clin Endocrinol Metab 36:207–214, 1973.
16. Simpson ER, Mahendroo MS, Means GD, et al. Aromatase cytochrome P450, the enzyme responsible for estrogen biosynthesis. Endocr Rev 15:342–355, 1994.
17. Buzdar AU, Hortobagyi G. Update on endocrine therapy for breast cancer. Clin Cancer Res 4:527–534, 1998.
18. Sasano H, Murakami H, Shizawa S, Satomi S, Nagura H, Harada N. Aromatase and sex steroid receptors in human vena cava. Endocr J 46:233–242, 1999.
19. Labrie F. Intracrinology. Mol Cell Endocrinol 78:C113–C118, 1991.
20. Bingham SA, Atkinson C, Liggins J, Bluck L, Coward A. Phyto-oestrogens: where are we now? Br J Nutr 79:393–406, 1998.
21. Martel C, Rhéaume E, Takahashi M, et al. Distribution of 17β-hydroxysteroid dehydrogenase gene expression and activity in rat and human tissues. J Steroid Biochem Mol Biol 41:597–603, 1992.
22. Pelletier G, Luu-The V, Tetu B, Labrie F. Immunocytochemical localization of type 5 17β-hydroxysteroid dehydrogenase in human reproductive tissues. J Histochem Cytochem 47:731–738, 1999.
23. Söderqvist G, Poutanen M, Wickman M, von Schoultz B, Skoog L, Vihko R. 17β-Hydroxysteroid dehydrogenase type 1 in normal breast tissue during the menstrual cycle and hormonal contraception. J Clin Endocrinol Metab 83:1190–1193, 1998.
24. Peltoketo H, Luu-The V, Simard J, Adamski J. 17β-Hydroxysteroid dehydrogenase (HSD)/17-ketosteroid reductase (KSR) family; nomenclature and main characteristics of the 17HSD/KSR enzymes. J Mol Endocrinol 23:1–11, 1999.
25. Li KX, Smith RE, Krozowski ZS. Cloning and expression of a novel tissue specific 17β-hydroxysteroid dehydrogenase. Endocr Res 24:663–667, 1998.
26. He XY, Merz G, Mehta P, Schulz H, Yang SY. Human brain short chain L-3-hydroxyacyl coenzyme A dehydrogenase is a single-domain multifunctional enzyme. Characterization of a novel 17β-hydroxysteroid dehydrogenase. J Biol Chem 274:15014–15019, 1999.
27. Lanisnik Rizner T, Möller G, Thole HH, Zakelj-Mavric M, Adamski J. A novel 17β-hydroxysteroid dehydrogenase in the fungus Cochliobolus lunatus: new insights into the evolution of steroid-hormone signalling. Biochem J 337:425–431, 1999.
28. Puranen T, Poutanen M, Ghosh D, Vihko R, Vihko P. Origin of substrate specificity of human and rat 17β-hydroxysteroid dehydrogenase type 1, using chimeric enzymes and site-directed substitutions. Endocrinology 138:3532–3539, 1997.
29. Poutanen M, Miettinen M, Vihko R. Differential estrogen substrate specificities for transiently expressed human placental 17β-hydroxysteroid dehydrogenase and an endogenous enzyme expressed in cultured COS-m6 cells. Endocrinology 133:2639–2644, 1993.
30. Jin JZ, Lin SX. Human estrogenic 17β-hydroxysteroid dehydrogenase: predominance of estrone reduction and its induction by NADPH. Biochem Biophys Res Commun 259:489–493, 1999.
31. Nokelainen P, Puranen T, Peltoketo H, Orava M, Vihko P, Vihko R. Molecular cloning of mouse 17β-hydroxysteroid dehydrogenase type 1 and characterization of enzyme activity. Eur J Biochem 236:482–490, 1996.
32. Wu L, Einstein M, Geissler WM, Chan HK, Elliston KO, Andersson S. Expression cloning and characterization of human 17β-hydroxysteroid

dehydrogenase type 2, a microsomal enzyme possessing 20α-hydroxysteroid dehydrogenase activity. J Biol Chem 268:12964–12969, 1993.
33. Labrie F, Luu-The V, Lin SX, et al. The key role of 17β-hydroxysteroid dehydrogenases in sex steroid biology. Steroids 62:148–158, 1997.
34. Sawetawan C, Milewich L, Word A, Carr BC, Rainey WE. Compartmentalization of type I 17β-hydroxysteroid oxidoreductase in the human ovary. Mol Cell Endocrinol 99:161–168, 1994.
35. Fournet-Dulguerov N, MacLusky NJ, Leranth CZ, et al. Immunohistochemical localization of aromatase cytochrome P-450 and estradiol dehydrogenase in the syncytiotrophoblast of the human placenta. J Clin Endocrinol Metab 65:757–764, 1987.
36. Mustonen MV, Poutanen MH, Kellokumpu S, et al. Mouse 17β-hydroxysteroid dehydrogenase type 2 mRNA is predominantly expressed in hepatocytes and in surface epithelial cells of the gastrointestinal and urinary tracts. J Mol Endocrinol 20:67–74, 1998.
37. English MA, Kane KF, Cruickshank N, Langman MJ, Stewart PM, Hewison M. Loss of estrogen inactivation in colonic cancer. J Clin Endocrinol Metab 84:2080–2085, 1999.
38. Mustonen MV, Isomaa VV, Vaskivuo T, et al. Human 17β-hydroxysteroid dehydrogenase type 2 messenger ribonucleic acid expression and localization in term placenta and in endometrium during the menstrual cycle. J Clin Endocrinol Metab 83:1319–1324, 1998.
39. Moghrabi N, Head JR, Andersson S. Cell type-specific expression of 17β-hydroxysteroid dehydrogenase type 2 in human placenta and fetal liver. J Clin Endocrinol Metab 82:3872–3878, 1997.
40. Adashi EY. The ovarian life cycle. In: Yen SSC, Jaffe RB (eds). Reproductive Endocrinology, Philadelphia: WB Saunders, 1991, pp 181–237.
41. Hillier SG, Whitelaw PF, Smyth CD. Follicular oestrogen synthesis: "two-cell, two-gonadotropin" model revisited. Mol Cell Endocrinol 100:51–54, 1994.
42. Richards JS, Hickey FJ, Chen S, et al. Hormonal regulation of estradiol biosynthesis, aromatase activity and aromatase mRNA in rat ovarian follicles and corpora lutea. Steroids 50:393–409, 1987.
43. Fortune JE. Ovarian follicular growth and development in mammals. Biol Reprod 50:225–232, 1994.
44. Sasano H, Suzuki T, Niikura H, et al. 17β-Hydroxysteroid dehydrogenase in common epithelial ovarian tumors. Mod Pathol 9:386–391, 1996.
45. Fomitcheva J, Baker ME, Anderson E, Lee GY, Aziz N. Characterization of Ke 6, a new 17β-hydroxysteroid dehydrogenase, and its expression in gonadal tissues. J Biol Chem 273:22664–22671, 1998.
46. Aziz N, Maxwell MM, St. Jacques B, Brenner BM. Downregulation of Ke 6, a novel gene encoded within the major histocompatibility complex, in murine polycystic kidney disease [published erratum appears in Mol Cell Biol 13:6614, 1993]. Mol Cell Biol 13:1847–1853, 1993.
47. Ramirez S, Fomitcheva I, Aziz N. Abnormal regulation of the Ke 6 gene, a new 17β-hydroxysteroid dehydrogenase in the *cpk* mouse kidney. Mol Cell Endocrinol 143:9–22, 1998.
48. Nokelainen P, Peltoketo H, Vihko R, Vihko P. Expression cloning of a novel estrogenic mouse 17β-hydroxysteroid dehydrogenase/17-ketosteroid reductase (m17HSD7), previously described as a prolactin receptor–associated protein (PRAP) in rat. Mol Endocrinol 12:1048–1059, 1998.
49. Duan WR, Parmer TG, Albarracin CT, Zhong L, Gibori G. PRAP, a prolactin receptor associated protein: its gene expression and regulation in the corpus luteum. Endocrinology 138:3216–3221, 1997.
50. Nokelainen P, Peltoketo H, Mustonen M, Vihko P. Expression of mouse 17β-hydroxysteroid dehydrogenase/17-ketosteroid reductase type 7 in ovary, uterus and placenta: localization from implantation to late pregnancy. Endocrinology 141:772–778, 2000.
51. Miettinen M, Mustonen M, Poutanen M, et al. 17β-Hydroxysteroid dehydrogenase in normal human mammary epithelial cells and breast tissue. Breast Cancer Res Treat 1438:1–8, 1999.
52. Speirs V, Green AR, Walton DS, et al. Short-term primary culture of epithelial cells derived from human breast tumours. Br J Cancer 78:1421–1429, 1998.
53. Speirs V, Green AR, Atkin SL. Activity and gene expression of 17β-hydroxysteroid dehydrogenase type I in primary cultures of epithelial and stromal cells derived from normal and tumourous human breast tissue: the role of IL-8. J Steroid Biochem Mol Biol 67:267–274, 1998.
54. Poutanen M, Isomaa V, Lehto V-P, Vihko R. Immunological analysis of 17β-hydroxysteroid dehydrogenase in benign and malignant human breast tissue. Int J Cancer 50:386–390, 1992.
55. Sasano H, Frost AR, Saitoh R, et al. Aromatase and 17β-hydroxysteroid dehydrogenase type 1 in human breast carcinoma. J Clin Endocrinol Metab 81:4042–4046, 1996.
56. Miettinen MM, Mustonen MV, Poutanen MH, Isomaa VV, Vihko RK. Human 17β-hydroxysteroid dehydrogenase type 1 and type 2 isoenzymes have opposite activities in cultured cells and characteristic cell- and tissue-specific expression. Biochem J 314:839–845, 1996.
57. Poutanen M, Isomaa V, Peltoketo H, Vihko R. Role of 17β-hydroxysteroid dehydrogenase type 1 in endocrine and intracrine estradiol biosynthesis. J Steroid Biochem Mol Biol 55:525–532, 1995.

58. Fishman J, Nisselbaum JS, Menendez-Botet CJ, Schwartz MK. Estrone and estradiol content in human breast tumors: relationship to estradiol receptors. J Steroid Biochem 8:893–896, 1977.
59. Vermeulen A, Deslypere JP, Paridaens R, Leclercq G, Roy F, Heuson C. Aromatase, 17β-hydroxysteroid dehydrogenase and intratissular sex hormone concentrations in cancerous and normal glandular breast tissue in postmenopausal women. Eur J Cancer Clin Oncol 22:515–525, 1986.
60. Blankenstein MA, van de Ven J, Maitimu-Smeele I, et al. Intratumoral levels of estrogens in breast cancer. J Steroid Biochem Mol Biol 69:293–297, 1999.
61. Miettinen MM, Poutanen MH, Vihko RK. Characterization of estrogen-dependent growth of cultured MCF-7 human breast-cancer cells expressing 17β-hydroxysteroid dehydrogenase type 1. Int J Cancer 68:600–604, 1996.
62. Mäentausta O, Sormunen R, Isomaa V, Lehto V-P, Jouppila P, Vihko R. Immunohistochemical localization of 17β-hydroxysteroid dehydrogenase in the human endometrium during the menstrual cycle. Lab Invest 65:582–587, 1991.
63. Mäentausta O, Svalander P, Danielsson KG, Bygdeman M, Vihko R. The effects of an antiprogestin, mifepristone, and an antiestrogen, tamoxifen, on endometrial 17β-hydroxysteroid dehydrogenase and progestin and estrogen receptors during the luteal phase of the menstrual cycle: an immunohistochemical study. J Clin Endocrinol Metab 77:913–918, 1993.
64. Tseng L, Gurpide E. Stimulation of various 17β- and 20α-hydroxysteroid dehydrogenase activities by progestins in human endometrium. Endocrinology 104:1745–1748, 1979.
65. Mäentausta O, Boman K, Isomaa V, Stendahl U, Bäckstrom T, Vihko R. Immunohistochemical study of the human 17β-hydroxysteroid dehydrogenase and steroid receptors in endometrial adenocarcinoma. Cancer 70:1551–1555, 1992.
66. Elo JP, Akinola LA, Poutanen M, et al. Characterization of 17β-hydroxysteroid dehydrogenase isoenzyme expression in benign and malignant human prostate. Int J Cancer 66:37–41, 1996.
67. Casey ML, MacDonald PC, Andersson S. 17β-Hydroxysteroid dehydrogenase type 2: chromosomal assignment and progestin regulation of gene expression in human endometrium. J Clin Invest 94:2135–2141, 1994.
68. Mustonen M, Poutanen M, Chotteau-Lelievre A, et al. Ontogeny of 17β-hydroxysteroid dehydrogenase type 2 mRNA expression in the developing mouse placenta and fetus. Mol Cell Endocrinol 134:33–40, 1997.
69. Mustonen MVJ, Poutanen MH, Isomaa VV, Vihko PT, Vihko RK. Cloning of mouse 17β-hydroxysteroid dehydrogenase type 2, and analysing expression of the mRNAs for types 1, 2, 3, 4 and 5 in mouse embryos and adult tissues. Biochem J 325:199–205, 1997.
70. Luu-The V, Labrie C, Simard J, et al. Structure of two in tandem human 17β-hydroxysteroid dehydrogenase genes. Mol Endocrinol 4:268–275, 1990.
71. Peltoketo H, Isomaa V, Vihko R. Genomic organization and DNA sequences of human 17β-hydroxysteroid dehydrogenase genes and flanking regions. Localization of multiple Alu sequences and putative cis-acting elements. Eur J Biochem 209:459–466, 1992.
72. Labrie Y, Durocher F, Lachance Y, et al. The human type II 17 β-hydroxysteroid dehydrogenase gene encodes two alternatively spliced mRNA species. DNA Cell Biol 14:849–861, 1995.
73. Luu The V, Labrie C, Zhao HF, et al. Characterization of cDNAs for human estradiol 17β-dehydrogenase and assignment of the gene to chromosome 17: evidence of two mRNA species with distinct 5'-termini in human placenta. Mol Endocrinol 3:1301–1309, 1989.
74. Poutanen M, Moncharmont B, Vihko R. 17β-Hydroxysteroid dehydrogenase gene expression in human breast cancer cells: regulation of expression by a progestin. Cancer Res 52:290–294, 1992.
75. Tremblay Y, Ringler GE, Morel Y, et al. Regulation of the gene for estrogenic 17-ketosteroid reductase lying on chromosome 17cen → q25. J Biol Chem 264:20458–20462, 1989.
76. Tremblay Y, Beaudoin C. Regulation of 3β-hydroxysteroid dehydrogenase and 17β-hydroxysteroid dehydrogenase messenger ribonucleic acid levels by cyclic adenosine 3',5'-monophosphate and phorbol myristate acetate in human choriocarcinoma cells. Mol Endocrinol 7:355–364, 1993.
77. Reed MJ, Rea D, Duncan LJ, Parker MG. Regulation of estradiol 17β-hydroxysteroid dehydrogenase expression and activity by retinoic acid in T47D breast cancer cells. Endocrinology 135:4–9, 1994.
78. Piao YS, Peltoketo H, Jouppila A, Vihko R. Retinoic acids increase 17β-hydroxysteroid dehydrogenase type 1 expression in JEG-3 and T47D cells, but the stimulation is potentiated by epidermal growth factor, 12-O-tetradecanoylphorbol-13-acetate, and cyclic adenosine 3',5'-monophosphate only in JEG-3 cells. Endocrinology 138:898–904, 1997.
79. Winqvist R, Peltoketo H, Isomaa V, Grzeschik KH, Mannermaa A, Vihko R. The gene for 17β-hydroxysteroid dehydrogenase maps to human chromosome 17, bands q12–q21, and shows an RFLP with ScaI. Hum Genet 85:473–476, 1990.
80. Simard J, Feunteun J, Lenoir G, et al. Genetic mapping of the breast–ovarian cancer syndrome to a small interval on chromosome 17q12–21:

exclusion of candidate genes EDH17B2 and RARA. Hum Mol Genet 2:1193–1199, 1993.
81. Durocher F, Morissette J, Labrie Y, Labrie F, Simard J. Mapping of the *HSD17B2* gene encoding type II 17β-hydroxysteroid dehydrogenase close to D16S422 on chromosome 16q24.1–q24.2. Genomics 25:724–726, 1995.
82. Sato T, Akiyama F, Sakamoto G, Kasumi F, Nakamura Y. Accumulation of genetic alterations and progression of primary breast cancer. Cancer Res 51:5794–5799, 1991.
83. Godfrey TE, Cher ML, Chhabra V, Jensen RH. Allelic imbalance mapping of chromosome 16 shows two regions of common deletion in prostate adenocarcinoma. Cancer Genet Cytogenet 98:36–42, 1997.
84. Chen T, Sahin A, Aldaz CM. Deletion map of chromosome 16q in ductal carcinoma in situ of the breast: refining a putative tumor suppressor gene region. Cancer Res 56:5605–5609, 1996.
85. Tsuda H, Callen DF, Fukutomi T, Nakamura Y, Hirohashi S. Allele loss on chromosome 16q24.2-qter occurs frequently in breast cancers irrespectively of differences in phenotype and extend of spread. Cancer Res 54:513–517, 1994.
86. Tirkkonen M, Tanner M, Karhu R, Kallioniemi A, Isola J, Kallioniemi OP. Molecular cytogenetics of primary breast cancer by CGH. Genes Chromosomes Cancer 21:177–184, 1998.
87. Carter BS, Ewing CM, Ward WS, et al. Allelic loss of chromosomes 16q and 10q in human prostate cancer. Proc Natl Acad Sci USA 87:8751–8755, 1990.
88. Bova GS, Isaacs WB. Review of allelic loss and gain in prostate cancer. World J Urol 14:338–346, 1996.
89. Verma RS, Manikal M, Conte RA, Godec CJ. Chromosomal basis of adenocarcinoma of the prostate. Cancer Invest 17:441–447, 1999.
90. Elo JP, Härkönen P, Kyllönen AP, Lukkarinen O, Vihko P. Three independently deleted regions at chromosome arm 16q in human prostate cancer: allelic loss at 16q24.1–q24.2 is associated with aggressive behaviour of the disease, recurrent growth, poor differentiation of the tumour and poor prognosis for the patient. Br J Cancer 79:156–160, 1999.
91. Visakorpi T. Molecular genetics of prostate cancer. Ann Chir Gynaecol 88:11–16, 1999.
92. Elo JP, Härkonen P, Kyllönen AP, et al. Loss of heterozygosity at 16q24.1–q24.2 is significantly associated with metastatic and aggressive behavior of prostate cancer. Cancer Res 57:3356–3359, 1997.
93. Suzuki H, Komiya A, Emi M, et al. Three distinct commonly deleted regions of chromosome arm 16q in human primary and metastatic prostate cancers. Genes Chromosomes Cancer 17:225–233, 1996.
94. Normand T, Narod S, Labrie F, Simard J. Detection of polymorphisms in the estradiol 17β-hydroxysteroid dehydrogenase II gene at the EDH17B2 locus on 17q11–q21. Hum Mol Genet 2:479–483, 1993.
95. Peltoketo H, Piao Y, Mannermaa A, et al. A point mutation in the putative TATA box, detected in nondiseased individuals and patients with hereditary breast cancer, decreases promoter activity of the 17β-hydroxysteroid dehydrogenase type 1 gene 2 (EDH17B2) *in vitro*. Genomics 23:250–252, 1994.
96. Kelsell DP, Black DM, Bishop DT, Spurr NK. Genetic analysis of the *BRCA1* region in a large breast/ovarian family: refinement of the minimal region containing *BRCA1*. Hum Mol Genet 2:1823–1828, 1993.
97. Puranen TJ, Poutanen MH, Peltoketo HE, Vihko PT, Vihko RK. Site-directed mutagenesis of the putative active site of human 17β-hydroxysteroid dehydrogenase type 1. Biochem J 304:289–293, 1994.
98. Jantus Lewintre E, Orava M, Peltoketo H, Vihko R. Characterization of 17β-hydroxysteroid dehydrogenase type 1 in choriocarcinoma cells: regulation by basic fibroblast growth factor. Mol Cell Endocrinol 104:1–9, 1994.
99. Richards JS. Hormonal control of gene expression in the ovary. Endocr Rev 15:725–751, 1994.
100. Ghersevich S, Poutanen M, Tapanainen J, Vihko R. Hormonal regulation of rat 17β-hydroxysteroid dehydrogenase type 1 in cultured rat granulosa cells: effects of recombinant follicle-stimulating hormone, estrogens, androgens, and epidermal growth factor. Endocrinology 135:1963–1971, 1994.
101. Ghersevich SA, Poutanen MH, Martikainen HK, Vihko RK. Expression of 17β-hydroxysteroid dehydrogenase in human granulosa cells: correlation with follicular size, cytochrome P450 aromatase activity and oestradiol production. J Endocrinol 143:139–150, 1994.
102. Kaminski T, Akinola L, Poutanen M, Vihko R, Vihko P. Growth factors and phorbol-12-myristate-13-acetate modulate the follicle-stimulating hormone- and cyclic adenosine-3′,5′-monophosphate–dependent regulation of 17β-hydroxysteroid dehydrogenase type 1 expression in rat granulosa cells. Mol Cell Endocrinol 136:47–56, 1997.
103. Ghersevich S, Nokelainen P, Poutanen M, et al. Rat 17β-hydroxysteroid dehydrogenase type 1: primary structure and regulation of enzyme expression in rat ovary by diethylstilbestrol and gonadotropins in vivo. Endocrinology 135:1477–1487, 1994.
104. Jantus Lewintre E, Orava M, Vihko R. Regulation of 17β-hydroxysteroid dehydrogenase type 1 by epidermal growth factor and transforming growth factor-α in choriocarcinoma cells. Endocrinology 135:2629–2634, 1994.
105. Ritvos O, Voutilainen R. Regulation of aro-

matase cytochrome P-450 and 17β-hydroxysteroid dehydrogenase messenger ribonucleic acid levels in choriocarcinoma cells. Endocrinology 130:61–67, 1992.
106. Piao YS, Peltoketo H, Oikarinen J, Vihko R. Coordination of transcription of the human 17β-hydroxysteroid dehydrogenase type 1 gene (EDH17B2) by a cell-specific enhancer and a silencer: identification of a retinoic acid response element. Mol Endocrinol 9:1633–1644, 1995.
107. Piao YS, Peltoketo H, Vihko P, Vihko R. The proximal promoter region of the gene encoding human 17β-hydroxysteroid dehydrogenase type 1 contains GATA, AP-2, and Sp1 response elements: analysis of promoter function in choriocarcinoma cells. Endocrinology 138:3417–3425, 1997.
108. Leivonen S, Piao YS, Peltoketo H, Numchaisrika P, Vihko R, Vihko P. Identification of essential subelements in the hHSD17B1 enhancer: difference in function of the enhancer and that of the hHSD17BP1 analog is due to −480C and −486G. Endocrinology 140:3478–3487, 1999.
109. Peltoketo H, Isomaa V, Poutanen M, Vihko R. Expression and regulation of 17β-hydroxysteroid dehydrogenase type 1. J Endocrinol 150(Suppl):S21–S30, 1996.
110. Hagen G, Müller S, Beato M, Suske G. Cloning by recognition site screening of two novel GT box binding proteins: a family of Sp1 related genes. Nucleic Acid Res 20:5519–5525, 1992.
111. Simon MC. Gotta have GATA. Nat Genet 11:9–11, 1995.
112. Ng Y-K, George KM, Engel JD, Linzer DIH. GATA factor activity is required for the trophoblast-specific transcriptional regulation of the mouse placental lactogen I gene. Development 120:3257–3266, 1994.
113. Lecointe N, Bernard O, Naert K, et al. GATA- and Sp1-binding sites are required for the full activity of the tissue-specific promoter of the tal-1 gene. Oncogene 9:2623–2632, 1994.
114. Li Y, Mak G, Franza RB. In vitro study of functional involvement of Sp1, NF-kappaB/Rel, and AP1 in phorbol 12-myristate 13-acetate mediated HIV-1 long terminal repeat activation. J Biol Chem 269:30616–30619, 1994.
115. Jiang S-W, Shepard AR, Eberhardt NL. An initiator element is required for maximal human chorionic somatomammotropin gene promoter and enhancer function. J Biol Chem 270:3628–3692, 1995.
116. Birnbaum MJ, van Wijnen AJ, Odgren PR, et al. Sp-1 trans-activation of cell cycle regulated promoters is selectively repressed by Sp3. Biochemistry 34:16503–16508, 1995.
117. Cowan PJ, Tsang D, Pedic CM, et al. The human ICAM-2 promoter is endothelial cell-specific in vitro and in vivo and contains critical Sp1 and GATA binding sites. J Biol Chem 273:11737–11744, 1998.
118. Braun H, Suske G. Combinatorial action of HNF3 and Sp family transcription factors in the activation of the rabbit uteroglobin/CC10 promoter. J Biol Chem 273:9821–9828, 1998.
119. Pena P, Reutens AT, Albanese C, et al. Activator protein-2 mediates transcriptional activation of the CYP11A1 gene by interaction with Sp1 rather than binding to DNA. Mol Endocrinol 13:1402–1416, 1999.
120. Langmann T, Buechler C, Ries S, et al. Transcriptional factors Sp1 and AP-2 mediate induction of acid sphingomyelinase during monocytic differentiation. J Lipid Res 40:870–880, 1999.
121. Taniguchi A, Matsumoto K. Epithelial-cell-specific transcriptional regulation of human Galbeta1,3GalNAc/Galbeta1,4GlcNAc α2,3-sialyltransferase (hST3Gal IV) gene. Biochem Biophys Res Commun 257:516–522, 1999.
122. Reed MJ, Beranek PA, Bonney RC, Ghilchik MW, James VH. The effect of ethynyloestradiol and medroxyprogesterone acetate on the in vivo uptake and metabolism of ^{3}H-oestradiol by breast tumour tissue in postmenopausal women. Anticancer Res 7:1265–1269, 1987.
123. Ghosh D, Pletnev VZ, Zhu DW, et al. Structure of human estrogenic 17β-hydroxysteroid dehydrogenase at 2.20 Å resolution. Structure 3:503–513, 1995.
124. Puranen T, Poutanen M, Ghosh D, Vihko P, Vihko R. Characterization of structural and functional properties of human 17β-hydroxysteroid dehydrogenase type 1 using recombinant enzymes and site-directed mutagenesis. Mol Endocrinol 11:77–86, 1997.
125. Azzi A, Rehse PH, Zhu DW, Campbell RL, Labrie F, Lin SX. Crystal structure of human estrogenic 17β-hydroxysteroid dehydrogenase complexed with 17β-estradiol [letter]. Nat Struct Biol 3:665–668, 1996.
126. Breton R, Housset D, Mazza C, Fontecilla-Camps JC. The structure of a complex of human 17β-hydroxysteroid dehydrogenase with estradiol and NADP$^+$ identifies two principal targets for the design of inhibitors. Structure 4:905–915, 1996.
127. Lin SX, Zhu DW, Azzi A, et al. Studies on the three-dimensional structure of estrogenic 17β-hydroxysteroid dehydrogenase [published erratum appears in J Endocrinol 151:151, 1996]. J Endocrinol 150(Suppl):S13–S20, 1996.
128. Mazza C, Breton R, Housset D, Fontecilla-Camps JC. Unusual charge stabilization of NADP$^+$ in 17β-hydroxysteroid dehydrogenase. J Biol Chem 273:8145–8152, 1998.
129. Jörnvall H, Persson B, Krook M, et al. Short-chain dehydrogenases/reductases (SDR). Biochemistry 34:6003–6013, 1995.
130. Ghosh D, Weeks CM, Grochulski P, Duax WL, Erman M, Orr JC. Three-dimensional structure of holo 3α,20β-hydroxysteroid dehydrogenase: a member of the short-chain dehydrogenase fam-

ily. Proc Natl Acad Sci USA 88:10064–10068, 1991.
131. Ghosh D, Wawrzak Z, Weeks CM, Duax WL, Erman M. The refined three-dimensional structure of 3α,20β-hydroxysteroid dehydrogenase and possible roles of the residues conserved in short-chain dehydrogenases. Structure 2:629–640, 1994.
132. Penning TM. 17β-Hydroxysteroid dehydrogenase: inhibitors and inhibitor design. Endocr-Relat Cancer 3:41–56, 1996.
133. Sawicki MW, Erman M, Puranen T, Vihko P, Ghosh D. Structure of the ternary complex of human 17β-hydroxysteroid dehydrogenase type 1 with 3-hydroxyestra-1,3,5,7-tetraen-17-one (equilin) and NADP$^+$. Proc Natl Acad Sci USA 96:840–845, 1999.
134. Tanenbaum DM, Wang Y, Williams SP, Sigler PB. Crystallographic comparison of the estrogen and progesterone receptor's ligand binding domains. Proc Natl Acad Sci USA 95:5998–6003, 1998.
135. Sawicki MW, Li N, Ghosh D. Equilin. Acta Crystallogr C 55:425–427, 1999.
136. Lin SX, Yang F, Jin JZ, et al. Subunit identity of the dimeric 17β-hydroxysteroid dehydrogenase from human placenta. J Biol Chem 267:16182–16187, 1992.
137. Peltoketo H, Isomaa V, Mäentausta O, Vihko R. Complete amino acid sequence of human placental 17β-hydroxysteroid dehydrogenase deduced from cDNA. FEBS Lett 239:73–77, 1988.
138. Dupont E, Labrie F, Luu-The V, Pelletier G. Localization of 17β-hydroxysteroid dehydrogenase throughout gestation in human placenta. J Histochem Cytochem 39:1403–1407, 1991.
139. Puranen TJ, Kurkela RM, Lakkakorpi JT, et al. Characterization of molecular and catalytic properties of intact and truncated human 17β-hydroxysteroid dehydrogenase type 2 enzymes: intracellular localization of the wild-type enzyme in the endoplasmic reticulum. Endocrinology 140:3334–3341, 1999.
140. Rommens JM, Durocher F, McArthur J, et al. Generation of a transcription map at the *HSD17B* locus centromeric to *BRCA1* at 17q21. Genomics 28:530–542, 1995.
141. Peltoketo H, Nokelainen P, Piao Y-S, Vihko R, Vihko P. Two 17β-hydroxysteroid dehydrogenases (17HSDs) of biosynthesis: 17HSD type 1 and type 7. J Steroid Biochem Mol Biol 69:431–439, 1999.

10

Breast Cancer: Intervention in *BRCA1* and *BRCA2* Families

FRANCES V. ELMSLIE
ROSALIND A. EELES

There are a number of strategies for intervention in families known to harbor a mutation in the breast cancer susceptibility genes *BRCA1* and *BRCA2*. The type of intervention considered will depend on the individual's status (whether affected or unaffected) and wishes. The options fall broadly into three groups: those aimed at primary prevention of cancer, those aimed at early detection, and those aimed at prevention of recurrence or the development of a second malignancy in individuals who have been previously affected. Research into the management of men and women who have a known genetic susceptibility to cancer is ongoing. Although most studies are retrospective, data are beginning to accrue that demonstrate benefit for some preventative strategies. More work is needed in this specific group of patients to determine the strategies for management. This is most optimal in the context of multicenter trials.

RISKS OF DEVELOPING CANCER

A number of studies have estimated the risk of developing cancer in carriers of *BRCA1/BRCA2* mutations. Easton et al.[1] investigated families linked to the *BRCA1* locus and estimated the risk of breast cancer to be 51% by age 50 and 85% by age 70. The risk of ovarian cancer was estimated to be 63% by age 70, based on the incidence of ovarian cancer in *BRCA1*-linked families, or 44%, based on the risk of ovarian cancer in carriers with a previous breast cancer. The incidence of contralateral breast cancers in women with a previous breast cancer from *BRCA1*-linked families is 64% by age 75.[2] However, studies of individuals found to carry *BRCA1/2* mutations ascertained from the general population have found a lower penetrance. A study of carriers of the common Ashkenazi mutations (185delAG and 5382insC in *BRCA1* and 6174delT in *BRCA2*) found the breast cancer risk to be 56% by age 70.[3] The ovarian cancer risk was 16% by this age. The authors found no significant difference in the incidence of these cancers between carriers of *BRCA1* mutations and those of *BRCA2* mutations. A further study investigated the carriers of a founder mutation in *BRCA2* (999del5), estimated to be present in 0.6% of the Icelandic population.[4] This provided a much lower estimate of the breast cancer risk, 37.2% by age 70. However, the authors were unable to draw conclusions about the ovarian cancer risk.

There is some evidence that different mutations in *BRCA1* and *BRCA2* are associated with different ovarian cancer risks,[1,5–7] although these figures are not yet used in counseling. In practice, most cancer geneticists quote a risk of ovarian cancer of between 40% and 60% for BRCA1 carriers and of 27% for BRCA2 carriers by age 80.[8]

Male carriers of mutations in BRCA1/2 are also at increased risk of developing cancers. The lifetime risk of breast cancer in a male carrier of a mutation in BRCA2 is about 5%; only a few cases of male breast cancer in BRCA1 families have been reported. Male carriers of BRCA1/2 have a higher incidence of cancer of the prostate; the lifetime risk has been estimated to be 6% in BRCA1 carriers[9] and 6%–14% in BRCA2 carriers by age 74 (the population risk being 2% at this age).[10,11] There is a 6% risk of colon cancer in BRCA1 carriers.[9]

GENETIC TESTING FOR BRCA1/2 MUTATIONS

The first step in genetic testing for BRCA1/2 is to identify a mutation in an affected family member. Mutations are widespread throughout the gene, and many are novel, i.e., specific to a family. It is important to establish that a mutation is definitely the cause of the disease in the family: mutation testing is then offered to unaffected relatives for the specific mutation previously found in their affected relative. This is accompanied by full counseling, which usually consists of at least two counseling sessions 1 month apart (the so-called cooling-off period). A negative result in this situation is truly negative since the mutation has already been identified in the family. When a mutation is not identified in the initial screen, this does not exclude the presence of a mutation. Current screening methods identify mutations only in the coding and splice site recognition sequences of the BRCA1 and BRCA2 genes. 10%–20% of mutations are thought to lie in the regulatory regions of the gene, for which analysis is not yet possible. Alternatively, a mutation may reside in another, currently unidentified gene. Founder mutations have been described in many populations, e.g. the Ashkenazi Jewish, Icelandic, and Scottish populations.[4,12–14] If an affected individual from one of these populations does not carry a founder mutation, the risk that their disease is due to a mutation in BRCA1/2 is substantially reduced.

The uptake of predictive genetic testing is higher in breast cancer families than in other genetic diseases, e.g., Huntington's disease, for which no preventive measures can be offered.[15] In research families, the uptake of BRCA1 testing is about 44% overall and higher in women than men.[16,17] One year after test results, Watson et al.[18] did not find any adverse psychological features. Similar findings were reported by Lerman et al.[19]

MANAGEMENT OF BRCA1/2 CARRIERS

There are currently several approaches to the management of men and women with an increased cancer risk due to a genetic predisposition. The options available depend on whether the individual is already affected or an unaffected carrier.

- Early detection through screening programs
- Chemoprevention
- Change in lifestyle
- Prophylactic surgery

Management of the Unaffected Carrier

Early Detection Through Screening

Screening for female carriers of BRCA1/2 mutations is currently under evaluation. Population mammographic screening is offered from the age of 50 until 65 in the United Kingdom and earlier in the United States as this has been shown to reduce mortality by at least 20%.[20] However, BRCA1/2 carriers are at risk of developing cancers before population screening starts. Current practice is to teach breast self-awareness and to offer annual breast surveillance with mammography from the age of 35, as recommended by the British Association of Surgical Oncologists and the U.K. Cancer Genetics Group (UKCGG).[21] If young cases of breast cancer have been observed in the family, some centers start breast surveillance 5 years before the age of the youngest case, although it is recommended that mammography is not introduced until after the age of 35. Lalloo et al.[22] showed that screening a high-risk population under the age of 50 detects as many cancers as the National Health Service Breast Screening Programme.

Alternative methods of imaging are being assessed in the high-risk population in the hope that they may provide an alternative to mammography. Mammography may be less effective in the younger population because breast tissue in premenopausal women is more dense. In addition, there is some concern that the radiation dose associated with mammography may cause tumor progression in gene carriers. The value of magnetic resonance imaging in screening for breast cancer is being evaluated in a U.K. study of known BRCA1/2 carriers or women at 50% genetic risk.[23]

Ovarian screening is more controversial because it has not been proven to be of benefit. It is currently offered on the basis of local availability, although the UKCGG has recommended that it should be offered only as part of a clinical trial.[21] When it is available, women are screened annually by transvaginal ultrasonography from the age of 35, usually together with measurement of serum CA125 antigen. In a pilot study of 21,935 postmenopausal women from the general population, Jacobs et al.[24] found that measurement of serum CA125 followed by transvaginal ultrasound in those that had a raised value had a positive predictive value of 20.7%. Although the median survival in women from the screened group was significantly greater than that in the control group, the authors could not conclude that this was definitely due to the screening. The number of deaths from an index cancer did not differ significantly, although the trial was not designed to show such an effect. They did demonstrate that ovarian screening was feasible and recommended that it should be evaluated further. A national study in the United Kingdom is ongoing.

Male carriers are at increased risk of developing cancer of the prostate, as previously mentioned. Targeted screening in first-degree relatives of brother pairs with the disease demonstrated a higher detection rate of prostate cancer than expected; however, whether there is a reduction in mortality is uncertain.[25] A study of screening in BRCA1/2 carriers by annual measurement of prostate-specific antigen and rectal examination from 50 to 69 years is being proposed (R.A. Eeles, personal communication).

Carriers of BRCA2 mutations are also at risk of developing melanoma. Patients who have a family history of melanoma should be examined annually and advised to report any suspicious skin lesions. Those with a previous personal history of melanoma should be followed up regularly by a dermatologist.

Chemoprevention

The antioestrogenic agent tamoxifen reduces mortality from breast cancer. Furthermore, women receiving tamoxifen following the diagnosis of a primary breast cancer have a significantly lower risk of developing a contralateral breast cancer.[26] It was therefore suggested that tamoxifen therapy may be effective in the primary prevention of breast cancer in women at high risk. The efficacy of chemoprevention in BRCA1/2 carriers has not been specifically investigated. However, the results from four trials of tamoxifen in healthy women at increased risk of developing breast cancer have been reported. The National Surgical Adjuvant Breast and Bowel Project P-1 (NSABP-1) study randomized 13,388 healthy women at increased risk of developing breast cancer on the basis of the Gail model to receive tamoxifen or placebo.[27] This study showed a 49% reduction in incidence of invasive breast cancer and a 50% reduction in the incidence of noninvasive breast cancer in the tamoxifen group. However, the rate of endometrial cancer was increased in the tamoxifen group (risk ratio = 2.53), although no deaths attributable to endometrial cancer were observed.

Two other trials have produced contradictory results. Interim analysis of a U.K. trial, has been reported. Investigators recruited 2494 women with a family history of breast cancer and randomized them to receive tamoxifen or placebo for up to 8 years, with 2471 followed up for a median of 70 months.[28] No protective effect of tamoxifen was demonstrated: the overall occurrence of breast cancer was the same in the group receiving tamoxifen as in the group receiving placebo. An increased number of endometrial cancers was observed in the tamoxifen group, although this was not significant. The authors suggested that the observed difference between this trial and the NSABPP-1 trial could be due to a difference in the study populations. Preliminary findings in an unselected group of healthy hysterectomized Italian women also failed to show a decrease in breast cancer in the group receiving tamoxifen.[29] The IBIS trial has shown intermediate results with about a third reduction in breast cancer risk, but with an increased risk of death on tamoxifen due to causes other than breast cancer.[30]

Clearly, further follow-up of these trials is warranted. If a protective effect is demonstrated, the effect of tamoxifen specifically in gene carriers should be assessed. The preliminary results of the U.K. trial suggest that it gives less protection when genetic factors are present.[31] BRCA1/2 mutation analysis of women who developed cancer in the NSABPP1 study has suggested tamoxifen may be more protective in BRCA2 than BRCA1 carriers, but the numbers are too small to dictate a change in clinical practice in our opinion.[32]

Lifestyle Factors

A number of lifestyle factors influence the development of breast cancer in the general population, although most have not been specifically investigated in those at increased risk because of a genetic predisposition. A national study in the United Kingdom (the EMBRACE EpideMiology of BRcA CarriErs study) is currently evaluating the role of these factors in cancer risk in known carriers. Lifestyle factors that will be discussed include the following:

- Diet
- Smoking and alcohol
- Exercise
- Oral contraception
- Reproductive factors
- Hormone-replacement therapy

DIET

A number of studies have investigated the role of dietary factors in the development of breast cancer in the general population. These have produced conflicting results. However, there is a consistent association with high consumption of red meat[33] and, in one report, well-done meat.[34] This effect may be related to the formation of heterocyclic amines, which are known carcinogens, when meat is cooked.[33] N-Acetyltransferase-1 (NAT1) is one of the major enzymes in human breast tissue that activates aromatic and heterocyclic amines. A possible association between two polymorphic variants of the NAT1 gene and breast cancer has been noted.[35] The positive association was more evident among smokers and those who consumed a high level of red meat.

SMOKING AND ALCOHOL

Smoking has not been found to be a risk factor for breast cancer in the general population.[36] However, smoking was found to be protective against breast cancer in BRCA1/2 carriers in one study.[37] The reduction in breast cancer incidence was significant for BRCA1 carriers who had smoked 4 or more pack-years (odds ratio = 0.47, 95% confidence interval 0.26–0.86). The reduction in BRCA2 carriers was greater (0.39) but did not reach statistical significance, possibly because of small sample size. The authors postulated that the protective effect may be due to the known antioestrogenic properties of smoking.[38] Some experimental data suggest that this antiestrogenic effect is due to increased hepatic metabolism of estrogens by estradiol-2 hydroxylation. Estrogens are converted by this enzyme to 2-hydroxyestrogens, which are less potent and rapidly cleared.[39]

It would, of course, be impossible to advocate smoking as protection against breast cancer in gene carriers because of the other known carcinogenic effects. However, the mechanism by which smoking acts should be further investigated with the aim of developing therapies that mimic this effect. Alcohol does increase the risk of breast cancer in the general population.

EXERCISE

Several studies have demonstrated an association between regular physical activity and reduced risk of breast cancer, although this has not been shown in all.[40,41] A large study of 25,624 Norwegian women with follow-up over 13.7 years demonstrated a reduction in the risk of breast cancer of 37% in women who reported exercising regularly at study base.[42] The reduction in risk was greatest in lean women, women under 45, and women who exercised regularly over a period of 3–5 years. Another study demonstrated a reduction in risk in postmenopausal women who exercised.[43] Several mechanisms by which exercise might reduce the risk of breast cancer have been postulated.[44] These include delay of the onset of menarche and reduction of the number of ovulatory cycles. However, very strenuous activity is required to reduce the number of ovulatory cycles. Obese women have an increased risk of breast cancer, possibly due to increased levels of endogenous estrogens. Reduction in body fat by exercise may reduce the aromatization of androgens to estrogens and therefore decrease the risk of breast cancer. Alternatively, the effect of exercise on the immune system, which includes increases in the number and activity of macrophages, natural killer cells, lymphokine-activated killer cells, and regulating cytokines, may also protect against cancer.

ORAL CONTRACEPTION

It has been well demonstrated that women who take the oral contraceptive pill or who are within 10 years of stopping it have a slightly increased risk of developing breast cancer.[45] A single study has investigated the effect of oral contraceptives on breast cancer risk in Ashkenazi BRCA1/2 carriers.[46] In a group of 50 breast cancer patients under the age of 40, women who reported long-term oral contraceptive use (>48 months) were significantly more likely to have mutations in BRCA1 or -2. However, the numbers in the study were small and must be viewed with caution.

The effect of oral contraceptive use on ovarian cancer risk in 207 BRCA1 and BRCA2 carriers has also been investigated.[47] Oral contraception had a protective effect against ovarian cancer in this group, with a reduction in risk of 20% for up to 3 years of use, increasing to 60% after 6 years of use. These results suggest that oral contraception has a similar effect on the reduction of ovarian cancer risk in those who have a genetic predisposition as in the general population. The authors suggest that oral contraception should be considered for chemoprevention in this group of patients. However, the effect on breast cancer risk needs to be clarified before such a recommendation can be made to patients.

REPRODUCTIVE FACTORS

It has been consistently shown in the general population that early age at first pregnancy and an increase in parity are associated with a reduction in the incidence of breast cancer.[48,49] In addition, multiparity is protective against ovarian cancer.[50] However, several studies of women at high risk because of a family history have produced conflicting results. Colditz et al.[51] showed an adverse effect of early first pregnancy on breast cancer risk, whereas other studies have shown no difference between these women and the general population.[52,53] A reduction in the risk of breast cancer in multiparous women with a family history has also been demonstrated.[54] The effect of reproductive factors has been investigated in women who are known gene carriers.

Risk modifiers have been examined in an historical cohort of BRCA1 carriers.[55] The authors found that early menarche, parity of <3, and year of birth after 1930 were associated with an increased risk of breast cancer. In contrast, the ovarian cancer risk increased with increasing parity but was also greater in the recent birth cohort. Both nulliparity and older age at last birth had a protective effect.

The effect of pregnancy on the risk of early-onset breast cancer in BRCA1 and BRCA2 carriers was examined in 189 BRCA1 carriers and 47 BRCA2 carriers who had developed breast cancer before the age of 40; these were matched to controls who had either not had breast cancer or had developed it after 40.[56] A higher proportion of carriers with children developed early-onset breast cancer compared with nulliparous carriers (odds ratio = 1.71). The risk increased with the number of births and did not diminish with time since the last pregnancy. In addition, an early first pregnancy did not confer protection against early-onset breast cancer. In a separate group of women, pregnancy was found to have no effect in those who developed breast cancer after the age of 40.

These findings have important implications for the counseling of BRCA1/2 carriers for whom early pregnancy appears to increase the risk of both breast and ovarian cancers.

HORMONE-REPLACEMENT THERAPY

An increased risk of breast cancer has been demonstrated among women on long-term (>5 years) hormone-replacement therapy (HRT). Beral et al.[57] found an excess incidence of six breast cancers in 1000 women who had used HRT for more than 10 years. In addition, the risks were greater when a combined estrogen and progesterone preparation was used compared with estrogen alone.[58,59] For the majority of women, the benefits in terms of protection against osteoporosis outweigh the breast cancer risk. However, one model has shown that, for women who have a >30% lifetime breast cancer risk and an average risk of cardiac events, life expectancy is no longer increased. In this group, HRT should be used with caution or for short periods.[60]

Prophylactic Surgery

Both prophylactic mastectomy and prophylactic oophorectomy are options available to a known carrier of a mutation in BRCA1 or BRCA2. The uptake of prophylactic mastectomy is greater in the United Kingdom and Holland[61,62] than in the United States.[63] The efficacy of this procedure has now been evaluated retrospectively in BRCA1/2 carriers. Two studies have found at least 90% reduction in the incidence of breast cancer in women with BRCA1 or BRCA2 mutations. Further evidence for a reduction in breast cancer risk following breast removal was found in a study that examined the incidence of breast cancer following surgical breast reduction.[65] Women who had breast reduction had a relative risk of 0.61 of developing breast cancer (95% confidence interval 0.42–0.86). The relative risk was related to the amount of breast tissue removed: those who had a greater amount of breast tissue removed had a lower relative risk of developing breast cancer. There is a report of a male having a prophylactic mastectomy; he was at a 1 in 2 risk in a BRCA2-like family in which no mutation had been identified.[66] His risk of developing breast cancer was only 2.5%, which is substantially less than the lifetime population risk in women. This illustrates the fact that individuals seeking prophylactic mastectomy are driven more by anxiety than the actual level of risk.[67] For this reason, women seeking prophylactic mastectomy are managed according to a strict protocol of cancer genetics risk assessment, a clinical psychological assessment, and consultations with the breast surgeon and clinical nurse specialist in breast care. This should include photographs of good and poor cosmetic results, and the breast surgeon should have experience of reconstructive surgery or close liaison with a plastic surgeon.

The type of mastectomy offered will depend on the woman's physique and preference. For prophylaxis, as much as possible of the breast tissue is removed and either an implant is inserted or a tram flap is used (the muscle of the flap replacing the breast tissue). There is no evidence that silicone implants are associated with increased cancer risk.[68] Prospective studies of these operations are needed in known gene carriers, and these are beginning in the United States and Europe.

Prophylactic oophorectomy has been shown to reduce the incidence of ovarian cancer in susceptible patients.[69–71] However, cancer can still occur in the peritoneum because the cells have the same embryological origin. The incidence of this is thought to be low, on the order of 2%–3%.[72] In addition, prophylactic oophorectomy reduced the risk of breast cancer in BRCA1 carriers (hazard ratio = 0.53, 95% confidence interval = 0.33–0.84).[73] The risk reduction increased with increasing number of years after surgery and was not affected by HRT.

Management of the Affected Carrier

A number of factors need to be considered in the management of an affected carrier.

- Is tumor pathology different, and does this impact on management?
- Do the chemosensitivity and radiosensitivity of the tumors differ?
- Is survival from the cancer different?
- Prophylactic surgery
- Screening for second cancers

Tumor Pathology, Management, and Survival

Tumors in BRCA1 carriers tend to be of higher grade than sporadic breast cancers from controls, and this is reflected in a higher mitotic rate. Carriers have a higher proportion of atypical medullary and medullary cancers and a lower rate of carcinoma in situ.[74,75] BRCA2 carriers have a lower rate of tubule formation, which is a bad prognostic feature.[75] However, the mitotic rate of BRCA2-associated tumors in this study was not increased compared with sporadic controls,[75] suggesting that cancers from BRCA2 heterozygotes do not share the increased proliferative rate seen in BRCA1-associated tumors. A study of both BRCA1 and BRCA2 carriers in the Jewish population showed that overall the breast cancers in known carriers were more likely to be of histological grade III, to have axillary involvement, and to be estrogen-receptor negative.[76] Verhoog et al.[77] studied 28 patients with BRCA2 mutations and found that the tumors were larger than those in sporadic patients and more likely to be steroid (particularly progesterone) receptor-positive. Many of these features suggest that survival of gene carriers should be worse, but studies of survival have produced conflicting results. Chappuis et al.[78] reviewed 31 studies of survival from breast cancer and familial factors reported between 1996 and 1999. These were divided into family history studies ($n = 18$), four showing better survival, two worse survival, and the rest no difference; linkage studies ($n = 3$), two families linked to BRCA1 showing better survival and one linked to BRCA2 showing worse survival; and mutation-based studies ($n = 10$), eight showing no difference and two showing a worse outcome in mutation carriers.

All studies of survival in gene carriers have an inherent bias in that women must be alive in order to undergo a genetic test. This can be circumvented in two ways. First, the proband can be excluded from the survival analysis. When Verhoog et al.[79] excluded probands from their survival analysis, they found that survival changed from no difference between BRCA1 carriers and controls to worse survival in carriers, but this difference was not statistically significant. Lee et al.[80] found no difference in the survival of affected carrier relatives of Ashkenazi Jewish probands with BRCA1/2 mutations and the affected relatives of noncarriers. Second, stored breast cancer tissue from all patients can be analyzed for mutations in BRCA1/2. This method is currently feasible only in populations in which founder mutations occur. Foulkes et al.[81] found that the presence of a BRCA1 mutation was an adverse prognostic factor, with a significantly worse 5-year disease-free survival.

If BRCA1-associated breast cancer does have a worse prognosis, it may be important to treat this group of patients more aggressively, even for early tumors. For example, this may be a group in which small (<1 centimeter) grade III tumors should be treated with adjuvant chemotherapy, an area in which there is currently controversy about adjuvant chemotherapy. In addition, the fact that most tumors are steroid receptor-negative may limit the efficacy of endocrine therapy and chemoprevention in BRCA1 carriers.

Robson et al.[82] showed that Jewish women with founder BRCA1/2 mutations are at increased risk for ipsilateral breast cancer–related events after breast conservation, although this was not statistically significant. Both this group and Verhoog et al.[79] found a statistically increased contralateral breast cancer risk among BRCA1/2 carriers or carriers of BRCA1 alone. In view of this fact, oncologists are debating whether contralateral prophylactic mastectomy should be offered at the time of initial surgery. Ongoing studies of follow-up of carriers should answer this question.

There are data that suggest that survival in familial cases of ovarian cancer is worse whether or not a BRCA1/2 mutation is present.[83-87] One study reported improved survival in this group.[88]

Radiosensitivity and Chemosensitivity

It is not known whether the tumors in human BRCA1/2 carriers have different sensitivity to radiation or chemotherapeutic agents. Mice null for brca1 or brca2 have increased radiosensitivity and chemosensitivity.[89-91]

Screening for Second Primary Cancers and Prophylactic Surgery

BRCA1/2 carriers are at risk of developing a second primary cancer after development of cancer. The lifetime risk of a second primary breast cancer in female BRCA1 carriers is 64%; for BRCA2 carriers, it is 56%.[2,92] This fact needs to be taken into account when treating the first cancer in these women. As previously discussed, prophylactic contralateral mastectomy may be of benefit. Prophylactic oophorectomy also has benefits in terms of reduction of the breast and ovarian cancer risk and improving survival from breast cancer.[93]

CONCLUSION

The cancer risks, management, and outcome in men and women with proven *BRCA1/BRCA2* mutations are areas of intensive research. More information is needed to counsel these families accurately. In addition, as we learn more about the natural history of cancer in this group of patients, it may be possible to individualize care of the affected carrier according to the precise mutation present. Prospective studies will yield the best and least biased results.

REFERENCES

1. Easton DF, Ford D, Bishop DT, Breast Cancer Linkage Consortium. Breast and ovarian cancer incidence in *BRCA1*-mutation carriers. Am J Hum Genet 56:265–271, 1995.
2. Ford D, Easton DF, Peto J. Estimates of the gene frequency of *BRCA1* and its contribution to breast and ovarian cancer incidence. Am J Hum Genet 57:1457–1462, 1995.
3. Struewing JP, Hartge P, Wacholder S, et al. The risk of cancer associated with specific mutations of *BRCA1* and *BRCA2* among Ashkenazi Jews. N Engl J Med 336:1401–1408, 1997.
4. Thorlacius S, Struewing JP, Hartge P, et al. Population-based study of the risk of breast cancer in carriers of *BRCA2* mutation. Lancet 352:1337–1339, 1998.
5. Holt JT, Thompson ME, Szabo C, et al. Growth retardation and tumour inhibition by *BRCA1*. Nat Genet 12:298–302, 1996.
6. Gayther SA, Warren W, Mazoyer S, et al. Germline mutations of the *BRCA1* gene in breast and ovarian cancer families provide evidence for a genotype–phenotype correlation. Nat Genet 11:428–433, 1995.
7. Gayther SA, Mangion J, Russell P, et al. Variation of risks of breast and ovarian cancer associated with different germline mutations in the *BRCA2* gene. Nat Genet 15:103–105, 1997.
8. Ford D, Easton DF, Stratton M, et al. Genetic heterogeneity and penetrance analysis of the *BRCA1* and *BRCA2* genes in breast cancer families. Am J Hum Genet 62:676–689, 1998.
9. Ford D, Easton DF, Bishop DT, Narod SA, Goldgar DE, Breast Cancer Linkage Consortium. Risks of cancer in *BRCA1*-mutation carriers. Lancet 343:692–695, 1994.
10. Phelan CM, Lancaster JM, Tonin P, et al. Mutation analysis of the *BRCA2* gene in 49 site-specific breast cancer families. Nat Genet 13:120–122, 1996.
11. Breast Cancer Linkage Consortium. Cancer risks in *BRCA2* mutation carriers. J Natl Cancer Inst 91:1310–1316, 1999.
12. Oddoux C, Struewing JP, Clayton CM, et al. The carrier frequency of *BRCA2* 6174delT mutation among Ashkenazi Jewish individuals is approximately 1%. Nat Genet 14:188–190, 1996.
13. Offit K, Gilewski T, McGuire P, et al. Germline *BRCA1* 185delAG mutations in Jewish women with breast cancer. Lancet 347:1643–1645, 1996.
14. Liede A, Cohen B, Black DM, et al. Evidence of a founder *BRCA1* mutation in Scotland. Br J Cancer 82:705–711, 2000.
15. Craufurd D, Dodge A, Kerzin-Storrar L, et al. Uptake of presymptomatic predictive testing for Huntington's disease. Lancet 2:603–605, 1989.
16. Craufurd D, Evans DGR, Binchy A. Response to *BRCA1* (linkage) testing. Poster presented to the Cancer Family Study Group Annual Meeting, Manchester U.K., 1995.
17. Watson M, Murday VA, Lloyd S, et al. Genetic testing in breast/ovarian cancer (*BRCA1*) families [letter]. Lancet 346:583, 1995.
18. Watson M, Lloyd SM, Eeles RA. Pyschosocial impact of testing (by linkage) for the *BRCA1* breast cancer gene: an investigation of two families in the research setting. Psycho-Oncology 5:233–239, 1996.
19. Lerman C, Narod S, Schulman K, et al. *BRCA1* testing in families with hereditary breast–ovarian cancer. A prospective study of patient decision-making and outcomes. JAMA 275:1885–1892, 1996.
20. Chamberlain J. Screening for breast cancer in high risk populations. In: Eeles RA, Ponder BAJ, Easton DF, Horwich A (eds). Genetic Predisposition to Cancer. London: Chapman and Hall, 1996, pp 253–266.
21. Eccles DM, Evans PG, Mackay J. Guidelines for a genetic risk–based approach to advising women with a family history of breast cancer. J Med Genet 37:203–209, 2000.
22. Lalloo F, Boggis CRM, Evans DGR, Shenton A, Threlfall AG, Howell A. Screening by mammography, women with a family history of breast cancer. Eur J Cancer 34:937–940, 1998.
23. MRI Breast Screening Study Advisory Group. National study of magnetic-resonance imaging to screen women at genetic risk of breast cancer. Lancet Interactive, Reviews Protocol 17/4, http://www.thelancet.com/newlancet/reg/author//protocol7_4.html, 1998.
24. Jacobs IJ, Skates SJ, MacDonald N, et al. Screening for ovarian cancer: a pilot randomised trial. Lancet 353:1207–1210, 1999.

25. McWhorter WP, Hernandez AO, Meikle AW, et al. A screening study for prostate cancer in high risk families. J Urol 148:826–828, 1992.
26. Early Breast Cancer Trialists Collaborative Group. Tamoxifen for early breast cancer: an overview of the randomised trials. Lancet 351:1451–1467, 1998.
27. Fisher B, Constantino JP, Wickerham DL, et al. Tamoxifen for prevention of breast cancer: report of the National Surgical Adjuvant Breast and Bowel Project P-1 Study. J Natl Cancer Inst 90:1371–1388, 1998.
28. Powles T, Eeles R, Ashley S, et al. Interim analysis of the incidence of breast cancer in the Royal Marsden Hospital tamoxifen randomised chemoprevention trial. Lancet 352:98–101, 1998.
29. Veronesi U, Maisonneuve P, Costa A, et al. Prevention of breast cancer with tamoxifen: preliminary findings from the Italian randomised trial among hysterectomised women. Lancet 352:93–97, 1998.
30. First results from the IBIS1 Breast Cancer Prevention Trial. The IBIS Investigators. Abstract presented at the Third European Breast Cancer Conference, Barcelona 2002.
31. Eeles RA, Powles TP, Ashley S, et al. BRCA1, BRCA2 and pedigree genetic analysis to determine genetic risk in the UK Royal Marsden Hospital tamoxifen prevention trial. Am Hum Genet 65:A124, 1999.
32. King M-C, Wieand S, Halek, et al. Tamoxifen and breast cancer incidence among women with inherited mutations in BRCA1 and BRCA2: NSABP1 Breast Cancer Prevention Trial. JAMA 286(18):2251–6, 2001.
33. Bingham SA. High meat diets and cancer risks. Proc Nutr Soc 58:243–248, 1999.
34. Zheng W, Gustafson DR, Sinha R, et al. Well-done meat intake and the risk of breast cancer. J Natl Cancer Inst 90:1687–1689, 1998.
35. Zheng W, Deitz AC, Campbell DR, et al. N-Acetyltransferase 1 genetic polymorphism, cigarette smoking, well-done meat intake, and breast cancer risk. Cancer Epidemiol Biomarkers Prev 8:233–239, 1999.
36. Baron JA, Newcomb PA, Longnecker MP, et al. Cigarette smoking and breast cancer. Cancer Epidemiol Biomarkers Prev 5:399–403, 1996.
37. Brunet JS, Ghadirian P, Rebbeck TR, et al. Effect of smoking on breast cancer in carriers of mutant BRCA1 or BRCA2 genes. J Natl Cancer Inst 90:761–765, 1998.
38. Baron JA, LaVecchia C, Levi F. The antioestrogenic effect of cigarette smoking in women. Am J Obstet Gynecol 162:502–514, 1990.
39. Michnovicz JJ, Hershcopf RJ, Naganuma H, Bradlow HL, Fishman J. Increased 2-hydroxylation of estradiol as a possible mechanism for the anti-estrogenic effect of cigarette smoking. N Engl J Med 315:1305–1309, 1986.
40. Rockhill B, Willett WC, Hunter DJ, et al. Physical activity and breast cancer risk in a cohort of young women. J Natl Cancer Inst 90:1155–1160, 1998.
41. Gammon MD, Schoenberg JB, Britton JA, et al. Recreational physical activity and breast cancer risk among women under age 45 years. Am J Epidemiol 147:273–280, 1998.
42. Thune I, Brenn T, Lund E, Gaard M. Physical activity and the risk of breast cancer. N Engl J Med 336:1269–1275, 1997.
43. Carpenter CL, Ross RK, Paganini-Hill A, Bernstein L. Lifetime exercise activity and breast cancer risk among post-menopausal women. Br J Cancer 80:1852–1858, 1999.
44. McTiernan A. Exercise and breast cancer-time to get moving? N Engl J Med 336:1311–1312, 1997.
45. Collaborative Group on Hormonal Factors and Breast Cancer. Breast cancer and hormonal contraceptives: collaborative reanalysis of individual data on 53,297 women with breast cancer and 100,239 women without breast cancer from 54 epidemiological studies. Lancet 347:1713–1727, 1996.
46. Ursin G, Henderson BE, Haile RW, et al. Does oral contraceptive use increase the risk of breast cancer in women with BRCA1/BRCA2 mutations more than in other women? Cancer Res 57:3678–3681, 1997.
47. Narod SA, Risch H, Moslehi R, et al. Oral contraceptives and the risk of hereditary ovarian cancer. N Engl J Med 339:424–428, 1998.
48. MacMahon B, Cole P, Lin TM, et al. Age at first birth and breast cancer risk. Bull World Health Organ 43:209–221, 1970.
49. Kelsey JL, Gammon MD, John EM. Reproductive factors and breast cancer. Epidemiol Rev 15:36–47, 1993.
50. Adami HO, Hsieh CC, Lambe M, et al. Parity, age at first childbirth, and risk of ovarian cancer. Lancet 40:1250–1254, 1994.
51. Colditz GA, Rosner BA, Speizer FE. Risk factors for breast cancer according to family history of breast cancer. J Natl Cancer Inst 88:365–371, 1996.
52. Magnusson C, Colditz GA, Rosner B, Berstrom R, Persson I. Association of family history and other risk factors with breast cancer risk (Sweden). Cancer Causes Control 9:259–267, 1998.
53. McCredie MR, Dite GS, Giles GG, Hopper JL. Family history and risk of breast cancer in New Zealand. Int J Cancer 73:503–507, 1997.
54. Egan KM, Stampfer MJ, Rosner BA, et al. Risk factors for breast cancer in women with a breast cancer family history. Cancer Epidemiol Biomarkers Prev 7:359–364, 1998.
55. Narod SA, Goldgar D, Cannon-Albright L, et al. Risk modifiers in carriers of BRCA1 mutations. Int J Cancer 64:394–398, 1995.
56. Jernstrom H, Lerman C, Ghadiran P, et al. Pregnancy and the risk of early breast cancer in carriers of BRCA1 and BRCA2. Lancet 354:1846–1849, 1999.

57. Beral V, Banks E, Reeves G, Appleby P. Use of HRT and the subsequent risk of cancer. J Epidemiol Biostat 4:191–210, 1999.
58. Ross RK, Paganini-Hill A, Wan PC, Pike MC. Effect of hormone replacement therapy on breast cancer risk: estrogen versus estrogen plus progestin. J Natl Cancer Inst 92:328–332, 2000.
59. Schairer C, Lubin J, Troisi R, Sturgeon S, Brinton L, Hoover R. Menopausal estrogen and estrogen–progestin replacement therapy and breast cancer risk. JAMA 283:485–491, 2000.
60. Armstrong K, Eisen A, Weber B. Assessing the risk of breast cancer. N Engl J Med 342:564–571, 2000.
61. Evans DGR, Anderson E, Lalloo F, et al. Utilisation of prophylactic mastectomy in 10 European centres. Dis Markers 15:148–151, 1999.
62. Meijers-Heijboer H, Verhoog L, Brekelmans C, et al. Prophylactic surgery in BRCA1/2 mutation carriers: predictive factors and follow up. Am J Hum Genet, 65:A22, 1999.
63. Lerman C, Narod S, Schulman K, et al. BRCA1 testing in families with hereditary breast-ovarian cancer. A prospective study of patient decision-making and outcomes. JAMA 275:1885–1892, 1996.
64. Hartmann LC, Schaid DJ, Woods JE, Crotty TP, Myers JL, et al. Efficacy of bilateral prophylactic mastectomy in women with a family history of breast cancer. N Engl J Med 340:77–84, 1999.
65. Baasch M, Nielsen SF, Engholm G, Lund K. Breast cancer incidence subsequent to surgical reduction of the female breast. Br J Cancer 73:961–963, 1996.
66. Daltrey IR, Eeles RA, Kissin MW. Bilateral prophylactic mastectomy: not just a woman's problem! Breast 7:236–237, 1998.
67. Stefanek M, Enger C, Bekendorf J, Flamm-Honig S, Lerman C. Bilateral prophylactic mastectomy decision making: a vignette study. Prev Med 29:216–221, 1999.
68. Brinton LA, Brown SL. Breast implants and cancer. J Natl Cancer Inst 89:1341–1349, 1997.
69. Kauff ND, Satagopan JM, Robson MG, et al. Risk reducing salpingo-oophorectomy in women with a BRCA1 or BRCA2 mutation. NEJM 346:1609–15, 2002.
70. Rebbeck TR, Lynch HT, Neuhausen SL et al. Prophylactic oophorectomy in cancers of BRCA1 or BRCA2 mutations. NEJM 346:1616–22, 2002.
71. Struewing JP, Watson P, Easton DF, Ponder BA, Lynch HT, Tucker MA. Prophylactic oophorectomy in inherited breast/ovarian families. J Natl Cancer Inst Monogr 17:33–35, 1995.
72. Piver MS, Jishi MF, Tsukada Y, Nava G. Primary peritoneal carcinoma after prophylactic oophorectomy in women with a family history of ovarian carcinoma. A report of the Gilda Radner Familial Ovarian Cancer Registry. Cancer 71:2751–2755, 1993.
73. Rebbeck TR, Levin AM, Eisen A, et al. Breast cancer risk after bilateral prophylactic oophorectomy in BRCA1 mutation carriers. J Natl Cancer Inst 91:1475–1479, 1999.
74. Lakhani S, Sloane JP, Gusterson BA, Anderson TJ, et al. A detailed analysis of the morphological features associated with breast cancer in patients harbouring mutations in BRCA1 and BRCA2 predisposition genes. J Natl Cancer Inst 90:1138–1145, 1998.
75. Breast Cancer Linkage Consortium. Pathology of familial breast cancer: differences in carriers of BRCA1 or BRCA2 mutations and sporadic cases. Lancet 349:1505–1510, 1997.
76. Robson M, Rajan P, Rosen PP, et al. BRCA-associated breast cancer: absence of a characteristic immunophenotype. Cancer Res 58:1839–1842, 1998.
77. Verhoog LC, Brekelmans CT, Seynaeve C, et al. Survival in hereditary breast cancer associated with germline mutations of BRCA2. J Clin Oncol 17:3396–3402, 1999.
78. Chappuis PO, Rosenblatt J, Foulkes WD. The influence of familial and hereditary factors on the prognosis of breast cancer. Ann Oncol 10:1163–1170, 1999.
79. Verhoog LC, Brekelmans CTM, Seynaeve C, et al. Survival and tumour characteristics of breast cancer patients with germline mutations of BRCA1. Lancet 351:316–321, 1998.
80. Lee JS, Wacholder S, Struewing JP, et al. Survival after breast cancer in Ashkenazi Jewish BRCA1 and BRCA2 mutation carriers. J Natl Cancer Inst 91:259–263, 1999.
81. Foulkes WD, Wong N, Brunet J-S, et al. Germline BRCA1 mutation is an adverse prognostic factor in Ashkenazi Jewish women with breast cancer. Clin Cancer Res 3:2465–2469, 1997.
82. Robson M, Gilewski T, Haas B, et al. BRCA-associated breast cancer in young women. J Clin Oncol 16:1642–1649, 1998.
83. Pharoah PD, Easton DF, Stockton DL, Gayther S, Ponder BA. Survival in familial BRCA1-associated, and BRCA2-associated epilethial ovarian cancer. United Kingdom Coordinating Committee for Cancer Research (UKCCCR) Familial Ovarian Cancer Study Group. Cancer Res 59:868–871, 1999.
84. Johansson OT, Ranstam J, Borg A, Olsson H. Survival of BRCA1 breast and ovarian cancer patients: a population-based study from southern Sweden. J Clin Oncol 16:397–404, 1998.
85. Cannistra SA. BRCA1 mutations and survival in women with ovarian cancer. N Engl J Med 336:1256–1257, 1997.
86. Whitmore SE. BRCA1 mutations and survival in women with ovarian cancer. N Engl J Med 336:1254–1257, 1997.
87. Modan B. BRCA1 mutations and survival in women with ovarian cancer. N Engl J Med 336:1255–1257.
88. Rubin SC, Benjamin I, Behbakht K, et al. Clini-

cal and pathological features of ovarian cancer in women with germ-line mutations of *BRCA1*. N Engl J Med 335:1413–1416, 1996.
89. Sharan SK, Morimatsu M, Albrecht U. Embryonic lethality and radiation hypersensitivity mediated by Rad 51 in mice lacking *BRCA2*. Nature 386:804–810, 1997.
90. Shen SX, Weaver Z, Xu X, et al. A targeted disruption of the murine *BRCA1* gene causes gamma irradiation hypersensitivity and genetic instability. Oncogene 17:3115–3124, 1997.
91. Coleman CN. Molecular biology in radiation oncology. Radiation oncology perspective of *BRCA1* and *BRCA2*. Acta Oncol 38(Suppl 13):55–59, 1999.
92. Breast Cancer Linkage Consortium. Cancer risks in *BRCA2* mutation carriers. J Natl Cancer Inst 91:1310–1316, 1999.
93. Early Breast Cancer Trialists Collaborative Group. Ovarian ablation in early breast cancer: an overview of the randomised trials. Lancet 348:1189–1196, 1996.

11

Implications of Hormones and Hormonal Risk Factors on Screening Strategies

GISKE URSIN

In the past few decades, much interest and controversy have surrounded the question of who will benefit from regular mammographic screening. A number of studies have been conducted, including several randomized trials.[1,2] Although some issues have been resolved, many remain. Mammographic screening overall appears to be associated with reduced mortality from breast cancer. However, it is not clear at what age the benefit begins and at what age the cost–benefit effect becomes reasonable. The beneficial effect is most unequivocal for women in the age group 50–69. To what extent there are survival benefits in women under the age of 50 is undetermined.

The issues briefly described above demonstrate one major problem with the screening debate so far; the data have been discussed in women stratified by age group only. This may simplify the guidelines but does not necessarily make sense given the complexity of breast cancer. One could argue that mammographic screening would become more beneficial in all women if the focus was shifted from the age at which screening should start to what can be done to improve the benefit of screening in the groups that currently do poorest.

A number of other issues are crucial to improve the benefit of mammographic screening for the population as a whole, a major one being improving access to screening and health care. These issues are discussed in detail elsewhere. This chapter summarizes the effect of hormones on mammograms and how hormonally based strategies could improve the benefit of mammographic screening.

WHAT THE CURRENT DATA TELL US ABOUT THE BENEFIT OF MAMMOGRAPHIC SCREENING

A report from 1993[1] summarized the data from the eight randomized trials. The authors argued that the Canadian study, which found the smallest benefit of screening, could not measure the effect of screening on mortality because the control women had received annual clinical breast examinations. After excluding this study, Fletcher et al. concluded that the randomized trials to date (the Swedish trials, the Edinburgh trial, and the Health Insurance Plan trial) found that breast cancer screening was associated with a substantial reduction in breast cancer mortality in women aged 50–69. The benefit was observed after a few years, and after about 10 years of follow-up, the relative risk of dying from breast cancer was reduced by about a third in the screened compared to the unscreened group. The reason the benefit was smaller in the Canadian study may have been because the control women received clinical breast exams annually.

Combining the data from the eight randomized trials, Hendrick et al.[2] estimated that the overall benefit of mammographic screening was smaller in the 40–49 age group. Including the women who were aged 40–49 at entry, the relative risk (RR) of dying was 0.82 ([95% confidence interval (CI) 0.70–0.95] after 10 years of follow-up. It has been argued that the mortality benefit was observed only in women after they turned age 50,[3] and that it is unclear whether the benefit exists if analyses are limited to women diagnosed before age 50. However, both the National Cancer Institute and the American Cancer Society currently recommend mammographic screening to women over age 40.

In 2000, two Danish Cochrane investigators raised major concerns regarding the validity of the eight randomized trials.[4] After collecting further data on individual studies, Olsen and Gøtzsche[5,6] concluded that six of the eight trials were of poor quality or flawed and that "there is no reliable evidence that screening for breast cancer reduces mortality."[5] This report caused an intense debate. Although it raised a number of important methodological issues, many investigators had concerns about how the quality assessment was conducted. As a result, neither the National Cancer Institute nor the American Cancer Society saw any reason to alter screening guidelines following this report.[7] Further, much of the criticisms raised about four of the trials were subsequently addressed in an updated meta-analysis of the Swedish trials.[8] In this report, Nyström et al.[8] provided a detailed description of the methods used in the Swedish trials and an update on mortality data after 15 years of follow-up. Including all age groups, there were 511 breast cancer deaths in 1,864,770 women-years in the group invited for screening and 584 breast cancer deaths among the 1,688,440 women-years in the control group. This represents a 21% reduction in breast cancer mortality (RR = 0.79, 95% CI 0.70–0.89) and a 2% reduction in total mortality (RR = 0.98, 95% CI 0.96–1.00). Although reductions in breast cancer mortality were seen in all age groups after 4–8 years of follow-up, they were only statistically significant in women over 55 years of age. The accompanying editorial[9] concluded that the data confirm the benefit of mammographic screening on breast cancer mortality, that the effect varies with age, and that it was "time to move on."

WHY IS SCREENING LESS EXPENSIVE AND MORE BENEFICIAL IN OLDER WOMEN?

Since cancer is less common in the younger group, a slight benefit of screening women under age 50 would mean that the cost of saving one life by mammographic screening in this age group could be substantial. Salzmann et al.[10] estimated that biennial screening of women aged 50 to 69 would increase life expectancy by 12 days, and cost $704 per woman or $21,400 per year of life saved. Extending screening to every 18 months for women aged 40 to 49 would only increase life expectancy by 2.5 days, and would increase the costs by $676 per woman or $105,000 per year of life saved.

Breast cancers may be less likely to be fast-growing in women older than women younger than 50 years,[11] thus improving the chances that early detection and treatment will cure the disease in older women. However, the benefit of screening is determined not only by the biology of the disease; characteristics of the screening test are also important.

In general, important parameters for evaluating the characteristics of the screening test are sensitivity, positive predictive value, and specificity. *Sensitivity* is defined as the probability that the mammogram will correctly classify cancerous breasts as positive and is calculated as the number of true positive mammograms over the true positives plus the false-negatives. *Positive predictive value* is estimated as the number of true positive mammograms over the true positives and false-positive mammograms. *Specificity* is defined as the number of true negative mammograms divided by the true negative and false-positive mammograms.

Mammographic sensitivity is substantially higher in women older than women younger than age 50.[11,12] Because breast cancers are less common in younger women, the number of true negatives would be expected to be higher in the younger group and, therefore, specificity may be as good as in the older group. The positive predictive value is, however, twice as low in women under age 50 than in older women.[13]

Thus the most obvious method by which we can improve the overall benefit of screening in women under age 50 would be to improve sensitivity and reduce the false positive rate.

ROLE OF MAMMOGRAPHIC DENSITY

Mammographic sensitivity is lower in women with mammographically dense breasts.[11,12] The differences in sensitivity and false-positive rate observed in women below and above age 50 parallel the decrease in mammographic density observed with increasing age between 35 and 55.[14,15] Very dense mammographic patterns especially appear to change to lower-risk patterns during the perimenopausal period (age 45–55).[14] There is, of course, no abrupt change in the positive predictive value of mammogram at age 50,[16] but, rather, a gradual increase. This decrease in mammographic density and increase in sensitivity is most likely an effect of menopause and not age per se.[17,18]

Thus, to improve the benefit of mammographic screening in the younger group, we must focus on reducing mammographic density. This can be done with a hormonal approach.

ROLE OF HORMONES ON MAMMOGRAPHIC DENSITY

The evidence for a role of female hormones on mammographic density comes from studies of the effects of hormone-replacement therapy (HRT), tamoxifen, and a gonadotropin-releasing hormone agonist–based regimen on mammographic density. The evidence linking each of these to mammographic density is summarized below and in Table 11.1.

ROLE OF POSTMENOPAUSAL HORMONE REPLACEMENT THERAPY

There is growing evidence for an association between postmenopausal HRT and mammo-

Table 11.1 Percent of Patients Undergoing Changes in Mammographic Density with Hormone Manipulations

	Increase (%)	Decrease (%)	Reference
ERT	3.5		51
EPRT	16–24		51
Tamoxifen		44–87	32, 33
GnRHA-based regimen		75–92	36
Menopause		Observed	18

ERT, estrogen-replacement therapy; EPRT, estrogen–progestin-replacement therapy; GnRHA, gonadotropin-releasing hormone agonist.

graphic density, although the regimen used may determine the size of the effect.

Kaufman et al.[19] found that postmenopausal women ($n = 113$) who were on HRT had higher mammographic density than women who had not taken HRT ($n = 50$; odds ratio = 2.0 of high risk Wolfe parenchymal patterns P2/DY vs. low-risk patterns (N1/P1)); no details of the HRT were given. McNicholas et al.[20] found that nine of 33 women on HRT, but none of the 31 controls, developed increased density; changes were seen with all preparations (six different preparations were used, four containing progestin). Berkowitz et al.[21] found no changes in mammographic density in 14 postmenopausal women who underwent mammography before and after 1–72 months of estrogen-replacement therapy (ERT), but five of the 16 women who received estrogen and progestin-replacement therapy (EPRT) showed an increase in mammographic density after 3–20 months. Stomper et al.[22] found that 10 of 38 women who received EPRT, but only two of 12 women who received ERT, developed increased densities; and Laya et al.[23] found that 30 of 41 women developed increased density after 1 year on EPRT. Two population-based studies did not find an association between ERT and Wolfe's parenchymal patterns.[24,25] Thus, there is substantial evidence that EPRT, and somewhat weaker evidence that ERT, increases mammographic density.

Results from the largest study to date on ERT, EPRT, and mammographic density were reported by Greendale et al.[26] They reported results on Breast Imaging Reporting and Data System (BI-RADS) categories from a randomized, placebo-controlled, double-blind clinical trial of ERT and EPRT. Women in the ERT group took 0.625 milligrams conjugated equine estrogens

(CEE) daily; those in the EPRT group took 0.625 milligrams CEE daily plus either 200 milligrams micronized progesterone per day for 12 days per month, 2.5 milligrams medroxyprogesterone acetate (MPA) daily, or 10 milligrams MPA per day for 12 days per month. Greendale and colleagues[26] found that the percentages of women with an increase in BI-RADS grade after 12 months in the trial were as follows: 0% in the placebo group, 3.5% in the ERT group, and 16.4%, 19.4%, and 23.5% in the different EPRT groups listed above. These are probably the strongest data to date of an effect on exogenous sex steroids on mammographic density.

Two large studies have assessed the effect of HRT on mammographic sensitivity.[12,27] As expected, sensitivity was lowest in women who used HRT. No data were provided on the effect of different HRT regimens.

One study addressed the effect of short-term cessation of HRT on mammographic density.[28] Of 47 women who had a new mass or density increase while on HRT, 35 (74.5%) experienced a reduction in density or resolution of the new mass after short-term cessation (10–30 days) of HRT. Somewhat more women experienced a reduction in the EPRT group (19/23, or 82.6%) then in the ERT group (15/23, or 65.2%).

Assuming the results above are true, if 1000 women start EPRT, up to 250 would be expected to undergo an increase in mammographic density. If these women stop their regimen a short period, one would expect 186 of them to experience a density reduction.

TAMOXIFEN

Tamoxifen is widely used in the treatment of breast cancer. Although the exact mechanisms for its effect are not known, it has also been found to protect against breast cancer.[29] As such, one would expect tamoxifen to result in reduced mammographic density and increased sensitivity of the mammograms.

In a pilot study of 19 premenopausal women, tamoxifen was associated with a substantial reduction in density.[30] Similarly, a change in Wolfe's parenchymal pattern to more lucid patterns was observed in 94 breast cancer patients who had been treated with tamoxifen, while no such change was observed in 188 controls.[31] In a study of 152 Korean breast cancer patients and 20 healthy controls, 87% of premenopausal patients and 30% of postmenopausal patients treated with tamoxifen had a density decrease, while only 10% of healthy controls had a decrease in density.[32] Among participants in a Canadian trial, 16 of 36 (44%) women randomized to receive 20 milligrams tamoxifen per day changed to a lower density pattern compared to five of 33 (15%) who received placebo.[33] In women under age 50, 67% of those randomized to receive tamoxifen developed a lower density pattern.

Thus, it appears that premenopausal women who take tamoxifen experience a substantial reduction in mammographic density.

To what extent other selective estrogen modulators such as raloxifene modify mammographic density is currently unknown. The effect of aromatase inhibitors on mammographic density is also unknown.

GONADOTROPIN-RELEASING HORMONE AGONIST–INDUCED TEMPORARY MENOPAUSE

Spicer et al.[34] described a gonadotropin-releasing hormone agonist (GnRHA)–based regimen designed to reduce endogenous levels of estrogen and progesterone. If used for a long enough period, this regimen would be expected to substantially reduce the risk of breast cancer. Over a 1-year period, we observed a marked decrease in mammographic densities in the 13 women treated with such a GnRH (which completely blocks ovarian function) plus low-dose add-back estrogen and intermittent progestogen.[35,36] This regimen would be expected to improve mammographic sensitivity, possibly in a substantial number of women.[37]

We are currently testing a revised version of this regimen in women with genetic alterations who are at very high risk of breast cancer.

ENDOGENOUS HORMONE LEVELS AND MAMMOGRAPHIC DENSITY

The data listed above suggest that endogenous hormones are important determinants of mammographic density. However, to what extent normal variations in hormone levels in premenopausal women predict mammographic density levels is largely unknown.

One study suggests that plasma levels of estrogen were the same but progesterone was 30% and 50% higher in women with the high-risk P2 or DY Wolfe pattern than in women with the lower risk N1 or N2 pattern.[38] Women who were moderately physically active were less likely to have high-risk mammographic patterns in one study;[39] however, whether this was due to lower levels of circulating steroids in these women is unknown.

In one report, women who obtained a mammogram in the luteal phase were more likely to have dense mammograms than women who had a mammogram in the follicular phase of the menstrual cycle.[40] In a pilot study of 11 women where mammograms were obtained sequentially over one menstrual cycle, six of 11 women had some (1%–7%) increase in mammographic density from the follicular to the luteal phase of the menstrual cycle.[41] These two studies suggest that to optimize mammographic sensitivity mammograms should be obtained in the follicular phase of the cycle.

We have also addressed the issue of normal variations in steroid hormones indirectly by looking at whether polymorphisms in genes involved in steroid hormone biosynthesis and metabolism, $CYP17$ ($T^{27}C$), $COMT$ ($Val^{158}Met$), $17HSDB1$ ($Ser^{312}Gly$), and $3HSDB1$ ($Asn^{367}Thr$), predict mammographic density. However, in a study of of 396 Caucasian and African-American women, we found no consistent evidence that any of these polymorphisms is associated with mammographic density.[42]

HOW OVARIAN STEROIDS AFFECT MAMMOGRAPHIC DENSITY

The association between ovarian steroids and mammographic density parallels what is known about the effects of ovarian steroids and breast cell proliferation.[43] Breast cell proliferation is higher in pre- than postmenopausal women and higher in women in the luteal than in the follicular phase of the menstrual cycle. In short, high estrogen/progesterone levels are associated with high amounts of breast cell proliferation.

A number of studies have examined what histological components are associated with mammographic densities. Although several studies have found an association between epithelial hyperplasia[15,44–46] and mammographic densities, others have not.[47–49] Oza and Boyd[50] concluded that it was predominantly stromal proliferation that was associated with increased mammographic density. They further hypothesized that the complex interaction between breast stroma and epithelial tissue could explain why stromal proliferation shown by the mammogram is associated with an increased risk of epithelial malignancy.

Thus, it is likely that mammographic density is, if not a direct measure, at least a marker for breast cell proliferation. This means that reducing mammographic density may not only improve mammographic sensitivity but also ultimately reduce breast cell proliferation and breast cancer risk.

IMPLICATIONS FOR WOMEN WITH FAMILY HISTORY

Sensitivity of mammography may be substantially lower in women with than without a family history of breast cancer. Further, sensitivity has been found to drop faster as the screening interval increased from 7 to 13 months in women with a family history (78.6% to 68.8%) compared to all women under age 50 (87.5% to 83.6%).[11] This suggests that these women are more likely to have fast-growing tumors. Whether hormonal manipulation is more or less likely to result in density changes in these women is unknown and remains an important area of study. Improving the sensitivity of mammograms in these women would be particularly beneficial.

USING HORMONAL STRATEGIES TO IMPROVE MAMMOGRAPHIC DENSITY AND SENSITIVITY OF SCREENING MAMMOGRAMS

Early oophorectomy, tamoxifen treatment, a GnRHA-based regimen, and cessation of HRT use would be expected to increase mammographic sensitivity and, possibly, reduce breast cancer risk. However, only certain women at very high risk of breast cancer are likely to opt for tamoxifen or early oophorectomy and then only after completing childbearing. Treatment with a GnRHA-based regimen could represent an option for young women who still would like to have children.[34]

It is possible that the benefit of such hormonal strategies varies substantially. It is clear that not everyone who takes a GnRHA regimen has a reduction in density,[36] just as not everyone who starts EPRT has an increase.[51] The biological or genetic predictors of mammographic density changes with hormonal alterations are currently unknown. Thus, in the absence of such predictive markers, the best clinical approach currently to improve the screening benefit in a young woman with dense breasts would be to simply test whether a temporary endogenous hormone reduction will result in a reduced amount of density on her mammogram. For premenopausal women, a reversible approach, such as a GnRHA-based regimen, ought to be attempted first. Women with a substantial reduction in mammographic density with a GnRHA-based regimen may be more likely to benefit from tamoxifen treatment or early oophorectomy than women who do not have density reduction with such hormonal manipulation. Postmenopausal women who have substantial increases in density when they start HRT may similarly benefit from temporary cessation of the therapy prior to their screening mammogram.

There may be women in whom reducing hormone levels will not result in reduced density. For such women and for women who due to medical (high cholesterol) or other reasons do not wish to reduce their hormone levels, alternative imaging techniques, currently being developed and tested, may be a good alternative.[52]

CONCLUSION AND FUTURE DIRECTIONS

A hormonal approach to reduce mammographic density and thereby improve sensitivity and reduce the number of false-positives could benefit many women. A reduction in mammographic density should improve mammographic sensitivity and, if sustained over time, could lead to not only earlier diagnosis but also reduced breast cancer risk. Currently, we cannot predict which women will benefit from such hormonal manipulations up front; further research in this area is clearly needed. To identify women in whom hormonal manipulations will be most helpful, we need to improve our understanding of the genetic determinants of mammographic density and mammographic density alterations. A number of candidate genes are currently being evaluated, and we should soon be able to better individualize recommendations on how to improve the benefit of mammographic screening.

REFERENCES

1. Fletcher SW, Black W, Harris R, Rimer B, Shapiro S. Report of the International Workshop on Screening for Breast Cancer. J Natl Cancer Inst 85:1644–1656, 1993.
2. Hendrick RE, Smith RA, Rutledge JHI, Smart CR. Benefit of screening mammography in women aged 40–49: a new meta-analysis of randomized controlled trials. J Natl Cancer Inst Monogr 22:87–92, 1997.
3. Fletcher SW. Breast cancer screening among women in their forties: an overview of the issues. J Natl Cancer Inst Monogr 22:5–9, 1997.
4. Gøtzsche PC, Olsen O. Is screening for breast cancer with mammography justifiable [see comments]? Lancet 355:129–134, 2000.
5. Olsen O, Gøtzsche PC. Cochrane review on screening for breast cancer with mammography. Lancet 358:1340–1342, 2001.
6. Olsen O, Gøtzsche PC. Screening for breast cancer with mammography "(Cochrane Review)." Cochrane Database Syst Rev 4:CD001877, 2001.
7. McLellan F. Independent US panel fans debate on mammography. Lancet 359:409, 2002.
8. Nyström L, Andersson I, Bjurstam N, Frisell J, Nordenskjöld B, Rutqvist LE. Long-term effects of mammography screening: updated overview of the Swedish randomised trials. Lancet 359:909–919, 2002.
9. Gelmon KA, Olivotto I. The mammography screening debate: time to move on. Lancet 359:904–905, 2002.
10. Salzmann P, Kerlikowske K, Phillips K. Cost-effectiveness of extending screening mammography guidelines to include women 40 to 49 years of age [erratum appears in Ann Intern Med 128:878, 1998] [see comments]. Ann Intern Med 127:955–965, 1997.
11. Kerlikowske K, Grady D, Barclay J, Sickles EA, Ernster V. Effect of age, breast density, and family history on the sensitivity of first screening

mammography [see comments]. JAMA 276:33–38, 1996.
12. Rosenberg RD, Hunt WC, Williamson MR, et al. Effects of age, breast density, ethnicity, and estrogen replacement therapy on screening mammographic sensitivity and cancer stage at diagnosis: review of 183,134 screening mammograms in Albuquerque, New Mexico. Radiology 209:511–518, 1998.
13. Kerlikowske K, Barclay J. Outcomes of modern screening mammography. J Natl Cancer Inst Monogr 22:105–111, 1997.
14. Wolfe J. Breast parenchymal patterns and their changes with age. Radiology 121:545–552, 1976.
15. Bartow SA, Pathak DR, Mettler FA. Radiographic microcalcification and parenchymal pattern as indicators of histologic "high-risk" benign breast disease. Cancer 66:1721–1725, 1990.
16. Kopans DB, Moore RH, McCarthy KA, et al. Positive predictive value of breast biopsy performed as a result of mammography: there is no abrupt change at age 50 years. Radiology 200:357–360, 1996.
17. Grove JS, Goodman MJ, Gilbert FI, Mi MP. Factors associated with mammographic pattern. Br J Radiol 58:21–25, 1985.
18. Grove JS, Goodman MJ, Gilbert F, Clyde D. Factors associated with breast structure in breast cancer patients. Cancer 43:1895–1899, 1979.
19. Kaufman Z, Garstin WIH, Hayes R, Michell MJ, Baum M. The mammographic parenchymal patterns of women on hormonal replacement therapy. Clin Radiol 48:389–392, 1991.
20. McNicholas MM, Heneghan JP, Milner MH, Tunney T, Hourihane JB, MacErlaine DP. Pain and increased mammographic density in women receiving hormone replacement therapy. AJR Am J Roentgenol 163:311–315, 1994.
21. Berkowitz JE, Gatewood OMB, Goldblum LE, Gayler BW. Hormonal replacement therapy: mammographic manifestations. Radiology 174:199–201, 1990.
22. Stomper PC, Van Voorhis BJ, Ravnikar VA, Meyer JE. Mammographic changes associated with postmenopausal hormone replacement therapy: a longitudinal study. Radiology 174:487–490, 1990.
23. Laya MB, Gallagher JC, Schreiman JS, Larson EB, Watson P, Weinstein L. Effect of postmenopausal hormonal replacement therapy on mammographic density and parenchymal pattern. Radiology 196:433–437, 1995.
24. Bergkvist L, Tabar L, Adami H-O, Persson I, Bergstrom R. Mammographic parenchymal patterns in women receiving noncontraceptive estrogen treatment. Am J Epidemiol 130:503–510, 1989.
25. Bland KI, Buchanan JB, Weisberg BF, Hagan TA, Gray LA. The effects of exogenous estrogen replacement therapy of the breasts: breast cancer risk and mammographic parenchymal patterns. Cancer 45:3027–3033, 1980.
26. Greendale GA, Reboussin BA, Sie A, et al. Effects of estrogen and estrogen–progestin on mammographic parenchymal density. Ann Intern Med 130:262–269, 1999.
27. Kavanagh AM, Mitchell H, Giles GG. Hormone replacement therapy and accuracy of mammographic screening. Lancet 355:270–274, 2000.
28. Harvey JA, Pinkerton JV, Herman CR. Short-term cessation of hormone replacement therapy and improvement in mammographic specificity. J Natl Cancer Inst 89:1623–1625, 1997.
29. Fisher B, Costantino JP, Wickerham DL, et al. Tamoxifen for prevention of breast cancer: report of the National Surgical Adjuvant Breast and Bowel Project P-1 Study. J Natl Cancer Inst 90:1371–1388, 1998.
30. Ursin G, Pike MC, Spicer DV, Porrath SA, Reitherman RW. Can mammographic densities predict effects of tamoxifen on the breast? J Natl Cancer Inst 88:128–129, 1996.
31. Atkinson C, Warren R, Bingham SA, Day NE. Mammographic patterns as a predictive biomarker of breast cancer risk: effect of tamoxifen. Cancer Epidemiol Biomarkers Prev 8:863–866, 1999.
32. Son HJ, Oh KK. Significance of follow-up mammography in estimating the effect of tamoxifen in breast cancer patients who have undergone surgery. AJR 173:905–909, 1999.
33. Brisson J, Brisson B, Cote G, Maunsell E, Berube S, Robert J. Tamoxifen and mammographic breast densities. Cancer Epidemiol Biomarkers Prev 9:911–915, 2000.
34. Spicer DV, Shoupe D, Pike MC. GnRH agonists as contraceptive agents: predicted significantly reduced risk of breast cancer. Contraception 44:289–310, 1991.
35. Spicer DV, Ursin G, Parisky YR, et al. Changes in mammographic densities induced by a hormonal contraceptive designed to reduce breast cancer risk. J Natl Cancer Inst 86:431–436, 1994.
36. Ursin G, Astrahan MA, Salane M, et al. The detection of changes in mammographic densities. Cancer Epidemiol Biomarkers Prev 7:43–47, 1998.
37. Feig SA. Hormonal reduction of mammographic densities: potential effect on breast cancer risk and performance of diagnostic and screening mammography. J Natl Cancer Inst 86:408–409, 1994.
38. Meyer F, Brisson J, Morrison AS, Brown JB. Endogenous sex hormones, prolactin, and mammographic features of breast tissue in premenopausal women. J Natl Cancer Inst 77:617–620, 1986.
39. Gram IT, Funkhouser E, Tabar L. Moderate physical activity in relation to mammographic patterns. Cancer Epidemiol Biomarkers Prev 8:117–122, 1999.
40. White E, Velentgas P, Mandelson MT, et al. Vari-

ation in mammographic breast density by time in menstrual cycle among women aged 40–49 years. J Natl Cancer Inst 90:906–910, 1998.
41. Ursin G, Parisky YR, Pike MC, Spicer DV. Mammographic density changes during the menstrual cycle. Cancer Epidemiol Biomarkers Prev 10:141–142, 2001.
42. Haiman CA, Bernstein L, Van Den Berg D, Ingles S, Salane M, Ursin G. Genetic determinants of mammographic density. Breast Cancer Res 4:R5, 2002.
43. Pike MC, Bernstein L, Spicer DV. Exogenous hormones and breast cancer risk. In: Niederhuber JE (ed). Current Therapy in Oncology. St. Louis: BC Decker-Mosby Year Book, pp 292–303, 1993.
44. Bright RA, Morrison AS, Brisson J, et al. Relationship between mammographic and histologic features of breast tissue in women with benign biopsies. Cancer 61:266–271, 1988.
45. Boyd NF, Jensen HM, Han HL. Relationship between mammographic and histological risk factors for breast cancer. J Natl Cancer Inst 84:1170–1179, 1992.
46. Urbanski S, Jensen HM, Cooke G, et al. The association of histological and radiological indicators of breast cancer risk. Br J Cancer 58:474–479, 1988.
47. Fisher ER, Palekar A, Kim WS, Redmond C. The histopathology of mammographic patterns. Am J Clin Pathol 69:421–426, 1978.
48. Moskowitz M, Gartside P, McLaughlin C. Mammographic patterns as markers for high-risk benign breast disease and incident cancers. Radiology 134:293–295, 1980.
49. Arthur JE, Ellis IO, Flowers C, Roebuck E, Elston CW, Blamey RW. The relationship of "high risk" mammographic patterns to histological risk factors for development of cancer in the human breast. Br J Radiol 63:845–849, 1990.
50. Oza AM, Boyd NF. Mammographic parenchymal patterns: a marker of breast cancer risk. Epidemiol Rev 15:196–208, 1993.
51. Greendale GA, Reboussin BA, Sie A, et al. Effects of estrogen and estrogen–progestin on mammographic parenchymal density. Ann Intern Med 130:262–269, 1999.
52. Buchberger W, DeKoekkoek-Doll P, Springer P, Obrist P, Dunser M. Incidental findings on sonography of the breast: clinical significance and diagnostic workup. AJR Am J Roentgenol 173:921–927, 1999.

Hormonal Chemoprevention with Tamoxifen and Selective Estrogen Receptor Modulators

DAVID J. BENTREM
V. CRAIG JORDAN

Prevention of disease is far superior to treating symptomatic illness. Until recently, the prevention of cancer focused on the identification of environmental carcinogens and the reduction of their exposure to the population. Education to avoid cigarette smoking is an example of primary prevention of lung cancer. Secondary prevention consists of screening individuals at increased risk for malignancy with the assumption that early detection and treatment will affect survival. Mammography and self-examination have been successful methods of secondary prevention, identifying early-stage breast cancers. A third mechanism of prevention, chemoprevention, involves intervention within the premalignant process with specific chemical agents to prevent the development of invasive cancer. Chemoprevention is designed to work as an adjunct to, rather than a replacement for, established modalities.

In 1896, George Beatson[1] demonstrated that the removal of ovaries from premenopausal women with metastatic breast cancer could cause regression of the disease and improve the prognosis; however, by 1900 Stanley Boyd[2] established that only one in three patients would respond and only for 1 year. Despite this disappointment, a link was established between an ovarian factor and the growth of some breast cancers, which was to become the foundation of the modern clinical practice of using antiestrogens to treat breast cancer.[3,4] Simultaneously, studies were being conducted in the laboratory to complement clinical efforts. Inbred strains of mice were being established for medical research, and certain strains had a high incidence of mammary tumors. Lathrop and Loeb[5] reported that an early ovariectomy could prevent the spontaneous development of tumors, but not until Allen and Doisey[6] identified the estrus-stimulating principle were these hormones linked to the development of breast cancer. By 1936, Antoine Lacassagne,[7] again working with high-incidence strains of mice, hypothesized that if breast cancer was caused by a special hereditary sensitivity to estrogen, then developing a therapeutic antagonist to estrogen action in the breast could prevent it.[7] There were no therapeutic antagonists of estrogen at the time or even a target to design drug molecules. Exciting developments in the discovery of nonsteroidal estrogens established the structural basis of carrier molecules, which resulted in the design of the two drugs tamoxifen and raloxifene (Fig. 12.1). Both were originally described as

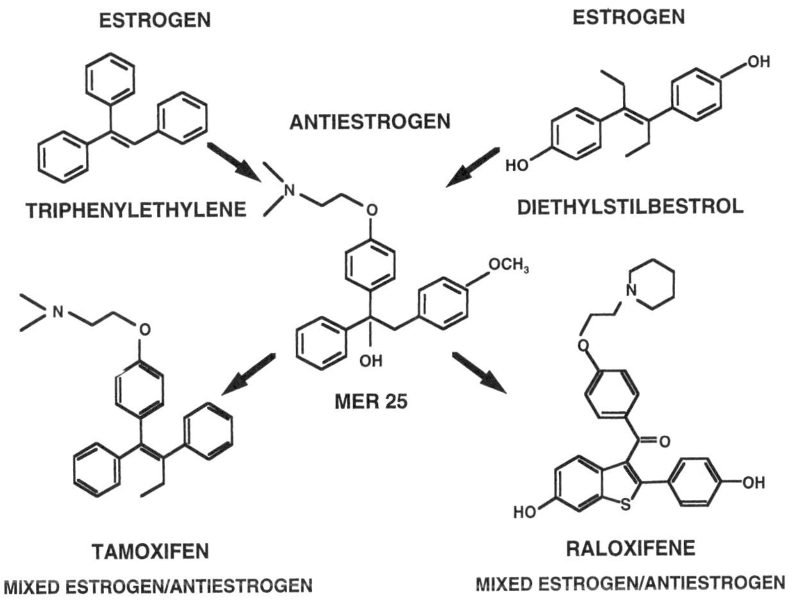

Figure 12.1 The discovery of MER-25 and the knowledge that a strategically placed alkylaminoethoxy side chain confers antiestrogenic properties was important in the development of the antiestrogens tamoxifen and raloxifene from the known nonsteroidal estrogens triphenylethylene and diethylstilbestrol.

antiestrogens and are being used today in clinical trials for the prevention of breast cancer in high-risk women.

In the 1950s and 1960s, it became clear that adrenalectomy, with glucocorticoid support, improves the prognosis of some postmenopausal women with advanced breast cancer.[8] In fact, about one-third of the women responded, approximately the same proportion as premenopausal women after oophorectomy. The reason for the apparently arbitrary responses was not resolved until the discovery of the estrogen receptor (ER),[9] and the subsequent application of this knowledge to predict the hormone responsiveness of a patient's tumor to endocrine ablation.[10] Only patients whose tumors had high levels of ER were likely to respond to endocrine ablative surgery.[11] This knowledge prevented patients with ER-negative tumors from having additional surgery with little hope of a response. However, in the early 1970s, there was no significant clinical experience with antiestrogens, and no large clinical studies had linked the efficacy of antiestrogens with the presence or absence of the ER. Tamoxifen was the first clinically useful antiestrogen for the treatment of advanced breast cancer, approved by the Food and Drug Administration in the United States in 1977.[12]

The discovery of the nonsteroidal antiestrogens was serendipitous and the result of an interest in contraception in the 1950s. Lerner et al.[13] described the first nonsteroidal antiestrogen, MER-25, in 1958. It was a blocking drug for estrogen action with almost no estrogenic properties. The drug failed in clinical trial because the low potency required large doses, causing serious central nervous system side effects.[14] Although this was disappointing, it must be stressed that a pure antiestrogen such as MER-25 would ultimately have been catastrophic as an agent to prevent breast cancer. Drug discovery switched to the triphenylethylene-based compounds with a high affinity for the ER that resulted first in clomiphene and then tamoxifen.[12]

The critical property of antiestrogens, which permitted their subsequent development as a chemopreventive for breast cancer, was antiestrogenic action at some sites, including the breast, but estrogenic properties at other sites, maintaining bone density and lowering circulating cholesterol.[15–22] The unusual target site-specific action as an estrogen or as an antiestrogen was true for both tamoxifen and raloxifene.[16,17] Lacassagne's[7] goal of developing an antagonist to estrogen action to prevent breast cancer in healthy women would not have been met if the available drugs increased the risk of osteoporosis and coronary heart disease.

LACASSAGNE'S PREVENTION PRINCIPLE: A TARGET AND AN ESTROGEN ANTAGONIST

In 1962, Jensen and Jacobson[9] demonstrated that [^{3}H]-estradiol bound to and was retained by estrogen target tissue, e.g., uterus, vagina, and pituitary gland, in the immature female rat. They proposed that an ER must be present in these tissues to capture circulating steroids and initiate the cascade of biochemical events. Toft and colleagues[23,24] first identified the ER as an extractable protein from rat uterus. Subsequently, Gorski et al.[25] and Jensen et al.[26] independently proposed subcellular models of estrogen action in target tissues. Jensen et al.[10] took the process one step further by proposing a clinical ER assay to predict hormone-responsive breast cancer. Thus, a link between estrogen action and the growth of breast cancer was established.

The MCF-7 breast cancer cell line is ER-positive,[27,28] and the cells have had applications in cancer research laboratories throughout the world.[29] Most importantly, access to these cells has resulted in a fundamental change in the understanding of hormone action, leading to the discovery of the steroid receptor superfamily of receptors.[30,31] Greene et al.[32,33] first developed monoclonal antibodies to the ER derived from MCF-7 cells and his colleagues established that the ER was a nuclear protein.[34] Immunocytochemistry is now standard for the determination of receptor status in breast biopsies.[35,36] The application of monoclonal antibodies as probes to clone and sequence the ER gene[37–39] was of fundamental significance for the understanding of the ER as a nuclear transcription factor.

Having found a target for drug discovery, investigation centered on the requirement of an antiestrogen to block estrogen action. Tamoxifen blocks the binding of [^{3}H]-estradiol to the ER derived from rat uterus[40–43] and human breast tumor.[44,45] Initial clinical studies with tamoxifen were conducted exclusively on postmenopausal women with advanced breast cancer,[3,4] and not until 1977 was it suspected that tamoxifen was more effective in ER-positive breast cancer.[46] Tamoxifen is currently used as a palliative therapy in the treatment of pre- and postmenopausal patients with ER-positive, advanced breast cancer. Adjuvant therapy revolutionized the treatment of breast cancer and is now used after breast surgery to destroy undetected micrometastases in a woman's body. The general principles derived from the use of tamoxifen as a therapy for breast cancer serve as the basis for consideration of tamoxifen as an estrogen antagonist in the prevention of breast cancer.

EVIDENCE FOR TESTING TAMOXIFEN AS A CHEMOPREVENTIVE

The 1998, Oxford overview analysis[47] combined 55 randomized trials that began before 1990, comparing adjuvant tamoxifen therapy to no tamoxifen prior to recurrence. The study population of 37,000 women with node-positive and node-negative breast cancer comprised 87% of the known, randomized clinical trials with tamoxifen. Of these women, fewer than 8000 had a very low or zero level of ER and 18,000 were classified as ER-positive. The ER status of the remaining nearly 12,000 women was unknown. Based on the normal distribution of ER in random populations, the authors estimated that two-thirds would be ER-positive.

This clinical trial database[47] is now used to answer the questions raised over the past two decades by laboratory results and hypotheses. In the 1970s, three laboratory observations emerged that merited evaluation in clinical trials: (1) tamoxifen blocks estrogen from binding to the ER, and patients with ER-positive disease are more likely to benefit from therapy than those with ER-negative disease;[48] (2) tamoxifen prevents mammary cancer in rats;[49,50] and (3) long-term treatment was more efficacious than short-term treatment for prevention of rat mammary carcinogenesis.[51–53]

Estrogen Receptor Status

The ER status of the patient is predictive of a treatment response to long-term tamoxifen ther-

Table 12.1 Comparison of the Proportional Risk Reduction of Adjuvant Tamoxifen Therapy Based on Estrogen Receptor (ER) Status

	Duration of Tamoxifen (years)		
	1	2	5
ER-poor			
Percent reduction in recurrence rates (±SD)	6 ± 8	13 ± 5	6 ± 11
Percent reduction in death rates (±SD)	6 ± 8	7 ± 5	−3 ± 11
ER-positive			
Percent reduction in recurrence rates (±SD)	21 ± 5	28 ± 3	50 ± 4
Percent reduction in death rates (±SD)	14 ± 5	18 ± 4	28 ± 5

Nearly 8000 patients were ER-poor and 18,000 patients ER-positive.
Source: Early Breast Cancer Trialists' Collaborative Group.[47]

apy. The treatment effect, based on receptor status, is summarized in Table 12.1. The reduced risk of recurrent disease following tamoxifen therapy in ER-positive patients is highly significant, and the difference between the various durations of tamoxifen administration is also significant ($\Pi^2 = 45.5$, $p < 0.00001$). In two trials of approximately 5 years of tamoxifen therapy, the reductions of recurrence were 43 ± 5% and 60 ± 6%. This translated to reductions in mortality of 23 ± 6% and 36 ± 7%, respectively. Clearly, the ER is a powerful predictor of tamoxifen response, a conclusion consistent with its proven mechanism of action as an estrogen antagonist in the breast.[54]

Duration of Therapy

The overview analysis also provides unequivocal proof of the laboratory principle that longer tamoxifen adjuvant therapy is more beneficial.[51–53] The duration of therapy is extremely important for the ER-positive premenopausal woman with large amounts of circulating estrogen that can rapidly reverse the effect of short-term tamoxifen treatment (Fig. 12.2). The benefit of 1 year of tamoxifen in the premenopausal woman is virtually nonexistent, with a 20-fold increase in effectiveness with 5 years of therapy (Fig. 12.2). In contrast, the effect of tamoxifen duration on women over the age of 60 is less dra-

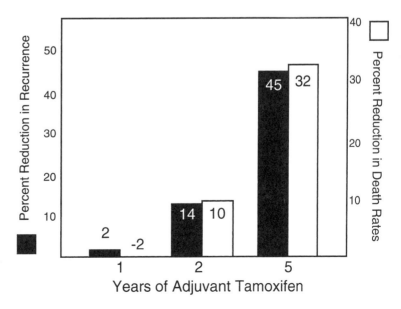

Figure 12.2 Relationship between duration of adjuvant tamoxifen therapy in estrogen receptor-positive premenopausal patients and reduction in recurrence and death rate. A longer duration of treatment has a dramatic effect on survival. (Adapted from Early Breast Cancer Trialists' Collaborative Group[47].)

Table 12.2 Proportional Risk Reductions in Women 60–69 Years Old when Estrogen Receptor-Poor Patients Are Excluded

	Duration of tamoxifen (years)		
	1	2	5
Percent reduction in recurrence rates (±SD)	26 ± 6	33 ± 4	54 ± 5
Percent reduction in death rates (±SD)	12 ± 6	12 ± 5	33 ± 6

Source: Early Breast Cancer Trialists' Collaborative Group.[47]

matic because 1 year of tamoxifen is much more effective in postmenopausal women. These data are illustrated in Table 12.2. The principle that longer therapy is more beneficial than shorter therapy is also demonstrated in the control of contralateral breast cancer (Fig. 12.3). Tamoxifen has no significant effect on the contralateral breast if 1 year of treatment is used, but extending therapy to 5 years produces a 47% reduction in breast cancer incidence. This is a powerful demonstration of the effect of tamoxifen as a chemopreventive.

Conclusions

Tamoxifen has been extensively tested in clinical trials of adjuvant therapy for 20 years.[55] Oxford overview analysis has established the laboratory concepts that tamoxifen is most effective in ER-positive disease, longer duration of use is more beneficial, and tamoxifen can prevent primary breast cancer, in this case contralateral disease.[48–53] The benefits of tamoxifen following 5 years of therapy increased throughout the first 10 years of the study. The patient clearly benefits from tamoxifen even after therapy has been discontinued. In other words, long-term benefits accrue despite the fact that therapy is not being taken. There is an accumulation of the tumoristatic/tumoricidal actions of tamoxifen for at least the first 5 years of treatment. This is also true for the reduction in contralateral breast cancer; the breast is protected even after therapy stops. This observation is extremely important for the application of tamoxifen as a preventive because a 5-year course would be expected to protect a woman from breast cancer for many years afterward.

SELECTIVE ESTROGEN RECEPTOR MODULATION

Nonsteroidal antiestrogens were originally defined as compounds that would inhibit estradiol-stimulated rat uterine weight. The compounds tamoxifen (ICI 46,474),[56] nafoxidine (U-11,100A),[57] nitromifene (C1628),[58] and clomiphene (MRL

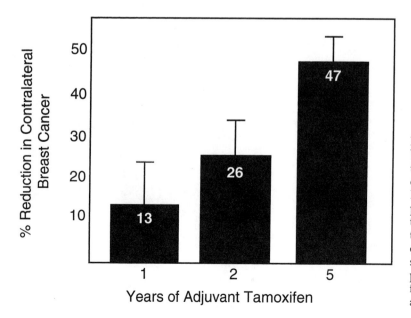

Figure 12.3 Relationship between duration of adjuvant tamoxifen and reduction in contralateral breast cancer. Longer duration is clearly superior, and the 5-year result that produces a 47% reduction in contralateral breast cancer is equivalent to the result observed in the tamoxifen prevention trial. (Adapted from Early Breast Cancer Trialists' Collaborative Group[47].)

41)[59] are partial estrogen agonists in the uterus that also inhibit the growth of dimethylbenz[a]-anthracene (DMBA)–induced rat mammary tumors[42,60,61] and ER-positive MCF-7 breast cancer cells in vitro.[62] Thus, in the 1960s and 1970s, antiestrogenicity was correlated with antitumor activity. However, findings that the compounds had increased estrogenic properties, i.e., vaginal cornification and increased uterine weight in the mouse,[56,63] raised questions about the reasons for the species specificity. One obvious possibility was species-specific metabolism; i.e., the mouse converts antiestrogen to estrogens via novel metabolic pathways. However, no species-specific metabolic routes to known estrogens have been identified.[64,65] The mouse model created a new dimension for study, which ultimately led to the recognition of the target site-specific actions of antiestrogens. This concept was subsequently referred to as selective estrogen receptor modulation (SERM) to describe the target site-specific effects of raloxifene. Now, all of the drugs in the class are known as SERMs.

The ER-positive breast cancer cell line MCF-7 can be heterotransplanted into immune-deficient athymic mice and will grow into tumors only with estrogen support. Paradoxically, tamoxifen, an estrogen in the mouse, does not support tumor growth[66] but stimulates mouse uterine growth with the same spectrum of metabolites present in both the uterus and the human tumor.[15] To explain the selective actions of tamoxifen in different targets of the same host, it was suggested that the ER complex could be interpreted as a stimulatory or inhibitory signal at different sites.[15] The concept was consolidated with experimental evidence from two further models. First, tamoxifen and raloxifene maintain bone density in the ovariectomized rat, but both compounds inhibit estradiol-stimulated uterine weight,[16] and prevent carcinogen-induced tumorigenesis.[17] Second, tamoxifen will partially stimulate the growth of human endometrial carcinoma transplanted into athymic mice,[67] allowing the simultaneous investigation of two human tumors bitransplanted into the same mouse to determine whether tamoxifen would inhibit estrogen-stimulated growth of two tumors in the same host equally.[68] Tamoxifen demonstrated target site specificity: breast tumor growth was controlled, but endometrial tumors continued growing. Again, the range of tamoxifen metabolites was consistent in all target tissues despite the contrasting biological responses, and it was concluded that the ER complexes must be interpreted differently in different target tissues.

During the past decade an intense effort has been made to discover the molecular basis for the target site-specific effects of antiestrogens. This knowledge will permit a rational application of tamoxifen and raloxifene for patients, and the discovery of new mechanisms for drug selectivity will open the door for innovations in drug discovery.

Antiestrogenic Activity at the Estrogen Receptor

Crystallization of the ligand-binding domain of the ER with estradiol and raloxifene has provided important insight into the conformational changes that occur in the receptor.[69] Estradiol causes helix 12 to seal the ligand inside the hydrophobic pocket of the ligand-binding domain (Fig. 12.4A). This causes receptor activation through the binding of coactivators on the surface of helix 12. In contrast, the binding of raloxifene prevents helix 12 from sealing the hydrophobic pocket (Fig. 12.4B), and gene transcription cannot occur because coactivators cannot bind. Unfortunately, the final shapes of the estrogen–ER and antiestrogen–ER complexes do not reveal the tertiary changes in protein structure. However, the crystal structure provides proof of the critical importance of the ER's AA351 (aspartate) for antiestrogen action. Removal of the side chain from the antiestrogen results in loss of all activity and exclusive estrogenic properties.[70] The side chain is thought to bind to an "antiestrogenic region" in the ligand-binding domain of the ER, neutralizing the estrogenic properties of the receptor.[71,72] Simply stated, the antiestrogen acts like a stick to prevent the jaws of the ER from closing around the ligand. An estrogenic complex is only created by the protein enveloping the ligand. The antiestrogenic region is now known to include AA351 on helix 3.[73–79] The interaction at the critical contact point of helix 3 and helix 12 prevents helix 12 from sealing the ligand into the binding pocket (Fig. 12.4B). The AA 351 has no role in the antiestrogenic action of the pure antiestrogen ICI

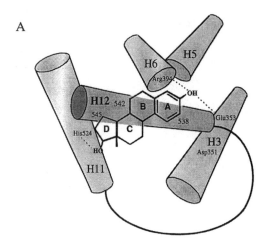

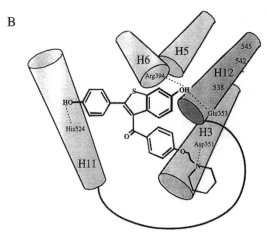

Figure 12.4 Comparison of the binding of estradiol (A) and raloxifene (B) in the ligand-binding domain of the human estrogen receptor. (Redrawn and based on data presented in Brzozowski et al.[69].)

182,780.[78] Recently, the X-ray crystallograph of the ICI 164,384 ER complex was resolved.[80] The steroid flips 180° so that the hydrophobic side chain now travels along the same route as the SERM side chain, however, the increased length permits the binding to the grove normally occupied by helix 12. This prevents helix 12 repositioning and results in rapid destruction of the complex. These data confirm a binding model for ICI 164,384 proposed in the 1980s.[81]

Coactivators for the Estrogen Receptor

A host of coactivator and corepressor proteins participate in the construction of a transcription complex in target cells.[82–88] The discovery that an antiestrogen–ER complex could become increasingly estrogenic in different cell contexts[89,90] suggested that the differential distribution of coactivators or corepressors is responsible for opposing effects in the breast and, e.g., bones.[91,92] The activating function region (AF-2) in the ligand-binding domain of the ER (Fig. 12.5) is repressed by tamoxifen and raloxifene, but the AF-1 region is unaffected by tamoxifen binding.[90] Clearly, the shape of the particular complex of ligand and ER will be different for different drugs.[93,94] Thus, coactivators could modulate estrogenicity differentially in different target sites. The candidate proteins may modify the antiestrogen–ER complex into an estrogenic complex. Alternatively, the antiestrogen–ER complex might recruit completely new proteins at a specific target site to induce or suppress gene transcription.

An Alternate Estrogen Receptor

The discovery of a second ER, ERβ,[95] has introduced a new dimension of the possible mechanisms of tamoxifen and raloxifene actions. Although ERβ has similar functional homology in the DNA-binding domain, there is only 55% homology between ERβ and ERα in the ligand-binding domain. One possible explanation of the target site specificity and altered estrogenicity of antiestrogens is a differential distribution of ERα and ERβ in different tissues.[96] Also, ERβ does not have an AF-1 site like ERα, so antiestrogen action may be modulated by a mix of ERα and ERβ in the same cell.[97–99] The differing mechanisms of action of ERα and ERβ may also involve different methods of gene activation. A novel signal-transduction pathway has been identified, a protein–protein interaction between antiestrogen–ERβ complexes and AP-1 (fos and jun)[100] that is capable of activating a reporter gene. Estradiol, however, does not activate the reporter. Interestingly, an ERα–tamoxifen complex will activate AP-1 reporter systems in the context of an endometrial cancer cell.[101] This has led to speculation that the target site specificity of antiestrogens could be both receptor- and context-selective. There is currently some confusion about the ERβ AP-1 signal-transduction pathway. Not only do raloxifene

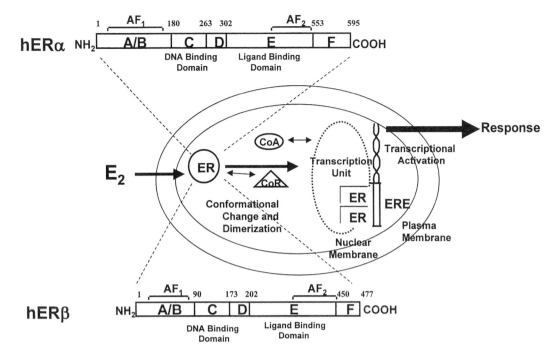

Figure 12.5 Regulation of estradiol (E_2) action through the estrogen receptor (ER) signal-transduction pathway. The respective ERα and ERβ are activated and dimerize (homo- or hetero-) to form a transcription unit. This is located appropriately at an estrogen response element (ERE) in the promoter region of an estrogen-responsive gene. The transcription unit is formed by coactivator (CoA) molecules but can be neutralized if antiestrogens are bound to the ER. The antiestrogen–ER complex does not bind CoA molecules but attracts corepressor (CoR) molecules. Selective estrogen receptor modulator action may depend on the ratio of ERα and ERβ or the ratio of CoA and CoR molecules in different target tissues.

and 4-hydroxytamoxifen increase engineered target gene transcription through an ERβ AP-1 site, but ICI 182,780 is also active.[102] No genes are known to be activated in nature by ICI 182,780, and this is an experimental paradox.

BIOLOGICAL BASIS FOR TAMOXIFEN AS A BREAST CANCER PREVENTIVE

Tamoxifen was selected for testing as a preventive based on (1) animal studies that demonstrated it could prevent carcinogenesis,[49–52,103] (2) extensive clinical experience that showed few serious side effects, and (3) a beneficial profile of estrogen-like action in maintaining bone density and reducing circulating cholesterol. Tamoxifen was already known to reduce the incidence of contralateral breast cancer[47] and to have a favorable toxicity profile, making the drug the primary agent to test in high-risk women.

Sporadic reports[19,104] and placebo-controlled randomized trials[20,105] demonstrated that tamoxifen can increase bone density in the lumbar spine, forearm, and neck of the femur by 1%–2%. Although the increases are modest compared to the results obtained with estrogen or biphosphonates (5% increase in bone density), tamoxifen produced a marginal decrease in hip and wrist fractures as a secondary end point in the breast cancer prevention trial.[106]

Tamoxifen reduces circulating cholesterol.[21,22] Low-density lipoprotein cholesterol is reduced by about 15%, and high-density lipoprotein cholesterol is maintained. It is hypothesized that this magnitude of fall in circulating cholesterol is a good surrogate marker for protection from coronary heart disease. There is evidence that women who have been treated with 5 years of adjuvant tamoxifen for breast cancer have a reduced incidence of fatal myocardial infarction.[107,108] Conversely, a large study in the

United States of 5 or more years of adjuvant tamoxifen found no statistically strong evidence for protection from coronary heart disease.[109] Nevertheless, the incidence of coronary heart disease increased once tamoxifen treatment was stopped, and there was no evidence for a detrimental effect of tamoxifen. Only a prospective, randomized trial in a population with cardiac risk factors would provide accurate data supporting cardioprotection from tamoxifen.

Tamoxifen produces partial agonist action in the rat uterus,[56] but until the late 1980s there was little information about its actions in the normal human uterus. It is now clear that a variety of endometrial changes occur in unselected populations of women.[110] The most significant finding is an increase in the stromal component, rather than endometrial hyperplasia.[111,112] Laboratory data suggesting that tamoxifen has the potential to encourage the growth of preexisting disease harbored in the uterus[67,68] provoked an intense investigation of the rates of endometrial cancer detection in women using adjuvant tamoxifen treatment for breast cancer. It is clear from the results of the tamoxifen prevention trial[106] that tamoxifen does not cause an excess of endometrial cancer in premenopausal women but does increase risk by three- to fourfold in postmenopausal women. This is consistent with the fact that women harbor four to five times the level of endometrial cancer than is detected clinically.[113] In other words, the increase in the detection of endometrial cancer from 1 per 1000 women per year to 3 per 1000 per year is consistent with the known rate of occult disease. Most importantly, the stage and grade of endometrial cancers observed in women taking tamoxifen are the same as those in the general population.[106,114] Bernstein and colleagues[115] analyzed the incidence of endometrial cancer associated with tamoxifen therapy with known risk factors for the disease. Women taking tamoxifen who developed endometrial cancer were obese or had previously taken estrogen-replacement therapy. This population of women would warrant careful evaluation prior to the use of tamoxifen.

RISK FACTORS FOR BREAST CANCER

If tamoxifen is the appropriate agent to test as a chemopreventive, then the issue becomes identification of women at risk for definitive clinical trials. Family history is probably the best-recognized risk factor for breast cancer. An inherited gene mutation is thought to account for 5%–10% of breast cancer cases.[116,117] Although infrequent, these mutations are significant since they are associated with a lifetime risk of breast cancer of 50%–80%,[118,119] beginning at a young age. At present, two predisposition genes, BRCA1, located on chromosome 17q21,[120] and BRCA2, located on chromosome 13q12–13,[121] have been identified, both of which are inherited in an autosomal dominant pattern.

Most women with a family history of breast cancer do not have the genetically transmitted form of the disease, and their risk is much less than that seen in women who have inherited a predisposing gene. The cumulative probability that a 30-year-old woman with a mother and sister with breast cancer will develop breast cancer by the age of 70 is between 7% and 18%.[122,123] While this risk increases as the number of relatives with breast cancer increases, the probability of cancer development if both a mother and a sister have bilateral breast cancer is only 25%.[123,124] The cumulative risk of breast cancer in women with a family history rarely exceeds 30%, making it critically important to distinguish those women with hereditary breast cancer from those with a family history of the disease.

Breast cancer is clearly related to endogenous hormones, and numerous studies have linked breast cancer risk to age at menarche, menopause, and first pregnancy. Although the absolute age-specific incidence of breast cancer is higher in postmenopausal than premenopausal women,[125] the absolute rate of rise of the curve is greatest up to the time of menopause, then slows to one-sixth of that seen in the premenopausal period. Further support for the promotional role of estrogen in breast cancer comes from observations that early menarche,[126] late menopause,[127] nulliparity, and late age at first birth[128] all increase the risk of breast cancer development. An increased number of ovulatory cycles is suggested to be the common mechanism of increased risk.

Other hormonal risk factors have been suggested but are not as well established. Abortion, whether spontaneous or induced, has been reported by some to increase risk,[129,130] while others have found no relationship between abortion and breast cancer.[131,132] Studies of the effect of lactation on breast cancer risk have also been inconclusive,[133,134] but recent studies have suggested that a long duration of lactation reduces breast cancer risk in premenopausal women.[135] Physical activity in adolescence is reported to decrease risk, perhaps due to a higher rate of anovulatory cycles;[136,137] but an increased level of physical activity later in life has not been shown to reduce breast cancer risk.[138] Postmenopausal obesity increases risk,[139] perhaps due to increased peripheral estrogen production; but this relationship between weight and risk is not observed in premenopausal women.

The effects of exogenous hormones in the form of oral contraceptives and hormone-replacement therapy (HRT) on breast cancer risk have been studied extensively. Overall, there is no convincing evidence of increased risk with use of oral contraceptives.[140] Some studies have suggested that long-term use of oral contraceptives in young women prior to first pregnancy may increase risk.[141,142] Two meta-analyses of the effect of HRT demonstrated small, but statistically significant, increases in risk.[143,144] The Iowa Women's Health Study found that the duration of HRT was directly associated with risk of invasive carcinoma with a favorable histology, with a relative risk (RR) of 1.81 for HRT taken at less than 5 years vs. an RR of 2.65 for HRT taken at greater than 5 years.[145] Although hormonal risk factors are clearly implicated in the pathogenesis of breast cancer, most of them are associated with an RR of less than 3; therefore, the presence of a single hormonal risk factor is insufficient to classify a woman as high-risk.

Environmental factors have also been linked to breast cancer. A large amount of attention has been directed to the role of diet in the etiology of breast cancer. This link is suggested by the large international variation in breast cancer incidence and the observation that national per capita fat consumption correlates with breast cancer incidence and mortality.[146] Prospective studies, however, of diet and breast cancer risk have failed to identify a relationship with 10 years of follow-up.[147] The lack of a relationship between dietary fat intake and cancer risk within the context of a Western diet is confirmed by a pooled analysis of seven cohort studies involving a total of 337,816 women, which demonstrated no difference in incidence in women in the lowest and highest quintiles of fat intake.[148]

Interactions among Risk Factors

A major problem in identifying the "high-risk woman" is lack of knowledge of the interactions among the various risk factors since the majority of studies have focused on defining individual risk factors. Most women have a combination of factors, which both increase and decrease risk, complicating the assessment of an individual's level of risk. In addition, it is unclear whether the risk conferred by multiple risk factors is additive, is multiplicative, or varies with the specific risk factor.

The interactions between a family history of breast cancer and other risk factors have been examined, often with conflicting results. Data from the Nurses Health Study[149] show that women with known risk factors, such as age at menarche or menopause, parity, age at first birth, alcohol use, the presence of benign breast disease, and a mother or sister with breast cancer, develop disease at rates equivalent to women with a family history alone. In contrast, Anderson and Badzioch[150] and Brinton et al.[151] reported that hormonal factors further modulate risk in women with a family history of breast cancer, although the effect varies with the factor under study. Studies of the interaction between HRT and other known risk factors also have variable results, depending on the risk factor under study. In a meta-analysis of 16 published studies, Steinberg et al.[143] found that the effect of HRT did not differ among parous and nulliparous women and those with or without benign breast disease; however, enhanced risk was observed in women with a family history of breast cancer.

Identification of Candidates for Chemoprevention

Women at increased risk for breast cancer would seem to be ideal candidates for chemoprevention initiatives; however, the problem of identification of the high-risk woman is far from solved. There is no consensus regarding the level of risk that is clinically relevant. The interactions among risk factors and their variability over time are poorly understood, and most of the data come from studies of white women; thus, little is known about the impact of ethnic diversity. Finally, with the exception of women with predisposing genetic mutations, the majority of women with risk factors will not develop breast carcinoma. A recent study of the fraction of breast cancer cases in the United States attributable to risk factors[152] found that fewer than 50% of women who develop the disease have any identifiable risk factors. Family history of breast cancer accounted for only 9% of cases, while relatively minor risk factors, such as later age at first birth and nulliparity, were seen in 29% of cases. In a similar study, Seidman et al.[153] noted that only 21% of breast cancer cases in women aged 30–54 and 29% of cases in women aged 55–84 occurred in women with at least one of 10 common breast cancer risk factors. The majority of women had minor risk factors, which increase the RR of breast cancer only twofold, and most had only a single risk factor. This level of increased risk would not meet the entry criteria for the trials of breast cancer prevention in high-risk women discussed below. These data suggest that even if women with a very small increase in breast cancer risk were targeted for prevention initiatives, a large number of cases would continue to be missed. Two strategies have been proposed to reduce the risk of breast cancer: (1) use of tamoxifen in high-risk pre- and postmenopausal women identified with the Gail model[154,155] and (2) use of a SERM for the prevention of osteoporosis, which will prevent breast cancer as a side effect in older postmenopausal women.[12]

PREVENTION OF BREAST CANCER WITH TAMOXIFEN

In 1986, Cuzick et al.[156] outlined a prevention trial with tamoxifen in women at high risk of breast cancer. The previous year, Cuzick and Baum[157] documented a marginally significant reduction in the risk of a second primary cancer in the contralateral breast of breast cancer patients after 2 years of tamoxifen. This initial observation was confirmed in larger randomized clinical trials. Fornander et al.[158] recruited 1846 postmenopausal patients for the Stockholm Adjuvant Tamoxifen Trial and randomized them to a trial of adjuvant tamoxifen or nothing for early breast cancers. In the tamoxifen patients, new primary breast cancers occurred less often (RR = 0.55). Similar results were obtained by the National Surgical Adjuvant Breast and Bowel Project (NSABP) B-14.[159] These results led to the obvious interest in tamoxifen as a possible preventive for primary disease. Carefully controlled clinical chemoprevention trials continue to establish the activity and toxicity of specific agents and combinations in breast cancer prevention. Two studies have addressed this question, the Royal Marsden Pilot Study[160] and the NSABBP Protocol P-1.[106] Additionally, an incomplete Italian[161] report of the efficacy of tamoxifen in low-risk women has been published. It is important to stress that only the NSABP P-1 trial was a prospective study that met all of the requirements for a chemoprevention study. The Royal Marsden Study[160] was established as a toxicity study prior to a national trial. This placebo-controlled trial is still ongoing under the title International Breast Intervention Study (IBIS). The recruitment goal of 7000 volunteers should be complete by 2002 (see note at the end of references).

Royal Marsden Pilot Study

Powles and coworkers[162] recruited high-risk women aged 30–70 to a placebo-controlled trial using 20 milligrams of tamoxifen daily for up to 8 years. Each participant had at least one first-degree relative with breast cancer under age 50, a first-degree relative affected at any age plus an additional affected first- or second-degree relative, or a first-degree relative with bilateral breast cancer. Women with a history of benign breast biopsy and an affected first-degree relative of any age were also eligible. Women with a history of venous thrombosis, any previous malignancy, or an estimated life expectancy of fewer than 10 years were excluded.[160,163] A total of 2494 women consented to participate in the study, and 23 were excluded from final analysis due to the presence of preexisting ductal carcinoma in situ (DCIS) or invasive breast carcinoma. The trial was undertaken to evaluate the problems of accrual, toxicity, compliance, and safety of tamoxifen prior to a large national, multicenter trail of chemoprevention. The stated goal was to evaluate the feasibility of a volunteer trial of 20,000 patients throughout the United Kingdom and Australia; however, breast cancer incidence was also analyzed.[160]

Acute toxicity was low for participants in the pilot study, and compliance remained correspondingly high: 77% of women on tamoxifen and 82% of women on placebo remained on medication at 5 years.[163] There was a significant increase in hot flashes (34% vs. 20%, $p < 0.005$), mostly in premenopausal women; vaginal discharge (16% vs. 4%, $p < 0.005$); and menstrual irregularities (14% vs. 9%, $p < 0.005$). At the most recent follow-up, 320 women had discontinued tamoxifen and 176 had discontinued placebo prior to the study's completion.[160]

At 70 months, no significant difference in the incidence of deep vein thrombosis or pulmonary embolism was observed between groups.[163] A detailed analysis of other coagulation parameters in a sequential subset of women found no significant changes in protein S, protein C, or crosslinked fibrinogen degradation products. A significant fall in total plasma cholesterol occurred within 3 months and was sustained over 5 years of treatment.[163] The decrease affected low-density lipoprotein, with no change in apolipoproteins A and B or high-density lipoprotein cholesterol.

In contrast, tamoxifen exerted antiestrogenic or estrogenic effects on bone density, depending on menopausal status. In premenopausal women, early findings demonstrated a small but significant ($p < 0.05$) loss of bone in both the lumbar spine and hip at 3 years. In contrast, postmenopausal women had increased bone mineral density in the spine ($p < 0.005$) and hip ($p < 0.001$) compared to untreated women.[164]

The Marsden group extensively studied gynecological complications associated with tamoxifen treatment in healthy women. Since ovarian and uterine assessment by transvaginal ultrasound became available some time after the trial's start, many subjects did not have a baseline evaluation. Ovarian screening demonstrated a significantly increased risk ($p < 0.005$) of benign ovarian cysts in premenopausal women who had received tamoxifen for more than 3 months compared to controls. There were no changes in ovarian appearance in postmenopausal women.[163] Careful examination of the uterus with transvaginal ultrasonography using color Doppler imaging in women taking tamoxifen showed that it was usually larger and that women with histological abnormalities had significantly thicker endometria.[165]

Analysis of breast cancer incidence was reported at a median follow-up of 70 months, when 42% of the participants had completed therapy or withdrawn.[160] During the study, 336 women receiving tamoxifen and 305 on placebo received HRT. No difference in the incidence of breast cancer was observed between the groups. There were 34 carcinomas in the tamoxifen group and 36 in the placebo group (RR = 0.98). Of the 70 cancers, only eight were DCIS. Analysis of the subset of women on HRT did not demonstrate an interaction with tamoxifen treatment. The authors suggested that they had a high population of BRCA1 and BRCA2 carriers who were hormone-unresponsive, accounting for a lack of response in the tamoxifen arm;[160] but this is unproven and the incidence of cancer is too low for it to be likely. Also, the recent finding that early oophorectomy in BRCA1 carriers reduces the risk of breast cancer[166] points to a role for hormones in breast carcinogenesis for gene mutation carriers. The fact that women in the Marsden trial used HRT is unlikely to be responsible for the outcome. We believe the study is underpowered. The Marsden group originally calculated a minimum of 5000 patients for significant results.

National Surgical Adjuvant Breast Project P-1 Study

The National Surgical Adjuvant Breast Project P-1 Study (NSABP/NCI) study opened in the United States and Canada in May 1992 with an accrual goal of 16,000 women from 100 North American sites.[106] The specific aim was to test the utility of tamoxifen as a preventive for breast cancer. It closed after accruing 13,338 women in 1997. The study design is illustrated in Figure 12.6. Women over the age of 60 or between the ages of 35 and 59 whose 5-year risk of developing breast cancer, as predicted by the Gail model,[154] was equal to that of a 60-year-old woman were eligible for inclusion. Additionally, women over age 35 with a diagnosis of lobular

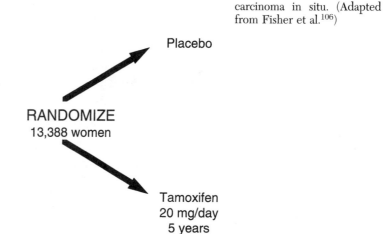

Figure 12.6 The eligibility and design of the National Surgical Adjuvant Breast and Bowel Project tamoxifen prevention trial. LCIS, lobular carcinoma in situ. (Adapted from Fisher et al.[106])

carcinoma in situ (LCIS) treated by biopsy alone were eligible for entry. In the absence of LCIS, the risk factors necessary to enter the study varied with age, such that a 35-year-old woman had to have an RR of 5.07, whereas the required RR for a 45-year-old woman was 1.79. Routine endometrial biopsies were performed in both arms of the study.

The breast cancer risk of women enrolled in the study was generally extremely high. Recruitment was also age-balanced, with about one-third younger than 50 years, one-third between 50 and 60 years, and one-third older than 60 years. Secondary end points of the study include the incidence of fractures and cardiovascular deaths. Most importantly, the study will provide information regarding tamoxifen's role in the treatment of women who carry somatic mutations in the BRCA1 gene. Laboratory results are not yet available.

The first results of the NSABP study were reported in September 1998, after a mean follow-up of 47.7 months.[106] There was a total of 363 invasive and noninvasive breast cancers in the participants, 124 in the tamoxifen group and 239 in the placebo group. A 49% reduction in the risk of invasive breast cancer and a 50% reduction in the risk of noninvasive breast cancer were seen in the tamoxifen arm. A subset analysis of women at risk due to prior LCIS demonstrated a 56% reduction of risk. The most dramatic risk reduction, 86%, was seen in women at risk due to atypical hyperplasia.

The benefits of tamoxifen were observed in all age groups, with a RRs of breast cancer ranging from 0.45 in women aged 60 and older to 0.49 for those in the 50–59 age group and 0.56 for women aged 49 and younger (Fig. 12.7). The benefit of tamoxifen was observed in all levels of breast cancer risk within the study and was not confined to a particular lower-risk or higher-risk subset.

As expected, tamoxifen affected the incidence of ER-positive tumors, which were reduced by 69% per year. The rate of ER-negative tumors in the tamoxifen group (1.46 per 1000 women) did not significantly differ from that in the placebo group (1.20 per 1000 women) (Fig. 12.8). Tamoxifen reduced the rate of invasive cancers of all sizes, but the greatest reduction was in the incidence of tumors 2.0 centimeters or less in size. Tamoxifen also reduced the incidence of both node-positive and node-negative breast cancer. The beneficial effects of tamoxifen were observed for each year of follow-up. After year 1, the risk was reduced by 33%, and in year 5, by 69%.

Tamoxifen also reduced the incidence of osteoporotic fractures of the hip, spine, and radius by 19% (Fig. 12.9),[106] approaching but not

Figure 12.7 Overall reduction in invasive breast cancer observed in the National Surgical Adjuvant Breast and Bowel Project tamoxifen prevention trial P-1 in high-risk women. (Adapted from Fisher et al.[106])

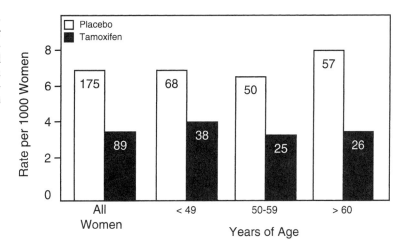

Figure 12.8 Incidence of estrogen receptor (ER)-positive and ER-negative breast cancer in the placebo- and tamoxifen-treated arms of the National Surgical Adjuvant Breast and Bowel Project tamoxifen prevention trial P-1. The antiestrogen reduces the risk of developing ER-positive breast cancer, but there is no change in the incidence of ER-negative breast cancer. (Adapted from Fisher et al.[106])

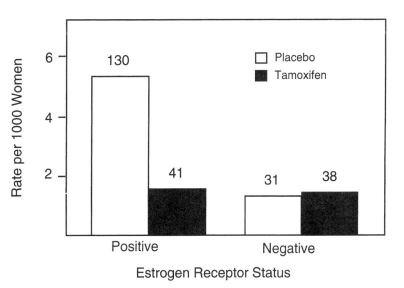

Figure 12.9 Incidence of osteoporotic fractures of the hip, wrist, and spine observed in the placebo- and tamoxifen-treated arms of the National Surgical Adjuvant Breast and Bowel Project tamoxifen prevention trial P-1. (Adapted from Fisher et al.[106])

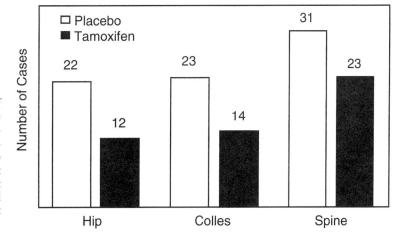

reaching statistical significance. This reduction was greatest in women who were 50 years and older at study entry. No differences in the risks of myocardial infarction, angina, coronary artery bypass grafting, or angioplasty were noted between groups.[106]

This study confirms the association between tamoxifen and endometrial carcinoma.[114,158] The RR of endometrial cancer in the tamoxifen group was 2.5, with women over age 50 having a RR of 4.01. All endometrial cancers in the tamoxifen group were grade 1, and none of the women receiving tamoxifen died of endometrial cancer. There was one endometrial cancer death in the placebo group. Although there is no doubt that tamoxifen increases the risk of endometrial cancer, it is important to recognize that this increase translates to a yearly incidence of 2.3 women per 1000.

More women in the tamoxifen group developed deep vein thrombosis[106] with the excess risk confined to women over 50 years. The RR of deep vein thrombosis in the older age group was 1.71 (95% confidence interval 0.85–3.58). An increase in pulmonary emboli was also seen in the older women taking tamoxifen, with an RR of approximately 3. Three deaths from pulmonary emboli occurred in the tamoxifen arm, all in women with significant comorbidities. An increased incidence of stroke (RR = 1.75) was also seen in the tamoxifen group, but this did not reach statistical significance.

Assessment of quality of life showed similar depression scores between groups. Hot flashes were noted in 81% of the women on tamoxifen compared to 69% of the placebo group, and the tamoxifen-associated hot flashes appeared to be of comparable severity. In the tamoxifen group, 29% of the women and 13% in the placebo group reported moderate or severe vaginal discharge. No differences in the occurrence of irregular menses, nausea, fluid retention, skin changes, or weight gain or loss were reported.

Italian Study

The third tamoxifen prevention study, performed in Italy, began in October 1992 and randomized 5408 women aged 35–70 to 20 milligrams of tamoxifen or placebo daily for 5 years.[161] Originally, 20,000 volunteers without risk factors were sought, but the study was stopped prematurely because of poor recruitment and compliance. Women were required to have had a hysterectomy for a nonneoplastic condition to obviate concerns about an increased risk of endometrial carcinoma. There was no requirement that participants be at risk for breast cancer development, and those who underwent premenopausal oophorectomy with hysterectomy (47%) actually had a slightly reduced risk of breast cancer development. Women with endometriosis, cardiac disease, and deep venous thrombosis were excluded from the study. Although 5408 women were randomized into this study, 1422 withdrew and only 149 completed 5 years of treatment.

The incidence of breast cancer did not differ between groups, with 19 cases in the tamoxifen group and 22 in the placebo group. Tumor characteristics, including size, grade, lymph node status, and receptor status, were similar between groups. The incidence of thrombophlebitis was increased in the tamoxifen group. A total of 64 events were reported in 56 women, 38 in the tamoxifen group and 18 in the placebo group ($p = 0.0053$); however, 42 of these were superficial phlebitis. No differences in the incidence of cerebrovascular ischemic events were observed.

Conclusions

Based on a single trial with a positive result and two with negative results, it may seem that the role of tamoxifen in breast cancer prevention remains unresolved; however, critical differences exist between these three studies. The negative findings in the Italian study[161] are readily explained by the relatively low risk of breast cancer development in the study population, the high dropout rate, and the small number of participants who completed 5 years of treatment. At present, the only conclusion that can be drawn from this study is that the benefits of tamoxifen are likely to be small in women with an average or decreased risk of breast cancer.

The Royal Marsden study was initially described as a pilot study to examine toxicity and compliance,[162,163,167] serving as a feasibility assessment for a larger trial to examine using tamoxifen for chemoprevention. In spite of being designed as a pilot study, the authors reported a

90% power to detect a 50% reduction in breast cancer incidence.[160] The authors of the Royal Marsden study suggest that the positive results of the NSABP trial at 3.5 years of follow-up are most likely due to the treatment of occult carcinoma, rather than the prevention of new breast cancers. Of the 368 total cancers in the NSABP study, 99 (28%) were DCIS[106] compared to 11% of the 70 cancers in the Royal Marsden study. The higher percentage of DCIS in the NSABP trial indicates that the detection of subclinical cancers was successful and that any treated occult cancer is not truly amenable to detection by current means. Whether occult carcinoma was treated or true prevention occurred, a significantly greater number of women were spared surgery, irradiation, and chemotherapy. The data from the overview analysis[47] do not support the contention that these cancers will become clinically evident when tamoxifen is stopped since the reduction in contralateral breast cancer persisted through 10 years even though tamoxifen treatment was stopped at 5 years.

Overall, the results of the NSABP trial,[106] with its large study population, clearly support the use of tamoxifen for breast cancer prevention in high-risk women. These findings are consistent with laboratory observations and with the contralateral breast cancer risk reduction seen with tamoxifen therapy (Fig. 12.3).

Tamoxifen was approved in 1998 for risk reduction in pre- and postmenopausal high-risk women. The results of the NSABP prevention trial have established tamoxifen as the current standard of care, opening the door for the evaluation of other agents that might have improved safety or efficacy profiles in clinical trial. An agent that does not promote liver tumors in rats[168,169] or is less estrogenic in the rodent uterus might have value for application as a breast cancer preventive.

BIOLOGICAL BASIS FOR RALOXIFENE AS A BREAST CANCER PREVENTIVE

Raloxifene, originally named keoxifene or LY156758,[170] was discovered in the breast cancer program at the laboratories of Eli Lilly (Indianapolis, IN). The drug has a high binding affinity for the ER,[171,172] primarily because it has strategically located phenolic groups (Fig. 12.1). Raloxifene and an analogue, LY117018,[59,173–175] are short-acting compounds with poor bioavailability due to rapid phase II metabolism.[176] Indeed, a concern in the early clinical trials was an inability to monitor blood levels. Although numerous assays are available to monitor tamoxifen and its metabolites,[177] the structure of raloxifene does not permit the use of similar chemical methods of detection. The analytical technique used to monitor raloxifene has not been published; therefore, there has been limited clinical experience for the treatment of breast cancer. The initial study, conducted at the M.D. Anderson Cancer Center in Houston, showed no responses in heavily pretreated patients with stage IV disease.[178] A second small study of 18 ER-positive patients with previously untreated metastatic disease showed a modest response rate of 30%, with a dose of 300 milligrams daily.[179] A key issue that has not been addressed is cross-resistance between raloxifene and tamoxifen. A clinical trial comparing adjuvant treatment with tamoxifen for 5 years to treatment with tamoxifen followed by raloxifene is needed to determine whether raloxifene-stimulated tumor growth is a clinical reality.

Antitumor Actions

Raloxifene inhibits the growth of DMBA-induced rat mammary carcinoma[180] and, more importantly for the proposed evaluation as a preventive, reduces the incidence of N-nitrosomethylurea-induced tumors if given after the carcinogen but before the appearance of palpable tumors[17,181] (Fig. 12.10). As would be anticipated with a drug that has a short biological half-life, raloxifene is not superior to tamoxifen at equivalent doses. There is no doubt that raloxifene and its analogues are effective and potent inhibitors of the growth of breast cancer cells in culture,[182,183] but the complication of first-pass metabolism in vivo reduces potency. For this reason, doses above 60 milligrams daily have been tested in clinical trials to prevent osteoporosis.

Based on the hypothesis that raloxifene reduces the incidence of breast cancer as a beneficial side effect of the prevention of osteoporosis,[12] placebo-controlled trials with raloxifene

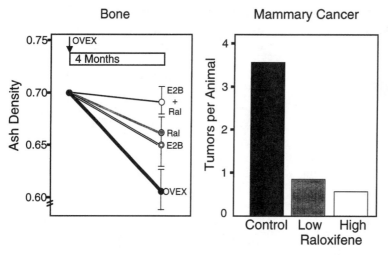

Figure 12.10 Comparison of the effects of raloxifene on femur ash density in ovariectomized (OVEX) rats and incidence of rat mammary tumors after administration of N-nitrosomethylurea. Data illustrate the target site specificity of a drug predicted to prevent osteoporosis and breast cancer simultaneously. (Adapted from Jordan et al.[16].)

are ongoing. The Multiple Outcomes of Raloxifene Evaluation (MORE) has randomized 7704 postmenopausal women, mean age 66.5 years, with osteoporosis defined by hip or spine bone density at least 2.5 SD below normal or vertebral fractures, and no history of breast or endometrial cancer to placebo or 60 or 120 milligrmas of raloxifene daily. Results at 3 years, with a total of 40 cases of breast cancer confirmed, indicate a 76% reduction in risk.[184] Another database pools all placebo-controlled trials and includes 10,553 women monitored for, on average, 3 years. In this group, there is a 54% reduction in the incidence of breast cancer among raloxifene-treated patients.[185,186] Similar to tamoxifen, raloxifene reduces the incidence of ER-positive breast cancer only. These are strong preliminary data for the Study of Tamoxifen and Raloxifene (STAR) in high-risk women.

Bones

Raloxifene maintains bone density in ovariectomized rats (Fig. 12.10).[15,187-193] It inhibits bone loss by reducing bone resorption,[188] as evidenced by a reduction in the elevated urinary pyridinoline and serum osteocalcin levels seen after ovariectomy. In general, raloxifene does not maintain bone density in the ovariectomized rat to the level observed in an intact animal,[187] but efficacy appears to be equivalent to estrogen treatment without an increase in uterine weight.

Preliminary studies in 251 postmenopausal women randomized to placebo, 200 milligrams raloxifene, 600 milligrams raloxifene, or 0.625 milligrams conjugated estrogens (Premarin; Wyeth-Ayerst, Philadelphia, PA) show equivalent decreases in serum alkaline phosphatase, serum osteocalcin, urinary pyridinoline, and urinary calcium excretion with raloxifene and estrogen.[194] These doses of raloxifene are far higher than the 60 milligrams currently recommended for the prevention of osteoporosis. Evaluation of 60 milligrams daily raloxifene on bone remodeling in early postmenopausal women using calcium tracer kinetic methods found that although remodeling suppression was greater for estrogen, the remodeling balance was the same for the two agents.[195] The authors concluded that raloxifene is an estrogen agonist in bone. Indeed, these results are consistent with the increased bone density of 2.4 ± 0.4% in the lumber spine and 2.4 ± 0.4% for the total hip seen with raloxifene.[196] Although the increases in bone density are not as high as would be anticipated with estrogen or bisphosphonates, it is now clear that raloxifene produces a 40% decrease in spine fractures.

Lipids

Raloxifene produces a significant decrease in low-density lipoprotein cholesterol, but high-density lipoprotein cholesterol remains the same.[194,196,197] Additionally, triglycerides do not rise during raloxifene treatment. Laboratory data on the rabbit strongly support that raloxifene prevents atherosclerosis;[198] however, data from primates fed high-cholesterol diets do not confirm this.[199] A prospective, randomized clinical trial is currently addressing the merit of raloxifene in the reduction of coronary heart disease in postmenopausal women with risk factors. The study, Raloxifene Use for the Heart (RUTH), has randomized 10,000 high-risk women to placebo or 60 milligrams daily raloxifene treatment for 5 years.

Uterus

Raloxifene and its analogues have limited estrogenic effects in the rat uterus.[170,172–174] The raloxifene analogue LY117018 blocks the estrogenic actions of tamoxifen in the rat uterus;[174] however, raloxifene and its analogues cannot be classified as pure antiestrogens. There is not a complete lack of uterotropic properties,[200,201] and estrogen-regulated genes, such as the progesterone receptor, are partially activated.[175]

Raloxifene's effect in the human uterus is currently being evaluated. A study in women without preexisting endometrial abnormalities shows that raloxifene, unlike estrogen, does not increase endometrial thickness.[202] Raloxifene is less estrogenic than tamoxifen and increases the growth of human endometrial carcinomas only under laboratory conditions by approximately 50% of tamoxifen.[203] This coupled with the preliminary data of raloxifene as a potential preventive for breast cancer in elderly women[184–186] is sufficient to propose testing against the current standard of care, tamoxifen.

STUDY OF TAMOXIFEN AND RALOXIFENE

The STAR trial is a phase III, double blind trial that is assigning eligible postmenopausal women to either 20 milligrams daily tamoxifen or 60 milligrams daily raloxifene therapy for 5 years. Trial participants will complete a minimum of 2 additional years of follow-up after therapy is stopped.

The STAR trial's primary aim is to determine if long-term raloxifene therapy is effective at preventing the occurrence of invasive breast cancer in high-risk postmenopausal women. It will additionally compare cardiovascular data, fracture data, and general toxicities for raloxifene and tamoxifen. It is clear that the activation or suppression of various target sites is similar for tamoxifen and raloxifene, but evaluation of the comparative benefits of the agents will provide an important new clinical database for raloxifene in postmenopausal women.

Premenopausal women at risk for breast cancer are currently ineligible for the STAR trial. Although there is extensive information about the efficacy of tamoxifen in premenopausal treatment and prevention of breast cancer,[47,106] clinical experience with raloxifene is confined to postmenopausal women. The National Cancer Institute is currently conducting a randomized study of 60 and 300 milligrams daily raloxifene in high-risk women to address its effect on bone density. Additionally, short-term raloxifene treatment up to 28 days causes elevation in circulating estradiol but does not prevent ovulation,[204] consistent with the known elevation of steroid hormones produced by tamoxifen in premenopausal breast cancer patients.[205] The changes in endocrine function produced by raloxifene will also be assessed as a prelude to the recruitment of premenopausal high-risk women to the STAR trial.

The results of the STAR trial are expected by 2006. Clearly, it will be invaluable to establish the overall benefits of the drugs with regard to breast cancer incidence, coronary heart disease, and osteoporosis. Comparisons of endometrial cancer will be especially interesting because the standard of care, i.e., self-reporting, will be employed in the STAR trial rather than routine screening with annual biopsies.

NEW AGENTS

Newer SERMs with slight modifications of chemical structure are being studied with the goal of increasing the spectrum of antitumor activity and reducing toxicity (Fig. 12.11). The ongoing investigation of several molecules may demonstrate applications as breast cancer and/or osteoporosis treatments. Most importantly, the drugs may be multifunctional medicines useful for the prevention of osteoporosis, coronary heart disease, and breast cancer. The ideal agent would: (1) not produce premature drug resist-

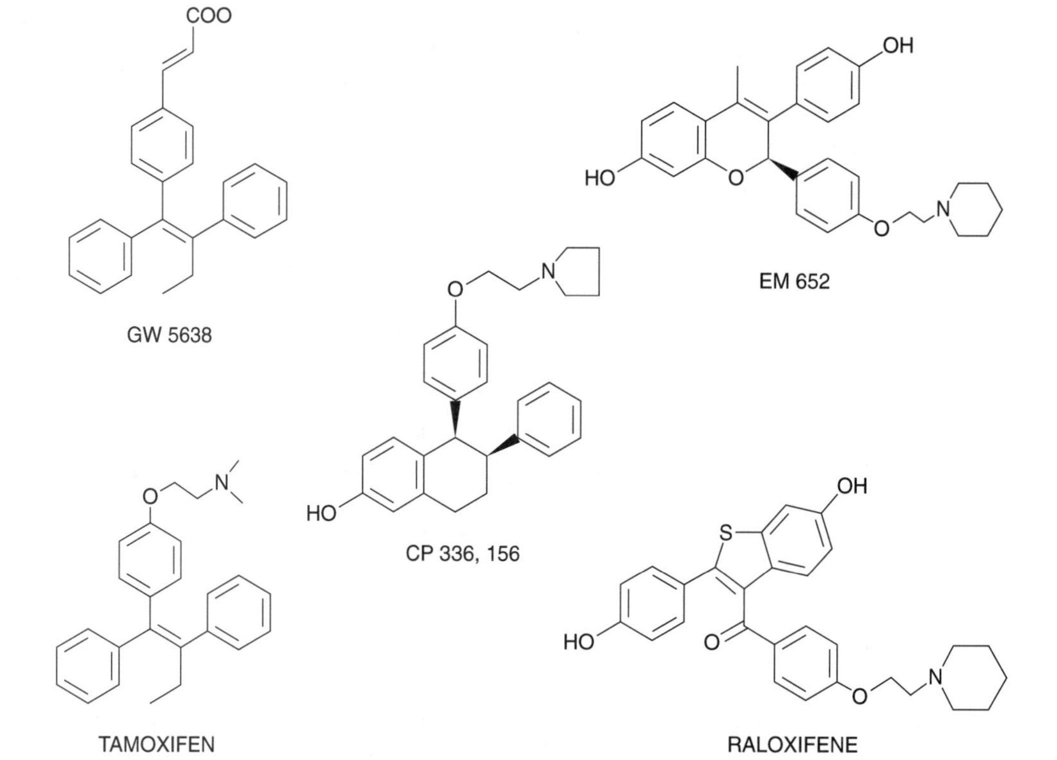

Figure 12.11 Comparison of the molecular structures of new and established selective estrogen receptor modulators.

ance, (2) not be cross-resistant with tamoxifen, and (3) have fewer side effects such as a lower incidence of endometrial cancer observed with tamoxifen.

Steroidal antiestrogens have been developed and have proven to be antiestrogenic in all target tissues[206] through premature destruction of the ER.[207,208] The "pure" antiestrogens are effective in laboratory models as second-line treatment for tamoxifen stimulated breast[209] and endometrial cancer.[12,203] Following successful phase II studies,[210] randomized trials were initiated comparing Faslodex with another hormonal therapy, the aromatase inhibitor anastrozole,* in patients with tamoxifen-refractory advanced breast cancer. Faslodex and anastrozole are equivalent treatments.[211,212] Although Falsodex may be studied as an adjunct therapy for node-negative ER positive breast cancer, there are no plans to use a pure antiestrogen as a preventative before rigorous evaluations of safety are complete.

LY 353,381.HCL (Arzoxifene) is a long-acting raloxifene analog,[213] which is being developed for breast cancer treatment and prevention. The compound is not completely cross-resistant with tamoxifen in laboratory models.[214] However, clinical data from phase II trials of advanced breast cancer have shown efficacy,[215] yet concern for uterine safety remains if used as a second line agent. Tamoxifen and Arzoxifene show an antitumor effect on the growth of Tam-naïve ECC-1 endometrial tumors in athymic mice. However, when tamoxifen resistant endometrial tumors were treated with Tam or Arzoxifene, a stimulatory effect of both compounds has been shown in these models.[216]

A diaryltetrahydronaphthalene derivative, CP 336,156, reportedly has a high affinity for the ER while preserving bone density in the rat.[217] Of the two diastereometric salts of CP 336,156, the *l* enantiomer has 20 times the binding affinity of the *d* enantiomer. Studies have demonstrated that the *l* enantiomer has twice the bioavailability of the *d* enantiomer.[218]

GW 5638, discovered by Willson et al.[204] in

1994 at Glaxo Wellcome (Research Triangle Park, NC), is effective at preserving bone density with minimal uterotropic activity in animal models.[219] The compound departs from the usual tertiary amino antiestrogenic side chain with a shorter allyl carboxylic group on a triphenylethylene molecule. The molecule induces a complex with properties similar to pure antiestrogens[220,221] yet does not degrade ER.[222] This interesting new SERM is currently being tested in animal models to establish a lack of cross-resistance to tamoxifen, without enhancement of endometrial cancer growth. As it is a tamoxifen analogue, laboratory research must also evaluate GW 5638 for any role in rat liver carcinogenesis before clinical trials in well women.

The research group at Wyeth-Ayerst has developed two new SERMs, ERA-923 and TSE-424 which are highly selective with nonestrogenic profiles on rat uterine tissue.[223] These drugs showed no uterine stimulation when dosed alone and were able to completely block the effect of estrogen. TSE-424 prevented bone loss and reduced total cholesterol in animal models. ERA-923 is in phase II clinical trials for the treatment of hormone-dependent metastatic breast cancer, and TSE-424 has completed phase II clinical trials for the treatment of postmenopausal osteoporosis.

The compound EM 800 and its active metabolite EM 652 are orally active agents with virtually no uterotropic activity. They could be described as pure antiestrogens because the molecule silences both AF-1 and AF-2 in ERα.[224] The antiestrogenic side chain is located similarly to the steroidal pure antiestrogens but appears to be too short for optimal activity. EM 800 is an antitumor agent in the DMBA model[225] and has antiestrogenic activity in mice with none of the estrogenic activity seen with tamoxifen.[226] The drug is extremely potent against breast cancer cells in culture and prevents the growth of estrogen-stimulated tumor xenografts in athymic mice.[227] However, unlike other pure antiestrogens, EM 800 does not decrease bone density in the rat. As an orally active pure antiestrogen, EM 800 could be used as a second-line therapy following tamoxifen failure. Based on the structural similarity with other raloxifene analogues, EM 652 appears to be a SERM with potential cross-resistance with tamoxifen. Both EM 652 and raloxifene have the antiestrogenic side chain interacting with AA351 in the ER. The D351Y mutant converts both EM 652 and raloxifene to an estrogenic complex, whereas a pure antiestrogen, ICI 182,780, is unaffected.[228] Therefore, EM 800 may fail as a second-line agent after tamoxifen and may be more beneficial as first-line therapy. EM 652 may have broad applications as a raloxifene-like drug.

At this point it is not possible to describe all the mechanisms of SERMs at a target site because the relative importance of co-regulator proteins and pathways are not known. It is possible to describe modulation of the ERα by SERMs through changes in the ligand antiestrogenic side chain and/or the docking amino acids of the ER. With the substitution of a noncharged amino acid at 351, it is possible to silence estrogen action of the 4-hydroxytamoxifen-ER complex.[229] Thus, the 4-hydroxytamoxifen-ER complex must bind co-activators at a novel site as the traditional AF-2 site is blocked in the 4-hydroxytamoxifen-ER complex.[229] The new activating site for co-activator binding on the SERM-ER complex is called AF-2b[229] in distinction to AF-2 used by estrogens in the ligand binding domain.[81] Additionally, changing the side chain of 4-hydroxytamoxifen to that of GW7604 with a carboxylic acid repels the aspartate at 351[222] or increasing the shielding of aspartate 351 by raloxifene[230] silences the estrogen action in the complex. Thus, the estrogen-like actions of SERMs are maintained through an interaction with a surface amino acid at 351.

Insight into how the compounds bind and affect the ER function allows for a molecular classification of new agents. Furthermore, drug discovery will proceed based on molecular function and not on clinical trial and error. Potentially new agents will be designed to give the ideal antitumor action with limited side effects.

THE FUTURE OF PREVENTION

Sporn and colleagues[231,232] proposed the concept of the chemoprevention of cancer in the mid-1970s. Effective implementation of the strategy requires a target to prevent either the initiation or the promotion of the cancer cell.

The key to success is a well-defined target so that a selective action can be applied without general toxicity. In the case of breast cancer the target is the ER,[9] but serendipitously the antiestrogens tamoxifen and raloxifene, which block breast cancer cell growth, were also found to modulate the physiological requirements for estrogen action selectively at other target sites.[15–17,233]

The major clinical question for the current application of tamoxifen as a chemopreventive is when the 5-year course should be taken and how long the effects will protect a woman at elevated risk for breast cancer. The simple answer to the first part of the question is that a woman who fits the elevated risk criteria for breast cancer will receive benefit through a 55% risk reduction whenever she takes tamoxifen. However, since there are no rules that define when a woman will develop breast cancer, earlier rather than later would be an appropriate strategy. In regard to the duration of therapy, the answer is less clear; but there are clues that 5 years of tamoxifen therapy results in protection for at least 5 years after the drug is stopped, based on contralateral breast cancer data from the overview analysis.[47] At this time, further follow-up is not available. It will be important to discover the mechanism for the long-term beneficial effects of tamoxifen as a chemopreventive as this could be exploited further. Similar questions about the optimal duration of raloxifene therapy will need to be addressed in the future. With raloxifene, this is not as important of an issue because long-term therapy for the prevention of osteoporosis is necessary.

The question of whether 5 years of tamoxifen will be sufficient for prolonged protection from breast cancer or whether longer durations of initial treatment will provide longer periods of protection remains unanswered. For example, after 10 or 15 years of treatment, can a woman get 20 years of risk reduction? There is reluctance to consider this type of clinical experiment because of concern over tamoxifen-stimulated primary breast cancer. Although 5 years of adjuvant tamoxifen appears to be as effective as 10 years,[234] this does not justify restriction of chemoprevention to 5 years. Drug resistance by metastatic breast cancer cells can probably develop much more rapidly than during the process of carcinogenesis in primary breast cancer. As a result, it may be important to evaluate longer durations of tamoxifen in clinical chemoprevention trials. The current challenge is to establish the efficacy of raloxifene compared with tamoxifen in the STAR trial and then to determine the optimal application of SERMs as a new drug group for the benefit of women's health.

Another unexplored aspect of chemoprevention with tamoxifen is evaluation of lower doses. This strategy may reduce side effects, while maintaining protection for the breast. Pilot studies have been completed in Italy,[235,236] and clinical trials with large populations are planned. Finally, a new generation of agents that specifically modulate ERα and ERβ selectively may become available for clinical testing within the next decade.[233] A number of postmenopausal diseases will be targeted, such as osteoporosis and coronary heart disease; but the beneficial side effects should include a reduction in uterine and breast cancers in the general population.[237] The present reduction of breast cancer in high-risk women by 50% is an important first step that has resulted from the rational application of translational research.

REFERENCES

1. Beatson GT. On the treatment of inoperable cases of carcinoma of the mamma: suggestions for a new method of treatment with illustrative cases. Lancet 2:104–107, 1896.
2. Boyd S. On oophorectomy in cancer of the breast. BMJ 2:1161–1167, 1900.
3. Cole MP, Jones CJ, Todd IDH. A new antiestrogenic agent in late breast cancer. An early appraisal of ICI 46,474. Br J Cancer 25:270–275, 1971.
4. Ward HWC. Antioestrogenic therapy for breast cancer: a trial of tamoxifen at two dose levels. BMJ 2:13–14, 1973.
5. Lathrop AEC, Loeb L. Further investigations on the origins of tumors in mice III. On the part played by internal secretions in the spontaneous development of tumors. J Cancer Res 1:1–16, 1916.
6. Allen E, Doisy EA. An ovarian hormone: preliminary report on its localization, extraction and partial purification and action in test animals. JAMA 81:819–821, 1923.
7. Lacassagne A. Hormonal pathogenesis of adenocarcinoma of the breast. Am J Cancer 27:217–225, 1936.
8. Kennedy BJ. Hormonal therapy of advanced breast cancer. Cancer 18:1551–1557, 1965.

9. Jensen EV, Jacobson HI. Basic guides to the mechanism of estrogen action. Recent Prog Horm Res 18:387–414, 1962.
10. Jensen EV, Block GE, Smith S, Kyser K, Desombre ER. Estrogen receptors and breast cancer response to adrenalectomy. NCI Monogr 34:55–70, 1971.
11. McGuire WL, Carbone PP, Volmer EP. Estrogen Receptors in Human Breast Cancer. New York: Raven Press, 1975.
12. Lerner LJ, Jordan VC. Development of antiestrogens and their use in breast cancer. Cancer Res 50:4177–4189, 1990.
13. Lerner LJ, Holthaus JF, Thompson CR. A nonsteroidal estrogen antagonist 1-(p-2diethylaminoethoxyphenyl)-1-phenyl-2-p-methoxyphenyl-ethanol. Endocrinology 63:295–318, 1958.
14. Lerner LJ. The first non-steroidal antioestrogen—MER 25. In: Sutherland RL, Jordan VC (eds). Non Steroidal Antioestrogens: Molecular Pharmacology and Antitumour Activity. Sydney: Academic Press, 1981, pp 1–6.
15. Jordan VC, Robinson SP. Species specific pharmacology of antiestrogens: role of metabolism. Fed Proc 46:1870–1874, 1987.
16. Jordan VC, Phelps E, Lindgren JU. Effects of antiestrogens on bone in castrated and intact female rats. Breast Cancer Res Treat 10:31–35, 1987.
17. Gottardis MM, Jordan VC. The antitumor actions of keoxifene (raloxifene) and tamoxifen in the N-nitrosomethylurea-induced rat mammary carcinoma model. Cancer Res 47:4020–4024, 1987.
18. Turner RT, Wakley GK, Hannon KS, Bell NH. Tamoxifen prevents the skeletal effects of ovarian hormone deficiency in rats. J Bone Miner Res 2:449–456, 1987.
19. Turken S, Siris E, Seldin D, Flaster E, Hyman G, Lindsay R. Effects of tamoxifen on spinal bone density in women with breast cancer. J Natl Cancer Inst 81:1086–1088, 1989.
20. Love RR, Mazess RB, Barden HS, et al. Effects of tamoxifen on bone mineral density in postmenopausal women with breast cancer. N Engl J Med 326:852–856, 1992.
21. Rossner S, Wallgren A. Serum lipoproteins and proteins after breast cancer surgery and effects of tamoxifen. Atherosclerosis 52:339–346, 1984.
22. Love RR, Wiebe DA, Newcomb PA, et al. Effects of tamoxifen on cardiovascular risk factors in postmenopausal women. Ann Intern Med 115:860–864, 1991.
23. Toft D, Gorski J. A receptor molecule for estrogens: isolation from the rat uterus and preliminary characterization. Proc Natl Acad Sci USA 55:1574–1581, 1966.
24. Toft D, Shyamala G, Gorski J. A receptor molecule for estrogen: studies using a cell-free system. Proc Natl Acad Sci USA 57:1740–1743, 1967.
25. Gorski J, Toft D, Shyamala G, Smith D, Notides A. Hormone receptors: studies on the interaction of estrogen with the uterus. Recent Prog Horm Res 24:45–80, 1968.
26. Jensen EV, Suzuki T, Karashima T, Stumpf WE, Jungblut PW, DeSombre ER: A two step mechanism for the interaction of estradiol with rat uterus. Proc Natl Acad Sci USA 59:632–638, 1968.
27. Soule HD, Vazquez J, Lang A, Albert S, Brennan M. A human cell line from a pleural effusion derived from a breast carcinoma. J Natl Cancer Inst 51:1409–1416, 1973.
28. Brooks SC, Locke ER, Soule HD. Estrogen receptors in a human cell line (MCF-7) from breast carcinoma. J Biol Chem 248:6251–6253, 1973.
29. Levenson AS, Jordan VC. MCF-7: the first hormone responsive breast cancer cell line. Cancer Res 57:3071–3078, 1997.
30. Tsai MJ, O'Malley BW. Mechanisms of action of steroid/thyroid receptor superfamily members. Annu Rev Biochem 63:451–486, 1994.
31. Beato M, Herrlich P, Schutz G. Steroid hormone receptor: many actors in search of a plot. Cell 82:851–857, 1994.
32. Greene GL, Fitch FW, Jensen EV. Monoclonal antibodies to estrophilim: probes for the study of estrogen receptors. Proc Natl Acad Sci USA 77:157–161, 1980.
33. Greene GL, Nolan C, Engler P, Jensen EV. Monoclonal antibodies to human estrogen receptor. Proc Natl Acad Sci USA 77:5115–5119, 1980.
34. King WJ, Greene GL. Monoclonal antibodies localize oestrogen receptor in the nuclei of target tissues. Nature 307:745–747, 1984.
35. King WJ, DeSombre ER, Jensen EV, Greene GL. Comparison of immunocytochemical and steroid binding assays for estrogen receptor in human breast tumors. Cancer Res 45:293–304, 1985.
36. DeSombre ER, Thorpe SM, Rose C, et al. Prognostic usefulness of estrogen receptor immunocytochemical assays (ERICA) for human breast cancer. Cancer Res 8:4256–4264, 1986.
37. Walter P, Green S, Greene GL, et al. Cloning of the human estrogen receptor cDNA. Proc Natl Acad Sci USA 82:7889–7893, 1985.
38. Green S, Walter P, Kumar V, et al. Human oestrogen receptor cDNA: sequence, expression and homology with v-erb A. Nature 320:134–139, 1986.
39. Greene GL, Gilna P, Waterfield M, Barker A, Hort Y, Shine J. Sequence and expressions of human estrogen receptor complementary DNA. Science 231:1150–1154, 1986.
40. Skidmore JR, Walpole AL, Woodburn J. Effect of some triphenylethylenes on oestradiol binding in vitro to macromolecules from uterus and anterior pituitary. J Endocrinol 52:289–298, 1972.

41. Nicholson RI, Golder MP. The effect of synthetic antioestrogens on the growth and biochemistry of rat mammary tumours. Eur J Cancer 11:571–579, 1975.
42. Jordan VC, Dowse LJ. Tamoxifen as an antitumour agent: effect on oestrogen binding. J Endocrinol 68:297–303, 1976.
43. Jordan VC, Prestwich G. Binding of [^{3}H] tamoxifen in rat uterine cytosols: a comparison of swinging bucket and vertical tube rotor sucrose density gradient analysis. Mol Cell Endocrinol 8:179–188, 1977.
44. Jordan VC, Koerner S. Tamoxifen (ICI 46,474) and the human carcinoma 8S oestrogen receptor. Eur J Cancer 11:205–206, 1975.
45. Coezy E, Borgna JL, Rochefort H. Tamoxifen and metabolites in MCF-7 cells: correlation between binding to estrogen receptor and cell growth inhibition. Cancer Res 42:317–323, 1982.
46. Kiang DT, Kennedy BJ. Tamoxifen (antiestrogen) therapy in advanced breast cancer. Ann Intern Med 87:687–690, 1977.
47. Early Breast Cancer Trialists' Collaborative Group. Tamoxifen for early breast cancer: an overview of the randomized trials. Lancet 351:1451–1467, 1998.
48. Jordan VC, Jaspan T. Tamoxifen as an antitumour agent: oestrogen binding as a predictive test for tumour response. J Endocrinol 68:453–460, 1976.
49. Jordan VC. Antitumour activity of the antioestrogen ICI 46,474 (tamoxifen) in the dimethylbenzanthracene (DMBA)–induced rat mammary carcinoma model. J Steroid Biochem 5:354, 1974.
50. Jordan VC. Effect of tamoxifen (ICI 46,474) on initiation and growth of DMBA-induced rat mammary carcinomata. Eur J Cancer 12:419–424, 1976.
51. Jordan VC, Dix CJ, Allen KE. The effectiveness of long term tamoxifen treatment in a laboratory model for adjuvant hormone therapy of breast cancer. In: Salmon SE, Jones SE (eds). Adjuvant Therapy of Cancer II. New York: Grune and Stratton, 1979, pp 19–26.
52. Jordan VC, Allen KE. Evaluation of the antitumour activity of the nonsteroidal antioestrogen monohydroxytamoxifen in the DMBA-induced rat mammary carcinoma model. Eur J Cancer 16:239–251, 1980.
53. Jordan VC. Laboratory studies to develop general principles for the adjuvant treatment of breast cancer with antiestrogens: problems and potential for future clinical applications. Breast Cancer Res Treat 3:73–86, 1983.
54. MacGregor JI, Jordan VC. Basic guide to the mechanisms of antiestrogen action. Pharmacol Rev 50:151–196, 1998.
55. Osborne CK. Tamoxifen in the treatment of breast cancer. N Engl J Med 339:1609–1618, 1998.
56. Harper MJK, Walpole AL. A new derivative of triphenylethylene: effect on implantation and mode of action in rats. J Reprod Fertil 13:101–119, 1967.
57. Duncan GW, Lyster SC, Clark JJ, Lednicer D. Antifertility activities of two diphenyl-dihydronaphthalene derivatives. Proc Soc Exp Biol Med 112:439–442, 1963.
58. Callantine MR, Humphrey RR, Lee SL, Windsor BL, Schottin NH, O'Brien OP. Action of an estrogen antagonist on reproductive mechanisms in the rat. Endocrinology 79:153–169, 1966.
59. Holtkamp DE, Greslin SC, Root CA, Lerner LJ. Gonadotropin inhibiting and antifecundity effects of chloramiphene. Proc Soc Exp Biol Med 105:197–201, 1960.
60. Terenius L. Antiestrogen and breast cancer. Eur J Cancer 7:57–64, 1971.
61. DeSombre ER, Arbogast LY. Effect of the antiestrogen CI628 on the growth of rat mammary tumors. Cancer Res 34:1971–1976, 1974.
62. Lippman ME, Bolan G, Huff K. The effects of estrogen and antiestrogens on hormone-responsive human breast cancer in long term tissue culture. Cancer Res 36:4595–4601, 1976.
63. Terenius L. Structure–activity relationships of antioestrogens with regard to interaction with 17β-oestradiol in the mouse uterus and vagina. Acta Endocrinol (Copenh) 66(Suppl):431–447, 1971.
64. Lyman SD, Jordan VC. Metabolism of nonsteroidal antiestrogens. In: Jordan VC (ed). Estrogen/Antiestrogen Action and Breast Cancer Therapy. Madison: University of Wisconsin Press, 1986, pp 191–219.
65. Robinson SP, Langan-Fahey SM, Jordan VC. Implications of tamoxifen metabolism in the athymic mouse for the study of antitumor effects upon human breast cancer xenografts. Eur J Cancer Clin Oncol 25:1769–1776, 1989.
66. Osborne CK, Hobbs K, Clark GM. Effects of estrogens and antiestrogens on growth of human breast cancer cells in athymic nude mice. Cancer Res 45:584–590, 1985.
67. Satyaswaroop PG, Zaino RJ, Mortel R. Estrogen-like effects of tamoxifen on endometrial carcinoma transplanted in nude mice. Cancer Res 44:4006–4010, 1984.
68. Gottardis MM, Robinson SP, Satyaswaroop PG, Jordan VC. Contrasting actions of tamoxifen on endometrial and breast tumor growth in the athymic mouse. Cancer Res 48:812–815, 1988.
69. Brzozowski AM, Pike ACW, Dauter Z, et al. Molecular basis of agonism and antagonism in the oestrogen receptor. Nature 389:753–758, 1997.
70. Jordan VC, Gosden B. Importance of the alkylamino-ethoxy side chain for the estrogenic and antiestrogenic actions of tamoxifen and trioxifene in the immature rat uterus. Mol Cell Endocrinol 27:291–306, 1982.

71. Lieberman ME, Gorski J, Jordan VC. An estrogen receptor model to describe the regulation of prolactin synthesis by antiestrogens in vitro. J Biol Chem 258:4741–4745, 1983.
72. Tate AC, Greene GL, DeSombre ER, Jensen EV, Jordan VC. Differences between estrogen and antiestrogen–estrogen receptor complexes identified with an antibody raised against the estrogen receptor. Cancer Res 44:1012–1018, 1984.
73. Wolf DM, Jordan VC. The estrogen receptor from a tamoxifen stimulated MCF-7 tumor variant contains a point mutation in the ligand binding domain. Breast Cancer Res Treat 31:129–138, 1994.
74. Catherino WH, Wolf DM, Jordan VC. A naturally occurring estrogen receptor mutation results in increased estrogenicity for a tamoxifen analog. Mol Endocrinol 9:1053–1063, 1995.
75. Levenson AS, Tonetti DA, Jordan VC. The oestrogen-like effect of 4-hydroxytamoxifen on induction of transforming growth factor α mRNA in MDA-MB-231 breast cancer cells stably expressing the oestrogen receptor. Br J Cancer 77:1812–1819, 1998.
76. Jordan VC, Collins MM, Rowsby L, Prestwich GA. Monohydroxylated metabolite of tamoxifen with potent antiestrogenic activity. J Endocrinol 75:305–306, 1977.
77. Levenson AS, Catherino WH, Jordan VC. Estrogenic activity is increased for an antiestrogen by a natural mutation of the estrogen receptor. J Steroid Biochem Mol Biol 60:261–268, 1997.
78. Levenson AS, Jordan VC. The key to the antiestrogenic mechanism of raloxifene is amino acid 351 (aspartate) in the estrogen receptor. Cancer Res 58:1872–1875, 1998.
79. Shiau AK, Barstad D, Loria PM, et al. The structural basis of estrogen receptor/co-activator recognition and the antagonism of this interaction by tamoxifen. Cell 95:927–937, 1998.
80. Pike AC, Brzozowski AM, Walton J, et al. Structural insights into the mode of action of a pure antiestrogen. Structure 9:145–153, 2001.
81. Jordan VC, Koch R. Regulation of prolactin synthesis in vitro by estrogenic and antiestrogenic derivatives of estradiol and estrone. Endocrinology 124:1717–1726, 1989.
82. Horwitz KB, Jackson TA, Bain DL, Richer JK, Takimoto GS, Tung L. Nuclear receptor co-activators and co-repressors. Mol Endocrinol 10:1167–1177, 1996.
83. Jenster G. Co-activators and co-repressors as mediators of nuclear receptor function: an update. Mol Cell Endocrinol 143:1–7, 1998.
84. Halamichi S, Marden E, Martin G, Mackey H, Abbondonza C, Brown M. Estrogen receptor associated proteins: possible mediators of hormone induced transcription. Science 264:1455–1458, 1993.
85. Hanstein B, Eckner R, DiRenzo J, et al. p300 is a component of an estrogen receptor p-coactivator complex. Proc Natl Acad Sci USA 93:11540–11545, 1996.
86. Baniahmad C, Nawaz Z, Banaihmad A, Gleeson MAG, Tsai MJ, O'Malley BW. Enhancement of human estrogen receptor activity by SPT6: a potential co-activator. Mol Endocrinol 9:34–43, 1995.
87. Onate SA, Tsai SY, Tsai MJ, O'Malley BW. Sequence and characterization of a coactivator for the steroid receptor superfamily. Science 270:1354–1357, 1995.
88. Katzenellenbogen JA, O'Malley BW, Katzenellenbogen BS. Tripartite steroid receptor pharmocology: interaction with multiple effector sites as a basis for the cell- and promoter-specific action of these hormones. Mol Endocrinol 10:119–131, 1996.
89. Berry M, Metzger D, Chambon P. Role of the two activating domains of the estrogen receptor in the cell type end promoter context dependent agonist activity of the antioestrogen 4 hydroxytamoxifen. EMBO J 9:2811–2818, 1990.
90. McDonnell DP, Clemm DL, Herman T, Goldman ME, Pike JW. Analysis of estrogen receptor function reveals three distinct classes of antiestrogens. Mol Endocrinol 9:659–669, 1995.
91. Smith CL, Nawaz Z, O'Malley BW. Co-activator and co-repressor regulation of the agonist/antagonist activity of the mixed antiestrogen, 4-hydroxytamoxifen. Mol Endocrinol 11:657–666, 1997.
92. Jackson TA, Richer JK, Bain DL, Takimoto GS, Tung L, Horwitz KB. The partial agonist activity of antagonist-occupied steroid receptors is controlled by a novel hinge domain-binding co-activator L7/SPA and the corepressors N-COR or SMRT. Mol Endocrinol 11:693–705, 1997.
93. Norris JD, Paige LA, Christensen DJ, et al. Peptide antagonists of the human estrogen receptor. Science 285:744–746, 1999.
94. Wijayaratne AL, Nagel SC, Paige LA, Christensen DJ, Norris JD, Fowlkes DM, McDonnell DP. Comparative analyses of mechanistic differences among antiestrogens. Endocrinology 140:5828–5840, 1999.
95. Kuiper GG, Enmark E, Pelto-Huikko M, Nilsson S, Gustafsson JA. Cloning of a novel estrogen receptor expressed in rat prostate and ovary. Proc Natl Acad Sci USA 93:5925–5930, 1996.
96. Kuiper GG, Carlsson B, Grnadien K, et al. Comparison of ligand binding specificity and transcript tissue distribution of estrogen receptor α and β. Endocrinology 138:863–870, 1997.
97. McInerney EM, Weis WE, Sun J, Mosselman S, Katzenellenbogen BS. Transcription activation by the human estrogen receptor subtype beta (ER beta) studied with ERbeta and ER-alpha receptor chimeras. Endocrinology 139:4513–4522, 1998.
98. Hall JM, McDonnell DP. The estrogen recep-

tor β-isoform (ERβ) of the human estrogen receptor modulates ERα transcriptional activity and is a key regulator of the cellular response to estrogens and antiestrogens. Endocrinology 140:5566–5578, 1999.
99. Barkhem T, Carlsson B, Nilsson Y, Enmark E, Gustafsson J, Nilsson S. Differential response of estrogen receptor alpha and estrogen receptor beta to partial estrogen agonists/antagonists. Mol Pharmacol 54:105–112, 1998.
100. Paech K, Webb P, Kuiper GG, et al. Differential ligand activation of estrogen receptor ERα and ERβ at AP-1 sites. Science 277:1508–1510, 1997.
101. Webb P, Lopez GN, Uht RM, Kusher PJ. Tamoxifen activation of the estrogen receptor/AP-1 pathway: potential origin for the cell specific estrogen-like effects of antiestrogen. Mol Endocrinol 9:443–456, 1995.
102. Webb P, Nguyen P, Valentine C. The estrogen receptor enhances AP-1 activity by two distinct mechanisms with different requirements for receptor transactivation functions. Mol Endocrinol. 13:1672–1685, 1999.
103. Welsch CW, Goodrich-Smith M, Brown CK, Clifton K. Effect of an estrogen antagonist (tamoxifen) on the initiation and progression of radiation-induced mammary tumors in female Sprague Dawley rats. Eur J Cancer 17:1255–1258, 1981.
104. Ward RL, Morgan G, Dalley D, Kelly PJ. Tamoxifen reduces bone turnover and prevents lumbar spine and proximal femoral bone loss in early postmenopausal women. Bone Miner 22: 87–94, 1993.
105. Kristensen B, Ejlertsen B, Dolgard P, et al. Tamoxifen and bone metabolism in postmenopausal low risk breast cancer patients: a randomized study. J Clin Oncol 12:992–997, 1994.
106. Fisher B, Costantino JP, Wickerham, et al. Tamoxifen for prevention of breast cancer: report of the National Surgical Adjuvant Breast and Bowel Project P-1 Study. J Natl Cancer Inst 90:1371–1388, 1998.
107. McDonald CC, Stewart HJ. Fatal myocardial infarction in the Scottish tamoxifen trial. BMJ 303:435–437, 1991.
108. McDonald CC, Alexander FE, Whyte BW, Forest AP, Steward HJ. Cardiac and vascular morbidity in women receiving adjuvant tamoxifen for breast cancer in a randomized trial. BMJ 311:977–980, 1995.
109. Costantino JP, Kuller LH, Ives DG, Fisher B, Dignam J. Coronary heart disease mortality and adjuvant tamoxifen therapy. J Natl Cancer Inst 89:776–782, 1997.
110. Assikis VJ, Jordan VC. Gynecological effects of tamoxifen and the association with endometrial cancer. Int J Gynecol Obstet 49:241–257, 1995.
111. Goldstein SR. Unusual ultrasonographic appearance of the uterus in patients receiving tamoxifen. Am J Obstet Gynecol 170:447–451, 1994.
112. Decensi A, Fontana V, Bruno S, Gustavino C, Gatteschi B, Costa A. Effect of tamoxifen on endometrial proliferation. J Clin Oncol 14:434–440, 1996.
113. Horwitz RI, Feinstein AR, Horwitz SM, Robboy SJ. Necropsy diagnosis of endometrial cancer and detection-bias in case/control studies. Lancet ii:66–68, 1981.
114. Fisher B, Costantino JP, Redmond CK, Fisher ER, Wickerham DL, Cronin WM. Endometrial cancer in tamoxifen treated breast cancer patients. Findings from the National Surgical Adjuvant Breast and Bowel Project (NSABP). J Natl Cancer Inst 86:527–537, 1994.
115. Bernstein L, Deapen D, Cerham JR, et al. Tamoxifen for breast cancer and endometrial cancer risk. J Natl Cancer Inst 91:1654, 1999.
116. Newman B, Austin MA, Lee M, King MC. Inheritance of human breast cancer: evidence for autosomal dominant transmission in high risk families. Proc Natl Acad Sci USA 85:3044–3048, 1988.
117. Claus EB, Risch N, Thompson WD. Genetic analysis of breast cancer in the cancer and steroid hormone study. Am J Hum Genet 48: 232–242, 1991.
118. Easton DF, Bishop DT, Ford D, Crockford GP, Breast Cancer Linkage Consortium. Genetic linkage analysis in familial breast and ovarian cancer results from 214 families. Am J Hum Genet 52:678–701, 1993.
119. Struewing JP, Hartge P, Wacholder S, et al. The risk of cancer associated with specific mutations of BRCA1 and BRCA2 among Ashkenazi Jews. N Engl J Med 336:1401–1408, 1997.
120. Miki Y, Swen J, Shattuck-Eidens D, et al. A strong candidate for the breast and ovarian cancer susceptibility gene, BRCA1. Science 266:66–71, 1994.
121. Wooster R, Neuhausen SL, Mangion J, et al. Localization of a breast cancer susceptibility gene, BRCA2, to chromosome 13q12–13. Science 265:2088–2090, 1994.
122. Ottman R, Pike MC, King MC, Henderson BE. Practical guide for estimating risk for familial breast cancer. Lancet 3:556–558, 1983.
123. Anderson DE, Badzioch MD. Risk of familial breast cancer. Cancer 56:383–387, 1985.
124. Sidrensky D, Tokino T, Helzlsouer K. Inherited p53 gene mutation in breast cancer. Cancer Res 52:2984–2989, 1992.
125. Ries LAG, Miller BA, Hankey BF. SEER Cancer Statistics Review, 1973–1991. NIH Publ 94. Bethesda, MD: National Cancer Institute, 1994, pp 2789–2803.
126. MacMahon B, Trichopoulos D, Brown J, et al. Age at menarche, probability of ovulation and breast cancer risk. Int J Cancer 29:13–16, 1982.

127. Trichopolos D, MacMahon B, Cole P. Menopause and breast cancer risk. J Natl Cancer Inst 48:605–613, 1972.
128. MacMahon B, Cole P, Lin TM, et al. Age at first birth and breast cancer risk. Bull WHO 43:209–221, 1970.
129. Daling JR, Malone KE, Voight LF, White E, Wiess NS. Risk of breast cancer among young women: relationship to induced abortion. J Natl Cancer Inst 86:1584–1592, 1994.
130. Newcomb PA, Storer BE, Longnecker MP, Mittendorf R, Greenberg ER, Willett WC. Pregnancy termination in relation to risk of breast cancer. JAMA 275:283–287, 1996.
131. Harris BM, Eklund G, Meririk O, Rutqvist LE, Wiklund K. Risk of cancer of the breast after legal abortion during the first trimester: a Swedish register study. BMJ 299:1430–1432, 1989.
132. Melbye M, Wohlfahrt J, Osen JH, et al. Induced abortion and the risk of breast cancer. N Engl J Med 336:81–85, 1997.
133. Layde PM, Webster LA, Baughman AL, Wingo PA, Rubin GL, Ory HW. The independent associations of parity, age at first full-term pregnancy, and duration of breast feeding with the risk of breast cancer. J Clin Epidemiol 42:963–973, 1989.
134. Kvale G, Heuch I. Lactation and cancer risk: is there a relation specific to breast cancer? J Epidemiol Community Health 42:30–37, 1988.
135. Newcomb PA, Storer BE, Longnecker MP, et al. Lactation and a reduced risk of premenopausal breast cancer. N Engl J Med 330:81–87, 1994.
136. Frisch RE, Gotz-Welbergen AV, McArthur JW, et al. Delayed menarche and amenorrhea of college athletes in relation to age of onset of training. JAMA 246:1559–1563, 1981.
137. Bernstein L, Henderson BE, Hanisch R, Sullivan-Halley J, Ross RK. Physical exercise and reduced risk of breast cancer in young women. J Natl Cancer Inst 86:1403–1408, 1994.
138. Dorgan JF, Brown C, Barrett M, et al. Physical activity and risk of breast cancer in the Framingham heart study. Am J Epidemiol 139:662–669, 1994.
139. De Waard F, Baanders-van Halecijn E. A prospective study in general practice on breast cancer risk in postmenopausal women. Int J Cancer 14:153–160, 1974.
140. Malone KE, Daling JR, Weiss NS. Oral contraceptives in relation to breast cancer. Epidemiol Rev 15:80–97, 1993.
141. Meirik O, Lund E, Adami HO, Bergstrom R, Christoffersen T, Bergsjo P. Oral contraceptive use and breast cancer risk in young women. A joint national case control study in Sweden and Norway. Lancet 2:650–654, 1986.
142. UK National Case-Control Study Group. Oral contraceptive use and breast cancer risk in young women. Lancet 1:976, 1989.
143. Steinberg KK, Thacker SB, Smith SJ, et al. A meta-analysis of the effect of estrogen replacement therapy on the risk of breast cancer. JAMA 265:1985–1990, 1991.
144. Sillero-Arenas M, Delgado-Rodriguez M, Rodigues-Canteras R, Bueno-Cavanillas A, Galvez-Vargas R. Menopausal hormone replacement therapy and breast cancer: a meta-analysis. Obstet Gynecol 79:286–294, 1992.
145. Gapstur S, Morrow M, Sellers T. Hormone replacement therapy and risk of breast cancer with a favorable histology. JAMA 22:2091–2097, 1999.
146. Armstrong B, Doll R. Environmental factors and cancer incidence and mortality in different countries with special reference to dietary practices. Int J Cancer 15:617, 1975.
147. Hunter DJ, Willett WC. Dietary factors. In: Harris JR, Lippman ME, Morrow M, Hellman S (eds). Diseases of the Breast. Philadelphia: Lippincott-Raven, 1996, pp 201–212.
148. Hunter DJ, Spiegelman D, Adami HO, et al. Cohort studies of fat intake and the risk of breast cancer—a pooled analysis. N Engl J Med 334:356–361, 1996.
149. Colditz GA, Willett WC, Hunter DJ, et al. Family history, age and risk of breast cancer. Prospective data from the Nurses' Health Study. JAMA 270:338–343, 1993.
150. Anderson DE, Badzioch MD. Combined effect of family history and reproductive factors on breast cancer risk. Cancer 63:349–353, 1989.
151. Brinton LA, William RR, Hoover RN, Stegens NL, Feinleib M, Fraumeni JF Jr. Breast cancer risk factors among screening program participants. J Natl Cancer Inst 62:37–44, 1979.
152. Madigan MP, Ziegler RG, Benichou J, Byrne C, Hoover RN. Proportion of breast cancer cases in the United States explained by well established risk factors. J Natl Cancer Inst 87:1681–1685, 1995.
153. Seidman H, Stellman SD, Mushinski MH. A different perspective on breast cancer risk factors: some implications of the non-attributable risk. CA Cancer J Clin 32:301–313, 1982.
154. Gail MH, Brinton LA, Byar DP, et al. Projecting individualized probabilities of developing breast cancer for white females who are being examined annually. J Natl Cancer Inst 81:1879–1886, 1989.
155. Costantino JP, Gail MH, Pee D, et al. Validation studies for models projecting the risk of invasive and total breast cancer incidence. J Natl Cancer Inst 91:1541, 1999.
156. Cuzick J, Wang DY, Bulbrook RD. The prevention of breast cancer. Lancet 12:83–86, 1986.
157. Cuzick J, Baum M. Tamoxifen and contralateral breast cancer. Lancet ii:282, 1985.
158. Fornander T, Cedermark B, Mattsson A, et al. Adjuvant tamoxifen in early breast cancer: occurrence of new primary cancers. Lancet ii:117–120, 1989.

159. Fisher B, Costantino J, Redmond C, et al. A randomized clinical trial evaluating tamoxifen in the treatment of patients with node-negative breast cancer who have estrogen-receptor-positive tumors. N Engl J Med 320:479–484, 1989.
160. Powles TJ, Eeles R, Ashley SE, et al. Interim analysis of the incident breast cancer in the Royal Marsden Hospital tamoxifen randomized chemoprevention trial. Lancet 362:98–101, 1998.
161. Veronesi U, Maisonneuve P, Costa A, et al. Prevention of breast cancer with tamoxifen: preliminary findings from the Italian randomized trial among hysterectomized women. Lancet 362:93–97, 1998.
162. Powles TJ, Hardy JR, Ashley SE, et al. A pilot trial to evaluate the acute toxicity and feasibility of tamoxifen for prevention of breast cancer. Br J Cancer 60:126–133, 1989.
163. Powles TJ, Jones AL, Ashley SE, et al. The Royal Marsden Hospital pilot tamoxifen chemoprevention trial. Breast Cancer Res Treat 31:73–82, 1994.
164. Powles TJ, Hickish T, Kanis JA, Tidy VA, Ashley S. Effect of tamoxifen on bone mineral density measured by dual energy X-ray absorptiometry in healthy premenopausal and postmenopausal women. J Clin Oncol 14:78–84, 1996.
165. Kedar RP, Bourne TH, Powles TJ, et al. Effects of tamoxifen on uterus and ovaries of postmenopausal women in a randomized breast cancer prevention trial. Lancet 342:1318–1321, 1994.
166. Rebbeck TR, Levin AM, Eisen A, et al. Breast cancer risk after bilateral prophylactic oophorectomy in BRCA1 mutation carriers. J Natl Cancer Inst 91:1475–1479, 1999.
167. Powles TJ, Tillyer CP, Jones AL, et al. Prevention of breast cancer with tamoxifen—an update on the Royal Marsden pilot program. Eur J Cancer 26:680–684, 1990.
168. Williams GM, Iatropoulos MJ, Djordjevic MV, Kaltenberg OP. The triphenylethylene drug tamoxifen is a strong liver carcinogen in the rat. Carcinogenesis 14:315–317, 1993.
169. Greaves P, Goonetilleke R, Nunn G, Topham J, Orton T. Two year carcinogenicity study of tamoxifen in Alderley Park Wister-derived rats. Cancer Res 53:3919–3924, 1993.
170. Jones CD, Jevnikar MG, Pike AJ, et al. Antiestrogens 2. Structure–activity studies in a series of 3 aroyl-2-arylbenzo[b]thiophene derivatives leading to [6-hydroxy-2-(4-hydrotyphenyl)-benzo[b]thiene-3-yl] [4-[2-(1-piperidinyl) ethoxy]-phenyl] methanone hydrochloride (LY156758), a remarkably effective estrogen antagonist with only minimal estrogenicity. J Med Chem 27:1057–1066, 1984.
171. Black LJ, Jones CD, Clark JH, Clemens JA. LY156758: a unique antiestrogen displaying high affinity for estrogen receptors, negligible estrogenic activity, and nearly total estrogen antagonism in vivo. Breast Cancer Res Treat 2:279, 1982.
172. Black LJ, Jones CD, Falcone JF. Antagonism of estrogen action with a new benzothiophene derived antiestrogen. Life Sci 32:1031–1036, 1983.
173. Black LJ, Goode RL. Uterine bioassay of tamoxifen, trioxifene and a new estrogen antagonist (LY117018) in rats and mice. Life Sci 26:1453–1458, 1980.
174. Jordan VC, Gosden B. Inhibition of the uterotropic activity of estrogens and antiestrogens by the short acting antiestrogen LY117018. Endocrinology 113:463–468, 1983.
175. Jordan VC, Gosden B. Differential antiestrogen action on the immature rat uterus: a comparison of hydroxylated antiestrogens with high affinity for the estrogen receptor. J Steroid Biochem 19:1249–1258, 1983.
176. Snyder KR, Sparano N, Malinowski JM. Raloxifene hydrochloride. Am J Health Syst Pharm 57:1669–1675, 2000.
177. Jordan VC. Biochemical pharmacology of antiestrogen action. Pharmacol Rev 36:245–276, 1984.
178. Buzdar AU, Marcus C, Holmes F, Hug V, Hortobagyi G. Phase II evaluation of LY156758 in metastatic breast cancer. Oncology 45:344–345, 1988.
179. Gradishar WJ, Glusman JE, Vogel CL, et al. Raloxifene HCl a new endocrine agent is active in estorgen receptor positive metastatic breast cancer. Breast Cancer Res Treat 46:53, 1997.
180. Clemens JA, Bennett DR, Black LJ, Jones CD. Effects of new antiestrogen keoxifene LY156758 on growth of carcinogen-induced mammary tumors and on LH and prolactin levels. Life Sci 32:2869–2875, 1983.
181. Anzano MA, Peer CW, Smith JM, et al. Chemoprevention of mammary carcinogenesis in the rat combined use of raloxifene and 9-cis-retinoic acid JNCI 88:123–125, 1996.
182. Poulin R, Merand Y, Poirier D, Levesque C, Dufor J-M, Labrie F. Antiestrogenic properties of keoxifene, trans 4-hydroxytamoxifen and ICI 164, 380, a new steroidal antiestrogen in ZR-75-1 human breast cancer cells. Breast Cancer Res Treat 14:65–76, 1989.
183. Jiang SY, Parker CJ, Jordan VC. A model to describe how a point mutation of the estrogen receptor alters the structure function relationship of antiestrogens. Breast Cancer Res Treat 26:139–147, 1993.
184. Cauley JA, Norton L, Lippman ME, et al. Continued breast cancer risk reduction in postmenopausal women treated with raloxifene: 4-year results from the MORE trial. Breast Cancer Res Treat 65:125–134, 2001.
185. Cummings SR, Norton L, Eckert S, et al. Raloxifene reduces the risk of breast cancer and may decrease the risk of endometrial cancer in post-

menopausal women. Two-year findings from the Multiple Outcomes of Raloxifene Evaluation (MORE) trial. Proc ASCO 2a 1998. Abstract nr 3.
186. Jordan VC, Glusman JE, Eckert S, et al. Raloxifene reduces incident primary breast cancers. Integrated data from multicenter double blind, placebo controlled randomized trials in postmenopausal women. Breast Cancer Res Treat 50:227, 1998. Abstract nr 2.
187. Black LJ, Sato M, Rowley ER, et al. Raloxifene (LY139,481 HCl) prevents bone loss and reduces serum cholesterol without causing uterine hypertrophy in ovariectomized rats. J Clin Invest 93:63–69, 1994.
188. Frolick CA, Bryant HU, Black EC, Magee DE, Chandrasekhar S. Time-dependent changes in biochemical bone markers and serum cholesterol in ovariectomized rats: effects of raloxifene HCl, tamoxifen, estrogen and alendronate. Bone 18:621–627, 1996.
189. Sato M, McClintock C, Kim J, et al. Dual-energy X-ray absorptiometry of raloxifene effects on the lumbar vertebrae and femora of ovariectomized rats. J Bone Miner Res 9:715–724, 1994.
190. Sato M, Kim J, Short LL, Slemenda CW, Bryant HU. Longitudinal and cross sectional analysis of raloxifene effects on tibiae from ovariectomized aged rats. J Pharmacol Exp Ther 272:1252–1259, 1995.
191. Sato M, Bryant HU, Iverson P, et al. Advantage of raloxifene over alendronate or estrogen on non-reproductive and reproductive tissues in the long term dosing of ovariectomized rats. J Pharmacol Exp Ther 279:298–305, 1996.
192. Sato M, Rippy MK, Bryant HU. Raloxifene, tamoxifen, nafoxidine and estrogen effects on reproductive and non-reproductive tissues in ovariectomized rats. FASEB J 10:905–912, 1996.
193. Turner CH, Sato M, Bryant HU. Raloxifene preserves bone strength and bone mass in ovariectomized rats. Endocrinology 135:2001–2005, 1994.
194. Draper MW, Flowers DE, Huster WJ, Neild JA. Effects of raloxifene (LY139,481 HCl) on biochemical markers of bone and lipid metabolism in healthy postmenopausal women. In: Christiansen C, Rii S (eds). Proceedings of the 4th International Symposium on Osteoporosis and Consensus Development Conference, Aalborg, Denmark, 1993, pp 119–121.
195. Heaney RP, Draper MW. Raloxifene and estrogen: comparative bone remodeling kinetics. J Clin Endocrinol Metab 82:3425–3429, 1997.
196. Delmas PD, Bjarnason NH, Mitlak BH, et al. Effects of raloxifene on bone mineral density, serum cholesterol concentrations and uterine endometrium in postmenopausal women. N Engl J Med 337:1641–1647, 1997.
197. Walsh BW, Kuller LH, Wild RA, et al. Effects of raloxifene on serum lipids and coagulation factors in healthy postmenopausal women. JAMA 279:1445–1451, 1998.
198. Bjarnason NH, Haarbo J, Byrjalsen I, Kauffman RF, Christiansen C. Raloxifene inhibits aortic accumulation of cholesterol in ovariectomized cholesterol fed rabbits. Circulation 96:1964–1969, 1997.
199. Clarkson TB, Anthony MS, Jerome CP. Lack of effect of raloxifene on coronary artery atherosclerosis of postmenopausal monkeys. J Clin Endocrinol Metab 83:721–726, 1998.
200. Grese TA, Cho S, Finley DR, et al. Structure–activity relationships of selective estrogen receptor modulators: modifications to the 2-arylbenzothiophene core of raloxifene. J Med Chem 40:146–167, 1997.
201. Grese TA, Sluka JP, Bryant HU, et al. Molecular determinants of tissue selectivity in estrogen receptor modulators. Proc Natl Acad Sci USA 94:14105–14110, 1997.
202. Boss SM, Huster WJ, Neild JA, Glant MD, Eisenhut CC, Draper MW. Effect of raloxifene hydrochloride on the endometrium of postmenopausal women. Am J Obstet Gynecol 177:1458–1464, 1997.
203. Gottardis MM, Ricchio ME, Satyaswaroop PG, Jordan VC. Effect of steroidal and non steroidal antiestrogens on the growth of a tamoxifen-stimulated human endometrial carcinoma (EnCa 101) in athymic mice. Cancer Res 50:3189–3192, 1990.
204. Baker VL, Draper M, Paul S, et al. Reproductive endocrine and endometrial effects of raloxifene hydrochloride, a selective estrogen receptor modulator in women with regular menstrual cycles. J Clin Endocrinol Metab 83:6–13, 1998.
205. Jordan VC, Fritz NF, Langan-Fahey S, Thompson M, Tormey DC. Alteration of endocrine parameters in premenopausal women with breast cancer during long term adjuvant therapy with tamoxifen as the single agent. J Natl Cancer Inst 83:1488–1491, 1991.
206. Wakeling AE, Bowler J. Steroidal pure antiestrogens. J Endocrinol 112:R7–R10, 1987.
207. Dauvois S, Danielian PS, White R, Parker MG. Antiestrogen ICI 164,384 reduces cellular estrogen receptor by increasing its turnover. Proc Natl Acad Sci USA 89:4037–4041, 1992.
208. Gibson MK, Nemmers LA, Beckman, et al. The mechanism of ICI 164,384 antiestrogenicity involves rapid loss of estrogen receptor in uterine tissue. Endocrinology 129:2000–2010, 1991.
209. Gottardis MM, Jiang SY, Jeng MH, et al. Inhibition of tamoxifen-stimulated growth of an MCF-7 tumor variant in athymic mice by novel steroidal antiestrogens. Cancer Res 49:4090–4093, 1989.
210. Howell A, DeFriend D, Robertson J, et al. Response of a specific antioestrogen (ICI182,780) in tamoxifen-resistant breast cancer. Lancet 345:29–30, 1995.

211. Howell A, Robertson JFR, Quaresma Albano J, et al. Comparison of the efficacy of Fulvestrant (Faslodex) with Anastrozole (Arimidex) in post menopausal women with advanced breast cancer. Breast Cancer Res Treat 64:27, 2000.
212. Osborne CK, on behalf of the North American Faslodex Investigator Group. A double blind randomized trial comparing the efficacy and tolerability of Faslodex (Fulvestrant) with Arimidex (Anastrozole) in postmenopausal women with advanced breast cancer. Breast Cancer Res Treat 64:27, 2000.
213. Sato M, Turner CH, Wang T, et al. LY 353,381.HCL: A novel raloxifene analog with improved SERM potency and efficacy in vivo. J Pharm Exper Ther 287:1–7, 1998.
214. MacGregor-Schafer J, Lee ES, Dardes R, Bentrem D, O'Regan RM, De Los Reyes A, and Jordan VC. Analysis of cross-resistance of the selective estrogen receptor modulators Arzoxifene (LY353381) and LY117018 in tamoxifen-stimulated breast cancer xenografts. Clin Cancer Res 7:2505–2512, 2001
215. Munster PN, Buzdar A, Dhingra K, et al. Phase I study of a third-generation selective estrogen receptor modulator, LY353381.HCL, in metastatic breast cancer. J Clin Oncol 19:2002–2009, 2001.
216. Dardes RC, Bentrem DJ, O'Regan RM, MacGregor Schafer J, Jordan VC. Effects of the new selective estrogen receptor modulator LY353381.HCl (Arzoxifene) on human endometrial cancer growth in athymic mice. Clin Cancer Res 7:4149–4155, 2001.
217. Ke HZ, Paralkar VM, Grasser WA. Effects of CP-336,156, a new, nonsteroidal estrogen agonist/antagonist, on bone, serum cholesterol, uterus, and body composition in rat models. Endocrinology 139:2068–2076, 1998.
218. Rosati RL, Da Silva JP, Cameron KO. Discovery and preclinical pharmacology of a novel, potent, nonsteroidal estrogen receptor agonist/antagonist, CP-336156, a diaryltetrahydronaphthalene. J Med Chem 41:2928–2931, 1998.
219. Willson TM, Henke BR, Momtahen TM, Charifson PS, Batchelor KW, Lubahn DB, Moore LB, Oliver BB, Sauls HR, Triantafillou JA. 3-[4-(1,2-Diphenylbut-1-enyl)phenyl]acrylic acid: a non-steroidal estrogen with functional selectivity for bone over uterus in rats. J Med Chem 37:1550–1552, 1994.
220. Willson TM, Norris JD, Wagner BL. Dissection of the molecular mechanism of action of GW5638, a novel estrogen receptor ligand, provides insights into the role of estrogen receptor in bone. Endocrinology 138:3901–3911, 1997.
221. Wijayaratne AL, Nagel SC, Paige LA, Christensen DJ, Norris JD, Fowlkes DM, McDonnell DP. Comparative analyses of mechanistic differences among antiestrogens. Endocrinology 140:5828–5840, 1999.
222. Bentrem DJ, Dardes RC, Liu H, MacGregor-Schafer J, Zapf JW, Jordan VC. Molecular mechanism of action at estrogen receptor α of a new clinically relevant antiestrogen (GW7604) related to tamoxifen. Endocrinology 142:838–846, 2001.
223. Miller CP, Collini MD, Tran BD, Harris HA, et al. Design, synthesis, and preclinical characterization of novel, highly selective indole estrogens. J Med Chem 44:1654–1657, 2001.
224. Gauthier S, Caron B, Cloutier J. (S)-(+)-[4-[7-(2,2-Dimethyl-1-oxopropoxy)-4-methyl-2-[4-[2-(1-piperidinyl)ethoxy]phenyl]-2H-1-benzopyran-3-yl]phenyl]-2,2-dimethylpropanoate (EM-800): a highly potent, specific and orally active non-steroidal antiestrogen. J Med Chem 40:2117–2122, 1997.
225. Luo S, Labrie C, Belanger A, Candas B, Labrie F. Prevention of development of dimethylbenz(a)anthracene (DMBA) induced–mammary tumors in the rat by the new nonsteroidal antiestrogen EM800 (Sch57050). Breast Cancer Res Treat 49:1–11, 1998.
226. Martel C, Labrie C, Belanger A, Gauthier S, Merand Y, Li X, Provencher L, Candas B, Labrie F. Comparison of the effects of the new orally active antiestrogen EM-800 with ICI 182,780 and toremifene on estrogen-sensitive parameters in the ovariectomized mouse. Endocrinology 139:2486–2492, 1998.
227. Couillad S, Gutman M, Labrie C, Belanger A, Landas B, Labrie F. Comparison of the effects of the antiestrogens EM-800 and tamoxifen on the growth of human breast ZR-75-1 cancer xenografts in nude mice. Cancer Res 58:60–64, 1998.
228. MacGregor JI, Liu H, Tonetti DA, Jordan VC. The interaction of raloxifene and the active metabolite of the antiestrogen EM-800 with the human estrogen receptor (ER). Cancer Res 59:4308–4313, 1999.
229. MacGregor Schafer J, Liu H, Bentrem DJ, Zapf JW, Jordan VC. Allosteric silencing of activating function 1 in the 4-hydroxytamoxifen estrogen receptor complex is induced by substituting glycine for aspartate at amino acid 351. Cancer Res 60:5097–5105, 2000.
230. Liu H, Lee ES, De Los Reyes A, Zapf JW, Jordan VC. Silencing and reactivation of the estrogen receptor modulator-estrogen receptor alpha complex. Cancer Res 61:3632–3639, 2001.
231. Sporn MB, Dunlop NM, Newton DL, Smith JM. Prevention of chemical carcinogenesis by vitamin A and its synthetic analogs (retinoids). Fed Proc 35:1332–1338, 1976.
232. Sporn MB. Approaches to prevention of epithelial cancer during the preneoplastic period. Cancer Res 36:2699–2702, 1976.
233. Jordan VC. Selective estrogen receptor modulation: A personal perspective. Cancer Res 61:5683–5687, 2001.

234. Fisher B, Dignam J, Bryant J, et al. Five years versus more than five years of tamoxifen therapy for breast cancer patients with negative lymph nodes and estrogen receptor-positive tumors. J Natl Cancer Inst 88:1529–1542, 1996.
235. Decensi A, Bonanni B, Guerrieri-Gonzaga A, et al. Biologic activity of tamoxifen at low doses in healthy women. J Natl Cancer Inst 90:1461–1467, 1998.
236. Costa A, Bonanni B, Manetti L, Guerrieri-Gonzaga A, Torrisi R, Decensi A. Prevention of breast cancer: focus on chemoprevention. Cancer Res 152:11–21, 1998.
237. Jordan VC, Gapstur S, Morrow M. Selective estrogen receptor modulation and reduction in risk of breast cancer, osteoporosis and coronary heart disease? JNCI 93:1449–1457, 2001.

NOTE ADDED IN PROOF

The results of IBIS I have recently been published and confirm the general conclusions of the NSABP P-1 study. (IBIS investigators. First Results from the International Breast Cancer Intervention Study: a randomised prevention trial. Lancet 360:817–824, 2002.)

13

Other Hormonal Prevention Strategies

DARCY V. SPICER
MALCOLM C. PIKE

A potential role for analogues of gonadotropin hormone–releasing hormone (GnRH) to prevent cancers of the breast, ovary, and endometrium was first suggested by Pike et al.[1] The rationale for considering GnRH analogues is their ability to suppress ovulation and sex-steroid production. The suppression of ovulation should prevent ovarian cancer, and the suppression of sex steroids should prevent cancers of the breast and endometrium. However, the GnRH analogue–induced hypoestrogenic state will have associated symptoms (hot flashes) and morbidity (urogenital atrophy and osteoporotic fractures). To minimize the deleterious effects of hypoestrogenemia, addition of other agents, including bisphosphonates, selective estrogen receptor modulators (SERMS), and low-dose add-back sex steroids, is under consideration for study. In this chapter, the epidemiological basis for considering GnRH analogues to prevent breast cancer and available preclinical and clinical data will be reviewed.

EPIDEMIOLOGICAL BASIS FOR CONSIDERING GONADOTROPIN-HORMONE–RELEASING HORMONE ANALOGUES

The key observations from epidemiological studies of importance in this context are the effects of early menopause or surgical oophorectomy on breast cancer risk. Epidemiological studies clearly demonstrate that early menopause, whether natural or artificial (bilateral oophorectomy), substantially reduces breast cancer risk. The large case-control study of Trichopoulos et al.[2] showed that artificial menopause below age 35 is associated with a breast cancer relative risk of 0.36 (a 64% reduction). Feinleib[3] noted in his large cohort study that among 1278 women with artificial menopause before age 40, six had breast cancer compared to an expected incidence of 24.0, a 75% reduction. Hirayama and Wynder's[4] epidemiological study found that the relative risk of breast cancer was 0.56 for women with bilateral oophorectomy; for women who were oophorectomized before age 37, the relative risk was 0.41 (i.e., a 59% reduction). What is of key importance is the magnitude of the benefit and the consistency of these findings. Menopause before age 35 is associated with a 60%–75% reduction in breast cancer risk.

The age–incidence curve for breast cancer is depicted in Figure 13.1. Age at menopause determines the transition point from the steeply rising premenopausal slope to the gentle postmenopausal slope. The effect of early oophorectomy on the age–incidence curve of breast cancer is shown.

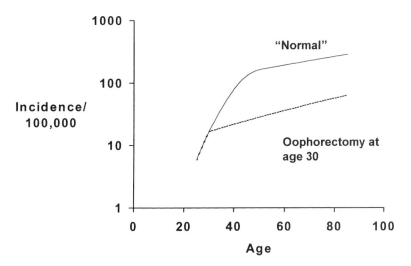

Figure 13.1 Effect of oophorectomy on the age–incidence curve for breast cancer. Calculations were made using the model described by Pike.[31]

Of significant interest is the finding of a protective effect of oophorectomy in women at high risk for breast cancer. In women who carry germline mutations in *BRCA1*, a protective effect of oophorectomy has been seen. In the study of Rebbeck et al.,[5] bilateral prophylactic oophorectomy, which was done to prevent ovarian cancer, was associated with a substantial reduction in breast cancer risk (hazard ratio = 0.53).

BREAST CELL PROLIFERATION AND THE ESTROGEN PLUS PROGESTERONE HYPOTHESIS

Breast cancers are thought to arise from the epithelial cells of the terminal duct lobular unit (TDLU). Studies of cell proliferation rates of the TDLU are therefore of substantial interest to our understanding of the factors that influence breast cancer risk. Repetitive cell proliferation is central to the risk of many common human cancers, and factors that increase cell proliferation in a tissue may result in malignant transformation by increasing the probability of converting DNA damage, however caused, into stable mutations.[6–10] In the postmenopausal human breast, the rate of TDLU cell proliferation is low compared to the premenopausal breast.[11,12] This low rate of cell proliferation is consistent with the small change in breast cancer risk seen during the postmenopausal years. The steeply rising premenopausal breast cancer rates are consistent with the measured higher breast epithelial cell proliferation rates and vary associated with the phase of the menstrual cycle, consistent with a direct effect of ovarian hormones on cell proliferation rates. Studies show low rates during the follicular phase of the menstrual cycle, increasing two- to threefold in the luteal phase.[12–17] This suggests that follicular-phase estrogen levels induces some breast cell proliferation but that the combined effect of estrogen and progesterone produces greater breast cell proliferation; i.e., both estrogen and progesterone are mitogens in the normal human breast epithelial cell (see Pike et al.[18] for review). Studies of the effects of estrogen- or combined estrogen and progesterone–replacement therapy in postmenopausal women have confirmed the greater cell proliferation with the combination regimen.[12] The estrogen plus progesterone hypothesis explains the effects of these hormones on breast cancer risk.

The effect of menopause or oophorectomy on breast cancer risk is thus predictable in light of these effects of estrogen and progesterone on breast epithelial cell proliferation. Cessation of ovarian function reduces cancer risk by eliminating the breast mitogen progesterone and reducing estrogen levels.

GONADOTROPIN-HORMONE–RELEASING HORMONE ANALOGUES

In women, GnRH analogues induce reversible inhibition of ovarian function and steroid production. Native GnRH produced by the hypothalamus controls the secretion of follicle-stimulating hormone (FSH) and luteinizing hormone (LH) by the pituitary and, thence, gonadal steroid hormone production. Inhibition of GnRH action can be achieved either with recently introduced antagonists or with agonists that desensitize GnRH receptors. Potent synthetic agonists of GnRH administered to premenopausal women produce a transient rise in FSH/LH release followed by a sustained suppression. Reduction in serum estradiol and serum progesterone to oophorectomized levels by GnRH agonists has been demonstrated in numerous reports and has led to their use in the treatment of hormone-responsive metastatic breast cancer in premenopausal women.[19]

The role of ovarian ablation in the adjuvant therapy of early breast cancer remains unsettled, and studies evaluating GnRH analogues continue. Adjuvant studies in hormone receptor-positive premenopausal breast cancer patients show a benefit similar to some chemotherapy regimens.[20] Of particular interest is the large multicenter trial evaluating the GnRH agonist goserelin in the adjuvant setting [Zolzdex in Premenopausal Patients (ZIPP) study] reported by Baum.[21] In the preliminary report of this trial, the incidence of contralateral breast tumors was reduced in patients receiving the goserelin [relative risk (RR) = 0.60, $p = 0.05$]. The demonstration of a protective effect of adjuvant GnRH analogue on contralateral breast tumors is analogous to that which was seen with adjuvant tamoxifen and provides support for the continued study of this GnRH analogues in breast cancer prevention.

The chemopreventive potential of GnRH analogues has been evaluated in animal models. In the dimethylbenz[a]anthracene (DMBA)–induced rat mammary carcinoma model, a GnRH agonist was as effective as surgical oophorectomy in the prevention of tumors.[22]

GONADOTROPIN-HORMONE–RELEASING HORMONE ANALOGUES IN COMBINATION

Use of GnRH analogues in premenopausal women is predictably associated with hypoestrogenic symptoms, including hot flushes, vaginal dryness, and sleep disturbances. In the majority of studies with protracted GnRH analogue treatment, loss of bone mineral density (BMD) has been evident. As oophorectomy at a young age is associated with an increased risk of cardiovascular disease, long-term use of GnRH analogues is of concern. While the side effects and risks associated with hypoestrogenemia are acceptable in the setting of metastatic breast cancer and in the adjuvant treatment of early breast cancer, such effects may not be acceptable to women only at risk for the development of the disease. While a GnRH analogue should achieve major reduction in a woman's lifetime breast cancer risk, the benefit will occur only if the agent were to be continued for prolonged periods of time. To permit such protracted use, methods for reducing the side effects and morbidity must be considered.

The effect of protracted GnRH analogue treatment on BMD has prompted consideration of several strategies to combat the loss of bone density. A logical choice is addition of a SERM such as tamoxifen, which is known to have protective effects on BMD.[23] The use of tamoxifen combined with a GnRH analogue is also of interest because of the known chemopreventive effects of tamoxifen.[24] Bisphosphonates are important agents in the management of osteopenia in postmenopausal women. A study testing the combination of a GnRH analogue and a bisphosphonate has also been proposed. While either approach may ultimately prove useful in reducing the loss of BMD, the tolerance of women to hypoestrogenic symptoms, which will not be alleviated by these approaches, remains to be evaluated. Analogues of GnRH are associated with greater symptoms than tamoxifen alone in the adjuvant setting,[25] and control of these symptoms may significantly impact on use in the preventive setting.

GONADOTROPIN-HORMONE–RELEASING ANALOGUES WITH ADD-BACK SEX STEROIDS

Low-dose hormone-replacement therapy, as proposed by Pike et al.,[1] will reduce the hypoestrogenic symptoms of a GnRH analogue.

While there is greater hormone exposure than would occur with a GnRH agonist alone, the overall reduction in hormone exposure compared to remaining premenopausal is substantial with reductions in estrogen exposure of 60% and in progestogen exposure of 75%. As the add-back low dose should permit long-term use, protracted reductions in hormone exposure would be possible. The predicted reductions in breast cancer incidence are less than with a GnRH analogue alone (Table 13.1) but remain substantial. As will be discussed below, the dose of add-back proposed is similar to that used as hormone-replacement therapy in postmenopausal women; however, the schedule of progestin administration differs. The add-back will, of course, have an effect on breast cancer risk. The effect of the add-back on breast cancer risk is consistent with that reported from studies of hormone-replacement therapy on breast cancer risk. In postmenopausal women, at the estrogen-replacement therapy (ERT) dose usually administered in the United States, breast cancer risk increases approximately 2% per year of use (RR ≈ 1.02 per year of use).[26,27] Longer use increases the risk proportionately. With add-back estrogen, it is also necessary to use add-back progestin to protect the endometrium from the estrogen treatment. Estrogen plus progestin–replacement therapy (EPRT) in postmenopausal women increases breast cancer risk to a greater extent than ERT.[28-30] The RR per year of EPRT use is approximately three times that associated with ERT use, approximately 1.06. To minimize progestin exposure while still providing protection of the endometrium, progestin is given for 14 days in every 3-month period. To estimate the effect on breast cancer risk of use of such a GnRH analogue plus add-back regimen, we have assumed that the GnRH analogue will induce medical oophorectomy and that the effect of the estrogen and progestin will be the same as that observed in postmenopausal hormone-replacement therapy users. Estimates for breast cancer are calculated using a mathematical model.[1,31,32] Lifetime breast cancer risk is predicted to be reduced by about 39% if the regimen is used for 5 years, by about 64% if used for 10 years, and by about 80% if used for 15 years. We conducted a pilot study to determine the effects of a GnRH analogue plus add-back on BMD, lipoprotein metabolism, the endometrium, and menopausal symptoms in women at high risk for breast cancer.[33,34] The regimen included a depot GnRH agonist administered monthly, low-dose estrogen with conjugated estrogens 0.625–0.9 milligrams for 6 days each week, and the progestogen medroxyprogesterone acetate 10 milligrams for 14 days every 112 days (4 months).

Subsequently, the effects of replacing ovarian androgen (which is suppressed by GnRH analogues) to the add-back hormone regimen were evaluated. Subjects were premenopausal women, aged 25–40, with one of the following breast cancer risk factors: lobular carcinoma in situ, mother and sister with breast cancer (at least one premenopausal), or a mother or sister with bilateral premenopausal breast cancer. Twenty-one subjects were entered and randomized on a 2:1 basis to an intervention group and a control group.

Overall, the regimen was well tolerated.[34] A questionnaire assessed frequency and intensity of possible symptoms of menopausal distress and premenstrual syndrome. Menopausal symptoms were infrequent and the few occurrences of hot flashes were eliminated by increasing the dose of conjugated estrogens to 0.9 milligrams. Treated subjects had a decrease in luteal phase or premenstrual syndrome symptoms of abdominal bloating or fullness; abdominal cramps or pain; breast swelling; breast pain or tenderness; anxious, tense, or nervous; irritable, angry, or impatient; and mood swings.[34]

In the first year of the study, a small but significant 2.9% reduction in lumbar spine BMD was noted in the treated group. As a result, the study was modified to replace ovarian androgens suppressed by the GnRH analogue using methyltestosterone. The mean change in BMD at the lumbar spine after the addition of the an-

Table 13.1 Predicted Reduction in Breast Cancer Risk with a Gonadotropin-Hormone–Releasing Hormone (GnRH) Analogue Alone or an Add-back Regimen

	Duration of Use (years)		
	5	10	15
GnRH analogue alone	47%	75%	89%
GnRH analogue + add-back	39%	64%	80%

drogen is depicted in Table 13.2. These results suggest that addition of the androgen may have an effect on maintenance of BMD.

Favorable effects on lipid profiles were seen with the initial regimen during the months when medroxyprogesterone acetate was not administered.[34] Addition of methyltestosterone, as expected, eliminated the beneficial effect of the regimen on lipoproteins. However, the changes in cholesterol (compared to baseline values) were not different from those of the controls. Oral methyltestosterone is not considered an optimal method of replacement of ovarian androgens. Scheduled bleeding occurred following most progestogen administrations and unscheduled bleeding or spotting occurred infrequently.[34] There was no evidence of endometrial hyperplasia identified by endometrial biopsies performed each year.

Epidemiological studies have consistently found that increased mammographic densities are associated with greater breast cancer risk independent of other risk factors.[35-38] In the pilot study, changes in mammographic densities were measured in women on the treatment regimen and in control women. The study directly measured changes in mammographic density because mammographic classification schemes are unable to distinguish fine changes that do not cause a change of category. The radiologists were masked both as to whether the mammograms were from treated or patients controls and as to whether they were baseline or follow-up studies.[39] Fibrous septae became more apparent, and there were less nodular densities and a smaller amount of confluent areas at 1 year than at baseline. Figure 13.2 shows the substantial im-

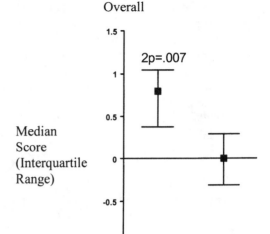

Figure 13.2 Mammographic changes in treated and control subjects.

provement overall in the treated group. The reduced estrogen and progestogen exposures achieved by the regimen resulted in significant reductions in follow-up mammographic densities, further supporting the evidence that a GnRH analogue and add-back may contribute significantly to breast cancer reduction. A second study in high-risk women is ongoing.

REFERENCES

1. Pike MC, Ross RK, Lobo RA, Key RJA, Potts M, Henderson BE. LHRH agonists and the prevention of breast and ovarian cancer. Br J Cancer 60: 142–148, 1980.
2. Trichopoulos D, MacMahon B, Cole P. Menopause and breast cancer risk. J Natl Cancer Inst 48:605–613, 1972.
3. Feinleib M. Breast cancer and artificial menopause: a cohort study. J Natl Cancer Inst 41: 315–329, 1968.
4. Hirayama T, Wynder EL. A study of the epidemiology of cancer of the breast II. The influence of hysterectomy. Cancer 15:28–38, 1962.
5. Rebbeck R, Levin A, Eisen A, et al. Breast cancer risk after bilateral prophylactic oophorectomy in BRCA1 mutation carriers. J Natl Cancer Inst 91:1475–1479, 1999.
6. Henderson B, Ross RK, Pike M, Casagrande J. Endogenous hormones as a major factor in human cancer. Cancer Res 42:3232–3239, 1982.
7. Ames BN, Gold LS. Too many rodent carcino-

Table 13.2 Annualized Change in Bone Mineral Density*

Group	Lumbar Spine	Femoral Neck
Control	0.4%	0.2%
Treated CE + MPA	-2.9%†	-2.2%‡
Treated CE + MT + MPA	0%	1.6%

*Assessed using quantitative digital radiography
†1p = .001.
‡1p = .006.
CE, conjugated estrogens; MPA, medroxyprogesterone acetate; MT, methyltestosterone.

gens: mitogenesis increases mutagenesis. Science 249:970–971, 1990.
8. Cohen SM, Ellwein L. Cell proliferation in carcinogenesis. Science 249:1007–1011, 1990.
9. Preston-Martin S, Pike MC, Ross RK, Jones A, Henderson BE. Increased cell division as a cause of human cancer. Cancer Res 50:7415–7421, 1990.
10. Butterworth B, Slaga T. Chemically induced cell proliferation: implications for risk assessment. New York: Wiley-Liss, 1991.
11. Meyer JS, Connor RE. Cell proliferation in fibrocystic disease and postmenopausal breast ducts measured by thymidine labeling. Cancer 50:746–751, 1982.
12. Hofseth LJ, Raafat AM, Osuch JR, Pathak DR, Slomski CA, Haslam SZ. Hormone replacement therapy with estrogen or estrogen plus medroxyprogesterone acetate is associated with increased epithelial proliferation in the normal postmenopausal breast. J Clin Endocrinol Metab 84:4559–4565, 1999.
13. Meyer JS. Cell proliferation in normal human breast ducts, fibroadenomas, and other duct hyperplasias, measured by nuclear labeling with tritiated thymidine. Hum Pathol 8:67–81, 1977.
14. Anderson TJ, Ferguson DJP, Raab GM. Cell turnover in the "resting" human breast: influence of parity, contraceptive pill, age and laterality. Br J Cancer 46:376–382, 1982.
15. Longacre TA, Bartow SA. A correlative morphologic study of human breast and endometrium in the menstrual cycle. Am J Surg Pathol 10:382–393, 1986.
16. Anderson TJ, Battersby S, King RJB, McPherson K, Going JJ. Oral contraceptive use influences resting breast proliferation. Hum Pathol 20:1139–1144, 1989.
17. Williams G, Anderson E, Howell A. Oral contraceptive (OCP) use increases proliferation and decreases oestrogen receptor content of epithelial cells in the normal human breast. Int J Cancer 48:206–210, 1991.
18. Pike MC, Spicer DV, Dahmoush L, Press MF. Estrogens, progestogens, normal breast cell proliferation and breast cancer risk. Epidemiol Rev 15:17–35, 1993.
19. Kaufman M, Jonat W, Kleeberg U, et al. Goserelin, a depot gonadotrophin-releasing hormone agonist in the treatment of premenopausal patients with metastatic breast cancer. J Clin Oncol 7:1113–1119, 1989.
20. Boccardo F, Rubagotti A, Amoroso D, et al. Cyclophosphamide, methotrexate, and fluorouracil versus tamoxifen plus ovarian suppression as adjuvant treatment of estrogen receptor-positive pre-/perimenopausal breast cancer patients: results of the Italian Breast Cancer Adjuvant Study Group 02 randomized trial. J Clin Oncol 18:2718–2727, 2000.
21. Baum M. Adjuvant treatment of premenopausal breast cancer with zoladex and tamoxifen. Breast Cancer Res Treat 57:30, 1999.
22. Jett EA, Lerner MR, Lightfoot SA, Hanas JS, Brackett DJ, Hollingsworth AB. Prevention of rat mammary carcinoma utilizing leuprolide as an equivalent to oophorectomy. Breast Cancer Res Treat 58:131–136, 1999.
23. Love RR, Mazess RB, Barden JS, et al. Effects of tamoxifen on bone mineral density in postmenopausal women with breast cancer. N Engl J Med 326:852–856, 1992.
24. Fisher B, Costantino JP, Wickerham DL, Redmond CK. Tamoxifen for prevention of breast cancer: report of the National Surgical Adjuvant Breast and Bowel Project P-1 Study. J Natl Cancer Inst 90:1371–1388, 1998.
25. Nystedt M, Berglund G, Bolund C, Brandberg Y, Fornander T, Rutqvist L. Randomized trial of adjuvant tamoxifen and/or goserelin in premenopausal breast cancer—self-rated physiological effects and symptoms. Acta Oncol 39:959–968, 2000.
26. Steinberg K, Thacker S, Smith S, Stroup D, Zack M, Flanders W. A meta-analysis of the effect of estrogen replacement therapy on the risk of breast cancer. JAMA 265:1985–1990, 1991.
27. Collaborative Group on Hormonal Factors in Breast Cancer. Breast cancer and hormone replacement therapy: collaborative reanalysis of data from 51 epidemiological studies of 52,705 women with breast cancer and 108,411 women without breast cancer. Lancet 350:1047–1059, 1997.
28. Magnusson C, Baron J, Correia N, Bergstrom R, Adami H-O, Persson I. Breast-cancer risk following long-term oestrogen- and oestrogen–progestin-replacement therapy. Int J Cancer 81:339–344, 1999.
29. Schairer C, Lubin J, Troisi R, Sturgeon S, Brinton L, Hoover R. Menopausal estrogen and estrogen—progestin replacement therapy and breast cancer risk. JAMA 283:485–491, 2000.
30. Ross R, Paganini-Hill A, Wan, Pike M. Estrogen versus estrogen–progestin hormone replacement therapy: effect on breast cancer risk. J Natl Cancer Inst 92:1475–1479, 2000.
31. Pike MC. Age-related factors in cancer of the breast, ovary and endometrium. J Chron Dis 40:59–69, 1987.
32. Spicer D, Shoupe D, Pike M. Gonadotropin-releasing hormone agonist plus add-back sex steroids to reduce risk of breast cancer [letter, comment]. J Natl Cancer Inst 83:23, 1991.
33. Spicer DV, Shoupe D, Pike M. GnRH agonists as contraceptive agents: predicted significantly reduced risk of breast cancer. Contraception 44:289–310, 1991.
34. Spicer DV, Pike MC, Pike A, Rude R, Shoupe D, Richardson J. Pilot trial of a gonadotropin hormone agonist with replacement hormones as a

prototype contraceptive to prevent breast cancer. Contraception 47:427–444, 1993.
35. Saftlas AF, Szklo M. Mammographic parenchymal patterns and breast cancer risk. Epidemiol Rev 9:146–174, 1987.
36. Brisson J, Morrison AS, Khalid N. Mammographic parenchymal features and breast cancer in the breast cancer detection demonstration project. J Natl Cancer Inst 80:1532–1540, 1988.
37. Warner E, Lockwood G, Trichler D, Boyd N. The risk of breast cancer associated with mammographic parenchymal patterns: a meta-analysis of the published literature to examine the effect of method of classification. Cancer Detect Prev 16:67–72, 1992.
38. Oza AM, Boyd NF. Mammographic parenchymal patterns: a marker of breast cancer risk. Epidemiol Rev 15:196–208, 1993.
39. Spicer D, Ursin G, Parisky Y, et al. Changes in mammographic densities induced by a hormonal contraceptive designed to reduce breast cancer risk. J Natl Cancer Inst 86:431–436, 1994.

14

Progression from Hormone-Dependent to Hormone-Independent Breast Cancer

ISABELL A. SCHMITT
MILANA DOLEZAL
MICHAEL F. PRESS

Breast cancer is the most common malignancy in women, and treatment is challenging. The potential role of hormonal therapy in breast cancer treatment was first demonstrated by George Beatson more than 100 years ago, when he showed that two of three premenopausal women with metastatic breast cancer responded to ovarian ablation.[1] Subsequently, Huggins and Bergenstal[2] demonstrated that postmenopausal women with metastatic breast cancer responded to ovariectomy and adrenalectomy. Despite the usefulness of hormonal ablation in some women, only approximately 30% of unselected women with metastatic breast cancer responded to the treatment.[3] Thus, there was a need to distinguish those women whose breast cancers were hormone-dependent from those whose cancers were hormone-independent. Jensen and collaborators[4] demonstrated that patient responsiveness to endocrine manipulative management was correlated with the uptake of estrogen in the breast cancer tissue. The first results showed that uptake of radioactive hormone was higher in breast cancer tissue compared to muscle tissue from four women who responded to hormonal treatment than from six women who did not respond. These findings were soon confirmed and extended by reports from other investigators.[5–8]

The initial in vivo human studies demonstrating a correlation between estrogen uptake by the tumor tissue and responsiveness to hormonal ablation therapy were followed by in vitro observations in primary breast cancer slices or homogenates. Procedures for cell fractionation and sucrose density gradient analysis were developed, which permitted the identification and quantification of estrogen binding in breast cancer samples. Using these techniques, the characteristics of an estrogen receptor (ER) were defined as a protein which bound its estrogen ligand in a highly specific fashion, with high affinity (K_d of 10^{-10} M) and with saturation by excess ligand. This interaction with the receptor protein was considered to mediate a variety of biological responses. Use of in vitro assays of tumor tissue samples improved the predictions of patient response to hormonal therapy. Approximately 50%–60% of women with ER-rich breast cancers responded to such therapy, while fewer than 10% with ER-poor breast cancers responded. Subsequently, an ER protein was isolated,[9,10] and monoclonal antibodies to ER were used to develop immunohistochemical assays that permitted identification of ER in tissue sections.[11,12] A second form of ER, ERβ, was identified using reverse transcriptase-polymerase chain reaction and cDNA cloning.[13,14] This form of the receptor has a normal tissue distribution, which overlaps with the classic ERα but is not identical.[15] There have been relatively few studies of ERβ in human breast cancers.[16–20] The potential role of ERβ in breast cancer responsiveness is not yet well established but may be significant.[16–18]

The use of ER assays alone predicted the response to hormonal manipulative management in the majority of breast cancer cases, although these assays probably measure only ERα. Most of the inaccuracy in predicting hormone responsiveness in breast cancer is due to unresponsive ER-positive tumors. This could be the result of heterogeneity in ER expression in different tumor cells, which does exist. However, it could also be due to an ER that was present but failed to have biological action in the target cell. Estrogens stimulate both cell division and the synthesis of specific hormone-responsive proteins in breast cells. One of these estrogen-induced proteins is the progesterone receptor (PR).[21] Since PR is an estrogen-inducible protein, assays for it have been suggested to identify tumors that are hormonally responsive to estrogen and to improve the overall predictive value of steroid hormone receptor assays.[21]

The use of in vitro assays for ER and PR in breast cancer tissue samples improves the prediction of patient responsiveness to hormonal therapy. While approximately three-fourths of women with ER-positive, PR-positive breast cancers respond to such therapy, fewer than 10% of women with ER-poor, PR-poor breast cancers respond (Table 14.1). Women with receptor-poor breast cancers are not likely to respond to hormonal manipulative treatment and, therefore, show hormone-independent growth from the initial diagnosis. Breast cancers can also progress from a hormone-dependent, antiestrogen-responsive phenotype to a hormone-independent, therapy-resistant phenotype.

Instead of surgical hormone ablation, as practiced in the past, currently antiestrogen compounds are used to induce clinical remissions. The most frequently used antiestrogen in clinical practice is tamoxifen. It is effective and has a relative lack of side effects. Anti-estrogenic compounds can be divided into purely antagonist drugs (type I), such as ICI 164,384, and mixed antagonist–agonist drugs (type II), such as tamoxifen. Tamoxifen induces ER dimerization and DNA binding, described below, but has a mixed agonist–antagonist transcriptional profile that is tissue- and promoter-specific.[22,23] Tamoxifen acts as an antagonist of ER action in the breast but as a weak ER agonist in the uterus and bone.[24] Such tissue-specific estrogenic ligands are termed selective estrogen receptor modulators (SERMs). Although new antiestrogenic compounds are being developed and tested in clinical trials, treatment with tamoxifen is currently the most common therapy for hormone-dependent breast cancers (tamoxifen and related compounds are extensively reviewed in Chapter 12).

Hormone-independent breast cancer can be encountered in various clinical settings. These can be briefly summarized as follows:

1. ER-positive breast cancers, initially responsive but subsequently unresponsive after long-term therapy with tamoxifen
2. ER-positive, tamoxifen-nonresponsive breast cancers at diagnosis

Table 14.1 Estrogen Receptor and Progesterone Receptor Content in Breast Cancer Tissue: Incidence, Prognosis, and Responsiveness to Hormonal Therapy

| Receptor Content | Incidence | Clinical Follow-up at 5 Years | | Response Rate to Endocrine Rx |
		DFS	Survival	
ER$^+$PR$^+$	50%	73%	91%	77%
ER$^+$PR$^-$	20%	75%	93%	27%
ER$^-$PR$^+$	5%	68%	88%	46%
ER$^-$PR$^-$	25%	64%	77%	7%

Sources: Picon et al.,[148] Osborne et al.,[149] Thorpe et al.,[150] Young et al.,[151] Fisher et al.[152]

3. ER-poor, tamoxifen-resistant breast cancers at diagnosis

The first represents acquired antiestrogen resistance or hormone independence, and the other two situations represent de novo antiestrogen resistance or hormone independence. The distinction is clinically useful because ER-poor breast cancers seldom respond to other second- or third-line antiestrogen agents, whereas approximately half of initially tamoxifen-responsive but subsequently tamoxifen-unresponsive breast cancers respond to second- or third-line agents.

Since hormone responsiveness and breast cancer hormone dependence are mediated through the ER pathway, alteration in any step of this pathway may lead to hormonal resistance. This resistance to endocrine therapy is a major problem that is likely mediated by various molecular mechanisms in the ER pathway. Therefore, understanding the biology of this process is important in the design of new, rational therapies.

ESTROGEN RECEPTOR ACTION

Cell Biology of Estrogen Receptor

Steroid-unoccupied forms of ER, although originally believed to be present in the cytoplasm,[4,25] are now known to be present in the nucleus of steroid-responsive cells.[12,26,27] Estrogen receptors α and β are nuclear, ligand-activated transcription factors. Estrogen traverses the cell membrane by passive or facilitated diffusion, enters the nucleus, binds to the unoccupied receptor, and causes receptor dimerization/activation through receptor phosphorylation.[28,29] The activated hormone–receptor complex interacts with nuclear chromatin by binding to estrogen response element (ERE) in the promoter/enhancer region of selected genes in such a way that a specific set of RNAs is expressed.[30] Some of the induced RNAs code for proteins involved in regulating the metabolism of the target cell. The PR is one such regulatory protein induced by activated ERα.

The nuclear distribution of the steroid-unoccupied form of ER requires continuous metabolic activity. Use of various inhibitors of energy synthesis in cultured cells expressing ERα demonstrated that the nuclear residence of the receptor reflects a dynamic state. In the presence of energy inhibitors, receptor diffuses from the nucleus to the cytoplasm. When the inhibitors are removed and glucose is returned to the culture medium, ERs are transported back to the nucleus.[31] Nuclear localization is also dependent on the presence of specific nuclear localization signals in the primary amino acid sequence of the receptors, as described below.

Molecular Biology of Estrogen Receptors α and β

Expression cloning of the human ER using monoclonal ER antibodies provided the first full-length sequence of ERα.[32,33] In common with the other members of the steroid/thyroid hormone receptor family of nuclear receptors, ERα has six main structural and functional domains (A–F): a variable or regulatory domain containing a transcriptional activation function (A/B), a DNA-binding domain (C), a hinge region (D), a ligand-binding domain (E), and a region that plays a role in distinguishing agonists vs. antagonists (F) (Fig. 14.1). Functions for these regions were initially proposed based on the coded amino acid sequence and hydrophobicity analyses of these sequences. In vitro site mutagenesis assays, deletion mutation studies, and domain swapping experiments have confirmed functional roles for these domains.[34–36] A portion of the N-terminal regulatory domain, amino acids 41–150 known as activation function-1 (AF-1), is associated with modulation of transcriptional activity. This domain contains three serine residues, which are phosphorylated during activation of ERα. The DNA-binding domain (C) contains two zinc fingers, which are responsible for direct interaction with the DNA helix. The hinge region (D) between the DNA-binding domain and the ligand-binding domain (E) contains sequences for receptor dimerization and nuclear localization.[37] A portion of the ligand-binding domain, amino acids 530–553 referred to as AF-2, is also associated with transcriptional regulation through association with coregulatory proteins (see below). Both ERα and ERβ have a similar domain structure (Fig. 14.1). The

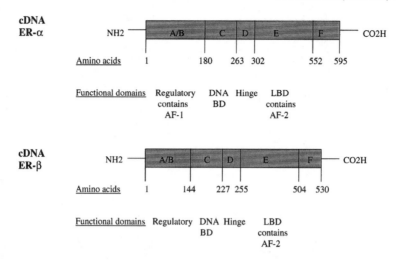

Figure 14.1 Schematic diagram of estrogen receptor (ER) α and ERβ domain structures. Both types have several functional domains (A–F). These domains are the variable or regulatory domain (A/B), containing the activation function-1 (AF-1), necessary for coactivator interaction; the DNA-binding domain (BD) (C); the hinge region (D), containing sequences for the nuclear localization signal (NLS) and ER dimerization; the ligand-binding domain (LBD) or steroid-binding domain (E), containing AF-2 for coactivator recruitment and transcriptional activation of target genes.

DNA-binding domain of ERβ is 96% and the ligand-binding domain is 58% homologous with the ERα sequence.[14] However, the regulatory (or A/B) domain, hinge region (D), and F region show little conservation between these receptor types.[14] Like ERα, ERβ binds 17β-estradiol with high affinity and can transactivate gene expression at EREs by an estrogen-dependent mechanism. However, at the AP-1 enhancer element, ERα and ERβ signal in opposite directions. With ERα, 17β-estradiol activates transcription, whereas with ERβ, 17β-estradiol inhibits transcription (see Color Figure 14.2A in separate color insert).[38] The antiestrogens tamoxifen, raloxifene, and ICI 164,384 (N-n-butyl-N-methyl-11-(3,17β-dihydroxyoestra-1,3,5(10)-trien-7α-yl) undecamide) are potent transcriptional activators with ERβ at an AP1 site (see Color Fig. 14.2B in separate color insert). Thus, the two ERs signal in different ways depending on the ligand and response element (see Color Fig. 14.2A,B in separate color insert).

The ERα gene is located on the long arm of chromosome 6 (band q24–27),[36] and the ERβ gene is on the long arm of chromosome 14 (band q22–24).[39] The ERα gene is composed of eight exons, which are, respectively, 684, 191, 117, 336, 139, 134, 184, and 4537 bp in size. Including the intervening introns, the ER gene spans at least 140 kb.[40] The amino-terminal hypervariable region is predominantly coded for by exon 1. Exons 2 and 3 each code for one zinc finger of the DNA-binding domain (Fig. 14.1).[40] The hinge region is coded for by exon 4. The large hydrophobic hormone-binding domain is encoded by five different exons, including part of exon 4, exons 5–7, and part of exon 8.[40] The genomic DNA structure of ERβ has not been described.

The ER recognizes the DNA promoter consensus sequence 5'-<u>AGGTCA</u>NNN<u>TGACCT</u>-3', which is known as the ERE.[30,41,42] These response elements are palindromic sequences; i.e. the underlined sequences are complementary to one another. The apparent dyadic symmetry of these elements indicates that they interact with receptor dimers, either homodimers or heterodimers, of ERα and ERβ. If fewer than three nucleotides separate the half-palindromes, then ER no longer recognizes its hormone response element sequence.

The three-dimensional structure of the ligand-binding domain for ERα has been determined using X-ray crystallography (see Color Fig. 14.3 in separate color insert).[43,44] The estradiol-binding cavity is completely partitioned from the external environment and occupies a relatively large portion of the ligand-binding domain hydrophobic core. This binding pocket is formed from portions of helix 3, helix 6, helix 8, helix 11, helix 12, and the S1/S2 hairpin (see Color Fig. 14.3 in separate color insert).[43] The ER-binding pocket binds a wide variety of nonsteroidal compounds. This is probably due to the size of the cavity, which has a probe-accessible volume nearly twice that of estradiol's molecular volume.[43] The length and breadth of the estradiol skeleton are well matched by the receptor, but there are large unoccupied cavities opposite the α face of the B ring and the β face of the C ring of estradiol.[43]

In estradiol-liganded ERα, helix 12 is positioned directly over the ligand-binding cavity (see Color Fig. 14.3A in separate color insert).[43,44] Although helix 12 of ER makes no direct contact with estradiol, it forms the "lid" of the binding cavity and projects its inner hydrophobic surface toward the bound hormone. The charged surface of helix 12 is directed away from the body of the ligand-binding domain, on the side of the molecule lying perpendicular to the dimerization surface.[43,44] This precise positioning of helix 12 appears to be a prerequisite for transcriptional activation since, by sealing the ligand-binding cavity, it generates a competent activation function (AF-2) that is capable of interacting with coactivators. The ligand-binding domain's transcriptional activation function (AF-2) can interact with a number of putative transcriptional coactivators in a ligand-dependent manner.[45] Helix 12 is essential for such transactivation as either loss or mutation in this region results in a receptor that is unresponsive to ligand.

In contrast to the above description of estradiol-bound ER, the alignment of helix 12 over the ligand-binding cavity is prevented by raloxifene and tamoxifen (see Color Fig. 14.3B,D in separate color insert).[43,44] These nonsteroidal antiestrogens possess large side chains that are too long to be contained within the binding cavity, and these displace helix 12 so that it protrudes from the pocket. This antagonist-induced repositioning of helix 12 involves a rotation of 130 degrees and is expected to be a general feature of both steroidal and nonsteroidal antiestrogens that possess bulky side chains. In addition, binding of tamoxifen by the ER results in an alternative packing arrangement of the ligand-binding pocket amino acid residues around tamoxifen (see Color Fig. 14.3D in separate color insert) and stabilizes a conformation of the ligand-binding domain that permits helix 12 to reach the static region of the AF-2 surface and to mimic bound coactivator.[44] This displacement of helix 12 by antiestrogens (especially tamoxifen) (see Color Fig. 14.3D in separate color insert) and alternative packing arrangement prevent the formation of AF-2 for interaction with coactivators such as GRIP1.[44]

Estrogen Receptor Coactivation Proteins

Estrogen receptor coregulatory proteins can be classified according to their ability to impart either activator or repressor functions on the DNA-bound receptor protein. These coregulatory proteins activate or repress transcription by different mechanisms. The coactivators associate with ligand-bound and DNA-bound receptor, while most corepressors bind the unliganded, DNA-bound receptors (see Color Fig. 14.2A in separate color insert).

A family of p160 coactivator proteins for steroid hormone receptor proteins have been isolated and characterized. These include steroid receptor coactivator-1 (SRC1)[46,47], SRC2, glucocorticoid receptor interacting protein-1 (GRIP1),[48–50] and SRC3 or amplified in breast cancer-1 (AIB1).[50–53] Using transient transfection assays, the p160 family members have been shown to potentiate transcription of ERα severalfold. In addition, secondary coactivators associated with the p160 family members further potentiate transcriptional activity. For example, coactivator-associated arginine methyltransferase-1 (CARM1), a secondary coactivator protein that binds to the carboxyl-terminal region of p160 coactivators, enhances transcriptional activation by ER but only when GRIP1 or SRC1a is coexpressed.[54] CARM1 can methylate histone H3. Thus, coactivator-mediated methylation of proteins in the transcription machinery may also contribute to transcriptional regulation.

Therefore, ER functions as a ligand-activated enhancer protein for mRNA transcription of the selected target genes (see Color Fig. 14.2A in separate color insert). This mechanism for stimulating transcriptional activity involves direct interactions between the target DNA and DNA-bound receptor as well as between ER and coactivators and integrator proteins such as BRCA1[55] for the entire protein complex to interact with RNA polymerase and establish a transcription complex at the promoter (see Color Fig. 14.2A in separate color insert). In addition, ER stimulates gene expression through an indirect mechanism where ER does not bind directly to DNA but interacts with another DNA-bound transcription factor in a way that stabilizes the DNA binding of that transcription factor and/or recruits coactivators to the complex (see Color Fig. 14.2A in separate color insert).[30]

MECHANISMS FOR LOSS OF HORMONE RESPONSIVENESS

Lack of tamoxifen responsiveness, especially when acquired after an initial period of responsiveness, is probably due to changes in the ER signaling pathway or the regulation of this pathway. The possible changes in the ER signaling pathway include (1) reduction in amount of tamoxifen available to interact with ER, (2) loss of ERα expression, (3) change in ERα mRNA sequence through selection of mutant tumor cell clones or expression of alternatively spliced expression products, (4) change in the ERE, (5) alteration of ER interaction with coregulator proteins, (6) activation of ERα through a ligand-independent alternative pathway and crosstalk between growth factors/growth factor receptors and ER pathways, and (7) increased expression of ERβ.[56] We will briefly address each of these possibilities.

Reduction in Amount of Tamoxifen Available to Interact with Estrogen Receptor

The pharmacology of tamoxifen is complicated because it can have either antagonistic or agonistic effects. Agonistic effects have been observed when the intracellular tamoxifen level is low.[57] This could be related to dimers of tamoxifen-occupied ER showing inhibition of ER function but heterodimers of tamoxifen-occupied ER–estrogen-occupied ER showing transcriptional activation (see Color Fig. 14.2B in separate color insert). Reduced concentrations of intratumoral tamoxifen have been observed in women whose breast cancers acquired tamoxifen resistance but not in women whose breast cancers were resistant from diagnosis.[58,59] Some patients who initially responded to tamoxifen and subsequently relapsed with tamoxifen-resistant breast cancer showed a 10-fold or greater reduction in intratumoral tamoxifen levels.[58] This change could be due to reduced uptake by the tumor, increased tamoxifen metabolism by the tumor cells, increased extrusion of tamoxifen, or decreased availability to ER. Neither reduced uptake nor increased metabolism is supported by the available studies.[60] Although increased efflux due to a multidrug resistance type of membrane pump is attractive, there is little evidence to support this mechanism in breast cancer. Investigations of the multidrug resistance efflux pump P-glycoprotein in breast cancer show that it is not expressed at significant levels in breast cancer cells but that it is expressed in the mononuclear lymphoid cells present in the tumor mass; therefore, it probably does not play a role in reducing tamoxifen levels in breast cancer cells.[61] If an efflux pump is involved in lowering intratumoral tamoxifen, it is not likely to be P-glycoprotein. While reduced levels of intratumoral tamoxifen have been observed, this mechanism does not account for loss of hormone responsiveness in the majority of breast cancers.

Loss of Estrogen Receptor Expression

Binding of tamoxifen to ER is critical for its antiestrogenic effect in breast tumor cells. Therefore, loss of ER expression is another potential mechanism that could lead to resistance to tamoxifen therapy. In de novo antiestrogenic resistance, the dominant form is lack of ER/PR expression.[62] Of tumors with acquired antiestrogenic resistance, 50%–70% remain ER-positive after tamoxifen therapy failure.[63,64] Among tumors that lose ER gene expression, different mechanisms for the loss of expression can be hypothesized, e.g., deletion of the entire ER gene,[65] population remodeling through selective pressures under tamoxifen therapy,[62] transcriptional repression of ER gene expression as in downregulation or silencing of the ER gene by DNA methylation, and, finally, false-negative ER assay results.[66,67]

While deletion of the entire ER gene is possible, it seems to be a rare event and there are not enough data to assess the number of breast cancer cases with the ER gene entirely deleted. One study investigated the loss of chromosome arms 6q (with the ER gene) and 11q (with the PR gene) in 95 breast cancers and found that in tumors with 6q deletions ER positivity was half of that in tumors without them. The investigators' conclusion was that gene dosage might play a secondary, although still important, role in the loss of ER expression in some breast cancers.[65] Some investigators describe cellular heterogeneity for ER-positive tumors.[68–70] Selection of ER-negative tumor cells during tamoxifen therapy could explain the switch in ER status from ER-positive to ER-negative tumor cells. This hypothesis was supported by a model established for the T47D human breast cancer cell line in which tamoxifen-resistant cells continued growing while tamoxifen-sensitive cells diminished from the heterogenous cell line.[71]

A more likely explanation for the loss of ER expression is transcriptional repression of ER expression, as seen with DNA methylation. Evidence over the past 10 years indicates that DNA methylation in normal cells regulates the organization of the human genome into active and inactive regions of gene transcription. Two types of methylation are known: de novo methylation during embryogenesis and at certain points in the differentiation of adult cells[72,73] and maintenance methylation. DNA methylation consists of recognition and complementary methylation of CpG sites in the genome by DNA (cytosine-5) methyltransferase (DMT), which catalyzes cytosine methylation. The same enzyme, DMT, appears to be responsible for both de novo methylation and maintenance of an established methylation pattern.

The role of methylation in the development and progression of human cancers has been investigated. Abnormal patterns of DNA methylation are still not very well understood. To date, several alterations are found in tumors, which play a role in different concepts of tumor progression. These include loss of methylation at normally methylated sites,[74] increased activity of DMT,[69] and increased regional hypermethylation.[75–77] In breast cancer cells, DNA methylation seems to be of importance in the loss of ER expression through transcriptional repression, enabling cancer cells to develop resistance to tamoxifen therapy.

Several patterns of hypermethylation can be linked to lack of ER expression. In studies with several cultured ER-positive and ER-negative breast cancer cell lines, a lack of ER mRNA, a higher capacity to methylate DNA, and an extensive methylation of the CpG island in the 5′ promoter region of the ER gene have been found in ER-negative breast cancer cell lines.[78] Hypermethylation of ER CpG islands is detected in 25% of ER-negative breast cancer specimens and PR hypermethylation in 40% of PR-negative breast cancer specimens.[79] Methylation of the P0 and P1 promoter regions of the ER gene is inversely correlated with ER expression.[80] However, no association could be established between the ER status of breast tumors and methylation at these sites in an earlier study.[81] The lack or loss of ER expression is due at least in part to transcriptional repression of the ER gene and can be reversed by inhibition of methylation at the ER CpG islands through nucleoside and nucleotide analogues (AzaC and deoxyC).[82]

Abnormal methylation patterns in ER-negative cell lines are the result of overexpression of DMT.[78] In ER-negative breast cancer cell lines, DMT mRNA levels as well as protein levels are elevated compared to ER-positive cell lines. These DMT mRNA expression levels have been studied in ER-positive and ER-negative breast cancer cells by Northern hybridization, with 2- to 10-fold increased levels of DMT reported for ER-negative cell lines.[78]

Normally, DMT is expressed during the S phase of the cell cycle. There are differences in DMT expression between ER-positive and ER-negative breast cancer cells during the cell cycle. In estrogen receptor-positive breast cancer cells, there is a significant correlation between the S-phase fraction and DMT protein expression, which is not found in ER-negative cells. In ER-negative breast cancer cells, the DMT protein is expressed throughout the cell cycle.[83] However, in ER-positive cells, there is no significant expression outside of the S phase. The difference in expression levels is most striking in the G_1 phase of the cell cycle, with only 20% of ER-positive breast cancer cells staining positive for DMT and 80% of ER-negative cells staining positive.[83] Therefore, regulatory control of DMT expression in ER-negative breast cancer cells is altered.[83]

These changes in DMT are reversible. Studies of DMT inhibitors in cultured cells demonstrate reversal of ER suppression and reestablishment of hormone responsiveness. Cultured breast cancer cell lines with low to intermediate levels of ER expression (MCF-7, MCF-7/Adriamycin-resistant, MDA-435, and ZR75-1) when treated with 5,6-dihydro-5′-azacytidine, a cytosine DMT inhibitor, and then exposed to tamoxifen or estradiol show growth stimulation in response to estrogen and growth suppression in response to tamoxifen.[84]

The last possibility, a false-negative result in immunohistochemistry staining assays, has been described as one of the major dilemmas in ER testing. An overall false-negative rate of 30%–60% as well as a rate of only 37% of laboratories being able to detect low levels of ER staining are reported by the U.K. National External Quality Assessment Scheme for Immunohistochemistry.[67] A third of all laboratories fail to detect any ER staining in the weaker estrogen-expressing cases at all. This failure rate most likely contributes to the suggested "loss" of ER since there is no uniform validation of staining results.[66] Improvements of antibodies used for staining and better validation will ensure a more accurate assessment of ER status in the future.

In conclusion, loss of ER can be found in approximately half of breast cancers which were previously ER-positive. Therefore, DNA methylation is a potential source of transcriptional silencing of the ER gene but can account for, at best, only a small proportion of breast cancers that become unresponsive to tamoxifen. A substantial proportion of those recurrent or metastatic breast cancers interpreted as ER-negative might be due to inadequate assessment of ER since, as stated above, there is a lack of consistency in current immunohistochemical staining procedures and assessments.

Change in Estrogen Receptor Sequence through Selection of Mutant Clones or Expression of Alternatively Spliced Expression Products

Changes in the ER mRNA sequence and, therefore, ER amino acid sequence might cause loss of ligand binding by ER agonists as well as antiestrogens like tamoxifen through conformational changes at the receptor level. This loss of ligand recognition by the ER represents another model of tamoxifen resistance in ER-positive breast cancer clones.

Variants of ER can be either mutant forms or alternatively spliced expression products. Some investigators find different ER variants in breast cancers: dominant positive receptors, transcriptionally active in the absence of estrogen, and transcriptionally inactive, so-called dominant negative ER. The former alternative splice products contain deletion of exon 3, 5, or 7 in the ER gene. The latter variants without transcriptional activity inhibit normal ER action. Both types may play a role as prognostic factors in ER-positive breast cancers.[85,86] This finding is confirmed by a group demonstrating a variant ER mRNA missing the coding sequence for exon 5. This variant is capable of constitutively inducing the expression of an estrogen-responsive reporter gene in vitro.[87] Anchorage-dependent growth of breast cancer cells stably transfected with this splice variant is not inhibited by the antiestrogens tamoxifen and 4-hydroxytamoxifen.[88] Two hypotheses for the development and expression of this splice variant exist: during antiestrogenic therapy, either breast cancer cells undergo genetic alterations and develop resistance to tamoxifen while exposed or there is a selective advantage for preexisting splice variants which outgrow other clones through selection.

The observation of an ER exon 5 deletion splice variant was confirmed in another study, and a significantly higher expression of this variant is found in ER-negative but PR- and pS2-expressing tumors compared to other phenotypes. Also, a higher percentage of exon 5 deletion variant/wild-type ER is detected in some tamoxifen-resistant tumors, particularly in ER-positive, PR-positive/pS2-positive tumors, which demonstrate a constitutively active exon 5 deletion splice variant. This ER-positive, PR-positive and ER-positive, pS2-positive state in relapsing patients on tamoxifen therapy is suggested as a possible future marker for these constitutively active splice variants.[89] A wide variety of ER splice variants exist in ER-positive as well as ER-negative breast cancer cell lines.[90] The function of most of these variants remains unknown and is currently under investigation. Splice variants most likely play a minor role in tamoxifen resistance overall but might be responsible for resistance in some tumors.

Another study shows that only 1% of breast cancers have point mutations in the ER coding region and that the ER-negative breast cancer phenotype most likely is not caused by mutations in the ER gene. Lack of ER expression has been assigned to a deficiency in expression caused by other mechanisms instead.[91] Two of 20 (10%) tamoxifen-resistant breast cancers have ER mutations.[92] An ER mutant recognizing tamoxifen only as an agonist instead of an antagonist has been isolated from tamoxifen-stimulated MCF-7 cells, but similar mutations have not been found in breast cancer biopsies.[93] While some breast cancers might acquire tamoxifen resistance through point mutations in the coding region of the ER, this appears to be relatively infrequent.

Change in the Estrogen Response Element

The ER acts by binding to specific target gene sequences in response to ligand binding (described above). The ERE is located in the 5′-flanking region of promoters of estrogen-regulated genes. The activated ERα or ERβ dimer binds to the ERE in the promoter/enhancer region of responsive genes. Since this DNA-binding site is highly conserved, there is considerable homology among the EREs of different genes.[30] However, only a limited number of estrogen-responsive genes have been identified, and these show variations in the recognition sequences.[30]

The binding affinity of ERα and ERβ to EREs and subsequent transcriptional activity is reduced by most substitution mutations within the ERE.[30,94,95] It is possible that any one of a number of genes that are usually activated by estradiol and inhibited by antiestrogens could lose responsiveness to tamoxifen by a change in the ERE. It is also possible that a single base substitution in the ERE could result in enhanced transcriptional activity.[37] The ERE sequence might act as an allosteric effector, which could affect the interaction with coregulatory proteins through conformational changes of the ER and, therefore, alter transcriptional activity. Such an ERE sequence has been identified, which increases the agonistic acitivity of tamoxifen and nafoxidine from <5% to approximately 10% of the level observed with estradiol, suggesting that alterations of the ERE at the consensus sequence could lead to tamoxifen resistance.[96] This was accomplished using yeast cells expressing the human ER, transformed with a random oligonucleotide library in a vector with expression of a selectable marker requiring insertion of an upstream activating sequence. Sixty-five hormone-dependent clones containing activating sequences have been screened for a consensus septamer (5′-GGTCAMV-3′, M=7A or C and V= not T) related to the 13 bp palindromic consensus sequence of estrogen-regulated genes. The clone showing increased tamoxifen binding and agonistic activity in mammalian cells has a septamer sequence 5′-GGTCACC-3′ in comparison to the ERE consensus sequence described above.

Overall, alterations of ERE sequences that potentially lead to tamoxifen resistance may exist, but their frequency in breast cancer–related genes has not been established.

Alteration of Estrogen Receptor Interaction with Coregulator Proteins

The ER is activated by ligand binding, which leads to ER dimerization, binding of the activated dimer to the promoters of target genes, and subsequently regulated expression of those genes. Expression of target genes therefore depends on the characteristics of the ligand bound to the ER and on specific coregulatory proteins that interact with the promoter-bound receptor–ligand complex in the nucleus.[97,98] There are two separate groups of coregulatory proteins: coactivators and corepressors.

Coactivators enhance transcription of target

genes. Several coactivators are described by different investigators: the members of the p160 family [SRC1, SRC2 (GRIP1), and SRC3 (AIB1)], the cyclic adenosine monophosphate (cAMP) response element binding protein (CBP), as well as the CBP/p300-associated factor and CARM1. There are two known functions of these complexes, protein binding involved in transcriptional activity and chromatin remodeling. Initially, pure agonists like estradiol precipitate coactivator binding to ER to form a complex resulting in transcriptional activation. Tamoxifen is different from pure agonists like estradiol in its agonist–antagonist properties. A novel coactivator, L7/SPA, has been described for mixed antagonists like tamoxifen.[99] L7/SPA is specific for mixed antagonists and does not act on estrogen-bound ERα or progesterone-bound PR. It increases the agonist activity of mixed antagonists by 3- to 10-fold.[100] L7/SPA as well as nuclear corepressor proteins interact with the hinge region of ERα. This interaction increases the partial agonist activity of tamoxifen–ER-α, RU486–PR, and RU486–glucocorticoid receptor.[99] The agonist activity leads to stimulation of breast cancer cells by tamoxifen instead of inhibition of their growth and could explain resistance to therapy in some patients.

The coactivator AIB1 (or SRC3) is amplified and overexpressed in approximately 5% of breast cancers and 7% of ovarian cancers.[51,101] AIB1 is correlated with ER and PR positivity and larger tumor size, but its potential role in tamoxifen resistance is not characterized. These breast cancers may show increased sensitivity to estrogens.

The corepressors silencing mediator of retinoid and thyroid receptors (SMRT) and nuclear receptor-corepressor (N-CoR) bind to ER in the absence of ligand and silence transcription. Unlike coactivators that have intrinsic histone acetylase activity, one of the functions of corepressors is to recruit a complex of proteins having histone deacetylase activity, to repress gene expression by maintaining chromatin in a more condensed state.[37,99] The inhibitory potency of antiestrogens is enhanced by recruitment of other coregulators, including an ER-selective repressor of ER activity in addition to SMRT and N-CoR.[102–104] Corepressor expression is reduced in some tamoxifen-resistant breast cancers.[104]

Activation of Estrogen Receptor through a Ligand-Independent Alternative Pathway and Crosstalk between Growth Factors and Estrogen Receptor Pathways

The ER is activated in a ligand-dependent way by estrogens. Ligand-independent activation of ER has been suggested. However, the complex signaling pathways of estrogens and certain growth factors seem to be linked in a network of phosphorylation cascades (see Color Fig. 14.2D in separate color insert). Growth factors like epidermal growth factor (EGF), insulin-like growth factor (IGF), and transforming growth factor α (TGF-α) are target genes of the activated ER, each having an upstream ERE.[30,105] However, ER can also be activated by several growth factors or growth factor receptors without involvement of estrogen in a ligand-independent manner.[106] This dual regulation of target genes through ligands as well as other molecules has been known for some time, but it has been assumed to be a separate mechanism of ER activation. Nuclear receptor proteins like ER are phosphoproteins, which gain their transcriptional function through phosphorylation.[107,108] Only in recent years has it been recognized that both the ligand-dependent and -independent pathways lead to ER phosphorylation. So far, all factors activating ER in the absence of ligands seem to alter the activity of kinases or phosphatases.

Antiestrogens like tamoxifen act as ER-binding modulators of transcription. Resistance to tamoxifen among certain breast cancers could be explained by ligand-independent stimulation of the ER despite antiestrogenic therapy. This ligand-independent pathway could also enhance tamoxifen's partial agonist activity.[109]

The effects of estrogens on the expression of members of the IGF-I signaling pathways have been described in the past, but ligand-independent regulation of ERα by IGF-I has been reported in recent years.[110–112] Ligand-independent activation of the ER through stimulation with cAMP and IGF-I is accompanied by altered phosphorylation of ER. Estrogen-stimulated as well as cAMP- and IGF-I-stimulated ER phosphorylation are blocked by antiestrogens.[110] Stimulation of IGF-I in MCF-7 cells results in increased activation of the ERα protein but decreased expression of ERα due to a putative negative response element in the proximal promoter of the ERα gene.

In transformed breast cells, TGF-α may induce hyperplastic changes.[113,114] Its expression in breast cancer cells has been positively correlated with de novo lack of endocrine responsiveness in ER-positive breast cancers.[115] In ER-negative breast cancer cells stably transfected with ERα cDNA, 4-hydroxytamoxifen is a full agonist at the TGF-α gene and increased expression of TGF-α mRNA and protein is detected in a drug concentration-dependent manner.[116] Some investigators suggest that diminished TGF-α expression in patients might correlate with better response and longer duration of response during tamoxifen treatment, but no data have been published on this subject.

Several other growth factors and growth factor receptors have been investigated to evaluate their relationship with the ER. An inverse correlation between the EGF receptor (EGFR) and ER expression is observed in breast cancers. There is also an inverse relationship between EGFR expression and lack of responsiveness to endocrine therapy.[117] Treatment with EGF activates the ER both in vivo and in vitro, detected by increased markers for ER activation/phosphorylation and DNA synthesis. These effects can be reversed by the antiestrogen ICI 164,384.[118,119] Decreased levels of DNA synthesis and PR mRNA are detected with EGF treatment in ER knockout mice with functional EGFR.[120] This implicates a connection between the EGFR and ER signaling pathways. For EGF-mediated induction of ER transcription factors, the amino terminus of the ER has to be present.[121,122] The EGF–EGFR interaction activates nuclear factor κB (NF-κB). This activation is inhibited by an EGFR antibody. The NF-κB transactivation of cyclin D subsequently causes increased phosphorylation of the ER. This activation is inhibited by an antibody to the EGFR.[123]

HER-2/neu, another member of the EGFR tyrosine kinase family, is also inversely correlated with ER expression and is a marker for lack of responsiveness to tamoxifen treatment. In addition, HER-2/neu and EGFR seem to have an additive effect in reducing the likelihood of response to therapy.[124,125] A number of investigations have demonstrated an inverse correlation between HER-2/neu expression and expression of ER and/or PR.[126–137] Introduction of HER-2/neu overexpression into breast cancer cells results in the development of estrogen-independent growth, which is insensitive to both estrogen and the antiestrogen tamoxifen. The peptide growth factor heregulin leads to phosphorylation and activation of ER with expression of PR. Ligand-independent downregulation of ER expression by HER-2/neu overexpression indicates a direct link between these two pathways.[138]

By stimulation of EGFR and/or HER-2/neu, the ligand-dependent pathway can be circumvented and growth of ER-positive cells can still be stimulated through the ER. For these cancer cells, antiestrogen therapy with tamoxifen has a higher rate of failure than for growth factor receptor-negative cells.

However, TGF-β1 is a growth-inhibitory factor. The role of TGF-β1 in breast cancer progression is not resolved. Antiestrogens induce growth inhibition in cultured human breast cancer cells by upregulating the expression and/or secretion of TGF-β1.[139] In ER-positive MCF-7 cells, antiestrogen treatment induces TGF-β1 secretion and increases TGF-β1 protein expression levels 8- to 27-fold.[139] Cell growth inhibition by TGF-β1 alone is demonstrated in ER-negative MDA-MB-231 cells. These TGF-β1 inhibitory effects can be blocked in both ER-positive and ER-negative TGF-β1-treated cells by specific anti-TGF-β1 antibodies.[139] However, blockade of all TGF-β1 isoforms does not alter tamoxifen-induced cell cycle arrest or inhibition of MCF-7 and T47D colony growth and, therefore, does not suggest a connection between pathways of growth inhibition by antiestrogens and TGF-β1. In addition, transfection of the TGF-β1 receptor into MCF-7 cells does not enhance sensitivity to tamoxifen therapy.[140] In breast cancer specimens, TGF-β1 immunohistochemical staining is correlated with the rate of breast cancer progression.[141] Among these breast cancers, tamoxifen-resistant tumors express higher levels of TGF-β1 than tamoxifen-sensitive breast tumors.

Tumor progression in these resistant cancers might be mediated by interaction of TGF-β1 with other factors, e.g., increased peritumoral angiogenesis and inhibition of the immune response.[142] The connections between the ER and TGF-β1 pathways are not understood.

Protein kinase A (PKA) has been suggested as a link between the proliferative pathways of IGF-I and ERα. Activation of PKA signaling pathways causes increased phosphorylation of the carboxy-terminal portion of the ER.[143] Stimulation of PKA signaling pathways in MCF-7 cells enhances the agonist and decreases the antagonist properties of tamoxifen and related antiestrogens. In these MCF-7 transfection studies, increased intracellular cAMP levels and expression of PKA catalytic subunits lead to increased agonistic properties of the tamoxifen–ER complex, with transcriptional activity levels between 20% and 75% of those in the estradiol-liganded ER.[144] However, the antiestrogen ICI 164,384 does not show the same activity when applied under the same conditions, suggesting that only tamoxifen-like antiestrogens activate ER transcription in the presence of cAMP. Assessment of the regulatory subunit Iα of PKA showed that mRNA levels of this subunit tend to rise in tumors developing tamoxifen resistance during tamoxifen exposure. Of 32 patients, eight showed tamoxifen resistance and six developed increased Iα subunit levels of mRNA.[145]

New studies indicate that additional genes are involved in ligand-independent activation of ER. BRCA1, a gene altered in hereditary breast cancers, mediates ligand-independent repression of ER.[146] In Brca1-null mouse fibroblasts, ER shows ligand-independent transcriptional activity. Expression of wild-type BRCA1, but not mutated BRCA1, in those cells restores the ligand-independent repression of ER. In addition, BRCA1 is a ligand-reversible barrier of transcriptional activation.[146] This suggests a possible mechanism by which functional inactivation of BRCA1 could promote tumorigenesis through inappropriate hormonal regulation. A possible role of this BRCA1 alteration in tamoxifen resistance should be examined in the future.

In summary, there is a ligand-independent pathway of ER activation that involves phosphorylation of ER by protein kinases/phosphatases. Estrogen-independent cancers can still be growth-stimulated through ER action despite antiestrogen therapy, and antiestrogens like tamoxifen can even gain agonist activities through different molecules involved in this network.

Increased Expression of Estrogen Receptor β

Estrogen receptor β is significantly upregulated in some tamoxifen-resistant breast cancers compared to tamoxifen-sensitive breast cancers.[18] In a recent study, the mean ERβ mRNA expression level was higher in the breast cancer tissue from nine tamoxifen-resistant women compared to eight tamoxifen-sensitive women.[18] In vitro studies with tamoxifen-sensitive and tamoxifen-resistant MCF-7 human breast cancer cell lines show a similar difference in ERβ expression levels. Formation of ERα–ERβ heterodimers is of potential importance in the responsiveness to estrogens and antiestrogens. Ligand-bound receptor signaling through the conventional ERE and activating protein-1 (AP-1) sites can show opposite effects. Antiestrogen–ERβ complexes are agonists of AP-1-driven reporter genes, whereas no transcriptional activity is observed when the same ligand–receptor complex signals from a conventional ERE.[147] These studies suggest that ERβ expression may be important in the acquisition of a tamoxifen-resistant phenotype. However, ERβ expression in breast cancer without ERα expression is infrequent (Table 14.2), and clinical trials are needed to evaluate the potential role of ERβ in antiestrogen-resistant breast cancer.

CONCLUSION

The development of hormone-independent breast cancer, either at diagnosis or during antiestrogen therapy, is the result of any of a variety of different pathogenetic mechanisms in differ-

Table 14.2 Estrogen Receptor (ER) α and ERβ Content in Normal and Breast Cancer Tissue*

Receptor Content	Normal Tissue Incidence	Breast Cancer Incidence	p†
ERα^+ERβ^+	13%	50%	0.0002
ERα^+ERβ^-	13%	27%	0.249
ERα^-ERβ^+	22%	0%	0.0011
ERα^-ERβ^-	52%	23%	0.017

*Summary of analysis of 60 breast cancers and 23 normal breast tissues by RT-PCR.
†Tumor vs. normal tissue.
Source: Speirs et al.[17]

ent breast cancers. Determination of the range and frequency of those alterations will require detailed molecular analysis of a large series of hormone-independent breast cancers. Until such a study is performed, we can only speculate about the relative role of each alteration in the emerging, increasingly complex biochemical pathways that regulate estrogen action. The ER and its ligands function in a variety of cellular environments regulated by growth factors and protein kinases as well as the phosphorylation state of coregulator proteins. As these critical components become better defined, they provide the foundation for the development of novel SERMs with optimal profiles of selectivity.

REFERENCES

1. Beatson G. On the treatment of inoperable cases of carcinoma of the mamma: Suggestions for a new method of treatment, with illustrative cases. Lancet 2:104–107, 1896.
2. Huggins C, Bergenstal D. Inhibition of mammary and prostatic cancer by adrenalectomy. Cancer Res 12:134–141, 1952.
3. Harris HJ, Spratt JJ. Bilateral adrenalectomy in metastatic mammary cancer: an analysis of sixty-four cases. Cancer 23:145–151, 1969.
4. Jensen E, Block G, Smith S, Kyser K, DeSombre E. Estrogen receptors and breast cancer response to adrenalectomy. Natl Cancer Inst Monogr 34:55–70, 1971.
5. Engelsman E, Persijn J, Korsten C, Cleton F. Oestrogen receptors in human breast cancer tissue and response to endocrine therapy. BMJ 2:750–752, 1973.
6. Leung B, Fletcher W, Lindell T, Wood D, Krippaehne W. Predictability of response to endocrine ablation in advanced breast carcinoma. Arch Surg 106:515–519, 1973.
7. Maass H, Engel B, Hohmeister H, Lehmann F, Trams G. Estrogen receptors in human breast cancer tissue. Am J Obstet Gynecol 113:377–382, 1972.
8. Savlov E, Wittliff J, Hilf R, Hall T. Correlations between certain biochemical properties of breast cancer and response to therapy: a preliminary report. Cancer 33:303–309, 1974.
9. Greene G, Nolan C, Engler J, Jensen E. Monoclonal antibodies to human estrogen receptor. Proc Natl Acad Sci USA 77:5115–5119, 1980.
10. Greene G, Fitch F, Jensen E. Monoclonal antibodies to estrophilin: probes for the study of estrogen receptors. Proc Natl Acad Sci USA 77:157–161, 1980.
11. Press M, Greene G. An immunocytochemical method for demonstrating estrogen receptor in human uterus using monoclonal antibodies to human estrophilin. Lab Invest 50:480–486, 1984.
12. King W, Greene G. Monoclonal antibodies localize oestrogen receptor in the nuclei of target cells. Nature 307:745–747, 1984.
13. Kuiper G, Enmark E, Pelto-Huikko M, et al. Cloning of a novel estrogen receptor expressed in rat prostate and ovary. Proc Natl Acad Sci USA 93:5925–5930, 1996.
14. Mosselman S, Polman J, Dijkema R. ERβ: identification and characterization of a novel human estrogen receptor. FEBS Lett 392:49–53, 1996.
15. Kuiper G, Carlsson B, Grandien K, Enmark E, Haggblad J, Nilsson S, Gustafsson J. Comparison of the ligand binding specificity and transcript tissue distribution of estrogen receptors alpha and beta. Endocrinology 138:863–870, 1997.
16. Dotzlaw H, Leygue E, Watson P, Murphy L. Expression of estrogen receptor-β in human breast tumors. J Clin Endocrinol Metab 82:2371–2374, 1997.
17. Speirs V, Parkes A, Kerin M, Walton D, Carleton P, Fox J, Atkin S. Coexpression of estrogen receptor alpha and beta: poor prognostic factors in human breast cancer? Cancer Res 59:525–528, 1999.
18. Speirs V, Malone C, Walton D, Kerin M, Atkin S. Increased expression of estrogen receptor β mRNA in tamoxifen-resistant breast cancer patients. Cancer Res 59:5421–5424, 2001.
19. Lazennec G, Bresson D, Lucas A, Chauveau C, Vignon F. ERbeta inhibits proliferation and invasion of breast cancer cells. Endocrinology. 142:4120–4130, 2001.
20. Cullen R, Maguire T, McDermott E, Hill A, O'Higgins N, Duffy M. Studies on oestrogen receptor-alpha and -beta mRNA in breast cancer. Eur J Cancer 37:1118–1122, 2001.
21. Horwitz K, McGuire W, Pearson O, Segaloff A. Predicting response to endocrine therapy in human breast cancer: a hypothesis. Science 189:726–727, 1975.
22. McDonnell D, Clemm D, Hermann T, Goldman M, Pike J. Analysis of estrogen receptor function in vitro reveals three distinct classes of antiestrogens. Mol Endocrinol 9:659–669, 1995.
23. Takimoto G, Graham J, Jackson T, Tung L, Powell R, Horwitz L, Horwitz K. Tamoxifen resistant breast cancer: coregulators determine the direction of transcription by antagonist-occupied steroid receptors. J Steroid Biochem Mol Biol 69:45–50, 1999.
24. Curtis R, Boice J, Shriner D, Hankey B, Fraumeni J. Second cancers after adjuvant tamoxifen therapy for breast cancer. J Natl Cancer Inst 88:832, 1996.
25. Jensen E, Suzuki T, Kawashima T, Stumpf W, Jungblut P, DeSombre E. A two-step mechanism for the interaction of estradiol with rat

uterus. Proc Natl Acad Sci USA 59:632–636, 1968.
26. Press M, Xu S.-H, Wang J.-D, Greene G. Subcellular distribution of estrogen receptor and progesterone receptor with and without specific ligand. Am J Pathol 135:857–864, 1989.
27. Welshons W, Lieberman M, Gorski J. Nuclear localization of unoccupied oestrogen receptors. Nature 307:747, 1984.
28. Arnold S, Vorojeikina D, Notides A. Phosphorylation of tyrosine 537 on the human estrogen receptor is required for binding to an estrogen response element. J Biol Chem 270:30205–30212, 1995.
29. Arnold S, Obourn J, Yudt M, Carter T, Notides A. In vivo and in vitro phosphorylation of the human estrogen receptor. J Steroid Biochem Mol Biol 52:159–171, 1995.
30. Klinge C. Estrogen receptor interaction with estrogen response elements. Nucleic Acids Res 29:2905–2919, 2001.
31. Guiochon-Mantel A, Lescop P, Chgristin-Maitre S, Losfelt H, Perrot-Applanat M, Milgrom E. Nucleocytoplasmic shuttling of the progesterone receptor. EMBO J 10:3851–3859, 1991.
32. Greene G, Gilna P, Waterfield M, Baker A, Hort Y, Shine J. Sequence and expression of human estrogen receptor complementary DNA. Science 231:1150–1154, 1986.
33. Green S, Walter P, Kumar V, Krust A, Bornert J.-M, Argos P, Chambon P. Human oestrogen receptor cDNA: sequence, expression and homology to v-*erb*-A. Nature 320:134–139, 1986.
34. Green S, Chambon P. Oestradiol induction of a glucocorticoid-responsive gene by a chimaeric receptor. Nature 325:75–78, 1987.
35. Kumar V, Green S, Staub A, Chambon P. Localisation of the oestradiol-binding and putative DNA-binding domains of the human oestrogen receptor. EMBO J 5:2231–2236, 1986.
36. Kumar V, Green S, Stack G, Berry M, Jin J, Chambon P. Functional domains of the human estrogen receptor. Cell 51:941, 1987.
37. Klinge C. Estrogen receptor interaction with coactivators and co-repressors. Steroids 65:227–251, 2000.
38. Paech K, Webb P, Kuiper G, Nilsson S, Gustafsson J, Kushner P, Scanlan T. Differential ligand activation of estrogen receptors ERalpha and ERbeta at AP1 sites. Science 277:1508–1510, 1997.
39. Enmark E, Pelto-Huikko M, Grandien K, Lagercrantz S, Lagercrantz J, Fried G, Nordenskjöld M, Gustaffson J. Human estrogen receptor β-gene structure, chromosomal localization, and expression pattern. J Clin Endocrinol Metab 82:4258–4265, 1997.
40. Ponglikitmongkol M, Green S, Chambon P. Genomic organization of the human oestrogen receptor gene. EMBO J 7:3385–3388, 1988.
41. Wahli W, Martinez E. Superfamily of steroid nuclear receptors: positive and negative regulators of gene expression. FASEB J 5:2243–2249, 1991.
42. Evans R. The steroid and thyroid hormone receptor superfamily. Science 240:889–895, 1988.
43. Brzozowski A, Pike A, Dauter Z, Hubbard R, Bonn T, Engstrom O, Ohman L, Greene G, Gustafsson J, Carlquist M. Molecular basis of agonism and antagonism in the oestrogen receptor. Nature 389:753–758, 1997.
44. Shiau A, Barstad D, Loria P, Cheng L, Kushner P, Agard D, Greene G. The structural basis of estrogen receptor/coactivator recognition and the antagonism of this interaction by tamoxifen. Cell 95:927–937, 1998.
45. Henttu P, Kalkhoven E, Parker M. AF-2 activity and recruitment of steroid receptor coactivator 1 to the estrogen receptor depend on a lysine residue conserved in nuclear receptors. Mol Cell Biol 17:1832–1839, 1997.
46. Kamei Y, Xu L, Heinzel T, Torchia J, Kurokawa R, Gloss B, Lin SHA, Heyman RA, Rose D, Glass C, Rosenfeld M. A CBP integrator complex mediates transcriptional activation and AP-1 inhibition by nuclear receptors. Cell 85: 403–414, 1996.
47. Onate S, Tsai S, Tsai M.-J, O'Malley B. Sequence and characterization of a coactivator for the steroid hormone receptor family. Science 270:1354–1357, 1995.
48. Hong H, Kohli K, Trivedi A, Johnson D, Stallcup M. GRIP1, a novel mouse protein that serves as a transcriptional coactivator in yeast for the hormone binding domains of steroid receptors, Proc Natl Acad Sci USA 93:4948–4952, 1996.
49. Voegel J, Heine M, Zechel C, Chambon P, Gronemeyer H. TIFS, a 160 kDa transcriptional mediator for the ligand-dependent activation function AF-2 of nuclear receptors. EMBO J 15:3667–3675, 1996.
50. Torchia J, Rose D, Inostroza J, Kamel Y, Westin S, Glass C, Rosenfeld M. The transcriptional coactivator p/CIP binds CBP and mediates nuclear-receptor function. Nature 387:677–684, 1997.
51. Anzick S, Kononen J, Walker R, Azorsa D, Tanner M, Guan X-Y, Sauter G, Kallioniemi O-P, Trent J, Meltzer P. AIB1, a steroid receptor coactivator amplified in breast and ovarian cancer. Science 277:965–968, 1997.
52. Chen H, Lin R, Schiltz R, Chakravarti D, Nash A, Nagy L, Privalsky M, Nakatani Y, Evans R. Nuclear receptor coactivator ACTR is a novel histone acetyltransferase and forms a multimeric activation complex with P/CAF and CBP/p300. Cell 90:569–580, 1997.
53. Li H, Gomes P, Chen J. RAC3, a steroid/nuclear receptor–associated coactivator that is related to SRC-1 and TIF2. Proc Natl Acad Sci USA 94:8479–8484, 1997.

54. Chen D, Ma H, Hong H, Koh S, Huang S, Schurter B, Aswad D, Stallcup M. Regulation of transcription by a protein methyltransferase. Science 284:2174–2177, 1999.
55. Park J, Irvine R, Buchanan G, Koh S, Park J, Tilley W, Stallcup M, Press M, Coetzee G. BRCA1 is a coactivator of the androgen receptor. Cancer Res 60:5946–5949, 2000.
56. Lykkesfeldt A. Mechanisms of tamoxifen resistance in the treatment of advanced breast cancer. Acta Oncol 35(Suppl 5):9–14, 1996.
57. Horwitz K, Koseki Y, McGuire W. Estrogen control of progesterone receptor in human breast cancer: role of estradiol and antiestrogen. Endocrinology 102:1742–1745, 1978.
58. Johnston S, Haynes B, Smith I, Jarman M, Sacks N, Ebbs S, Dowsett M. Acquired tamoxifen resistance in human breast cancer and reduced intratumoural drug concentration. Lancet 342:1521–1522, 1993.
59. Osborne C, Wiebe V, McGuire W, Ciocca D, DeGregorio M. Tamoxifen and the isomers of 4-hydroxytamoxifen in tamoxifen-resistant tumors from breast cancer patients. J Clin Oncol 10:304–310, 1992.
60. Dowsett M. Endocrine resistance in advanced breast cancer. Acta Oncol 35(Suppl 5):91–95, 1996.
61. Yang X, Uziely B, Groshen S, Lukas J, Israel V, Russell C, Dunnington G, Formenti S, Muggia F, Press M. MDR1 gene expression in primary and advanced breast cancer. Lab Invest 79:271–280, 1999.
62. Clarke R, Skaar T, Bouker K, Davis N, Lee Y, Welch J, Leonessa F. Molecular and pharmacological aspects of antiestrogen resistance. J Steroid Biochem Mol Biol 76:71–84, 2001.
63. Encarnacion C, Ciocca D, McGuire W, Clark G, Fuqua S, Osborne C. Measurement of steroid hormone receptors in breast cancer patients on tamoxifen. Breast Cancer Res Treat 26:237–246, 1993.
64. Johnston S, Saccanti-Jotti G, Smith I, Newby J, Dowsett M. Change in oestrogen receptor expression and function in tamoxifen-resistant breast cancer. Endocr Relat Cancer 2:105–110, 1995.
65. Magdelenat H, Gerbault-Seureau M, Dutrillaux B. Relationship between loss of estrogen and progesterone receptor expression and 6q and 11q chromosome arms in breast cancer. Int J Cancer 57:63–66, 1994.
66. McCann J. Better assays needed for hormone receptor status, experts say. J Natl Cancer Inst 93:579–580, 2001.
67. Rhodes A, Jasani B, Barnes D, Brobow L, Miller K. Reliability of immunohistochemical demonstration of oestrogen receptors in routine practice: interlaboratory variance in the sensitivity of detection and evaluation of scoring systems. J Clin Pathol 53:125–130, 2000.
68. Walker K, McClelland R, Candlish W, Blamey R, Nicholson R. Heterogeneity of oestrogen receptor expression in normal and malignant breast tissue. Eur J Cancer 28:34–37, 1992.
69. Van Netten J, Algard F, Coy P. Heterogenous estrogen receptor levels detected via multiple microsamples from individual breast tumors. Cancer 56:2019–2024, 1985.
70. Van Netten J, Armstrong J, Carlyle S, Goodchild N, Thornton I, Brigden M, Coy P, Fletcher C. Estrogen receptor distribution in the peripheral, intermediate and central regions of breast cancers. Eur J Cancer Clin Oncol 24:1885–1889, 1988.
71. Graham M, Smith J, Jewett P, Horwitz K. Heterogeneity of progesterone receptor content and remodelling by tamoxifen characterize subpopulations of cultured human breast cancer cells: analysis by quantitative dual parameter flow cytometry. Cancer Res 52:593–602, 1992.
72. Razin A, Cedar H. DNA Methylation: Molecular Biology and Biological Significance. Basel: Birkhauser-Verlag, 1993, pp 341–357.
73. Razin A, Shemer R. DNA methylation in early development. Hum Mol Genet 4:1751–1755, 1995.
74. Feinberg A, Vogelstein B. Hypomethylation distinguishes genes of some human cancers from their normal counterparts. Nature 301:89–92, 1983.
75. Baylin S, Makos M, Wu J, Yen R, deBustros A, Vertino P, Nelkin B. Abnormal patterns of DNA methylation in human neoplasia: potential consequences for tumor progression. Cancer Cells 3:383–390, 1991.
76. Jones P. DNA methylation errors and cancer. Cancer Res 56:2463–2467, 1996.
77. Laird P, Jaenisch R. DNA methylation and cancer. Hum Mol Genet 3:1487–1495, 1994.
78. Ottaviano Y, Issa J-P, Parl F, Smith H, Baylin S, Davidson N. Methylation of the estrogen receptor gene CpG island marks loss of estrogen receptor expression in human breast cancer cells. Cancer Res 54:2552–2555, 1994.
79. Lapidus R, Ferguson A, Ottaviano Y, Parl F, Smith H, Wetizman S, Baylin S, Issa J-P, Davidson N. Methylation of estrogen and progesterone receptor gene 5′ CpG islands correlates with lack of estrogen and progesterone receptor gene expression in breast tumors. Clin Cancer Res 2:805–810, 1996.
80. Iwase H, Omoto Y, Iwata H, Toyama T, Hara Y, Ando Y, Ito Y, Fujii Y, Kobayashi S. DNA methylation analysis at distal and proximal promoter regions of the oestrogen receptor gene in breast cancers. Br J Cancer 80:1982–1986, 1999.
81. Falette N, Fuqua S, Chamness G, Cheah M, Greene G, McGuire W. Estrogen receptor gene methylation in human breast tumors. Cancer Res 50:3974–3978, 1990.
82. Ferguson A, Lapidus R, Baylin S, Davidson N.

Demethylation of the estrogen receptor gene in estrogen receptor-negative breast cancer cells can reactivate estrogen receptor gene expression. Cancer Res 55:2279–2283, 1995.
83. Nass S, Ferguson A, El-Ashry D, Nelson W, Davidson N. Expression of DNA methyl-transferase (DMT) and the cell cycle in human breast cancer cells. Oncogene 18:7453–7461, 1999.
84. Izbicka E, Davidson K, Lawrence R, MacDonald J, Von Hoff D. 5,6-Dihydro-5′-azacytidine (DHAC) affects estrogen sensitivity in estrogen-refractory human breast carcinoma cell lines. Anticancer Res 19:1293–1298, 1999.
85. McGuire W, Chamness G, Fuqua S. Estrogen receptor variants in clinical breast cancer. Mol Endocrinol 5:1571–1577, 1991.
86. Sluyser M. Role of estrogen receptor variants in the development of hormone resistance in breast cancer. Clin Biochem 25:405–414, 1992.
87. Fuqua S, Fitzgerald S, Chamness G, Tandon A, McDonnell D, Nawaz Z, O'Malley B, McGuire W. Variant human breast tumor estrogen receptor with constitutive transcriptional activity. Cancer Res 51:105–109, 1991.
88. Fuqua S, Wolf D. Molecular aspects of estrogen receptor variants in breast cancer. Breast Cancer Res Treat 35:233–241, 1995.
89. Daffada A, Johnston S, Smith I, Detre S, King N, Dowsett M. Exon 5 deletion variant estrogen receptor messenger RNA expression in relation to tamoxifen resistance and progesterone receptor/pS2 status in human breast cancer. Cancer Res 55:288–293, 1995.
90. Poola I, Koduri S, Chatra S, Clarke R. Identification of twenty alternatively spliced estrogen receptor alpha mRNAs in breast cancer cell lines and tumors using splice targeted primer approach. J Steroid Biochem Mol Biol 72:249–258, 2000.
91. Roodi N, Bailey L, Kao W, Verrier C, Yee C, Dupont W, Parl F. Estrogen receptor gene analysis in estrogen receptor-positive and receptor-negative primary breast cancer. J Natl Cancer Inst 87:446–451, 1995.
92. Karnik P, Kulkarni S, Liu X, Budd G, Bukowski R. Estrogen receptor mutations in tamoxifen-resistant breast cancer. Cancer Res 54:349–353, 1994.
93. Jiang S, Langan-Fahey S, Stella A, McCague R, Jordan V. Point mutation of estrogen receptor (ER) in the ligand-binding domain changes the pharmacology of antiestrogens in ER-negative breast cancer cells stably expressing complementary DNAs for ER. Mol Endocrinol 6:2167–2174, 1992.
94. Klinge C. Role of estrogen receptor ligand and estrogen response element sequence on interaction with chicken ovalbumin upstream promoter transcription factor (COUP-TF). J Steroid Biochem Mol Biol 71:1–19, 1999.
95. Driscoll M, Sathya G, Muyan M, Klinge C, Hilf R, Bambara R. Sequence requirements for estrogen receptor binding to estrogen response elements. J Biol Chem 273:29321–29330, 1998.
96. Dana S, Hoener P, Wheeler D, Lawrence C, McDonnell D. Novel estrogen response elements identified by genetic selection in yeast are differentially responsive to estrogens and antiestrogens in mammalian cells. Mol Endocrinol 8:1193–1207, 1994.
97. Glass C, Rose D, Rosenfeld M. Nuclear receptor coactivators. Curr Opin Cell Biol 9:222–232, 1997.
98. Horwitz K, Jackson T, Bain D, Richer J, Takimoto G, Tung L. Nuclear receptor coactivators and corepressors. Mol Endocrinol 10:1167–1177, 1996.
99. Jackson T, Richer J, Bain D, Takimoto G, Tung L, Horwitz K. The partial agonist activity of antagonist-occupied steroid receptors is controlled by a novel hinge domain-binding coactivator L7/SPA and the corepressors N-CoR or SMRT. Mol Endocrinol 11:693–705, 1997.
100. Graham J, Bain D, Richer J, Jackson T, Tung L, Horwitz K. Nuclear receptor conformation, coregulators, and tamoxifen-resistant breast cancer. Steroids 65:579–584, 2000.
101. Bautista S, Valles H, Walker R, et al. In breast cancer, amplification of the steroid receptor coactivator gene AIB1 is correlated with estrogen and progesterone receptor positivity. Clin Cancer Res 4:2925–2929, 1998.
102. Montano M, Ekena K, Delage-Mourroux R, Chang W, Martini P, Katzenellenbogen B. An estrogen receptor-selective coregulator that potentiates the effectiveness of antiestrogens and represses the activity of estrogens. Proc Natl Acad Sci USA 96:6947–6952, 1999.
103. Smith C, Nawaz Z, O'Malley B. Coactivator and corepressor regulation of the agonist/antagonist activity of the mixed antiestrogen, 4-hydroxytamoxifen. Mol Endocrinol 11:657–666, 1997.
104. Lavinsky RJ, Jepsen K, Heinzel T, Torchia J, Mullen T-M, Schiff R, Del-Rio A, Ricote M, Ngo S, Gemsch J, Hilsenbeck S, Osborne C, Glass C, Rosenfel M, Rose D. Diverse signaling pathways modulate nuclear receptor recruitment of N-CoR and SMRT complexes. Proc Natl Acad Sci USA 95:2920–2925, 1998.
105. Westley B, May, F. Role of insulin-like growth factors in steroid-modulated proliferation. J Steroid Biochem 51:1–9, 1994.
106. Chalbos D, Philips A, Rochefort H. Genomic cross-talk between the estrogen receptor and growth factor regulatory pathways in estrogen target tissues. Semin Cancer Biol 5:361–368, 1994.
107. Ali S, Metzger D, Bornert J-M, Chambon P. Modulation of transcriptional activation by ligand-dependent phosphorylation of the human oestrogen receptor A/B region. EMBO J 12:1153–1160, 1993.

108. Weigel N. Steroid hormone receptors and their regulation by phosphorylation. Biochem J 319:657–667, 1996.
109. Katzenellenbogen B, Montano M, Le Groff P. Antiestrogens: mechanisms and actions in target cells. J Steroid Biochem Molec Biol 53:387–393, 1995.
110. Aronica S, Katzenellenbogen B. Stimulation of estrogen receptor-mediated transcription and alteration in the phosphorylation state of the rat uterine estrogen receptor by estrogen, cyclic adenosine monophosphate, and insulin-like growth factor-1. Mol Endocrinol 7:743–752, 1993.
111. Smith C. Cross-talk between peptide growth factor and estrogen receptor signaling pathways. Biol Reprod 58:627–632, 1998.
112. Stoica A, Saceda M, Fakhro A, Joyner M, Martin M. Role of insulin-like growth factor-I in regulating estrogen receptor-alpha gene expression. J Cell Biochem 76:605–614, 2000.
113. Sandgren E, Luetteke N, Palmiter R, Brinster R, Lee D. Overexpression of TGFalpha in transgenic mice: induction of epithelial hyperplasia, pancreatic metaplasia, and carcinoma of the breast. Cell 61:1121–1135, 1990.
114. Matsui Y, Halter S, Holt J, Hogan B, Coffrey R. Development of mammary hyperplasia and neoplasia in MMTV-TGF alpha transgenic mice. Cell 61:1147–1155, 1990.
115. Nicholson R, McClelland R, Gee J, DL, M, Cannon P, Robertson J, Ellis I, Blamey R. Transforming growth factor-alpha and endocrine sensitivity in breast cancer. Cancer Res 54:1684–1689, 1994.
116. Levenson A, Tonetti D, Jordan V. The oestrogen-like effect of 4-hydroxytamoxifen on induction of transforming growth factor alpha mRNA in MDA-MB-231 breast cancer cells stably expressing the oestrogen receptor. Br J Cancer 77:1812–1819, 1998.
117. Klijn J, Berns P, Schmitz P, Foekens J. The clinical significance of epidermal growth factor receptor (EGF-R) in human breast cancer: a review on 5232 patients. Endocr Rev 13:3–17, 1992.
118. Ignar-Trowbridge D, Nelson K, Bidwell M, Curtis S, Washburn T, Machlachlan J, Korach K. Coupling of dual signaling pathways: epidermal growth factor action involves the estrogen receptor. Proc Natl Acad Sci USA 89:4658–4662, 1992.
119. Ignar-Trowbridge D, Teng C, Ross K, Parker M, Korach K, McLachlan J. Peptide growth factors elicit estrogen receptor-dependent transcriptional activation of an estrogen responsive element. Mol Endicrinol 7:992–998, 1993.
120. Curtis S, Washburn T, Sewall C, DiAugustine R, Lindzey J, Couse J, Korach K. Physiological coupling of growth factor and steroid receptor signaling pathways: estrogen receptor knockout mice lack estrogen-like response to epidermal growth factor. Proc Natl Acad Sci USA 93:12626–12630, 1996.
121. Bunone G, Briand P-A, Miksicek R, Picard D. Activation of the unliganded estrogen receptor by EGF involves the MAP kinase pathway and direct phosphorylation. EMBO J 15:2174–2183, 1996.
122. Ignar-Trowbridge D, Pimentel M, Parker M, McLachlan J, Korach K. Peptide growth factor cross-talk with the estrogen receptor requires the A/B domain and occurs independently of protein kinase C or estradiol. Endocrinology 137:1735–1744, 1996.
123. Biswas D, Cruz A, Gansberger E, Pardee A. Epidermal growth factor–induced nuclear factor κB activation: a major pathway of cell-cycle progression in estrogen-receptor negative breast cancer cells. Proc Natl Acad Sci USA 97:8542–8547, 2000.
124. Carlomagno C, Perrone F, Gallo C, DeLaurentiis M, Lauria R, Morabito A, Pettinato G, Panico L, D'Antonoi A, Bianco A, DePlacido S. c-erbB2 overexpression decreases the benefit of adjuvant tamoxifen in early-stage breast cancer without axillary lymph node metastases. J Clin Oncol 14:2702–2708, 1996.
125. Wright C, Nicholson S, Angus B, Sainsbury J, Farndon J, Cairns J, Harris A, Horne C. Relationship between c-erbB-2 protein product expression and response to endocrine therapy in advanced breast cancer. Br J Cancer 65:118–121, 1992.
126. Anbazhagan R, Gelber R, Bettelheim R, Goldhirsch K, Gusterson B. Association of c-erbB-2 expression and S-phase fraction in the prognosis of node positive breast cancer. Ann Oncol 2:47–53, 1991.
127. Bacus S, Chin D, Yarden Y, Zelnick C, Stern D. Type 1 receptor tyrosine kinases are differentially phosphorylated in mammary carcinoma and differentially associated with steroid receptors. Am J Pathol 148:549–558, 1996.
128. Ciocca D, Fujimura F, Tandon A, Clark G, Mark C, Lee-Chen G, Pounds G, Vendely P, Owens M, Pandian M. Correlation of HER-2/neu amplification with expression and with other prognostic factors in 1103 breast cancers. J Natl Cancer Inst 84:1279–1282, 1992.
129. De Potter C, Beghin C, Makar A, Vandekerckhove D, Roels H. The neu-oncogene protein as a predictive factor for haematogenous metastases in breast cancer patients. Int J Cancer 45:55–58, 1990.
130. DiGiovanna M, Carter D, Flynn S, Stern D. Functional assay for HER-2/neu demonstrates active signalling in a minority of HER-2/neu-overexpressing invasive human breast tumours. Br J Cancer 74:802–806, 1996.
131. Gusterson B, Gelber R, Goldhirsch A, Price K, Save-Soderborgh J, Anbazhagan R, Styles J,

Rudenstam C, Golouh R, Reed R. Prognostic importance of c-erbB-2 expression in breast cancer. J Clin Oncol 10:1034–1036, 1992.

132. Hartmann L, Ingle J, Wold L, Farr GJ, Grill J, Su J, Maihle N, Krook J, Witzig T, Roche P. Prognostic value of c-erbB2 overexpression in axillary lymph node positive breast cancer. Results from a randomized adjuvant treatment protocol. Cancer 74:2956–2963, 1994.

133. Heintz N, Leslie K, Rogers L, Howard P. Amplification of the c-erbB-2 oncogene and prognosis of breast adenocarcinoma. Arch Pathol Lab Med 114:160–163, 1990.

134. Kallioniemi O, Holli K, Visakorpi T, Koivula T, Helin H, Isola J. Association of c-erbB-2 protein over-expression with high rate of cell proliferation, increased risk of visceral metastasis and poor long-term survival in breast cancer. Int J Cancer 49:650–655, 1991.

135. Quenel N, Wafflart J, Bonichon F, de Mascarel I, Trojani M, Durand M, Avril A, Coindre J. The prognostic value of c-erbB2 in primary breast carcinomas: a study on 942 cases. Breast Cancer Res Treat 35:283–91, 1995.

136. Tetu B, Brisson J. Prognostic significance of HER-2/neu oncoprotein expression in node-positive breast cancer. The influence of the pattern of immunostaining and adjuvant therapy. Cancer 73:2359–2365, 1994.

137. Wright C, Angus B, Nicholson S, Sainsbury J, Cairns J, Gullick W, Kelly P, Harris A, Horne C. Expression of c-erbB-2 oncoprotein: a prognostic indicator in human breast cancer. Cancer Res 49:2087–2090, 1989.

138. Pietras R, Arboleda J, Reese D, Wongvipat N, Pegram M, Ramos L, Gorman C, Parker M, Sliwkowski M, Slamon D. Her-2 tyrosine kinase pathway targets estrogen receptor and promotes hormone-independent growth in human breast cancer cells. Oncogene 10:2435–2446, 1995.

139. Knabbe C, Lippman M, Wakefield L, Flanders K, Derynck R, Dickson R. Evidence that transforming growth factor-beta is a hormonally regulated negative growth factor in human breast cancer cells. Cell 48:417–428, 1987.

140. Koli K, Ramsey T, Ko Y, Dugger T, Brattain M, Arteaga C. Blockade of transforming growth factor-β signaling does not abrogate antiestrogen-induced growth inhibition of human breast carcinoma cells. J Biol Chem 272:8296–8302, 1997.

141. Gorsch S, Memoli V, Stukel T, Gold L, Arrick B. Immunohistochemical staining for transforming growth factor beta 1 associates with disease progression in human breast cancer. Cancer Res 52:6949–6952, 1992.

142. Koli K, Arteaga C. Complex role of tumor cell TGF-betas on breast carcinoma progression. J Mammary Gland Biol Neopl 1:373–380, 1996.

143. LeGoff P, Montano M, Schodin D, Katzenellenbogen B. Phosphorylation of the human estrogen receptor. Identification of hormone-regulated sites and examination of their influence on transcriptional activity. J Biol Chem 269:4458–4466, 1994.

144. Fujimoto N, Katzenellenbogen B. Alteration in the agonist/antagonist balance of antiestrogens by activation of protein kinase A signaling pathways in breast cancer cells: antiestrogen selectivity and promoter dependence. Mol Endocrinol 8:296–304, 1994.

145. Miller W, Hulme M, Bartlett J, MacCallum J, Dixon J. Changes in messenger RNA expression of protein kinase A regulatory subunit alpha in breast cancer patients treated with tamoxifen. Clin Cancer Res 3:2399–2404, 1997.

146. Zheng L, Annab L, Afshari C, Lee W-H, Boyer T. BRCA1 mediates ligand-independent transcriptional repression of the estrogen receptor. Proc Natl Acad Sci USA 98:9587–9592, 2001.

147. Tremblay G, Tremblay A, Copeland N, Gilbert D, Jenkins N, Labrie F, Giguere V. Cloning, chromosomal localization, and functional analysis of the murine estrogen receptor beta. Mol Endocrinol 11:353–365, 1997.

148. Pichon M, Broet P, Magdelenat H, Delarue J, Spyratos F, Basuyau J, Saez S, Rallet A, Courriere P, Millon R, Asselain B. Prognostic value of steroid receptors after long-term follow-up of 2257 operable breast cancers. Br J Cancer 75:1545–1551, 1997.

149. Osborne C, Yochmowitz M, Knight W III, McGuire W. The value of estrogen and progesterone receptors in the treatment of breast cancer. Cancer 46:2884–2888, 1980.

150. Thorpe S, Rose C, Rasmussen B, Mouridsen H, Bayer T, Keiding N. Prognostic value of steroid hormone receptors: multivariate analysis of systemically untreated patients with node negative primary breast cancer. Cancer Res 47:6125–6133, 1987.

151. Young P, Ehrlich C, Einhorn L. Relationship between steroid receptors and response to endocrine therapy and cytotoxic chemotherapy in metastatic breast cancer. Cancer 46:2961–2963, 1980.

152. Fisher B, Redmond C, Fisher E, et al. Relative worth of estrogen or progesterone receptor and pathologic characteristics of differentiation as indicators of prognosis in node negative breast cancer patients: findings from National Surgical Adjuvant Breast and Bowel Project Protocol B-06. J Clin Oncol 6:1076–1087, 1988.

153. McKenna N, Lanz R, O'Malley B. Nuclear receptor coregulators: cellular and molecular biology. Endocr Rev 20:321–344, 1999.

154. Weigel N, Zhang Y. Ligand-independent activation of steroid hormone receptors. J Mol Med 76:469–479, 1998.

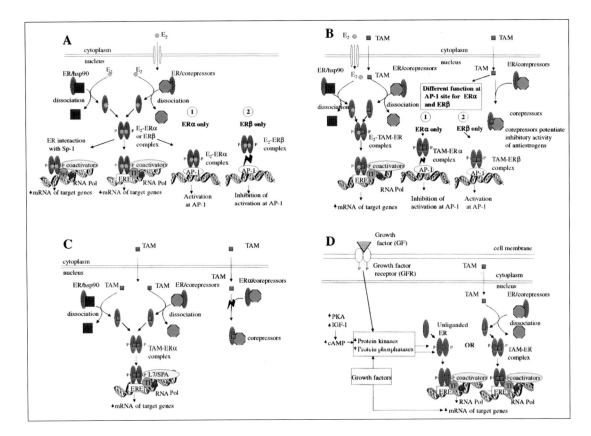

Color Figure 14.2 Schematic model of direct and indirect activation of target genes by estrogen receptor (ER). (A) Interaction of the full agonist 17β-estradiol (E_2) with ERα and ERβ. After entry to the nucleus, E_2 binds to ER monomers, which are complexed with either the heat shock protein (hsp) complex or corepressors while in their inactive state. Estradiol-liganded ER monomers dimerize. Activated E_2–ER dimers bind to estrogen response elements (EREs) in the promoter region of target genes as indicated. The activated complex not only interacts with EREs but can also cause a protein–protein interaction with transcription factors like Sp-1 and, therefore, enhance transcription of Sp-1 target genes. Also, ER enhances transcription at activating protein-1 (AP-1) sites, where ERα and ERβ show different mechanisms of action. Dimers of E_2–ERα bind at AP-1 sites and activate transcription of the target genes, whereas E_2–ERβ dimers inhibit activation at AP-1 sites and, therefore, transcription of the target genes. This difference in activation of target genes plays an important role in different mechanisms of tamoxifen action as indicated in (B). In (B), the mechanisms of action of tamoxifen (TAM) are shown. At low levels and in tissues where it acts as an agonist, tamoxifen activates ER monomers and forms heterodimers with E_2-liganded ERs. In this form, tamoxifen can activate the same target genes as E_2 by binding to EREs (*illustration on the left*). At high levels and in tissues where it mainly acts as an antagonist, tamoxifen stabilizes ER binding to corepressors like REA (repressor of ER activity) (*illustration on the right*). As with E_2 binding, there is a difference in mechanism of action at AP-1 sites for tamoxifen-bound ERα and ERβ dimers. The tamoxifen–ERα dimer inhibits activation of transcription at AP-1 sites and, therefore, inhibits subsequent target gene expression. However, the tamoxifen–ERβ dimer activates transcription at AP-1 sites and, therefore, acts as an agonist (*illustration in the middle*). This may be one mechanism for the development of antiestrogen resistance in some breast cancers. (C) Interaction of tamoxifen-bound ER with specific coregulators. A coactivator, L7/SPA, interacts with tamoxifen-liganded ER, recruiting transcription factors and increasing mRNA of target genes. However, decreases in corepressors like REA can decrease the stabilization of corepressor-bound ER through tamoxifen, leading to increased tamoxifen resistance. (D) Activation of ER through crosstalk with growth factors and/or growth factor receptors is shown as a model. In the absence of E_2 and/or the presence of antiestrogen, ER is phosphorylated and activated by signal-transduction cascades from membrane receptors or growth factors. These cascades lead to an increase in protein kinase or a decrease in protein phosphatase activity, increasing the level of phosphorylation at ERs. These phosphorylated, unliganded, or tamoxifen-liganded ER dimers can activate transcription of target genes by binding to EREs and stimulate cell growth in the absence of E_2. PKA, protein kinase A; IGF-I, insulin-like growth factor I; cAMP, cyclic adenosine monophosphate.

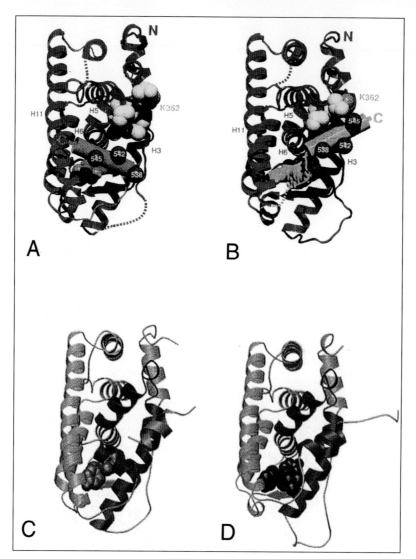

Color Figure 14.3 Comparison of the three-dimensional structures of the estrogen receptor (ER) α ligand-binding domain (LBD) in a complex with either agonists or antagonists. *(A)* Three-dimensional schematic illustration of estradiol bound to the LBD of ERα. Helix 12 (H12, blue cylinder) covers the ligand-binding cavity and interacts with helices H3, H5/6, and H11 (red). The H12 surface charged with amino acids Asp^{538}, Asp^{545}, and Glu^{542} (brown spheres on the blue H12 cylinder) is directed away from the ligand-binding cavity. Estradiol (blue space-filling molecule below the blue H12 cylinder) occupies the central hydrophobic core binding pocket of the LBD. Hydrophobic residues in the groove between H3 and H5 (yellow) and Lys^{362} (K^{362}) are illustrated in space-filling form. In this conformation, the crucial residue Lys^{362} is not blocked by H12 and can be bound by coactivators. *(B)* Three-dimensional schematic illustration of raloxifene bound to the LBD of ERα. H12 (green cylinder) is repositioned upon raloxifene binding (raloxifene illustrated as green molecule below H6) to the LBD and its alignment over the hydrophobic cavity prevented. H12 lies in a groove formed by H5 and H3 (red) instead. Lys^{362} (K^{362}, in pink space-filling form), required for recruitment of coactivators, is partially blocked by the repositioned H12. *(C)* Three-dimensional schematic illustration of diethylstilbestrol (DES) bound to the LBD of ERα. The LBD is depicted as a ribbon drawing. H12 is illustrated as a magenta helix; helices 3, 4, and 5 are blue; and DES is represented as a green space-filling molecule in the central hydrophobic core binding pocket of the LBD, comparable to the conformation shown in *(A)*. In addition, the LBD nuclear receptor box II peptide is illustrated (golden helix), which is required for recognition by coactivators. *(D)* Three-dimensional schematic illustration of 4-hydroxytamoxifen, the active form of tamoxifen bound to the LBD of ERα. The LBD is shown as a ribbon drawing. As in *(C)*, H12 is illustrated as a magenta helix; helices 3, 4, and 5 are blue. 4-Hydroxytamoxifen is illustrated as a brown space-filling molecule. Repositioning of H12 prevents binding of coactivators to the LBD nuclear receptor box II and, therefore, prevents transcription of target genes. (From Brzozowski et al.[43] and Shiau et al.[6] with permission.)

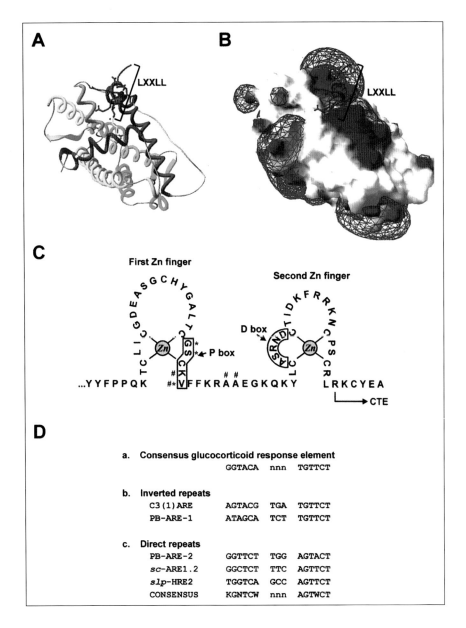

Color Figure 16.2 (A) Ribbon diagram of the androgen receptor (AR) ligand-binding domain (LBD) based on the published crystal structure showing the α helix and β sheet with the LXXLL peptide of the p160 coactivator glucocorticoid receptor interacting protein-2 (GRIP1), modeled into the hydrophobic cleft formed by activation function 2 (AF-2). (B) Topology and electrostatic surface potential diagram of the AR LBD showing position of the bound LXXLL peptide of GRIP1. Homology modeling of the AR (A, B) was performed by Dr. Jonathan Harris (Queensland Institute of Technology). (C) The AR DNA-binding domain (DBD) consists of a highly conserved cysteine-rich sequence that contains two zinc finger motifs and a short C-terminal extension (CTE), which forms part of the hinge. The first zinc finger mediates DNA recognition via the highly conserved P box and binds to the major groove of DNA, while conserved amino acids in the D box of the second zinc finger mediate dimerization between steroid receptor monomers. Sequences in the CTE bind to the minor groove of DNA on the opposite side of the helix to the first zinc finger and mediate sequence specificity of steroid receptors. Indicated are amino acids that determine the specificity of DNA binding (°) and those likely to make base-pair contacts in the hormone response element (HRE) half-site (#). (D) Known androgen response elements (AREs). a: Consensus glucocorticoid response element. b: AREs that have an inverted repeat structure consisting of imperfect palindromes bind both AR and glucocorticoid receptor. c: AREs that bind AR selectively have a partial direct repeat structure. PB, rat probasin gene; sc, human secretory component gene; slp, mouse sex limited protein gene.

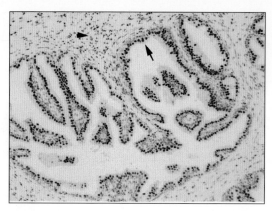

Color Figure 18.1 The androgen receptor is localized by immunohistochemistry to both the stromal (*arrowhead*) and glandular (*arrow*) cells in the normal prostate.

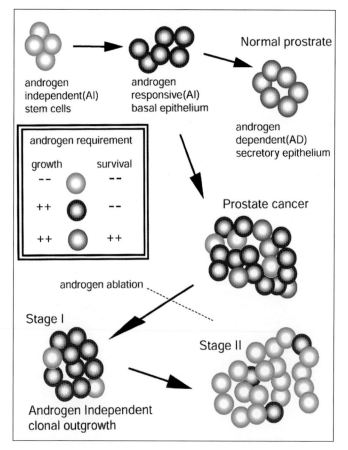

Color Figure 18.2 Model for development of androgen-independent (AI) prostate cancer through two distinct stages. A two-step model for progression to AI through clonal selection is shown. Three types of cell are postulated, which vary in their requirement for androgen as a growth and survival factor. Cells that require androgen for both growth and survival are represented by blue and correspond to the secretory epithelium in normal prostate. Cells that require androgen for growth but not survival (androgen-responsive, AR) are red and may correspond to the basal epithelial cells in the normal prostate. Cells that do not require androgen for growth or survival (AI) are represented by green. It is unknown if a counterpart for these cells exists in the normal human prostate, but it may correspond to the putative prostate stem cell. The first step in the progression to AI (stage I) is enrichment for AR (red) cells as a consequence of apoptotic death of the androgen-dependent (AD) (blue) cells after androgen-ablation therapy. The second step (stage II) is clonal outgrowth of AI (green) cells. (Modified from Craft et al.[34])

15

Prostate Cancer: Epidemiology and Molecular Endocrinology

RONALD K. ROSS
NICK M. MAKRIDAKIS
JUERGEN K. V. REICHARDT

The epidemiology and etiology of prostate cancer are probably the least understood of any of the numerically major cancers. Although androgens have long been thought to play an important role in prostate cancer development, this remains unproven. While there is substantial indirect evidence for such involvement (reviewed below), direct evidence is still lacking.

The epidemiology of prostate cancer is controlled by two risk factors, age and race/ethnicity; but there is no firm understanding of why these two factors play such a crucial role in modifying risk.

AGE

Among adult cancers of epithelial origin, prostate cancer is by far the most age-related. Before the advent of prostate-specific antigen (PSA) as a method to detect occult disease, prostate cancer was extremely rare before age 40 and very uncommon even up to age 50. However, after age 50 the rate of increase with aging is greater than for any other cancer.[1] When age is plotted against incidence on a log–log scale for most common adult epithelial cancers (e.g., lung, stomach, bladder, and colon but with female breast, ovarian, and endometrial cancers representing exceptions), one observes a straight-line relationship, although the slope varies substantially from cancer to cancer. The steepest slope (age raised approximately to the 8.5 power) is observed for cancer of the prostate. Although it is difficult to observe given the steepness of the curve, it is noteworthy that for prostate cancer the relationship is not quite linear: the slope is actually steepest in the early part of the age range in which prostate cancer occurs and then tapers somewhat with further aging.[2]

RACE/ETHNICITY

Probably the most interesting epidemiological feature of prostate cancer is its rather enormous international, or more precisely racial/ethnic, variation in incidence. It is among the most internationally variant of all cancers.[3] African-American men have the highest rates of prostate cancer in the world, exceeding the rate of any other racial/ethnic group and any other countrywide incidence by almost twofold. Although there has been speculation that black Africans also have high rates of prostate cancer, convincing data are currently lacking, due to an absence of comprehensive cancer registries in black African countries and to the relatively short average life span still sustained by most African men. Such data would prove most useful in helping resolve whether the high risk in African Americans is largely due to genetic predisposition, lifestyle characteristics, or a combination of these factors. Asian men indigenous to China, Japan, and Korea and probably other countries as well have the lowest reported prostate cancer rates worldwide.[3] Although historically the difference in incidence between African-American men, on the one hand, and Asian men, on the other, has been reported to be as much as 100-fold or more, much of this is due to differences in diagnostic strategies in different populations.[4] For example in the United States, prostate tissue from transurethral resections of the prostate for benign disease is systematically evaluated, and given the frequency of this procedure, this becomes an important source for prostate cancer diagnosis but with uncertain clinical significance. Comparisons of incidence rates across populations have been further complicated in the past decade by the increasing use of PSA blood test screening for prostate cancer detection in some populations. As such use results in uneven detection of both clinically relevant and, especially, occult prostate cancer of uncertain clinical relevance, comparisons across populations have become largely meaningless. Mortality rate comparisons are somewhat more reliable, although still subject to disparate methods between populations, in assigning cause of death.[3] Nonetheless differences across countries and racial/ethnic groups in mortality from prostate cancer tend to be substantially less than differences in incidence, although patterns tend to be comparable (e.g., African Americans > Caucasian Americans > indigenous Asian populations).

FAMILY RISK

Other than age and race/ethnicity, the only firmly established risk factor for prostate cancer is family history. Men with a first-degree relative (father, sibling) with a history of prostate cancer have roughly two to three times the risk of prostate cancer compared to men with no such history. Risk is increasingly elevated as the number of first-degree relatives with prostate cancer increases; men with three or more such first-degree relatives have an 11-fold increase in risk over men with no such history.[5] Although there have been surprisingly few studies of this relationship, having such a relative with prostate cancer at a young age (<65) conveys a higher risk than if such relatives develop prostate cancer at a later age. Although having either a brother or a father with prostate cancer conveys higher risk than having no relative with the disease, several studies have shown that the level of risk so conveyed is not comparable. Risk tends to be higher with a brother with prostate cancer than with a father. This pattern of familial risk with a preferentially higher risk with an affected brother than with an affected father extends across multiple racial/ethnic groups (African Americans, Japanese, Latinos, and Caucasians).[6] This distinctive pattern of familial risk suggests an X-linked or recessive gene mode of inheritance, although linkage studies in large multiplex prostate cancer families support an autosomal dominant inheritance pattern in these highly selected families.[5]

The strong and consistent relationship between family history and prostate cancer risk has led to a massive scientific effort to identify one or more single-locus, high-penetrance genes responsible for this familial risk. Several national and international registries of multiplex prostate cancer families have been established to support this endeavor. Linkage analyses have identified a series of possible candidate loci (see Chapter 17 for details), but confirmatory studies for each of these have produced highly inconsistent results.[7] Thus, it appears that the identification and cloning of one or several such high-penetrance genes for prostate cancer, comparable to *BRCA1* and *BRCA2* for breast cancer (i.e., that convey very high lifetime risk in their mutated form, explain a readily measurable fraction of familial risk, and are reproducibly found in diverse populations), are unlikely in the near future. This realization has led to alternative approaches to understanding the molecular genetics of prostate cancer risk, including especially the search for common variants of low-penetrance genes using a pathway-driven approach and the search for possible environmental or genetic modifiers of risk either for the high-penetrance candidate loci or for polygenic low-penetrance, risk-modifying variants. Chapters 16 and 17 in this book discuss in detail the current status in the search for both single-locus, high-penetrance genes for prostate cancer and for certain low-penetrance candidate genes, especially in the androgen transactivation pathway. We summarize below this and other molecular genetic etiological pathways currently being explored to better understand prostate cancer risk.

Many potential risk factors for prostate cancer have been evaluated by epidemiologists over the past several decades, but none is a fully established, widely accepted determinant of risk. These have been the subject of extensive reviews.[2,8] Among the most promising general areas still being extensively evaluated are diet and sexual/reproductive risk factors. We briefly review these and other possible risk factors by way of introduction to the molecular genetic pathway summaries which follow.

DIETARY PROSTATE CANCER RISK FACTORS

There are at least seven major dietary macro- or micronutrients that are under intense scrutiny currently as dietary risk or protective factors (Table 15.1). Prominent among these is dietary fat or some component of fat (e.g., saturated fat), which first received attention as a mechanism to possibly explain the low risk in native Asian populations and the apparent rapid shift in risk upon migration of Asian populations to the United States. There are suggestive data that increased fat consumption is associated with higher circulating testosterone levels, providing a possible mechanism for a fat–prostate cancer relationship.[9] Both case-control and cohort data tend to support a relationship between fat consumption overall and prostate cancer risk;[10,11] however, there are sufficient inconsistencies in the data, and the magnitude of risk even between extreme categories of estimated fat intake is sufficiently modest that fat is still not considered a proven prostate cancer risk factor.

Antioxidant vitamins and minerals are of great current interest, not only as antiprostate cancer agents but also as anticarcinogenic agents in a broad sense. Antioxidants are potential anticancer agents because they bind *free radicals*, chemical entities that can damage DNA, create mutations, and lead to malignant transformation.[12] Two antioxidants are of special interest in the context of prostate cancer. Lycopene is a member of the carotenoid family of vitamins and is the major antioxidant in tomatoes and tomato products. It is concentrated in the prostate and may suppress prostate cancer growth.[13] Observational studies have demonstrated inverse relationships between lycopene and/or frequency of consumption of tomato products and prostate cancer risk.[14] Moreover, prospective serological studies have shown that high circulating levels of lycopene predict low prostate cancer risk.[15] Additional prospective dietary and serological studies are ongoing to confirm this relationship and evaluate lycopene as a potential protective agent in greater detail. Vitamin E is an antioxidant vitamin that has been tested in a clinical trial setting as a chemopreventive agent for lung cancer in heavy smokers. Although vitamin E was ineffective against lung cancer in that setting, prostate cancer risk was substantially reduced in the trial arm that received vitamin E.[16] While this observation is highly encouraging, it is unclear why there would be specificity to the prostate in terms of cancer reduction. This fac-

Table 15.1 Suggested Dietary Risk or Protective Factors for Prostate Cancer and Their Possible Mechanisms of Action

Dietary Factor	Direction of Risk	Possible Mechanism
Fat	Increased	Androgen activity/cell proliferation
Fiber	Decreased	Androgen activity/reduced cell proliferation
Lycopene	Decreased	Antioxidant
Phytoestrogens	Decreased	Androgen activity/reduced cell proliferation
Selenium	Decreased	Antioxidant, proapoptosis
Vitamin D	Decreased	Vitamin D activity/reduced cell proliferation/pro-differentiation
Vitamin E	Decreased	Antioxidant

tor plus the incidental nature of the finding require that additional confirmatory studies be conducted.

Selenium is a trace mineral that has generated considerable interest as an anticancer agent. Like the carotenoids and vitamin E, selenium is a potent antioxidant, but it also has additional anticancer properties experimentally, including especially proapoptotic activity, although other mechanisms may also be involved.[17] As with vitamin E, selenium was observed to reduce prostate cancer risk as an incidental finding in a clinical trial of secondary skin cancer prevention.[18] In part because of its proven efficacy as an anticancer agent experimentally and because there are epidemiological biomarker data supporting selenium as a prostate cancer chemopreventive agent, there are ongoing short- and long-term clinical trials to directly test the efficacy of selenium to prevent prostate cancer. Results of the short-term trials, which will evaluate the impact of selenium supplementation on the prostate cancer precursor lesion prostate intraepithelial neoplasia (PIN) and on possible biomarkers of prostate cancer risk (e.g., rate of epithelial cell proliferation) should be available within 2 years. However, results of full-scale trials to determine the impact of selenium supplementation on clinical prostate cancer risk per se in healthy men will not be available for at least 7–10 years.

Phytoestrogens, i.e., plant-derived estrogens such as from soy products, have been hypothesized to protect against prostate cancer, possibly by antagonist actions of estrogens on androgen-mediated growth.[19] There are some experimental animal data supporting an inhibitory effect of phytoestrogen intake on prostate tumor growth.[19,20] Phytoestrogen intake has been hypothesized to play a role in the reduced prostate cancer rates observed in Asian populations. Cross-sectional population studies have shown urinary phytoestrogen biomarkers to be substantially higher in native Japanese than other racial/ethnic groups, but there is little direct evidence linking high phytoestrogen consumption to low prostate cancer risk.[21] One study found tofu consumption, a soy source of phytoestrogens, to be associated with reduced risk; but there is great need for further epidemiological investigations.[22]

Fiber can bind steroid hormones that are being recirculated through the enterohepatic circulation. Fiber binds these hormones, causing them to be excreted rather than reabsorbed, which, in theory, will reduce circulating levels of steroid hormones. Thus, fiber intake might be expected to result in some reduction in prostate cancer risk by reducing androgen stimulation of the prostate, but this hypothesis is also largely untested.

Vitamin D is a promising potential chemopreventive agent for prostate cancer. 1,25-Dihydroxyvtamin D, the active form of vitamin D, can reduce cell proliferation and induce differentiation of prostate cancer cells in vitro.[23,24] Both of these actions would be anticipated to reduce prostate cancer risk or at least reduce progression from occult to clinically relevant disease. Epidemiological studies are generally supportive. Low circulating levels of 1,25-dihydroxyvitamin D have been associated with high prostate cancer risk, either overall or in the subgroup of men with low 25-hydroxyvitamin D levels.[25,26] No chemoprevention trials are under way as hypercalcemia is an unacceptable side effect of supplementation with the natural form of active vitamin D. A large number of vitamin D analogues have been developed, but as yet none exists with absolutely no hypercalcemic potential. The molecular genetic epidemiology of vitamin D and prostate cancer risk is described below.

NONDIETARY PROSTATE CANCER RISK FACTORS

Although most attention has focused on possible dietary risk or protective factors for prostate cancer as the most likely environmental risk factor category to explain the racial/ethnic variation in incidence and the impact of migration on risk modification, other factors have also been evaluated over the past few decades. Among these, cigarette smoking and a history of any type of sexually transmitted disease are among the most reproducible. As there are no highly suspected carcinogens to the prostate found in cigarette smoke or any direct evidence of an infectious etiology of prostate cancer, it has been proposed that both of these risk factors might be indices of an androgenic profile associated with prostate cancer development.[27] Smokers have higher circulating testosterone levels than nonsmokers.[28] While there is no direct link between androgen levels and sexually transmitted diseases, androgen levels may be modestly correlated with indices of sexual activity.[29]

The substantial literature on smoking and prostate cancer has been reviewed,[30] with the overall conclusion that smokers have only a very modest, if any, increase in risk but that risk may be somewhat higher for fatal than incident disease. The reasons for this possible risk differential remain obscure.

Among the various parameters of sexual activity indices and prostate cancer risk that have been evaluated (e.g., age at first intercourse, frequency of intercourse, marital status), the only one that has been consistently linked to prostate cancer risk is a history of a sexually transmitted disease.[31] While this has suggested to some a possible infectious etiology, the epidemiology does not strongly support this premise. For example, Ross and colleagues[32] showed that celibate priests do not have a low risk of prostate cancer; and unlike cervical cancer (known to be caused by a sexually transmitted infectious agent), prostate cancer is not inversely related to socioeconomic status, nor is risk elevated in Latinos. An alternative hypothesis is that such a history is an index of sexual activity, which in turn is correlated with underlying androgen levels.

Other factors have been suggested to increase prostate cancer risk but with highly inconsistent results. Alcohol consumption, which is of interest primarily because of its impact on circulatory steroid hormone levels, was recently the subject of a meta-analysis involving 33 epidemiological studies that had assessed alcohol consumption and prostate cancer risk.[33] Alcohol consumers overall had a risk of prostate cancer of 1.05 compared to non-drinkers, but risk increased to 1.21 among men consuming at least four drinks per day.

Obesity and indices of body size have also been of interest in terms of prostate cancer risk. Body mass index and height are of interest mechanistically in terms of both possible impact on steroid hormone levels and possible relationships with insulin-like growth factors (see below, Insulin-like Growth Factor Signaling Pathways). No consistent pattern has emerged from these studies in terms of either height, weight, body mass index, or other indices of body size.[34]

In addition to these possible nondietary risk factors, two nondietary preventive factors have been suggested for prostate cancer. Lee et al.[35] summarized the 23 studies to date that have examined a role of physical activity in prostate cancer risk.[35] They calculated a summary relative risk of 0.8 for the most active versus the least active physical activity categories across this range of studies. The quality of physical activity assessment varied substantially across these studies, and none was designed specifically to assess this relationship. One suggested mechanism for a protective effect of physical activity is via modulation of androgen levels.[36]

Several recent studies have suggested that regular use of nonsteroidal antiinflammatory drugs may reduce risk of prostate cancer, but no definitive study has been done.[37] The probable mechanism for any preventive efficacy is via inhibition of cyclooxygenase 2, an inducible enzyme with proangiogenic, prometastatic, and antiapoptotic properties.[38]

RISK PATHWAYS FOR PROSTATE CANCER

Several molecular etiological pathways have been suggested for prostate cancer. Although androgen transactivation pathways clearly have received the greatest attention, others have been the focus of increasing research activity. Among the most prominent of these additional pathways are vitamin D metabolism, insulin-like growth factor (IGF) signaling pathways, and chemical carcinogenic pathways (Fig. 15.1). Hsing and Devesa[39] developed an etiological model that integrates these pathways and incorporates many of the possible risk factors suggested for prostate cancer.[39] With the availability in the past decade of polymerase chain reaction technology to study polymorphic variants in genes on a large-scale population basis, epidemiologists and molecular biologists have been afforded an opportunity to begin detailed evaluation of these pathways with special attention to the contribution of polygenic variation in prostate cancer development within and between populations. In the section below,

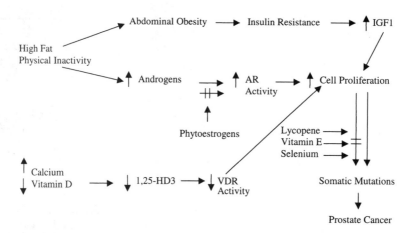

Figure 15.1 A simple model of prostate cancer development: possible interfaces in etiological pathways. AR, androgen receptor; IGF, insulin-like growth factor; VDR, vitamin D receptor; 1,25-HD3, 1,25-dihydroxyvitamin D_3.

we overview the state of the art related to understanding the contribution of these pathways to prostate cancer incidence. We have chosen not to review the current state of knowledge regarding polymorphisms in metabolic genes and prostate cancer as we believe this is premature in the absence of any clearly established environmental risk exposure. These have been reviewed by Chen.[40] More detailed descriptions of our current understanding of the genes involved in the androgen-activation pathway are provided in the chapters which follow.

ANDROGEN TRANSACTIVATION PATHWAYS

Androgens play a critical role in normal and abnormal prostate development. Studies of androgens and prostate cancer go back nearly 60 years. The pioneering work of Huggins in this area was rewarded with a Nobel prize. Substantial epidemiological data support a critical role for androgens in prostate cancer etiology.[2,3] Prostate cancer development is absent in men with marked androgen deficiency, such as eunuchs, or men with markedly reduced androgenization of the prostate, such as those with constitutional absence of 5α-reductase enzyme activity in whom the prostate remains a vestigial organ. Normal prostate development is induced by dehydrotestosterone (DHT), which is formed from testosterone (T) by the enzyme steroid 5α-reductase.[41] In men, T is produced in large amounts primarily by the testes. It is then irreversibly metabolized intracellulary to DHT. The DHT (or, much less efficiently, T) is bound by an intracellular cytosolic receptor, the androgen receptor (AR). This complex is then translocated to the cell nucleus, where it activates transcription of genes with androgen-responsive elements (AREs) in their promoter regions.

Ross and colleagues[3] introduced the concept of a polygenic etiology of prostate cancer related to multiple functional polymorphic variants in genes involved in androgen biosynthesis, transport, activation, and metabolism. They proposed that each such variant might have only a minor impact on androgen transactivation and, therefore, on prostate cancer risk but that multiple such variants in combination might have a more substantial impact. This group also introduced the notion that there might exist multiple polymorphic variants in the same gene, which might either act jointly in their impact on androgen transactivation activity or even "cancel" each other out.

To date, only a few genes in this pathway have been studied in relationship to prostate cancer risk (Table 15.2). Two of these have been particularly well studied, the 5α-reductase type II (*SRD5A2*) gene, which encodes the enzyme that

PROSTATE CANCER: EPIDEMIOLOGY AND MOLECULAR ENDOCRINOLOGY

Table 15.2. Characteristics of Selected Androgen-Metabolic Genes

Gene Symbol	Gene Product	Chromosomal Location	Gene Size (kb)
AR	Androgen receptor	Xq11–12	>54
CYP17	Cytochrome P-450c17	10q24.3	8.7
HSD3B2	3β-Hydroxysteroid dehydrogenase	1p13	7.8
HSD17B3	17β-Hydroxysteroid dehydrogenase	9q22	>60
SRD5A2	Steroid 5α-reductase	2p23	>40

activates T in the prostate to its reduced and more biologically active form, DHT, and the AR, which binds DHT and then transports the ligand–receptor complex to the nucleus for DNA binding and transactivation of androgen-responsive genes.

The *SRD5A2* gene is located on chromosome 2, spanning 40 kb with five exons (Fig. 15.2).[42] Steroid 5α-reductase is a membrane-bound enzyme that catalyzes the irreversible conversion of T to DHT with reduced nicotinamide adenine dinucleotide phosphate (NADPH) as a cofactor.[41] Two isozymes exist: the type I enzyme with an alkaline pH optimum, encoded by the *SRD5A1* gene, and the type II isozyme with an acidic pH optimum, encoded by the *SRD5A2* gene.[43] Thigpen et al.[43] reported immunological studies showing that the type I enzyme is expressed primarily in newborn scalp and in skin and liver. The type II isozyme protein is expressed primarily in genital skin, liver, and prostate. Molecular genetic studies have shown that a rare disorder of male sexual differentiation, male pseudohermaphroditism, is due to 5α-reductase deficiency and inactivating germline mutations in the *SRD5A2* gene.[42] Affected males exhibit genital ambiguity and external female phenotype until puberty, at which point there is some development of secondary sex characteristics but the prostate remains highly underdeveloped.[44] Thus, normal prostate development requires normal function of the *SRD5A2* gene.

Reichardt and colleagues[45] initially screened the gene for polymorphic variants among a population with either high or low circulatory levels of androstanediol glucuronide, a DHT metabolite and a circulating correlate of 5α-reductase activity in the prostate. This process initially yielded eight single-nucleotide substitution polymorphisms, i.e., variants resulting in an amino acid change in the protein product, as well as several silent polymorphisms, i.e., single-nucleotide variants resulting in no amino acid change.[45] This group also described a series of previously unknown variants of a dinucleotide (TA) repeat sequence in the 3'-untranslated region (UTR) of the gene.[46] Reichardt's group proceeded to characterize the eight substitution polymorphisms in vitro using transfection assays to assess enzyme activity versus wild-type.[47] Although a few of these were true polymorphisms, resulting in no detectable change in enzyme activity despite the amino acid change in the pro-

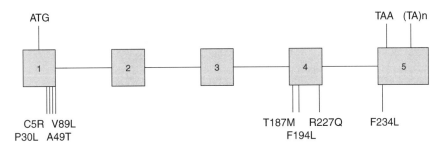

Figure 15.2 Schematic of the *SRD5A2* locus encoding the prostatic or type II steroid 5α-reductase. The translation-start (ATG) and stop (TAA) codons are indicated along with the (TA)$_n$ dinucleotide repeat polymorphism in the 3'-untranslated region. Constitutional DNA missense substitutions are identified below the gene in the single-letter amino acid code.

tein, others resulted in either increases or decreases in enzyme kinetics. One in particular, an alanine-to-threonine substitution at codon 49 (A49T), resulted in a substantial increase in enzyme activity. Although uncommon in healthy control men, the Reichardt group assessed its possible role in prostate cancer etiology in a multiracial nested case-control study because of its substantial impact in vitro. In both African-American and Latino men, the A49T polymorphism was associated with substantial increases in prostate cancer risk, especially for advanced disease at presentation [relative risk (RR) = 1.5 for localized and 7.1 for advanced disease, respectively, in African Americans, $p = 0.001$ for the latter association; RR = 1.7 for localized and 3.6 for advanced disease, respectively, among Latinos, $p = 0.043$ for the latter association]. The contribution of the A49T mutation in prostate cancer predisposition worldwide is uncertain; increased incidence of the A49T mutation has been reported in Italian prostate cancer patients,[48] while no difference between prostate cancer and control patients was reported in Finland.[49]

This same group has investigated the frequency of somatic mutations in SRD5A2 in tissue from prostate cancer patients.[50] By far, the most common de novo mutation in prostate cancer tissue in the SRD5A2 gene is the same A49T variant associated with such a substantial increase in prostate cancer risk in constitutional DNA.

Finally, the work by Reichardt, Ross, and their colleagues on 5α-reductase hormonal indices and genotypes led directly to an ongoing national trial to evaluate chemopreventive efficacy of the 5α-reductase inhibitor finasteride on prostate cancer risk in 18,000 healthy men with normal PSA levels.[51] Importantly, Reichardt et al.[47] demonstrated that inhibition of 5α-reductase activity in vitro depends strongly on SRD5A2 genotype. For example, finasteride displays 11- to 7-fold lower affinity for the A49T mutant enzyme than for the wild-type protein in vitro. Therefore, if finasteride proves efficacious in preventing prostate cancer in this trial, the degree of protection may vary substantially according to constitutional polymorphic variation in the SRD5A2 gene, which may eventually allow better targeting for such chemopreventive interventions.

The SRD5A2 gene association with prostate cancer risk has supplied a useful paradigm to systematically investigate other candidate genes for prostate cancer, i.e., beginning with rigorous sequencing of the gene among healthy individuals with extreme levels of biochemical correlates of the protein product of the gene, to identify polymorphic variants and their frequencies across racial/ethnic groups; then, proceeding to assess the functional significance of each genetic variant in vitro; for those of interest, to then conduct an appropriate epidemiological study to determine the strength of any association between specific variants and disease risk; to determine whether somatic mutations of this variant occur in the course of disease progression; and, if possible, to use this information to design and implement prevention intervention strategies.

The AR gene is reviewed in Chapter 16. The gene is located on the X chromosome and has a structure similar to other steroid hormone receptor genes, with a DNA-binding domain, a ligand-binding domain, and a transcription-modulatory domain.[52] Given the multiple functions of the receptor in androgen activity in the prostate, it is a strong candidate gene in prostate cancer etiology. Coetzee and Ross[53] noted that expansion of a trinucleotide $(CAG)_n$ repeat polymorphism in exon 1 (the transactivation domain) of the gene caused a rare adult-onset motor neuron disease, spinal and bulbar muscular atrophy, or Kennedy's disease.[53] Men with this genetic disorder have evidence of reduced androgen activity, even though androgen binding by their ARs appears to be normal.[54] Coetzee and Ross[53] hypothesized that if there is suboptimal transactivation in the expanded CAG repeat range (>36 CAGs) associated with Kennedy's disease, there might also be a range of androgen transactivation within the normal range of repeats (≈8–33 CAGs); thus, men in the higher part of the range might have less androgen transactivation than men in the lower part of the range. Their hypothesis further predicted that men with relatively short CAG repeats would be at higher risk of prostate cancer than men with relatively long CAG repeats and that African-American men would have shorter repeats on average than whites, who in turn would have shorter repeats than Asians.[53]

The hypothesis first received support from in vitro transfection assays demonstrating normal androgen binding but reduced AR-mediated transactivation activity, first in the abnormal range associated with Kennedy's disease[54] and then in the normal range with transactivation being negatively correlated with CAG length.[55] We and others have shown, as predicted, that African Americans have shorter CAG repeats on average than whites, who in turn have shorter repeats than Asians.[53,56] Ingles and colleagues[57] showed that, in whites, CAG length predicted prostate cancer risk, with risk increasing as length decreases. This effect was particularly pronounced for men presenting with advanced disease clinically. This finding, or a variant of it, has been the most reproducible to date in the molecular genetic epidemiology of prostate cancer. Studies done in diverse populations have rather consistently found that CAG repeat length is related either to prostate cancer risk overall, to advanced stage of prostate cancer if not prostate cancer risk overall, or age at onset of prostate cancer (see Table 16.1, Chapter 16).

Coetzee and colleagues have been attempting to better understand the mechanism behind the relationship of CAG repeat length and androgen transactivation efficiency. They have shown, e.g., that CAG modulation of transcription activity is dependent on the presence of coactivator proteins, particularly of the p160 family, which bind to the N-terminal domain of the AR, just downstream from the polyglutamine tract encoded by the CAG repeat.[58,59]

Work initiated on the CAG repeat and prostate cancer has led to additional research demonstrating the value of polygenic approaches to etiology. Xue and colleagues[60] studied one of the "downstream" genes, i.e., a gene that is transactivated by the AR in conjunction with polymorphic variation in the AR CAG repeat. Initially, they demonstrated a correlation between PSA levels and CAG repeat length; the PSA gene is one of only a relatively few genes that are known to be controlled by the AR. The PSA gene contains a single-nucleotide polymorphism (SNP) in the ARE of its promoter region.[61] This polymorphic variant predicted risk of prostate cancer, especially advanced prostate cancer (RR = 2.0 and 2.4, respectively, for prostate cancer overall and for advanced disease) in a population-based pilot case-control study.[62] CAG repeat length alone predicted risk, but the greatest risk levels were observed for the CAG repeat size and the PSA variant in combination (RR = 4.6 overall and 9.6 for advanced disease). This work may demonstrate in principle the value of polygenic pathway-driven approaches to better understand prostate cancer etiology. This work may also connect two major potential etiological pathways in that the PSA gene encodes a protease that participates in cleavage of IGF-I from its main binding protein (see below, Insulin-like Growth Factor Signaling Pathways).[63]

Several other androgen-signaling genes have been preliminarily studied in the context of prostate cancer risk. Testosterone is synthesized from cholesterol in a series of enzymatic steps involving several of the cytochrome P-450 enzymes.[64] The enzyme cytochrome P-450c17 catalyzes two sequential reactions of the biosynthesis of T, in both the gonads and the adrenals. The first step is the conversion of pregnenolone to 17-hydroxypregnenolone (hydroxylase activity), and the second is its subsequent conversion to C19 steroid dehydroepiandrosterone (lyase activity), a steroid with androgenic activity.[64] The *CYP17* gene on chromosome 10 encodes the P-450c17 enzyme involved in these two sequential reactions in T biosynthesis.[65] A T-to-C transition SNP exists in the 5′-UTR of the *CYP17* gene (A2 allele).[66] While the functional relevance of this polymorphism is in dispute, it has been linked to polycystic ovarian cancer risk in women, male pattern baldness in men,[66] various estrogen metabolic parameters,[67] breast cancer risk factors,[68] and breast cancer risk per se.[68] One U.S. study found the *CYP17* A2 allele to be associated with higher risk of prostate cancer,[69] whereas a Swedish study found the opposite.[70] Obviously, additional epidemiological and basic research is needed to fully understand the role of this gene in prostate cancer development.

As discussed above, DHT is the most active intraprostatic androgen. It is synthesized from T by the enzyme steroid 5α-reductase and inactivated through a reductive reaction catalyzed by 3α- or 3β-hydroxysteroid dehydrogenase. Thus, the dehydrogenase reactions initiate the irreversible inactivation of DHT in the prostate, and these enzymes are critical for the regulation of intraprostatic DHT steady-state levels by controlling its degradation rate. Activity of the two 3-hydroxysteroid dehydrogenase enzymes is significantly lower in abnormal compared to healthy prostatic tissue.[71] Thus, DHT might accumulate because of slowed degradation. 3β-Hydroxysteroid dehydrogenase activity in humans is encoded by two closely linked yet distinct loci: the *HSD3B1* and *HSD3B2* genes, both located on chromosome band 1p13.[72] The type 1 gene (*HSD3B1*) encodes the isoform present in placenta and peripheral tissue such as skin and mammary gland, while expression of the type 2 enzyme is restricted to adrenals and reproductive organs.[73] Thus, the type 2 enzyme encoded by the *HSD3B2* gene would regulate DHT levels by initiating the inactivation of this potent androgen in the prostate. Cloning and characterization of the human *HSD3B2* gene have revealed that it spans about 7.8 kb of genomic DNA in four exons (Fig. 15.3).[74]

The *HSD3B2* gene may play dual competing roles in prostate cancer etiology: the enzyme product (noted above), is one of the two enzymes that irreversibly inactivates DHT in the prostate; however, it is also responsible for production of the adrenal androgen androstenedione, which can undergo further conversion to T.[41] A complex dinucleotide repeat polymorphism exists in intron 3, for which multiple alleles have been described with substantial variation in frequency across race/ethnicity.[75] In preliminary studies, several of these occur more frequently in prostate cancer patients than in healthy men.[75]

The number of candidate genes and their polymorphic variants in this pathway will continue to grow. The androgen-signaling pathway is highly intricate and complex, involving not just genes involved in androgen biosynthesis, transport, activation, and detoxification but also genes encoding coactivator proteins and a whole series of downstream genes, most of which have yet to be identified.

INSULIN-LIKE GROWTH FACTOR SIGNALING PATHWAYS

There is growing interest in the role of growth factors in prostate cancer development as these proteins can play important roles in regulating cell proliferation in the prostate. Particular attention has been focused on IGF-I and its binding proteins as prospective studies of circulating levels of IGF-I have suggested that individuals with high levels, especially when combined with low levels of binding proteins, may be at high risk of prostate cancer.[76,77] Moreover, IGF-I activity may be one mechanism by which multiple possible etiological pathways are joined, including vitamin D pathways and androgen signaling, as both vitamin D analogues and castration can upregulate IGF-I binding proteins.[78,79] As described above, an important androgen-regulated gene, the *PSA* gene, encodes a protease that can cleave IGF-I from its binding protein, leading to increased IGF-1 bioavailability, thus offering another possible connection between etiological pathways.

Animal models are consistent with an etiological role of IGF signaling in prostate cancer. In rat prostates, IGF-I induces prostate growth but deficiency reduces cell proliferation.[80] Prostate cancer progression in the transgenic adenocarcinoma of the mouse prostate (TRAMP) model is associated with increased IGF-I expression.[81] Administration of the 5α-reductase inhibitor finasteride in animal models reduces IGF-I levels, suggesting a further tie between androgen and IGF signaling pathways in prostate cancer development.[82] The genes encoding IGF-1 and its growth factors are polymorphic and may in-

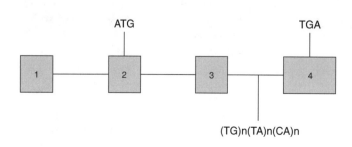

Figure 15.3 Schematic representation of the *HSD3B2* gene encoding the type II 3β-hydroxysteroid dehydrogenase enzyme. The translation-start and -stop codons are indicated above the gene. The complex $(TG)_n(TA)_n(CA)_n$ tandem repeat polymorphism in intron 3 is marked below the gene.

fluence circulatory levels of these proteins.[83,84] Thus, there exist opportunities for polygenic studies of this pathway alone or in combination with the androgen or vitamin D signaling pathways.

VITAMIN D METABOLISM PATHWAYS

Another molecular pathway currently receiving substantial interest in terms of prostate cancer development is the vitamin D metabolic pathway. 1,25-Dihydroxyvitamin D, the bioactive vitamin D metabolite, is a potent antiproliferation agent in the prostate as well as a prodifferentiation agent for prostate cells in vitro.[85] Vitamin D also has inhibitory effects on prostate cancer growth in animal models.[86] As prostate cells themselves metabolize vitamin D compounds to 1,25-dihydroxyvitamin D, vitamin D stimulation of the prostate is under both local and systemic control.

As noted above, the first prospective study of the relationship between circulating 1,25-dihydroxyvitamin D levels and subsequent risk of prostate cancer found that men who developed prostate cancer had substantially and significantly lower levels of 1,25-dihydroxyvitamin D than men who did not develop prostate cancer.[25] However, another prospective study, the Physician's Health Study, found no overall relationship between 1,25-dihydroxyvitamin D levels and prostate cancer risk.[26] In a subsequent report from that study, a positive association was observed between estimated dietary intake of calcium and prostate cancer risk, which the authors suggested might be due to a suppressive effect of calcium on conversion of 25-dihydroxyvitamin D to 1,25-dihydroxyvitamin D.[87] In fact, there has been a consistently observed association between intake of dairy products and prostate cancer risk.[39] Although this relationship has been generally attributed to fat content, calcium intake may be an alternate explanation.

Vitamin D activity is mediated by the vitamin D receptor (VDR) encoded by the VDR gene. A number of polymorphisms with common allelic variants have been identified in and around the 3′-UTR of the VDR gene, some of which have biological correlates, such as associations with indices of bone mineral density. Ingles et al.[57] found that a bimodal polyA microsatellite in the 3′-UTR was strongly related to prostate cancer risk in whites. In that study, men with at least one "long" A allele had a 4.6-fold increase in prostate cancer risk compared to men homozygous for "short" polyA alleles. Moreover, of the 26 patients tested who presented with advanced disease, all had at least one long A allele versus only 36/169 healthy control men of similar age. Taylor and colleagues[88] found an association between prostate cancer risk and a second polymorphic marker, a TaqI restriction fragment length polymorphism (RFLP). Individuals who had a TaqI allele had a three-fold higher prostate cancer risk than men who were tt homozygotes.[88] A third study found the BsmI RFLP associated with an increase in prostate cancer risk only among men with low circulating levels of 25-hydroxyvitamin D.[89]

Ingles et al.[90] explored the relationship between these polymorphic markers of the VDR gene and prostate cancer risk in African Americans.[90] As there is only very weak linkage disequilibrium in the 3′-UTR area of the VDR gene in African Americans, these investigators devised a haplotype assay to look directly at two of these markers in combination (the BsmI RFLP in intron 8, which is in linkage disequilibrium with the polyA microsatellite in whites, and the polyA microsatellite itself). They found that the BsmIB/long polyA haplotype was associated with a twofold increase in risk of advanced prostate cancer compared to individuals with no B/long A haplotypes, as predicted from results in whites.[90]

THE FUTURE

The immediate future of molecular epidemiology will undoubtedly center around four main areas. Despite the limited reproducibility of loci identified to date as candidates for single-locus, high-penetrance susceptibility genes for prostate cancer, we can expect continued linkage analyses, identification of additional candidate loci, and undoubtedly the eventual identification and cloning of one or more of these genes.

We can expect continued emphasis on the search for candidate susceptibility genes. These searches will increasingly be pathway-driven, involving an increasing number of polymorphic variants, and will utilize high-throughput technology for large-scale analyses. Laboratory technology is moving toward utilization of haplotypes (multiple linked SNPs) to characterize genetic variants. Whether haplotype strategies will prove more useful than association studies involving multiple functional alleles to determine polygenic models remains to be seen. As phenotype is ultimately of greater interest than genotype, the molecular epidemiology of prostate cancer will likely turn increasingly to proteomics as this field evolves.

While epidemiologists have mainly focused on polymorphic variants in genes to study etiology, changes in gene expression can also be controlled epigenetically through DNA methylation. We can expect to see molecular epidemiological studies of prostate cancer increasingly incorporate DNA methylation analyses as part of a more general strategy to fully understand etiology.

Finally, an important part of the molecular epidemiology of prostate cancer, not fully discussed in this chapter, is to understand why only a small proportion of prostate cancers progress from occult, clinically nondetectable lesions to clinically meaningful ones and why this rate of progression varies substantially across racial/ethnic groups. To fully understand this phenomenon genetically will require tissue analyses looking for somatic genetic changes across different stages of prostate cancer and between racial/ethnic groups.

ACKNOWLEDGMENTS

This work was supported by a Department of Defense Prostate Cancer Center Institution Grant DAMD 17-00-1-0102.

REFERENCES

1. Henderson BE, Ross RK, Pike MC, Casagrande JT. Endogenous hormones as a major factor in human cancer. Cancer Res 42:3232–3239, 1982.
2. Ross RK, Schottenfeld D. Prostate cancer. In: Schottenfeld D, Fraumeni JF (eds). Cancer Epidemiology and Prevention, 2nd ed. New York: Oxford University Press, 1996, pp 1180–1206.
3. Ross RK, Pike MC, Coetzee GA, Reichardt JKV, Yu MC, Feigelson H, Stanczyk FZ, Kolonel LN, Henderson BE. Androgen metabolism and prostate cancer: establishing a model of genetic susceptibility. Cancer Res 58:4497–4504, 1998.
4. Shimizu H, Ross RK, Bernstein L. Underestimation of the incidence rate of prostate cancer in Japan. Jpn J Cancer Res 82:483–485, 1991.
5. Carter BS, Beaty TH, Steinberg GD, Childs B, Walsh PC. Mendelian inheritance of familial prostate cancer. Proc Natl Acad Sci USA 89:3367–3371, 1992.
6. Monroe KR, Yu MC, Kolonel LN, Coetzee GA, Wilkens LR, Ross RK, Henderson BE. Evidence of an X-linked genetic component to prostate cancer risk. Nat Med 1:827–829, 1995.
7. Stanford JL, Ostrander EA. Familial prostate cancer. Epidemiol Rev 23:19–23, 2001.
8. Hsing AW, Chang L, Nomura AM, Isaacs WB, Armenian HK. A glimpse into the future. Epidemiol Rev 23:2, 2001.
9. Ross RK, Coetzee GA, Reichardt J, Skinner E, Henderson BE. Does the racial–ethnic variation in prostate cancer risk have a hormonal basis? Cancer 75:1778–1782, 1995.
10. Kolonel LN, Yoshizawa CN, Hankin JH. Diet and prostatic cancer: a case-control study in Hawaii. Am J Epidemiol 127:999–1012, 1988.
11. Giovannucci E, Rimm EB, Colditz GA, Stampfer MJ, Ascherio A, Chute CC, Willett WC. A prospective study of dietary fat and risk of prostate cancer. J Natl Cancer Inst 85:1571–1579, 1993.
12. Brody T. Vitamin A. In: Nutritional Biochemistry. Boston: Academic Press, 1999, pp 554–565.
13. Giovannucci E. Tomatoes, tomato-based products, lycopene, and cancer: review of the epidemiologic literature. J Natl Cancer Inst 91:317–331, 1999.
14. Giovannucci E, Ascherio A, Rimm EB, Stampfer MJ, Colditz GA, Willett WC. Intake of carotenoids and retinol in relation to risk of prostate cancer. J Natl Cancer Inst 87:1767–1776, 1995.
15. Gann PH, Ma J, Giovannucci E, Willett W, Sacks FM, Hennekens CH, Stampfer MJ. Lower prostate cancer risk in men with elevated plasma lycopene levels: results of a prospective analysis. Cancer Res 59:1225–1230, 1999.
16. Heinonen OP, Albanes D, Virtamo J, Taylor PR, Huttunen JK, Hartman AM, Haapakoski J, Malila N, Rautalahti M, Ripatti S, Mäenpää H, Teerenhovi L, Koss L, Virolainen M, Edwards BK. Prostate cancer and supplementation with α-tocopherol and β-carotene: incidence and mortality in a controlled trial. J Natl Cancer Inst 90:440–446, 1998.
17. Menter DG, Sabichi AL, Lippman SM. Selenium effects on prostate cell growth. Cancer Epidemiol Biomarkers Prev 9:1171–1182, 2000.
18. Clark LC, Dalkin B, Krongrad A, Combs GFJ, Turnbull BW, Slate EH, Witherington R, Her-

long JH, Janosko E, Carpenter D, Borosso C, Falk S, Rounder J. Decreased incidence of prostate cancer with selenium supplementation: results of a double-blind cancer prevention trial. Br J Urol 81:730–734, 1998.
19. Onozawa M, Kawamori T, Baba M, Fukuda K, Toda T, Sato H, Ohtani M, Akaza H, Sugimura T, Wakabayashi K. Effects of a soybean isoflavone mixture on carcinogenesis in prostate and seminal vesicles of F344 rats. Jpn J Cancer Res 90:393–398, 1999.
20. Kato K, Takahashi S, Cui L, Toda T, Suzuki S, Futakuchi M, Sugiura S, Shirai T. Suppressive effects of dietary genistein and daidzin on rat prostate carcinogenesis. Jpn J Cancer Res 91:786–791, 2000.
21. Barnes S. Role of phytochemicals in prevention and treatment of prostate cancer. Epidemiol Rev 23:102–105, 2001.
22. Severson RK, Nomura AM, Grove JS, Stemmermann GN. A prospective study of demographics, diet, and prostate cancer among men of Japanese ancestry in Hawaii. Cancer Res 49:1857–1860, 1989.
23. Schwartz GG, Oeler TA, Uskokovic MR, Bahnson RR. Human prostate cancer cells: inhibition of proliferation by vitamin D analogs. Anticancer Res 14:1077–1081, 1994.
24. Esquenet M, Swinnen JV, Heyns W, Verhoeven G. Control of LNCaP proliferation and differentiation: actions and interactions of androgens, 1α,25-dihydroxycholecalciferol, all-*trans* retinoic acid, 9-*cis* retinoic acid, and phenylacetate. Prostate 28:182–194, 1996.
25. Corder EH, Guess HA, Hulka BS, Friedman GD, Sadler M, Vollmer RT, Lobaugh B, Drezner MK, Vogelman JH, Orentreich N. Vitamin D and prostate cancer: a prediagnostic study with stored sera. Cancer Epidemiol Biomarkers Prev 2:467–472, 1993.
26. Gann PH, Ma J, Hennekens CH, Hollis BW, Haddad JG, Stampfer MJ. Circulating vitamin D metabolites in relation to subsequent development of prostate cancer. Cancer Epidemiol Biomarkers Prev 5:121–126, 1996.
27. Honda GD, Bernstein L, Ross RK, Greenland S, Gerkins V, Henderson BE. Vasectomy, cigarette smoking, and age at first sexual intercourse as risk factors for prostate cancer in middle-aged men. Br J Cancer 57:326–331, 1988.
28. Zmuda JM, Cauley JA, Kriska A, Glynn NW, Gutai JP, Kuller LH. Longitudinal relation between endogenous testosterone and cardiovascular disease risk factors in middle-aged men. A 13-year follow-up of former Multiple Risk Factor Intervention Trial participants. Am J Epidemiol 146:609–617, 1997.
29. Tsitouras PD, Martin CE, Harman SM. Relationship of serum testosterone to sexual activity in healthy elderly men. J Gerontol 37:288–293, 1982.
30. Hickey K, Do KA, Green A. Smoking and prostate cancer. Epidemiol Rev 23:115–125, 2001.
31. Strickler HD, Goedert JJ. Sexual behavior and evidence for an infectious cause of prostate cancer. Epidemiol Rev 23:144–151, 2001.
32. Ross RK, Deapen DM, Casagrande JT, Paganini-Hill A, Henderson BE. A cohort study of mortality from cancer of the prostate in Catholic priests. Br J Cancer 43:233–235, 1981.
33. Dennis LK, Hayes RB. Alcohol and prostate cancer. Epidemiol Rev 23:110–114, 2001.
34. Nomura AM. Body size and prostate cancer. Epidemiol Rev 23:126–131, 2001.
35. Lee IM, Sesso HD, Chen JJ, Paffenbarger RS Jr. Does physical activity play a role in the prevention of prostate cancer? Epidemiol Rev 23:132–137, 2001.
36. Wheeler GD, Wall SR, Belcastro AN, Cumming DC. Reduced serum testosterone and prolactin levels in male distance runners. JAMA 252:514–516, 1984.
37. Nelson JE, Harris RE. Inverse association of prostate cancer and non-steroidal anti-inflammatory drugs (NSAIDs): results of a case-control study. Oncol Rep 7:169–170, 2000.
38. Lim JT, Piazza GA, Han EK, Delohery TM, Li H, Finn TS, Buttyan R, Yamamoto H, Sperl GJ, Brendel K, Gross PH, Pamukcu R, Weinstein IB. Sulindac derivatives inhibit growth and induce apoptosis in human prostate cancer cell lines. Biochem Pharmacol 58:1097–1107, 1999.
39. Hsing AW, Devesa SS. Trends and patterns of prostate cancer: what do they suggest? Epidemiol Rev 23:3–13, 2001.
40. Chen C. Risk of prostate cancer in relation to polymorphisms of metabolic genes. Epidemiol Rev 23:30–35, 2001.
41. Cheng E, Lee C, Grayhack J. Endocrinology of the prostate. In: Lepor H, Lawson RK (eds). Prostate Diseases. Philadelphia: WB Saunders, 1993, pp 57–71.
42. Russell DW, Wilson JD. Steroid 5α-reductase: two genes/two enzymes. Annu Rev Biochem 63:25–61, 1994.
43. Thigpen AE, Silver RI, Guileyardo JM, Casey ML, McConnell JD, Russell DW. Tissue distribution and ontogeny of steroid 5α-reductase isozyme expression. J Clin Invest 92:903–910, 1993.
44. Wilson JD, Griffin JE, Russell DW. Steroid 5α-reductase 2 deficiency. Endocrinol Rev 14:577–593, 1993.
45. Reichardt JKV, Makridakis N, Henderson BE, Yu MC, Pike MC, Ross RK. Genetic variability of the human *SRD5A2* gene: implications for prostate cancer risk. Cancer Res 55:3973–3975, 1995.
46. Makridakis N, Ross RK, Pike MC, Chang L, Stanczyk FZ, Kolonel LN, Shi C-Y, Yu MC, Henderson BE, Reichardt JKV. A prevalent missense substitution that modulates activity of prostatic steroid 5α-reductase. Cancer Res 57:1020–1022, 1997.

47. Makridakis NM, di Salle E, Reichardt JK. Biochemical and pharmacogenetic dissection of human steroid 5alpha-reductase type II. Pharmacogenetics 10:407–413, 2000.
48. Margiotti K, Sangiuolo F, De Luca A, Froio F, Pearce CL, Ricci-Barbini V, Micali F, Bonafe M, Franceschi C, Dallapiccola B, Novelli G, Reichardt JK. Evidence for an association between the SRD5A2 (type II steroid 5alpha-reductase) locus and prostate cancer in Italian patients. Dis Markers 16:147–150, 2000.
49. Mononen N, Ikonen T, Syrjakoski K, Matikainen M, Schleutker J, Tammela TL, Koivisto PA, Kallioniemi OP. A missense substitution A49T in the steroid 5-alpha-reductase gene (SRD5A2) is not associated with prostate cancer in Finland. Br J Cancer 84:1344–1347, 2001.
50. Makridakis NM, Ross RK, Pike MC, Crocitto LE, Kolonel LN, Pearce CL, Henderson BE, Reichardt JK. Association of mis-sense substitution in SRD5A2 gene with prostate cancer in African-American and Hispanic men in Los Angeles, USA. Lancet 354:975–978, 1999.
51. Thompson IM, Coltman CA Jr, Crowley J. Chemoprevention of prostate cancer: the Prostate Cancer Prevention Trial. Prostate 33:217–221, 1997.
52. Irvine RA, Yu MC, Ross RK, Coetzee GA. The CAG and GGC microsatellites of the androgen receptor gene are in linkage disequilibrium in men with prostate cancer. Cancer Res 55:1937–1940, 1995.
53. Coetzee GA, Ross RK. Prostate cancer and the androgen receptor. J Natl Cancer Inst 86:872–873, 1994.
54. Mhatre AN, Trifiro MA, Kaufman M, Kazemi-Esfarjani P, Figlewicz D, Rouleau G, Pinsky L. Reduced transcriptional regulatory competence of the androgen receptor in X-linked spinal and bulbar muscular atrophy. Nat Genet 5:184–188, 1993.
55. Chamberlain NL, Driver ED, Miesfeld RL. The length and location of CAG trinucleotide repeats in the androgen receptor N-terminal domain affect transactivation function. Nucleic Acids Res 22:3181–3186, 1994.
56. Edwards A, Hammond HA, Jin L, Caskey CT, Chakraborty R. Genetic variation at five trimeric and tetrameric tandem repeat loci in four human population groups. Genomics 12:241–253, 1992.
57. Ingles SA, Ross RK, Yu MC, Irvine RA, La Pera G, Haile RW, Coetzee GA. Association of prostate cancer risk with genetic polymorphisms in vitamin D receptor and androgen receptor. J Natl Cancer Inst 89:166–170, 1997.
58. Ma H, Hong H, Huang SM, Irvine RA, Webb P, Kushner PJ, Coetzee GA, Stallcup MR. Multiple signal input and output domains of the 160-kilodalton nuclear receptor coactivator proteins. Mol Cell Biol 19:6164–6173, 1999.
59. Irvine RA, Ma H, Yu MC, Ross RK, Stallcup MR, Coetzee GA. Inhibition of p160-mediated coactivation with increasing androgen receptor polyglutamine length. Hum Mol Genet 9:267–274, 2000.
60. Xue WM, Coetzee GA, Ross RK, Irvine R, Kolonel L, Henderson BE, Ingles SA. Genetic determinants of serum prostate-specific antigen levels in healthy men from a multiethnic cohort. Cancer Epidemiol Biomarkers Prev 10:575–579, 2001.
61. Rao A, Cramer SD. Identification of a polymorphism in the ARE I region of the PSA promoter. Proc Am Assoc Cancer Res 40:65, 1999.
62. Xue W, Irvine RA, Yu MC, Ross RK, Coetzee GA, Ingles SA. Susceptibility to prostate cancer: interaction between genotypes at the androgen receptor and prostate-specific antigen loci. Cancer Res 60:839–841, 2000.
63. Cohen P, Peehl DM, Graves HC, Rosenfeld RG. Biological effects of prostate specific antigen as an insulin-like growth factor binding protein-3 protease. J Endocrinol 142:407–415, 1994.
64. Waterman MR, Keeney DS. Genes involved in androgen biosynthesis and the male phenotype. Horm Res 38:217–221, 1992.
65. Picado-Leonard J, Miller WL. Cloning and sequence of the human gene for P450c17 (steroid 17alpha-hydroxylase/17,20-lyase): similarity with the gene for P450c21. DNA 6:439–448, 1987.
66. Carey AH, Waterworth D, Patel K, White D, Little J, Novelli P, Franks S, Williamson R. Polycystic ovaries and premature male pattern baldness are associated with one allele of the steroid metabolism gene CYP17. Hum Mol Genet 3:1873–1876, 1994.
67. Haiman CA, Hankinson SE, Spiegelman D, Colditz GA, Willett WC, Speizer FE, Kelsey KT, Hunter DJ. The relationship between a polymorphism in CYP17 with plasma hormone levels and breast cancer. Cancer Res 59:1015–1020, 1999.
68. Feigelson HS, Coetzee GA, Kolonel LN, Ross RK, Henderson BE. A polymorphism in the CYP17 gene increases the risk of breast cancer. Cancer Res 57:1063–1065, 1997.
69. Lunn RM, Bell DA, Mohler JL, Taylor JA. Prostate cancer risk and polymorphism in 17 hydroxylase (CYP17) and steroid reductase (SRD5A2). Carcinogenesis 20:1727–1731, 1999.
70. Wadelius M, Andersson AO, Johansson JE, Wadelius C, Rane E. Prostate cancer associated with CYP17 genotype. Pharmacogenetics 9:635–639, 1999.
71. Bartsch W, Klein H, Schiemann U, Bauer HW, Voigt KD. Enzymes of androgen formation and degradation in the human prostate. Ann NY Acad Sci 595:53–66, 1990.
72. Berube D, Luu The V, Lachance Y, Gagne R, Labrie F. Assignment of the human 3beta-hydroxysteroid dehydrogenase gene (HSDB3) to the p13 band of chromosome 1. Cytogenet Cell Genet 52:199–200, 1989.

73. Rheaume E, Simard J, Morel Y, Mebarki F, Zachmann M, Forest MG, New MI, Labrie F. Congenital adrenal hyperplasia due to point mutations in the type II 3β-hydroxysteroid dehydrogenase gene. Nat Genet 1:239–245, 1992.
74. Lachance Y, Luu-The V, Verreault H, Dumont M, Rheaume E, Leblanc G, Labrie F. Structure of the human type II 3beta-hydroxysteroid dehydrogenase/delta5-delta4 isomerase (3beta-HSD) gene: adrenal and gonadal specificity. DNA Cell Biol 10:701–711, 1991.
75. Devgan SA, Henderson BE, Yu MC, Shi CY, Pike MC, Ross RK, Reichardt JK. Genetic variation of 3β-hydroxysteroid dehydrogenase type II in three racial/ethnic groups: implications for prostate cancer risk. Prostate 33:9–12, 1997.
76. Chan JM, Stampfer MJ, Giovannucci E, Gann PH, Ma J, Wilkinson P, Hennekens CH, Pollak M. Plasma insulin-like growth factor-I and prostate cancer risk: a prospective study. Science 279:563–566, 1998.
77. Harman SM, Metter EJ, Blackman MR, Landis PK, Carter HB. Serum levels of insulin-like growth factor I (IGF-I), IGF-II, IGF-binding protein-3, and prostate-specific antigen as predictors of clinical prostate cancer. J Clin Endocrinol Metab 85:4258–4265, 2000.
78. Nickerson T, Pollak M, Huynh H. Castration-induced apoptosis in the rat ventral prostate is associated with increased expression of genes encoding insulin-like growth factor binding proteins 2, 3, 4 and 5. Endocrinology 139:807–810, 1998.
79. Nickerson T, Huynh H. Vitamin D analogue EB1089-induced prostate regression is associated with increased gene expression of insulin-like growth factor binding proteins. J Endocrinol 160:223–229, 1999.
80. Pollak M, Beamer W, Zhang JC. Insulin-like growth factors and prostate cancer. Cancer Metastasis Rev 17:383–390, 1998.
81. Kaplan PJ, Mohan S, Cohen P, Foster BA, Greenberg NM. The insulin-like growth factor axis and prostate cancer: lessons from the transgenic adenocarcinoma of mouse prostate (TRAMP) model. Cancer Res 59:2203–2209, 1999.
82. Huynh H, Seyam RM, Brock GB. Reduction of ventral prostate weight by finasteride is associated with suppression of insulin-like growth factor I (IGF-I) and IGF-I receptor genes and with an increase in IGF binding protein 3. Cancer Res 58:215–218, 1998.
83. Rosen CJ, Kurland ES, Vereault D, Adler RA, Rackoff PJ, Craig WY, Witte S, Rogers J, Bilezikian JP. Association between serum insulin growth factor-I (IGF-I) and a simple sequence repeat in IGF-I gene: implications for genetic studies of bone mineral density. J Clin Endocrinol Metab 83:2286–2290, 1998.
84. Deal C, Ma J, Wilkin F, Paquette J, Rozen F, Ge B, Hudson T, Stampfer M, Pollak M. Novel promoter polymorphism in insulin-like growth factor-binding protein-3: correlation with serum levels and interaction with known regulators. J Clin Endocrinol Metab 86:1274–1280, 2001.
85. Feldman D, Skowronski RJ, Peehl DM. Vitamin D and prostate cancer. In: American Institute for Cancer Research (ed). Diet and Cancer: Molecular Mechanisms of Interactions. New York: Plenum Press, 1995, pp 53–63.
86. Schwartz GG, Hill CC, Oeler TA, Becich MJ, Bahnson RR. 1,25-Dihydroxy-16-ene-23-yne-vitamin D_3 and prostate cancer cell proliferation in vivo. Urology 46:365–369, 1995.
87. Giovannucci E, Rimm EB, Wolk A, Ascherio A, Stampfer MJ, Colditz GA, Willett WC. Calcium and fructose intake in relation to risk of prostate cancer. Cancer Res 58:442–447, 1998.
88. Taylor JA, Hirvonen A, Watson M, Pittman G, Mohler JL, Bell DA. Association of prostate cancer with vitamin D receptor gene polymorphism. Cancer Res 56:4108–4110, 1996.
89. Ma J, Stampfer MJ, Gann PH, Hough HL, Giovannucci E, Kelsey KT, Hennekens CH, Hunter DJ. Vitamin D receptor polymorphisms, circulating vitamin D metabolites, and risk of prostate cancer in United States physicians. Cancer Epidemiol Biomarkers Prev 7:385–390, 1998.
90. Ingles SA, Coetzee GA, Ross RK, Henderson BE, Kolonel LN, Crocitto L, Wang W, Haile RW. Association of prostate cancer with vitamin D receptor haplotypes in African-Americans. Cancer Res 58:1620–1623, 1998.

16

Androgen Receptor Signaling in Prostate Cancer

WAYNE D. TILLEY
GRANT BUCHANAN
GERHARD A. COETZEE

The androgen-signaling axis is the principal regulator of the development, function, and growth of the prostate gland and plays a vital role in prostate cancer predisposition and progression. The major components of this axis include the biosynthesis and transport of testosterone to target tissues, where it is converted to the more active metabolite 5α-dihydrotestosterone (DHT); maturation of the androgen receptor (AR) to a ligand-binding-competent form; nuclear import; and the subsequent transcriptional regulation of AR target genes. The AR, which is the pivotal component of androgen signaling, is a member of the superfamily of nuclear transcription factors that regulate a diverse range of cellular functions by providing a direct link between signaling molecules and gene transcription.[1–3] The AR is unique among nuclear receptors (NRs) in that a strong constitutive transactivation function involving at least three overlapping regions of the N-terminal domain (NTD) is responsible for most, if not all, of its transactivation activity. The AR NTD also contains two polymorphic regions (polyQ and polyG stretches) that are associated with prostate cancer predisposition. In particular, there is an inverse correlation between polyQ size and both receptor activity and risk of developing prostate cancer.

The status of the androgen-signaling axis is especially important in patients who either are diagnosed with or subsequently develop metastatic prostate cancer. Currently, the success of the only treatment option for metastatic disease, androgen ablation [i.e., orchidectomy, treatment with luteinizing hormone–releasing hormone (LHRH) agonists/antagonists and/or AR antagonists[4,5]], is dependent on a functional androgen-signaling axis. Despite an initial response to androgen ablation in 80%–90% of patients, these hormonal therapies are essentially palliative and disease progression eventually ensues.[5,6] Resistance to androgen ablation is not necessarily due to loss of androgen sensitivity but may develop as a consequence of a deregulated androgen-signaling axis resulting from increased levels of, or gain of function mutations in, the AR gene, altered interaction between the receptor and coregulatory molecules, or ligand-independent activation of the AR by growth factors and cytokines.[7–12] In this chapter, we review the contribution of molecular aberrations in the AR gene to genetic predisposition, tumor progression, and the development of resistance to androgen-ablation therapies. These new insights into AR signaling in prostate cancer provide a unifying conceptual framework for understanding the development of resistance to androgen-ablation therapies.

ANDROGEN RECEPTOR STRUCTURE AND FUNCTION

The AR is a member of the superfamily of nuclear transcription factors, which consists of more than 300 members across vertebrates, arthropods, and nematodes and includes receptors for steroid hormones, vitamin D, retinoic acids, and thyroid hormones, as well as a number of "orphan" receptors for which no ligand has yet been identified.[1] Specifically, the AR is a member of the class III NR family, including receptors for estrogen (ER), progestin, glucocorticoids (GR), mineralocorticoids (MR), and the orphan estrogen-related receptors, which represent the terminal derivatives of cholesterol biosynthesis.[3] Like many other NRs [e.g., GRs, progesterone receptors (PRs), and ERs], the AR contains a ligand-dependent activation function (AF-2) in helix 12 of the ligand-binding domain (LBD), which interacts predominantly with LxxLL motifs of the p160 coactivators. A key functional role for AF-2 is to recruit, in a ligand-dependent manner, chromatin-remodeling factors, which may bridge or stabilize the interaction between the DNA-bound AR and proteins in the basal transcription complex.

Gene and Protein Structure

The human AR is encoded by a large single-copy gene (>90 kb), located on the long arm of the X chromosome at Xq11–12, that is organized into eight exons divided by long intronic sequences (Fig. 16.1).[13–15] Although the AR promoter lacks classic TATA or CAAT box elements, there are more than 22 potential binding sites for known transcription factors, including Sp1, the cyclic adenosine monophosphate (cAMP) response element binding protein factor, sex determining region Y (SRY), nuclear factor κB (NF-κB), and a critical homopurine/homopyrimidine stretch adjacent to the functional GC box that directs efficient recruitment of transcription factor IID (TFIID).[16] Although the AR gene appears to be ubiquitously expressed, the extent of expression in various tissues differs by two to three orders of magnitude.[17]

Two AR mRNA species, of approximately 10.5 and 7.5 kb, which arise from alternative splicing of the AR transcript, have been identified in human tissues, with the 10.5 kb form observed at significantly greater abundance (Fig. 16.1).[18] The AR coding sequence of approximately 3.0 kb gives rise to two naturally occurring AR isoforms by translation initiation at two methionine residues.[19] The predominant isoform, AR-B, is an approximately 917–amino acid protein with a molecular weight of 98.9 kDa (Fig. 16.1).[14,15,20] The smaller AR-A isoform lacks the N-terminal 187 amino acids of AR-B and has a molecular weight of 80 kDa. Although both isoforms coexist in a variety of adult tissues, AR-A is present at about one-tenth the level of AR-B.[19] Post-translational modifications, including phosphorylation, account for the observed size of AR A (~87 kDa) and AR B (~110 kDa) on sodium dodecyl sulfate polyacrylamide gel electrophoresis.

The AR protein can be defined broadly in terms of three major functional domains: a large NTD encoded entirely by exon 1, a DNA-binding domain (DBD) encoded by exons 2 and 3, and a carboxy-terminal LBD encoded by exons 4–8 that contains a highly conserved ligand-dependent transactivation function (AF-2) (Fig. 16.1).[21] The DBD and LBD are highly conserved among the steroid receptors, while the NTD and the hinge, which is a small region separating the DBD from the LBD encoded by the 5′ portion of exon 4, show poor homology with the corresponding domains of other receptors.

Amino-Terminal Transactivation Domain

The large NTD of the AR is unique among NRs in that it is responsible for almost the entire transactivation potential of the protein. It contains at least three overlapping constitutive transactivation domains, which, in the context of the full-length receptor, are activated or inhibited by conformational changes resulting from hormone binding at the LBD or by other mechanisms. In particular, two activation functions, AF-1 (aa 51–210) and AF-5 (aa 369–492), which contain LxxLL-like motifs, appear to cooperate to facilitate ligand-induced N and C interactions, the latter with an interaction surface that overlaps with AF-2 in helix 12 of the LBD (Fig. 16.1).

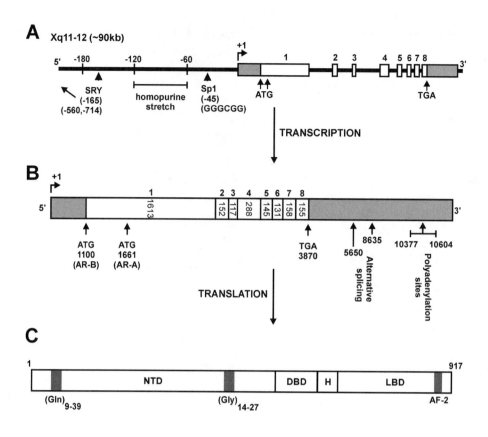
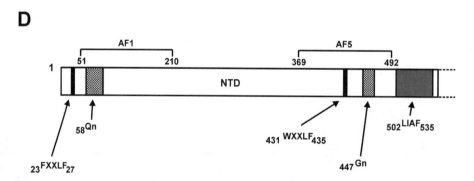

Figure 16.1 Androgen receptor (AR) gene and protein structure. **A:** Schematic representation of the AR gene on chromosome Xq11–12 showing important binding sites for sex determining region Y (SRY) and SP1 transcription factors. Individual exons are separated by up to 16 kb of intronic sequence. **B:** AR mRNA transcript showing alternative splice and polyadenylation sites. Translation is primarily directed from the first of two initiating methionine residues. **C:** Structure of the predominant (AR-B) form of the AR. Indicated are the N-terminal transactivation domain (NTD), DNA-binding domain (DBD), hinge region (H), ligand-binding domain (LBD), and ligand-dependent activation function (AF-2). **D:** The structure of the NTD showing the position of the two polymeric amino acid stretches (polyglutamine, Qn; polyglycine, Gn), ligand-independent activation function (LIAF), constitutive activation functions (AF1 and AF5), and the position of the two LxxLL-like motifs implicated in the interaction between the NTD and LBD. Structures are not to scale.

Access to the hydrophobic binding surface of LxxLL-like peptides of coactivators also is blocked by helix 12 of the LBD. An N–C interaction can occur intra- or intermolecularly and is associated with increased affinity of receptor binding to androgen response elements of target gene promoters and stabilization of the receptor complex. It is not clear if there is functional redundancy between AF-1 and AF-5 in the AR NTD and AF-2 in the LBD or whether both NTD activation functions are absolutely required for recruitment of coregulatory factors and transcription initiation. We and others have shown that p160 cofactors can form a bridge to tether the AR N and C termini via interaction between a glutamine-rich region of the cofactor and the LxxLL-like motifs in the AR NTD and between LxxLL motifs of the cofactor and AF-2 in the AR.[22,23] These interactions may be promoted by the cofactor to stabilize the receptor for maximal activity.

The transactivation potential of subdomains in the AR NTD is modulated by two polymorphic trinucleotide microsatellite repeats, CAG and GGN, in the coding sequence. The normal size distribution of these microsatellites in the population is 9–39 CAG and 14–27 GGN repeats, with a modal repeat size of 21 and 23, respectively.[24,25] These repeats encode variable-length polyglutamine (polyQ) and polyglycine (polyG) tracts, respectively, in the receptor (Fig. 16.1). The NTD also contains a strong ligand-independent activation function (LIAF, aa 502–535; see "Multiple Pathways of Androgen Receptor Activation" below) located downstream of AF-5 (Fig. 16.1), which appears to be silenced by interaction with an inhibitory subdomain (ISD) in the NTD containing the polyQ region.[22]

Expansion of the CAG microsatellite to 40 or more repeats causes a rare, X-linked, adult-onset, neurodegenerative disorder called spinal and bulbar muscular atrophy (SBMA), or Kennedy's disease.[26,27] In addition to progressive muscle weakness and atrophy due to loss of brain stem and spinal cord motor neurons, men with this disorder frequently present with symptoms of partial androgen insensitivity (i.e., gynecomastia and testicular atrophy), indicative of aberrant AR function.[28,29] Receptor proteins encoded by SBMA AR alleles have normal androgen-binding affinities but reduced transactivation capacity compared to wild-type AR.[30,31] Indeed, an inverse relationship between AR transactivation activity and CAG repeat length has been established over a CAG size range encompassing normal AR alleles.[22,32–34]

Ligand-Binding Domain

The ability of steroid receptors to respond rapidly to hormone-signaling events in a highly sensitive and specific manner resides in their ability to discriminate between low circulating levels of structurally similar hormones. Remarkably, this facility is bestowed by the highly homologous LBDs of these receptors. The recently resolved crystal structure determined that the AR LBD, like those of other steroid receptors, consists of 11α helices [as with the progesterone receptor (PR), the AR LBD has no helix 2) and four short β sheets arranged in a "helical sandwich" around the protected, relatively compact ligand-binding pocket (see Color Fig. 16.2 in separate color insert).[35–37] High-affinity ligands such as DHT, testosterone, and the synthetic androgen R1881 make direct contact with 18 individual amino acids in helices 3, 4, 5, 7, and 11 and with one of the β sheets. Although the crystal structure of the apo-AR has not been determined, by inference from the unliganded crystal structures of other steroid receptors, it appears that high-affinity ligand binding to the AR causes significant conformational changes resulting in the repositioning of the critical AF-2 region of helix 12 across the ligand-binding pocket onto the scaffold formed by helices 3, 4, and 5.[36–39] This interaction results in the formation of a distinct hydrophobic cleft in the surface of the receptor, which is critical for receptor activity and can bind short hydrophobic LxxLL-like motifs found in coregulatory proteins and the AR NTD (see Color Fig. 16.2 in separate color insert).[40–42]

Conformational Maturation

Following translation of the AR, conformational maturation of steroid receptors by a multiprotein chaperone heterocomplex is essential for the acquisition of ligand-binding competence.[43] The specific details of this process are poorly defined, but in general, it requires at least three heat-shock proteins, Hsp40, Hsp70, and Hsp90, and the cochaperones p23 and Hop.[43] In the final stages of receptor maturation, Hsp90 becomes directly associated with the receptor LBD in a process stabilized by p23 and with one of the tetratricopeptide repeat (TPR)–containing proteins, which include the immunophilins FKBP51/52 and CyP40 and the protein-serine phosphatase PP5.[43] Interaction of the cochaperone p23 with Hsp90 is an absolute requirement for heterocomplex stabilization of unliganded NRs.[43] Although many TPR-containing proteins appear to have a similar affinity for Hsp90, they exhibit specific preferences for different steroid receptors and may play a role in hormone action by altering the affinity and specificity for ligand. It is thought that the Hsp90-containing heterocomplex dynamically associates with steroid receptors to maintain them in a conformation that, although unstable, has a high affinity for ligand binding.[43] Following hormone binding in the cytoplasm, the Hsp90-containing heterocomplex is dissociated and the steroid receptor rapidly translocated into the nucleus.

DNA Binding

In the nucleus, the AR dimerizes and binds to the DNA double helix at specific DNA sequences called androgen-response elements (AREs). The DBDs of steroid receptors are highly conserved, cysteine-rich sequences that contain two zinc finger motifs and a short C-terminal extension that forms part of the hinge (see Color Fig. 16.2 in separate color insert).[44,45] By inference from studies of the GR, the first zinc finger mediates DNA recognition and binds to the major groove of DNA,[46,47] while conserved amino acids in the DBD (D box) of the second zinc finger mediate dimerization between steroid receptor monomers. Recognition of specific DNA sequences by steroid receptors is determined by conserved amino acids in the P box of the first zinc finger, which contact specific base pairs in consensus steroid-response elements.[44] However, despite their diverse biological effects, the P boxes of the closely related steroid receptors AR, GR, MR and PR are almost identical and the core DBD, which includes both zinc fingers, shares up to 73% identity.[45]

Response elements for steroid receptors generally consist of hexameric half-sites, separated by three nucleotides, arranged as either inverted repeats (symmetrical palindromes) or direct repeats (see Color Fig. 16.2 in separate color insert). Recent evidence suggests that half-sites arranged as inverted repeats may induce head-to-head dimerization of the AR, while the polarity of direct repeats may lead to head-to-tail dimerization.[47–49] This is in contrast to the GR, which is able to bind only to AREs arranged as inverted repeats. Receptor selectivity for particular target sequences and alternative modes of DNA binding may be determined by sequence differences in the C-terminal extension of the DBD (see Color Fig. 16.2 in separate color insert), which exhibits only ~30% identity between steroid receptors.[50] Following binding of the first zinc finger to the major grove of DNA at the hexameric recognition site, an α helix formed by residues in the C-terminal extension enters the minor grove of DNA, binding to the hexameric half-site from the opposite side.[48] Together, these various sequences mediate specific AR binding and the unique regulation of target genes by androgens.[51] Following ligand dissociation, the AR is shuttled back to the cytoplasm, where it can reassociate with Hsp90 and ligand, subsequently undergoing multiple rounds of nucleocytoplasmic recycling and gene activation.[52]

Recruitment of Cofactors

The DNA-bound AR dimer recruits a multiprotein complex containing members of the basal transcription machinery (e.g., TFIIF) and additional essential proteins, termed cofactors, which upregulate (*coactivators*) or inhibit (*corepressors*) target gene expression.[53] The p160 family of coactivators, consisting of three related proteins, steroid receptor coactivator-1 (SRC-1), glucocorticoid receptor interacting protein-1 (GRIP1) and amplified in breast cancer-1 (AIB1) are the best-characterized of the NR-interacting proteins. The p160 coactivators are recruited to the ligand-bound receptor, where they interact directly with the hydrophobic cleft formed by AF-2 residues in the LBD via a set of conserved LxxLL-like motifs termed *NR boxes*.[54] The p160 coactivators enhance steroid receptor transcriptional activity by actively recruiting secondary coactivators, such as p300/cAMP response element binding protein (CBP) and p300/CBP-associated factor (PCAF), resulting in chromatin remodeling via targeted histone acetylation and stable assembly of the preinitiation transcriptional complex, leading to enhanced rates of transcription initiation by RNA polymerase II.[53] In contrast, corepressors such as the related proteins silencing mediator of retinoic and thyroid hormone receptors (SMRT) and NR-corepressor (N-CoR) are thought to interact with only the apo-LBD of the steroid receptors. However, recent studies in our laboratories (Buchanan, Coetzee, and Tilley, unpublished data) suggest that SMRT may interact with multiple domains of the AR by ligand-dependent and ligand-independent mechanisms and may play an important role in AR signaling. The following is a summary of the key AR coregulatory molecules.

Androgen Receptor Coactivators

P160 COACTIVATORS

Three related proteins (SRC-1, GRIP1/TIF2, pCIP/ACTR/RAC3/AIB1/TRAM1) belong to the p160 coactivator family and collectively are the best characterized of the NR coactivators.[55] Each has three (or in the case of SRC-1a, isoform, four) LxxLL motifs called NR boxes, which bind to the AF-2 region of all NRs. While this binding interaction is conserved among all NRs, the specific binding interaction varies. Most NRs can bind to small regions of p160 proteins containing just the NR boxes. However, the AR hormone-binding domain requires the NR boxes plus an additional region of the p160 coactivator located far downstream near the CBP binding site.[56] The AR and GR hormone-binding domains also prefer to bind NR box 3 of GRIP1 or NR box 4 of SRC-1a, which is located at the extreme C terminus of one alternatively spliced isoform of SRC-1.[42] The AR can also bind to p160 coactivators by a second mechanism: the NTD of AR (LIAF region) can bind the C-terminal region of all three p160 proteins.[22,23,57] In this way, the p160 cofactors can function as a bridge between and can stabilize the N and C termini of the AR to facilitate dimerization. Phosphorylation of SRC1 by mitogen-activated protein kinase (MAPK) has been reported to enhance its interaction with NRs, suggesting that this may also be an important step to achieve maximal target gene transcription.[58] The p160 proteins furthermore facilitate AR-mediated transcriptional activation by recruiting additional coactivators, including p300/CBP, coactivator-associated arginine methyltransferase-1 (CARM1), and possibly PCAF. p300/CBP and PCAF can acetylate histones to help remodel nucleosomes and can acetylate other proteins in the transcription-initiation complex;[59] CARM1 is a histone methyltransferase, suggesting that methylation of histones or other proteins in the transcriptional machinery may also contribute to transcriptional activation.[60]

BRCA1

Mutations in *BRCA1* have been associated with familial breast cancer susceptibility. While functions for *BRCA1* in DNA repair have been proposed, it was shown that *BRCA1* represses the activity of the ER.[61] However, we have recently demonstrated that *BRCA1* functions as a coactivator for AR (and for ER under most conditions).[62] Additionally, *BRCA1* interacts with the AR, primarily through the LIAF region of the NTD.[62] While the mechanism is unknown, this function of *BRCA1* suggests that the AR may mediate the tumor-suppressor or growth-regulatory effects of *BRCA1*. Somatic mutations in the LIAF region of the AR potentially could abrogate such regulation.

P300/CBP AND CARM1

p300/CBP and CARM1 are considered secondary coactivators normally recruited to the preinitiation transcription complex by primary coactivators like the p160 family members.[55] They provide necessary histone acetylation (p300/CBP) or methylation (CARM1) activities for the conversion of chromatin to transcriptional permissive local structures. As stated above, evidence exists for an additional binding of p300/CBP to the LIAF subdomain of the AR. Recruitment of these secondary coactivators might be affected by structural changes in the AR NTD.

ANDROGEN RECEPTOR-ASSOCIATED PROTEINS

Chang and colleagues[63] identified several AR-associated (ARA) proteins that interact with either the AR LBD (ARA70/RFG/ELEI, ARA55, ARA54) or the AR NTD (ARA24, ARA160/TMF). The AR N-terminal interacting protein (ARA160), also known as TATA element modulatory factor, cooperates with ARA70 to enhance AR activity. Another AR NTD interacting protein, ARA24, interacts with the poly-Q tract. Binding of ARA24 is decreased by expanding the polyQ length within the AR NTD, the latter being inversely correlated with the transcriptional activity of AR. Additional studies have demonstrated that AR and some select AR coactivators, such as ARA70 and ARA54, can interact with CBP and PCAFs that have histone acetyltransferase activity for assisting in chromatin remodeling. Wang et al.[64] identified a new ARA protein, ARA267-α, that interacts with both the AR DBD and LBD. Unlike other coregulators, such as CBP, ARA267-α exerts little influence on AR N–C interactions but enhances AR transactivation in a ligand-dependent manner in prostate cancer cells. Also, ARA267-α enhances AR transactivation with other coregulators, such as ARA24 and PCAF in an additive manner. Collectively, these observations suggest that optimal AR transactivation in prostate cancer cells requires interaction of AR with an appropriate assembly of ARA coregulators.

Androgen Receptor Corepressors

SMRT
The corepressor SMRT was discovered in the Evans lab at the Salk Institute and inhibits NR-mediated transactivation activity.[54] It also acts as a powerful inhibitor of AR signaling (data not shown). Gain-of-function mutations in the AR might abolish/inhibit interaction with SMRT, thereby increasing AR transactivation activity.

CYCLIN D1
Cyclin D1 is a required component of the cyclin-dependent kinase-4 (CDK4) complex, which plays a role in cell cycle control via phosphorylation of the retinoblastoma tumor-suppressor gene.[65] Cyclin D1 possibly binds directly to the AR NTD and inhibits AR transactivation activity.[66] It is not known exactly which regions or subdomains in the AR NTD are necessary for cyclin D1 binding or how structural alterations in the AR might affect the binding of cyclin D1 and its AR inhibitory activity.

SHP
The protein short heterodimer partner (SHP) is an orphan receptor that lacks a DBD[67] and interacts with the AR NTD, causing dramatic inhibition of AR-mediated transactivation activity due to competition with AR coactivators like GRIP1/TIF2.[68]

MULTIPLE PATHWAYS OF ANDROGEN RECEPTOR ACTIVATION

Ligand-Dependent Activation

The sequence of events in the ligand-dependent activation of the AR and regulation of androgen-responsive genes is shown in Figure 16.3 and can be summarized as follows:

1. In the absence of native ligand (DHT), immature AR is complexed in the cytoplasm to a multiprotein chaperone complex, which is essential for receptor maturation and the acquisition of ligand-binding competence.[43]
2. Following hormone binding, the complex dissociates and the receptor is rapidly translocated into the nucleus.[52]
3. In the nucleus, the AR dimerizes and binds in the major groove of the DNA double helix at specific AREs. One can speculate that at this stage the receptor dimer remains transcriptionally repressed because the large NTD is folded in such a manner that protein-interaction surfaces in the AF subdomains are not assessable for binding to other proteins; this main inhibitory function is achieved by the interaction of AF-1 with AF-5 plus the silenced LIAF. Access to the hydrophobic binding surface of LxxLL-like peptides of coactivators is blocked by helix 12 of the

LBD. This structure is further stabilized by tight intra- or intermolecular interactions, potentially between multiple regions of the NTD with AF-2 in helix 12 and/or with other regions of the LBD.[41,69,70]

4. The subsequent transactivation reaction depends on the relative competitive recruitment of coactivators vs. corepressors to interaction surfaces in the LBD and NTD. The binding of both types of coregulators is a consequence of, or results in, conformational changes that open protein–protein interaction motifs in the AR. Coactivators additionally recruit histone acetylases and methylases, while corepressors recruit histone deacetylases to the complex, resulting in chromatin decondensation (activation) and condensation (inhibition), respectively.[71]

5. Dissociation of the AR from promoters and recycling between cytoplasm and nucleus occur multiple times.[52]

Ligand-Independent Activation

Binding of high-affinity androgenic ligands is not the only mechanism by which the AR can activate target gene sequences. There is an accumulating body of evidence that the AR can be activated in in vitro systems in the absence of native ligand by growth factors (keratinocyte growth factor, insulin-like growth factor I, and epidermal growth factor), cytokines [interleukin-6 (IL-6)], protein kinase A, and overexpression of the tyrosine kinase receptor HER2/neu.[9,72–75] The mechanism(s) that causes LIA of the AR is best understood in the case of HER2/neu overexpression. HER2/neu is a transmembrane glycoprotein member of the epidermal growth factor receptor family and is overexpressed in carcinomas of the breast, ovary, stomach, and prostate.[76] Overexpression of HER2/neu in androgen-responsive prostate cancer cell lines enhances AR transactivation of androgen-regulated genes such as *PSA* in a ligand-independent manner and increases cell survival during androgen deprivation.[75] HER2/neu potentially affects two phosphorylation pathways leading to LIA of AR activity. One pathway involves activation of Akt and phosphorylation of the AR at two serine residues (213, 793) and results in suppression of androgen-induced apoptosis.[55] The other pathway involves MAPK. Inhibitors of HER2/neu and MAPK or mutation of AR-S513 (S513A),[77] which is located in a consensus MAPK phosphorylation site in the LIAF region of the AR NTD, abrogate AR transactivation activity in transient transfection assays in the presence of low concentrations of androgen.[9,74] We have identified multiple AR gene mutations in the LIAF region of the receptor in hormone-refractory prostate cancer patients. One of these mutations results in substitution of a glycine for a serine (S513G) in the same codon of the MAPK phosphorylation site that was mutated to an alanine in earlier studies.[9] Collectively, these observations suggest that structural alterations in this region of the AR NTD may impact on LIA of the receptor by MAPK. A significant difference between LIA of AR by growth factors and by the MAPK pathway is that the latter is not inhibited by casodex, suggesting that activation of the AR is independent of the LBD. Ueda et al.[75] showed that LIA of the AR by IL-6 involves the MAPK/signal transducer and activator of transcription 3 (STAT3) signaling pathway in LNCaP cells, and that this activity is mediated by the AR NTD. Thus, the MAPK pathway appears to be a key mediator of LIA of the AR by both IL-6 and HER2/neu.

ANDROGEN RECEPTOR POLYMORPHISMS AND PROSTATE CANCER RISK

Androgen Receptor CAG Repeat

In 1992, Edwards et al.[24] reported the allelic frequency distribution of AR CAG repeat size in different U.S. racial/ethnic populations as part of a larger survey of genetic variation in a series of different trimeric and tetrameric tandem repeats. Among African Americans, the frequency of AR alleles with fewer than 22 CAG repeats was 65% compared to 53% in Caucasians and 34% in Asian Americans. On the basis of these observations, Coetzee and Ross[78] hypothesized that AR CAG repeat length might be associated with the higher risk of prostate cancer in African Americans and the intermediate and low risks in

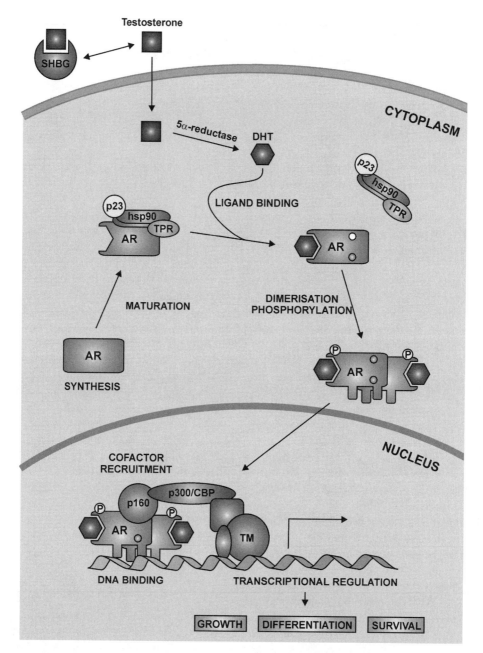

Figure 16.3 Mechanism of androgen receptor (AR) activation by ligand-dependent pathways. Following synthesis, the AR exists in dynamic equilibrium between an immature state and an active form capable of binding high-affinity androgenic ligands via association/dissociation with a complex that includes heat-shock proteins (hsp), p23, and a tetratricopeptide (TPR) containing protein. Ligand binding results in the dissociation of this complex, receptor dimerization and phosphorylation, nuclear transport, DNA binding, recruitment of components of the transcription machinery (TM) and other cofactor molecules (such as the p160 coactivators), and ultimately activation of androgen-regulated gene pathways. SHBG, sex hormone–binding globulin; DHT, dihydrotestosterone; CBP, cyclic adenosine monophosphate response element binding protein.

Caucasians and Asian Americans, respectively, and that enhanced transcriptional activity of receptors with a shorter AR CAG allele could promote tumorigenesis by enhancing prostatic epithelial cell turnover.

In 1995, the same investigators directly tested this hypothesis in a pilot case-control study comprising 68 prostate cancer patients and 123 control subjects.[79] In agreement with Edwards et al.,[24] there was a prevalence of short AR CAG alleles in African-American vs. Caucasian and Asian-American controls. In addition, modest, though not statistically significant, enrichment of short AR CAG alleles was observed in the Caucasian prostate cancer patients. These findings were extended in an expanded follow-up study, which showed a significantly higher prevalence of short AR CAG alleles among prostate cancer patients, especially those with advanced disease (Table 16.1).[80] In addition to our studies, Hakimi et al.[81] identified a subgroup of patients diagnosed with advanced prostate cancer who had shorter AR CAG repeats. Hardy et al.,[82] furthermore, demonstrated an association between age at onset and AR CAG repeat length.

Subsequently, several well-designed, matched case-control studies demonstrated an approximately twofold increased prostate cancer risk, decreased age at onset, and/or increased risk of advanced disease for reduced AR CAG repeat length (Table 16.1). Giovannucci et al.[25] used a population selected from the Physicians Health Study that included 587 prostate cancer cases and 588 matched controls. The large sample size allowed the stratification of cases by tumor grade and stage. A highly significant inverse correlation between AR CAG repeat length and risk of developing prostate cancer was observed when repeat size was analyzed as a semicontinuous variable. Short AR CAG alleles also correlated with increased risk of having advanced disease, defined as a high stage or high-grade tumor at diagnosis.[25] Stanford et al.[83] analyzed AR CAG repeat length and prostate cancer risk in 301 prostate cancer cases and 277 matched controls. They noted only a small increase in the frequency of AR CAG alleles with fewer than 22 repeats in cancer patients compared to controls. Nevertheless, when AR CAG repeat size was examined as a continuous variable, an overall age-adjusted relative odds of developing prostate cancer of 0.97 was observed for each additional CAG. Hsing et al.[84] reported that AR CAG alleles were significantly shorter in prostate cancer patients compared to controls among Shanghai Chinese. This study was the first to demonstrate an association in a non-Caucasian population. In a case-control study in an Australian Caucasian population, no association was observed between AR CAG repeat length and prostate cancer risk, but a significant effect of age at onset was observed.[85] In other studies (Table 16.1), associations between AR CAG repeat length and prostate cancer risk were not consistently observed, possibly due to small sample sizes, population differences, and/or failure to appropriately match cases and controls.[86–96]

While the epidemiological studies discussed above consistently provided evidence for an association between AR CAG repeat length and prostate cancer risk, they did not address the molecular mechanisms underlying changes in receptor activity. As stated above, in vitro transient cotransfection studies have shown that ARs with longer polyQ tracts (encoded by the polymorphic CAG repeat) have normal ligand-binding affinity but lower transactivation activity.[30–33,85] Protein expression levels are unlikely to account for this effect since they have been found to be similar for ARs containing 9–42 polyQ repeats.[22] However, two studies have reported that AR constructs with longer repeat lengths (CAG-50 to CAG-52) are unstable and undergo accelerated degradation, potentially in a ligand-dependent manner.[22,97] The polyQ size effect in AR transactivation activity observed in most in vitro studies is thought to be mediated, at least in part, through altered functional interactions with cofactors. In transient cotransfection experiments, the p160 coactivators GRIP1, AIB1, and SRC-1 exaggerate the relative difference in AR transactivation activity with altered polyQ length.[22] As the p160 coactivators bind to regions of the AR distinct from the polyQ tract, this effect may be mediated by steric hindrance of p160–receptor interactions when polyQ length is increased.[22] The RAS-related G protein Ran/ARA24, which binds to the AR NTD in the polyQ region, is an AR cofactor that ap-

Table 16.1 Associations between Androgen Receptor (AR)-CAG and/or -GGN Microsatellites and Prostate Cancer Risk, Nature of Disease at Diagnosis, and Age at Onset

		AR-CAG Repeat Associations			AR-GGN Repeat Associations		
Reference	Subjects	Risk	State/Grade	Age at Onset	Risk	Stage/Grade	Age at Onset
Pilot studies							
Irvine et al., 1995[79]	US Caucasian	Yes	N/A	N/A	Yes	N/A	N/A
Hardy et al., 1996[82]	US Caucasian	N/A	No	Yes	N/A	N/A	N/A
Ingles et al., 1997[80]	US Caucasian	Yes	Yes	N/A	N/A	N/A	N/A
Hakimi et al., 1997[81]	US Caucasian	Yes	Yes	No	Yes	No	No
Matched case-control studies							
Giovannucci et al., 1997[25]	US Caucasian	Yes	Yes	No	N/A	N/A	N/A
Stanford et al., 1997[83]	US Caucasian	Yes	No	Yes	Yes	No	Yes
Platz et al., 1998[102]	US Caucasian	N/A	N/A	N/A	Yes	N/A	N/A
Hsing et al., 2000[84]	Chinese	Yes	No	No	Yes	No	No
Beilin et al., 2001[85]	Australian Caucasian	No	No	Yes	N/A	N/A	N/A
Other studies							
Ekman et al., 1999[86]	Swedish Caucasian	Yes	N/A	N/A	N/A	N/A	N/A
Edwards et al., 1999[87]	British Caucasian	No	No	N/A	Yes	No	N/A
Correa-Cerro et al., 1999[88]	French/German Caucasian	No	No	No	No	No	No
Bratt et al., 1999[89]	Swedish Caucasian	No	Yes	Yes	N/A	N/A	N/A
Lange et al., 2000[90]	US Caucasian (high risk)	No	No	No	N/A	N/A	N/A
Nam et al., 2001[92]	Canadian	N/A	Yes	No	N/A	N/A	N/A
Latil et al., 2001[92]	French Caucasian	No	No	Yes	N/A	N/A	N/A
Modugno et al., 2001[93]	US Caucasian	Yes	N/A	N/A	N/A	N/A	N/A
Miller et al., 2001[94]	US Caucasian	No	N/A	N/A	No	N/A	N/A
Panz et al., 2001[95]	S. Africans (black and white)	Yes	Yes	N/A	N/A	N/A	N/A
Cude et al., 2002[96]	Unknown US population	N/A	Yes	No	N/A	N/A	N/A

N/A, not applicable or not assessed; Yes, association between polymorphism and listed parameter; No, no significant association detected between polymorphism and listed parameter.

pears to enhance AR activity in a polyQ size-dependent manner.[98] Given the well-described role for Ran in protein nuclear transport, it is possible that larger polyQ tracts inhibit the efficiency of Ran-directed AR nuclear import.[99] Moreover, recent observations from our laboratories (Buchanan, Coetzee, and Tilley, unpublished data) indicate that the corepressor SMRT largely abolishes the differences in transactivation activity mediated by different Q_n lengths. Together, these results suggest that net AR NTD transactivation activity may be a function of the relative amounts of coactivators vs. corepressors and that the penetrance of the polyQ variations in the AR NTD can be affected by the ratios of different transcriptional coregulators. Clearly, more studies are required to determine whether the effects of other cofactors that act in a cell-, promoter, and/or AR specific manner can be directly influenced by polyQ length and to determine how variation in AR polyQ length can influence prostate cancer cell growth.

Androgen Receptor GGN Repeat

Allelic distributions of the GGN microsatellite are significantly different among racial/ethnic groups,[79] with the 23-repeat GGN allele being least prevalent among high-risk African Americans (i.e., 20%) and most prevalent in low-risk Asians Americans (i.e., 70%). This is suggestive of a protective role for this allele in prostate cancer risk. It is possible that the 23-repeat GGN allele encodes an AR containing a polyG tract of optimal length for normal receptor function in prostatic epithelial cells. While this is speculative, as it is not known whether variation in polyG length modulates AR activity, a weak, though non-significant, paucity of the 23-repeat GGN allele was observed among Caucasian prostate cancer patients compared to control subjects, suggesting that there is enrichment of putative risk alleles (i.e., non-23-repeat GGN alleles) among cases.[79]

Because the AR gene is X-linked, with each male inheriting a single maternal copy, it is possible to define a putative AR prostate cancer risk allelotype of short CAG (i.e., <22 repeats) and non-23-repeat GGN. Indeed, the distribution of this allelotype has been shown to be significantly different among control subjects, with African Americans and Asians having the highest and lowest prevalence, respectively.[79] Among Caucasian prostate cancer patients, the <22 CAG/non-23-repeat GGN haplotype conferred a twofold increase in risk of prostate cancer, although statistical significance was not reached.[79] Among prostate cancer patients, a nonrandom distribution of CAG and GGN alleles was also observed: 66% of patients with a short CAG allele also had a non-23-repeat GGN allele, while only 25% of patients with long CAG alleles had a non-23-repeat GGN allele. As the CAG and GGN microsatellites are in close proximity at the AR locus, it was not surprising to find evidence of linkage disequilibrium between the intragenic markers in patient samples. In contrast, there was no evidence of linkage disequilibrium between control samples when assessed either together or by ethnicity. This indicates that in normal men either one or both of the microsatellites are hypermutable, resulting in a random distribution of CAG and GGN alleles at the AR locus. Indeed, when the rate of mutation at the CAG microsatellite was measured using single-cell assays of sperm, an exceptionally high rate of 1%–4% was observed.[100] Collectively, these data suggest that a nonrandom subset of CAG and GGN AR alleles occurs in men with prostate cancer.

In three matched case-control studies (Table 16.1), a positive association between AR GGN repeat length variation and prostate cancer risk was found.[84,101,102] The failure to consistently demonstrate this association in other studies (Table 16.1) might be due to a lack of statistical power and/or failure to appropriately match cases with controls. A more detailed assessment of the effects of the AR GGN repeat on prostate cancer risk awaits elucidation of the effects of alterations in polyG tract length on AR function.

ANDROGEN RECEPTOR AND LOCALIZED DISEASE

The role of AR in the progression of clinically localized prostate cancer has only recently been addressed.[103,104] Henshall et al.[103] reported that AR was expressed in more than 70% of the tumor cells in localized prostate cancer but that there was loss of AR immunoreactivity in the adjacent peritumoral stroma, which was associated with earlier relapse after radical prostatectomy. Sweat et al.[104] found no association between AR expression and disease progression in a highly selected cohort of tumors with a Gleason score of 6–9. In our laboratory, the level of AR protein in tumor foci determined by video image analysis was a strong predictor of the risk of relapse following radical prostatectomy (unpublished observations). While further studies are necessary to determine how AR influences disease progression in clinically localized prostate cancer, a number of mechanisms have been identified in prostatic tumors that potentially explain the increase in levels of AR immunostaining observed in tumor cells in our study. These mechanisms include amplification of the AR gene,[106] changes in the methylation status of the AR promoter and hence transcription of the AR gene,[107,108] altered stability of AR mRNA,[109] and ligand-independent activation of the receptor.[9,110] Irrespective of the mechanism, increased AR levels likely result in altered expression profiles of androgen-regulated proteins, including angiogenic factors, cell adhesion molecules, and cell cycle regulators (e.g., vascular endothelial growth factor, integrins, and CDKs and their inhibitors,[111–113]) which could collectively contribute to disease progression.

HORMONAL THERAPIES FOR METASTATIC PROSTATE CANCER

Due to the advent of screening with serum prostate-specific antigen (PSA) in the 1990s, the majority of prostate cancer cases are now diagnosed with clinically organ-confined disease. However, the assumption that earlier diagnosis would result in improved survival has not been ratified. In particular, approximately 30% of patients with organ-confined disease and 60%–80% of those with extraprostatic disease who receive potentially curative local therapy will eventually relapse.[114] These data suggest that micrometastases are already present at the time of diagnosis.

Although it is more than 60 years since Huggins et al.[115] established that surgical castration could provide effective symptomatic relief for men with prostate cancer, targeting the androgen-signaling pathway remains the predominant form of treatment for patients who are either diagnosed with or subsequently develop metastatic disease (Fig. 16.4).[4,5] A myriad of therapeutic agents that target the androgen-signaling pathway have been developed since the initial observation of Huggins and colleagues.[115] The majority of these agents reduce circulating levels of androgens (LHRH agonists/antagonists such as goserelin and leuprorelin or the estrogens diethylstilbestrol and estramustine phosphate) or block the action of androgens through their primary cellular target, the AR (the steroidal antiandrogen cyproterone acetate and pure antiandrogens such as hydroxyflutamide, bicalutamide, and nilutamide; Fig. 16.4). However, despite an initial response to androgen ablation in 80%–90% of patients with metastatic disease, androgen ablation is essentially palliative and disease progression eventually ensues.[5,6] Few patients who receive androgen-ablative therapies survive for more than 5 years following initiation of hormonal therapies for metastatic disease.[5,6]

As adrenal androgens also have the potential to contribute to prostate cell growth, pure antiandrogens have been used in combination with an LHRH agonist/antagonist, i.e., combined androgen blockade (CAB). In the late 1980s, Labrie and colleagues[116] reported that CAB could prolong survival of patients with metastatic prostate cancer by about 17 months. However, subsequent reports were conflicting, and currently most patients initially are treated with monotherapy, usually LHRH agonists. Indeed, a recent systematic review of CAB, encompassing 20 individual trials and more than 6000 patients, concluded that it results in only a modest increase in survival compared to monotherapy alone but is more likely to be associated with adverse events and reduced quality of life.[117] Recent evidence regarding the mechanisms contributing to therapy failure (see below, Androgen Receptor and the Failure of Androgen-Ablation Therapies) suggests that combinational approaches with LHRH agonists and receptor antagonists cannot completely abrogate androgen action and may select for cells with a growth state permissive for a particular hormonal envi-

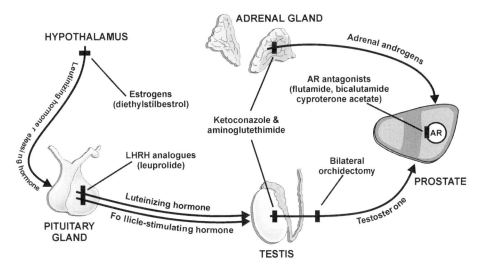

Figure 16.4 Androgen signaling axis in prostate cancer: androgen ablation therapies. The growth and development of the normal prostate require a functioning androgen-signaling pathway, which originates with the hypothalamus–pituitary axis. The continued reliance of metastatic prostate cancer on androgen signaling for growth is exploited in androgen ablation therapies, which aim to either reduce the circulating level of androgens or block their action in the prostate. Lines represent the path of the normal signaling molecules, while dark bars represent the molecular or cellular target of agents that have been used to treat prostate cancer. Combined androgen blockade, which uses a combination of luteinizing hormone–releasing hormone (LHRH) analogues and androgen receptor (AR) antagonists, is often used for metastatic disease that has progressed on monotherapy.

ronment. The potential clinical importance of a therapy-mediated selective pressure is illustrated by the syndrome of steroid-hormone and anti-antiandrogen withdrawal, which is characterized by tumor regression and decreasing serum levels of PSA when treatment with an antiandrogen, progestational agent, or estrogen is selectively discontinued at a time of clinical progression.[118] A withdrawal response has been observed in up to 30% of patients with hormone-refractory prostate cancer when treatment with the antiandrogen hydroxyflutamide is terminated[118–120] and has also been documented following withdrawal of the AR antagonists nilutamide and bicalutamide, the estrogens diethylstilbestrol and megestrol acetate, and the progestational agent chlormadinone acetate.[121–126] Withdrawal responses have been reported at a higher incidence following combined therapy, consisting of castration or LHRH agonists in combination with an AR antagonist, compared to antagonist alone, leading to the conclusion that prolonged exposure to antiandrogens was the predominant factor in the withdrawal response rather than a low level of androgens.[122,127] In one study, inhibition of adrenal steroid production with ketoconazole following discontinuation of antiandrogen therapy resulted in a higher proportion of patients (55%) exhibiting a withdrawal response and an increased duration of response[128] than reported for withdrawal of the antiandrogen alone.[127]

Resistance to androgen-ablation therapies may not necessarily be due to loss of androgen sensitivity but may develop as a consequence of a deregulated androgen-signaling axis resulting from (1) amplification and/or altered expression of the AR, (2) mutation of the AR gene, (3) inappropriate interaction with AR coregulatory molecules (co-activators, corepressors), or (4) ligand-independent activation of the AR by growth factors and cytokines.[7–12] In addition, many somatic genetic alterations implicated in prostate cancer initiation and progression may have a direct effect on androgen signaling in prostate cells (Buchanan, Tilley, and Coetzee, unpublished observations).

ANDROGEN RECEPTOR AND THE FAILURE OF ANDROGEN-ABLATION THERAPIES

Androgen Receptor Levels

Initial studies using both androgen-unresponsive Dunning rat adenocarcinoma and human prostate cancer cell lines suggested that loss of AR mRNA and protein could be a mechanism to explain the failure of androgen-ablation therapies.[129,130] Subsequent immunohistochemical studies of clinical prostate cancer demonstrated that the AR is expressed in essentially all metastatic tumors, including those that continue to grow following androgen ablation.[131–137] Other studies have shown that the AR in recurrent, hormone-refractory prostate cancer is expressed at levels similar to those in androgen-dependent prostate tumors.[138] Recent studies using the recurrent CWR22 xenograft model and its derived cell line, CWR-R1, as well as the LNCaP C4-2 cell line derived from LNCaP cells after prolonged periods of culture in the absence of androgen suggest that the AR is expressed at similar levels in both androgen-dependent and recurrent tumors but is more stable in recurrent tumors in the absence of androgen.[139] In addition, the concentration of androgen required for stimulation of the CWR-R1 and LNCaP C4-2 cell lines is fourfold lower than that required for androgen-dependent LNCaP cells. This concentration of DHT is comparable to the levels in prostates of men treated with androgen ablation. The observations that AR is expressed at relatively high levels in recurrent CWR22 human tumor xenografts and cell lines and is hypersensitive to low levels of androgens suggest that androgen signaling and the associated activation of androgen-regulated genes in human prostate tumors is sufficient to maintain tumor growth following androgen ablation. Kim et al.[140] reported marked reduction in AR expression in CWR22 tumors 2 days post-castration but subsequent reexpression of AR and androgen-regulated genes in recurrent tumors 150 days post-castration to levels comparable to those observed in tumors from intact mice. Reactivation of AR expression was associated with renewed proliferation of tumor cells at 120–150 days postcastration, suggesting that this is a critical factor contributing to renewed tumor growth.

Amplification of the AR gene, which also has the potential to contribute to increased AR protein levels, has been reported in 22% of prostate cancer metastases[141] and in 23%–28% of primary tumors following androgen deprivation.[106,142] An average twofold increase in the levels of both AR and PSA proteins has been reported in prostate tumor samples with AR gene amplification compared to samples where no AR amplification was found.[7,143] Increased AR protein levels in metastatic prostate tumors may also augment the sensitivity of the androgen-signaling axis, thereby contributing to disease progression during the course of androgen ablation. In addition, AR gene amplification and overexpression of the receptor have been reported in hormone-refractory prostate cancer following monotherapy but not in primary prostate tumors,[7] suggesting that hormonal therapy could select for cells with an ability to maintain growth during treatment.

Androgen Receptor Gene Mutations

The first indication that AR gene mutations might contribute to the failure of androgen-ablation therapies came from studies of the androgen-responsive human prostate cancer cell line LNCaP. The AR in LNCaP cells contains a single–amino acid substitution (Thr-Ala^{877}), which facilitates inappropriate activation by glucocorticoids, progestins, adrenal androgens, estradiol, and the antiandrogen hydroxyflutamide.[144,145] Subsequently, somatic missense mutations have been detected throughout the AR coding sequence at frequencies of up to 50% in advanced primary tumors and metastatic deposits.[10,146–149] These mutations consistently result in receptors that exhibit decreased specificity of ligand binding and enhanced receptor activation by androgens and non-classical ligands compared to wild-type AR (wtAR).[150,151] More recently, in collaboration with Dr. Norman Greenberg (Baylor College of Medicine, Houston, TX), we reported the identification of AR gene mutations in the autochthonous transgenic adenocarcinoma of mouse prostate (TRAMP) model.[152] Analogous to the findings in clinical prostate cancer, AR gene mutations detected in TRAMP tumors also result in receptors that contribute to altered androgen signaling.[152] In the TRAMP model, different hormonal environments result in the selection of AR variants with mutations in distinctly different regions of the receptor.[152] All of the missense AR gene mutations (6/6) identified in tumors derived from intact TRAMP mice at 24–28 weeks of age were located in the LBD (Fig. 16.5).[152] In contrast, most of the missense mutations (7/9) identified in mice castrated at 12 weeks of age were located in the AR NTD. Four of the mutations identified in castrated TRAMP mice resulted in receptors with increased transactivation function in the absence of ligand. Selection for AR gene mutations with a phenotype permissive for growth is also evident from studies of clinical prostate cancer. In one study, AR gene missense mutations were detected more frequently in patients treated with CAB using hydroxyflutamide (5/16) compared to monotherapy (1/17) with androgen ablation.[148] The same mutation in codon 877 (T877A) was found in all five of the patients treated with CAB. A different mutation resulting in a D890N substitution in the AR was identified in the patient treated with monotherapy. Whereas no difference in activation of the D890N variant and wtAR was observed, functional studies demonstrated that the T877A mutation resulted in receptors exhibiting a marked increase in activity in response to hydroxyflutamide but not to DHT or other androgenic ligands, suggesting that flutamide treatment selects for tumor cells expressing a hydroxyflutamide-inducible AR variant.[148] Haapala and colleagues[153] reported the identification of AR gene mutations in 36% (4/11) of tumors from

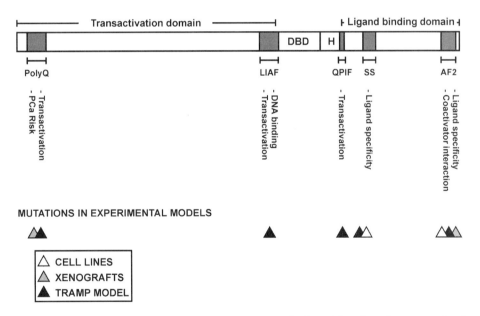

Figure 16.5 Collocation of androgen receptor (AR) gene mutations in prostate cancer (PCa). The majority (80%) of AR gene mutations identified in clinical prostate cancer collocate to discrete regions within the receptor N-terminal domain (NTD) and ligand-binding domain (LBD) as indicated. Mutations in each of these regions (PolyQ, polyglutamine repeat; LIAF, ligand-independent activation function; QPIF, a tetrapeptide at the boundary of the hinge and LBD; SS, signature sequence; AF2, activation function 2) have been shown to result in changes to receptor activity, which could explain the outgrowth of prostate cancer cells containing these mutations during androgen ablation therapies. The position of mutations identified in human prostate cancer cell lines, xenografts, and the transgenic adenocarcinoma of the mouse prostate (TRAMP) model are indicated. Hormonal treatment of TRAMP mice drives the selection of AR gene mutations to either the NTD (7/9 castrate mice vs. 0/6 intact animals) or the LBD (6/6 intact mice vs. 2/9 castrate).

patients treated with CAB. These mutations are in AR regions different from those reported in codon 877 by Taplin et al.,[148] suggesting that different types of AR variant may be specifically selected for by different treatments. These data support the hypothesis that AR gene mutations identified in prostate cancer provide a selective growth advantage given the appropriate hormonal environment, resulting in the re-emergence of tumor growth during the course of hormone-ablation therapy.

Structural and Functional Collocation of Androgen Receptor Variants

Nearly 80% of missense AR gene mutations identified in clinical prostate cancer cluster to discrete regions of the receptor that collectively span less than 15% of the coding sequence (Fig. 16.5).[151]

LIGAND-BINDING DOMAIN VARIANTS

In the LBD, mutations collocate to *(1)* the "signature sequence," a conserved 20–amino acid region of NRs involved in ligand recognition and specificity;[154] *(2)* AF-2, a binding site for the p160 cofactors; and *(3)* a region at the boundary of the hinge and LBD containing a 4–amino acid tetrapeptide (^{668}QPIF^{671}), which may define a protein–protein interaction surface (Fig. 16.5). Many of the AR gene mutations identified in the LBD of the AR in the TRAMP model, xenografts, and cell lines occur in the same three regions as mutations in clinical prostate cancer (Fig. 16.5). For example, a Phe-Ile^{671} mutation identified in an intact TRAMP mouse colocates to the ^{668}QPIF^{671} tetrapeptide with mutations identified in human prostate cancer.[8] Mutations of the AR gene identified in both clinical prostate cancer and TRAMP tumors in this region exhibit a two- to fourfold greater transactivation activity in response to DHT, nonclassical ligands, and hydroxyflutamide compared to wtAR,[8] without altering ligand-binding kinetics, receptor levels, or DNA-binding capacity. Homology modeling reveals that the ^{668}QPIF^{671} tetrapeptide residues form a potential protein–protein interaction surface, which is markedly disrupted by the naturally occurring mutations, providing a mechanism that could explain the observed gain in transcriptional activity.[8]

Another AR missense mutation identified in the TRAMP model, Phe-Ser^{697}, is located adjacent to the signature sequence. The Ser697 AR variant exhibits markedly reduced transactivation responses to progesterone and 17β-estradiol but enhanced response to R1881 compared to wtAR.[152] These results are consistent with the role of the signature sequence in ligand recognition and specificity and with previous reports of mutations in this region in clinical disease.[10,151] Analysis of the Thr-Ala^{877} AR variant, identified in a significant proportion of clinical prostate tumors and in the human prostate cancer cell line LNCaP has confirmed that this mutation exhibits increased transactivation activity in response to progesterone, 17β-estradiol, adrenal androgens, and hydroxyflutamide compared to wtAR.[155] The recently determined AR LBD crystal structure[36,37] allowed us to use homology modeling to demonstrate that this mutation results in changes to the shape and volume of the ligand-binding pocket such that bulkier ligands like progesterone can be accommodated.[152]

DNA-BINDING DOMAIN VARIANTS

In addition to the above observations, five somatic missense mutations have been identified in clinical prostate tumors that collocate to a 14–amino acid region at the carboxyl-terminal end of the first zinc finger motif in the DBD of the AR.[146,149] The effect of each of these mutations is unknown, but none of the codons in which they occur has been reported to contain mutations that cause receptor inactivation in the clinical syndrome of androgen insensitivity. Mutations in the AR DBD selectively affect the transactivation and transrepression functions of the AR on different promoters despite reduced DNA-binding ability[156,157] and may represent a predisposing factor for male breast cancer.[158] Due to the high homology of the DBD across members of the NR superfamily, the cell and promoter specificity of different receptors is, in part, mediated by only a few changes in DBD sequence.[159] Mutations in the DBD may result in AR variants that bind to response elements normally specific for other NRs,[158] leading to inappropriate activation or repression of growth-regulatory pathways. In an analogous manner, mutations in AREs increase the sensitivity of the enhancer for GR.[160] Modeling experiments suggest that residues in the AR DBD could form a protein interaction surface,[158] and several AR coactivators that interact with the DBD in a ligand-dependent manner[161–163] are predicted to alter receptor activity via local chromatin remodeling,[161] interaction with components of the transcriptional machinery,[162] or inhibition of nuclear export.[164] It is also possible that mutations in the DBD of the AR gene identified in prostate cancer could alter the affinity of receptor binding to response elements, resulting in altered expression of a range of target genes regulated by the AR.

AMINO-TERMINAL TRANSACTIVATION DOMAIN VARIANTS

Nearly half of the AR gene mutations identified in clinical prostate cancer are located in the NTD of the receptor. Mutations of the AR gene in this domain have also been identified in the TRAMP model, and analogous to the observations for the LBD, these mutations cluster with those identified in clinical prostate cancer to discrete regions within the NTD that are implicated in receptor function. The main regions of colocation in the NTD are (1) within and adjacent to the polyQ tract (codons 54–78), which, as discussed above, has been implicated in modulating receptor activity and prostate cancer risk, and (2) a region amino-terminal to the DBD (codons 502–535) known to modulate the transactivation capacity of the receptor in both a ligand-dependent and ligand-independent manner (i.e., LIAF) (Fig. 16.5).[165,166]

Somatic contractions in the CAG repeat of the AR gene, which potentially increase AR activity in a subpopulation of cells and thereby contribute to disease progression, have been identified in three independent studies of clinical prostate tumors.[167–169] In addition, we have identified a somatic mutation within the CAG repeat of the AR gene in a primary prostate tumor that results in interruption of the polyQ repeat by two leucine residues. This AR variant has a two- to fourfold greater ability to transactivate target genes compared to wtAR in the presence of physiological concentrations of DHT.[170] Four additional somatic mutations have been identified in or adjacent to the CAG repeat region of the AR gene in human prostate cancer[146] but have not yet been characterized. Further analysis of inherited and somatic alterations in this region of the AR gene in prostate cancer is warranted to determine the contribution of this motif to AR activity and its potential to influence the development and/or progression of the disease.

A second region of the NTD where amino acid substitutions have been identified in hormone-refractory prostate tumors is the LIAF. We examined the complete coding sequence of the AR gene for mutations in metastatic tissue biopsies from 12 patients who exhibited the clinical syndrome of steroid-hormone and anti-androgen withdrawal response to hydroxyflutamide.[151] Five mutations identified in this study colocated with a previously identified mutation[146] to a small carboxyl-terminal portion (aa 502–535) of the NTD of the AR. An additional mutation (Asp-Gly^{526}) previously identified in our studies of primary prostate tumors[146] and another identified in the TRAMP model[152] colocate to this region of the AR. This carboxyl-terminal region of the AR NTD modulates the transactivation capacity of the receptor in both a ligand-dependent and ligand-independent manner[165,166] and is involved in direct interactions with the p160 coactivators and the transcription regulator p300/CBP.[22,23,171] This region of the NTD also contains the binding site for the receptor accessory factor, which enhances the specific DNA binding of rat AR.[172] Further evidence in support of the LIAF region being important in AR transactivation has been provided by studies demonstrating that mutation of S513, which is located in a consensus MAPK phosphorylation site, partially inhibits ligand-independent activation of the receptor by *HER-2/neu*.[9] Collectively, these observations suggest that the LIAF region may contain an interaction surface for accessory proteins that promote ligand-independent transactivation. Mutations in this region may alter the ability of the receptor to respond to these and other coregulatory proteins, thereby altering the transactivation capacity of the AR in a manner that could provide a growth advantage to prostate cancer cells in an appropriate hormonal environment.

Altered Interaction with Coregulatory Molecules

Gregory et al.[173] showed that overexpression of the p160 coactivators TIF2/GRIP1 and SRC-1, observed in recurrent tumors from CWR22 human prostate xenografts and clinical prostate cancer, increases AR transactivation capacity at physiological concentrations of nonclassical ligands (adrenal androgens, estradiol, and progesterone). Overexpression of p160 coactivators, especially TIF2, favors an interaction between the coactivator and AF-2 in preference to formation of an N–C interaction (between an LxxLL-like motif in the NTD and a hydrophobic interaction surface that overlaps with AF-2). Overexpression of TIF2 could therefore result in the recruitment of TIF2 to AF-2 by low-affinity steroids, resulting in inhibition of N–C interaction and derepression of LIAF in the NTD.

Characterization of the functional relationship between the AR and ARA coregulators by Yeh and colleagues[174,175] has shown that ARA70 and ARA55 can enhance the androgenic effects of 17β-estradiol and hydroxyflutamide, the latter an antiandrogen commonly used in the treatment of metastatic prostate cancer. ARA70 can enhance the androgenic activity of 17β-estradiol and antiandrogens toward AR, suggesting that specificity of sex hormones and agonist activity of antiandrogens can be modulated by different coregulators. Their studies also demonstrated that ARA70, ARA55, and ARA54, but not SRC-1 and Rb, could significantly enhance AR-mediated transactivation by androst-5-ene-3,17β-diol, a precursor to testosterone. Thus, it is possible that the specificity and sensitivity of sex hormones and antiandrogens for activating the AR can be selectively modulated by AR coactivators, thereby contributing to the growth of prostate tumors in the presence of low concentrations of endogenous androgenic steroids or nonclassical ligands.

Characterization of a Glu-Gly[236] substitution identified in a TRAMP tumor revealed that the variant receptor had increased transactivation function compared to wtAR in response to R1881 and 17β-estradiol only in the presence of the coactivator ARA70 and increased response to R1881, but not 17β-estradiol, in the presence of the coactivator ARA160.[152] In addition, phosphorylation of cofactors (e.g., SRC-1) by MAPK could alter the transactivation activity of the AR by more effectively recruiting the cofactor to the basal transcription complex[58] or by bridging AF-1/AF-5 in the NTD with AF-2, thereby promoting an N–C interaction and a maximal functional response. Collectively, these findings suggest that the phenotype of some AR gene mutations may be apparent only in the presence of the appropriate milieu of coregulators. All of the mechanisms discussed above theoretically could result in enhanced AR transactivation activity in vivo in the presence of low concentrations of DHT or in the absence of circulating levels of native ligand, thereby contributing to the development of resistance to androgen-ablation therapies used in the treatment of metastatic prostate cancer. These findings also suggest that disruption of the interaction between the AR and AR-specific cofactors may serve as a novel therapeutic target in the treatment of metastatic prostate cancer.

Ligand-Independent Activation of the Androgen Receptor

Another potentially important mechanism contributing to the failure of androgen ablation is LIA of the AR. Aberrant expression of growth factor receptors also contributes to the development and progression of prostate cancer.[10,151] HER2/*neu* (c-*erb*B-2), a transmembrane glycoprotein member of the epidermal growth factor receptor family, is overexpressed in carcinomas of the breast, ovary, stomach, and prostate.[176–179] Unlike other epidermal growth factor receptor members, HER2/*neu* has intrinsic tyrosine kinase activity and mediates signal transduction in the absence of ligand.[180] HER2/*neu* expression is increased in hormone-refractory prostate tumors compared to earlier stages of disease.[76,181] Overexpression of HER2/*neu* in androgen-responsive prostate cancer cell lines enhances AR transactivation of androgen-regulated genes such as PSA in a ligand-independent manner and increases cell survival during androgen deprivation.[9,74] Although the mechanism involved in HER2/*neu* modulation of AR transactivation has not been fully characterized, HER2/*neu* expression is associated with activation of the MAPK and Akt (protein kinase B) pathways, which have been implicated in LIA of the AR.[9,110] Tumor cells with increased HER2/*neu* expression and high AR may have a selective growth advantage.

For example, HER2/*neu* activation of an Akt–AR pathway[9,74,110] may confer a clonal advantage by promoting cancer cell survival via the androgen-signaling axis[110] or by induction of Akt-dependent pathways.[180] In a study using prostate cancer xenograft models, monotherapy with a monoclonal antibody directed against activated HER2/*neu* resulted in antiproliferative activity in androgen-dependent LNCaP and CWR22 tumors, but no significant growth inhibition was observed in androgen-independent CWR22 tumors.[182] These observations suggest that signaling through the AR is a requisite for a response to Herceptin in prostate cancer.[182] Thus, patients with elevated levels of both HER2/*neu* and AR immunostaining may benefit from early treatment targeting both the AR and HER2/*neu* signaling cascades.

CONCLUSIONS

Androgen receptor signaling plays a key role in many phases of prostate cancer biology. Carefully matched case-control and other studies suggest that length variation of the AR CAG and GGN polymorphic microsatellite repeats contributes to prostate cancer risk and may influence age at onset and tumor pathology by altering AR transcriptional activity and/or interaction with coregulators. In addition, the consistent findings that AR expression is maintained in virtually all advanced prostate tumors and that increased expression of and/or gain-of-function mutations in the AR gene occur frequently in clinical disease have provided compelling evidence that the failure of androgen-ablation therapy for prostate cancer does not result from loss of androgen signaling but, rather, acquisition of genetic changes that lead to aberrant activation of the androgen-signaling axis. Other mechanisms that potentially contribute to alterations in AR signaling in prostate cancer include inappropriate interaction with coregulatory proteins (e.g., between p160 cofactors and AF-2 to alter the specificity of activation of the receptor) and ligand-independent activation of the AR by growth factors, receptor tyrosine kinases, and cytokines. Thus, resistance to androgen ablation and survival of prostate cancer cells is not necessarily due to the evolution of a growth state that circumvents the androgen-signaling axis but could be explained in part by increased activity of the AR in the presence of either native or alternative ligands or cross-talk with other signaling pathways. This provides a unifying hypothesis for the failure of conventional androgen-ablation therapies and has important implications for the development of novel therapeutic interventions.

ACKNOWLEDGMENTS

This work was supported by grants from the National Health and Medical Research Council of Australia, the Cancer Council of South Australia and the NIH/NCI (R01CA84890).

REFERENCES

1. Laudet V. Evolution of the nuclear receptor superfamily: early diversification from an ancestral orphan receptor. J Mol Endocrinol 19:207–226, 1997.
2. Baker ME. Evolution of 17beta-hydroxysteroid dehydrogenases and their role in androgen, estrogen and retinoid action. Mol Cell Endocrinol 171: 211–215, 2001.
3. Whitfield GK, Jurutka PW, Haussler CA, Haussler MR. Steroid hormone receptors: evolution, ligands, and molecular basis of biologic function. J Cell Biochem Suppl 32–33:110–122, 1999.
4. Santen RJ. Clinical review 37. Endocrine treatment of prostate cancer. J Clin Endocrinol Metab 75:685–689, 1992.
5. Hellerstedt BA, Pienta KJ. The current state of hormonal therapy for prostate cancer. CA Cancer J Clin 52:154–179, 2002.
6. Kozlowski JM, Ellis WJ, Grayhack JT. Advanced prostatic carcinoma: early vs. late endocrine therapy. Urol Clin North Am 18:15–24, 1991.
7. Linja MJ, Savinainen KJ, Saramaki OR, Tammela TL, Vessella RL, Visakorpi T. Amplification and overexpression of androgen receptor gene in hormone-refractory prostate cancer. Cancer Res 61:3550–3555, 2001.
8. Buchanan G, Yang M, Nahm SJ, Han G, Moore N, Bentel JM, Matusik RJ, Horsfall DJ, Marshall VR, Greenberg NM, Tilley WD. Mutations at the boundary of the hinge and ligand binding domain of the androgen receptor confer increased transactivation function. Mol Endocrinol 15:46–56, 2000.
9. Yeh S, Lin HK, Kang HY, Thin TH, Lin MF, Chang C. From HER2/Neu signal cascade to androgen receptor and its coactivators: a novel pathway by induction of androgen target genes through MAP kinase in prostate can-

cer cells. Proc Natl Acad Sci USA 96:5458–5463, 1999.
10. Bentel JM, Tilley WD. Androgen receptors in prostate cancer. J Endocrinol 151:1–11, 1996.
11. Sadar MD, Hussain M, Bruchovsky N. Prostate cancer: molecular biology of early progression to androgen independence. Endocr Relat Cancer 6:487–502, 1999.
12. Jenster G. The role of the androgen receptor in the development and progression of prostate cancer. Semin Oncol 26:407–421, 1999.
13. Brown CJ, Goss SJ, Lubahn DB, Joseph DR, Wilson EM, French FS, Willard HF. Androgen receptor locus on the human X chromosome: regional localization to Xq11–12 and description of a DNA polymorphism. Am J Hum Genet 44:264–269, 1989.
14. Tilley WD, Marcelli M, Wilson JD, McPhaul MJ. Characterization and expression of a cDNA encoding the human androgen receptor. Proc Natl Acad Sci USA 86:327–331, 1989.
15. Lubahn DB, Joseph DR, Sullivan PM, Willard HF, French FS, Wilson EM. Cloning of human androgen receptor complementary DNA and localization to the X chromosome. Science 240:327–330, 1988.
16. Chen S, Supakar PC, Vellanoweth RL, Song CS, Chatterjee B, Roy AK. Functional role of a conformationally flexible homopurine/homopyrimidine domain of the androgen receptor gene promoter interacting with Sp1 and a pyrimidine single strand DNA-binding protein. Mol Endocrinol 11:3–15, 1997.
17. Roy AK, Chatterjee B. Androgen action. Crit Rev Eukaryot Gene Expr 5:157–176, 1995.
18. Lubahn DB, Joseph DR, Sar M, Tan J, Higgs HN, Larson RE, French FS, Wilson EM. The human androgen receptor: complementary deoxyribonucleic acid cloning, sequence analysis and gene expression in prostate. Mol Endocrinol 2:1265–1275, 1988.
19. Wilson CM, McPhaul MJ. A and B forms of the androgen receptor are present in human genital skin fibroblasts. Proc Natl Acad Sci USA 91:1234–1238, 1994.
20. Chang CS, Kokontis J, Liao ST. Molecular cloning of human and rat complementary DNA encoding androgen receptors. Science 240:324–326, 1988.
21. Kallio PJ, Pakvimo JJ, Janne OA. Genetic regulation of androgen action. Prostate Suppl 6:45–51, 1996.
22. Irvine RA, Ma H, Yu MC, Ross RK, Stallcup MR, Coetzee GA. Inhibition of p160-mediated coactivation with increasing androgen receptor polyglutamine length. Hum Mol Genet 9:267–274, 2000.
23. Ma H, Hong H, Huang SM, Irvine RA, Webb P, Kushner PJ, Coetzee GA, Stallcup MR. Multiple signal input and output domains of the 160-kilodalton nuclear receptor coactivator proteins. Mol Cell Biol 19:6164–6173, 1999.
24. Edwards A, Hammond HA, Jin L, Caskey CT, Chakraborty R. Genetic variation at five trimeric and tetrameric tandem repeat loci in four human population groups. Genomics 12:241–253, 1992.
25. Giovannucci E, Stampfer MJ, Krithivas K, Brown M, Dahl D, Brufsky A, Talcott J, Hennekens CH, Kantoff PW. The CAG repeat within the androgen receptor gene and its relationship to prostate cancer. Proc Natl Acad Sci USA 94:3320–3323, 1997.
26. La Spada AR, Wilson EM, Lubahn DB, Harding AE, Fischbeck KH. Androgen receptor gene mutations in X-linked spinal and bulbar muscular atrophy. Nature 352:77–79, 1991.
27. La Spada AR, Roling DB, Harding AE, Warner CL, Spiegel R, Hausmanowa-Petrusewicz I, Yee WC, Fischbeck KH. Meiotic stability and genotype–phenotype correlation of the trinucleotide repeat in X-linked spinal and bulbar muscular atrophy. Nat Genet 2:301–304, 1992.
28. Arbizu T, Santamaria J, Gomez JM, Quilez A, Serra JP. A family with adult spinal and bulbar muscular atrophy, X-linked inheritance and associated testicular failure. J Neurol Sci 59:371–382, 1983.
29. Nagashima T, Seko K, Hirose K, Mannen T, Yoshimura S, Arima R, Nagashima K, Morimatsu Y. Familial bulbo-spinal muscular atrophy associated with testicular atrophy and sensory neuropathy (Kennedy-Alter-Sung syndrome). Autopsy case report of two brothers. J Neurol Sci 87:141–152, 1988.
30. Mhatre AN, Trifiro MA, Kaufman M, Kazemi-Esfarjani P, Figlewicz D, Rouleau G, Pinsky L. Reduced transcriptional regulatory competence of the androgen receptor in X-linked spinal and bulbar muscular atrophy. Nat Genet 5:184–188, 1993.
31. Chamberlain NL, Driver ED, Miesfeld RL. The length and location of CAG trinucleotide repeats in the androgen receptor N-terminal domain affect transactivation function. Nucleic Acids Res 22:3181–3186, 1994.
32. Kazemi-Esfarjani P, Trifiro MA, Pinsky L. Evidence for a repressive function of the long polyglutamine tract in the human androgen receptor: possible pathogenetic relevance for the $(CAG)_n$-expanded neuronopathies. Hum Mol Genet 4:523–527, 1995.
33. Tut TG, Ghadessy FJ, Trifiro MA, Pinsky L, Yong EL. Long polyglutamine tracts in the androgen receptor are associated with reduced *trans*-activation, impaired sperm production, and male infertility. J Clin Endocrinol Metab 82:3777–3782, 1997.
34. Beilin J, Ball EM, Favaloro JM, Zajac JD. Effect of the androgen receptor CAG repeat poly-

morphism on transcriptional activity: specificity in prostate and non-prostate cell lines. J Mol Endocrinol 25:85–96, 2000.
35. Bourguet W, Ruff M, Chambon P, Gronemeyer H, Moras D. Crystal structure of the ligand-binding domain of the human nuclear receptor RXR-alpha. Nature 375:377–382, 1995.
36. Sack JS, Kish KF, Wang C, Attar RM, Kiefer SE, An Y, Wu GY, Scheffler JE, Salvati ME, Krystek SR, Weinmann R, Einspahr HM. Crystallographic structures of the ligand-binding domains of the androgen receptor and its T877A mutant complexed with the natural agonist dihydrotestosterone. Proc Natl Acad Sci USA 98:4904–4909, 2001.
37. Matias PM, Donner P, Coelho R, Thomaz M, Peixoto C, Macedo S, Otto N, Joschko S, Scholz P, Wegg A, Basler S, Schafer M, Egner U, Carrondo MA. Structural evidence for ligand specificity in the binding domain of the human androgen receptor: implications for pathogenic gene mutations. J Biol Chem 275:26164–26171, 2000.
38. Tanenbaum DM, Wang Y, Williams SP, Sigler PB. Crystallographic comparison of the estrogen and progesterone receptor's ligand binding domains. Proc Natl Acad Sci USA 95:5998–6003, 1998.
39. Ribeiro RC, Apriletti JW, Wagner RL, Feng W, Kushner PJ, Nilsson S, Scanlan TS, West BL, Fletterick RJ, Baxter JD. X-ray crystallographic and functional studies of thyroid hormone receptor. J Steroid Biochem Mol Biol 65:133–141, 1998.
40. Heery DM, Kalkhoven E, Hoare S, Parker MG. A signature motif in transcriptional co-activators mediates binding to nuclear receptors. Nature 387:733–736, 1997.
41. He B, Kemppainen JA, Wilson EM. FXXLF and WXXLF sequences mediate the NH_2-terminal interaction with the ligand binding domain of the androgen receptor. J Biol Chem 275:22986–22994, 2000.
42. Ding XF, Anderson CM, Ma H, Hong H, Uht RM, Kushner PJ, Stallcup MR. Nuclear receptor-binding sites of coactivators glucocorticoid receptor interacting protein 1 (GRIP1) and steroid receptor coactivator 1 (SRC-1): multiple motifs with different binding specificities. Mol Endocrinol 12:302–313, 1998.
43. Pratt WB, Toft DO. Steroid receptor interactions with heat shock protein and immunophilin chaperones. Endocr Rev 18:306–360, 1997.
44. Mader S, Leroy P, Chen JY, Chambon P. Multiple parameters control the selectivity of nuclear receptors for their response elements. Selectivity and promiscuity in response element recognition by retinoic acid receptors and retinoid X receptors. J Biol Chem 268:591–600, 1993.
45. Schoenmakers E, Alen P, Verrijdt G, Peeters B, Verhoeven G, Rombauts W, Claessens F. Differential DNA binding by the androgen and glucocorticoid receptors involves the second Zn-finger and a C-terminal extension of the DNA-binding domains. Biochem J 341:515–521, 1999.
46. Baumann H, Paulsen K, Kovacs H, Berglund H, Wright AP, Gustafsson JA, Hard T. Refined solution structure of the glucocorticoid receptor DNA-binding domain. Biochemistry 32:13463–13471, 1993.
47. Luisi BF, Xu WX, Otwinowski Z, Freedman LP, Yamamoto KR, Sigler PB. Crystallographic analysis of the interaction of the glucocorticoid receptor with DNA. Nature 352:497–505, 1991.
48. Rastinejad F, Perlmann T, Evans RM, Sigler PB. Structural determinants of nuclear receptor assembly on DNA direct repeats. Nature 375:203–211, 1995.
49. Gronemeyer H, Moras D. Nuclear receptors. How to finger DNA. Nature 375:190–191, 1995.
50. Schoenmakers E, Verrijdt G, Peeters B, Verhoeven G, Rombauts W, Claessens F. Differences in DNA binding characteristics of the androgen and glucocorticoid receptors can determine hormone-specific responses. J Biol Chem 275:12290–12297, 2000.
51. Reid KJ, Hendy SC, Saito JL, Sorensen P, Nelson CC. Two classes of androgen receptor elements mediate cooperativity through allosteric interactions. J Biol Chem 276:2943–2952, 2001.
52. Tyagi RK, Lavrovsky Y, Ahn SC, Song CS, Chatterjee B, Roy AK. Dynamics of intracellular movement and nucleocytoplasmic recycling of the ligand-activated androgen receptor in living cells. Mol Endocrinol 14:1162–1174, 2000.
53. Jenster G, Spencer TE, Burcin MM, Tsai SY, Tsai MJ, O'Malley BW. Steroid receptor induction of gene transcription: a two-step model. Proc Natl Acad Sci USA 94:7879–7884, 1997.
54. Glass CK, Rosenfeld MG. The coregulator exchange in transcriptional functions of nuclear receptors. Genes Dev 14:121–141, 2000.
55. Lin HK, Yeh S, Kang HY, Chang C. Akt suppresses androgen-induced apoptosis by phosphorylating and inhibiting androgen receptor. Proc Natl Acad Sci USA 98:7200–7205, 2001.
56. Hong H, Yang L, Stallcup MR. Hormone-independent transcriptional activation and coactivator binding by novel orphan nuclear receptor ERR3. J Biol Chem 274:22618–22626, 1999.
57. Bevan CL, Hoare S, Claessens F, Heery DM, Parker MG. The AF1 and AF2 domains of the androgen receptor interact with distinct regions of SRC1. Mol Cell Biol 19:8383–8392, 1999.
58. Rowan BG, Garrison N, Weigel NL, O'Malley BW. 8-Bromo-cyclic AMP induces phosphorylation of two sites in SRC-1 that facilitate ligand-independent activation of the chicken progesterone receptor and are critical for functional

cooperation between SRC-1 and CREB binding protein. Mol Cell Biol 20:8720–8730, 2000.
59. Cheung WL, Briggs SD, Allis CD. Acetylation and chromosomal functions. Curr Opin Cell Biol 12:326–333, 2000.
60. Chen D, Ma H, Hong H, Koh SS, Huang SM, Schurter BT, Aswad DW, Stallcup MR. Regulation of transcription by a protein methyltransferase. Science 284:2174–2177, 1999.
61. Fan S, Wang J, Yuan R, Ma Y, Meng Q, Erdos MR, Pestell RG, Yuan F, Auborn KJ, Goldberg ID, Rosen EM. BRCA1 inhibition of estrogen receptor signaling in transfected cells. Science 284:1354–1356, 1999.
62. Park JJ, Irvine RA, Buchanan G, Koh SS, Park JM, Tilley WD, Stallcup MR, Press MF, Coetzee GA. Breast cancer susceptibility gene 1 (BRCA1) is a coactivator of the androgen receptor. Cancer Res 60:5946–5949, 2000.
63. Sampson ER, Yeh SY, Chang HC, Tsai MY, Wang X, Ting HJ, Chang C. Identification and characterization of androgen receptor associated coregulators in prostate cancer cells. J Biol Regul Homeost Agents 15:123–129, 2001.
64. Wang X, Yeh S, Wu G, Hsu CL, Wang L, Chiang T, Yang Y, Guo Y, Chang C. Identification and characterization of a novel androgen receptor coregulator ARA267-alpha in prostate cancer cells. J Biol Chem 276:40417–40423, 2001.
65. Harbour JW, Dean DC. Rb function in cell-cycle regulation and apoptosis. Nat Cell Biol 2:E65–E67, 2000.
66. Petre CE, Wetherill YB, Danielsen M, Knudsen KE. Cyclin D1: mechanism and consequence of androgen receptor co-repressor activity. J Biol Chem 277:2207–2215, 2002.
67. Seol W, Chung M, Moore DD. Novel receptor interaction and repression domains in the orphan receptor SHP. Mol Cell Biol 17:7126–7131, 1997.
68. Gobinet J, Auzou G, Nicolas JC, Sultan C, Jalaguier S. Characterization of the interaction between androgen receptor and a new transcriptional inhibitor, SHP. Biochemistry 40:15369–15377, 2001.
69. He B, Kemppainen JA, Voegel JJ, Gronemeyer H, Wilson EM. Activation function 2 in the human androgen receptor ligand binding domain mediates interdomain communication with the NH$_2$-terminal domain. J Biol Chem 274:37219–37225, 1999.
70. He B, Bowen NT, Minges JT, Wilson EM. Androgen-induced NH$_2$- and COOH-terminal interaction inhibits p160 coactivator recruitment by activation function 2. J Biol Chem 276:42293–42301, 2001.
71. Rosenfeld MG, Glass CK. Coregulator codes of transcriptional regulation by nuclear receptors. J Biol Chem 276:36865–36868, 2001.
72. Wang X, Yeh S, Wu G, Hsu CL, Wang L, Chiang T, Yang Y, Guo Y, Chang C. Identification and characterization of a novel androgen receptor coregulator ARA267-alpha in prostate cancer cells. J Biol Chem 276:40417–40423, 2001.
73. Culig Z, Hobisch A, Cronauer MV, Radmayr C, Trapman J, Hittmair A, Bartsch G, Klocker H. Androgen receptor activation in prostatic tumor cell lines by insulin-like growth factor-I, keratinocyte growth factor, and epidermal growth factor. Cancer Res 54:5474–5478, 1994.
74. Craft N, Shostak Y, Carey M, Sawyers CL. A mechanism for hormone-independent prostate cancer through modulation of androgen receptor signaling by the HER-2/neu tyrosine kinase. Nat Med 5:280–285, 1999.
75. Ueda T, Bruchovsky N, Sadar MD. Activation of the androgen receptor N-terminal domain by IL-6 via MAPK and STAT3 signal transduction pathways. J Biol Chem 277:7076–7085, 2002.
76. Osman I, Scher HI, Drobnjak M, Verbel D, Morris M, Agus D, Ross JS, Cordon-Cardo C. HER-2/neu (p185neu) protein expression in the natural or treated history of prostate cancer. Clin Cancer Res 7:2643–2647, 2001.
77. Sadar MD, Gleave ME. Ligand-independent activation of the androgen receptor by the differentiation agent butyrate in human prostate cancer cells. Cancer Res 60:5825–5831, 2000.
78. Coetzee GA, Ross RK. Re: Prostate cancer and the androgen receptor. J Natl Cancer Inst 86:872–873, 1994.
79. Irvine RA, Yu MC, Ross RK, Coetzee GA. The CAG and GGC microsatellites of the androgen receptor gene are in linkage disequilibrium in men with prostate cancer. Cancer Res 55:1937–1940, 1995.
80. Ingles SA, Ross RK, Yu MC, Irvine RA, La Pera G, Haile RW, Coetzee GA. Association of prostate cancer risk with genetic polymorphisms in vitamin D receptor and androgen receptor. J Natl Cancer Inst 89:166–170, 1997.
81. Hakimi JM, Schoenberg MP, Rondinelli RH, Piantadosi S, Barrack ER. Androgen receptor variants with short glutamine or glycine repeats may identify unique subpopulations of men with prostate cancer. Clin Cancer Res 3:1599–1608, 1997.
82. Hardy DO, Scher HI, Bogenrieder T, Sabbatini P, Zhang ZF, Nanus DM, Catterall JF. Androgen receptor CAG repeat lengths in prostate cancer: correlation with age of onset. J Clin Endocrinol Metab 81:4400–4405, 1996.
83. Stanford JL, Just JJ, Gibbs M, Wicklund KG, Neal CL, Blumenstein BA, Ostrander EA. Polymorphic repeats in the androgen receptor gene: molecular markers of prostate cancer risk. Cancer Res 57:1194–1198, 1997.
84. Hsing AW, Gao YT, Wu G, Wang X, Deng J, Chen YL, Sesterhenn IA, Mostofi FK, Benichou J, Chang C. Polymorphic CAG and GGN repeat lengths in the androgen receptor gene and prostate cancer risk: a population-based case-

control study in China. Cancer Res 60:5111–5116, 2000.
85. Beilin J, Harewood L, Frydenberg M, Mamegham H, Martyres RF, Farish SJ, Olin C, Deam DR, Byron KA, Zajac JD. A case-control study of the androgen receptor gene CAG repeat polymorphism in Australian prostate carcinoma subjects. Cancer 92:941–949, 2001.
86. Ekman P, Gronberg H, Matsuyama H, Kivineva M, Bergerheim US, Li C. Links between genetic and environmental factors and prostate cancer risk. Prostate 39:262–268, 1999.
87. Edwards SM, Badzioch MD, Minter R, Hamoudi R, Collins N, Ardern-Jones A, Dowe A, Osborne S, Kelly J, Shearer R, Easton DF, Saunders GF, Dearnaley DP, Eeles RA. Androgen receptor polymorphisms: association with prostate cancer risk, relapse and overall survival. Int J Cancer 84:458–465, 1999.
88. Correa-Cerro L, Wohr G, Haussler J, Berthon P, Drelon E, Mangin P, Fournier G, Cussenot O, Kraus P, Just W, Paiss T, Cantu JM, Vogel W. (CAG)$_n$CAA and GGN repeats in the human androgen receptor gene are not associated with prostate cancer in a French–German population. Eur J Hum Genet 7:357–362, 1999.
89. Bratt O, Borg A, Kristoffersson U, Lundgren R, Zhang QX, Olsson H. CAG repeat length in the androgen receptor gene is related to age at diagnosis of prostate cancer and response to endocrine therapy, but not to prostate cancer risk. Br J Cancer 81:672–676, 1999.
90. Lange EM, Chen H, Brierley K, Livermore H, Wojno KJ, Langefeld CD, Lange K, Cooney KA. The polymorphic exon 1 androgen receptor CAG repeat in men with a potential inherited predisposition to prostate cancer. Cancer Epidemiol Biomarkers Prev 9:439–442, 2000.
91. Nam RK, Elhaji Y, Krahn MD, Hakimi J, Ho M, Chu W, Sweet J, Trachtenberg J, Jewett MA, Narod SA. Significance of the cag repeat polymorphism of the androgen receptor gene in prostate cancer progression. J Urol 164:567–572, 2000.
92. Latil AG, Azzouzi R, Cancel GS, Guillaume EC, Cochan-Priollet B, Berthon PL, Cussenot O. Prostate carcinoma risk and allelic variants of genes involved in androgen biosynthesis and metabolism pathways. Cancer 92:1130–1137, 2001.
93. Modugno F, Weissfeld JL, Trump DL, Zmuda JM, Shea P, Cauley JA, Ferrell RE. Allelic variants of aromatase and the androgen and estrogen receptors: toward a multigenic model of prostate cancer risk. Clin Cancer Res 7:3092–3096, 2001.
94. Miller EA, Stanford JL, Hsu L, Noonan E, Ostrander EA. Polymorphic repeats in the androgen receptor gene in high-risk sibships. Prostate 48:200–205, 2001.
95. Panz VR, Joffe BI, Spitz I, Lindenberg T, Farkas A, Haffejee M. Tandem CAG repeats of the androgen receptor gene and prostate cancer risk in black and white men. Endocrine 15:213–216, 2001.
96. Cude KJ, Montgomery JS, Price DK, Dixon SC, Kincaid RL, Kovacs KF, Venzon DJ, Liewehr DJ, Johnson ME, Reed E, Figg WD. The role of an androgen receptor polymorphism in the clinical outcome of patients with metastatic prostate cancer. Urol Int 68:16–23, 2002.
97. Butler R, Leigh PN, McPhaul MJ, Gallo JM. Truncated forms of the androgen receptor are associated with polyglutamine expansion in X-linked spinal and bulbar muscular atrophy. Hum Mol Genet 7:121–127, 1998.
98. Hsiao PW, Lin DL, Nakao R, Chang C. The linkage of Kennedy's neuron disease to ARA24, the first identified androgen receptor polyglutamine region–associated coactivator. J Biol Chem 274:20229–20234, 1999.
99. Rush MG, Drivas G, D'Eustachio P. The small nuclear GTPase Ran: how much does it run? Bioessays 18:103–112, 1996.
100. Zhang L, Leeflang EP, Yu J, Arnheim N. Studying human mutations by sperm typing: instability of CAG trinucleotide repeats in the human androgen receptor gene. Nat Genet 7:531–535, 1994.
101. Lu S, Tsai SY, Tsai MJ. Molecular mechanisms of androgen-independent growth of human prostate cancer LNCaP-AI cells. Endocrinology 140:5054–5059, 1999.
102. Platz EA, Giovannucci E, Dahl DM, Krithivas K, Hennekens CH, Brown M, Stampfer MJ, Kantoff PW. The androgen receptor gene GGN microsatellite and prostate cancer risk. Cancer Epidemiol Biomarkers Prev 7:379–384, 1998.
103. Henshall SM, Quinn DI, Lee CS, Head DR, Golovsky D, Brenner PC, Delprado W, Stricker PD, Grygiel JJ, Sutherland RL. Altered expression of androgen receptor in the malignant epithelium and adjacent stroma is associated with early relapse in prostate cancer. Cancer Res 61:423–427, 2001.
104. Sweat SD, Pacelli A, Bergstralh EJ, Slezak JM, Bostwick DG. Androgen receptor expression in prostatic intraepithelial neoplasia and cancer. J Urol 161:1229–1232, 1999.
105. Hall RE, Vivekanandan S, Ricciardelli C, Stapleton A, Scardino P, Stahl J, Haagensen DE, Marshall VR, Tilley WD. Androgen receptor and apolipoprotein-D: Diagnostic and prognostic significance in prostate cancer. Proc Am Assoc Cancer Res 40:408, 1999. (Abstract nr 2700).
106. Koivisto P, Kononen J, Palmberg C, Tammela T, Hyytinen E, Isola J, Trapman J, Cleutjens K, Noordzij A, Visakorpi T, Kallioniemi OP. Androgen receptor gene amplification: a possible molecular mechanism for androgen deprivation therapy failure in prostate cancer. Cancer Res 57:314–319, 1997.
107. Jarrard DF, Kinoshita H, Shi Y, Sandefur C,

Hoff D, Meisner LF, Chang C, Herman JG, Isaacs WB, Nassif N. Methylation of the androgen receptor promoter CpG island is associated with loss of androgen receptor expression in prostate cancer cells. Cancer Res 58:5310–5314, 1998.
108. Kinoshita H, Shi Y, Sandefur C, Meisner LF, Chang C, Choon A, Reznikoff CR, Bova GS, Friedl A, Jarrard DF. Methylation of the androgen receptor minimal promoter silences transcription in human prostate cancer. Cancer Res 60:3623–3630, 2000.
109. Yeap BB, Krueger RG, Leedman PJ. Differential posttranscriptional regulation of androgen receptor gene expression by androgen in prostate and breast cancer cells. Endocrinology 140:3282–3291, 1999.
110. Wen Y, Hu MC, Makino K, Spohn B, Bartholomeusz G, Yan DH, Hung M. C. HER-2/neu promotes androgen-independent survival and growth of prostate cancer cells through the Akt pathway. Cancer Res 60:6841–6845, 2000.
111. Ruohola JK, Valve EM, Karkkainen MJ, Joukov V, Alitalo K, Harkonen PL. Vascular endothelial growth factors are differentially regulated by steroid hormones and antiestrogens in breast cancer cells. Mol Cell Endocrinol 149:29–40, 1999.
112. Goldberg YP, Kalchman MA, Metzler M, Nasir J, Zeisler J, Graham R, Koide HB, O'Kusky J, Sharp AH, Ross CA, Jirik F, Hayden MR. Absence of disease phenotype and intergenerational stability of the CAG repeat in transgenic mice expressing the human Huntington disease transcript. Hum Mol Genet 5:177–185, 1996.
113. Lu S, Tsai SY, Tsai MJ. Regulation of androgen-dependent prostatic cancer cell growth: androgen regulation of CDK2, CDK4, and CKI $p16$ genes. Cancer Res 57:4511–4516, 1997.
114. Hussain M. The biology of metastatic prostate cancer. In: Peeling WB (ed). Questions and Uncertainties about Prostate Cancer. Oxford: Blackwell, 1996, pp 67–77.
115. Huggins C, Stevens RE, Hodges CV. Studies on prostatic cancer. II. The effect of castration on advanced carcinoma of the prostate gland. Arch Surg 43:209–228, 1941.
116. Labrie F, Dupont A, Cusan L, Gomez J, Emond J, Monfette G. Combination therapy with flutamide and medical (LHRH agonist) or surgical castration in advanced prostate cancer: 7-year clinical experience. J Steroid Biochem Mol Biol 37:943–950, 1990.
117. Schmitt B, Bennett C, Seidenfeld J, Samson D, Wilt T. Maximal androgen blockade for advanced prostate cancer. Cochrane Database Syst Rev ⟨www.update-software.com⟩ Article CD001526, 2002.
118. Kelly WK, Slovin S, Scher HI. Steroid hormone withdrawal syndromes. Pathophysiology and clinical significance. Urol Clin North Am 24:421–431, 1997.

119. Kelly WK, Scher HI. Prostate specific antigen decline after antiandrogen withdrawal: the flutamide withdrawal syndrome. J Urol 149:607–609, 1993.
120. Small EJ, Srinivas S. The antiandrogen withdrawal syndrome. Experience in a large cohort of unselected patients with advanced prostate cancer. Cancer 76:1428–1434, 1995.
121. Huan SD, Gerridzen RG, Yau JC, Stewart DJ. Antiandrogen withdrawal syndrome with nilutamide. Urology 49:632–634, 1997.
122. Nieh PT. Withdrawal phenomenon with the antiandrogen casodex. J Urol 153:1070–1072, 1995.
123. Small EJ, Carroll PR. Prostate-specific antigen decline after casodex withdrawal: evidence for an antiandrogen withdrawal syndrome. Urology 43:408–410, 1994.
124. Bissada NK, Kaczmarek AT. Complete remission of hormone refractory adenocarcinoma of the prostate in response to withdrawal of diethylstilbestrol. J Urol 153:1944–1945, 1995.
125. Dawson NA, McLeod DG. Dramatic prostate specific antigen decrease in response to discontinuation of megestrol acetate in advanced prostate cancer: expansion of the antiandrogen withdrawal syndrome. J Urol 153:1946–1947, 1995.
126. Akakura K, Akimoto S, Furuya Y, Ito H. Incidence and characteristics of antiandrogen withdrawal syndrome in prostate cancer after treatment with chlormadinone acetate. Eur Urol 33:567–571, 1998.
127. Scher HI, Kelly WK. Flutamide withdrawal syndrome: its impact on clinical trials in hormone-refractory prostate cancer. J Clin Oncol 11:1566–1572, 1993.
128. Small EJ, Baron A, Bok R. Simultaneous antiandrogen withdrawal and treatment with ketoconazole and hydrocortisone in patients with advanced prostate carcinoma. Cancer 80:1755–1759, 1997.
129. Quarmby VE, Beckman WCJ, Cooke DB, Lubahn DB, Joseph DR, Wilson EM, French FS. Expression and localization of androgen receptor in the R-3327 Dunning rat prostatic adenocarcinoma. Cancer Res 50:735–739, 1990.
130. Tilley WD, Wilson CM, Marcelli M, McPhaul MJ. Androgen receptor gene expression in human prostate carcinoma cell lines. Cancer Res 50:5382–5386, 1990.
131. Culig Z, Hobisch A, Hittmair A, Peterziel H, Cato AC, Bartsch G, Klocker H. Expression, structure, and function of androgen receptor in advanced prostatic carcinoma. Prostate 35:63–70, 1998.
132. Tilley WD, Lim-Tio SS, Horsfall DJ, Aspinall JO, Marshall VR, Skinner JM. Detection of discrete androgen receptor epitopes in prostate cancer by immunostaining: measurement by color video image analysis. Cancer Res 54:4096–4102, 1994.

133. Prins GS, Sklarew RJ, Pertschuk LP. Image analysis of androgen receptor immunostaining in prostate cancer accurately predicts response to hormonal therapy. J Urol 159:641–649, 1998.
134. Sadi MV, Barrack ER. Image analysis of androgen receptor immunostaining in metastatic prostate cancer. Heterogeneity as a predictor of response to hormonal therapy. Cancer 71:2574–2580, 1993.
135. Takeda H, Akakura K, Masai M, Akimoto S, Yatani R, Shimazaki J. Androgen receptor content of prostate carcinoma cells estimated by immunohistochemistry is related to prognosis of patients with stage D2 prostate carcinoma. Cancer 77:934–940, 1996.
136. Pertschuk LP, Schaeffer H, Feldman JG, Macchia RJ, Kim YD, Eisenberg K, Braithwaite LV, Axiotis CA, Prins G, Green GL. Immunostaining for prostate cancer androgen receptor in paraffin identifies a subset of men with a poor prognosis. Lab Invest 73:302–305, 1995.
137. Pertschuk LP, Maccha RJ, Feldman JG, Brady KA, Levine M, Kim DS, Eisenberg KB, Rainford E, Prins GS, Greene GL. Immunocytochemical assay for androgen receptors in prostate cancer: a prospective study of 63 cases with long-term follow-up. Ann Surg Oncol 1:495–503, 1994.
138. De Vere White RW, Meyers F, Chi SG, Chamberlain S, Siders D, Lee F, Stewart S, Gumerlock PH. Human androgen receptor expression in prostate cancer following androgen ablation. Eur Urol 31:1–6, 1997.
139. Gregory CW, Johnson RT, Mohler JL, French FS, Wilson EM. Androgen receptor stabilization in recurrent prostate cancer is associated with hypersensitivity to low androgen. Cancer Res 61:2892–2898, 2001.
140. Kim D, Gregory CW, French FS, Smith GJ, Mohler JL. Androgen receptor expression and cellular proliferation during transition from androgen-dependent to recurrent growth after castration in the CWR22 prostate cancer xenograft. Am J Pathol 160:219–226, 2002.
141. Bubendorf L, Kononen J, Koivisto P, Schraml P, Moch H, Gasser TC, Willi N, Mihatsch MJ, Sauter G, Kallioniemi OP. Survey of gene amplifications during prostate cancer progression by high-throughput fluorescence in situ hybridization on tissue microarrays. Cancer Res 59:803–806, 1999.
142. Koivisto P, Visakorpi T, Kallioniemi OP. Androgen receptor gene amplification: a novel molecular mechanism for endocrine therapy resistance in human prostate cancer. Scand J Clin Lab Invest Suppl 226:57–63, 1996.
143. Koivisto PA, Helin HJ. Androgen receptor gene amplification increases tissue PSA protein expression in hormone-refractory prostate carcinoma. J Pathol 189:219–223, 1999.
144. Veldscholte J, Ris-Stalpers C, Kuiper GG, Jenster G, Berrevoets C, Claassen E, van Rooij HC, Trapman J, Brinkmann AO, Mulder E. A mutation in the ligand binding domain of the androgen receptor of human LNCaP cells affects steroid binding characteristics and response to anti-androgens. Biochem Biophys Res Commun 173:534–540, 1990.
145. Zhao XY, Malloy PJ, Krishnan AV, Swami S, Navone NM, Peehl DM, Feldman D. Glucocorticoids can promote androgen-independent growth of prostate cancer cells through a mutated androgen receptor. Nat Med 6:703–706, 2000.
146. Tilley WD, Buchanan G, Hickey TE, Bentel JM. Mutations in the androgen receptor gene are associated with progression of human prostate cancer to androgen independence. Clin Cancer Res 2:277–285, 1996.
147. Taplin ME, Bubley GJ, Shuster TD, Frantz ME, Spooner AE, Ogata GK, Keer HN, Balk SP. Mutation of the androgen-receptor gene in metastatic androgen-independent prostate cancer. N Engl J Med 332:1393–1398, 1995.
148. Taplin ME, Bubley GJ, Ko YJ, Small EJ, Upton M, Rajeshkumar B, Balk SP. Selection for androgen receptor mutations in prostate cancers treated with androgen antagonist. Cancer Res 59:2511–2515, 1999.
149. Marcelli M, Ittmann M, Mariani S, Sutherland R, Nigam R, Murthy L, Zhao Y, DiConcini D, Puxeddu E, Esen A, Eastham J, Weigel NL, Lamb DJ. Androgen receptor mutations in prostate cancer. Cancer Res 60:944–949, 2000.
150. Gelmann EP. Androgen receptor mutations in prostate cancer. Cancer Treat Res 87:285–302, 1996.
151. Buchanan G, Greenberg NM, Scher HI, Harris JM, Marshall VR, Tilley WD. Collocation of androgen receptor gene mutations in prostate cancer. Clin Cancer Res 7:1273–1281, 2001.
152. Han G, Foster BA, Mistry S, Buchanan G, Harris JM, Tilley WD, Greenberg NM. Hormone status selects for spontaneous somatic androgen receptor variants that demonstrate specific ligand and cofactor dependent activities in autochthonous prostate cancer. J Biol Chem 276:11204–11213, 2001.
153. Haapala K, Hyytinen ER, Roiha M, Laurila M, Rantala I, Helin HJ, Koivisto PA. Androgen receptor alterations in prostate cancer relapsed during a combined androgen blockade by orchiectomy and bicalutamide. Lab Invest 81:1647–1651, 2001.
154. Wurtz JM, Bourguet W, Renaud JP, Vivat V, Chambon P, Moras D, Gronemeyer H. A canonical structure for the ligand-binding domain of nuclear receptors. Nat Struct Biol 3:206, 1996.
155. Veldscholte J, Berrevoets CA, Ris-Stalpers C, Kuiper GG, Jenster G, Trapman J, Brinkmann AO, Mulder E. The androgen receptor in LNCaP cells contains a mutation in the ligand binding domain which affects steroid binding characteristics and response to antiandrogens. J Steroid Biochem Mol Biol 41:665–669, 1992.

156. Aarnisalo P, Santti H, Poukka H, Palvimo JJ, Janne OA. Transcription activating and repressing functions of the androgen receptor are differentially influenced by mutations in the deoxyribonucleic acid–binding domain. Endocrinology 140:3097–3105, 1999.
157. Bruggenwirth HT, Boehmer AL, Lobaccaro JM, Chiche L, Sultan C, Trapman J, Brinkmann AO. Substitution of Ala^{564} in the first zinc cluster of the deoxyribonucleic acid (DNA)–binding domain of the androgen receptor by Asp, Asn, or Leu exerts differential effects on DNA binding. Endocrinology 139:103–110, 1998.
158. Poujol N, Lobaccaro JM, Chiche L, Lumbroso S, Sultan C. Functional and structural analysis of R607Q and R608K androgen receptor substitutions associated with male breast cancer. Mol Cell Endocrinol 130:43–51, 1997.
159. Rundlett SE, Miesfeld RL. Quantitative differences in androgen and glucocorticoid receptor DNA binding properties contribute to receptor-selective transcriptional regulation. Mol Cell Endocrinol 109:1–10, 1995.
160. Verrijdt G, Schoenmakers E, Haelens A, Peeters B, Verhoeven G, Rombauts W, Claessens F. Change of specificity mutations in androgen-selective enhancers. Evidence for a role of differential DNA binding by the androgen receptor. J Biol Chem 275:12298–12305, 2000.
161. Blanco JG, Minucci S, Lu J, Yang XJ, Walker KK, Chen H, Evans RM, Nakatani Y, Ozato K. The histone acetylase PCAF is a nuclear receptor coactivator. Genes Dev 12:1638–1651, 1998.
162. Moilanen AM, Karvonen U, Poukka H, Yan W, Toppari J, Janne OA, Palvimo JJ. A testis-specific androgen receptor coregulator that belongs to a novel family of nuclear proteins. J Biol Chem 274:3700–3704, 1999.
163. Moilanen AM, Poukka H, Karvonen U, Hakli M, Janne OA, Palvimo JJ. Identification of a novel RING finger protein as a coregulator in steroid receptor–mediated gene transcription. Mol Cell Biol 18:5128–5139, 1998.
164. Poukka H, Karvonen U, Yoshikawa N, Tanaka H, Palvimo JJ, Janne OA. The RING finger protein SNURF modulates nuclear trafficking of the androgen receptor. J Cell Sci 113:2991–3001, 2000.
165. Chamberlain NL, Whitacre DC, Miesfeld RL. Delineation of two distinct type 1 activation functions in the androgen receptor amino-terminal domain. J Biol Chem 271:26772–25678, 1996.
166. Gao T, Marcelli M, McPhaul MJ. Transcriptional activation and transient expression of the human androgen receptor. J Steroid Biochem Mol Biol 59:9–20, 1996.
167. Schoenberg MP, Hakimi JM, Wang S, Bova GS, Epstein JI, Fischbeck KH, Isaacs WB, Walsh PC, Barrack ER. Microsatellite mutation (CAG24 → 18) in the androgen receptor gene in human prostate cancer. Biochem Biophys Res Commun 198:74–80, 1994.
168. Watanabe M, Ushijima T, Shiraishi T, Yatani R, Shimazaki J, Kotake T, Sugimura T, Nagao M. Genetic alterations of androgen receptor gene in Japanese human prostate cancer. Jpn J Clin Oncol 27:389–393, 1997.
169. Wallen MJ, Linja M, Kaartinen K, Schleutker J, Visakorpi T. Androgen receptor gene mutations in hormone-refractory prostate cancer. J Pathol 189:559–563, 1999.
170. Yang M, Raynor M, Neufing PJ, Buchanan G, Tilley WD. Disruption of the polyglutamine tract results in increased ligand-induced transcriptional activity of the androgen receptor. Proc Am Assoc Cancer Res 40:408, 1999. Abstract nr 2699.
171. Fronsdal K, Engedal N, Slagsvold T, Saatcioglu F. CREB binding protein is a coactivator for the androgen receptor and mediates cross-talk with AP-1. J Biol Chem 273:31853–31859, 1998.
172. Kupfer SR, Marschke KB, Wilson EM, French FS. Receptor accessory factor enhances specific DNA binding of androgen and glucocorticoid receptors. J Biol Chem 268:17519–17527, 1993.
173. Gregory CW, He B, Johnson RT, Ford OH, Mohler JL, French FS, Wilson EM. A mechanism for androgen receptor-mediated prostate cancer recurrence after androgen deprivation therapy. Cancer Res 61:4315–4319, 2001.
174. Yeh S, Miyamoto H, Shima H, Chang C. From estrogen to androgen receptor: a new pathway for sex hormones in prostate. Proc Natl Acad Sci USA 95:5527–5532, 1998.
175. Yeh S, Kang HY, Miyamoto H, Nishimura K, Chang HC, Ting HJ, Rahman M, Lin HK, Fujimoto N, Hu YC, Mizokami A, Huang KE, Chang C. Differential induction of androgen receptor transactivation by different androgen receptor coactivators in human prostate cancer DU145 cells. Endocrine 11:195–202, 1999.
176. Park JB, Rhim JS, Park SC, Kimm SW, Kraus MH. Amplification, overexpression, and rearrangement of the *erbB-2* protooncogene in primary human stomach carcinomas. Cancer Res 49:6605–6609, 1989.
177. Slamon DJ, Godolphin W, Jones LA, Holt JA, Wong SG, Keith DE, Levin WJ, Stuart SG, Udove J, Ullrich A. Studies of the *HER-2/neu* proto-oncogene in human breast and ovarian cancer. Science 244:707–712, 1989.
178. Pegram MD, Pauletti G, Slamon DJ. HER-2/neu as a predictive marker of response to breast cancer therapy. Breast Cancer Res Treat 52:65–77, 1998.
179. Kuhn EJ, Kurnot RA, Sesterhenn IA, Chang EH, Moul JW. Expression of the c-erbB-2 (HER-2/neu) oncoprotein in human prostatic carcinoma. J Urol 150:1427–1433, 1993.

180. Zhou BP, Hu MC, Miller SA, Yu Z, Xia W, Lin SY, Hung MC. HER-2/neu blocks tumor necrosis factor–induced apoptosis via the Akt/NF-kappaB pathway. J Biol Chem 275:8027–8031, 2000.
181. Sadasivan R, Morgan R, Jennings S, Austenfeld M, Van Veldhuizen P, Stephens R, Noble M. Overexpression of Her-2/neu may be an indicator of poor prognosis in prostate cancer. J Urol 150:126–131, 1993.
182. Agus DB, Scher HI, Higgins B, Fox WD, Heller G, Fazzari M, Cordon-Cardo C, Golde DW. Response of prostate cancer to anti-Her-2/neu antibody in androgen-dependent and -independent human xenograft models. Cancer Res 59:4761–4764, 1999.

Hereditary Prostate Cancer: The Search for Major Genes and the Role of Genes Involved in Androgen Action

BAO-LI CHANG
AUBREY R. TURNER
WILLIAM B. ISAACS
JIANFENG XU

Prostate cancer is the most frequently diagnosed noncutaneous cancer in men in Western countries. Both genetic and environmental factors have been implicated in the etiology of this disease. A genetic component of prostate cancer has been suggested by both twin and segregation studies.[1–6] About 9% of all prostate cancer cases have been proposed to be due to mutations in prostate cancer susceptibility genes,[1] and the overall effect of heritable factors has been estimated to be higher for prostate cancer than for other common cancers.[6] So far, linkage studies of prostate cancer families have provided evidence for several major prostate cancer susceptibility loci.[7–12] In addition to major gene effects, common polymorphisms that result in minor quantitative and qualitative functional differences in gene products most likely modify the risk of prostate cancer. As a hormone-related cancer, genes involved in the androgen-metabolism pathway are particularly compelling candidates for prostate cancer risk factors and have been the focus of many case-control studies. However, few studies have investigated the role of androgen-pathway genes in the etiology of hereditary prostate cancer (HPC), in which genetic factors may play more important roles compared to sporadic cases. This chapter reviews the search for HPC genes and recent efforts to understand the interrelationship of HPC and the androgen-action pathway.

FAMILIAL AGGREGATION: CLUES FOR GENETIC INFLUENCES IN THE ETIOLOGY OF PROSTATE CANCER

Evidence of familial clustering of prostate carcinoma is well documented in many case-control studies.[13–24] Three major conclusions can be drawn from these studies. First, estimates of the relative risk of developing prostate cancer are higher in first-degree relatives of prostate cancer cases compared to the general population. This increase of relative risk among relatives of prostate cancer patients has been demonstrated in several ethnic and racial groups, including African Americans, Caucasians, Hispanics, and Asians. Second, the risk of prostate cancer increases with increasing number of affected relatives. For example, in the study conducted by Lesko et al.,[23] the odds ratio (OR) was 2.2 for subjects with a family history of prostate cancer in one relative and went up to 3.9 for subjects with a family history of two or more affected relatives. Third, the risk of developing prostate cancer among first- and second-degree relatives is negatively correlated with a decrease in the age at diagnosis of index cases. In one study for men with a relative diagnosed before 65 years of age, the OR of prostate cancer was 4.1 compared to 0.76 for those with a relative diagnosed after age 74.[23] Several cohort studies, designed to decrease the recall bias and detection bias inherent in case-control studies have also shown the same trend as seen in case control studies.[25-28]

TWIN STUDIES: DISSECTION OF GENETIC FACTORS FROM ENVIRONMENTAL INFLUENCES

Familial aggregation may be due to genetic or shared environmental factors. Twin studies can be used to estimate the contribution of inheritance to the familial aggregation of a trait or disease by comparing the similarities (concordance rate) of a trait or disease in monozygotic (MZ, genetically identical) and dizygotic (DZ, sharing on average 50% of genetic material) twins. In a registry of 4840 male twin pairs in Sweden, 458 cases of prostate cancer were identified. There were 16 concordant pairs among 1649 MZ twin pairs (1%) but only six concordant pairs among 2983 DZ twin pairs (0.2%). The greater concordance rate seen in MZ twins compared to DZ twins is attributed to a greater degree of shared genes.[2] A second twin study, conducted in the United States, showed similar results.[4] Among 1009 twin pairs identified from a national twin registry, a significantly higher concordance rate was observed among MZ twins (27.1%) compared to DZ twins (7.1%). It was estimated that genetic influences account for approximately 57% and environmental influences for 43% of the variability in twin liability for prostate cancer. In the twin study conducted by Lichtenstein et al.,[6] a 21% concordance rate was found in MZ twins but only a 6% concordance rate in DZ twins. It was estimated that 42% [95% confidence interval (CI) = 29%–50%] of prostate cancer risk was due to heritable factors, and this genetic contribution was the highest among all neoplasms studied.

SEGREGATION ANALYSIS: EVIDENCE OF MAJOR SUSCEPTIBILITY GENES

The evidence from twin studies and family studies supports the importance of genetic factors in the development of prostate cancer. However, these studies cannot infer the genetic model of inheritance. To identify a specific inheritance model of hereditary prostate cancer, complex segregation analysis can be used. So far, there are four complex segregation analyses of prostate cancer that support an autosomal dominant inheritance model for prostate cancer susceptibility, especially for early-onset disease.[1,3,5,29] A prostate cancer susceptibility gene was estimated to be rare in the general population, with a frequency of about 0.003[1] to 0.006.[5] The penetrance of such a rare susceptibility gene was estimated to be ~88% by age 85 for carriers and 3%–5% for noncarriers. However, other inheritance models have been suggested for prostate cancer. The segregation analyses by Schaid et al.[5] confirmed that dominant inheritance was the best-fitting model for families with early-onset prostate cancer cases. However, this model did not adequately explain family data when the index cases were diagnosed at an older age (>70 years). Cui et al.[29] also suggested that two-locus models, combining autosomal dominant with either autosomal recessive or X-linked inheritance, fit their family data better than any single-locus model. The results from segregation analyses begin to illustrate the heterogeneous nature of prostate cancer.

LINKAGE STUDIES: IDENTIFYING MAJOR PROSTATE CANCER SUSCEPTIBILITY LOCI

Linkage analyses test for association between genetic markers at known chromosomal locations and disease phenotypes, to help identify and define specific chromosomal regions associated with prostate cancer. Genomewide screens for linkage have revealed several chromosomal regions that are likely to harbor prostate cancer susceptibility genes. Smith et al.[7] reported the first genomewide screen for prostate cancer susceptibility genes and identified the first prostate cancer susceptibility locus, *HPC1*, at chromosomal region 1q24–25. Significant evidence of locus heterogeneity was revealed by an admixture test, with an estimated 34% of families linked to this region. An additional analysis of these data showed that the positive evidence of linkage came from the subset of families with mean age at diagnosis of <65 years. Higher-grade cancers were also more common in these potentially linked families compared to potentially unlinked families.[30] The linkage to *HPC1* was replicated in four other independent studies.[31–34] However, other studies[8,35–40] failed to replicate the linkage results. In a combined analysis of 772 HPC families by the International Consortium for Prostate Cancer Genetics,[12] weak evidence of linkage to *HPC1* was confirmed, with a small proportion (6%) of families linked to this locus.

A second prostate cancer susceptibility locus was reported to be at 1q42–43 (*PCaP*) in a genomewide screen by Berthon et al.[8] Evidence of linkage to *PCaP* was found in 47 families of French and German origin, with a maximal two-point LOD score of 2.7. This locus is about 60 cM away from the *HPC1* locus. Three other studies failed to replicate the linkage to *PCaP*.[40–42] Some evidence of linkage to *PCaP* was found in our collection of 159 HPC families. However, conditional analyses revealed that the evidence of linkage to regions at 1q24–25 and 1q42–43 was likely related.[43]

The greater risk of prostate cancer for brothers of cases compared to sons or fathers of cases observed in case-control and cohort studies is consistent with an X-linked or autosomal recessive component to prostate cancer susceptibility. Evidence for a susceptibility locus on Xq27–28 (termed *HPCX*) was demonstrated by Xu et al.[9] in a panel of 360 prostate cancer families collected from North America, Finland, and Sweden. A peak two-point LOD score of 4.6 was obtained at marker DXS1113. The proportion of families linked to *HPCX* was estimated to be 16% in the combined study population and similar in each separate family collection. The linkage studies reported by Lange et al.[44] and Peters et al.[45] support linkage to *HPCX*. However, Bergthorsson et al.[46] and Hsieh et al.[40] failed to replicate these results.

Linkage to chromosomal region 1p36 was found in HPC families with a history of primary brain cancer.[47] This result was confirmed in one study;[43] however, two other studies[37,40] were not able to confirm the linkage. Another candidate locus, *HPC20* at chromosomal region 20q13, was identified.[11] As is the recurring theme, this linkage was confirmed in two other studies[48,49] but not by the study of Hsieh et al.[40]

In a genomewide screen of extended high-risk pedigrees from Utah, evidence of linkage was found on the short arm of chromosome 17.[50] *ELAC2/HPC2* was identified through positional cloning and mutation screening within the defined linkage region. Mutations in the *ELAC2/HPC2* gene were found to segregate with prostate cancer in two pedigrees. In addition, two common missense variants in this gene were associated with risk of prostate cancer.[50–52] However, this association was not confirmed by other studies.[53–55]

Many other chromosomal regions identified through genomewide screens have been suggested to harbor prostate cancer susceptibility genes. These regions include 1q33–42, 4q26–27, 5p12–13, 7p21, 13q31–33 (with dominant inheritance model), 10q, 12q, 14q (with dominant inheritance model), 1q, 8q, 10q, 16p (with recessive inheritance model), 2q, 12p, 15q, 16p, 16q, 5p13.3–5q13.1, 12p13.3–12.3, 19p13.3, 4q, and Xq12–13 (model-free).[7,10,34,40,56] Evidence of linkage has been also observed at 8p22–23 in multiple studies.[10,57]

The results from linkage studies fully reveals the complex nature of prostate cancer. Large collaborative studies will be necessary to begin to decipher and separate the signals produced by actual prostate cancer susceptibility genes from the noise arising from the vagaries of analyzing a complex disease like prostate cancer by linkage analysis. In all likelihood, multiple prostate cancer susceptibility genes are involved in the etiology of hereditary prostate cancer, with each gene responsible for a subset of families. In addition to rare, high-penetrant susceptibility genes, more common genes with lower penetrance may contribute to the development or manifestation of HPC. These genes may work collectively in the carcinogenesis process or act as modifiers for major susceptibility genes. The extensive variability observed in linkage analysis results for prostate cancer may reflect differing proportions of disease in different study populations due to the actions of both types of susceptibility gene.

ANDROGEN LEVELS AND RISK OF PROSTATE CANCER

The role of the androgens in the process of carcinogenesis to prostate cancer has been hypothesized and examined in many studies. Given that exogenous androgens can initiate and/or promote prostate cancer in rodents and humans and that eunuchs rarely develop prostate cancer,[58–61] circumstantial evidence implicates androgen levels in the etiology of prostate cancer. Further, a reduced risk of prostate cancer has been associated with certain hyperestrogenic–hypoandrogenic states,[62] and estrogen therapy has a palliative effect in advanced cases.[61] However, studies comparing hormone levels in men with and without prostate cancer have produced widely varying results.[63,64] Several possibilities may explain these variations: small sample sizes, analyses of hormone levels in blood collected after the diagnosis of cancer, nonrepresentative control subjects, no adjustment for other related hormone levels and hormone-binding protein levels, and no adjustment for other potential confounding factors such as obesity. To date, one of the largest and most extensive case-control studies in this area was a prospective nested study done by Gann et al.[65] to investigate sex hormone levels and the risk of prostate cancer. No clear association was found between the risk of prostate cancer and either unadjusted levels of sex hormones or sex hormone–binding globulin (SHBG). However, positive results were observed when hormone and SHBG levels were adjusted simultaneously. Specifically, a strong trend of increasing prostate cancer risk was observed with increasing levels of plasma testosterone (95% CI 1.34–5.02, $p = 0.004$), and an inverse trend in risk was observed with increasing levels of both estradiol (95% CI 0.32–0.98, $p = 0.03$) and SHBG (95% CI 0.24–0.89, $p = 0.01$). The meta-analysis of Shaneyfelt et al.[66] further confirmed the association of serum testosterone and SHBG levels with the risk of prostate cancer. The meta-analysis included three prospective cohort studies with data on all three sex hormones of interest.[65,67,68] After multivariate adjustment, including the levels of other hormones, age, and body mass index, men with total serum testosterone in the highest quartile were 2.34 times more likely to develop prostate cancer (95% CI 1.30–4.20). Men with the highest SHBG levels were less likely to develop prostate cancer compared to those with the lowest SHBG levels (OR 0.46, 95% CI 0.24–0.89). These studies provide us with strong evidence that androgen and estrogen levels can alter an individuals' risk of prostate cancer.

IMPORTANCE OF ANDROGENS IN THE CARCINOGENESIS OF PROSTATE CANCER

The prostate is an androgen-dependent organ. Testosterone and androgen precursors freely diffuse into prostate cells. Through a coordinated network of enzymes, they can be rapidly and irreversibly converted to their reduced and more potent metabolic form, dihydrotestosterone (DHT). Both DHT and, to a lesser extent, testosterone bind to and induce the conformational change and activation of the androgen receptor (AR). The activated androgen–receptor complex then binds to the androgen response element (ARE) of androgen-responsive genes and initiates or inhibits their transcription.[69–72] After fulfilling their role in transcriptional regulation, DHT and testosterone are oxidized to biologically inactive derivatives by another set of enzymes. The intracellular steady-state active androgen level, which is balanced by the availability of testosterone and dehydroepiandrosterone (DHEA), the formation of DHT, and the degradation of DHT and other androgens, is thought to be an important determinant in the control of cell proliferation and its countering process, cell death. The deregulation of the balance between cell proliferation and cell death is a common denominator in the pathogenesis of most human cancers,[73] and any genetic or environmental factor that can alter the hormonal environment, and thus alter this balance in the target organs, could participate in the process of malignant transformation in these target tissues.[74] Individual genetic variation that affects androgen metabolism and availability in target organs could contribute, by regulation of androgen levels, to individual susceptibility to prostate cancer.

GENES IN THE ANDROGEN PATHWAY AS CANDIDATES FOR HEREDITARY PROSTATE CANCER

A variety of genes are involved in the metabolic pathway of androgens and the mediation of the effects of androgens. Functionally important polymorphisms in genes that encode enzymes involved in androgen metabolism and transport may lead to differences in individual susceptibility to prostate cancer by altering the levels and effects of androgens, as proposed by Ross et al.[75] (see Chapter 15) Thus far, studies of androgen-pathway genes and the risk of prostate cancer have focused mainly on sporadic prostate cancer. In addition, studies of HPC have primarily focused on the identification of novel major susceptibility genes. However, we cannot exclude the possibility that mutations (or polymorphisms) in androgen-pathway genes are involved in the etiology of HPC for the following reasons. First, given the biological relevance of androgens in the etiology of prostate cancer in general, it is possible that mutations in the androgen-pathway genes that significantly affect androgen metabolism may be inherited as high-penetrance susceptibility genes in some families. Second, the androgen-pathway genes may, instead of being major genes, act as modifier genes for major susceptibility gene(s) segregating in high-risk families. Mutations in major susceptibility genes might not reveal their full potential and lead to clinically identifiable prostate cancer (incomplete penetrance) without mutations (or polymorphims) in androgen-pathway genes. Third, before major prostate cancer genes are identified and their roles in the genetics of HPC families fully understood, the possibility that multiple, minor genes are responsible for HPC should not be excluded.

FAMILY-BASED STUDY DESIGN: COMPENSATING FOR THE POPULATION-BASED STUDY DESIGN IN THE INVESTIGATION OF CANDIDATE GENES IN PROSTATE CANCER SUSCEPTIBILITY

The ability of any study design to detect a relationship between a mutation (or a polymorphism) and disease susceptibility is dependent on the frequency and penetrance of the mutation (or polymorphism). Because the underlying influence of DNA sequence variants in the androgen-pathway genes is unknown, approaches utilizing multiple study designs are needed to evaluate these genes in the etiology of prostate cancer. Several methodologies, including linkage analysis, the family-based association test, and the population-based association test, can be applied to thoroughly examine the role of candidate genes in complex diseases such as prostate cancer.

A genetic linkage study is one important design to evaluate candidate genes in complex diseases such as prostate cancer. A significant feature of the linkage approach is that it is less sensitive to allelic heterogeneity. If a mutation has a large effect (i.e., high penetrance) and there are multiple such mutations within a gene, a linkage study is likely to detect the effect of such a gene, while population-based association approaches are likely to fail. Information regarding specific sequence variants within a gene is not necessary for linkage studies, while it is essential for association studies. At this stage, it is technically difficult to identify and assess multiple sequence variants due to limited sequence data from large numbers of prostate cancer patients and unaffected control subjects.

To date, most studies have utilized population-based association tests to evaluate the relationship between genetic variants in androgen-pathway genes and the risk of prostate cancer, especially sporadic prostate cancer. A population-based association study is the method of choice to detect mutations with a high frequency but low penetrance. However, as mentioned above, genetic linkage studies and family-based association studies in families with multiple affected members are better designs for detecting mutations with a lower frequency but higher penetrance because the frequency of gene carriers is likely to be higher in these families and there is a higher likelihood that a rare mutation cosegregates with disease in family members carrying the disease. In addition, it is difficult to rule out the effect of population stratification in population-based association studies. A family-based test is free from population stratification and can compensate for the shortcomings inherent to the population-based study design.

Only a limited number of studies have investigated the potential impact of androgen-pathway genes on the risk of HPC. Four genes, AR, 3β-hydroxysteroid dehydrogenase types I and II (HSD3B1, HSD3B2), and cytochrome P-450c17α (CYP17), will be reviewed in the following sections.

Androgen Receptor

The AR gene, located on Xq11–12 (~50 cM centromeric to HPCX), is a compelling candidate for prostate cancer. The AR gene encodes for a transcription factor within the steroid receptor superfamily.[76,77] One critical function of the AR gene product is to activate the expression of other genes. The transactivation activity resides in the N-terminal domain of the protein, encoded by exon 1. Two polymorphic microsatellites are located approximately 1.1 kb apart in exon 1: a highly polymorphic CAG repeat and a less polymorphic GGC repeat.[78,79] The CAG repeat encodes a polyglutamine tract and usually contains 9–29 repeats.[80] Alleles of the GGC repeat code for a polyglycine tract and contain 4–21 repeats, with 16 repeats being the most common allele. An inverse correlation between the length of the CAG repeat and the transactivation activities of AR was demonstrated by several in vitro assays.[81–83] However, there is no report of an association between the length of GGC repeats and functional changes of the AR. The hypothesis that shorter alleles of CAG and (or) GGC repeats in the AR are associated with an increased risk for prostate cancer has also been tested in several association studies. Because these studies will be detailed in Chapters 15 and 18, we here summarize three inferences drawn from them. First, the association between AR repeats and prostate cancer is inconclusive. While some studies have reported a marginally increased risk for individuals with short CAG repeats and/or short GGC repeats, an almost equal number of studies have not found a significant association. This inconsistency stresses the importance of applying other study designs in the candidate gene approach. Second, among the studies that tested both CAG and GGC repeats, the association with prostate cancer risk was stronger with GGC repeats or combinations of GGC and CAG repeats.[84–86] The less-studied GGC repeats might play a more important role if AR is involved in the etiology of prostate cancer. Third, few studies have evaluated and compared the risk of CAG repeats in HPC,[87] and no study has evaluated GGC repeats in HPC. It is unclear whether the AR repeats impose a higher or lower cancer risk in HPC compared to the sporadic form.

Several linkage studies have tested the hypothesis that AR is a prostate susceptibility gene. Results from two large-scale prostate cancer genomewide screens have provided evidence for linkage to the AR region. In a study by Goddard et al.,[34] a LOD of 3.06 ($p = 0.0005$) at the AR region was reported in 254 families after Gleason score was included as a covariate. In the study by Hsieh et al.,[40] a multipoint nonparametric linkage (NPL) Z-score of 1.5 was observed in 98 multiple affected families. However, results from two other prostate cancer genomewide screens failed to provide evidence for linkage at the AR region.[7,10] The inconsistent results from linkage studies may be due to several factors, e.g., sample size, informativeness of the families, resolution of markers, choice of study design, and choice of analytical methods. Therefore, additional linkage studies using multiple markers at the AR in large and well-characterized prostate cancer families are warranted. Furthermore, because CAG and GGC repeats themselves, in addition to serving as polymorphic markers for linkage analysis, could be associated with prostate cancer risk, transmission/disequilibrium tests, which have greater power to detect linkage in the presence of association, would be more appropriate in this case.[88–90] We used this approach in a study of 159 HPC families, each with at least three first-degree relatives affected with prostate cancer,[91] to detect linkage between the AR repeats and a hypothetical rare prostate cancer susceptibility gene with X-linked inheritance. We detected positive, but not statistically significant, linkage scores across the AR region, with a maximum LOD under heterogeneity (HLOD) of 0.49 ($p = 0.12$). However, when the male-limited, X-linked transmission/disequilibrium test (XLRC-TDT) was applied, a preferential transmission of short GGC alleles (≤16 repeats) from heterozygous mothers to their affected sons ($z' = 2.65$, $p = 0.008$) was observed. A similar result was observed when the 16 GGC repeat was tested ($z' = 3.17, p = 0.001$). These results strongly suggest that GGC repeats are linked to a prostate cancer susceptibility gene. No significant overtransmission of CAG repeat alleles was observed, which is in agreement with the results of Lange et al.[87] When the allele frequencies of GGC and CAG repeats were compared among HPC probands, sporadic cases, and unaffected controls, significantly increased frequencies of the ≤16 GGC repeat alleles in 159 independent hereditary cases (71%) and 245 sporadic cases (68%) cases compared with 211 controls (59%) suggested that GGC repeats are associated with prostate cancer risk ($p = 0.02$). The evidence for the association between the ≤16 GGC repeats and prostate cancer risk was stronger using the XLRC-TDT ($z' = 2.66, p = 0.007$). No evidence for association between the CAG repeats and prostate cancer risk was observed. The consistent results from both linkage and association studies strongly implicate the GGC repeats in the AR as a risk factor for prostate cancer susceptibility. In addition, this experience clearly demonstrates the impact of the study population and the importance of methodology on the results of candidate gene studies.

HSD3B1 and HSD3B2

The enzyme 3β-hydroxysteroid dehydrogenase is a critical component of the androgen-metabolism pathway because it catalyzes androstendione production in steroidogenic tissues and converts active DHT into inactive metabolites in steroid target tissues. The HSD3B gene family has two genes and five pseudogenes, all of which map to chromosome 1p13.[92–94] The HSD3B1 gene encodes the type I enzyme, which is exclusively expressed in the placenta and peripheral tissues, such as prostate, breast, and skin. The HSD3B2 gene encodes the type II enzyme, which is predominantly expressed in classical steroidogenic tissues, namely, the adrenals, testis, and ovary.[93,95–98] A number of mutations in HSD3B2 have been found to cause congenital adrenal hyperplasia, a rare mendelian disease manifested by salt wasting and incomplete masculinization in males.[99]

Linkage findings at 1p13 significantly increase the likelihood that HSD3B genes play an important role in prostate cancer susceptibility. In a chromosomewide linkage study to evaluate different prostate cancer susceptibility loci on chromosome 1 in 159 HPC families, our group reported evidence for linkage in a broad region from 1p13 to 1q32.[43] The HLOD score was 1.31 ($p = 0.01$) and the allele sharing LOD score was 1.34 ($p = 0.01$) at HSD3B2. The evidence for linkage was stronger in families with five or more affected men (allele sharing LOD = 2.22, $p = 0.001$) and in families with mean age at onset >65 years (allele sharing LOD = 1.45, $p = 0.01$). In another genomewide scan for prostate cancer susceptibility loci, Goddard et al.[34] reported a LOD score of 3.25 ($p = 0.0001$) at 1p13, near markers D1S534 and D1S1653, when the Gleason score was included as a covariate. There are only a few studies on the sequence variants of HSD3B2 in prostate cancer. A complex $(TG)_n(TA)_n(CA)_n$ repeat has been described and studied in intron 3 of HSD3B2.[100,101] However, very few published studies have evaluated the association between prostate cancer risk and either this repeat or other sequence variants in HSD3B1.

Considering the biological importance of the HSD3B genes and the evidence that they are located in a chromosomal region that is likely to contain prostate cancer susceptibility genes, a systematic study to evaluate their roles in relation to prostate cancer was carried out by Chang et al.[102] In this study, sequence variants in the HSD3B1 and HSD3B2 genes were identified by directly sequencing the polymerase chain reaction products from the 500 bp promoter region, all exons, exon–intron junctions, and the 3'-untranslated region (UTR) of both genes in 96 subjects. A total of 11 single-nucleotide polymorphisms (SNPs) were identified, and four of these were frequent and informative for further analyses. The four frequent SNPs identified in HSD3B genes, B1-c7062t and B1-N367T in HSD3B1, and B2-c7474t and B2-c7519g in HSD3B2, were then tested for their association with prostate cancer risk by comparing their allele and genotype distributions in 159 HPC probands, 245 sporadic prostate cancer cases, and 222 unaffected controls. The frequency of allele C of B1-N367T was higher in the HPC probands (34%) and sporadic cases (33%) compared to the unaffected controls (26%). The differences were significant between HPC probands and controls ($p = 0.03$), between sporadic cases and controls ($p = 0.04$), and between either type of prostate cancer and controls ($p = 0.02$). The allele frequencies of the other three SNPs were higher in cases compared to controls, although none reached statistical significance. When the genotype frequencies of the four SNPs were compared, similar findings were observed. The frequencies of the variant genotypes of *B1-N367T* were higher in both HPC cases (55%) and sporadic cases (54%) than in controls (43%). Because of the similarity in genetic structure and biological function of the protein products encoded by the *HSD3B1* and *HSD3B2* genes, the joint effect of the two genes on prostate cancer risk was also tested. Proportions of men with the variant genotypes at either *B1-N367T* or *B2-c7519g* were 74% in HPC probands, 68% in sporadic cases, and 57% in unaffected controls. After adjustment for age, the differences were statistically significant between HPC probands and controls ($p = 0.004$), between sporadic cases and controls ($p = 0.02$), and between either type of prostate cancer and controls ($p = 0.003$). Compared to men with wild-type genotypes at both *B1-N367T* and *B2-c7519g*, the age-adjusted point estimates of relative risk for HPC, sporadic prostate cancer, and either type of prostate cancer were 2.17 (95% CI 1.29–3.65), 1.61 (95% CI 1.07–2.42), and 1.76 (95% CI 1.21–2.57), respectively, for men with the variant genotypes at either *B1-N367T* (C/A or C/C) or *B2-c7519g* (C/G or G/G).

To examine the relationship between the evidence for linkage at 1p13 and variants of the *HSD3B* genes, the association between *HSD3B* genes and the risk of prostate cancer was tested in unrelated HPC probands stratified by whether their families provided evidence for linkage at 1p13 (LOD > 0). Proportions of men with the variant genotypes at either *B1-N367T* or *B2-c7519g* were 78%, 71%, and 57% in the 66 HPC probands whose families linked to 1p13, the remaining 67 probands, and controls, respectively. The difference between the 66 HPC probands from linked families and controls was significant ($p = 0.008$). Thus, the subset of HPC probands whose families provided evidence for linkage at 1p13 predominantly contributed to the observed association. In addition, because the evidence for linkage at 1p13 was provided primarily by families with older mean age at onset, stratifying analyses were also performed in subjects aged 60 years or older. Larger differences in the proportion of men with either variant genotype of the two SNPs were observed among HPC probands (76%), sporadic cases (74%), and unaffected controls (51%). After adjustment for age, the differences were statistically significant between HPC probands and controls ($p = 0.002$), between sporadic cases and controls ($p = 0.005$), and between all cases and controls ($p = 0.0005$).

The exact mechanism by which sequence variants in either *HSD3B1* or *HSD3B2* increase prostate cancer susceptibility is not defined. The *HSD3B* genes encode membrane-bound microsomal proteins with two predicted transmembrane domains: one is a 16-residue segment between residues 75 and 91, and the other is a C-terminal 26-residue segment between residues 283 and 308. The *B1-N367T* variant is located in the C-terminal extramembrane domain. This SNP results in an amino acid change from Asn to Thr and may have an effect on conformation, enzymatic activity, stability, or regulation of HSD3B1 protein. This amino acid change creates a new putative protein kinase C (PKC) phosphorylation site (phosphorylation site pattern [ST][RK], http://maple.bioc.columbia.edu/predictprotein/). Isozymes of PKC are a family of kinases in the signal-transduction cascade involved in cell proliferation, antitumor resistance, and apoptosis. Expression of the *HSD3B1* gene is specifically induced by interleukin (IL)-4 and IL-13 in both human prostate cancer cell lines and primary prostatic epithelial cells.[96] In addition, the PKC activator phorbol-12-myristate-13-acetate (PMA) further enhances the stimulatory effect of *IL-4* on *HSD3B* activity.[103] It is possible that HSD3B proteins are regulated through phosphorylation by PKC, and it is worth exploring whether the new PKC phosphorylation site in a variant HSD3B1 protein alters the expression of HSD3B1 protein. Because SNP *B2-c7519g* is located in the 3'-UTR of *HSD3B2*, it has no effect on the amino acid sequence of HSD3B2 protein. However, the nucleotide change may result in a conformational change in the 3'-UTR of *HSD3B2* mRNA and may affect the stability of this mRNA. Posttranscriptional regulation of mRNA stability can have a significant impact on mRNA abundance and subsequent protein expression. Several elements in the 3′-UTR that are important to the stability of a variety of mRNA species have been identified, including the poly(A) site, AU-rich elements, iron-responsive element, 3′-terminal stem loop, long-range stem loop, exoribonuclease cleavage site, and endoribonuclease cleavage site. It is possible that the nucleotide change in the 3′-UTR of *HSD3B2* mRNA alters the structure of a protein binding site and, hence, alters the stability of the mRNA and the quantity of the protein produced. With differential expression of the *HSD3B1* and *HSD3B2* genes in different tissues, functional studies that assess the biological impact of genetic variants in *HSD3B* genes will be difficult, although necessary, for a better understanding of their roles in prostate cancer etiology.

CYP17

The enzyme P-450c17α mediates both 17α-hydroxylase and 17,20-lyase activity. It catalyzes both 17α-hydroxylation of pregnenolone and progesterone and 17,20-lysis of 17α-hydroxypregnenolone and 17α-hydroxyprogesterone, which are the key reactions for both sex steroid and cortisol biosynthesis. The *CYP17* gene maps to 10q24.3 and consists of eight exons.[104] Mutations in the *CYP17* gene result in disruption in testosterone synthesis and subsequently pseudohermaphroditism in males and impaired sex steroid hormone synthesis and absence of sexual maturation in females. Because of its importance in androgen synthesis, *CYP17* has been proposed and tested as a candidate gene for prostate cancer in many case-control studies. The single-base polymorphism (a T-to-C transition) in the 5′-UTR of the *CYP17* gene[105] has been the subject of most *CYP17* association studies. This single-base change creates an additional SP1-type (CCACC box) transcriptional factor binding site, which was hypothesized to result in increased expression of the gene. It also creates a recognition site for the restriction enzyme MspA1. Following MspA1 digestion of a polymerase chain reaction fragment, the A1 allele (representing the wild-type allele) and the A2 allele (representing the variant allele with C transition) were designated. Contradictory results regarding the *CYP17* A1/A2 polymorphism and the risk of prostate cancer have been reported. Lunn et al.,[106] Gsur et al.,[107] and Yamada et al.[108] reported positive associations between the A2 allele and increased risk of prostate cancer. However, two other studies[109,110] showed that A1 is the risk allele for prostate cancer. The association between risk of prostate cancer and the *CYP17* genotype remains unclear, even in age-stratified subgroups. Two groups observed an association between the *CYP17* genotype and prostate cancer risk in younger prostate cancer cases.[106,108] However, the association between the *CYP17* genotype and prostate cancer was documented in an older subgroup in two other publications.[107,110]

With the mixed results from case-control studies in mind, a study design that combined family-based and population-based approaches was employed to better understand the role of the *CYP17* gene in the risk of prostate cancer.[111] Among 159 HPC families, a linkage study was used to test whether genes in the region of *CYP17* increase prostate cancer susceptibility, and a family-based association test examined whether the 5′ promoter polymorphism in the *CYP17* gene is associated with prostate cancer. In addition, a population-based association study, which included the HPC probands, sporadic prostate cancer cases, and unaffected controls, was performed to evaluate whether the polymorphism increases the risk for both sporadic and HPC cases. Multipoint linkage analysis provided evidence for linkage with a peak HLOD of 1.30 ($p = 0.014$) at the *CYP17* region. In stratified linkage analyses, which are based on family characteristics such as mean age at diagnosis, number of affected members in the family, and ethnicity, the evidence for linkage at the *CYP17* gene region came mainly from families with mean age at diagnosis of ≥65 years ($n = 80$, HLOD = 1.21, $p = 0.018$), families with five or more affected members ($n = 90$, HLOD = 1.13, $p = 0.022$), and Caucasian families ($n = 133$, HLOD = 1.22, $p = 0.018$). However, the family-based association test did not confirm an association between the *CYP17* A1/A2 polymorphism and the risk for prostate cancer; neither did the population-based association test, which included HPC probands, sporadic cases, and unaffected controls. These results imply that

CYP17 or other gene(s) in this chromosomal region could be involved in the etiology of prostate cancer. However, it is also possible that the A1/A2 polymorphism in CYP17 is not associated with prostate cancer risk, while other polymorphisms in the gene might still be important.

Two other genomewide linkage studies also showed moderate evidence of linkage to the 10q chromosomal region.[10,56] However, the linkage signals in both studies were at 10q25-qter regions, which is approximately 30 cM telomeric to the CYP17 locus, where evidence of linkage was observed by Chang et al.[111] Interestingly, loss of heterozygosity in the region of 10q23–25, which covers the region with evidence of linkage in this study, was frequent in prostate carcinoma (~50% of tumors studied) as well as in other tumors, including glioblastoma multiforme, endometrial carcinoma, breast carcinoma, and melanoma.[112–118] A tumor-suppressor gene(s) located in this chromosomal region, including PTEN, has been proposed to be involved in the development of tumors.[119–121] Therefore, it is also possible that the linkage signal observed in and near the CYP17 locus actually came from the nearby tumor-suppressor gene. Further study of this chromosomal region is necessary to clarify these possibilities.

CONCLUSION

The role of androgens in normal prostate physiology is undisputed, and androgen blockade has been, and remains, one of the most widespread treatment strategies for advanced prostate cancer. Accordingly, a complete understanding of the role of androgens in prostate cancer has been pursued for quite some time. Based on new evidence that several genes in the androgen pathway, in addition to their roles in sporadic prostate cancer, may be involved in the etiology of HPC, further studies are warranted. By accessing candidate genes in the androgen-metabolism pathway from multiple angles, the collective knowledge gained may allow for not only better family risk assessment, which has been a direct focus of the HPC studies, but also improved screening for prostate cancer, which has been a target of many population-based studies. Improved use of existing hormonal therapies and the development of designer therapies are promising goals of all efforts to understand the role of the androgen-action pathway in prostate cancer. It would be extremely useful to identify men who will develop progressive disease in general and, specifically, androgen-refractory prostate cancer more quickly than others, thus warranting more aggressive therapies. In this hormonally promoted cancer, it is clear that we still have much to learn.

REFERENCES

1. Carter BS, Beaty TH, Steinberg GD, Childs B, Walsh PC. Mendelian inheritance of familial prostate cancer. Proc Natl Acad Sci USA 89(8):3367–3371, 1992.
2. Gronberg H, Damber L, Damber JE. Studies of genetic factors in prostate cancer in a twin population. J Urol 152:1484–1487, 1994.
3. Gronberg H, Damber L, Damber JE, Iselius L. Segregation analysis of prostate cancer in Sweden: support for dominant inheritance. Am J Epidemiol 146:552–557, 1997.
4. Page WF, Braun MM, Partin AW, Caporaso N, Walsh P. Hereditary prostate cancer: a study of World War II veteran twins. Prostate 33:240–245, 1997.
5. Schaid DJ, McDonnell SK, Blute ML, Thibodeau SN. Evidence for autosomal dominant inheritance of prostate cancer. Am J Hum Genet 62:1425–1438, 1998.
6. Lichtenstein P, Holm NV, Verkasalo PK, Iliadou A, Kaprio J, Koskenvuo M, Pukkala E, Skytthe A, Hemminki K. Environmental and heritable factors in the causation of cancer—analyses of cohorts of twins from Sweden, Denmark, and Finland. N Engl J Med 343:78–85, 2000.
7. Smith JR, Freije D, Carpten JD, Gronberg H, Xu J, Isaacs SD, Brownstein MJ, Bova GS, Guo H, Bujnovszky P, Nusskern DR, Damber JE, Bergh A, Emanuelsson M, Kallioniemi OP, Walker-Daniels J, Bailey-Wilson JE, Beaty TH, Meyers DA, Walsh PC, Collins FS, Trent JM, Isaacs WB. Major susceptibility locus for prostate cancer on chromosome 1 suggested by a genome-wide search. Science 274:1371–1374, 1996.
8. Berthon P, Valeri A, Cohen-Akenine A, et al. Predisposing gene for early-onset prostate cancer, localized on chromosome 1q42.2–43. Am J Hum Genet 62:1416–1424, 1998.
9. Xu J, Meyers D, Freije D, et al. Evidence for a prostate cancer susceptibility locus on the X chromosome. Nat Genet 20:175–179, 1998.
10. Gibbs M, Stanford JL, Jarvik GP, Janer M, Badzioch M, Peters MA, Goode EL, Kolb S,

Chakrabarti L, Shook M, Basom R, Ostrander EA, Hood L. A genomic scan of families with prostate cancer identifies multiple regions of interest. Am J Hum Genet 67:100–109, 2000.
11. Berry R, Schroeder JJ, French AJ, McDonnell SK, Peterson BJ, Cunningham JM, Thibodeau SN, Schaid DJ. Evidence for a prostate cancer-susceptibility locus on chromosome 20. Am J Hum Genet 67:82–91, 2000.
12. Xu J. Combined analysis of hereditary prostate cancer linkage to 1q24–25: results from 772 hereditary prostate cancer families from the International Consortium for Prostate Cancer Genetics. Am J Hum Genet 66:945–957, 2000.
13. Woolf CM. An investigation of familial aspects of carcinoma of the prostate. Cancer 13:739–744, 1960.
14. Cannon L, Bishop DT, Skolnick M, et al. Genetic epidemiology of prostate cancer in the Utah Mormon genealogy. Cancer Surv 1:47–69, 1982.
15. Meikle AW, Stanish WM. Familial prostatic cancer risk and low testosterone. J Clin Endocrinol Metab 54:1104–1108, 1982.
16. Steinberg GD, Carter BS, Beaty TH, Childs B, Walsh PC. Family history and the risk of prostate cancer. Prostate 17:337–347, 1990.
17. Spitz MR, Currier RD, Fueger JJ, et al. Familial patterns of prostate cancer: a case-control analysis. J Urol 146:1305–1307, 1991.
18. Keetch DW, Rice JP, Suarez BK, et al. Familial aspects of prostate cancer: a case control study. J Urol 154:2100–2102, 1995.
19. Narod SA, Dupont A, Cusan L, Diamond P, Gomez JL, Suburu R, Labrie F. The impact of family history on early detection of prostate cancer [letter]. Nat Med 1:99–101, 1995.
20. Monroe KR, Yu MC, Kolonel LN, et al. Evidence of an X-linked or recessive genetic component to prostate cancer risk. Nat Med 1:827–829, 1995.
21. Hayes RB, Liff JM, Pottern LM, et al. Prostate cancer risk in U.S. blacks and whites with a family history of cancer. Int J Cancer 60:361–364, 1995.
22. Whittemore AS, Wu AH, Kolonel LN, John EM, Gallagher RP, Howe GR, West DW, Teh CZ, Stamey T. Family history and prostate cancer risk in black, white, and Asian men in the United States and Canada. Am J Epidemiol 141:732–740, 1995.
23. Lesko SM, Rosenberg L, Shapiro S. Family history and prostate cancer risk. Am J Epidemiol 144:1041–1047, 1996.
24. Ghadirian P, Howe GR, Hislop TG, et al. Family history of prostate cancer: a multi-center case-control study in Canada. Int J Cancer 70:679–681, 1997.
25. Goldgar DE, Easton DF, Cannon-Albright LA, et al. Systematic population-based assessment of cancer risk in first-degree relatives of cancer probands. J Natl Cancer Inst 86:1600–1608, 1994.
26. Gronberg H, Damber L, Damber JE. Familial prostate cancer in Sweden. A nationwide register cohort study. Cancer 77:138–143, 1996.
27. Cerhan JR, Parker AS, Putnam SD, et al. Family history and prostate cancer risk in a population-based cohort of Iowa men. Cancer Epidemiol Biomarkers Prev 8:53–60, 1999.
28. Schuurman AG, Zeegers MP, Goldbohm RA, van den Brandt PA. A case-cohort study on prostate cancer risk in relation to family history of prostate cancer. Epidemiology 10:192–195, 1999.
29. Cui J, Staples MP, Hopper JL, English DR, McCredie MR, Giles GG. Segregation analyses of 1,476 population-based Australian families affected by prostate cancer. Am J Hum Genet 68:1207–1218, 2001.
30. Gronberg H, Isaacs SD, Smith JR, Carpten JD, Bova GS, Freije D, Xu J, Meyers DA, Collins FS, Trent JM, Walsh PC, Isaacs WB. Characteristics of prostate cancer in families potentially linked to the hereditary prostate cancer 1 (HPC1) locus. JAMA 278:1251–1255, 1997.
31. Cooney KA, McCarthy JD, Lange E, Huang L, Miesfeldt S, Montie JE, Oesterling JE, Sandler HM, Lange K. Prostate cancer susceptibility locus on chromosome 1q: a confirmatory study. J Natl Cancer Inst 89:955–959, 1997.
32. Hsieh CL, Oakley-Girvan I, Gallagher RP, Wu AH, Kolonel LN, Teh CZ, Halpern J, West DW, Paffenbarger RS Jr, Whittemore AS. Re: Prostate cancer susceptibility locus on chromosome 1q: a confirmatory study. J Natl Cancer Inst 89:1893–1894, 1997.
33. Neuhausen SL, Farnham JM, Kort E, Tavtigian SV, Skolnick MH, Cannon-Albright LA. Prostate cancer susceptibility locus HPC1 in Utah high-risk pedigrees. Hum Mol Genet 8:2437–2442, 1999.
34. Goddard KA, Witte JS, Suarez BK, Catalona WJ, Olson JM. Model-free linkage analysis with covariates confirms linkage of prostate cancer to chromosomes 1 and 4. Am J Hum Genet 68:1197–1206, 2001.
35. McIndoe RA, Stanford JL, Gibbs M, et al. Linkage analysis of 49 high-risk families does not support a common familial prostate cancer-susceptibility gene at 1q24–25. Am J Hum Genet 61:347–353, 1997.
36. Eeles RA, Durocher F, Edwards S, Teare D, Badzioch M, Hamoudi R, Gill S, Biggs P, Dearnaley D, Ardern-Jones A, Dowe A, Shearer R, McLennan DL, Norman RL, Ghadirian P, Aprikian A, Ford D, Amos C, King TM, Labrie F, Simard J, Narod SA, Easton D, Foulkes WD. Linkage analysis of chromosome 1q markers in 136 prostate cancer families. The Cancer Research Campaign/British Prostate Group U.K. Familial Prostate

Cancer Study Collaborators. Am J Hum Genet 62:653–658, 1998.
37. Berry R, Schaid DJ, Smith JR, French AJ, Schroeder JJ, McDonnell SK, Peterson BJ, Wang ZY, Carpten JD, Roberts SG, Tester DJ, Blute ML, Trent JM, Thibodeau SN. Linkage analyses at the chromosome 1 loci 1q24–25 (HPC1), 1q42.2–43 (PCAP), and 1p36 (CAPB) in families with hereditary prostate cancer. Am J Hum Genet 66:539–546, 2001.
38. Goode EL, Stanford JL, Chakrabarti L, Gibbs M, Kolb S, McIndoe RA, Buckley VA, Schuster EF, Neal CL, Miller EL, Brandzel S, Hood L, Ostrander EA, Jarvik GP. Linkage analysis of 150 high-risk prostate cancer families at 1q24–25. Genet Epidemiol 18:251–275, 2000.
39. Suarez BK, Lin J, Witte JS, Conti DV, Resnick MI, Klein EA, Burmester JK, Vaske DA, Banerjee TK, Catalona WJ. Replication linkage study for prostate cancer susceptibility genes. Prostate 45:106–114, 2000.
40. Hsieh CL, Oakley-Girvan I, Balise RR, Halpern J, Gallagher RP, Wu AH, Kolonel LN, O'Brien LE, Lin IG, Van Den Berg DJ, Teh CZ, West DW, Whittemore AS. A genome screen of families with multiple cases of prostate cancer: evidence of genetic heterogeneity. Am J Hum Genet 69:148–158, 2001.
41. Gibbs M, Chakrabarti L, Stanford JL, Goode EL, Kolb S, Schuster EF, Buckley VA, Shook M, Hood L, Jarvik GP, Ostrander EA. Analysis of chromosome 1q42.2–43 in 152 families with high risk of prostate cancer. Am J Hum Genet 64:1087–1095, 1999.
42. Whittemore AS, Lin IG, Oakley-Girvan I, Gallagher RP, Halpern J, Kolonel LN, Wu AH, Hsieh CL. No evidence of linkage for chromosome 1q42.2–43 in prostate cancer. Am J Hum Genet 65:254–256, 1999.
43. Xu J, Zheng SL, Chang B, Smith JR, Carpten JD, Stine OC, Isaacs SD, Wiley KE, Henning L, Ewing C, Bujnovszky P, Bleeker ER, Walsh PC, Trent JM, Meyers DA, Isaacs WB. Linkage of prostate cancer susceptibility loci to chromosome 1. Hum Genet 108:335–345, 2001.
44. Lange EM, Chen H, Brierley K, Perrone EE, Bock CH, Gillanders E, Ray ME, Cooney KA. Linkage analysis of 153 prostate cancer families over a 30-cM region containing the putative susceptibility locus HPCX. Clin Cancer Res 5:4013–4020, 1999.
45. Peters MA, Jarvik GP, Janer M, Chakrabarti L, Kolb S, Goode EL, Gibbs M, DuBois CC, Schuster EF, Hood L, Ostrander EA, Stanford JL. Genetic linkage analysis of prostate cancer families to Xq27–28. Hum Hered 51:107–113, 2001.
46. Bergthorsson JT, Johannesdottir G, Arason A, Benediktsdottir KR, Agnarsson BA, Bailey-Wilson JE, Gillanders E, Smith J, Trent J, Barkardottir RB. Analysis of HPC1, HPCX, and PCaP in Icelandic hereditary prostate cancer. Hum Genet 107:372–375, 2000.
47. Gibbs M, Stanford JL, McIndoe RA, Jarvik GP, Kolb S, Goode EL, Chakrabarti L, Schuster EF, Buckley VA, Miller EL, Brandzel S, Li S, Hood L, Ostrander EA. Evidence for a rare prostate cancer-susceptibility locus at chromosome 1p36. Am J Hum Genet 64:776–787, 1999.
48. Bock CH, Cunningham JM, McDonnell SK, Schaid DJ, Peterson BJ, Pavlic RJ, Schroeder JJ, Klein J, French AJ, Marks A, Thibodeau SN, Lange EM, Cooney KA. Analysis of the prostate cancer-susceptibility locus HPC20 in 172 families affected by prostate cancer. Am J Hum Genet 68:795–801, 2001.
49. Zheng SL, Xu J, Isaacs SD, Wiley K, Chang B, Bleecker ER, Walsh PC, Trent JM, Meyers DA, Isaacs WB. Evidence for a prostate cancer linkage to chromosome 20 in 159 hereditary prostate cancer families. Hum Genet 108:430–435, 2001.
50. Tavtigian SV, Simard J, Teng DH, Abtin V, Baumgard M, Beck A, Camp NJ, Carillo AR, Chen Y, Dayananth P, Desrochers M, Dumont M, Farnham JM, Frank D, Frye C, Ghaffari S, Gupte JS, Hu R, Iliev D, Janecki T, Kort EN, Laity KE, Leavitt A, Leblanc G, McArthur-Morrison J, Pederson A, Penn B, Peterson KT, Reid JE, Richards S, Schroeder M, Smith R, Snyder SC, Swedlund B, Swensen J, Thomas A, Tranchant M, Woodland AM, Labrie F, Skolnick MH, Neuhausen S, Rommens J, Cannon-Albright LA. A candidate prostate cancer susceptibility gene at chromosome 17p. Nat Genet 27:172–180, 2001.
51. Rebbeck TR, Walker AH, Zeigler-Johnson C, Weisburg S, Martin AM, Nathanson KL, Wein AJ, Malkowicz SB. Association of HPC2/ELAC2 genotypes and prostate cancer. Am J Hum Genet 67:1014–1019, 2000.
52. Suarez BK, Gerhard DS, Lin J, Haberer B, Nguyen L, Kesterson NK, Catalona WJ. Polymorphisms in the prostate cancer susceptibility gene HPC2/ELAC2 in multiplex families and healthy controls. Cancer Res 61:4982–4984, 2001.
53. Xu J, Zheng SL, Carpten JD, Nupponen NN, Robbins CM, Mestre J, Moses TY, Faith DA, Kelly BD, Isaacs SD, Wiley KE, Ewing CM, Bujnovszky P, Chang B, Bailey-Wilson J, Bleeker ER, Walsh PC, Trent JM, Meyers DA, Isaacs WB. Evaluation of linkage and association of HPC2/ELAC2 in patients with familial or sporadic prostate cancer. Am J Hum Genet 68:901–911, 2001.
54. Vesprini D, Nam RK, Trachtenberg J, Jewett MA, Tavtigian SV, Emami M, Ho M, Toi A, Narod SA. HPC2 variants and screen-detected prostate cancer. Am J Hum Genet 68:912–917, 2001.
55. Wang L, McDonnell SK, Elkins DA, Slager SL, Christensen E, Marks AF, Cunningham JM, Pe-

terson BJ, Jacobsen SJ, Cerhan JR, Blute ML, Schaid DJ, Thibodeau SN. Role of *HPC2/ELAC2* in hereditary prostate cancer. Cancer Res 61:6494–6499, 2001.
56. Suarez BK, Lin J, Burmester JK, Broman KW, Weber JL, Banerjee TK, Goddard KA, Witte JS, Elston RC, Catalona WJ. A genome screen of multiplex sibships with prostate cancer. Am J Hum Genet 66:933–944, 2000.
57. Xu J, Zheng SL, Hawkins GA, Faith DA, Kelly B, Isaacs SD, Wiley KE, Chang Bl, Ewing CM, Bujnovszky P, Carpten JD, Bleecker ER, Walsh PC, Trent JM, Meyers DA, Isaacs WB. Linkage and association studies of prostate cancer susceptibility: evidence for linkage at 8p22–23. Am J Hum Genet 69:341–350, 2001.
58. Henderson BE, Ross RK, Pike MC, Casagrande JT. Endogenous hormones as a major factor in human cancer. Cancer Res 42:3232–3239, 1982.
59. Noble RL. The development of prostatic adenocarcinoma in Nb rats following prolonged sex hormone administration. Cancer Res 37:1929–1933, 1977.
60. Noble RL. Sex steroids as a cause of adenocarcinoma of the dorsal prostate in Nb rats, and their influence on the growth of transplants. Oncology 34:138–141, 1977.
61. Huggins C, Hodges CV. Studies on prostatic cancer: effect of castration, of estrogen, and of androgen injection on serum phosphatases in metastatic carcinoma of the prostate. Cancer Res 1:293–297, 1941.
62. Glantz GM. Cirrhosis and carcinoma of the prostate gland. J Urol 92:291–293, 1964.
63. Nomura AM, Kolonel LN. Prostate cancer: a current perspective. Epidemiol Rev 13:200–227, 1991.
64. Hsing AW, Chua S Jr, Gao YT, Gentzschein E, Chang L, Deng J, Stanczyk FZ. Prostate cancer risk and serum levels of insulin and leptin: a population-based study. J Natl Cancer Inst 93:783–789, 2001.
65. Gann PH, Hennekens CH, Ma J, Longcope C, Stampfer MJ. Prospective study of sex hormone levels and risk of prostate cancer. J Natl Cancer Inst 88:1118–1126, 1996.
66. Shaneyfelt T, Husein R, Bubley G, Mantzoros CS. Hormonal predictors of prostate cancer: a meta-analysis. J Clin Oncol 18:847–853, 2000.
67. Hsing AW, Comstock GW. Serological precursors of cancer: serum hormones and risk of subsequent prostate cancer. Cancer Epidemiol Biomarkers Prev 2:27–32, 1993.
68. Nomura AM, Stemmermann GN, Chyou PH, Henderson BE, Stanczyk FZ. Serum androgens and prostate cancer. Cancer Epidemiol Biomarkers Prev 5:621–625, 1996.
69. Luke MC, Coffey DS. Human androgen receptor binding to the androgen response element of prostate specific antigen. J Androl 15:41–51, 1994.
70. Coffey DS. The molecular biology of the prostate. In: Prostate Disease. Philadelphia: WB Saunders, 1993, pp 28–56.
71. Culig Z, Hobisch A, Bartsch G, Klocker H. Androgen receptor—an update of mechanisms of action in prostate cancer. Urol Res 28:211–219, 2000.
72. Culig Z, Hobisch A, Bartsch G, Klocker H. Expression and function of androgen receptor in carcinoma of the prostate. Microsc Res Tech 51:447–455, 2000.
73. Perston-Martin S, Pike MC, Ross RK, Henderson BE. Epidemiologic evidence for the increased cell proliferation model of carcinogenesis. Environ Health Perspect 101(Suppl 5):137–138, 1993.
74. Henderson BE, Ross RK, Bernstein L. Estrogens as a cause of human cancer: the Richard and Hinda Rosenthal Foundation Award lecture. Cancer Res 48:246–253, 1988.
75. Ross RK, Pike MC, Coetzee GA, Reichardt JK, Yu MC, Feigelson H, Stanczyk FZ, Kolonel LN, Henderson BE. Androgen metabolism and prostate cancer: establishing a model of genetic susceptibility. Cancer Res 58:4497–4504, 1998.
76. Chang C, Kokontis J, Liao S. Molecular cloning of human and rat complementary DNA encoding androgen receptors. Science 240:324–326, 1988.
77. Lubahn DB, Joseph DR, Sar M, Tan J, Higgs HN, Larson RE, French FS, Wilson EM. The human androgen receptor: complementary deoxyribonucleic acid cloning, sequence analysis and gene expression in prostate. Mol Endocrinol 2:1265–1275, 1988.
78. Edwards A, Hammond HA, Jin L, Caskey CT, Chakraborty R. Genetic variation at five trimeric and tetrameric tandem repeat loci in four human population groups. Genomics 12:241–253, 1992.
79. Sleddens HF, Oostra BA, Brinkmann AO, Trapman J. Trinucleotide (GGN) repeat polymorphism in the human androgen receptor (AR) gene. Hum Mol Genet 2:493, 1993.
80. Irvine RA, Yu MC, Ross RK, Coetzee GA. The CAG and GGC microsatellites of the androgen receptor gene are in linkage disequilibrium in men with prostate cancer. Cancer Res 55:1937–1940, 1995.
81. Mhatre AN, Trifiro MA, Kaufman M, Kazemi-Esfarjani P, Figlewicz D, Rouleau G, Pinsky L. Reduced transcriptional regulatory competence of the androgen receptor in X-linked spinal and bulbar muscular atrophy. Nat Genet 5:184–188, 1993.
82. Chamberlain NL, Driver ED, Miesfeld RL. The length and location of CAG trinucleotide repeats in the androgen receptor N-terminal domain affect transactivation function. Nucleic Acids Res 22:3181–3186, 1994.
83. Beilin J, Ball EM, Favaloro JM, Zajac JD. Ef-

fect of the androgen receptor CAG repeat polymorphism on transcriptional activity: specificity in prostate and non-prostate cell lines. J Mol Endocrinol 25:85–96, 2000.
84. Hakimi JM, Schoenberg MP, Rondinelli RH, Piantadosi S, Barrack ER. Androgen receptor variants with short glutamine or glycine repeats may identify unique subpopulations of men with prostate cancer. Clin Cancer Res 3:1599–1608, 1997.
85. Stanford JL, Just JJ, Gibbs M, Wicklund KG, Neal CL, Blumenstein BA, Ostrander EA. Polymorphic repeats in the androgen receptor gene: molecular markers of prostate cancer risk. Cancer Res 57:1194–1198, 1997.
86. Platz EA, Giovannucci E, Dahl DM, Krithivas K, Hennekens CH, Brown M, Stampfer MJ, Kantoff PW. The androgen receptor gene GGN microsatellite and prostate cancer risk. Cancer Epidemiol Biomarkers Prev 7:379–384, 1998.
87. Lange EM, Chen H, Brierley K, Livermore H, Wojno KJ, Langcfcld CD, Lange K, Cooncy KA. The polymorphic exon 1 androgen receptor CAG repeat in men with a potential inherited predisposition to prostate cancer. Cancer Epidemiol Biomarkers Prev 9:439–442, 2000.
88. Speilman RS, McGinnis RE, Ewens WJ. Transmission test for linkage disequilibrium: the insulin gene region and insulin-dependent diabetes mellitus (IDDM). Am J Hum Genet 52:506–516, 1993.
89. Knapp M. A note on the application of the transmission disequilibrium test when a parent is missing. Am J Hum Genet 56:811–812, 1999.
90. Knapp M. The transmission/disequilibrium test and parental-genotype reconstruction: the reconstruction-combined transmission/disequilibrium test. 64:861–870, 1999.
91. Chang B, Zheng S, Hawkins G, Isaacs S, Wiley K, Turner A, Carpten J, Bleecker E, Walsh P, Trent J, Meyers D, Isaacs W, Xu J. Polymorphic GGC repeats in the androgen receptor gene are associated with hereditary and sporadic prostate cancer risk. Hum Genet 110:122–129, 2002.
92. Luu-The V, Lachance Y, Labrie C, Leblanc G, Thomas JL, Strickler RC, Labrie F. Full length cDNA structure and deduced amino acid sequence of human 3β-hydroxy-5-ene-steroid dehydrogenase. Mol Endocrinol 3:1310–1312, 1989.
93. Rheaume E, Lachance Y, Zhao HF, Breton N, Dumont M, Delaunoit Y, Trudel C, Luu-The V, Simard J, Labrie F. Structure and expression of a new complementary-DNA encoding the almost exclusive 3-beta-hydroxysteroid dehydrogenase delta-5-delta-4-isomerase in human adrenals and gonads. Mol Endocrinol 5:1147–1157, 1991.
94. McBride MW, McVie AJ, Burridge SM, Brintnell B, Craig N, Wallace AM, Wilson RH, Varley J, Sutcliffe RG. Cloning, expression, and physical mapping of the 3beta-hydroxysteroid dehydrogenase gene cluster (HSD3BP1–HSD3BP5) in human. Genomics 61:277–284, 1999.
95. Simard J, Durocher F, Mebarki F, Turgeon C, Sanchez R, Labrie Y, Couet J, Trudel C, Rheaume E, Morel Y, Luu-The V, Labrie F. Molecular biology and genetics of the 3beta-hydroxysteroid dehydrogenase/delta5-delta4 isomerase gene family. J Endocrinol 150(Suppl): S189–S207, 1996.
96. Gingras S, Simard J. Induction of 3beta-hydroxysteroid dehydrogenase/isomerase type 1 expression by interleukin-4 in human normal prostate epithelial cells, immortalized keratinocytes, colon, and cervix cancer cell lines. Endocrinology 140:4573–4584, 1999.
97. Gingras S, Moriggl R, Groner B, Simard J. Induction of 3beta-hydroxysteroid dehydrogenase/delta5-delta4 isomerase type 1 gene transcription in human breast cancer cell lines and in normal mammary epithelial cells by interleukin-4 and interleukin-13. Mol Endocrinol 13:66–81, 1999.
98. El-Alfy M, Luu-The V, Huang XF, Berger L, Labrie F, Pelletier G. Localization of type 5 17beta-hydroxysteroid dehydrogenase, 3beta-hydroxysteroid dehydrogenase, and androgen receptor in the human prostate by in situ hybridization and immunocytochemistry. Endocrinology 140:1481–1491, 1999.
99. Rheaume E, Simard J, Morel Y, Mebarki F, Zachmann M, Forest MG, New MI, Labrie F. Congenital adrenal hyperplasia due to point mutation in the type II 3β-hydroxysteroid dehydrogenase gene. Nat Genet 1:239–245, 1992.
100. Verreault H, Dufort I, Simard J, Labrie F, Luu-The V. Dinucleotide repeat polymorphism in the HSD3B2 gene. Hum Mol Genet 3:384, 1994.
101. Devgan SA, Henderson BE, Yu MC, Shi C-Y, Pike MC, Ross RK, Reichardt JK. Genetic variation of 3β-hydroxysteroid dehydrogenase type II in three racial/ethnic groups: implications for prostate cancer risk. Prostate 33:9–12, 1997.
102. Chang B, Zheng S, Hawkins G, Isaacs S, Wiley K, Turner A, Carpten J, Bleecker E, Walsh P, Trent J, Meyers D, Isaacs W, Xu J. Joint effect of HSD3B1 and HSD3B2 genes is associated with hereditary and sporadic prostate cancer susceptibility Cancer Res 62:1784–1789, 2002.
103. Gingras S, Cote S, Simard J. Multiple signal transduction pathways mediate interleukin-4-induced 3beta-hydroxysteroid dehydrogenase/delta5-delta4 isomerase in normal and tumoral target tissues. J Steroid Biochem Mol Biol 76:213–225, 2001.
104. Picado-Leonard J, Miller WL. Cloning and sequence of the human gene for P450c17 (steroid 17α-hydroxylase/17,20 lyase): similarity with the gene for P450c21. DNA 6:439–448, 1987.
105. Carey AH, Waterworth D, Patel K, White D,

Little J, Novelli P, Franks S, Williamson R. Polycystic ovaries and premature male pattern baldness are associated with one allele of the steroid metabolism gene *CYP17*. Hum Mol Genet 3:1873–1876, 1994.
106. Lunn RM, Bell DA, Mohler JL, Taylor JA. Prostate cancer risk and polymorphism in 17 hydroxylase (*CYP17*) and steroid reductase (*SRD5A2*). Carcinogenesis 20:1727–1731, 1999.
107. Gsur A, Bernhofer G, Hinteregger S, Haidinger G, Schatzl G, Madersbacher S, Marberger M, Vutus C, Micksche M. A polymorphism in the *CYP17* gene is asssociated with prostate cancer risk. Int J Cancer 87:434–437, 2000.
108. Yamada Y, Watanabe M, Murata M, Yamanaka M, Kubota Y, Ito H, Katoh T, Kawamura J, Yatani R, Shiraishi T. Impact of genetic polymorphisms of 17-hydroxylase cytochrome P-450 (*CYP17*) and steroid 5alpha-reductase type II (*SRD5A2*) genes on prostate-cancer risk among the Japanese population. Int J Cancer 92:683–686, 2001.
109. Wadelius M, Anderson AO, Johanson JE, Wadelius C, Rane E. Prostate cancer associated with *CYP17* genotype. Pharmacogenetics 9:635–639, 1999.
110. Habuchi T, Liqing Z, Suzuki T, Sasaki R, Tsuchiya N, Tachiki H, Shimoda N, Satoh S, Sato K, Kakehi Y, Kamoto T, Ogawa O, Kato T. Increased risk of prostate cancer and benign prostate hyperplasia associated with a *CYP17* gene polymorphism with a gene dosage effect. Cancer Res 60:5710–5713, 2000.
111. Chang B, Zheng SL, Isaacs SD, Wiley KE, Carpten JD, Hawkins GA, Bleecker ER, Walsh PC, Trent JM, Meyers DA, Isaacs WB, Xu J. Linkage and association of *CYP17* gene in hereditary and sporadic prostate cancer. Int J Cancer 95:354–359, 2001.
112. Isaacs WB, Carter BS. Genetic changes associated with prostate cancer in humans. Cancer Surv 11:15–24, 1991.
113. Gray IC, Phillips SM, Lee SJ, Neoptolemos JP, Weissenbach J, Spurr NK. Loss of the chromosomal region 10q23–25 in prostate cancer. Cancer Res 55:4800–4803, 1995.
114. Ittmann M. Allelic loss on chromosome 10 in prostate adenocarcinoma. Cancer Res 56:2143–2147, 1996.
115. Komiya A, Suzuki H, Ueda T, Yatani R, Emi M, Ito H, Shimazaki J. Allelic losses at loci on chromosome 10 are associated with metastasis and progression of human prostate cancer. Genes Chromosomes Cancer 17:245–253, 1996.
116. Trybus TM, Burgess AC, Wojno KJ, Glover TW, Macoska JA. Distinct areas of allelic loss on chromosomal regions 10p and 10q in human prostate cancer. Cancer Res 56:2263–2267, 1996.
117. Bose S, Wang SI, Terry MB, Hibshoosh H, Parsons R. Allelic loss of chromosome 10q23 is associated with tumor progression in breast carcinomas. Oncogene 17:123–127, 1998.
118. Rasheed BK, Wiltshire RN, Bigner SH, Bigner DD. Molecular pathogenesis of malignant gliomas. Curr Opin Oncol 11:162–167, 1999.
119. Li DM, Sun H. TEP1, encoded by a candidate tumor suppressor locus, is a novel protein tyrosine phosphatase regulated by transforming growth factor beta. Cancer Res 57:2124–2129, 1997.
120. Li J, Yen C, Liaw D, Podsypanina K, Bose S, Wang SI, Puc J, Miliaresis C, Rodgers L, McCombie R, Bigner SH, Giovanella BC, Ittmann M, Tycko B, Hibshoosh H, Wigler MH, Parsons R. *PTEN*, a putative protein tyrosine phosphatase gene mutated in human brain, breast, and prostate cancer. Science 275:1943–1947, 1997.
121. Steck PA, Pershouse MA, Jasser SA, Yung WK, Lin H, Ligon AH, Langford LA, Baumgard ML, Hattier T, Davis T, Frye C, Hu R, Swedlund B, Teng DH, Tavtigian SV. Identification of a candidate tumour suppressor gene, *MMAC1*, at chromosome 10q23.3 that is mutated in multiple advanced cancers. Nat Genet 15:356–362, 1997.

18

Androgen-Independent Prostate Cancer Progression: Mechanistic Insights

GEORGE V. THOMAS
CHARLES L. SAWYERS

Androgen-independent prostate cancer is incurable and is responsible for the majority of cancer-related deaths. The intriguing question is how a cancer that is initially exquisitely sensitive to androgen-ablation therapy becomes universally refractory to all available forms of such therapy.

Both normal and cancerous prostate cells are dependent on androgens, mainly in the form of testosterone and dihydrotestosterone (DHT), to promote proliferation and inhibit apoptosis.[1–3] Androgen mediates its biological actions through the androgen receptor (AR), a ligand-activated transcription factor of the steroid/nuclear receptor superfamily (see Color Fig. 18.1 in separate color insert).[4–6] Androgen-ablative treatment modalities rely on either surgical or medical castration, the latter usually administered via luteinizing hormone–releasing hormone (LHRH) analogs (termed *monotherapy* when used singly) and AR antagonists (termed *total androgen-ablative therapy* when used in combination).[7,8] Counter intuitively, AR continues to be expressed even as the cancer progresses from the androgen-dependent (AD) to an androgen-independent (AI) stage. Immunohistochemical studies have shown the presence of AR in both AD and AI tumors.[9,10] A rise in serum prostatic-specific antigen (PSA), an androgen-regulated gene, is usually the first marker of relapse/development of AI cancer. This suggests that AI prostate cancers rely on expression of AR-dependent genes.[11,12]

In this chapter, we critically discuss the current mechanistic concepts involved in the development of AI prostate cancer. These mechanisms can be broadly categorized as those known to signal through the AR and those whose mechanism for promoting AI growth remains unclear.

ANDROGENS AND THE ANDROGEN RECEPTOR

The testis, which produces testosterone, and adrenal glands, which produce androstenedione, dehydroepiandrostene, and dehydroepiandrostene sulfate, contribute to the bulk of circulating androgens. These are converted in the prostate or peripherally by 5α-reductase to DHT, which is approximately 10 times more active than testosterone.[13]

Androgens affect both the epithelial and mesenchymal components of the prostate, regulating growth and differentiation and inhibiting apoptosis. In murine prostates, castration results in glandular atrophy and involution, which is reversed by androgen replacement. The histological hallmarks of androgen-ablation therapy in the human prostate are atrophy, vacuolation, squamous and transitional cell hyperplasia, and basal cell hyperplasia.[14,15] Basal cells are considered to harbor the stem cells of the prostate, not requiring androgens for survival but at the same time able to proliferate and differentiate in response to androgen exposure. In contrast, the terminally differentiated luminal/secretory cells are dependent on androgens for survival and undergo apoptosis when androgens are removed.[2,16,17]

The AR-signaling cascade is initiated by the binding of androgen (ligand) to the AR.[6] The AR is a member of the steroid/nuclear receptor superfamily of transcription factors (hereafter designated as NR).[18,19] Members of the NR superfamily share three major functional domains: a central DNA-binding domain, an NH_3 terminal transactivation domain, and a COOH terminal ligand-binding domain.[11,20] The DNA-binding domain has two zinc fingers, one to bind DNA and the other to facilitate receptor dimerization as well as binding to coactivators. The NH_3 terminal contains one of the two transactivational domains found in the AR, i.e., activator function-1 (AF-1). This region accounts for the greatest diversity that exists between NR members. The AR has polymorphic glutamine and glycine repeats in this region. The COOH ligand-binding domain has a hydrophobic pocket in which the ligand binds. Members of the NR superfamily share a tertiary antiparallel α-helical sandwich with 12 helices, which undergoes conformational changes when the ligand is bound, facilitating coactivator–receptor interactions.[21–24] The COOH terminus also has a transcription-activating domain, AF-2, which is active only in the presence of ligand.[25] In the inactive state, the COOH terminus of AR is bound to heat shock proteins 70 and 90. Androgen binding leads to dissociation of AR from these heat shock proteins with subsequent phosphorylation, ligand–receptor stabilization, and receptor dimerization. The ligand–receptor complex can then bind to androgen response elements in the promoter regions of target genes. The AR recognizes a consensus glucocorticoid response element (5′-AGAACA-3′).[26] This activated DNA-bound AR homodimer complex is then able to recruit a plethora of coactivators and corepressors, which interact with the transcriptional machinery, resulting in stimulation or inhibition of target gene transcription. The prototypic androgen-responsive gene in the prostate is *PSA*.[27–30] Other examples of androgen-responsive genes include kallikriens, prostatic acid phosphatase (*PSAP*), and *p21*.[31,32]

Until recently, model systems to study the transition from AD to AI that closely recapitulate the human disease have been lacking. We utilized two human prostate xenografts created in our laboratory, LAPC-4 and LAPC-9, both of which grow and passage successfully in severe combined immunodeficiency disease mice as well as continuously passaged cell lines.[33] Both xenografts require androgen for growth, possess a nonmutant AR, synthesize PSA, and progress from AD to AI in response to castration. We demonstrated that a small fraction of cells in these xenografts die by apoptosis in response to castration, whereas the majority withdraw from the cell cycle. These cells remain in a dormant yet viable state and respond rapidly when re-exposed to androgen by re-entering the cell cycle and resuming tumor growth, even 6 months after androgen deprivation. After longer intervals, some LAPC-9 tumors resume growth as AI cancers. Our results suggest that AI progression occurs in two distinct stages.[34] At the time of initial diagnosis, a fraction of the cells in a prostate cancer tumor are dependent on androgen for survival and undergo apoptosis in response to androgen-ablation therapy. Clinical evidence for this conclusion has been well documented in studies of prostate cancer patients who receive neoadjuvant hormone-ablation therapy prior to radical prostatectomy. The first step in AI progression is a transition stage in which tumor cells remain androgen-responsive yet no longer require androgen for survival. The second stage involves the outgrowth of a tumor that is independent of androgen for both growth and survival, as observed clinically with hormone-refractory cancers that progress despite androgen blockade. Through serial dilutions and fluctuation analysis of the LAPC-4 and LAPC-9 cell lines, we showed that this second stage results from clonal expansion of a small number of AI cells present in the AD xenografts under selective pressure of androgen ablative therapy (see Color Figure 18.2 in separate color section). Our findings are thus consistent with the hypothesis that hormone-refractory cancer evolves through clonal outgrowth of a small number of AI cells that are preexisting or develop at a low frequency due to secondary genetic mutations. A critical next step is to identify the molecular basis for AI survival as opposed to AD growth.

LIGAND-DEPENDENT, ANDROGEN RECEPTOR–MEDIATED SIGNALING

Prostate cancer cells utilize components of a functional AR-signaling pathway in their progression from an AD to an AI state. Here, we review the evidence implicating *AR* gene amplification and mutations in disease progression.

Androgen Receptor Overexpression

Can the initiation, development, and progression of prostate cancer be attributed to *AR* overexpression by virtue of its positive effects on proliferation and its antiapoptotic effects? Stanbrough et al.,[35] using the rat probasin promoter to overexpress the murine *AR* transgene in mice, reported increased proliferation rates in the ventral and dorsolateral prostates of transgenic mice compared to wild-type littermates.[35] In addition, older mice developed prostatic intraepithelial neoplasia, a precursor lesion to prostate cancer. Hence, at least in mice, it appears that *AR* overexpression can initiate the development of precursor lesions but does not cause invasive cancer.

In human prostate cancer, the implication of *AR* overexpression in AI progression has come from the finding of *AR* gene amplification in tumors. Interestingly, *AR* amplification appears to be a feature of tumors that have been previously treated with androgen-ablation therapies and is very rarely found in untreated prostate cancers. Koivisto et al.[36] reported *AR* amplification in approximately 30% of AI tumors but none in matched tumors from the same patients prior to treatment. These data are consistent with the hypothesis that the amplification was a result of clonal selection of cells that were able to survive despite low levels of ligand. In an associated study from the same group, *AR* gene amplification (detected using fluorescence in situ hybridization, FISH) in prostate cancers progressing during androgen-blockade monotherapy was associated with better response to second-line combined androgen blockade (i.e., LHRH agonists combined with AR antagonists, e.g., biclutamide).[37] These results allow speculation into the possible mechanisms underlying progression while on androgen-ablative therapy; e.g., perhaps *AR* amplification results in increased *AR* responsiveness to low levels of androgen. Alternatively, *AR* amplification may result in increased expression of androgen-regulated genes, especially cell cycle genes and growth factors (see below, Growth Factors), which may enhance cellular proliferation in relative androgen absence.

Androgen Receptor Mutations

The *AR* gene maps to the X chromosome, thus rendering all mutations dominant in prostate cancer. Germ-line inactivating mutations of *AR* are well documented in the androgen-insensitivity syndrome.[6,38]

Similar to *AR* amplification, the frequency of *AR* mutations is rare in untreated prostate cancer. In contrast, *AR* mutations have been detected in as high as 50% of advanced and metastatic tumors, indicating that mutations are much more common in AI cancers.[39,40] These mutations may be present from the onset of the cancer (but remain undetectable due to the low sensitivity of the available assays) or may be acquired. In either case, it appears that these mutations gain prominence under the selective pressure of androgen-ablative therapy, as seen with *AR* amplification.

Evidence for *AR* mutations in prostate cancer first surfaced when *AR* was studied in the metastatic prostate cancer cell line LNCaP. Sequencing results uncovered a single missense point mutation in codon 877 of *AR*, resulting in a threonine-to-alanine substitution (T877A).[41,42] This mutation is located in the ligand-binding site of *AR*, and functional studies have revealed a change in the ligand-binding properties of the mutant *AR*. This mutation allows *AR* to bind to androgens, estrogens, progestagens, adrenal androgens, as well as many antiandrogens.[43]

This lack of specificity may also explain the flutamide withdrawal syndrome, in which patients show clinical deterioration with rising serum PSA levels while on the AR antagonist flutamide but then improve when the drug is removed. In other words, the antagonist behaves as an agonist. Taplin et al.[40] showed the presence of the T877A mutation in 5/16 patients with bone metastases while on flutamide. However, this mutation is not unique as several other mutations, e.g., T877S, are able to convert flutamide to an agonist as well. Functional analysis revealed strong stimulation of these mutant ARs by flutamide. Interestingly, these mutants were inhibited by bicalutamide (Casodex), a related nonsteroidal AR antagonist. Since the initial publications, numerous mutations of AR have been discovered (e.g., V715M, V730M, H874Y, and L701H).[44] Functional characterization of many of these mutants is in progress, and an updated list can be accessed at the Androgen Receptor Mutations Database (www.mcgill.ca/androgendb). Mutations of AR (including T877A) have been found in cell lines from patients not previously treated with flutamide (i.e., LNCaP and CRW22).[40] These AR mutants may allow prostate cancer cells to escape decreased androgen levels as well as conferring them with other as yet uncharacterized survival mechanisms. While neither AR amplification or AR mutations are likely to account for the majority of AI prostate cancers, they do illustrate the central role of altered AR function in late-stage disease.

LIGAND-INDEPENDENT, ANDROGEN RECEPTOR–MEDIATED SIGNALING PATHWAYS

Growth Factors

An alternate mechanism that plays a role in the progression of AI tumors involves the recruitment of nonsteroid receptor signal-transduction pathways that activate AR in the clinical setting of androgen deprivation. Studies in breast cancer have shown that both the estrogen receptor (ER) and progesterone receptor (also members of the NR superfamily) can be activated by growth factors such as epidermal growth factor (EGF) and insulin-like growth factor I (IGF-I) as well as stimuli which activate cyclic adenosine monophosphate. This ligand-independent activation is promoted by EGF, which phosphorylates the serine-118 residue of ER, through activation of the EGF receptor (EGFR)–Ras-Erk signaling pathway.[45,46] Culig et al.[47] showed that EGF, IGF-I, and keratinocyte growth factor were able to activate AR and subsequently induce AR target genes in the absence of ligand.[47] However, the mechanism is not yet defined.

There is growing evidence that activation of AR through these receptor tyrosine kinase pathways may play a functional role in human prostate cancer. Constitutive activation of EGFR has been seen with the expression of a mutant receptor, EGFRVIII, which lacks much of its extracellular domain. This mutant is amplified in glioblastomas.[48] In prostate cancer, one group has reported increased protein expression of EGFRVIII in a large fraction of advanced-stage disease, but confirmatory studies are needed.[49]

Serum IGF-I levels are increased in prostate cancer and have been correlated with more aggressive tumors.[50,51] However, unlike EGFR, no somatic mutations or amplifications of this growth factor gene and its receptor have been reported in prostate cancer. The receptor for IGF (IGF-R) signals through the phosphoinositide 3-kinase (PI3K) pathway (see below, PI3K-AKT Pathway), a parallel pathway to the Erk signal-transduction cascade, with ample opportunities for crosstalk along the way. Using human prostate cancer xenograft models, we observed that progression to AI growth was associated with a 60-fold increase in IGF-I mRNA expression in LAPC-9 xenografts and a 28-fold increase in LNCaP xenografts relative to initial AD neoplasms. This rise in IGF-I was mirrored by an increase in IGF-IR mRNA expression.[52]

Bicalutamide is able to completely block the activation of AR by EGF and IGF-I, thus confirming the central role that AR plays in these triggered signal-transduction pathways as well as alluding to other escape mechanisms the cancer cells have developed to evade complete androgen-ablative therapy.[47] However, the very presence of these growth factors points to the larger role of receptor tyrosine kinases in AI progression.

Her2/Neu Receptor Tyrosine Kinase

The *Her2/Neu* oncogene encodes a transmembrane glycoprotein with a tyrosine kinase domain that is structurally related to the EGFR family of receptors. It has intrinsic kinase activity and is able to activate receptor-mediated signal transduction when overexpressed. *Her2/Neu* is amplified and overexpressed in 20%–30% of breast and ovarian cancers.[53,54] Overexpression in breast cancer is associated with poor prognosis. With the development of humanized monoclonal anti-Her2/Neu antibody as a viable treatment option, the significance of targeting specific receptor tyrosine kinases has gained clinical prominence.[55] However, studies of Her2/Neu expression in human prostate cancer have revealed conflicting results, with different groups reporting differing levels of gene copy amplification and protein expression.[56–61] Her2/Neu is expressed in normal prostatic epithelial cells, and its putative ligand, heregulin, is expressed in the stroma, suggesting a paracrine mode of growth factor receptor stimulation. In one of the larger cohorts examined, Signoretti et al.[62] showed that Her2/Neu protein expression (as detected by immunohistochemistry) was present in a small number of early-stage, untreated prostate cancers and that the proportion of Her2/Neu-positive tumors was significantly higher in tumors that had received androgen-ablation therapy as well as in AI tumors.[62] In addition, while Her2/Neu mRNA was detected in all tumors, no amplification of the gene was detected by FISH. Hence, in contrast to breast cancer, in prostate cancer Her2/Neu is not commonly overexpressed by gene amplification. These findings have been confirmed by two separate studies.[63,64] How does increased Her2/Neu expression mediate AI progression? Functional studies of Her2/Neu in established AD and AI xenografts (LAPC-4 AD and LAPC-4 AI, respectively) revealed that forced expression of Her2/Neu was able to convert the AD cell line to an AI cell line in vitro and accelerate the progression to AI in castrate animals.[65] In addition, Her2/Neu activated the AR-signaling pathway in the absence of ligand and enhanced the magnitude of the AR response in the presence of low levels of androgen. This mirrors the clinical situation of men receiving medical castration. Furthermore, reconstitution experiments with AR established that the effects of Her2/Neu on the AR pathway required AR expression. However, in striking contrast to the effects mediated by IGF-I, the effect mediated by Her2/Neu could not be blocked by bicalutamide, indicating that Her2/Neu acts on the AR pathway distal to the interaction between ligand and receptor. Therefore, activation of the Her2/Neu pathway is able to rescue prostate cancer cells from the growth arrest imposed by androgen-ablative therapies. While the exact mechanism for the crosstalk between Her2/Neu and AR has yet to be defined, several studies have uncovered possible candidates.

Focusing on the mitogen-activated protein kinase (MAPK) pathway, Yeh et al.[66] reported that MAPK directly phosphorylated AR and that inhibitors of MAPK decreased the Her2/Neu-mediated activation of AR.[66] Another candidate pathway involves the interaction between Her2/Neu and the PI3K/AKT pathway (see below, PI3K/AKT).

Alternatively, Her2/Neu may directly or indirectly interact with AR and/or its coactivators/co-repressors.

Phosphoinositide 3-Kinase–AKT Pathway

The receptor for EGF, IGF-R, and Her2/Neu recruit PI3K to the receptor. In turn, PI3K can phosphorylate membrane lipids, specifically phosphotidylinositol (3,4,5)-triphosphate [PtdIns (3,4,5)P3], which activates the serine/threonine kinase AKT.[67] Recruitment of AKT to the cell membrane results in a conformational change of AKT, enabling residues Thr^{308} and Ser^{473} to be phosphorylated by upstream kinases phosphoinositide-dependent protein kinase-1 (PDK-1) and phosphoinositide-dependent protein kinase-2 (PDK-2) or integrin-linked kinase (ILK), respectively. Activated AKT phosphorylates specific targets such as mammalian target of Rapamycin (mTOR), glycogen synthase kinase-3 (GSK3), BAD, procaspase-9, and the forkhead transcription factor FKHRL1, which promote cell survival, block apoptosis, and control translational and transcriptional as well as cellular metabolism.[68] Thus, PI3K-AKT is an important component of multiple oncogenic signaling pathways. The discovery of the tumor-suppressor gene *PTEN* has focused a high degree of interest on this pathway. The *PTEN* gene possesses both lipid and protein phosphatase activity. In the context of the PI3K pathway, *PTEN* removes the 3-phosphate from PtdIns(3,4,5)P3, thus blocking or dampening the PI3K–AKT pathway. The *PTEN* gene localizes to chromosome 10q23. Loss of heterozygosity of this region has been found in 20%–60% of prostate cancers, with more advanced/metastatic tumors exhibiting higher rates.[69–71] Loss of protein expression (as seen by lack of immunohistochemical staining for *PTEN*) is found predominantly in tumors of higher Gleason grade and stage, thus correlating *PTEN* loss with more aggressive tumors.[72] Mice heterozygous for *PTEN* as well as conditional *PTEN* knockouts (pure *PTEN* knockouts result in embryonic lethality) develop diffuse prostatic hyperplasia and eventually cancer.[73–75]

Graff and coworkers,[76] using AI LNCaP cells (which are *PTEN* null), found increased Akt activity in the propagated AI sublines compared to parental AD lines. Overexpression of AKT resulted in accelerated tumor growth. Interestingly, this phenotype was accompanied by a concurrent decrease in p27 levels. This is in keeping with previously published data that *PTEN*-deficient cell lines exhibit accelerated entry into S phase, accompanied by downregulation of p27.[77]

Wen et al.[78] found that Her2/Neu can activate the AKT pathway, to promote survival and growth. These effects were correlated with AKT phosphorylation of Ser^{213} and Ser^{791} of AR and abrogated by the expression of a dominant negative AKT.

Coactivators

Studies in the NR field have identified families of proteins known as coactivators and corepressors, which modulate the function of transcription factors.[79,80] Coactivators act by binding to the receptor in a ligand-dependent fashion and enhance ligand-dependent gene transcription. Overexpression of these coactivators can enhance the transcription of androgen-responsive genes in vitro. Conversely, downregulation of the corepressors is postulated to achieve the same effect. Many NR coactivators contain one or multiple copies of the α-helical LXXLL signature motif (where L is leucine and X is any amino acid). This motif mediates ligand-dependent coactivator NR or coactivator–coactivator interactions. Examples of coactivators known to directly interact with NRs include steroid receptor coactivator-1 (SRC-1), transcriptional intermediary factor 2 (TIF2), and amplified in breast cancer-1 (AIB-1). The latter is amplified in breast and ovarian cancers.[81] Gregory et al.[82] reported higher levels of AR, TIF2 and SRC-1 in recurrent human prostate cancer as well as in the CRW22 xenograft model. The AR-associated protein ARA70 is a prototype AR coactivator whose expression not only activates the transcription of androgen-responsive genes but also alters the ligand specificity of AR such that antagonists function as agonist.[83]

Two known corepressors of the NR superfamily are the NR corepressor (N-CoR) and silencing mediator of retinoic and thyroid hormone receptors. Decreased levels of N-CoR correlate with acquisition of resistance to the ER tamoxifen in a breast cancer xenograft model.[84] Evidence on AR corepressors was lacking until a recent report detailing the role of members of the protein inhibitor of activated signal transducer and activator of transcription protein inhibitor of activated STAT (PIAS) family. Four members of the PIAS family have been identified: PIAS1, PIAS3, PIASx, and PIASy. The latter acts as a potent inhibitor of AR transcriptional activity.[85] In contrast, PIAS1 and PIAS3 act as coactivators. Even more interestingly, PIASy possesses an LXXLL signature motif, which appears to be essential for its transrepression activity. Clearly, our knowledge of the interactions between AR and its coactivators and corepressors is still limited and needs further investigation.

ANDROGEN RECEPTOR–INDEPENDENT PATHWAYS

Modulating Apoptosis

The mechanisms that determine AD to AI progression can be broadly divided into those that signal through AR and those that do not. Survival signals utilized by cancer cells include the activation of pathways that inhibit apoptosis. Androgen-ablative therapy initially results in growth arrest, followed by proliferation, all while successfully evading apoptosis. Pathways that regulate apoptosis contribute to tumorigenesis and resistance to therapy. Bcl-2 is a potent modulator of the antiapoptotic pathway, which does not function through the AR pathway in prostate cancer cells.

Bcl-2 is normally expressed in the basal cells of the prostate, which are resistant to the effects of androgen withdrawal. Basal cells are also thought to house the putative stem cells of the prostate (see below, Stem Cells). Bcl-2-negative secretory/luminal cells undergo apoptosis in response to androgen deprivation.

High levels of bcl-2 protein expression (as measured by immunohistochemistry) are seen with greater frequency as prostate cancers progress from localized (7%) to AI tumors (67%).[86,87] Thus, it appears that bcl-2 enables prostate cancer cells to remain viable despite low levels of androgen and that androgen ablation may select for bcl-2-positive cells that fail to undergo apoptosis. Apoptosis can also be modulated via the AKT pathway. AKT can phosphorylate Bad, a proapoptotic peptide. Phosphorylated Bad binds the cytosolic 14-3-3 protein, whereas dephosphorylated Bad binds and sequesters bcl-2 family proteins, such as bcl-X_L, to promote apoptosis.[88] Thus, phosphorylation of Bad through AKT will enhance the activity of the anti-apoptotic kinase bcl-2.

Stem Cells

We have alluded to the basal cell layer as the site of the putative stem cells of the prostate. These stem cells differentiate into secretory epithelial cells of the prostate. Basal cells do not express PSA and are dependent on androgen for survival but not growth.[17] A pathognomic histological feature of androgen-ablation therapy is basal cell hyperplasia.[14] Under the selective pressure of androgen ablation, we may select for subsets of AI basal cells, which serve as seeds for tumor progression. Ablation of these cells may be a desirable therapeutic goal, but a better understanding of these cells is required. Studies have begun to identify candidate genes that define this stem cell population.

Prostate stem cell antigen (*PSCA*) is one such candidate prostate stem cell gene.[89] It shares 30% nucleotide homology with stem cell antigen 2, a member of the Thy-1/Ly-6 superfamily of glycosyl-phosphatidylinositol-anchored cell surface antigens. Interestingly, signaling through stem cell antigen 2 can prevent apoptosis in immature thymocytes. Immunohistochemical studies in human prostate cancer showed increased PSCA protein expression in higher-stage prostate cancer, especially bone metastases, suggesting a role in prostate cancer progression.[90] In addition, *PSCA* is coamplified with c-*myc*. C-*myc* amplification has been identified in more than 20% of recurrent and metastatic prostate cancers.[91] Anti-PSCA monoclonal antibodies inhibit tumor formation in xenograft mice models as well as slow the growth of established orthotopic tumors.[92] These findings suggest its utility as a novel immunotherapeutic modality for the treatment of prostate cancer.

p63 is a homolog of the *p53* tumor-suppressor gene, is expressed in the basal cells of many epithelial organs, and may play a role in the maintenance of the stem cell compartment in these organs.[93,94] *p63* knockout mice ($p63^{-/-}$) do not develop prostates, providing further evidence that the basal cells may indeed represent prostatic stem cells.[95] In human prostates, positive *p63* immunohistochemical staining is seen in basal cells of all normal prostatic glands, benign hyperplastic lesions, and prostatic intraepithelial neoplasia. In contrast to PSCA, almost no prostatic cancers stain with *p63*. The differences in protein expression between PSCA and p63 suggest that these genes represent different subpopulations of intermediately differentiated stem cells.

SUMMARY

We have presented possible pathways that may lead to AI prostate cancer. An important factor in this progression involves the ability of prostate cancer cells to survive and grow in an androgen-depleted milieu. Cancer cells may gain this ability through alterations in AR function, which include amplification, mutations, and perturbations in signaling pathway components, which involve ligands, kinases, and coactivator–corepressor complexes. Alternatively, parallel pathways not involving AR signaling may also contribute to progression. Due to the heterogeneity of prostate cancer, several pathways, acting singly or in concert, may ultimately lead to AI progression. We have highlighted the role of receptor tyrosine kinase signaling pathways since this is the arena in which many of the innovative clinical candidates are being developed, e.g., STI-571 and CCI-779, in addition to antitumor monoclonal antibodies, e.g., Trastuzumab and mAB-C225.[96–98] The paradigm on which these agents have been developed is founded on the hypothesis that cancers have specific molecular signatures that, once identified, define that pathway and may serve as suitable molecular targets. While it may be argued that we have progressed very little since the discovery that orchiectomy slows the progression of prostate cancer, the last few years have certainly delivered the hope of new therapeutic options.

ACKNOWLEDGMENTS

This work was supported by Department of Defense (DOD) grant PC-010383 and a UCLA Jonsson Cancer Center Foundation Fellowship grant to G.V.T. C.L.S. received grant support from the National Cancer Institute, DOD, CAPCure, and the Leukemia and Lymphoma Society. We thank Ms. Moya Costello for the artwork.

REFERENCES

1. Ross RK, Pike MC, Coetzee GA, Reichardt JK, Yu MC, Feigelson H, Stanczyk FZ, Kolonel LN, Henderson BE. Androgen metabolism and prostate cancer: establishing a model of genetic susceptibility. Cancer Res 58:4497–4504, 1998.
2. Denmeade SR, Lin XS, Isaacs JT. Role of programmed (apoptotic) cell death during the progression and therapy for prostate cancer. Prostate 28:251–265, 1996.
3. Colombel M, Gil Diez S, Radvanyi F, Buttyan R, Thiery JP, Chopin D. Apoptosis in prostate cancer. Molecular basis to study hormone refractory mechanisms. Ann NY Acad Sci 784:63–69, 1996.
4. Brinkmann AO, Blok LJ, de Ruiter PE, Doesburg P, Steketee K, Berrevoets CA, Trapman J. Mechanisms of androgen receptor activation and function. J Steroid Biochem Mol Biol 69:307–313, 1999.
5. Brinkmann AO, Trapman J. Genetic analysis of androgen receptors in development and disease. Adv Pharmacol 47:317–341, 2000.
6. Quigley CA, De Bellis A, Marschke KB, el-Awady MK, Wilson EM, French FS. Androgen receptor defects: historical, clinical, and molecular perspectives. Endocr Rev 16:271–321, 1995.
7. Labrie F, Belanger A, Dupont A, Luu-The V, Simard J, Labrie C. Science behind total androgen blockade: from gene to combination therapy. Clin Invest Med 16:475–492, 1993.
8. Labrie F, Dupont A, Belanger A, St-Arnaud R, Giguere M, Lacourciere Y, Emond J, Monfette G. Treatment of prostate cancer with gonadotropin-releasing hormone agonists. Endocr Rev 7:67–74, 1986.
9. Culig Z, Hobisch A, Hittmair A, Peterziel H, Cato AC, Bartsch G, Klocker H. Expression, structure, and function of androgen receptor in advanced prostatic carcinoma. Prostate 35:63–70, 1998.
10. van der Kwast TH, Schalken J, Ruizeveld de Winter JA, van Vroonhoven CC, Mulder E, Boersma W, Trapman J. Androgen receptors in endocrine-therapy-resistant human prostate cancer. Int J Cancer 48:189–193, 1991.
11. Gnanapragasam VJ, Robson CN, Leung HY, Neal DE. Androgen receptor signalling in the prostate. BJU Int 86:1001–1013, 2000.
12. Jenster G. The role of the androgen receptor in the development and progression of prostate cancer. Semin Oncol 26:407–421, 1999.
13. Avila DM, Zoppi S, McPhaul MJ. The androgen receptor (AR) in syndromes of androgen insensitivity and in prostate cancer. J Steroid Biochem Mol Biol 76:135–142, 2001.
14. Reuter VE. Pathological changes in benign and malignant prostatic tissue following androgen deprivation therapy. Urology 49:16–22, 1997.
15. Westin P, Bergh A, Damber JE. Castration rapidly results in a major reduction in epithelial cell numbers in the rat prostate, but not in the highly differentiated Dunning R3327 prostatic adenocarcinoma. Prostate 22:65–74, 1993.
16. Isaacs JT. The biology of hormone refractory prostate cancer. Why does it develop? Urol Clin North Am 26:263–273, 1999.
17. Bui M, Reiter RE. Stem cell genes in androgen-independent prostate cancer. Cancer Metastasis Rev 17:391–399, 1998.
18. Beato M. Gene regulation by steroid hormones. Cell 56:335–344, 1989.
19. Evans RM. The steroid and thyroid hormone receptor superfamily. Science 240:889–895, 1988.
20. Culig Z, Hobisch A, Bartsch G, Klocker H. Expression and function of androgen receptor in carcinoma of the prostate. Microsc Res Tech 51:447–455, 2000.
21. Giguere V, Hollenberg SM, Rosenfeld MG, Evans RM. Functional domains of the human glucocorticoid receptor. Cell 46:645–652, 1986.
22. Fawell SE, Lees JA, White R, Parker MG. Characterization and colocalization of steroid binding and dimerization activities in the mouse estrogen receptor. Cell 60:953–962, 1990.

23. Wagner RL, Apriletti JW, McGrath ME, West BL, Baxter JD, Fletterick RJ. A structural role for hormone in the thyroid hormone receptor. Nature 378:690–697, 1995.
24. Bourguet W, Ruff M, Chambon P, Gronemeyer H, Moras D. Crystal structure of the ligand-binding domain of the human nuclear receptor RXR-alpha. Nature 375:377–382, 1995.
25. Tora L, White J, Brou C, Tasset D, Webster N, Scheer E, Chambon P. The human estrogen receptor has two independent nonacidic transcriptional activation functions. Cell 59:477–487, 1989.
26. Kumar V, Chambon P. The estrogen receptor binds tightly to its responsive element as a ligand-induced homodimer. Cell 55:145–156, 1988.
27. Riegman PH, Vlietstra RJ, van der Korput JA, Brinkmann AO, and Trapman J. The promoter of the prostate-specific antigen gene contains a functional androgen responsive element. Mol Endocrinol 5:1921–1930, 1991.
28. Wolf DA, Schulz P, Fittler F. Transcriptional regulation of prostate kallikrein-like genes by androgen. Mol Endocrinol 6:753–762, 1992.
29. Young CY, Montgomery BT, Andrews PE, Qui SD, Bilhartz DL, Tindall DJ. Hormonal regulation of prostate-specific antigen messenger RNA in human prostatic adenocarcinoma cell line LNCaP. Cancer Res 51:3748–3752, 1991.
30. Shan JD, Porvari K, Ruokonen M, Karhu A, Launonen V, Hedberg P, Oikarinen J, Vihko P. Steroid-involved transcriptional regulation of human genes encoding prostatic acid phosphatase, prostate-specific antigen, and prostate-specific glandular kallikrein. Endocrinology 138:3764–3770, 1997.
31. Henttu P, Vihko P. Steroids inversely affect the biosynthesis and secretion of human prostatic acid phosphatase and prostate-specific antigen in the LNCaP cell line. J Steroid Biochem Mol Biol 41:349–360, 1992.
32. Lu S, Liu M, Epner DE, Tsai SY, Tsai MJ. Androgen regulation of the cyclin-dependent kinase inhibitor p21 gene through an androgen response element in the proximal promoter. Mol Endocrinol 13:376–384, 1999.
33. Klein KA, Reiter RE, Redula J, Moradi H, Zhu XL, Brothman AR, Lamb DJ, Marcelli M, Belldegrun A, Witte ON, Sawyers CL. Progression of metastatic human prostate cancer to androgen independence in immunodeficient SCID mice. Nat Med 3:402–408, 1997.
34. Craft N, Chhor C, Tran C, Belldegrun A, DeKernion J, Witte ON, Said J, Reiter RE, Sawyers CL. Evidence for clonal outgrowth of androgen-independent prostate cancer cells from androgen-dependent tumors through a two-step process. Cancer Res 59:5030–5036, 1999.
35. Stanbrough M, Leav I, Kwan PW, Bubley GJ, Balk SP. Prostatic intraepithelial neoplasia in mice expressing an androgen receptor transgene in prostate epithelium. Proc Natl Acad Sci USA 98:10823–10828, 2001.
36. Koivisto P, Kononen J, Palmberg C, Tammela T, Hyytinen E, Isola J, Trapman J, Cleutjens K, Noordzij A, Visakorpi T, Kallioniemi OP. Androgen receptor gene amplification: a possible molecular mechanism for androgen deprivation therapy failure in prostate cancer. Cancer Res 57:314–319, 1997.
37. Palmberg C, Koivisto P, Kakkola L, Tammela TL, Kallioniemi OP, Visakorpi T. Androgen receptor gene amplification at primary progression predicts response to combined androgen blockade as second line therapy for advanced prostate cancer. J Urol 164:1992–1995, 2000.
38. Quigley CA, French FS. Androgen insensitivity syndromes. Curr Ther Endocrinol Metab 5:342–351, 1994.
39. Taplin ME, Bubley GJ, Shuster TD, Frantz ME, Spooner AE, Ogata GK, Keer HN, Balk SP. Mutation of the androgen-receptor gene in metastatic androgen-independent prostate cancer. N Engl J Med 332:1393–1398, 1995.
40. Taplin ME, Bubley GJ, Ko YJ, Small EJ, Upton M, Rajeshkumar B, Balk SP. Selection for androgen receptor mutations in prostate cancers treated with androgen antagonist. Cancer Res 59:2511–2515, 1999.
41. Veldscholte J, Berrevoets CA, Ris-Stalpers C, Kuiper GG, Jenster G, Trapman J, Brinkmann AO, Mulder E. The androgen receptor in LNCaP cells contains a mutation in the ligand binding domain which affects steroid binding characteristics and response to antiandrogens. J Steroid Biochem Mol Biol 41:665–669, 1992.
42. Gaddipati JP, McLeod DG, Heidenberg HB, Sesterhenn IA, Finger MJ, Moul JW, Srivastava S. Frequent detection of codon 877 mutation in the androgen receptor gene in advanced prostate cancers. Cancer Res 54:2861–2864, 1994.
43. Culig Z, Hobisch A, Cronauer MV, Cato AC, Hittmair A, Radmayr C, Eberle J, Bartsch G, Klocker H. Mutant androgen receptor detected in an advanced-stage prostatic carcinoma is activated by adrenal androgens and progesterone. Mol Endocrinol 7:1541–1550, 1993.
44. Zhao XY, Malloy PJ, Krishnan AV, Swami S, Navone NM, Peehl DM, Feldman D. Glucocorticoids can promote androgen-independent growth of prostate cancer cells through a mutated androgen receptor. Nat Med 6:703–706, 2000.
45. Kato S, Endoh H, Masuhiro Y, et al. Activation of the estrogen receptor through phosphorylation by mitogen-activated protein kinase. Science 270:1491–1494, 1995.
46. Kato S. Estrogen receptor-mediated cross-talk with growth factor signaling pathways. Breast Cancer 8:3–9, 2001.
47. Culig Z, Hobisch A, Cronauer MV, Radmayr C, Trapman J, Hittmair A, Bartsch G, Klocker H. Androgen receptor activation in prostatic tumor

cell lines by insulin-like growth factor-I, keratinocyte growth factor, and epidermal growth factor. Cancer Res 54:5474–5478, 1994.
48. Steck PA, Lee P, Hung MC, Yung WK. Expression of an altered epidermal growth factor receptor by human glioblastoma cells. Cancer Res 48:5433–5439, 1988.
49. Olapade-Olaopa EO, Moscatello DK, MacKay EH, Horsburgh T, Sandhu DP, Terry TR, Wong AJ, Habib FK. Evidence for the differential expression of a variant EGF receptor protein in human prostate cancer. Br J Cancer 82:186–194, 2000.
50. Wolk A, Mantzoros CS, Andersson SO, Bergstrom R, Signorello LB, Lagiou P, Adami HO, Trichopoulos D. Insulin-like growth factor 1 and prostate cancer risk: a population-based, case-control study. J Natl Cancer Inst 90:911–915, 1998.
51. Signorello LB, Brismar K, Bergstrom R, Andersson SO, Wolk A, Trichopoulos D, Adami HO. Insulin-like growth factor-binding protein-1 and prostate cancer. J Natl Cancer Inst 91:1965–1967, 1999.
52. Nickerson T, Chang F, Lorimer D, Smeekens SP, Sawyers CL, Pollak M. In vivo progression of LAPC-9 and LNCaP prostate cancer models to androgen independence is associated with increased expression of insulin-like growth factor I (IGF-I) and IGF-I receptor (IGF-IR). Cancer Res 61:6276–80, 2001.
53. Slamon DJ, Godolphin W, Jones LA, et al. Studies of the HER-2/*neu* proto-oncogene in human breast and ovarian cancer. Science 244:707–712, 1989.
54. Venter DJ, Tuzi NL, Kumar S, Gullick WJ. Overexpression of the c-erbB-2 oncoprotein in human breast carcinomas: immunohistological assessment correlates with gene amplification. Lancet 2:69–72, 1987.
55. Cobleigh MA, Vogel CL, Tripathy D, Robert NJ, Scholl S, Fehrenbacher L, Wolter JM, Paton V, Shak S, Lieberman G, Slamon DJ. Multinational study of the efficacy and safety of humanized anti-HER2 monoclonal antibody in women who have HER2-overexpressing metastatic breast cancer that has progressed after chemotherapy for metastatic disease. J Clin Oncol 17:2639–2648, 1999.
56. Bubendorf L, Kononen J, Koivisto P, Schraml P, Moch H, Gasser TC, Willi N, Mihatsch MJ, Sauter G, Kallioniemi OP. Survey of gene amplifications during prostate cancer progression by high-throughout fluorescence in situ hybridization on tissue microarrays. Cancer Res 59:803–806, 1999.
57. Mark HF, Feldman D, Das S, Kye H, Mark S, Sun CL, Samy M. Fluorescence in situ hybridization study of HER-2/*neu* oncogene amplification in prostate cancer. Exp Mol Pathol 66:170–178, 1999.
58. Fournier G, Latil A, Amet Y, Abalain JH, Volant A, Mangin P, Floch HH, Lidereau R. Gene amplifications in advanced-stage human prostate cancer. Urol Res 22:343–347, 1995.
59. Latil A, Baron JC, Cussenot O, Fournier G, Boccon-Gibod L, Le Duc A, Lidereau R. Oncogene amplifications in early-stage human prostate carcinomas. Int J Cancer 59:637–638, 1994.
60. Ross JS, Sheehan CE, Hayner-Buchan AM, Ambros RA, Kallakury BV, Kaufman RP Jr, Fisher HA, Rifkin MD, Muraca PJ. Prognostic significance of HER-2/*neu* gene amplification status by fluorescence in situ hybridization of prostate carcinoma. Cancer 79:2162–2170, 1997.
61. Kaltz-Wittmer C, Klenk U, Glaessgen A, Aust DE, Diebold J, Lohrs U, Baretton GB. FISH analysis of gene aberrations (*MYC*, *CCND1*, *ERBB2*, *RB*, and *AR*) in advanced prostatic carcinomas before and after androgen deprivation therapy. Lab Invest 80:1455–1464, 2000.
62. Signoretti S, Montironi R, Manola J, Altimari A, Tam C, Bubley G, Balk S, Thomas G, Kaplan I, Hlatky L, Hahnfeldt P, Kantoff P, Loda M. Her-2-neu expression and progression toward androgen independence in human prostate cancer. J Natl Cancer Inst 92:1918–1925, 2000.
63. Osman I, Scher HI, Drobnjak M, Verbel D, Morris M, Agus D, Ross JS, Cordon-Cardo C. HER-2/neu (p185neu) protein expression in the natural or treated history of prostate cancer. Clin Cancer Res 7:2643–2647, 2001.
64. Shi Y, Brands FH, Chatterjee S, Feng AC, Groshen S, Schewe J, Lieskovsky G, Cote RJ. Her-2/neu expression in prostate cancer: high level of expression associated with exposure to hormone therapy and androgen independent disease. J Urol 166:1514–1519, 2001.
65. Craft N, Shostak Y, Carey M, Sawyers CL. A mechanism for hormone-independent prostate cancer through modulation of androgen receptor signaling by the HER-2/neu tyrosine kinase. Nat Med 5:280–285, 1999.
66. Yeh S, Lin HK, Kang HY, Thin TH, Lin MF, Chang C. From HER2/neu signal cascade to androgen receptor and its coactivators: a novel pathway by induction of androgen target genes through MAP kinase in prostate cancer cells. Proc Natl Acad Sci USA 96:5458–5463, 1999.
67. Wu X, Senechal K, Neshat MS, Whang YE, Sawyers CL. The PTEN/MMAC1 tumor suppressor phosphatase functions as a negative regulator of the phosphoinositide 3-kinase/Akt pathway. Proc Natl Acad Sci USA 95:15587–15591, 1998.
68. Vazquez F, Sellers WR. The PTEN tumor suppressor protein: an antagonist of phosphoinositide 3-kinase signaling. Biochim Biophys Acta 1470:M21–M35, 2000.
69. Gray IC, Phillips SM, Lee SJ, Neoptolemos JP, Weissenbach J, Spurr NK. Loss of the chromosomal region 10q23–25 in prostate cancer. Cancer Res 55:4800–4803, 1995.

70. Carter BS, Ewing CM, Ward WS, Treiger BF, Aalders TW, Schalken JA, Epstein JI, Isaacs WB. Allelic loss of chromosomes 16q and 10q in human prostate cancer. Proc Natl Acad Sci USA 87:8751–8755, 1990.
71. Komiya A, Suzuki H, Ueda T, Yatani R, Emi M, Ito H, Shimazaki J. Allelic losses at loci on chromosome 10 are associated with metastasis and progression of human prostate cancer. Genes Chromosomes Cancer 17:245–253, 1996.
72. McMenamin ME, Soung P, Perera S, Kaplan I, Loda M, Sellers WR. Loss of PTEN expression in paraffin-embedded primary prostate cancer correlates with high Gleason score and advanced stage. Cancer Res 59:4291–4296, 1999.
73. Di Cristofano A, Pandolfi PP. The multiple roles of PTEN in tumor suppression. Cell 100:387–390, 2000.
74. Di Cristofano A, De Acetis M, Koff A, Cordon-Cardo C, Pandolfi PP. Pten and p27^{KIP1} cooperate in prostate cancer tumor suppression in the mouse. Nat Genet 27:222–224, 2001.
75. Di Cristofano A, Pesce B, Cordon-Cardo C, Pandolfi PP. Pten is essential for embryonic development and tumour suppression. Nat Genet 19:348–355, 1998.
76. Graff JR, Konicek BW, McNulty AM, Wang Z, Houck K, Allen S, Paul JD, Hbaiu A, Goode RG, Sandusky GE, Vessella RL, Neubauer BL. Increased AKT activity contributes to prostate cancer progression by dramatically accelerating prostate tumor growth and diminishing p27^{Kip1} expression. J Biol Chem 275:24500–24505, 2000.
77. Sun H, Lesche R, Li DM, Liliental J, Zhang H, Gao J, Gavrilova N, Mueller B, Liu X, Wu H. PTEN modulates cell cycle progression and cell survival by regulating phosphatidylinositol 3,4,5,-trisphosphate and Akt/protein kinase B signaling pathway. Proc Natl Acad Sci USA 96:6199–6204, 1999.
78. Wen Y, Hu MC, Makino K, Spohn B, Bartholomeusz G, Yan DH, Hung MC. HER-2/neu promotes androgen-independent survival and growth of prostate cancer cells through the Akt pathway. Cancer Res 60:6841–6845, 2000.
79. Torchia J, Rose DW, Inostroza J, Kamei Y, Westin S, Glass CK, Rosenfeld MG. The transcriptional co-activator p/CIP binds CBP and mediates nuclear-receptor function. Nature 387:677–684, 1997.
80. Alland L, Muhle R, Hou H Jr, Potes J, Chin L, Schreiber-Agus N, DePinho RA. Role for N-CoR and histone deacetylase in Sin3-mediated transcriptional repression. Nature 387:49–55, 1997.
81. Anzick SL, Kononen J, Walker RL, Azorsa DO, Tanner MM, Guan XY, Sauter G, Kallioniemi OP, Trent JM, Meltzer PS. AIB1, a steroid receptor coactivator amplified in breast and ovarian cancer. Science 277:965–968, 1997.
82. Gregory CW, He B, Johnson RT, Ford OH, Mohler JL, French FS, Wilson EM. A mechanism for androgen receptor–mediated prostate cancer recurrence after androgen deprivation therapy. Cancer Res 61:4315–4319, 2001.
83. Yeh S, Kang HY, Miyamoto H, Nishimura K, Chang HC, Ting HJ, Rahman M, Lin HK, Fujimoto N, Hu YC, Mizokami A, Huang KE, Chang C. Differential induction of androgen receptor transactivation by different androgen receptor coactivators in human prostate cancer DU145 cells. Endocrine 11:195–202, 1999.
84. Lavinsky RM, Jepsen K, Heinzel T, Torchia J, Mullen TM, Schiff R, Del-Rio AL, Ricote M, Ngo S, Gemsch J, Hilsenbeck SG, Osborne CK, Glass CK, Rosenfeld MG, Rose DW. Diverse signaling pathways modulate nuclear receptor recruitment of N-CoR and SMRT complexes. Proc Natl Acad Sci USA 95:2920–2925, 1998.
85. Gross M, Liu B, Tan J, French FS, Carey M, Shuai K. Distinct effects of PIAS proteins on androgen-mediated gene activation in prostate cancer cells. Oncogene 20:3880–3887, 2001.
86. Furuya Y, Krajewski S, Epstein JI, Reed JC, Isaacs JT. Expression of bcl-2 and the progression of human and rodent prostatic cancers. Clin Cancer Res 2:389–398, 1996.
87. Oh WK, Kantoff PW. Treatment of locally advanced prostate cancer: is chemotherapy the next step? J Clin Oncol 17:3664–3675, 1999.
88. Zha J, Harada H, Yang E, Jockel J, Korsmeyer SJ. Serine phosphorylation of death agonist BAD in response to survival factor results in binding to 14-3-3 not BCL-X_L. Cell 87:619–628, 1996.
89. Reiter RE, Gu Z, Watabe T, Thomas G, Szigeti K, Davis E, Wahl M, Nisitani S, Yamashiro J, Le Beau MM, Loda M, Witte ON. Prostate stem cell antigen: a cell surface marker overexpressed in prostate cancer. Proc Natl Acad Sci USA 95:1735–1740, 1998.
90. Gu Z, Thomas G, Yamashiro J, Shintaku IP, Dorey F, Raitano A, Witte ON, Said JW, Loda M, Reiter RE. Prostate stem cell antigen (PSCA) expression increases with high Gleason score, advanced stage and bone metastasis in prostate cancer. Oncogene 19:1288–1296, 2000.
91. Reiter RE, Sato I, Thomas G, Qian J, Gu Z, Watabe T, Loda M, Jenkins RB. Coamplification of prostate stem cell antigen (PSCA) and MYC in locally advanced prostate cancer. Genes Chromosomes Cancer 27:95–103, 2000.
92. Saffran DC, Raitano AB, Hubert RS, Witte ON, Reiter RE, Jakobovits A. Anti-PSCA mAbs inhibit tumor growth and metastasis formation and prolong the survival of mice bearing human prostate cancer xenografts. Proc Natl Acad Sci USA 98:2658–2663, 2001.
93. Yang A, Kaghad M, Wang Y, Gillett E, Fleming MD, Dotsch V, Andrews NC, Caput D, McKeon F. p63, a p53 homolog at 3q27–29, encodes multiple products with transactivating, death-inducing, and dominant-negative activities. Mol Cells 2:305–316, 1998.

94. Yang A, Schweitzer R, Sun D, Kaghad M, Walker N, Bronson RT, Tabin C, Sharpe A, Caput D, Crum C, McKeon F. p63 is essential for regenerative proliferation in limb, craniofacial and epithelial development. Nature 398:714–718, 1999.
95. Signoretti S, Waltregny D, Dilks J, Isaac B, Lin D, Garraway L, Yang A, Montironi R, McKeon F, Loda M. p63 is a prostate basal cell marker and is required for prostate development. Am J Pathol 157:1769–1775, 2000.
96. Neshat MS, Mellinghoff IK, Tran C, Stiles B, Thomas G, Petersen R, Frost P, Gibbons JJ, Wu H, Sawyers CL. Enhanced sensitivity of PTEN-deficient tumors to inhibition of FRAP/mTOR. Proc Natl Acad Sci USA 98:10314–10319, 2001.
97. Podsypanina K, Lee RT, Politis C, Hennessy I, Crane A, Puc J, Neshat M, Wang H, Yang L, Gibbons J, Frost P, Dreisbach V, Blenis J, Gaciong Z, Fisher P, Sawyers C, Hedrick-Ellenson L, Parsons R. An inhibitor of mTOR reduces neoplasia and normalizes p70/S6 kinase activity in Pten$^{+/-}$ mice. Proc Natl Acad Sci USA 98:10320–10325, 2001.
98. Sawyers CL. Chronic myeloid leukemia. N Engl J Med 340:1330–1340, 1999.

19

Hormonal Therapies for Prostate Cancer

DAVID I. QUINN
DEREK RAGHAVAN

Prostate cancer is the commonest malignancy among men in developed countries and the second commonest cause of cancer death.[1] As populations age, prostate cancer becomes an even greater demographic problem. Hormonal therapy has been the mainstay of therapy for prostate cancer for the past century. New applications of such treatment to earlier stages of the disease may be altering its natural history.

CLINICAL BIOLOGY

Normal prostate development and growth as well as the development of prostatic neoplasia are dependent on the action of the testicular androgens testosterone (T) and dihydrotestosterone (DHT).[2] Testosterone, secreted by the testicular Leydig cells, is taken up by a variety of tissues, including normal prostate and prostate cancer. The type II 5α-reductase enzyme acts within the prostate to metabolize T into the more potent androgen DHT.[3] Dihydrotestosterone and T form complexes with the androgen receptor (AR), which then interact with DNA via coactivator recruitment and RNA polymerase II to induce protein synthesis and cell replication. The major mechanism for regulation of serum and tissue androgen concentrations is control of testicular Leydig cell T secretion via the hypothalamic–pituitary–gonadal (HPG) axis. The testes produce T in response to the serum luteinizing hormone (LH) concentration. This hormone is secreted from the anterior pituitary based on pulsatile and phasic production of LH-releasing hormone (LHRH) by the hypothalamus. In turn, and to complete the loop, LHRH secretion is suppressed by feedback at the hypothalamic level directly either by sex steroid hormones or by hormones released by their end-organ effect, such as inhibin and activin.

The testes are not the only source of androgens, but they are responsible for 90%–95% of the circulating T present in healthy males. Adrenal androgens [dihydro-epiandrostenedione (DHEA) and androstenedione] may be converted to T and DHT by hydroxysteroid dehydrogenases. Adrenal androgens may also activate mutant ARs in the absence of T or DHT.[4–6] This effect may be particularly important in patients with prostate cancer treated with medical or surgical castration but who develop progressive disease despite this. However, it appears that while adrenal androgens may be important in prostate cancer progression, they do not play a critical role in prostate development or carcinogenesis. Despite the presence of elevated serum concentrations of adrenal androgens, patients with hypogonadotrophic hypogonadism have abnormal prostate development but do not develop prostate neoplasia. Males castrated at or prior to puberty seemingly never develop prostate cancer.

The natural clinical history of prostate cancer is long, with projected transition from the first malignant cell in the prostate to distant metastatic disease taking more than 15 years in some cases and not occurring within the life span of a significant proportion of men.[7] Important points at which the clinical course of the disease appears to accelerate include the development of metastases and the onset of resistance to primary hormone therapy. Local extension from primary prostate cancer may involve the bladder and seminal vesicles. This tumor also metastasizes through lymphatics to regional lymph nodes in the pelvis.[8] Distant metastases commonly involve bone, especially the spine, but may involve distant lymph nodes and visceral organs, including the lung, liver, and adrenal glands.[9]

In advanced or metastatic prostate cancer, tumor response (measured by a variety of criteria) is reported in between 50% and 100% of cases with bilateral orchiectomy, exogenous estrogen therapy, or hormone therapy directed at the HPG axis and/or the intracellular AR pathway.[10–14] This response is characterized by an improvement in clinical symptoms (usually bone pain) and a fall in serum prostate-specific antigen (PSA).[15] Survival is better in individuals who have no bone pain, lower serum alkaline phosphatase, better performance status, and higher serum T levels prior to medical or surgical castration.[16,17] Following initial treatment with castration, patients with advanced prostate cancer typically experience clinical and biochemical quiescence of their disease for an average of 18–36 months. A small number of patients survive more than 10 years following castration for advanced prostate cancer.[18] However, the vast majority of patients go on to develop hormone-refractory prostate cancer characterized by worsening clinical symptoms, including bone pain, coupled with rising serum PSA. An increase in serum PSA subsequent to the hormone-induced nadir may precede clinical recurrence by several months.[19] Despite significant developments in the use of chemotherapy for patients with hormone-refractory prostate cancer in recent years, the median survival from the development of hormone-refractory prostate cancer ranges 10–18 months in most patients, depending on the extent of disease present at the initiation of salvage therapy. The biological correlate of this transient response to hormonal manipulation is triphasic:[20,21]

- Initial androgen withdrawal results in a variable proportion of prostate cancer cells undergoing apoptosis.[22] The remaining cells enter an androgen ablation–induced cell cycle arrest rather than undergoing apoptosis.
- Surviving cells undergo a variable quiescent or remission phase.
- Cell cycle induction and increased proliferation occur at the onset of hormone independence in a subset of cells, which correlates with biochemical and clinical progression.

Several key questions regarding the response of prostate cancer to hormone therapy remain to be resolved: Does combined androgen blockade provide an advantage over single therapy? Can therapy be modified to delay or alleviate the onset of hormone-refractory disease? Once prostate cancer becomes refractory to hormones, can further hormone therapy produce a meaningful disease response? Will a better understanding of the mechanisms underlying hormone resistance facilitate the development of drugs that delay or reverse it?

THERAPEUTIC MODALITIES

The various modalities of hormonal therapy used in prostate cancer are summarized in Table 19.1.

Orchiectomy

Surgical castration, most commonly achieved with bilateral subcapsular orchiectomy, was one of the first forms of therapy available for prostate cancer. Its continued use in the twenty-first century identifies it as the clinical "former standard" against which newer hormonal therapies must be compared.[23] Surgical castration results in a rapid fall in serum T concentration to 0%–10%

Table 19.1 Hormonal Therapies of Proven or Potential Use in Prostate Cancer

Surgical therapy
  Orchiectomy
  Hypophysectomy
Medical therapy
  Centrally acting inhibitors of LH secretion
    LHRH agonists
      Leuprolide, Goserelin, Buserelin
    LHRH antagonists
      Aborelix, Cetrorelix
  Peripheral androgen antagonists
    Nonsteroidal
      Flutamide, Nilutamide, Biclutamide
    Steroidal
      Cyproterone acetate
  Agents with estrogenic action
    Diethylstilbestrol, other synthetic estrogens
    PC-SPES
    Estramustine
  Agents with combined estrogenic and cytoxic action
    Estramustine
  Antagonists of adrenal androgen effect
    Ketoconazole
    Aminoglutethimide with cortisone acetate
    Prednisone and other corticosteroids

LH, luteinizing hormone; LHRH, LH-releasing hormone.

of its original level with a corresponding rise in serum LH levels. Around 80% of patients with advanced metastatic prostate cancer obtain improvement in clinical symptoms and extent of disease on imaging modalities as well as a fall in PSA in response to surgical castration. The mean duration of response to such hormonal manipulation is on the order of 12–18 months, after which the cancer progresses in a manner that is often resistant to further hormonal therapy as well as most other treatments.

Hypophysectomy and Adrenalectomy

Surgical hypophysectomy and adrenalectomy have been used to treat prostate cancer. However, they provide no clear therapeutic advantage over castration as first-line therapy. In patients relapsing after castration, hypophysectomy and adrenalectomy are reported to result in symptomatic improvement in around 20% of patients but are not clearly superior to second-line hormonal strategies.[24] They have the disadvantage of surgical morbidity. In particular, an added potential side effect of hypophysectomy is central abolition of adrenocorticotropin and thyroid-stimulating hormone secretion with a requirement for the patient to take exogenous corticosteroid and thyroid hormones. Patients undergoing bilateral adrenalectomy require lifelong corticosteroid replacement.

Luteinizing Hormone–Releasing Hormone Agonists and Antagonists

The physiological requirement of pulsatile and phasic hypothalamic LHRH secretion for basal LH production and secretion by the pituitary means that LH production may be suppressed either by blockade via LHRH antagonists or by agents that result in the abolition of pulsatile/phasic LHRH secretion by the hypothalamus, i.e., LHRH agonists.

Agonists

The first LH-suppressive agents in broad clinical use were LHRH agonists (leuprolide,[25] goserelin acetate,[12] buserelin[26]). They are the preferred form of hormonal therapy for prostate cancer in developed countries, despite being more expensive than surgical orchiectomy.[27–29] Agonists of LHRH suppress both serum LH and T concentrations by 3–4 weeks after commencement.[26,30,31] Prior to this, there is a transient increase in serum LH and T concentrations in the first few days that correlates with early response to increased LHRH effect on the HPG axis before the suppressive effect on LH secretion intervenes.[32] This is of clinical importance because patients may experience a flare in disease activity in the first few days of therapy with the potential result of increased bone pain, urinary tract obstruction, and/or spinal cord compression before disease regression occurs.[14,33] In practice, the best way to avoid this is with administration of a peripheral AR blocker for several days prior to and for several weeks after commencement of the LHRH agonist. In the setting of metastatic prostate cancer, LHRH agonists have similar clinical efficacy to estrogens[14] and to surgical castration in randomized clinical trials.[34]

Administration of LHRH agonists has evolved significantly in recent years. Initial trials required administration of multiple doses of drug by either daily subcutaneous injection or intranasal insufflation.[14,31,35,36] Subsequently, delayed-release depot formulations of LHRH agonists were developed, allowing monthly intramuscular or subcutaneous injection with better compliance and suppression of T.[37,38] This development saw this mode of therapy become more accepted by patients, with improved compliance as a result. More recently, depot preparations that provide castrate levels of serum T for at least 3 months have been developed, allowing longer intervals between injections.

Antagonists

Several LHRH antagonists (aborelix,[39,40] cetrorelix[41]) have been developed and are now in trials for a variety of indications, including prostate cancer. These trials show that antagonists produce earlier castrate levels of T than LHRH agonists and are not associated with an early surge in serum T concentrations.[40,42] The available preparations of LHRH antagonists require monthly injection, but depot formulation development to allow a longer interdose interval is under way. The role of LHRH antagonists in the treatment of prostate cancer remains to be determined, but they may provide an alternative to combined androgen blockade (CAB) when rapid castrate levels of T are required, such as in patients with spinal cord compression, bone pain, and/or potential for bladder outlet obstruction. Delineation of the role of LHRH antagonists as acceptable alternatives to castration and CAB will require carefully designed randomized, stratified studies that evaluate long-term efficacy, side effects, cost, and net quality-of-life (QOL) effects of each of the therapies.

Antiandrogens

In the setting of metastatic prostate cancer, peripheral androgen blockers as monotherapy are traditionally considered inferior to estrogens,[43] castration, or CAB.[44] However, such monotherapy may be associated with early preservation of sexual function compared to CAB (see below, Combined Androgen Ablation).[45,46]

Some patients with prostate cancer who develop disease progression as determined by increasing symptoms, increase in tumor volume, and/or increase in serum PSA while on antiandrogen therapy have improvement of symptoms, decrease in disease dimensions, and/or a fall in serum PSA when the antiandrogen is stopped. This phenomenon, which occurs in around 30% of patients on flutamide, is known as the *antiandrogen withdrawal response*.[47–49] Initially, it was thought that this response occurred only in patients on flutamide, but more recent reports describe withdrawal phenomena after cessation of biclutamide or nilutamide therapy.[50] The duration of response in these patients averages 3.5 months. Thereafter, the prostate cancer usually progresses again.[51] The postulated underlying mechanism for this response is selective survival advantage of cells with particular AR mutations, where the mutations result in antiandrogen AR activation rather than inhibition,[52,53] or possibly upregulation of AR expression. Other hormones, such as estrogens, progesterones, and DHEA, may also activate the so-called promiscuous AR.[5] It has been suggested that the occurrence of a withdrawal response correlates with a good response to second-line hormonal therapy, such as aminoglutethimide,[54] or, if associated with use of flutamide, with response to subsequent biclutamide therapy.[53]

Estrogens

Administration of therapeutic estrogen to postpubertal men results in a marked reduction in serum T by inhibiting LH production in the anterior pituitary.[55,56] Also, estrogens may have direct cytotoxic effects on prostate cancer cells in vitro. The efficacy of estrogen therapy, usually given as oral diethylstilbestrol, is limited by its side effects, such as painful gynecomastia and increased propensity for venous thromboembolic, cardiovascular, and cerebrovascular disease.[14,43,57,58]

PC-SPES

Herbal or "alternative" medicines are commonly used by patients with cancer.[59] Many such medicines contain hormones, hormone analogues, or tissue extracts with the potential to affect hormonal pathways. PC-SPES is a widely used herbal medicine that contains several different compounds, including a phytoestrogen with proestrogenic properties that inhibits prostate cancer cell proliferation in experimental models.[60] Therapeutic experience with PC-SPES suggests that its efficacy and side effect spectrum are similar to those of estrogens in hormone-naive patients.[61–63] It does appear to have an effect in patients with hormone-refractory prostate cancer, although this may be transient, rarely lasting more than several months.[63,64] The place of PC-SPES in hormonal therapy of prostate cancer remains to be determined, but initial results support further study of its action, efficacy, and toxicity profile. The University of California San Francisco is currently coordinating a randomized trial to compare the efficacy and toxicity of synthetic estrogens with PC-SPES.

5α-Reductase Enzyme Inhibitors

Given the pivotal intracellular role of 5α reductase in androgen metabolism, it has been postulated that inhibition of this enzyme with agents such as finasteride has the potential to inhibit prostate carcinogenesis and/or induce regression in established prostate cancer.[65,66] The recent demonstration that individuals with a germline 5α-reductase missense mutation that results in increased enzyme activity had a higher chance of developing prostate cancer adds weight to this hypothesis.[67,68] In addition, the tumor expression of 5α-reductase is increased in high-grade and androgen-insensitive prostate cancer.[69] Despite this, clinical evidence suggests that finasteride therapy does not inhibit the development or progression of prostate cancer[70] and may result in unexpected toxicity when given with other hormonal therapies.[71] Cote et al.,[72] in a study conducted at the University of Southern California, showed that finasteride did not inhibit the development of prostate cancer and noted that some patients demonstrated dedifferentiation on finasteride treatment. Interestingly, it appears that polymorphisms in the 5α-reductase gene and aromatase-metabolizing genes vary in incidence between ethnic groups, with different polymorphisms being most predictive of risk of prostate cancer and more advanced disease at presentation within individual groups.[73–76] Whether further studies will demonstrate a predictive effect of finasteride or other similar drugs in individuals based on 5α-reductase gene polymorphisms or independent of these factors remains to be seen.[67]

COMBINED ANDROGEN ABLATION

Combined, or "maximal," androgen ablation (CAB) has traditionally involved the combination of an LHRH antagonist (e.g., goserelin, leuprolide, or buserelin) or orchiectomy and a peripheral inhibitor of androgen action in target tissues (e.g., flutamide, nilutamide, bicalutamide, or cyproterone acetate). Labrie et al.[33,77] suggested that nearly all prostate cancer patients experience an objective response to CAB. Subsequent studies have failed to replicate fully these initial observations. However, several important studies have come from the desire to compare CAB with single-modality hormonal therapy.

A North American Intergroup trial evaluated 603 men with metastatic prostate cancer in a prospectively randomized placebo-controlled study of LHRH agonist therapy (leuprolide) with or without the peripheral androgen blocker flutamide.[78] There was a statistically significant benefit of leuprolide with flutamide over leuprolide with placebo in time to cancer progression (16.9 and 13.8 months, respectively) and overall survival (35.6 and 28.3 months, respectively). However, in the clinical context of prolonged survival after a diagnosis of metastatic prostate cancer, the differences between the two arms were less impressive. Several other randomized studies conducted in the same time frame as the first Intergroup trial failed to show a difference between CAB and LHRH monotherapy.

Interestingly, a retrospective subgroup analysis in the first North American Intergroup trial[78] demonstrated a greater advantage for CAB in patients with good performance status and minimal metastatic disease, defined as metastases confined to the axial skeleton or lymph nodes. Some clinicians interpreted this result as favoring CAB in minimal disease or good performance status patients. However, the retrospective nature of the analysis and small subgroup size limit extrapolation.

To test further the hypothesis that CAB may be superior to castration, a second Intergroup trial examined the role of flutamide in addition to bilateral orchiectomy.[79] This trial was stratified for extent of disease and accrued 1387 patients. It failed to demonstrate a difference in overall survival between the arms either in the group as a whole or in patients with limited extent of metastases (Fig. 19.1). This study reproduced almost exactly the results of an earlier randomized trial from Australia that compared orchiectomy and placebo with orchiectomy and flutamide.[80]

To evaluate the available evidence for differential efficacy with added peripheral androgen blockade, the Prostate Cancer Trialists' Collaborative Group[81] undertook a meta-analysis of all registered randomized trials comparing CAB with orchiectomy or LHRH agonist therapy alone. The meta-analysis encompassed trials in which more than 5000 prostate cancer deaths had occurred. There were minimal differences in outcome overall (Fig. 19.2). This emphasizes the fact that only three trials[78,82–84] have suggested an overall survival benefit of CAB while at least 15 more have been negative.[36,85–87] The results of time to progression analyses may be complicated by variation in the methods used to determine progression (biochemical, imaging, clinical) and the clinical course of the disease after progression occurs.

Differential efficacy is not the only issue in determining superiority for CAB against castration alone. Toxicity and net impact on QOL are important measures of outcome. Additional medication increases the risk of adverse events, particularly in an older population. In the case of flutamide, potential side effects include diarrhea, nausea, and increased liver transaminases.

A QOL analysis undertaken in one prospective trial suggested that gastrointestinal toxicity has an adverse impact in patients given antiandrogens as part of their regimen.[88] Additional drug also comes at additional cost, and the cost–effectiveness of the addition of an agent with marginal efficacy and potential toxicity has been questioned.

Trials evaluating the effect of newer nonsteroidal antiandrogens, such as biclutamide, in combination with or compared to castration are needed. Biclutamide is better tolerated than flutamide[89] and has similar efficacy when combined with castration. A randomized trial of 813 patients with metastatic prostate cancer compared LHRH agonist with either flutamide or biclutamide and found no difference in efficacy or outcome but did not evaluate QOL differences between the two arms.[90] Early evidence suggested that standard doses of biclutamide (50 milligrams per day) alone may be less effective than medical or surgical castration. However, high-dose biclutamide (150 milligrams per day) may provide equivalent cancer control with less diminution in QOL particularly with regard to sexual function.[45,91–93] In an Italian Prostate Cancer Project study,[46] 220 patients with stage C or D prostate cancer were randomly allocated to receive either biclutamide monotherapy (150 milligrams per day) or goserelin plus flutamide.[46] There was no difference in either progression-free survival or overall survival between the two arms. The biclutamide arm had fewer

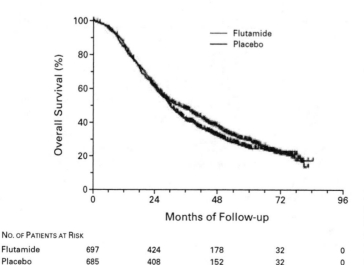

Figure 19.1 Overall survival among eligible patients with follow-up, according to treatment with flutamide or placebo. (From Eisenberger et al.[79] with permission.)

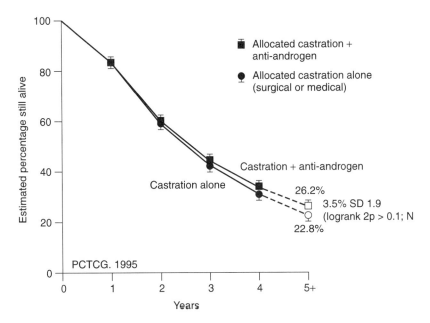

Figure 19.2 Estimated effect on survival of the addition of anti-androgen to castration for the treatment of metastatic prostate cancer. (From Prostate Cancer Trialists' Collaborative Group[81] with permission.)

treatment discontinuations, less erectile dysfunction, and better QOL than the goserelin plus flutamide arm. While this study might be interpreted as favoring use of high-dose biclutamide over CAB, there are some concerns related to the conduct, analysis, and interpretation of the data. Potentially significant differences in pretreatment features such as baseline PSA and tumor grade are not addressed and make efficacy data difficult to evaluate. The trial mixed stage C and D patients and suggested an early differential efficacy in stage C patients in favor of biclutamide and in stage D patients for CAB. As a result, there was a change in accrual criteria to exclude stage D patients after a given date. When analysis was complete, neither of the early trends proved significant. While QOL data were collected using a validated questionnaire, libido and erectile dysfunction were assessed in a nonstandard way, with individual physicians asked for a patient assessment at each office visit. Evidence from other investigators suggests that the onset of impotence or reduced sexuality-related QOL occurs soon after castration, whereas that associated with peripheral antiandrogen therapy is increasingly evident with longer follow-up.[94] The QOL assessments were initially undertaken by 60% of participants, and this number decreased with time on study. Thus, the differences represent a comparison of the subsets that completed questionnaires 6 months after initiation of therapy, when one might have expected biclutamide to be superior to CAB. Long-term QOL and efficacy data comparing high-dose biclutamide with castration in stage D disease are needed before biclutamide monotherapy is advocated outside of clinical trials.

DIFFERENCES BETWEEN SINGLE-MODALITY HORMONAL THERAPIES

When single-treatment modalities were compared in a systematic literature review and meta-analysis of 6600 patients,[44] there was little difference in outcome between orchiectomy and LHRH agonist therapy and between different types of LHRH agonist. This reaffirms the findings of several large randomized trials.[95,96] However, treatment with nonsteroidal antiandrogens alone was associated with a poorer outcome compared to either orchiectomy or LHRH agonists, although this failed to reach statistical significance. Previous studies have demonstrated that the response to LHRH agonist therapy is equivalent to that to diethylstilbestrol but that diethylstilbestrol has a much poorer toxicity profile.[14,34,58]

TIMING OF HORMONAL THERAPY: EARLY VERSUS DELAYED TREATMENT

In detecting and monitoring advanced disease, the progressive introduction and widespread availability of nucleotide bone scanning, serum PSA estimation, high-resolution computerized tomography, magnetic resonance imaging, and positron emission tomography have increasingly resulted in earlier diagnosis of asymptomatic patients with minimal recurrent or metastatic disease. The duration between the first abnormality in one of these tests and the development of clinically symptomatic prostate cancer can be many years.[97] On this basis, clinicians are confronted with a major stage shift of patients to an earlier time point in their cancer development, the phenomenon of *stage migration*. Patients with only a biochemical or minor imaging abnormality and no symptoms have a lower disease burden, longer natural disease history, and longer survival than patients with clinical symptoms. Earlier detection raises the question of whether earlier treatment improves outcome sufficiently to outweigh the side effects.

The British Medical Research Council undertook a trial designed to examine the role of early institution of hormonal therapy in patients with locally advanced or asymptomatic metastatic prostate cancer compared to similar treatment instigated at disease progression.[98] The study involved randomization of 934 men who had follow-up and management according to the participating clinician's normal practice. Information was collected annually on survival, local and distant progression, and major complications. Patients in the deferred arm developed progression and major complications statistically more often than those given early therapy. Deferred patients were also more likely to die from prostate cancer, but the benefit in survival was limited to patients with locally advanced prostate cancer. While the study provided support for early hormonal therapy in these patient groups, it has been criticized for relative immaturity of data in the locally advanced group as well as for inadequate staging, inconsistent follow-up data, and suboptimal data acquisition.[99] The study also failed to provide data on QOL in each group. Hence, further randomized studies are needed in this area but, given the results of this trial, may be faced with accrual difficulties.

ADJUVANT AND NEOADJUVANT HORMONAL STRATEGIES

The widespread availability and use of serum PSA testing and outpatient prostate biopsy in the past decade have resulted in many more men being diagnosed with clinically localized or locally advanced prostate cancer. One result of this is that these men are undergoing therapies such as radical prostatectomy and radiation for early disease. While the efficacy of such interventions in altering outcome is debated, it is clear that monomodality local therapies yield intermediate cancer control in only a proportion of patients. The reason for this limited effect is complex but may relate to undetected micrometastatic disease beyond the surgical specimen or radiation field and/or failure of local treatment due to inadequate surgical clearance (macroscopic surgical margin involvement) or cancer resistance to radiation. On this basis, a variety of strategies have been developed and tested with the aim of improving cancer control in patients undergoing therapy for clinically localized disease.

Neoadjuvant Hormonal Therapy

Neoadjuvant hormonal therapy (NHT) prior to radical prostatectomy has not been shown to improve disease-free survival despite a 30% reduction in surgical margin involvement on pathological evaluation compared to those not receiving such therapy before proceeding to surgery.[100–105] The degree of fall in serum PSA seen in these patients does not predict the likelihood of positive margins at surgery.[106] Patients with either clinical or pathological extension of prostate cancer through the prostatic capsule (stage T3) do poorly with NHT and radical prostatectomy.[107–110] Conversely, neoadjuvant and/or concurrent hormonal therapy improved outcome in patients undergoing radiation therapy in two randomized trials (Fig. 19.3).[111,112] Studies comparing cancer control and morbid-

Figure 19.3 *A:* Kaplan–Meier estimate of overall survival. The overall survival rate at 5 years was 79% (95% confidence interval 72%–86%) for the combined-treatment group and 62% (95% confidence interval 52%–72%) for the group treated only with radiotherapy. *B:* Kaplan–Meier estimate of the disease-free interval. This curve shows the proportion of surviving patients who were free of disease at each time point. The method takes the censoring process into account. The number of patients who are at risk for the event at each time point is the total of patients minus the number in whom disease progressed or who were lost to follow-up. (From Bolla et al.[112] with permission.)

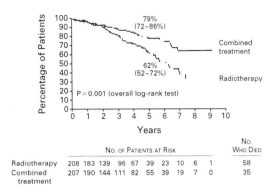

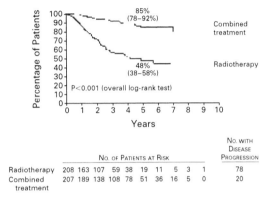

ity for radiation therapy combined with NHT and that observed for other treatments used in clinically localized prostate cancer are urgently needed. At present, NHT prior to planned surgery is not recommended outside a trial setting. Several accruing trials are examining the optimal type, timing, and duration of hormonal therapy combined with radiation for clinically localized disease.

Adjuvant Hormonal Therapy

For patients with lymph node involvement at radical prostatectomy, continuous hormonal therapy is considered standard care in some centers. A series of non-randomized comparisons supported the use of indefinite therapy in the form of either LHRH agonists or orchidectomy.[113–117] More recently, a randomized trial has demonstrated benefit from long-term LHRH agonist therapy for patients with lymph node involvement delineated at radical prostatectomy (Fig. 19.4).[118] This study has been criticized for low power and early termination. Whether patients with pelvic lymph node involvement can be treated with transient adjuvant therapy instead of indefinite hormonal therapy has not been determined. Whether long-term hormonal therapy is justified in patients with other adverse pathological findings at radical prostatectomy, such as seminal vesicle involvement or extensive extracapsular disease, remains to be determined. Despite widespread use in patients with these adverse pathological findings, there is little evidence in the literature to support therapy in this setting despite high rates of early biochemical (PSA) relapse. These issues are the subject of trials currently accruing.

Within the group of 945 patients in the Radiation Therapy Oncology Group (RTOG) 85-31 trial treated with primary radiation therapy for clinically localized prostate cancer, adjuvant LHRH antagonist resulted in a 5-year disease-free survival rate of 53% compared to 20% in the arm not given adjuvant therapy.[119] At a mean of 54 months follow-up, adjuvant therapy resulted in significantly improved overall survival in patients with a prostate biopsy Gleason score of 8

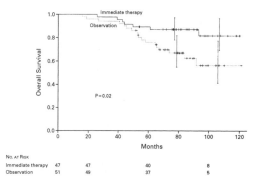

Figure 19.4 Kaplan–Meier estimates of overall survival based on immediate castration or observation for cases with pelvic lymph node metastases at radical prostatectomy for clinically localized prostate cancer. Vertical bars are 95% confidence intervals. The log-rank test was used to calculate p values. (From Messing et al.[118] with permission.)

or more. Patients with better-differentiated tumors did not achieve improvement in overall survival from adjunctive hormonal therapy, although longer follow-up in other subgroups will help to determine the broader applicability of these data. For patients in the RTOG 85-31 trial with biopsy-proven pelvic lymph node involvement, continuous hormonal therapy from the time of radiation therapy resulted in better biochemical progression-free survival and absolute survival than hormonal therapy given at relapse.[120] However, overall survival for the whole randomized trial did not reveal significant differences between the groups. Retrospective data from one case series suggest that continuous adjuvant antiandrogen therapy may be of benefit in patients with clinical stage T2B, Gleason score ≥ 7, and/or PSA >15 ng/ml notwithstanding the potential biases of the study design.[121] The role of further short-term adjuvant hormonal therapy following radiation therapy given subsequent to NHT is to be determined.[122] In one trial, which tested the utility of 6 months of adjuvant hormone after completion of NHT and radiation T, an early analysis suggested improved PSA-free survival from adjuvant therapy. However, this was not confirmed by later analysis (at 2 years).[122]

One criticism made of trials designed to assess the effect of combining castration and radiation therapy is that there are limited data to compare combined-modality treatment with castration alone. The British Medical Research Council addressed this in an early trial of orchiectomy, radiation therapy alone, and orchiectomy plus radiation therapy in patients with clinical T2–4 prostate cancer.[123] The net result was that those arms that included orchiectomy had significant delay in time to clinical progression but there was no difference in overall survival or local disease control. There was a non-significant trend of improved survival in the radiation plus orchiectomy group (5.2 years) vs. the orchiectomy alone group (4.5 years). The trial has been criticized because of the small number of patients, the techniques and doses of radiation therapy given, and for the large proportion of patients with T2 and/or low-grade tumors. However, it does suggest potential for combined therapy that needs further examination. At this time, adjuvant hormonal therapy following radiation should be considered in patients with a high probability of early relapse as predicted by Gleason score, PSA, and stage; but more data from randomized studies are required to define the fine details of this principle. The use of long-term hormonal therapy in patients with proven lymph node involvement improves survival in both the postradical prostatectomy and postradiation therapy settings.

SIDE EFFECTS OF HORMONAL THERAPY: SIMILARITIES AND DIFFERENCES

Hormonal therapy for prostate cancer eventually produces decreases in libido and potency in virtually all patients regardless of the modality used.[34,124] Additional side effects include lethargy, depression, anorexia, breast swelling with or without tenderness, hot flashes, anemia, and osteoporosis with potential for pathological fracture.[14,125–130] Most side effects, including impotence and infertility, are slowly reversible with cessation of therapy. However, reduced bone mineral density often does not reverse after prolonged hormonal suppression. There is a consensus that irreversible changes occur more often after suppression of longer than 18–24 months.

When different agents are compared, castration and cyproterone acetate are tolerated better than estrogens. Castration or cyproterone acetate produces less gynecomastia, diarrhea, nausea, and liver function abnormality than flutamide; but cyproterone is associated with a higher incidence of thrombotic problems.[131] Hot flashes are more common with castration than flutamide or biclutamide alone.[45]

INTERMITTENT VERSUS CONTINUOUS THERAPY

Intermittent cessation of hormonal therapy may improve long-term outcome by preventing the development of mutations in an environment of androgen deprivation or by impeding the evolution to AR overexpression that is a feature of prostate cancer progression.[132] To attenuate the side effects of hormonal suppression, recent investigative strategies have been designed to determine whether intermittent cessation of CAB improves QOL without compromising cancer control. The answer to this question awaits the outcome of a number of prospective randomized studies currently in accrual phase. Experimental and some clinical evidence suggests that tumors that respond initially and then progress when hormonal therapy is stopped will usually respond to further androgen deprivation.[133,134] Preliminary evidence suggests that interruption of hormonal therapy can improve QOL.[135] However, studies initiated in Canada in 1995 have largely been undertaken in patients with PSA relapse rather than patients with symptomatic advanced disease.[132] Recent phase II studies in patients with symptomatic advanced prostate cancer show variable responses to intermittent cessation of therapy;[136,137] therefore extrapolation of early data from patients with elevated PSA as the only evidence of disease to symptomatic patients requires randomized studies incorporating QOL assessment with survival analysis.

CONTINUATION OF HORMONE-ABLATIVE THERAPY IN HORMONE-REFRACTORY PROSTATE CANCER

When prostate cancer becomes hormone-refractory, the decision to continue or stop LHRH agonist therapy is a difficult one. Evidence in the literature is contradictory. A review of Southwest Oncology Group cytotoxic chemotherapy trials in patients with hormone-refractory prostate cancer failed to define a difference in outcome based on continued LHRH therapy or previous orchiectomy opposed to cessation of prior LHRH-based treatment.[138] However, another sizeable retrospective analysis (of patients treated with estrogens) suggested a modest benefit for continued hormonal therapy.[139] Given the evidence in support of both cessation and continuation of LHRH agonist therapy once hormone-refractory prostate cancer is present, many clinicians opt to continue it after commencing chemotherapy. However, a case can be made for discontinuation of hormonal therapy in individual patients, depending on the side effects of the therapy balanced against a modest chance of disease acceleration after treatment is stopped.

SECOND-LINE HORMONAL THERAPY

In considering therapy for patients progressing despite medical or surgical castration, options include adrenal steroid metabolism inhibitors, addition of an antiandrogen, and cytotoxic chemotherapy.

SECOND-LINE ANTIANDROGEN THERAPY

The European Organization for Research and Treatment of Cancer randomized 201 patients to either prednisolone 5 milligrams four times per day or flutamide 250 milligrams three times per day following symptomatic progression of prostate cancer treated with initial medical or surgical castration.[140] The investigators concluded that treatment with prednisolone or flutamide leads to similar rates of progression and overall survival and no difference in subjective or biochemical response. A U.S. National Prostatic Cancer Project trial randomized 220 men with hormone-refractory prostate cancer after initial surgical castration to either flutamide or estramustine therapy and found no difference in response or survival between the two drugs.[141]

ADRENAL STEROID METABOLISM INHIBITORS

Adrenal steroids contribute to prostatic growth, especially in the T-deficient environment. On this basis, drugs that act by altering adrenal androgen production are used as second-line hormonal therapy in patients failing primary testicular androgen suppression.

Glucocorticosteroids

Glucocorticosteroids have activity against prostate cancer. Their use as sole therapy is limited by long-term side effects including proximal myopathy, osteoporosis, and diabetes mellitus. While one study suggests they are as efficacious as flutamide in patients failing castration,[140] in practice glucocorticosteroids are most often used as an adjunct to other therapy, such as aminoglutethimide or mitoxantrone. When used with aminoglutethimide, their primary role is to compensate for suppression of essential glucocorticoid production by the adrenals.

Ketoconazole

Ketoconazole is an imidazole antifungal agent that inhibits both testicular and adrenal androgen production.[142,143] It produces a rapid fall in serum T to castrate levels[144,145] and a marked fall in androstenedione and DHEA sulfate concentrations.[146] Optimal suppression of adrenal androgens occurs at a dose of 400 milligrams three times daily, but a difference in cancer response between this dose level and 200 milligrams three times per day has not been demonstrated.[147] Ketoconazole in patients who have progressed despite previous castration or CAB has been reported to induce a response rate of more than 50%.[148,149] However, it can be accompanied by significant side effects, commonly including nausea and abdominal bloating, fatigue, liver function abnormalities, skin changes, and gynecomastia.[147,150] Gastrointestinal side effects are a common cause of cessation of therapy and noncompliance.[151] Rarer but more serious side effects include adrenal crisis and acute confusional states.[152–154] The recognition that ketoconazole absorption is dependent on an acidic gastric pH and is therefore decreased by H_2 antagonists and proton pump inhibitors has improved the efficacy of treatment.[155–157] Patients on antiulcer medication can ingest their ketoconazole with a cola drink or a similar acidic beverage to ensure absorption without compromising ulcer control.[157]

Aminoglutethimide

Aminoglutethimide inhibits several enzymes involved in the synthesis of corticosteroids as well as the aromatase enzyme that converts androgens to estrogens.[142] Given the clinical side effects of profound corticosteroid synthesis blockade, aminoglutethimide is administered with replacement therapy, usually in the form of oral cortisone acetate or hydrocortisone. Suppression of serum adrenal and testicular androgen concentration may vary in response to aminoglutethimide therapy, suggesting that individual variation may account for differences in response seen across different cohorts studied. Therapeutic monitoring of serum androgens has not been demonstrated to assist with the clinical management of this therapy in one previous study, although this study was not prospectively designed to assess the utility of dose modification based on serum androgen concentrations.[158] However, response to aminoglutethimide is reported to correlate with suppression of serum LH and follicle-stimulating hormone levels.[159] This suggests that adrenal androgens were elevated, causing feedback inhibition, and were amenable to suppression by aminoglutethimide. Assessment of data from studies correlating serum gonadotrophins and/or androgens with response to aminoglutethimide is made difficult by the potential effects of prior estrogen therapy on these parameters and response.

In patients refractory to first-line hormonal therapy with castration or estrogens, aminoglutethimide with hydrocortisone produces a symptomatic and biochemical response up to 48% of the time, with responding patients often experiencing clinical improvement for some months and occasionally up to 3–4 years.[159–162] Side effects include adrenal insufficiency (in the absence of corticosteroid-replacement therapy), skin rash, hypothyroidism, and less commonly thrombocytopenia.[159,163,164] Another troublesome side effect is sedation or depression. This is due to the fact that aminoglutethimide is an analogue of glutethimide, which was an early sedative.

NEW HORMONAL STRATEGIES

New Adrenal Aromatase Inhibitors

Given the efficacy but significant side effects of aminoglutethimide in prostate cancer progressing after castration, evaluation of newer aromatase inhibitors with more specific sites of action and proven activity in breast cancer seems logical. Unfortunately, anastrozole,[165] letrozole, and exemestane have produced negligible response rates in early-phase pharmaceutical company testing. The reasons for this are complex but may result from the more selective inhibition of adrenal aromatase relative to other steroid pathway–metabolizing enzymes compared to aminoglutethimide. Specific aromatase inhibition will block production of estradiol and estrone from T and androstenedione, respectively, with decreased levels of circulating estrogens but potential for increased androgen production. Interestingly, the very lack of specificity that produces toxicity with aminoglutethimide may also account for its activity in prostate cancer because aminoglutethimide inhibits the generation of androgen (and corticosteroids) through its inhibitory effect early in the steroidogenic pathway.

Liarazole

Liarazole inhibits both aromatase and retinoid catabolism. The result of the latter effect is that retinoic acid accumulates in cells. Retinoic acid and other vitamin A derivatives have been identified as differentiating agents and inducers of apoptosis in many in vitro settings and are in clinical use as differentiating agents in acute promyelocytic leukemia.[166,167] Early trials in hormone-refractory prostate cancer suggest the drug has limited activity in 18%–50% of patients failing castration and that most of its action may relate to its effects on retinoid metabolism rather than aromatase activity.[168–170] The side effects are mainly cutaneous, with drying and exfoliation of the skin, hair, and nails, as well as nausea, fatigue, somnolence, and mild alopecia. The limited efficacy and profile of toxicity have reduced its use in routine clinical practice.

Strategies Incorporating Hormonal Therapy and Cytotoxic Chemotherapy

The benefit of cytotoxic chemotherapy for patients with hormone-refractory prostate cancer is now firmly established.[171,172] Several attempts have been made to determine whether response to and duration of clinical remission from hormonal therapy can be improved by the early addition of cytotoxic chemotherapy. Generally, with the notable exception of a combination of estramustine with selected cytotoxic drugs (see below, Estramustine in Combination with Cytotoxic Drugs), these attempts have met with limited success. The Southwest Oncology Group studied the combination of endocrine therapy (estrogens or orchiectomy) with doxorubicin and cyclophosphamide randomized against endocrine therapy alone with addition of the same chemotherapy regimen at progression.[173] This trial accrued between September 1982 and October 1986. Patients on the combined chemo–endocrine therapy arm had a slightly higher response rate (63%) compared to those on endocrine therapy alone (48%), but this was not statistically significant ($p = 0.059$). The response rate for patients failing endocrine therapy and then receiving chemotherapy was low, with only 3/27 patients having a response. Despite differences in the initial response, time to disease progression and overall survival were the same in each group and randomization was not a factor predictive of outcome when recognized prognostic factors were placed in a multivariate model.

The combination of suramin, hydrocortisone, and androgen deprivation in patients with hormone-naive metastatic prostate cancer demonstrated high rates of major toxicity in a phase II Southwest Oncology Group study.[174] An evaluation of suramin and aminoglutethimide in patients progressing despite castration suggested that the combination added little to responses seen with aminoglutethimide alone.[175] A phase III trial comparing suramin and hydrocortisone with hydrocortisone alone in symptomatic hormone-refractory prostate cancer reported high symptomatic response in the suramin arm with a median 8-month duration of response and tolerable side effects.[176] The reasons for these differences in reported toxicity are unclear.

Attempts at combining mitomycin and aminoglutethimide in hormone-refractory prostate cancer resulted in high rates of toxicity and limited, short-lived responses in a small proportion of patients.[177] The combination of mitoxantrone with LHRH antagonist and flutamide provided no benefit in patients with metastatic disease over CAB alone.[178] However, this comparative study did suggest that there may be benefit of such a combination in patients with locally advanced prostate cancer. While this study suffered from methodological problems related to balance of known prognostic factors between the two groups compared, its results provide impetus for prospective randomized controlled studies to test the potential benefit of chemotherapy in combination with CAB in patients with locally advanced disease.

In the phase III Southwest Oncology Group study 9921, CAB with and without mitoxantrone as adjuvant therapy will be assessed in patients treated with radical prostatectomy who have poor prognostic features such as high Gleason score, extracapsular extension, seminal vesicle involvement, and/or lymph node involvement in the surgical specimen. A phase III study (RTOG 9902) will evaluate CAB with or without paclitaxel, estramustine, and etoposide in patients with clinical extracapsular extension, high biopsy Gleason score, and/or serum PSA between 20 and 100 ng/ml, who are undergoing definitive radiation therapy. Randomized studies to prospectively evaluate castration with and without chemotherapy in locally advanced disease will be forthcoming soon.

Estramustine in Combination with Cytotoxic Drugs

Estramustine is constituted by the carbamate linkage of estradiol and nornitrogen mustard molecules and is therefore best considered a combination hormonal–cytotoxic therapy. However, the mechanism by which estramustine exerts its antineoplastic effect is unclear. Following oral ingestion, estramustine is preferentially taken up by and retained in prostate tissue and prostate cancer cells. As a single agent, estramustine has not demonstrated benefit over continued or alternate hormonal therapy in hormone-refractory prostate cancer.[179] However, in combination with cytotoxic agents, estramustine appears to contribute to response.[180–183] While the mechanism for this is unclear, the effect of estramustine may occur through a phase activation of prostate cancer cells so that they are more sensitive to subsequent or concurrent cytotoxic effects. Estramustine alters cellular microtubular configuration and may have synergy with other drugs that act on microtubules, such as taxanes (paclitaxel, docetaxel) and vinca alkaloids (vincristine, vinblastine).[181,184–186] Reports from phase I and phase II trials suggest that the combination of estramustine and docetaxel is well tolerated and produces a decrease of over 50% in serum PSA in around 50% of hormone-refractory prostate cancer cases treated.[187–193] Given the significant estrogenic side effects related to estramustine therapy, its place in combination therapy with cytotoxic agents for hormone-refractory prostate cancer still requires proof of efficacy over the cytotoxic agents alone as well as dose scheduling to minimize toxicity. In addition to trials in patients with evidence of metastatic hormone-refractory prostate cancer, the Eastern Cooperative Oncology Group is currently accruing to a phase III randomized comparison of safety and efficacy of ketoconazole and hydrocortisone with the chemohormonal combination of estramustine and paclitaxel for patients with rising serum PSA values on androgen-suppression therapy. The results of this trial may assist in clinical decision making for patients progressing toward hormone-refractory prostate cancer who have only biochemical evidence of their disease.

FUTURE POSSIBILITIES

The molecular biology of prostate cancer progression is characterized by aberrant activity of several regulatory pathways in both the prostate cells and the surrounding milieu. These pathways can be broadly grouped into apoptosis (programmed cell death), AR signaling, cell cycle regulation, cellular adhesion/cohesion, and angiogenesis (blood vessel formation). The major apoptotic regulators p53[194] and bcl-2[195] demonstrate abnormal function and expression as prostate cancer progresses and are mechanistically implicated in hormone resistance.[196–201] Progressive heterogeneity of AR expression characterizes prostate cancer progression with a net increase in epithelial expression[202–205] and possible aberrant stromal signaling through a paracrine mechanism.[206] Aberrant AR signaling occurs in several different ways in advanced prostate cancer, including activation of AR by other hormones and antiandrogens, AR gene amplification, and replication of AR action by other pathways.[6,207,208] c-myc is consistently amplified in prostate cancer and may contribute, with other cell cycle abnormalities, to increase cellular proliferation as well as metastatic potential in ways yet to be characterized.[209–211] As prostate cancer progresses, cellular adhesion is altered, with reduced expression of a number of molecules, including E-cadherin[212–215] α-catenin,[216,217] and metalloproteinases[218–221] as well as altered expression of molecules in the prostatic stroma including chondroitin sulfate[222] within primary cancers known to be locally advanced or to have metastasized. Increased microvessel formation is a feature of many cancers including prostate cancer.[223–226] Blood vessel formation is regulated by molecules involved in adhesion as well as vascular endothelial growth factor,[220] nitric oxide, and cyclooxygenases. Aberrant blood vessel formation is associated with anomalies in pathways involved in apoptosis, AR signaling, and cellular adhesion.[227–230]

The sequential anomalies that occur in these pathways with prostate cancer progression are rarely isolated events but rather accompany each other in a manner that suggests that one anomaly may be causative or permissive of another in a different pathway. It is becoming increasingly evident that aberrant expression of molecules in one pathway is more likely in the presence of abnormalities in other pathways.[227–229,231] This emphasizes the complexity of molecular anomalies within prostate cancer as well as the potential for interdependence and redundancy within and between pathways responsible for apparently disparate biological functions.[232] Thus, targeting the processes responsible for hormonal resistance presents an immense challenge, with the attendant risk of little reward where the new drug works but does not improve outcome. Each of these pathways has been the target of in vitro and in vivo therapeutic maneuvers designed to improve the efficacy of hormonal therapy or to reverse hormone resistance. A number of agents that impact on putative key molecular targets are in early-phase trials, and a selection are discussed below.

Apoptosis Regulators

In vitro evidence suggests that bcl-2 overexpression may be induced by androgen deprivation and may result in resistance to hormone-induced apoptosis.[197,233] For this reason, inhibitors of bcl-2 expression or effect are potential candidates in the quest for partners to hormonal therapy.[234,235] Animal models have demonstrated a delay in time to androgen-independent progression with bcl-2 antisense therapy that blocks bcl-2 transcription.[235–237] Delivery to and effect of antisense therapy at disparate sites of metastatic cancer within the body is a major therapeutic hurdle to the use of such treatment in a clinical setting. Experience with bcl-2 antisense therapy in other tumors shows that while bcl-2 becomes undetectable in circulating lymphocytes, the reduction in tumor bcl-2 levels is much less dramatic.[238] Whether attainable reductions in bcl-2 expression in prostate cancer metastases will translate to therapeutic benefit needs to be tested. Early-phase clinical trials examining the safety and therapeutic potential of bcl-2 antisense therapy in prostate cancer have commenced and are ongoing.[239,240] However, the taxanes phosphorylate bcl-2 and induce apoptosis even in the presence of bcl-2 overexpression, suggesting a potential role for these agents concurrent with or subsequent to hormonal therapy.[241] Attempts at reinstating p53 function to prostate cancer cells involve delivery of normal or wild-type p53 via viral vector gene therapy.[242] Wild-type p53 can be delivered to cancer cells in the prostate by direct injection.[243] However, delivery to metastatic sites is more problematic. The clinical response to p53 gene therapy is still to be delineated. An alternative to gene therapy may be the development of other agents that induce p53-independent apoptosis.[229] If therapy directed at reinstating the normal apoptotic mechanism or bypassing a defective one in prostate cancer cells proves promising, trials testing the efficacy of such therapy with hormonal therapy and in hormone-resistant prostate cancer will be needed. At present, several novel therapeutic agents targeting a variety of apoptotic regulators are entering phase I clinical trials.

Angiogenesis Inhibitors

Angiogenesis inhibitors are now in clinical trials for a number of cancers. The clinical experience with these agents has been that toxicity is acceptable and a proportion of patients experience tumor shrinkage. However, a larger proportion of patients experience disease plateau with little change in the size of tumor deposits while on therapy.[244] Complete response to angiogenesis inhibitors as single agents is rare. Early trials using antiangiogenesis agents are under way in prostate cancer with interesting early results in heavily pretreated patients.[245,246] Given the "static" response to these agents in other cancers,[244] further clinical trials are needed to determine whether this response is additive or synergistic with other therapies in different phases of prostate cancer progression.

Metalloproteinase Inhibitors

Prostate cancer progression involves increased expression of selected matrix metalloproteinases (MMPs) with decreased expression of tissue inhibitors of metalloproteinases (TIMP). In vitro experiments demonstrate that neoangiogenesis by transformed prostate cells occurs concurrently with a decrease in TIMP-1 levels and an increase in MMP-2 and MMP-9 levels. This process is inhibited by the action of interleukin-10 and the bisphosphonate aledronate[247,248] and stimulated by interleukin-8. In addition, synthetic metalloproteinase inhibitors have been shown to inhibit prostate cancer growth in model systems,[249] while metalloproteinase secretion is important in the formation of bone metastases.[221] On this basis, exogenous metalloproteinase inhibitors may have the potential to improve the duration of response to hormonal therapy. Currently, MMP inhibitors being studied clinically include collagen peptidomimetics (matrilysin,[250] batimastat, marimastat) and nonpeptidomimetic inhibitors of MMP (AG3340,[251] BB-94[252]), interleukins, tetracycline derivatives (Col-3), and bisphosphonates.[253,254] Many other trials are under way using an array of other agents with MMP-inhibitory activity.[255] Bis-phosphonates have demonstrated a significant palliative effect in hormone-refractory prostate cancer patients with symptomatic bone metastases and the ability to diminish bone loss when combined with androgen ablation,[254,256] suggesting that they may be a suitable adjunct to androgen ablation in certain settings. Metalloproteinase inhibitors are likely to be cytostatic rather than cytotoxic.[257] Their place in prostate and other cancer therapy awaits the results of phase II and III trials currently under way.

CONCLUSION

Castration is the cornerstone of treatment for metastatic prostate cancer and has a role in selected patients with localized disease. Addition of other hormonal agents to castration has not consistently demonstrated improved outcome. However, newer approaches involving the concurrent use of chemotherapy or of agents directed at mechanisms involved in hormone resistance have the potential to increase the duration of the response to castration. In addition, these agents may provide therapeutic options for patients with progressive prostate cancer in the face of primary hormonal therapy. The biggest problem in the development of novel management strategies remains the paucity of patients with prostate cancer who are entered into well-designed, structured clinical trials.

REFERENCES

1. Greenlee RT, Murray T, Bolden S, Wingo PA. Cancer statistics, 2000. CA Cancer J Clin 50:7–33, 2000.
2. Fowler JE, Whitmore WF. Considerations for the use of testosterone with systemic chemotherapy in prostatic cancer. Cancer 49:1373–1377, 1982.

3. Thigpen AE, Silver RI, Guileyardo JM, Casey ML, McConnell JD, Russell DW. Tissue distribution and ontogeny of steroid 5alpha-reductase isozyme expression. J Clin Invest 92:903–910, 1993.
4. Veldscholte J, Ris-Stalpers C, Kuiper GG, Jenster G, Berrevoets C, Claassen E, van Rooij HC, Trapman J, Brinkmann AO, Mulder E. A mutation in the ligand binding domain of the androgen receptor of human LNCaP cells affects steroid binding characteristics and response to anti-androgens. Biochem Biophys Res Commun 173:534–540, 1990.
5. Tan J, Sharief Y, Hamil KG, Gregory CW, Zang DY, Sar M, Gumerlock PH, deVere White RW, Pretlow TG, Harris SE, Wilson EM, Mohler JL, French FS. Dehydroepiandrosterone activates mutant androgen receptors expressed in the androgen-dependent human prostate cancer xenograft CWR22 and LNCaP cells. Mol Endocrinol 11:450–459, 1997.
6. Jenster G. The role of the androgen receptor in the development and progression of prostate cancer. Semin Oncol 26:407–421, 1999.
7. Albertsen PC, Hanley JA, Gleason DF, Barry MJ. Competing risk analysis of men aged 55 to 74 years at diagnosis managed conservatively for clinically localized prostate cancer. JAMA 280:975–980, 1998.
8. Stamey TA, McNeal JE, Freiha FS, Redwine E. Morphometric and clinical studies on 68 consecutive radical prostatectomies. J Urol 139:1235–1241, 1988.
9. Bubendorf L, Schopfer A, Wagner U, Sauter G, Moch H, Willi N, Gasser TC, Mihatsch MJ. Metastatic patterns of prostate cancer: an autopsy study of 1,589 patients. Hum Pathol 31:578–583, 2000.
10. Grayhack JT, Keeler TC, Kozlowski JM. Carcinoma of the prostate. Hormonal therapy. Cancer 60:589–601, 1987.
11. Fowler JE Jr, Pandey P, Seaver LE, Feliz TP, Braswell NT. Prostate specific antigen regression and progression after androgen deprivation for localized and metastatic prostate cancer. J Urol 153:1860–1865, 1995.
12. Ahmann FR, Citrin DL, deHaan HA, Guinan P, Jordan VC, Kreis W, Scott M, Trump DL. Zoladex: a sustained-release, monthly luteinizing hormone–releasing hormone analogue for the treatment of advanced prostate cancer. J Clin Oncol 5:912–917, 1987.
13. Conn PM, Crowley WF Jr. Gonadotropin-releasing hormone and its analogues. N Engl J Med 324:93–103, 1991.
14. The Leuprolide Study Group. Leuprolide versus diethylstilbestrol for metastatic prostate cancer. N Engl J Med 311:1281–1286, 1984.
15. Matzkin H, Eber P, Todd B, van der Zwaag R, Soloway MS. Prognostic significance of changes in prostate-specific markers after endocrine treatment of stage D2 prostatic cancer. Cancer 70:2302–2309, 1992.
16. Kreis W, Ahmann FR, Lesser M, Scott M, Caplan R, Gau T, Vinciguerra V. Predictive initial parameters for response of stage D prostate cancer to treatment with the luteinizing hormone–releasing hormone agonist goserelin. J Clin Oncol 8:870–874, 1990.
17. Chodak GW, Vogelzang NJ, Caplan RJ, Soloway M, Smith JA. Independent prognostic factors in patients with metastatic (stage D2) prostate cancer. The Zoladex Study Group. JAMA 265:618–621, 1991.
18. Reiner WG, Scott WW, Eggleston JC, Walsh PC. Long-term survival after hormonal therapy for stage D prostatic cancer. J Urol 122:183–184, 1979.
19. Stamey TA, Kabalin JN, Ferrari M, Yang N. Prostate specific antigen in the diagnosis and treatment of adenocarcinoma of the prostate. IV. Anti-androgen treated patients. J Urol 141:1088–1090, 1989.
20. Westin P, Stattin P, Damber JE, Bergh A. Castration therapy rapidly induces apoptosis in a minority and decreases cell proliferation in a majority of human prostatic tumors. Am J Pathol 146:1368–1375, 1995.
21. Agus DB, Cordon-Cardo C, Fox W, Drobnjak M, Koff A, Golde DW, Scher HI. Prostate cancer cell cycle regulators: response to androgen withdrawal and development of androgen independence. J Natl Cancer Inst 91:1869–1876, 1999.
22. Matsushima H, Goto T, Hosaka Y, Kitamura T, Kawabe K. Correlation between proliferation, apoptosis, and angiogenesis in prostate carcinoma and their relation to androgen ablation. Cancer 85:1822–1827, 1999.
23. Chon JK, Jacobs SC, Naslund MJ. The cost value of medical versus surgical hormonal therapy for metastatic prostate cancer. J Urol 164:735–737, 2000.
24. Worgul TJ, Santen RJ, Samojlik E, Veldhuis JD, Lipton A, Harvey HA, Drago JR, Rohner TJ. Clinical and biochemical effect of aminoglutethimide in the treatment of advanced prostatic carcinoma. J Urol 129:51–55, 1983.
25. Ahmed SR, Brooman PJ, Shalet SM, Howell A, Blacklock NJ, Rickards D. Treatment of advanced prostatic cancer with LHRH analogue ICI 118630: clinical response and hormonal mechanisms. Lancet 2:415–419, 1983.
26. Borgmann V, Hardt W, Schmidt-Gollwitzer M, Adenauer H, Nagel R. Sustained suppression of testosterone production by the luteinising-hormone releasing-hormone agonist buserelin in patients with advanced prostate carcinoma. A new therapeutic approach? Lancet 1:1097–1099, 1982.

27. Nicol DL, Heathcote PS, Kateley GD, Lloyd S. Advanced prostate cancer. The role of high priced hormone therapy. Med J Aust 159:16–19, 1993.
28. Bayoumi AM, Brown AD, Garber AM. Cost-effectiveness of androgen suppression therapies in advanced prostate cancer. J Natl Cancer Inst 92:1731–1739, 2000.
29. Mariani AJ, Glover M, Arita S. Medical versus surgical androgen suppression therapy for prostate cancer: a 10-year longitudinal cost study. J Urol 165:104–107, 2001.
30. Grant JB, Ahmed SR, Shalet SM, Costello CB, Howell A, Blacklock NJ. Testosterone and gonadotrophin profiles in patients on daily or monthly LHRH analogue ICI 118630 (Zoladex) compared with orchiectomy. Br J Urol 58:539–544, 1986.
31. Eisenberger MA, O'Dwyer PJ, Friedman MA. Gonadotropin hormone–releasing hormone analogues: a new therapeutic approach for prostatic carcinoma. J Clin Oncol 4:414–424, 1986.
32. Sandow J. Clinical applications of LHRH and its analogues. Clin Endocrinol (Oxf) 18:571–592, 1983.
33. Labrie F, Dupont A, Giguere M, Borsanyi JP, Lacourciere Y, Belanger A, Lachance R, Emond J, Monfette G. Combination therapy with flutamide and castration (orchiectomy or LHRH agonist): the minimal endocrine therapy in both untreated and previously treated patients. J Steroid Biochem 27:525–532, 1987.
34. Peeling WB. Phase III studies to compare goserelin (Zoladex) with orchiectomy and with diethylstilbestrol in treatment of prostatic carcinoma. Urology 33:45–52, 1989.
35. Faure N, Labrie F, Lemay A, Belanger A, Gourdeau Y, Laroche B, Robert G. Inhibition of serum androgen levels by chronic intranasal and subcutaneous administration of a potent luteinizing hormone–releasing hormone (LHRH) agonist in adult men. Fertil Steril 37:416–424, 1982.
36. de Voogt HJ, Studer U, Schroder FH, Klijn JG, de Pauw M, Sylvester R. Maximum androgen blockade using LHRH agonist buserelin in combination with short-term (two weeks) or long-term (continuous) cyproterone acetate is not superior to standard androgen deprivation in the treatment of advanced prostate cancer. Final analysis of EORTC GU Group Trial 30843. European Organization for Research and Treatment of Cancer (EORTC) Genito-Urinary Tract Cancer Cooperative Group. Eur Urol 33:152–158, 1998.
37. Chrisp P, Sorkin EM. Leuprorelin. A review of its pharmacology and therapeutic use in prostatic disorders. Drugs Aging 1:487–509, 1991.
38. Plosker GL, Brogden RN. Leuprorelin. A review of its pharmacology and therapeutic use in prostatic cancer, endometriosis and other sex hormone–related disorders. Drugs 48:930–967, 1994.
39. Garnick MB, Campion M. Abarelix depot, a GnRH antagonist, v LHRH superagonists in prostate cancer: differential effects on follicle-stimulating hormone. Abarelix Depot Study Group. Mol Urol 4:275–277, 2000.
40. Cook T, Sheridan WP. Development of GnRH antagonists for prostate cancer: new approaches to treatment. Oncologist 5:162–168, 2000.
41. Reissmann T, Schally AV, Bouchard P, Riethmüller H, Engel J. The LHRH antagonist cetrorelix: a review. Hum Reprod Update 6:322–331, 2000.
42. Pechstein B, Nagaraja NV, Hermann R, Romeis P, Locher M, Derendorf H. Pharmacokinetic–pharmacodynamic modeling of testosterone and luteinizing hormone suppression by cetrorelix in healthy volunteers. J Clin Pharmacol 40:266–274, 2000.
43. Chang A, Yeap B, Davis T, Blum R, Hahn R, Khanna O, Fisher H, Rosenthal J, Witte R, Schinella R, Trump D. Double-blind, randomized study of primary hormonal treatment of stage D2 prostate carcinoma: flutamide versus diethylstilbestrol. J Clin Oncol 14:2250–2257, 1996.
44. Seidenfeld J, Samson DJ, Hasselblad V, Aronson N, Albertsen PC, Bennett CL, Wilt TJ. Single-therapy androgen suppression in men with advanced prostate cancer: a systematic review and meta-analysis. Ann Intern Med 132:566–577, 2000.
45. Iversen P, Tyrrell CJ, Kaisary AV, Anderson JB, Van Poppel H, Tammela TL, Chamberlain M, Carroll K, Melezinek I. Bicalutamide monotherapy compared with castration in patients with nonmetastatic locally advanced prostate cancer: 6.3 years of followup. J Urol 164:1579–1582, 2000.
46. Boccardo F, Rubagotti A, Barichello M, Battaglia M, Carmignani G, Comeri G, Conti G, Cruciani G, Dammino S, Delliponti U, Ditonno P, Ferraris V, Lilliu S, Montefiore F, Portoghese F, Spano G. Bicalutamide monotherapy versus flutamide plus goserelin in prostate cancer patients: results of an Italian Prostate Cancer Project study. J Clin Oncol 17:2027–2038, 1999.
47. Kelly WK, Slovin S, Scher HI. Steroid hormone withdrawal syndromes. Pathophysiology and clinical significance. Urol Clin North Am 24:421–431, 1997.
48. Kelly WK, Scher HI. Prostate specific antigen decline after antiandrogen withdrawal: the flutamide withdrawal syndrome. J Urol 149:607–609, 1993.
49. Paul R, Breul J. Antiandrogen withdrawal syndrome associated with prostate cancer therapies: incidence and clinical significance. Drug Saf 23:381–390, 2000.
50. Laufer M, Sinibaldi VJ, Carducci MA, Eisen-

berger MA. Rapid disease progression after the administration of bicalutamide in patients with metastatic prostate cancer. Urology 54:745, 1999.
51. Small EJ, Srinivas S. The antiandrogen withdrawal syndrome. Experience in a large cohort of unselected patients with advanced prostate cancer. Cancer 76:1428–1434, 1995.
52. Suzuki H, Akakura K, Komiya A, Aida S, Akimoto S, Shimazaki J. Codon 877 mutation in the androgen receptor gene in advanced prostate cancer: relation to antiandrogen withdrawal syndrome. Prostate 29:153–158, 1996.
53. Taplin ME, Bubley GJ, Ko YJ, Small EJ, Upton M, Rajeshkumar B, Balk SP. Selection for androgen receptor mutations in prostate cancers treated with androgen antagonist. Cancer Res 59:2511–2515, 1999.
54. Sartor O, Cooper M, Weinberger M, Headlee D, Thibault A, Tompkins A, Steinberg S, Figg WD, Linehan WM, Myers CE. Surprising activity of flutamide withdrawal, when combined with aminoglutethimide, in treatment of "hormone-refractory" prostate cancer [published erratum appears in J Natl Cancer Inst 86:463, 1994.] J Natl Cancer Inst 86:222–227, 1994.
55. Tomic R. Pituitary function after orchiectomy in patients with or without earlier estrogen treatment for prostatic carcinoma. J Endocrinol Invest 10:479–482, 1987.
56. Bishop MC, Selby C, Taylor M. Plasma hormone levels in patients with prostatic carcinoma treated with diethylstilboestrol and estramustine. Br J Urol 57:542–547, 1985.
57. Scott WW, Menon M, Walsh PC. Hormonal therapy of prostatic cancer. Cancer 45:1929–1936, 1980.
58. Garnick MB. Leuprolide versus diethylstilbestrol for previously untreated stage D2 prostate cancer. Results of a prospectively randomized trial. Urology 27:21–28, 1986.
59. Moyad MA. Alternative therapies for advanced prostate cancer. What should I tell my patients? Urol Clin North Am 26:413–417, 1999.
60. Kubota T, Hisatake J, Hisatake Y, Said JW, Chen SS, Holden S, Taguchi H, Koeffler HP. PC-SPES: a unique inhibitor of proliferation of prostate cancer cells in vitro and in vivo. Prostate 42:163–171, 2000.
61. de la Taille A, Buttyan R, Hayek O, Bagiella E, Shabsigh A, Burchardt M, Burchardt T, Chopin DK, Katz AE. Herbal therapy PC-SPES: in vitro effects and evaluation of its efficacy in 69 patients with prostate cancer. J Urol 164:1229–1234, 2000.
62. DiPaola RS, Zhang H, Lambert GH, Meeker R, Licitra E, Rafi MM, Zhu BT, Spaulding H, Goodin S, Toledano MB, Hait WN, Gallo MA. Clinical and biologic activity of an estrogenic herbal combination (PC-SPES) in prostate cancer. N Engl J Med 339:785–791, 1998.
63. Small EJ, Frohlich MW, Bok R, Shinohara K, Grossfeld G, Rozenblat Z, Kelly WK, Corry M, Reese DM. Prospective trial of the herbal supplement PC-SPES in patients with progressive prostate cancer. J Clin Oncol 18:3595–3603, 2000.
64. Pfeifer BL, Pirani JF, Hamann SR, Klippel KF. PC-SPES, a dietary supplement for the treatment of hormone-refractory prostate cancer. BJU Int 85:481–485, 2000.
65. Brawley OW, Ford LG, Thompson I, Perlman JA, Kramer BS. 5α-reductase inhibition and prostate cancer prevention. Cancer Epidemiol Biomarkers Prev 3:177–182, 1994.
66. Homma Y, Kaneko M, Kondo Y, Kawabe K, Kakizoe T. Inhibition of rat prostate carcinogenesis by a 5alpha-reductase inhibitor, FK143. J Natl Cancer Inst 89:803–807, 1997.
67. Makridakis NM, Ross RK, Pike MC, Crocitto LE, Kolonel LN, Pearce CL, Henderson BE, Reichardt JK. Association of mis-sense substitution in SRD5A2 gene with prostate cancer in African-American and Hispanic men in Los Angeles, USA. Lancet 354:975–978, 1999.
68. Makridakis N, Ross RK, Pike MC, Chang L, Stanczyk FZ, Kolonel LN, Shi CY, Yu MC, Henderson BE, Reichardt JK. A prevalent missense substitution that modulates activity of prostatic steroid 5α-reductase. Cancer Res 57:1020–1022, 1997.
69. Bonkhoff H, Stein U, Aumuller G, Remberger K. Differential expression of 5alpha-reductase isoenzymes in the human prostate and prostatic carcinomas. Prostate 29:261–267, 1996.
70. Kirby R, Robertson C, Turkes A, Griffiths K, Denis LJ, Boyle P, Altwein J, Schroder F. Finasteride in association with either flutamide or goserelin as combination hormonal therapy in patients with stage M1 carcinoma of the prostate gland. International Prostate Health Council (IPHC) Trial Study Group. Prostate 40:105–114, 1999.
71. Ornstein DK, Beiser JA, Andriole GL. Anaemia in men receiving combined finasteride and flutamide therapy for advanced prostate cancer. BJU Int 83:43–46, 1999.
72. Cote RJ, Skinner EC, Salem CE, Mertes SJ, Stanczyk FZ, Henderson BE, Pike MC, Ross RK. The effect of finasteride on the prostate gland in men with elevated serum prostate-specific antigen levels. Br J Cancer 78:413–418, 1998.
73. Latil AG, Azzouzi R, Cancel GS, Guillaume EC, Cochan-Priollet B, Berthon PL, Cussenot O. Prostate carcinoma risk and allelic variants of genes involved in androgen biosynthesis and metabolism pathways. Cancer 92:1130–1137, 2001.
74. Margiotti K, Sangiuolo F, De Luca A, Froio F, Pearce CL, Ricci-Barbini V, Micali F, Bonafe M, Franceschi C, Dallapiccola B, Novelli G, Reichardt JK. Evidence for an association between

the *SRD5A2* (type II steroid 5alpha-reductase) locus and prostate cancer in Italian patients. Dis Markers. 16:147–150, 2000.

75. Yamada Y, Watanabe M, Murata M, Yamanaka M, Kubota Y, Ito H, Katoh T, Kawamura J, Yatani R, Shiraishi T. Impact of genetic polymorphisms of 17-hydroxylase cytochrome P-450 (*CYP17*) and steroid 5alpha-reductase type II (*SRD5A2*) genes on prostate-cancer risk among the Japanese population. Int J Cancer 92:683–686, 2001.

76. Nam RK, Toi A, Vesprini D, Ho M, Chu W, Harvie S, Sweet J, Trachtenberg J, Jewett MA, Narod SA. V89L polymorphism of type-2,5-α-reductase enzyme gene predicts prostate cancer presence and progression. Urology 57:199–204, 2001.

77. Labrie F, Dupont A, Belanger A, Giguere M, Lacoursiere Y, Emond J, Monfette G, Bergeron V. Combination therapy with flutamide and castration (LHRH agonist or orchiectomy) in advanced prostate cancer: a marked improvement in response and survival. J Steroid Biochem 23:833–841, 1985.

78. Crawford ED, Eisenberger MA, McLeod DG, Spaulding JT, Benson R, Dorr FA, Blumenstein BA, Davis MA, Goodman PJ. A controlled trial of leuprolide with and without flutamide in prostatic carcinoma. N Engl J Med 321:419–424, 1989.

79. Eisenberger MA, Blumenstein BA, Crawford ED, Miller G, McLeod DG, Loehrer PJ, Wilding G, Sears K, Culkin DJ, Thompson IM Jr, Bueschen AJ, Lowe BA. Bilateral orchiectomy with or without flutamide for metastatic prostate cancer. N Engl J Med 339:1036–1042, 1998.

80. Zalcberg JR, Raghavan D, Marshall V, Thompson PJ. Bilateral orchidectomy and flutamide versus orchidectomy alone in newly diagnosed patients with metastatic carcinoma of the prostate—an Australian multicentre trial. Br J Urol 77:865–869, 1996.

81. Prostate Cancer Trialists' Collaborative Group. Maximum androgen blockade in advanced prostate cancer: an overview of 22 randomised trials with 3283 deaths in 5710 patients. Lancet 346:265–269, 1995.

82. Keuppens F, Whelan P, Carneiro de Moura JL, et al. Orchidectomy versus goserelin plus flutamide in patients with metastatic prostate cancer (EORTC 30853). European Organization for Research and Treatment of Cancer—Genitourinary Group. Cancer 72:3863–3869, 1993.

83. Denis LJ, Keuppens F, Smith PH, Whelan P, de Moura JL, Newling D, Bono A, Sylvester R. Maximal androgen blockade: final analysis of EORTC phase III trial 30853. EORTC Genito-Urinary Tract Cancer Cooperative Group and the EORTC Data Center. Eur Urol 33:144–151, 1998.

84. Denis LJ, Carnelro de Moura JL, Bono A, Sylvester R, Whelan P, Newling D, Depauw M. Goserelin acetate and flutamide versus bilateral orchiectomy: a phase III EORTC trial (30853). EORTC GU Group and EORTC Data Center. Urology 42:119–130, 1993.

85. Iversen P. Zoladex plus flutamide vs. orchidectomy for advanced prostatic cancer. Danish Prostatic Cancer Group (DAPROCA). Eur Urol 18(Suppl 3):41–44, 1990.

86. Iversen P, Rasmussen F, Klarskov P, Christensen IJ. Long-term results of Danish Prostatic Cancer Group trial 86. Goserelin acetate plus flutamide versus orchiectomy in advanced prostate cancer. Cancer 72:3851–3854, 1993.

87. Iversen P, Christensen MG, Friis E, et al. A phase III trial of Zoladex and flutamide versus orchiectomy in the treatment of patients with advanced carcinoma of the prostate. Cancer 66:1058–1066, 1990.

88. Moinpour CM, Savage MJ, Troxel A, Lovato LC, Eisenberger M, Veith RW, Higgins B, Skeel R, Yee M, Blumenstein BA, Crawford ED, Meyskens FL. Quality of life in advanced prostate cancer: results of a randomized therapeutic trial. J Natl Cancer Inst 90:1537–1544, 1998.

89. Boccon-Gibod L. Are non-steroidal anti-androgens appropriate as monotherapy in advanced prostate cancer? Eur Urol 33:159–164, 1998.

90. Schellhammer PF, Sharifi R, Block NL, Soloway MS, Venner PM, Patterson AL, Sarosdy MF, Vogelzang NJ, Chen Y, Kolvenbag GJ. A controlled trial of bicalutamide versus flutamide, each in combination with luteinizing hormone–releasing hormone analogue therapy, in patients with advanced prostate carcinoma. Analysis of time to progression. Casodex Combination Study Group. Cancer 78:2164–2169, 1996.

91. Kaisary AV, Tyrrell CJ, Beacock C, Lunglmayr G, Debruyne F. A randomised comparison of monotherapy with Casodex 50 mg daily and castration in the treatment of metastatic prostate carcinoma. Casodex Study Group. Eur Urol 28:215–222, 1995.

92. Chodak G, Sharifi R, Kasimis B, Block NL, Macramalla E, Kennealey GT. Single-agent therapy with bicalutamide: a comparison with medical or surgical castration in the treatment of advanced prostate carcinoma. Urology 46:849–855, 1995.

93. Iversen P, Tyrrell CJ, Kaisary AV, Anderson JB, Baert L, Tammela T, Chamberlain M, Carroll K, Gotting-Smith K, Blackledge GR. Casodex (bicalutamide) 150-mg monotherapy compared with castration in patients with previously untreated nonmetastatic prostate cancer: results from two multicenter randomized trials at a median follow-up of 4 years. Urology 51:389–396, 1998.

94. Schroder FH, Collette L, de Reijke TM, Whe-

lan P. Prostate cancer treated by anti-androgens: is sexual function preserved? EORTC Genitourinary Group. European Organization for Research and Treatment of Cancer. Br J Cancer 82:283–290, 2000.

95. Kaisary AV, Tyrrell CJ, Peeling WB, Griffiths K. Comparison of LHRH analogue (Zoladex) with orchiectomy in patients with metastatic prostatic carcinoma. Br J Urol 67:502–508, 1991.

96. Vogelzang NJ, Chodak GW, Soloway MS, Block NL, Schellhammer PF, Smith JA, Caplan RJ, Kennealey GT. Goserelin versus orchiectomy in the treatment of advanced prostate cancer: final results of a randomized trial. Zoladex Prostate Study Group. Urology 46:220–226, 1995.

97. Pound CR, Partin AW, Eisenberger MA, Chan DW, Walsh PC. Natural history of progression after PSA elevation following radical prostatectomy. JAMA 281:1591–1597, 1999.

98. The Medical Research Council Prostate Cancer Working Party Investigators Group. Immediate versus deferred treatment for advanced prostatic cancer: initial results of the Medical Research Council Trial. BJU Int 79:235–246, 1997.

99. Raghavan D. Prostate cancer management under scrutiny: one man's meta-analysis is another man's Poisson. J Clin Oncol 17:3371–3373, 1999.

100. Van Poppel H, De Ridder D, Elgamal AA, Van de Voorde W, Werbrouck P, Ackaert K, Oyen R, Pittomvils G, Baert L. Neoadjuvant hormonal therapy before radical prostatectomy decreases the number of positive surgical margins in stage T2 prostate cancer: interim results of a prospective randomized trial. The Belgian Uro-Oncological Study Group. J Urol 154:429–434, 1995.

101. Soloway MS, Sharifi R, Wajsman Z, McLeod D, Wood DP Jr, Puras-Baez A. Randomized prospective study comparing radical prostatectomy alone versus radical prostatectomy preceded by androgen blockade in clinical stage B2 (T2bNxM0) prostate cancer. The Lupron Depot Neoadjuvant Prostate Cancer Study Group. J Urol 154:424–428, 1995.

102. Aus G, Abrahamsson PA, Ahlgren G, Hugosson J, Lundberg S, Schain M, Schelin S, Pedersen K. Hormonal treatment before radical prostatectomy: a 3-year followup. J Urol 159:2013–2017, 1998.

103. Civantos F, Marcial MA, Banks ER, Ho CK, Speights VO, Drew PA, Murphy WM, Soloway MS. Pathology of androgen deprivation therapy in prostate carcinoma. A comparative study of 173 patients. Cancer 75:1634–1641, 1995.

104. Bonney WW, Schned AR, Timberlake DS. Neoadjuvant androgen ablation for localized prostate cancer. J Urol 160:1754–1760, 1996.

105. Scolieri MJ, Altman A, Resnick MI. Neoadjuvant hormonal ablative therapy before radical prostatectomy: a review. Is it indicated? J Urol 164:1465–1472, 2000.

106. McLeod DG, Johnson CF, Klein E, Peabody JO, Coffield S, Soloway M. PSA levels and the rate of positive surgical margins in radical prostatectomy specimens preceded by androgen blockade in clinical B2 (T2bNxMo) prostate cancer. The Lupron Depot Neoadjuvant Study Group. Urology 49:70–73, 1997.

107. Narayan P, Lowe BA, Carroll PR, Thompson IM. Neoadjuvant hormonal therapy and radical prostatectomy for clinical stage C carcinoma of the prostate. Br J Urol 73:544–548, 1994.

108. Cher ML, Shinohara K, Breslin S, Vapnek J, Carroll PR. High failure rate associated with long-term follow-up of neoadjuvant androgen deprivation followed by radical prostatectomy for stage C prostatic cancer. Br J Urol 75:771–777, 1995.

109. Gomella LG, Liberman SN, Mulholland SG, Petersen RO, Hyslop T, Corn BW. Induction androgen deprivation plus prostatectomy for stage T3 disease: failure to achieve prostate-specific antigen–based freedom from disease status in a phase II trial. Urology 47:870–877, 1996.

110. Fair WR, Cookson MS, Stroumbakis N, Cohen D, Aprikian AG, Wang Y, Russo P, Soloway SM, Sogani P, Sheinfeld J, Herr H, Dalgabni G, Begg CB, Heston WD, Reuter VE. The indications, rationale, and results of neoadjuvant androgen deprivation in the treatment of prostatic cancer: Memorial Sloan-Kettering Cancer Center results. Urology 49:46–55, 1997.

111. Grignon DJ, Caplan R, Sarkar FH, Lawton CA, Hammond EH, Pilepich MV, Forman JD, Mesic J, Fu KK, Abrams RA, Pajak TF, Shipley WU, Cox JD. p53 status and prognosis of locally advanced prostatic adenocarcinoma: a study based on RTOG 8610. J Natl Cancer Inst 89:158–165, 1997.

112. Bolla M, Gonzalez D, Warde P, Dubois JB, Mirimanoff RO, Storme G, Bernier J, Kuten A, Sternberg C, Gil T, Collette L, Pierart M. Improved survival in patients with locally advanced prostate cancer treated with radiotherapy and goserelin. N Engl J Med 337:295–300, 1997.

113. van Aubel OG, Hoekstra WJ, Schroder FH. Early orchiectomy for patients with stage D1 prostatic carcinoma. J Urol 134:292–294, 1985.

114. Myers RP, Larson-Keller JJ, Bergstralh EJ, Zincke H, Oesterling JE, Lieber MM. Hormonal treatment at time of radical retropubic prostatectomy for stage D1 prostate cancer: results of long-term followup. J Urol 147:910–915, 1992.

115. Zincke H, Bergstralh EJ, Larson-Keller JJ, Farrow GM, Myers RP, Lieber MM, Barrett DM, Rife CC, Gonchoroff NJ. Stage D1 prostate cancer treated by radical prostatectomy and adjuvant hormonal treatment. Evidence for favorable survival in patients with DNA diploid tumors. Cancer 70:311–323, 1992.

116. deKernion JB, Neuwirth H, Stein A, Dorey F, Stenzl A, Hannah J, Blyth B. Prognosis of pa-

tients with stage D1 prostate carcinoma following radical prostatectomy with and without early endocrine therapy. J Urol 144:700–703, 1990.
117. Kozlowski JM, Ellis WJ, Grayhack JT. Advanced prostatic carcinoma. Early versus late endocrine therapy. Urol Clin North Am 18:15–24, 1991.
118. Messing EM, Manola J, Sarosdy M, Wilding G, Crawford ED, Trump D. Immediate hormonal therapy compared with observation after radical prostatectomy and pelvic lymphadenectomy in men with node-positive prostate cancer. N Engl J Med 341:1781–1788, 1999.
119. Pilepich MV, Caplan R, Byhardt RW, Lawton CA, Gallagher MJ, Mesic JB, Hanks GE, Coughlin CT, Porter A, Shipley WU, Grignon D. Phase III trial of androgen suppression using goserelin in unfavorable-prognosis carcinoma of the prostate treated with definitive radiotherapy: report of Radiation Therapy Oncology Group Protocol 85-31. J Clin Oncol 15:1013–1021, 1997.
120. Lawton CA, Winter K, Byhardt R, Sause WT, Hanks GE, Russell AH, Rotman M Porter A, McGowan DG, DelRowe JD, Pilepich MV. Androgen suppression plus radiation versus radiation alone for patients with D1 (pN+) adenocarcinoma of the prostate (results based on a national prospective randomized trial, RTOG 85-31). Radiation Therapy Oncology Group. Int J Radiat Oncol Biol Phys. 38:931–939, 1997.
121. Anderson PR, Hanlon AL, Movsas B, Hanks GE. Prostate cancer patient subsets showing improved bNED control with adjuvant androgen deprivation. Int J Radiat Oncol Biol Phys. 39:1025–1030, 1997.
122. Laverdiere J, Gomez JL, Cusan L, Suburu ER, Diamond P, Lemay M, Candas B, Fortin A, Labrie F. Beneficial effect of combination hormonal therapy administered prior and following external beam radiation therapy in localized prostate cancer. Int J Radiat Oncol Biol Phys 37:247–252, 1997.
123. Fellows GJ, Clark PB, Beynon LL, Boreham J, Keen C, Parkinson MC, Peto R, Webb JN. Treatment of advanced localised prostatic cancer by orchiectomy, radiotherapy, or combined treatment. A Medical Research Council Study. Urological Cancer Working Party—Subgroup on Prostatic Cancer. Br J Urol 70:304–309, 1992.
124. Linde R, Doelle GC, Alexander N, Kirchner F, Vale W, Rivier J, Rabin D. Reversible inhibition of testicular steroidogenesis and spermatogenesis by a potent gonadotropin-releasing hormone agonist in normal men: an approach toward the development of a male contraceptive. N Engl J Med 305:663–667, 1981.
125. Karling P, Hammar M, Varenhorst E. Prevalence and duration of hot flushes after surgical or medical castration in men with prostatic carcinoma. J Urol 152:1170–1173, 1994.
126. Stege R. Potential side-effects of endocrine treatment of long duration in prostate cancer. Prostate Suppl 10:38–42, 2000.
127. Roux C, Pelissier C, Listrat V, Kolta S, Simonetta C, Guignard M, Dougados M, Amor B. Bone loss during gonadotropin releasing hormone agonist treatment and use of nasal calcitonin. Osteoporos Int 5:185–190, 1995.
128. Townsend MF, Sanders WH, Northway RO, Graham SD Jr. Bone fractures associated with luteinizing hormone–releasing hormone agonists used in the treatment of prostate carcinoma. Cancer 79:545–550, 1997.
129. Daniell HW. Osteoporosis after orchiectomy for prostate cancer. J Urol 157:439–444, 1997.
130. Daniell HW, Dunn SR, Ferguson DW, Lomas G, Niazi Z, Stratte PT. Progressive osteoporosis during androgen deprivation therapy for prostate cancer. J Urol 163:181–186, 2000.
131. Schroder FH. Antiandrogens as monotherapy for prostate cancer. Eur Urol 34(Suppl 3):12–17, 1998.
132. Bruchovsky N, Klotz LH, Sadar M, Crook JM, Hoffart D, Godwin L, Warkentin M, Gleave ME, Goldenberg SL. Intermittent androgen suppression for prostate cancer: Canadian Prospective Trial and related observations. Mol Urol 4:191–201, 2000.
133. Akakura K, Bruchovsky N, Goldenberg SL, Rennie PS, Buckley AR, Sullivan LD. Effects of intermittent androgen suppression on androgen-dependent tumors. Apoptosis and serum prostate-specific antigen. Cancer 71:2782–2790, 1993.
134. Higano CS, Ellis W, Russell K, Lange PH. Intermittent androgen suppression with leuprolide and flutamide for prostate cancer: a pilot study. Urology 48:800–804, 1996.
135. Goldenberg SL, Bruchovsky N, Gleave ME, Sullivan LD, Akakura K. Intermittent androgen suppression in the treatment of prostate cancer: a preliminary report. Urology 45:839–844, 1995.
136. Strum SB, Scholz MC, McDermed JE. Intermittent androgen deprivation in prostate cancer patients: factors predictive of prolonged time off therapy. Oncologist 5:45–52, 2000.
137. Bouchot O, Lenormand L, Karam G, Prunet D, Gaschignard N, Malinovsky JM, Buzelin JM. Intermittent androgen suppression in the treatment of metastatic prostate cancer. Eur Urol 38:543–549, 2000.
138. Hussain M, Wolf M, Marshall E, Crawford ED, Eisenberger M. Effects of continued androgen-deprivation therapy and other prognostic factors on response and survival in phase II chemotherapy trials for hormone-refractory prostate cancer: a Southwest Oncology Group report. J Clin Oncol 12:1868–1875, 1994.
139. Taylor CD, Elson P, Trump DL. Importance of continued testicular suppression in hormone-refractory prostate cancer. J Clin Oncol 11:2167–2172, 1993.

140. Fossa SD, Slee PH, Brausi M, Horenblas S, Hall RR, Hetherington JW, Aaronson N, de Prijck L, Collette L. Flutamide versus prednisone in patients with prostate cancer symptomatically progressing after androgen-ablative therapy: a phase III study of the European Organization for Research and Treatment of Cancer Genitourinary Group. J Clin Oncol 19:62–71, 2001.
141. de Kernion JN, Murphy GP, Priore R. Comparison of flutamide and Emcyt in hormone-refractory metastatic prostatic cancer. Urology 31:312–317, 1988.
142. Shaw MA, Nicholls PJ, Smith HJ. Aminoglutethimide and ketoconazole: historical perspectives and future prospects. J Steroid Biochem 31:137–146, 1988.
143. Trachtenberg J, Zadra J. Steroid synthesis inhibition by ketoconazole: sites of action. Clin Invest Med 11:1–5, 1988.
144. Trachtenberg J, Halpern N, Pont A. Ketoconazole: a novel and rapid treatment for advanced prostatic cancer. J Urol 130:152–153, 1983.
145. Lowe FC, Bamberger MH. Indications for use of ketoconazole in management of metastatic prostate cancer. Urology 36:541–545, 1990.
146. Trump DL, Havlin KH, Messing EM, Cummings KB, Lange PH, Jordan VC. High-dose ketoconazole in advanced hormone-refractory prostate cancer: endocrinologic and clinical effects. J Clin Oncol 7:1093–1098, 1989.
147. Williams G, Kerle DJ, Ware H, Doble A, Dunlop H, Smith C, Allen J, Yeo T, Bloom SR. Objective responses to ketoconazole therapy in patients with relapsed progressive prostatic cancer. Br J Urol 58:45–51, 1986.
148. Witjes FJ, Debruyne FM, Fernandez del Moral P, Geboers AD. Ketoconazole high dose in management of hormonally pretreated patients with progressive metastatic prostate cancer. Dutch South-Eastern Urological Cooperative Group. Urology 33:411–415, 1989.
149. Small EJ, Baron AD, Fippin L, Apodaca D. Ketoconazole retains activity in advanced prostate cancer patients with progression despite flutamide withdrawal. J Urol 157:1204–1207, 1997.
150. Bok RA, Small EJ. The treatment of advanced prostate cancer with ketoconazole: safety issues. Drug Saf 20:451–458, 1999.
151. Pont A. Long-term experience with high dose ketoconazole therapy in patients with stage D2 prostatic carcinoma. J Urol 137:902–904, 1987.
152. White MC, Kendall-Taylor P. Adrenal hypofunction in patients taking ketoconazole. Lancet 1:44–45, 1985.
153. Sarver RG, Dalkin BL, Ahmann FR. Ketoconazole-induced adrenal crisis in a patient with metastatic prostatic adenocarcinoma: case report and review of the literature. Urology 49:781–785, 1997.
154. Hanash KA. Neurologic complications of ketoconazole therapy for advanced prostatic cancer. Urology 33:466–467, 1989.
155. Quinn DI, Day RO. Drug interactions of clinical importance. An updated guide. Drug Saf 12:393–452, 1995.
156. Blum RA, D'Andrea DT, Florentino BM, Wilton JH, Hilligoss DM, Gardner MJ, Henry EB, Goldstein H, Schentag JJ. Increased gastric pH and the bioavailability of fluconazole and ketoconazole. Ann Intern Med 114:755–757, 1991.
157. Chin TW, Loeb M, Fong IW. Effects of an acidic beverage (Coca-Cola) on absorption of ketoconazole. Antimicrob Agents Chemother 39:1671–1675, 1995.
158. Ahmann FR, Crawford ED, Kreis W, Levasseur Y. Adrenal steroid levels in castrated men with prostatic carcinoma treated with aminoglutethimide plus hydrocortisone. Cancer Res 47:4736–4739, 1987.
159. Harnett PR, Raghavan D, Caterson I, Pearson B, Watt H, Teriana N, Coates A, Coorey G. Aminoglutethimide in advanced prostatic carcinoma. Br J Urol 59:323–327, 1987.
160. Ponder BA, Shearer RJ, Pocock RD, Miller J, Easton D, Chilvers CE, Dowsett M, Jeffcoate SL. Response to aminoglutethimide and cortisone acetate in advanced prostatic disease. Br J Cancer 50:757–763, 1984.
161. Labrie F, Dupont A, Belanger A, Cusan L, Brochu M, Turina E, Pinault S, Lacourciere Y, Emond J. Anti-hormone treatment for prostate cancer relapsing after treatment with flutamide and castration. Addition of aminoglutethimide and low dose hydrocortisone to combination therapy. Br J Urol 63:634–638, 1989.
162. Bezwoda WR. Treatment of stage D2 prostatic cancer refractory to or relapsed following castration plus oestrogens. Comparison of aminoglutethimide plus hydrocortisone with medroxyprogesterone acetate plus hydrocortisone. Br J Urol 66:196–201, 1990.
163. Figg WD, Thibault A, Sartor AO, Mays D, Headlee D, Calis KA, Cooper MR. Hypothyroidism associated with aminoglutethimide in patients with prostate cancer. Arch Intern Med 154:1023–1025, 1994.
164. Messeih AA, Lipton A, Santen RJ, Harvey HA, Boucher AE, Murray R, Ragaz J, Buzdar AU, Nagel GA, Henderson IC. Aminoglutethimide-induced hematologic toxicity: worldwide experience. Cancer Treat Rep 69:1003–1004, 1985.
165. Dukes M, Edwards PN, Large M, Smith IK, Boyle T. The preclinical pharmacology of "Arimidex" (anastrozole; ZD1033)—a potent, selective aromatase inhibitor. J Steroid Biochem Mol Biol 58:439–445, 1996.
166. Hall AK. Liarozole amplifies retinoid-induced apoptosis in human prostate cancer cells. Anticancer Drugs 7:312–320, 1996.
167. Trump DL. Retinoids in bladder, testis and

prostate cancer: epidemiologic, pre-clinical and clinical observations. Leukemia 8:S50–S54, 1994.
168. Dijkman GA, Fernandez del Moral P, Bruynseels J, de Porre P, Denis L, Debruyne FM. Liarozole (R75251) in hormone-resistant prostate cancer patients. Prostate 33:26–31, 1997.
169. Denis L, Debruyne F, De Porre P, Bruynseels, J. Early clinical experience with liarozole (Liazal) in patients with progressive prostate cancer. Eur J Cancer 34:469–475, 1998.
170. Debruyne FJ, Murray R, Fradet Y, Johansson JE, Tyrrell C, Boccardo F, Denis L, Marberger JM, Brune D, Rassweiler J, Vangeneugden T, Bruynseels J, Janssens M, De Porre P. Liarozole—a novel treatment approach for advanced prostate cancer: results of a large randomized trial versus cyproterone acetate. Liarozole Study Group. Urology 52:72–81, 1998.
171. Tannock IF, Osoba D, Stockler MR, Ernst DS, Neville AJ, Moore MJ, Armitage GR, Wilson JJ, Venner PM, Coppin CM, Murphy KC. Chemotherapy with mitoxantrone plus prednisone or prednisone alone for symptomatic hormone-resistant prostate cancer: a Canadian randomized trial with palliative end points. J. Clin Oncol 14:1756–1764, 1996.
172. Kantoff PW, Halabi S, Conaway M, Picus J, Kirshner J, Hars V, Trump D, Winer EP, Vogelzang NJ. Hydrocortisone with or without mitoxantrone in men with hormone-refractory prostate cancer: results of the cancer and leukemia group B 9182 study [see comments]. J Clin Oncol 17:2506–2513, 1999.
173. Osborne CK, Blumenstein B, Crawford ED, Coltman CA Jr, Smith AY, Lambuth BW, Chapman RA. Combined versus sequential chemoendocrine therapy in advanced prostate cancer: final results of a randomized Southwest Oncology Group study. J Clin Oncol 8:1675–1682, 1990.
174. Hussain M, Fisher EI, Petrylak DP, O'Connor J, Wood DP, Small EJ, Eisenberger MA, Crawford ED. Androgen deprivation and four courses of fixed-schedule suramin treatment in patients with newly diagnosed metastatic prostate cancer: a Southwest Oncology Group study. J Clin Oncol 18:1043–1049, 2000.
175. Dawson N, Figg WD, Brawley OW, Bergan R, Cooper MR, Senderowicz A, Headlee D, Steinberg SM, Sutherland M, Patronas N, Sausville E, Linehan WM, Reed E, Sartor O. Phase II study of suramin plus aminoglutethimide in two cohorts of patients with androgen-independent prostate cancer: simultaneous antiandrogen withdrawal and prior antiandrogen withdrawal. Clin Cancer Res 4:37–44, 1998.
176. Small EJ, Meyer M, Marshall ME, Reyno LM, Meyers FJ, Natale RB, Lenehan PF, Chen L, Slichenmyer WJ, Eisenberger M. Suramin therapy for patients with symptomatic hormone-refractory prostate cancer: results of a randomized phase III trial comparing suramin plus hydrocortisone to placebo plus hydrocortisone. J Clin Oncol 18:1440–1450, 2000.
177. Dik P, Blom JH, Schroder FH. Mitomycin C and aminoglutethimide in the treatment of metastatic prostatic cancer: a phase II study. Br J Urol 70:542–545, 1992.
178. Wang J, Halford S, Rigg A, Roylance R, Lynch M, Waxman J. Adjuvant mitozantrone chemotherapy in advanced prostate cancer. BJU Int 86:675–680, 2000.
179. Smith PH, Suciu S, Robinson MR, et al. A comparison of the effect of diethylstilbestrol with low dose estramustine phosphate in the treatment of advanced prostatic cancer: final analysis of a phase III trial of the European Organization for Research on Treatment of Cancer. J Urol 136:619–623, 1986.
180. Carles J, Domenech M, Gelabert-Mas A, Nogue M, Tabernero JM, Arcusa A, Guasch I, Miguel A, Ballesteros JJ, Fabregat X. Phase II study of estramustine and vinorelbine in hormone-refractory prostate carcinoma patients. Acta Oncol 37:187–191, 1998.
181. Hudes G, Einhorn L, Ross E, Balsham A, Loehrer P, Ramsey H, Sprandio J, Entmacher M, Dugan W, Ansari R, Monaco F, Hanna M, Roth B. Vinblastine versus vinblastine plus oral estramustine phosphate for patients with hormone-refractory prostate cancer: a Hoosier Oncology Group and Fox Chase Network phase III trial. J Clin Oncol 17:3160–3166, 1999.
182. Bracarda S, Tonato M, Rosi P, De Angelis V, Mearini E, Cesaroni S, Fornetti P, Porena M. Oral estramustine and cyclophosphamide in patients with metastatic hormone refractory prostate carcinoma: a phase II study. Cancer 88:1438–1444, 2000.
183. Sumiyoshi Y, Hashine K, Nakatsuzi H, Yamashita Y, Karashima T. Oral estramustine phosphate and oral etoposide for the treatment of hormone-refractory prostate cancer. Int J Urol 7:243–247, 2000.
184. Sangrajrang S, Denoulet P, Millot G, Tatoud R, Podgorniak MP, Tew KD, Calvo F, Fellous A. Estramustine resistance correlates with tau over-expression in human prostatic carcinoma cells. Int J Cancer 77:626–631, 1998.
185. Williams JF, Muenchen HJ, Kamradt JM, Korenchuk S, Pienta KJ. Treatment of androgen-independent prostate cancer using antimicrotubule agents docetaxel and estramustine in combination: an experimental study. Prostate 44:275–278, 2000.
186. Stein CA. Mechanisms of action of taxanes in prostate cancer. Semin Oncol 26:3–7, 1999.
187. Petrylak DP, Macarthur RB, O'Connor J, Shelton G, Judge T, Balog J, Pfaff C, Bagiella E,

Heitjan D, Fine R, Zuech N, Sawczuk I, Benson M, Olsson CA. Phase I trial of docetaxel with estramustine in androgen-independent prostate cancer. J Clin Oncol 17:958–967, 1999.
188. Weitzman A, Shelton G, Zuech N, England-Owen CJ N, Bagiella E, Katz A, Sawczuk I, Benson M, Olsson CA, Petrylak DP. Phase II study of estramustine combined with docetaxel in patients with androgen-independent prostate cancer. In: Proceedings of the American Society of Clinical Oncology, May 14–18, 1999, p A1369.
189. Savarese D, Taplin ME, Halabi S, Hars V, Kreis W, Vogelzang N. A phase II study of docetaxel (Taxotere), estramustine, and low-dose hydrocortisone in men with hormone-refractory prostate cancer: preliminary results of cancer and leukemia group B trial 9780. Semin Oncol 26:39–44, 1999.
190. Kreis W, Budman D. Daily oral estramustine and intermittent intravenous docetaxel (Taxotere) as chemotherapeutic treatment for metastatic, hormone-refractory prostate cancer. Semin Oncol 26:34–38, 1999.
191. Smith DC, Esper P, Strawderman M, Redman B, Pienta KJ. Phase II trial of oral estramustine, oral etoposide, and intravenous paclitaxel in hormone-refractory prostate cancer. J Clin Oncol 17:1664–1671, 1999.
192. Weitzman AL, Shelton G, Zuech N, Owen CE, Judge T, Benson M, Sawczuk I, Katz A, Olsson CA, Bagiella E, Pfaff C, Newhouse JH, Petrylak DP. Dexamethasone does not significantly contribute to the response rate of docetaxel and estramustine in androgen independent prostate cancer. J Urol 163:834–837, 2000.
193. Kosty MP, Ferreira A, Bryntesen T, Grossman J. Weekly docetaxel and low-dose estramustine phosphate in hormone refractory prostate cancer: a phase II study. In: Proc Am Soc Clin Oncol, New Orleans, LA, May 18–24, 2000, p A1442.
194. Kirsch DG, Kastan MB. Tumor-suppressor p53: implications for tumor development and prognosis. J Clin Oncol 16:3158–3168, 1998.
195. Reed JC. Bcl-2 and the regulation of programmed cell death. J Cell Biol 124:1–6, 1994.
196. Visakorpi T, Kallioniemi OP, Heikkinen A, Koivula T, Isola J. Small subgroup of aggressive, highly proliferative prostatic carcinomas defined by p53 accumulation. J Natl Cancer Inst 84:883–887, 1992.
197. Raffo AJ, Perlman H, Chen MW, Day ML, Streitman JS, Buttyan R. Overexpression of bcl-2 protects prostate cancer cells from apoptosis in vitro and confers resistance to androgen depletion in vivo. Cancer Res 55:4438–4445, 1995.
198. Bauer JJ, Sesterhenn IA, Mostofi FK, McLeod DG, Srivastava S, Moul JW. Elevated levels of apoptosis regulator proteins p53 and bcl-2 are independent prognostic biomarkers in surgically treated clinically localized prostate cancer. J Urol 156:1511–1516, 1996.
199. Apakama I, Robinson MC, Walter NM, Charlton RG, Royds JA, Fuller CE, Neal DE, Hamdy FC. Bcl-2 overexpression combined with p53 protein accumulation correlates with hormone-refractory prostate cancer. Br J Cancer 74:1258–1262, 1996.
200. Meyers FJ, Gumerlock PH, Chi SG, Borchers H, Deitch AD, deVere White RW. Very frequent p53 mutations in metastatic prostate carcinoma and in matched primary tumors. Cancer 83:2534–2539, 1998.
201. Quinn DI, Henshall SM, Head DR, Golovsky D, Wilson JD, Brenner PC, Turner JJ, Delprado W, Finlayson JF, Grygiel JJ, Stricker PD, Sutherland RL. Prognostic significance of p53 nuclear accumulation in localized prostate cancer treated with radical prostatectomy. Cancer Res 60:1585–1594, 2000.
202. Kirdani RY, Emrich LJ, Pontes EJ, Priore RL, Murphy GP. A comparison of estrogen and androgen receptor levels in human prostatic tissue from patients with non-metastatic and metastatic carcinoma and benign prostatic hyperplasia. J Steroid Biochem 22:569–575, 1985.
203. Prins GS, Sklarew RJ, Pertschuk LP. Image analysis of androgen receptor immunostaining in prostate cancer accurately predicts response to hormonal therapy. J Urol 159:641–649, 1998.
204. Taplin ME, Bubley GJ, Shuster TD, Frantz ME, Spooner AE, Ogata GK, Keer HN, Balk SP. Mutation of the androgen-receptor gene in metastatic androgen-independent prostate cancer. N Engl J Med 332:1393–98, 1995.
205. Sweat SD, Pacelli A, Bergstralh EJ, Slezak JM, Cheng L, Bostwick DG. Androgen receptor expression in prostate cancer lymph node metastases is predictive of outcome after surgery. J Urol 161:1233–1237, 1999.
206. Henshall SM, Quinn DI, Lee CS, Head DR, Golovsky D, Brenner PC, Delprado W, Stricker PD, Grygiel JJ, Sutherland RL. Altered expression of androgen receptor in the malignant epithelium and adjacent stroma is associated with early relapse in prostate cancer. Cancer Res 61:423–427, 2001.
207. Craft N, Shostak Y, Carey M, Sawyers CL. A mechanism for hormone-independent prostate cancer through modulation of androgen receptor signaling by the HER-2/neu tyrosine kinase. Nat Med 5:280–285, 1999.
208. Yeh S, Lin HK, Kang HY, Thin TH, Lin MF, Chang C. From HER2/Neu signal cascade to androgen receptor and its coactivators: a novel pathway by induction of androgen target genes through MAP kinase in prostate cancer cells. Proc Natl Acad Sci USA 96:5458–5463, 1999.
209. Jenkins RB, Qian J, Lieber MM, Bostwick DG. Detection of c-*myc* oncogene amplification and

chromosomal anomalies in metastatic prostatic carcinoma by fluorescence in situ hybridization. Cancer Res 57:524–531, 1997.
210. Sato K, Qian J, Slezak JM, Lieber MM, Bostwick DG, Bergstralh EJ, Jenkins RB. Clinical significance of alterations of chromosome 8 in high-grade, advanced, nonmetastatic prostate carcinoma. J Natl Cancer Inst 91:1574–1580, 1999.
211. Henshall SM, Quinn DI, Lee CS, Head DR, Golovsky D, Brenner PC, Delprado W, Stricker PD, Grygiel JJ, Sutherland RL. Overexpression of the cell cycle inhibitor p16^{INK4A} in high-grade prostatic intraepithelial neoplasia predicts early relapse in prostate cancer patients. Clin Cancer Res 7:544–550, 2001.
212. Umbas R, Schalken JA, Aalders TW, Carter BS, Karthaus HF, Schaafsma HE, Debruyne FM, Isaacs WB. Expression of the cellular adhesion molecule E-cadherin is reduced or absent in high-grade prostate cancer. Cancer Res 52:5104–5109, 1992.
213. Umbas R, Isaacs WB, Bringuier PP, Schaafsma HE, Karthaus HF, Oosterhof GO, Debruyne FM, Schalken JA. Decreased E-cadherin expression is associated with poor prognosis in patients with prostate cancer. Cancer Res 54:3929–3933, 1994.
214. Cheng L, Nagabhushan M, Pretlow TP, Amini SB, Pretlow TG. Expression of E-cadherin in primary and metastatic prostate cancer. Am J Pathol 148:1375–1380, 1996.
215. Morita N, Uemura H, Tsumatani K, Cho M, Hirao Y, Okajima E, Konishi N, Hiasa Y. E-cadherin and alpha-, beta- and gamma-catenin expression in prostate cancers: correlation with tumour invasion. Br J Cancer 79:1879–1883, 1999.
216. Richmond PJ, Karayiannakis AJ, Nagafuchi A, Kaisary AV, Pignatelli M. Aberrant E-cadherin and alpha-catenin expression in prostate cancer: correlation with patient survival. Cancer Res 57:3189–3193, 1997.
217. Aaltomaa S, Lipponen P, Ala-Opas M, Eskelinen M, Kosma VM. Alpha-catenin expression has prognostic value in local and locally advanced prostate cancer. Br J Cancer 80:477–482, 1999.
218. Wood M, Fudge K, Mohler JL, Frost AR, Garcia F, Wang M, Stearns ME. In situ hybridization studies of metalloproteinases 2 and 9 and TIMP-1 and TIMP-2 expression in human prostate cancer. Clin Exp Metastasis 15:246–258, 1997.
219. Luo J, Lubaroff DM, Hendrix MJ. Suppression of prostate cancer invasive potential and matrix metalloproteinase activity by E-cadherin transfection. Cancer Res 59:3552–3556, 1999.
220. Kuniyasu H, Troncoso P, Johnston D, Bucana CD, Tahara E, Fidler IJ, Pettaway CA. Relative expression of type IV collagenase, E-cadherin, and vascular endothelial growth factor/vascular permeability factor in prostatectomy specimens distinguishes organ-confined from pathologically advanced prostate cancers. Clin Cancer Res 6:2295–2308, 2000.
221. Sanchez-Sweatman OH, Orr FW, Singh G. Human metastatic prostate PC3 cell lines degrade bone using matrix metalloproteinases. Invasion Metastasis 18:297–305, 1998.
222. Ricciardelli C, Quinn DI, Raymond WA, McCaul K, Sutherland PD, Stricker PD, Grygiel JJ, Sutherland RL, Marshall VR, Tilley WD, Horsfall DJ. Elevated levels of peritumoral chondroitin sulfate are predictive of poor prognosis in patients treated by radical prostatectomy for early-stage prostate cancer. Cancer Res 59:2324–2328, 1999.
223. Gasparini G, Weidner N, Bevilacqua P, Maluta S, Dalla Palma P, Caffo O, Barbareschi M, Boracchi P, Marubini E, Pozza F. Tumor microvessel density, p53 expression, tumor size, and peritumoral lymphatic vessel invasion are relevant prognostic markers in node-negative breast carcinoma. J Clin Oncol 12:454–466, 1994.
224. Jackson MW, Bentel JM, Tilley WD. Vascular endothelial growth factor (VEGF) expression in prostate cancer and benign prostatic hyperplasia. J Urol 157:2323–2328, 1997.
225. Silberman MA, Partin AW, Veltri RW, Epstein JI. Tumor angiogenesis correlates with progression after radical prostatectomy but not with pathologic stage in Gleason sum 5 to 7 adenocarcinoma of the prostate. Cancer 79:772–779, 1997.
226. Mydlo JH, Kral JG, Volpe M, Axotis C, Macchia RJ, Pertschuk LP. An analysis of microvessel density, androgen receptor, p53 and HER-2/neu expression and Gleason score in prostate cancer. Preliminary results and therapeutic implications. Eur Urol 34:426–432, 1998.
227. Koivisto PA, Rantala I. Amplification of the androgen receptor gene is associated with p53 mutation in hormone-refractory recurrent prostate cancer. J Pathol 187:237–241, 1999.
228. Yu ED, Yu E, Meyer GE, Brawer MK. The relation of p53 protein nuclear accumulation and angiogenesis in human prostate cancer. Prostate Cancer Prostate Dis 1:39–44, 1997.
229. Strohmeyer D, Rossing C, Bauerfeind A, Kaufmann O, Schlechte H, Bartsch G, Loening S. Vascular endothelial growth factor and its correlation with angiogenesis and p53 expression in prostate cancer. Prostate 45:216–224, 2000.
230. Fernandez A, Udagawa T, Schwesinger C, Beecken W, Achilles-Gerte E, McDonnell T, D'Amato R. Angiogenic potential of prostate carcinoma cells overexpressing bcl-2. J Natl Cancer Inst 93:208–213, 2001.
231. Kokontis J, Takakura K, Hay N, Liao S. Increased androgen receptor activity and altered

c-myc expression in prostate cancer cells after long-term androgen deprivation. Cancer Res 54:1566–1573, 1994.
232. Schumacher G, Bruckheimer EM, Beham AW, Honda T, Brisbay S, Roth JA, Logothetis C, McDonnell TJ. Molecular determinants of cell death induction following adenovirus-mediated gene transfer of wild-type p53 in prostate cancer cells. Int J Cancer 91:159–166, 2001.
233. Cardillo M, Berchem G, Tarkington MA, Krajewski S, Krajewski M, Reed JC, Tehan T, Ortega L, Lage J, Gelmann EP. Resistance to apoptosis and upregulation of Bcl-2 in benign prostatic hyperplasia after androgen deprivation. J Urol 158:212–216, 1997.
234. DiPaola RS, Aisner J. Overcoming bcl-2- and p53-mediated resistance in prostate cancer. Semin Oncol 26:112–116, 1999.
235. Miayake H, Tolcher A, Gleave ME. Chemosensitization and delayed androgen-independent recurrence of prostate cancer with the use of antisense Bcl-2 oligodeoxynucleotides. J Natl Cancer Inst 92:34–41, 2000.
236. Gleave M, Tolcher A, Miyake H, Nelson C, Brown B, Beraldi E, Goldie J. Progression to androgen independence is delayed by adjuvant treatment with antisense Bcl-2 oligodeoxynucleotides after castration in the LNCaP prostate tumor model. Clin Cancer Res 5:2891–2898, 1999.
237. Miyake H, Tolcher A, Gleave ME. Antisense Bcl-2 oligodeoxynucleotides inhibit progression to androgen-independence after castration in the Shionogi tumor model. Cancer Res 59:4030–4034, 1999.
238. Waters JS, Webb A, Cunningham D, Clarke PA, Raynaud F, di Stefano F, Cotter FE. Phase I clinical and pharmacokinetic study of bcl-2 antisense oligonucleotide therapy in patients with non-Hodgkin's lymphoma. J Clin Oncol 18:1812–1823, 2000.
239. Scher HI, Morris MJ, Tong WP, Cordon-Cardo C, Drobnjak M, Kelly WM, Slovin SF, Terry KL, DiPaola RS, Rafi M, Rosen N. A phase I trial of G3139, a BCL2 antisense drug, by continuous infusion as a single agent and with weekly Taxol. In: Proc Annu Meet Am Soc Clin Oncol, New Orleans, LA, May 18–24, 2000, p A774.
240. Banerjee D. Genasense (Genta Inc). Curr Opin Investig Drugs 2:574–580, 2001.
241. Haldar S, Basu A, Croce CM. Bcl2 is the guardian of microtubule integrity. Cancer Res 57:229–233, 1997.
242. Eastham JA, Grafton W, Martin CM, Williams BJ. Suppression of primary tumor growth and the progression to metastasis with p53 adenovirus in human prostate cancer. J Urol 164:814–819, 2000.
243. Cowen D, Salem N, Ashoori F, Meyn R, Meistrich ML, Roth JA, Pollack A. Prostate cancer radiosensitization in vivo with adenovirus-mediated p53 gene therapy. Clin Cancer Res 6:4402–4408, 2000.
244. Herbst RS, Lee AT, Tran HT, Abbruzzese JL. Clinical studies of angiogenesis inhibitors: the University of Texas MD Anderson Center Trial of Human Endostatin. Curr Oncol Rep 3:131–140, 2001.
245. Logothetis CJ, Wu KK, Finn LD, Daliani D, Figg W, Ghaddar H, Gutterman JU. Phase I trial of the angiogenesis inhibitor TNP-470 for progressive androgen-independent prostate cancer. Clin Cancer Res 7:1198–1203, 2001.
246. Figg WD, Dahut W, Duray P, Hamilton M, Tompkins A, Steinberg SM, Jones E, Premkumar A, Linehan WM, Floeter MK, Chen CC, Dixon S, Kohler DR, Kruger EA, Gubish E, Pluda JM, Reed E. A randomized phase II trial of thalidomide, an angiogenesis inhibitor, in patients with androgen-independent prostate cancer. Clin Cancer Res 7:1888–1893, 2001.
247. Stearns ME, Wang M. Alendronate blocks metalloproteinase secretion and bone collagen I release by PC-3 ML cells in SCID mice. Clin Exp Metastasis 16:693–702, 1998.
248. Stearns ME, Rhim J, Wang M. Interleukin 10 (IL-10) inhibition of primary human prostate cell-induced angiogenesis: IL-10 stimulation of tissue inhibitor of metalloproteinase-1 and inhibition of matrix metalloproteinase (MMP)-2/MMP-9 secretion. Clin Cancer Res 5:189–196, 1999.
249. Lein M, Jung K, Le DK, Hasan T, Ortel B, Borchert D, Winkelmann B, Schnorr D, Loenings SA. Synthetic inhibitor of matrix metalloproteinases (batimastat) reduces prostate cancer growth in an orthotopic rat model. Prostate 43:77–82, 2000.
250. Miyazaki K, Koshikawa N, Hasegawa S, Momiyama N, Nagashima Y, Moriyama K, Ichikawa Y, Ishikawa T, Mitsuhashi M, Shimada H. Matrilysin as a target for chemotherapy for colon cancer: use of antisense oligonucleotides as antimetastatic agents. Cancer Chemother Pharmacol 43(Suppl):S52–S55, 1999.
251. Shalinsky DR, Brekken J, Zou H, McDermott CD, Forsyth P, Edwards D, Margosiak S, Bender S, Truitt G, Wood A, Varki NM, Appelt K. Broad antitumor and antiangiogenic activities of AG3340, a potent and selective MMP inhibitor undergoing advanced oncology clinical trials. Ann NY Acad Sci 878:236–270, 1999.
252. Knox JD, Bretton L, Lynch T, Bowden GT, Nagle RB. Synthetic matrix metalloproteinase inhibitor, BB-94, inhibits the invasion of neoplastic human prostate cells in a mouse model. Prostate 35:248–254, 1998.
253. Hidalgo M, Eckhardt SG. Development of matrix metalloproteinase inhibitors in cancer therapy. J Natl Cancer Inst 93:178–193, 2001.
254. Heidenreich A, Hofmann R, Engelmann UH.

The use of bisphosphonate for the palliative treatment of painful bone metastasis due to hormone refractory prostate cancer. J Urol 165:136–140, 2001.
255. Brown PD. Ongoing trials with matrix metalloproteinase inhibitors. Expert Opin Investig Drugs 9:2167–2177, 2000.
256. Smith MR, McGovern FJ, Zietman AL, Fallon MA, Hayden DL, Schoenfeld DA, Kantoff PW, Finkelstein JS. Pamidronate to prevent bone loss during androgen-deprivation therapy for prostate cancer. N Engl J Med 345:948–955, 2001.
257. Yip D, Ahmad A, Karapetis CS, Hawkins CA, Harper PG. Matrix metalloproteinase inhibitors: applications in oncology. Invest New Drugs 17:387–399, 1999.

20

Endometrial Cancer: Epidemiology and Molecular Endocrinology

LINDA S. COOK
JENNIFER A. DOHERTY
NOEL S. WEISS
CHU CHEN

Current evidence indicates that exposure of the endometrium to high circulating levels of estrogens increases the likelihood of developing endometrial cancer. Conversely, there is evidence that progestogens (both endogenous and exogenous) have a beneficial effect in terms of reducing the occurrence of endometrial cancer. The actions of many other known or suspected factors that can alter endometrial cancer risk, such as obesity, reproductive characteristics, certain medical conditions, and cigarette smoking, may be explained at least in part by their influence on estrogen and progestogen activity. Less studied are possible genetic factors, including germline allelic variants in enzymes involved in hormone synthesis and metabolism, which may also influence the level and activity of circulating hormones. In this chapter, we summarize the evidence bearing on a role of female gonadal sex hormones in the genesis of endometrial cancer. More comprehensive and detailed reviews of endometrial cancer occurrence and risk factors can be found elsewhere.[1,2] We then present detailed study results regarding the evidence for a genetic component in endometrial cancer etiology and discussions of candidate susceptibility genes involved in estrogen biosynthesis or catabolism.

HORMONAL RISK FACTORS

Endogenous Estrogen and Progesterone

Medical Conditions

Medical conditions that are known to affect hormone levels, such as those that result in relatively high endogenous estrogen levels and/or an atypically high ratio of endogenous estrogen to progesterone, are associated with an elevated risk of endometrial cancer.[3–9] For example, women with polycystic ovary syndrome (Stein-Leventhal syndrome) secrete abnormally large quantities of androstenedione due to chronically elevated levels of luteinizing hormone. This is converted to estrone (a potent estrogen) by the enzyme aromatase and results in levels of estrone similar to those found at the peak of the normal ovulatory cycle.[10] Women with this syndrome also have low progesterone levels due to a lack of cyclic progesterone secretion.[8,11,12]

Obesity

Epidemiological studies have consistently found that both pre- and postmenopausal women with endometrial cancer are more likely to be overweight than other women.[1,2] In postmenopausal women, obesity can lead to a net increase in the amount of endogenous estrogens, both through increased conversion of androstenedione to estrone in adipose tissue[13–15] and through decreased circulating levels of sex hormone–binding globulin.[16–18] In premenopausal women, obesity may increase risk through lower progesterone levels. A study of six obese oligomenorrheic premenopausal women found consistently subnormal levels of serum progesterone relative to 10 nonobese controls, even though the luteal phase in all subjects lasted at least 10 days.[19] Estrogen levels did not differ between the two groups.

Cigarette Smoking

A reduced risk of endometrial cancer has been noted among cigarette smokers in most epidemiological studies.[1,2] The reduced risk is especially notable among postmenopausal women, suggesting that cigarette smoking may reduce endometrial cancer occurrence in ways other than by reducing ovarian production of estrogens. There is some evidence that the negative association is particularly strong in recent smokers.[20–23] The reduced endometrial cancer risk among smokers may be due to differential metabolism of estrogens among smokers that favors the 2-hydroxylation pathway, producing a metabolite of low estrogenic activity (see Cytochrome P-450 1A1 below).[24,25] Similarly, higher levels of circulating progesterone, as well as a higher ratio of progesterone to estrogen among smokers,[26] may also act to diminish net estrogenic effects.

Events of Reproductive Life

Generally, women who experience a natural menopause at a relatively late age are at greater risk of endometrial cancer than other women.[1,2] This elevation in risk could be due to a more prolonged menopausal transition, when limited estrogen production continues (through peripheral conversion of adrenal androgens) without any cyclic progesterone production or substantial peripheral production of progesterone.

Nulliparity consistently has been associated with increased risk of endometrial cancer.[1,2] It is unclear at present if fertility issues explain the elevation in risk among nulliparous women. Impaired fertility, usually measured by self-report, has been associated with an increased risk of endometrial cancer in most studies after accounting for parity.[27–31] Additionally, the occurrence of endometrial cancer among cohorts of women either referred for fertility treatment or diagnosed with infertility was roughly three to five times that expected based on the rate in the general population.[31,32]

Exogenous Estrogen and Progesterone

Hormone-Replacement Therapy

Women who take estrogens unopposed by progestogens are more likely to develop adenomatous hyperplasia of the uterus[33–35] and endometrial cancer[1,2,36,37] than women who do not use hormone-replacement therapy (HRT). Briefly, the risk of endometrial cancer appears to be elevated with all commonly prescribed dosages of conjugated estrogens (0.3–1.25 milligrams per day or the equivalent amount of other estrogens). Endometrial cancer risk appears to increase with increasing duration of use and to decrease after cessation of use. A causal role of exogenous estrogens is supported by the sharp rise and fall in U.S. endometrial cancer incidence in the early 1970s that followed an increase and then a decrease in the use of unopposed estrogens among postmenopausal women.[38–40]

Exogenous progestogens counter the effects of exogenous estrogens by mimicking the actions of luteal progesterone in promoting differentiation and arresting proliferation of endometrial tissue. Several studies have found that estrogen-stimulated hyperplasia reverts to normal endometrium after administration of exogenous progestogens.[41,42] As predicted from these studies of hyperplasia, all studies assessing the impact of progestogen supplementation of postmenopausal estrogens on the risk of endometrial cancer have observed a lower risk of endometrial cancer associated with the use of combined therapy than with unopposed estrogens. Relative to hormone nonusers, modest elevations in risk have been associated with a short duration of cyclic progestogen use (i.e., for 10 or fewer days each month),[43–46] particularly when this regimen was used for 5 years or longer.[43,44,46] Women who take cyclic progestogen for more than 10 days per month appear to have a risk of endometrial cancer that is similar to that of hormone nonusers, except possibly if the total duration of use exceeds 5 years. Two studies have reported that women using continous progestogen (i.e., taken with estrogen every day) had a reduced risk compared to women who never used hormones,[47,48] although another study did not.[44]

Oral Contraceptives

Endometrial proliferative abnormalities and an increased risk of endometrial cancer were associated with the use of a particular type of sequential oral contraceptive pill (OCP), Oracon, which featured a strong estrogenic component and a weak progestogen.[1,2] In contrast, women who have taken the newer monophasic combination OCPs (with a fixed dosage of estrogen and progesterone during a cycle and a stronger progestogenic component) have about one-half the risk of endometrial cancer as nonusers.[1,2] Some studies report that the reduction in risk may be greatest with OCPs in which progestogen effects predominate[49] or that contain higher dosages of progestogen,[50] but one study found that a longer duration of use (>5 years), and not progestogen dosage, was most predictive of a reduced risk.[51] These results suggest that the duration of pre- and perimenopausal combination OCP use is most predictive of a persistent reduction in postmenopausal endometrial cancer risk.

Selective Estrogen Receptor Modulators

Tamoxifen, a nonsteroidal hormone used in the treatment and prevention of breast cancer, has been shown to be estrogenic in the human uterus[52,53] and appears to increase the risk of endometrial cancer, particularly when given at a relatively high daily dosage (30–40 milligrams)[54,55] or for relatively long periods of time (≥5 years).[56–59] Raloxifene, a non-steroidal hormone used to treat osteoporosis in postmenopausal women,[60] does not appear to stimulate proliferation of the endometrium[61–63] or to increase endometrial cancer risk.[64]

GENETIC INFLUENCES

As outlined above, there is strong evidence for a causal relationship between high levels of estrogen exposure, particularly when coupled with low progestogen exposure, and endometrial cancer risk. Although less well studied, there is increasing evidence to suggest that genetic factors are also important in the etiology of endometrial cancer. Such genetic factors include both familial cancer syndromes and potential susceptibility genes influencing estrogen biosynthesis and catabolism.

Familial Aggregation

Endometrial cancer risk is reported to be elevated in women with a first-degree family history of endometrial cancer and among those with certain familial cancer syndromes. Elevations in risk of 50%–90% were noted in several studies among predominantly postmenopausal women with a first-degree family history of endometrial cancer compared to women without such a history,[65,66] whereas a large cohort study among women 55–69 years of age found no association.[67] Larger increases in risk associated with a positive family history have been found among younger women. For example, elevated risks were present among women less than 45 years of age (five cases had a positive family history when only one case would have been expected based on the frequency of a positive family history in the controls),[27] among women less than 50 years of age [odds ratio (OR) = 2.1, 95% confidence interval (CI) 1.1–3.8],[68] and among women less than 55 years of age (OR = 2.8, 95% CI 1.9–4.2).[69] In a large Swedish study, the endometrial cancer risk for the daughters of women diagnosed with endometrial cancer at less than 50 years of age was increased by 6.8-fold, whereas the risk for the daughters of women diagnosed at 50 years of age or greater was increased by 2.9-fold.[70]

In studies of families with hereditary nonpolyposis colorectal cancer (HNPCC), endometrial cancer is one of several cancers observed to occur in excess of general population rates and at ages about two decades younger than that in the general population.[71] An autosomal dominant inheritance of mutations in several DNA mismatch repair genes characterizes HNPCC.[71] Among women who carry the mutations associated with HNPCC, the cumulative incidence of endometrial cancer by age 70 is about 60%.[72] However, these mutations are relatively rare, and the syndrome is thought to account for only 5% of endometrial cancer cases in the general population.[73] In addition to HNPCC, there may be a cancer syndrome specific to endometrial cancer that is consistent with autosomal dominant inheritance[74] and may account for another 5% of cases.[73]

Genetic Factors Related to Estrogen Biosynthesis and Catabolism

Only a small proportion of endometrial cancer cases, even those that run in families, can be attributed to the above autosomal dominant models of inheritance. Given that estrogen exposure is an important risk factor for endometrial cancer, interindividual variation in genes that govern the structure and/or activity of enzymes involved in estrogen biosynthesis and catabolism may play a role in susceptibility to endometrial cancer. Many of the genes involved in these pathways are known to be polymorphic (i.e., to have two or more variant subtypes present in 1% or more of the population). Although such variants may be associated with a small absolute increase in risk of disease for the carrier, they might account for a large proportion of cases due to their relatively high prevalence.

The relative importance of each gene in estrogen biosynthesis and catabolism may change over a woman's life. Beginning with menarche and continuing through perimenopause, women are exposed to large, cyclical fluctuations in the levels of ovarian steroid hormones, with only a small proportion of estrogen produced peripherally. At the time of menopause, however, a woman's hormonal milieu changes. Ovarian hormone production drops sharply, and adipose tissue becomes the primary source of estrogen biosynthesis through the conversion of androstenedione (produced primarily in the adrenal gland) to estrone.[75,76]

Genes involved in the many steps in the pathways that ultimately convert cholesterol to estrone and estradiol, genes involved in the subsequent catabolism of estrone and estradiol, and some of the genes that influence estrogen activity are discussed below (Fig. 20.1). The body of epidemiological work exploring the association between polymorphisms in these genes and cancer in women has focused on breast cancer. Given that the breast and endometrium are both hormonally sensitive tissues in which risk of cancer is related to estrogen exposure, results from breast cancer studies may help guide hypotheses concerning potential associations of polymorphic genes and endometrial cancer. Thus, results from relevant breast cancer studies and the few studies exploring the association between endometrial cancer and candidate susceptibility genes are presented below (for a more comprehensive review of hormone-related genes and their relation to breast cancer risk, see Chapter 5 and Dunning et al.[77]).

Estrogen Biosynthesis

cytochrome p-450 11A1

The first step in the biosynthesis of estrogen (and all other steroid hormones) is the formation of pregnenolone from cholesterol,[78,79] which is catalyzed by cholesterol side chain cleavage enzyme (P450scc, CYP11a, or CYP11α) encoded by *CYP11A1* (Fig. 20.1).[80] A pentanucleotide repeat $(TTTTA)_n$ microsatellite polymorphism at position −528 from the ATG start site of translation in the promotor region of the *CYP11A1* gene has four known alleles (216, 226, 236, and 241).[81] Studies of *CYP11A1* genotype and hormone levels in women have produced inconsistent results.[81–83] To date, there have been no studies of the association of *CYP11A1* variants and endometrial cancer risk.

3B-HYDROXYSTEROID DEHYDROGENASE, TYPES 1 AND 2

3β-Hydroxysteroid dehydrogenase (HSD3β) types 1 and 2 catalyze three separate reactions in estrogen biosynthesis, converting pregnenolone into progesterone, 17α-hydroxypregnenolone into 17α-hydroxyprogesterone, and dehydroepiandrosterone (DHEA) into androstenedione (Fig. 20.1). In addition to the contribution of HSD3β to estrogen biosynthesis, the conversion of pregnenolone to progesterone is a potentially important step in relation to endometrial cancer risk because of the effect of

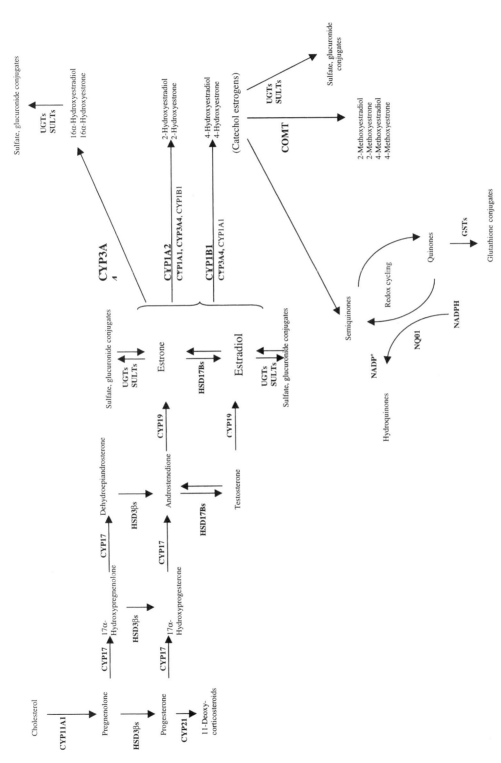

Figure 20.1 Genes involved in estrogen biosynthesis and catabolism. UGT, uridine diphosphate-glucuronosyltransferase; SULT, sulfotransferase; COMT, catechol-*O*-methyltransferase.

progesterone in countering estrogen-induced proliferation. A complex $(TG)_n(TA)_n(CA)_n$ repeat in the third intron of the *HSD3β2* gene with at least 25 variant alleles has been reported,[84,85] but no studies have examined the functional significance of these variants or their association with the occurrence of endometrial cancer.

CYTOCHROME P-450 17

Cytochrome P-450 17 (*CYP17*) mediates both 17α-hydroxylase and 17,20-lyase activities in the biosynthesis of the steroid hormones that are precursors for estradiol. CYP17 also influences progesterone levels by converting progesterone to androstenedione. A polymorphism 34 bases upstream from the transcription-start site in the 5' promotor region has been identified. The variant *CYP17* allele (*A2*) contains a T-to-C transition. This creates a new putative binding site for the transcription factor Sp-1, perhaps increasing the rate of *CYP17* transcription,[86] although a recent report found no evidence that the Sp-1 transcription factor binds to the polymorphic region.[87]

The *A2/A2* genotype may be associated with modest increases in the level of some hormones. In premenopausal women, *A2* homozygotes had 30%–57% higher serum levels of estradiol and progesterone than *A1* homozygotes.[88] In postmenopausal women, the *A2/A2* genotype was associated with only slight differences in levels of these hormones relative to the *A1/A1* genotype.[89,90] Other evidence supporting a possible functional role of the *A2* variant includes the observation that women with the *A2/A2* genotype are half as likely as women with the *A1/A1* genotype to take HRT.[91] However, this observation was not corroborated by the data of Haiman et al.[90]

The *CYP17* polymorphism does not appear to be related to the incidence of breast cancer.[86,87,89,92–101] For endometrial cancer, a small study (50 cases and 51 controls) reported an OR of 1.9 (95% CI 0.5–6.9) for *A2* homozygotes compared to women carrying one or none of the *A2* alleles (Table 20.1).[102] In contrast, in a larger study of 184 cases and 554 controls, women with one or more *A2* alleles were at decreased risk of endometrial cancer (OR = 0.89, 95% CI 0.62–1.27 for heterozygotes; OR = 0.43, 95% CI 0.23–0.80 for *A2/A2* homozygotes compared to *A1/A1* homozygotes) (Table 20.1).[90] This inverse relationship was strongest among never-users of HRT.[90] A multiethnic study (51 cases and 391 controls) observed that, among women who had ever taken estrogen-replacement therapy, those who carried the *A2* allele were at decreased risk of endometrial cancer compared to those with the *A1/A1* genotype (OR = 0.39, 95% CI 0.14–1.09 for *A1/A2*; and OR = 0 for *A2/A2*) (Table 20.1).[90,103]

CYTOCHROME P-450 19

The final and rate-limiting enzymatic reaction in the conversion of androgens to estradiol and estrone is catalyzed by the cytochrome P-450 aromatase (CYP19). CYP19 converts androstenedione to estrone and testosterone to estradiol. While active in the ovaries of pre-menopausal women, aromatase is also active in the adipose tissue of pre- and postmenopausal women[104] suggesting that aromatase may be of particular interest in relation to postmenopausal endometrial cancer risk.

Several sequence variants of the aromatase gene have been identified.[105–107] A C-to-T point mutation in exon 7 results in an arginine-to-cysteine change at codon 264.[108] The variant *Cys264* allele does not affect aromatase activity in vitro[108] and was not associated with breast cancer risk among Japanese women[107,108] or in a population of white, black, Hispanic, and Japanese women.[109] Another variant, a tryptophan-to-arginine exchange at codon 39, was associated with a decreased risk of breast cancer among Japanese women who carried at least one copy of the variant *Arg* allele relative to the *Trp/Trp* genotype.[107]

The most widely studied polymorphism is a variable tetranucleotide repeat $(TTTA)_n$ in intron 4. The varying numbers of repeats (7–13) are thought to affect the determination of DNA splice sites. Decreased plasma concentrations of estrone sulfate, estrone, and estradiol and a lower estrone-to-androstenedione ratio were reported for women with one or more of the seven repeat alleles (including the seven repeat alleles with a 3 bp deletion) compared to non-carriers of the seven-repeat allele.[110] Carriers of at least one eight-repeat allele had elevated plasma estrogen levels and a higher estrone-to-androstenedione ratio compared to noncarriers.[110] No differences in plasma estrogen and androgen levels were noted with the 12-repeat allele.[110]

Table 20.1 Summary of Case-Control Studies of Polymorphic Genes Involved in Estrogen Biosynthesis, Catabolism, and Response and Endometrial Cancer

Gene	Polymorphism	Number of Cases	Number of Controls	Comparison	Odds Ratio (95% CI)
CYP17	T-to-C change in promoter region, A2	50	51	A2/A2 vs. A1/A1 or A1/A2	1.9 (0.5–6.9)[102]
		184	554	A1/A2 vs. A1/A1	0.9 (0.6–1.3)[90]
				A2/A2 vs. A1/A1	0.4 (0.2–0.8)[90]
		51	391	A1/A2 vs. A1/A1	0.4 (0.1–1.1)[90,103]
				A2/A2 vs. A1/A1 (among ever-users of estrogens)	0[90,103]
CYP19	(TTTA)$_n$ in intron 4	85	110	≥1 12-repeat alleles vs. none	5.6 (1.1–39.4)*[114]
				≥1 11-repeat alleles vs. none	1.9 (1.0–3.6)*[114]
CYP1A1 m1	T6235C in 3' non-coding region, MspI	80	60	≥1 m1 variant vs. no m1 variant	3.7 (1.2–13.3)[185]
CYP1A1 m2	Ile462Val in exon 7	80	60	≥1 m2 variant vs. no m2 variant	3.7 (1.2–13.3)[185]
		43	36	≥1 m2 variant vs. no m2 variant	0.5 (0.1–2.0)[187]
CYP1A1 m4	Thr461Asn in exon 7	80	60	≥1 m4 variant vs. no m4 variant	6.4 (2.0–26.5)[186]
COMT	Val158Met in exon 4	49	29	Met/Val and Met/Met vs. Val/Val	1.6 (0.6–4.2)[102]
ERαTA repeat	(TA)$_n$ in promoter region	261	380	<19/≥19 repeats vs. ≥19/≥19 repeats	1.2 (0.8–2.0)[302]
				<19/<19 repeats vs. ≥19/≥19 repeats	1.6 (0.9–2.7)[302]
ERαPvuII	PvuII RFLP, p allele contains restriction site	261	380	Pp vs. pp	1.0 (0.6–1.4)[302]
				PP vs. pp	0.7 (0.4–1.2)[302]
ERα XbaI	XbaI RFLP, x allele contains restriction site	261	380	Xx vs. xx	0.8 (0.5–1.1)[302]
				XX vs. xx	0.6 (0.3–1.1)[302]

*Calculated by the present authors.

Another study did not observe different estrone-to-androstenedione ratios associated with any of the various genotypes.[109]

Studies of aromatase tetranucleotide repeat genotypes in relation to risk of breast cancer have produced conflicting results.[106,107,109–113] With respect to endometrial cancer, a Russian case-control study of 85 cases and 110 controls observed that the longer alleles (with 11 or 12 repeats) were more prevalent among cases (71%) than controls (50%). Specifically, at least one 12-repeat allele was present in 9% of the cases and 2% of the controls, with an unadjusted OR of 5.6 (95% CI 1.1–39.4) (calculated from the paper by the present authors), relative to all other women. At least one 11-repeat allele was present in 65% of cases and 49% of controls, with an unadjusted OR of 1.9 (95% CI 1.0–3.6) (again, calculated by the present authors), relative to all other women (Table 20.1).[114]

There appears to be strong linkage disequilibrium between the tetranucleotide repeat in intron 4 and both an exon 10 polymorphism (a C-to-T change in the 3′-untranslated region)[115] and an intron 6 polymorphism (a G-to-T change).[113] Although the intron 6 polymorphism was not associated with breast cancer risk,[113] the exon 10 *TT* genotype was associated with a twofold increased risk of breast cancer compared to *CC* homozygotes.[115] Therefore, an increased risk of endometrial cancer with the presence of longer aromatase tetranucleotide repeat variants, even if confirmed, may not be independent of other polymorphic sites that may influence endometrial cancer risk.

17β-HYDROXYSTEROID DEHYDROGENASE, TYPES 1, 2, AND 4

Type 1 of the enzyme 17β-hydroxysteroid dehydrogenase, encoded by *HSD17B1*, catalyzes the conversion of estrone to the more biologically active estradiol,[116–118] while types 2 and 4 are involved in the progesterone-stimulated conversion of estradiol to estrone.[117,119–123] *HSD17B1*, *HSD17B2*, and *HSD17B4* are expressed in the endometrium and adipose tissue, as well as in other tissues.[123–129]

No relevant polymorphisms in *HSD17B2* or *HSD17B4* have been reported, but there are various polymorphic sites in *HSD17B1*. An A-to-C change at position −27 that decreases promoter activity by 45% was equally prevalent among breast cancer cases and controls.[130] An A-to-G transition in exon 6 leads to the substitution of glycine for serine at codon 312.[131,132] The *Ser312* allele does not appear to affect the catalytic properties of the enzyme.[133] The *Ser312/Ser312* genotype was marginally associated with increased risk of breast cancer in one study[132] but not in another.[101] In the latter study women with the combined genotype of *HSD17B1 Ser312/Ser312* and *CYP17 A2/A2* had an increased risk of advanced breast cancer compared to women without either allele.[101]

Estrogen Catabolism

Estradiol and estrone undergo oxidative metabolism through hydroxylation at various sites, but the major pathways are 2-, 4-, and 16α-hydroxylation (Fig. 20.1). Of these, 2-hydroxylation predominates.[134] Estradiol, estrone, and their hydrophobic hydroxylated metabolites can be conjugated with glucuronides and sulfates or *O*-methylated to become less lipophilic and more easily excretable.

The most active conjugative pathway for the 2- and 4-hydroxylated catechol estrogens is *O*-methylation, forming methoxyestrogens.[135] The majority of methoxy compounds are excreted, and the 2-methylethers are the most common metabolites of estradiol and estrone in urine.[136–138] The 2- and 4-hydroxy catechol estrogens can also be oxidized to quinones and semiquinones.[139] These semiquinones and quinones can form DNA adducts[139] and undergo redox cycling, producing reactive oxygen species, which may cause oxidative stress, lipid peroxidation,[140] and DNA damage.[135,141–143] Catechol estrogen quinones can be deactivated by conjugation with glutathione[144] or converted to hydroquinones, which protect against redox cycling.[145,146]

Several hypotheses have been proposed concerning the carcinogenic potential of 2-, 4-, and 16α-hydroxylated products.[135] At present, there is limited human evidence supporting any of these hypotheses; almost all of the evidence comes from in vitro studies and animal models. This evidence is based on the ability of a metabolite to activate and/or covalently bind with the estrogen receptor (ER) and the potential of metabolites to induce estrogen-related carcinogenesis in the Syrian hamster kidney tumorigenesis model, which results in a cumulative tumor incidence of almost 100% with exposure to estradiol.[147,148]

Higher levels of 16α-hydroxylated estrogens may increase the risk of hormone-related cancer,[149] in part because they activate the ER[141,150] and are uterotropic.[150] Women with breast and endometrial cancer had a higher percentage of 16α-hydroxylated metabolites than women without cancer in one study,[151] and two cohort studies reported that a higher (vs. lower) 2:16α-hydroxyestrone ratio was associated with a modestly reduced risk of subsequent breast cancer.[152,153] However, 16α-hydroxyestrogens are only weakly carcinogenic in the Syrian hamster kidney tumor model.[148]

Higher 2-hydroxyestrogen levels and higher 2:16α-hydroxyestrogen ratios were reported among Finnish women relative to Asian women, who have relatively high and low risks of breast/endometrial cancer, respectively, after adjustment for age and body mass index.[154] However, 2-hydroxyestrogens do not induce tumors in the Syrian hamster kidney,[148,155] and they have a reduced binding affinity for the ER.[137,156] Additionally, O-methylation of 2-hydroxyestradiol forms 2-methoxyestradiol, which has been suggested to be an anticancer agent because it is a potent inhibitor of tumor cell proliferation and has antiangiogenic effects.[135]

Although 4-hydroxylated estrogens are formed in much smaller amounts than 2-hydroxylated estrogens in the liver, there are several reasons why 4-hydroxyestradiol might be important in carcinogenesis. Unlike 2-hydroxyestradiol, 4-hydroxyestradiol is as carcinogenic as estradiol in the Syrian hamster kidney model.[148,155] Additionally, 4-hydroxyestradiol retains potent hormonal activity, as evidenced by its activation of the ER.[137,156-158] 4-Hydroxyestradiol is rapidly formed and can accumulate in target tissues when high levels of 2-hydroxyestradiol inhibit the O-methylation of 4-hydroxyestradiol to a less active form.[135,159-161] The ratio of estradiol 4- to 2-hydroxylation increases during tumorigenesis in the Syrian hamster kidney model,[162] and this pattern was also observed in human uterine leiomyomata compared to normal myometrium.[160] Also, 4-hydroxy catechol estrogen quinones form depurinating DNA adducts, whereas 2-hydroxy catechol estrogen quinones form stable DNA adducts.[139,163] The depurinating adducts could potentially be involved in tumor initiation, by producing mutations in critical genes.[139]

CYTOCHROME P-450 1A1

CYP1A1 codes for aryl hydrocarbon hydroxylase (AHH), one of the enzymes that catalyzes the irreversible hydroxylation of estradiol and estrone into 2-hydroxyestrogens and, to a much lesser extent, 4-hydroxyestrogens.[164] Expression of AHH is primarily extrahepatic, and it is constitutively expressed in the endometrium.[165] *CYP1A1* gene transcription is highly inducible by various substances, including polycyclic aromatic hydrocarbons (PAHs), polyhalogenated aromatic hydrocarbons, and flavones.[166,167] An increase in 2-hydroxylation due to induction of the AHH-related pathway by components in cigarette smoke might explain, in part, why smokers have a reduced risk of endometrial cancer.[24,25]

There are several known variant alleles in the *CYP1A1* gene, including *m1*, a T-to-C exchange at position 6235 in the 3' noncoding region;[168] *m2*, an A-to-G exchange at position 4889 that results in an isoleucine-to-valine substitution in exon 7 in the heme-binding region;[169] *m3*, an African American-specific T-to-C exchange at position 5639 in the 3' noncoding region;[170] and *m4*, a C-to-A exchange at position 4887 that results in a threonine-to-asparagine substitution in exon 7 adjacent to *m2*.[171] It is unclear whether the *m1* variant gene is associated with a high-inducibility phenotype,[170,172-175] but the *m2* variant is associated with increased enzyme activity as well as increased *CYP1A1* gene inducibility.[170,175] The *m4* allele codes for a protein that has reduced activity toward estradiol compared to the wild type.[176] The functional significance of the *m3* allele has not yet been described.

The association between polymorphisms in *CYP1A1* and breast cancer risk has been inconsistent.[96,177-184] For endometrial cancer, a case-control study of 80 cases and 60 controls reported that women with at least one variant allele were at increased risk of endometrial cancer compared to women without the allele of interest. Odds ratios associated with the presence of the *m1*, *m2*, and *m4* alleles were, respectively, 3.67 (95% CI 1.21–13.26), 3.67 (95% CI 1.21–13.26), and 6.36 (95% CI 1.99–26.5) (Table 20.1).[185,186] In contrast, in a study of 43 cases and 36 controls, possessing at least one *m2* allele was associated with a decreased risk of endometrial cancer (OR = 0.51, 95% CI 0.13–1.97) (Table 20.1).[187]

CYTOCHROME P-450 1A2

While CYP1A2 is closely homologous to CYP1A1, it is expressed almost exclusively in liver.[188] CYP1A2 catalyzes 2-hydroxylation and, to a lesser extent, 4-hydroxylation of estradiol and estrone.[164,189–191] CYP1A2 is the most active enzyme in the 2-hydroxylation of estradiol and estrone in liver[190,191] and is involved in 16α-hydroxylation as well.[191] CYP1A2 activity is induced by PAHs, cigarette smoke, charred meats, cruciferous vegetables, and various drugs.[188] There is also evidence for gene–gene interactions between *CYP1A2*, *CYP1A1*, and *GSTM1*. The *CYP1A1 m2* variant was correlated with increased CYP1A2 activity after exposure to cigarette smoke or charred meat in one study.[192] Additionally, in that study, CYP1A2 activity was relatively higher among individuals with the *GSTM1* null allele.[192]

There is a large variation in human hepatic expression of CYP1A2.[188] Four polymorphisms in *CYP1A2* have been described. The first is a G-to-A transition at position −2964 in the 5′-flanking region, called *CYP1A2/A*. The adenine allele is associated with decreased inducibility.[193] The second polymorphism is an A-to-C substitution at position 734 in intron 1, called *CYP1A2/D*. The homozygous *CYP1A2/D* genotype is associated with low inducibility.[194,195] The other two polymorphisms are a T-deletion polymorphism in the 5′-flanking region (*CYP1A2/B*) and a T-to-G substitution in intron 1 (*CYP1A2/C*).[196] None of these variants has been investigated in relation to the incidence of endometrial cancer.

CYTOCHROME P-450 1B1

CYP1B1 is the main enzyme that mediates the 4-hydroxylation of estradiol. It is the most efficient estradiol hydroxylase that has been characterized.[197] Although the 4-catechol estrogens are short-lived, and thus circulating levels may have a limited impact on the endometrium, CYP1B1 is one of the major P-450 mRNAs found in the endometrium,[165,198] suggesting an important tissue-level effect. CYP1B1 is also expressed at low levels in the liver and various other tissues.[198] Like CYP1A1, CYP1B1 is involved in activation of several environmental carcinogens, including PAHs, heterocyclic and aryl amines, and nitroaromatic hydrocarbons.[199] It is inducible by dioxins and indole 3-carbinols.[197,200,201] While polymorphisms in *CYP1B1* may be relevant to endometrial cancer risk, the variants discussed below have not been investigated with respect to endometrial cancer.

Four known common polymorphisms in *CYP1B1* result in amino acid substitutions. The first is a leucine-to-valine exchange at codon 432 in exon 3.[202] Based on studies in human lung microsomes[203] and *Escherichia coli* recombinant systems,[204,205] the presence of the *Val432* allele results in increased 4-hydroxy metabolite formation compared to the *Leu432* allele, but these findings were not replicated in an insect cell expression system.[206] No association between the *Val432* allele and breast cancer was seen in Japanese[207] or Caucasian and African-American populations,[208] whereas the *Leu432/Leu432* genotype was associated with an increased risk of breast cancer compared to the *Val432/Val432* genotype in Chinese women.[209]

The second polymorphism is a G-to-T transition at nucleotide 701, resulting in an alanine-to-serine substitution at codon 119 in exon 2.[202] In *E. coli* recombinant systems, the *Ser119* variant exhibits higher 4-hydroxylation activity for 17β-estradiol compared to the wild-type enzyme.[204,205] The *Ser119* allele was associated with an increased risk of breast cancer in a Japanese population, while individuals who simultaneously carried the *Ala119* and *Leu432* alleles were at reduced risk of breast cancer compared to women who carried any other combination of alleles, even though the *Leu432* allele alone was not associated with breast cancer.[207] When this combined allele was expressed in a recombinant system, it had the lowest estradiol 4-hydroxylase activity compared to the other three possible combinations of these two polymorphisms.[207]

The third polymorphism is an arginine-to-glycine substitution at codon 48 in exon 2,[202] which is linked to the *Ala119Ser* variant.[210] Although the *Gly48* allele had higher 4-hydroxylation activity than the wild type in a recombinant system,[205] there was no difference between a combined *Gly48/Ser119* allele and the common allele with respect to catalytic activity and protein stability when expressed in yeast and mammalian COS-1 cells.[210]

The fourth *CYP1B1* polymorphism is an asparagine-to-serine substitution at codon 453 resulting from an A-to-G change at nucleotide 1358 in exon 3.[202] The variant is reported to have higher 4-hydroxylation activity than the wild type in an *E. coli* recombinant system,[205] but these findings were not replicated in an insect cell expression system.[206] No association between the *Ser453* allele and breast cancer was found.[208]

CYTOCHROME P-450 3A4

CYT3A4 has strong catalytic activity for estrone 16α-hydroxylation.[190,191,211,212] It is also involved in estradiol 16α-hydroxylation[191] and, to a lesser degree than CYP1A2, hepatic 2- and 4-hydroxylation of estradiol.[189,191,213] CYP3A4 is one of the major P-450 enzymes in human liver.[214,215] Unlike CYP1A2, CYP3A4 is also present in several extrahepatic tissues and may contribute substantially to 2-hydroxylation of estrogens in these tissues.[135] This enzyme is also highly inducible and is involved in the oxidative metabolism of over half of the drugs used clinically.[216]

Like CYP1A2, there is a large variation in human hepatic expression of CYP3A4.[215] A polymorphism in the promoter region of the gene, called *CYP3A4-V* (or *CYP3A4*1B*) is characterized by an A-to-G mutation in the 5′ nifedipine-specific element, which is −290 bp from the transcription-start site of the *CYP3A4* gene.[217] The variant does not appear to affect the properties of the enzyme,[218–220] although testosterone metabolism may be increased.[221] Another variant allele results from a serine-to-proline change at codon 222 in exon 7. Compared to the wild-type enzyme, this variant enzyme did not exhibit altered testosterone hydroxylation.[222] Neither of the two *CYP3A4* variants has been investigated in relation to risk of endometrial cancer.

CATECHOL-*O*-METHYLTRANSFERASE

The enzyme catechol-*O*-methyltransferase (COMT), which transforms catechol estrogens into inactive metabolites and prevents them from entering into redox cycling,[137] has received considerable attention. Its activity is present in both liver and endometrium.[223] In the Syrian hamster kidney model, inhibition of COMT activity resulted in increased kidney tumorigenesis.[224,225]

A valine-to-methionine substitution due to a G-to-A transition in codon 158 in exon 4 of the *COMT* gene results in a heat-labile enzyme that is 4- to 5-fold less effective at methylating catechol substrates in vitro.[226] The *Met* allele also leads to 2- to 3-fold lower levels of methoxyestrogen metabolite formation.[227] The relation of this low-activity allele to the incidence of breast cancer is unclear.[96,228–231] One small study (49 cases and 29 controls) reported a modestly increased risk of endometrial cancer associated with carrying one or two low-activity alleles, but the result was statistically imprecise (OR = 1.6, 95% CI 0.6–4.2) (Table 20.1).[102]

URIDINE DIPHOSPHATE-GLUCURONOSYLTRANSFERASES

The uridine diphosphate-glucuronosyltransferases (UGTs) are a family of enzymes that catalyze glucuronidation, making estrogens less lipophilic and more easily excreted in urine and bile. The UGTs are present in liver and various other tissues and have overlapping substrate specificity. They are of considerable interest in relation to endometrial cancer because of their direct role in the deactivation of estrogens and catechol estrogens and their potential to regulate steroid hormone levels in target tissues.[232–234] Several UGTs, including 1A1, 1A3, 1A4, 1A7, 1A8, 1A9, 1A10, 2B4, and 2B7, exhibit activity toward estrone, estradiol, and their catechol derivatives.[235] While polymorphic alleles have been described for *UGT1A1*, *UGT2B4*, and *UGT2B7*, there are no studies of any of the *UGT* genes in relation to endometrial cancer.

UGT1A1 is involved in the glucuronidation of estradiol, estriol, and 2-hydroxyestrogens,[236,237] as well as several other substrates including bilirubin. Numerous polymorphisms have been identified in *UGT1A1*, most of which are related to a decreased ability to glucuronidate bilirubin, as seen in Gilbert's syndrome and the rarer Crigler-Najjar syndrome. Common single-nucleotide polymorphisms in *UGT1A1* include a G-to-A transition at position 211 in exon 1 that results in a glycine-to-arginine change at codon 71 (UGT1A1*6)[238] and a C-to-A transversion at position 686 at codon 229 in exon 1 that changes proline to glutamine (UGT1A1*27).[239] Little is known about the functional consequences of these variants. A third polymorphism is a TA dinucleotide repeat in the promoter, 41 bp upstream of the translation-start site.[240] The common repeat allele has six repeats and is designated UGT1A1*1; the putative high-risk allele has seven repeats and is designated UGT1A1*28. Alleles with five (UGT1A1*33) and eight (UGT1A1*34) repeats have been found only among African Americans.[241] While gene transcription appears to be inversely associated with the number of repeats,[240–242] repeat number was not associated with steroid hormone levels among postmenopausal women.[243] Genotypes in African-American women that included seven or eight repeats were associated with an increased risk of premenopausal breast cancer compared to genotypes containing five or six repeats.[242] In contrast, no association between the TA repeat polymorphism and breast cancer was found among mostly Caucasian women.[243]

UGT2B4 glucuronidates 2-, 4-, and 16α-hydroxylated estrogens, with the highest efficiency toward 4-hydroxyestrone.[244,245] An aspartic acid-to-glutamic acid substitution at codon 458 does not affect the substrate specificity or catalytic activity of the protein.[246]

UGT2B7 glucuronidates estradiol and 2- and 16α-hydroxylated estrogens.[237,247] A polymorphism consisting of a histidine-to-tyrosine change at codon 268[244] does not seem to be related to altered function.[237,248]

SULFOTRANSFERASES

Sulfation by sulfotransferases (SULTs) is one of the main routes of conjugative metabolism and deactivation of estrogen and estrogen metabolites. Estrone sulfate is the most abundant circulating estrogen, and estrogen sulfates have a longer half-life than non-conjugated estrogens.[235] Estrogen-related SULT activity is present primarily in the liver[235] but is also present in other tissues, including the endometrium.[249,250] Progesterone stimulates estrogen sulfotransferase.[251,252] SULT1A1 (or TS PST1), SULT1A3 (or TL PST), SULT2A1 (or DHEA ST), and SULT1E1 (or EST2) may be of particular interest in endometrial carcinoma since they are able to conjugate estrogens and catechol estrogens. Both SULT1A3 (which can conjugate estradiol and 2-hydroxyestrogens[253]) and SULT1E1 (which can conjugate estrone and estradiol) exhibit considerable interindividual variation, but no genetic polymorphisms have been reported.[235,254] Additionally, none of the SULTs polymorphisms described below has been examined in relation to endometrial cancer risk.

SULT1A1 is able to conjugate estrone, estradiol, 4-hydroxyestrone and estradiol, 2-hydroxyestradiol,[255] and 2-methoxyestradiol.[256] A G-to-A substitution at position 638 in exon 7 results in an arginine-to-histidine exchange at codon 213 (also referred to as SULT1A1*2) and is associated with an approximately 10-fold decrease in enzyme activity among individuals homozygous for this variant.[255,257] One study reported an increased risk of breast cancer among women with the Arg/Arg genotype compared to women homozygous for the His allele,[258] while another study found that SULT1A1*2 genotype did not affect breast cancer risk.[259]

SULT1A2 has an A-to-C transversion that causes an asparagine-to-threonine exchange at codon 235. The Thr allele exhibits less enzyme activity than the Asn allele.[260] The alleles in SULT1A1 and 1A2 are in linkage disequilibrium: SULT1A1 Arg with SULT1A2 Asn (the high-activity alleles) and SULT1A1 His with SULT1A2 Thr (the low-activity alleles).[261]

SULT2A1 conjugates estrone, estradiol and 4-hydroxyestradiol, but not 4-hydroxyestrone or 2-hydroxyestradiol.[235] Two variant alleles have been reported: a T-to-C exchange at nucleotide 170, resulting in a methionine-to-threonine exchange at codon 57, and an A-to-T transversion at nucleotide 557 in exon 4, resulting in a glutamic acid-to-valine substitution at codon 186. When expressed in an in vitro recombinant system, these variants have reduced enzyme activity, although the variants were not correlated with activity in human liver samples.[262]

NAD(P)H:QUINONE OXIDOREDUCTASE 1

NAD(P)H:quinone oxidoreductase 1 (NQO1), previously referred to as DT-diaphorase, is a flavoprotein that catalyzes the reduction of quinones to hydroquinones using reduced nicotinamide adenine dinucleotide (NADH) or reduced NAD phosphate [NAD(P)H] as electron donors.[263,264] Since it prevents the formation of semiquinones and highly reactive oxygen species, it protects cells from the adverse effects of redox cycling.[145,146] The enzyme is induced by PAHs, indoles, antioxidants, oxidants, and a wide variety of other compounds,[265] including selective ER modulators.[266,267]

A C-to-T transition results in a proline-to-serine exchange in codon 187 in exon 6.[268] The *Ser187* allele, when expressed in a recombinant *E. coli* system, results in a protein with only 2% of the activity of the wild type.[269] Individuals homozygous for the *Ser187* allele have no detectable NQO1 protein,[270] and heterozygotes have NQO1 activity that is significantly lower than that of wild-type individuals.[270,271] A much less common allele,[272] a C-to-T substitution in exon 4 that results in an arginine-to-tryptophan exchange in codon 139, shows reduced activity toward several substrates.[273] Neither of these putative high-risk alleles has been examined in relation to endometrial cancer risk.

Other Factors Influencing Estrogen Activity

LUTEINIZING HORMONE

Luteinizing hormone (LH) is a glycopeptide hormone released from the anterior pituitary under the control of gonadotropin-releasing hormone. It regulates estrogen biosynthesis in the ovary by initiating the secretion of androstenedione and testosterone in the dominant follicles. It also stimulates progesterone secretion.[274] The hormone is a heterodimer composed of α and β subunits. The α subunit is highly conserved, but several polymorphic regions have been identified in the β subunit. A relatively common variant is associated with two missense mutations that result in the following amino acid changes in exon 2: a T-to-C change at nucleotide 22 that results in a tryptophan-to-arginine substitution at codon 8 and a T-to-C change at nucleotide 44 that results in an isoleucine-to-threonine substitution at codon 15,[275,276] creating a new glycosylation site.[277] The two nucleotide changes appear to always occur together.[278]

In vitro studies suggest that there is relatively increased bioactivity in individuals homozygous for the LH allele compared to individuals with the homozygous common genotype, although variant LH has a relatively short half-life compared to wild-type LH in vivo.[277,279] It is therefore unclear whether the variant is associated with overall higher or lower bioactivity compared to the wild type. Premenopausal women heterozygous for the variant allele had increased serum estradiol, sex hormone–binding globulin (SHBG), and testosterone concentrations compared to homozygotes for the more common allele,[280] although one small study did not observe such a difference.[281] The presence of one or more LH variant alleles was not associated with risk of postmenopausal breast cancer[282] or breast carcinoma in situ.[283] The variant has not been studied in relation to endometrial cancer risk.

SEX HORMONE–BINDING GLOBULIN

Estrogen, whether produced in the ovary or peripherally, is transported to target cells throughout the body by SHBG. While it appears that the proportion of circulating levels of bound and unbound estrogens are related to the serum concentration of SHBG,[284] the influence SHBG might have in the carcinogenic process remains unclear.[285–287] It is expressed in the endometrium[286,287] as well as in the liver and various other tissues.

A G-to-A mutation in exon 8 of the SHBG gene results in substitution of asparagine for aspartic acid at codon 327 introducing an additional N-glycosylation site.[288] Although the variant protein does not have different steroid-binding properties compared to the wild type,[288] it exhibits decreased clearance.[289] One study reported that the variant allele was present in a higher proportion of breast cancer cases than controls,[285] but no studies have examined the variant in relation to endometrial cancer risk.

ESTROGEN RECEPTOR

The ER mediates hormonal responses in estrogen-sensitive tissues at the cellular level and is therefore critical for hormone regulation. There are two forms of the ER, ERα and ERβ. Both types of ER are expressed in human endometrium.[290]

One polymorphism has been reported for the *ERβ* gene, a CA repeat in intron 5,[291] with at least 14 alleles. The *short/short* genotype (with eight or fewer repeats) is associated with higher levels of androstenedione and lower levels of SHBG than the *long/long* genotype.[292] This polymorphism has not been examined in relation to endometrial cancer risk.

A number of polymorphisms exist in the *ERα* gene. There may be an elevated breast cancer risk among women homozygous for a *Xba*I restriction fragment length polymorphism (RFLP) that results from an A-to-G exchange in intron 1 relative to wild-type homozygotes.[293] Other polymorphisms have an unclear relation to breast cancer risk, such as a silent C-to-G change at nucleotide 975 in codon 325[294–297] and a G-to-T change at nucleotide 478 resulting in a glycine-to-cysteine substitution at codon 160.[297,298] Still others appear to have no relation to breast cancer risk, including a silent G-to-C mutation in nucleotide 261 in codon 87 of exon 1 that results in a *Bst*UI RFLP,[299] an A-to-G substitution at nucleotide 908 resulting in a lysine-to-arginine exchange at codon 303 in exon 4, and a T-to-C point mutation that results in a *Pvu*II RFLP in intron 1.[293,297,300]

A common TA repeat polymorphism in the promoter region −1174 bp upstream of exon 1,[301] one unrelated to hormone levels in premenopausal women,[292] has been examined along with the *Pvu*II and *Xba*I RFLPs in a population-based endometrial cancer study with 261 cases and 380 controls.[302] Women who carried the short (<19 repeats) TA repeat genotype were at moderately increased risk of endometrial cancer (*short/short* compared to *long/long*, OR = 1.57, 95% CI 0.92–2.68; *short/long* compared to *long/long*, OR = 1.23, 95% CI 0.76–1.98).[302] The *Pvu*II PP and *Xba*I XX genotypes were modestly associated with a decreased risk of endometrial cancer, and there was a suggestion of a trend with the number of variant alleles carried for the *Xba*I polymorphism [ORs and 95% CIs are, respectively, for *PP* compared to *pp* 0.72 (0.43–1.22), for *Pp* compared to *pp* 0.96 (0.64–1.44), for *XX* compared to *xx* 0.60 (0.33–1.10), and for *Xx* compared to *xx* 0.75 (0.51–1.09)] (Table 20.1).

PROGESTERONE RECEPTOR

Progesterone is important in relation to endometrial cancer risk because it counters the proliferative effects of estrogen by inducing glandular and stromal differentiation of the endometrium. The progesterone receptor (PR) mediates the action of progesterone. The *PR* gene has two isoforms, *PRA* and *PRB*. The latter acts as a transcriptional activator, whereas *PRA* can repress both the transcriptional activity of *PRB* and *ER* gene activation. Both isoforms are present in the endometrium.[303] A variant allele, called PROGINS, has a 306 bp insertion in intron G of the *PR* gene.[304] Results in two published abstracts suggest that the variant allele exhibits increased transcriptional activity and stability compared to the wild-type allele.[305,306] There is conflicting evidence regarding the relation of breast cancer risk to the presence of the PROGINS variant allele.[307–309] This polymorphism has not been examined in relation to endometrial cancer risk.

SUMMARY

In this chapter, we have summarized the evidence concerning *(1)* hormonal risk factors for endometrial cancer; *(2)* familial aggregation of endometrial cancer; and *(3)* major enzymes in estrogen metabolism and catabolism, polymorphisms in genes that code for these enzymes, and the relationship between some of these variants and endometrial cancer occurrence. Support for a causal role of circulating estrogen levels in the etiology of endometrial cancer is quite strong, particularly when coupled with low levels of circulating progesterone. However, the etiological role in endometrial cancer of estrogen biosynthesis/catabolism enzymes and the genes that bear on their structure and function remains unclear.

Nonetheless, there are good biological reasons to think that a comprehensive causal model for this disease would include an imbalance of estrogen and progesterone exposure as well as increased activation or decreased detoxification/excretion of these compounds that, in one or more combinations, leads to a carcinogenic cascade. It is also reasonable to hypothesize that the action of other risk factors, such as exogenous hormone exposures, may be modulated by the enzymes and their variants involved in estrogen biosynthesis and catabolism. The study of genetic factors in endometrial cancer will continue and has the potential to provide both etiological insights and a means to identify women at greatest risk of this disease.

REFERENCES

1. Grady D, Ernster VL. Endometrial cancer. In: Schottenfeld D, Fraumeni JF (eds). Cancer Epidemiology and Prevention. New York: Oxford University Press, 1996, pp 1058–1089.
2. Cook LS, Weiss NS. Endometrial cancer. In: Goldman MB, Hatch MC (eds). Women and Health. London: Academic Press, 1999, pp 916–931.
3. Diddle AW. Granulosa and theca cell ovarian tumors: prognosis. Cancer 5:215–228, 1952.
4. Larson JA. Estrogens and endometrial cancer. Obstet Gynecol 3:551–572, 1954.
5. Salerno LJ. Feminizing mesenchymomas of the ovary—an analysis of 28 granulosa-theca cell tumors and their relationship to coexistent carcinoma. Am J Obstet Gynecol 84:731–738, 1962.
6. Gusberg SB, Kardon P. Proliferative endometrial response to theca-granulosa cell tumors. Am J Obstet Gynecol 111:633–643, 1971.
7. Dockerty MB, Lovelady SB, Foust GT Jr. Carcinoma of the corpus uteri in young women. Am J Obstet Gynecol 61:966–981, 1951.
8. Farhi DC, Nosanchuk J, Silverberg SG. Endometrial adenocarcinoma in women under 25 years of age. Obstet Gynecol 68:741–745, 1986.
9. Coulam CB, Annegers JF, Kranz JS. Chronic anovulation syndrome and associated neoplasia. Obstet Gynecol 61:403–407, 1983.
10. Siiteri PK, MacDonald PC. The role of extraglandular estrogen in human endocrinology. In: Handbook of Physiology. Endocrinology, Section 7, Volume 2, Washington, DC: American Physiological Society, 1973, p 615–629.
11. Lucas WE. Causal relationships between endocrine-metabolic variables in patients with endometrial carcinoma. Obstet Gynecol Surv 29:507–528, 1974.
12. Yen SS. The polycystic ovary syndrome. Clin Endocrinol (Oxf) 12:177–207, 1980.
13. MacDonald PC, Siiteri PK. The relationship between the extraglandular production of estrone and the occurrence of endometrial neoplasia. Gynecol Oncol 2:259–263, 1974.
14. MacDonald PC, Edman CD, Hemsell DL, Porter JC, Siiteri PK. Effect of obesity on conversion of plasma androstenedione to estrone in postmenopausal women with and without endometrial cancer. Am J Obstet Gynecol 130:448–455, 1978.
15. Edman CD, MacDonald PC. Effect of obesity on conversion of plasma androstenedione to estrone in ovulatory and anovulatory young women. Am J Obstet Gynecol 130:456–61, 1978.
16. Davidson BJ, Gambone JC, Lagasse LD, Castaldo TW, Hammond GL, Siiteri PK, Judd HL. Free estradiol in postmenopausal women with and without endometrial cancer. J Clin Endocrinol Metab 52:404–408, 1981.
17. Kaye SA, Folsom AR, Soler JT, Prineas RJ, Potter JD. Association of body mass and fat distribution with sex hormone concentrations in postmenopausal women. Int J Epidemiol 20:151–156, 1991.
18. Nyholm HC, Nielsen AL, Lyndrup J, Dreisler A, Hagen C, Haug E. Plasma oestrogens in postmenopausal women with endometrial cancer. Br J Obstet Gynaecol 100:1115–1119, 1993.
19. Sherman BM, Korenman SG. Measurement of serum LH, FSH, estradiol and progesterone in disorders of the human menstrual cycle: the inadequate luteal phase. J Clin Endocrinol Metab 39:145–149, 1974.
20. Folsom AR, Kaye SA, Potter JD, Prineas RJ. Association on incident carcinoma of the endometrium with body weight and fat distribution in older women: early findings of the Iowa Women's Health Study. Cancer Res 49:6828–6831, 1989.
21. Austin H, Drews C, Partridge EE. A case-control study of endometrial cancer in relation to cigarette smoking, serum estrogen levels, and alcohol use. Am J Obstet Gynecol 169:1086–1091, 1993.
22. Elliott EA, Matanoski GM, Rosenshein NB, Grumbine FC, Diamond EL. Body fat patterning in women with endometrial cancer. Gynecol Oncol 39:253–258, 1990.
23. Brinton LA, Barrett RJ, Berman ML, Mortel R, Twiggs LB, Wilbanks GD. Cigarette smoking and the risk of endometrial cancer. Am J Epidemiol 137:281–291, 1993.
24. Michnovicz JJ, Hershcopf RJ, Naganuma H, Bradlow HL, Fishman J. Increased 2-hydroxylation of estradiol as a possible mechanism for the anti-estrogenic effect of cigarette smoking. N Engl J Med 315:1305–1309, 1986.
25. Key TJA, Pike MC, Brown JB, Hermon C, Allen DS, Wang DY. Cigarette smoking and urinary oestrogen excretion in premenopausal and postmenopausal women. Br J Cancer 74:1313–1316, 1996.
26. Friedman AJ, Ravnikar VA, Barbieri RL. Serum steroid hormone profiles in postmenopausal smokers and nonsmokers. Fertil Steril 47:398–401, 1987.
27. Henderson BE, Casagrande JT, Pike MC, Mack T, Rosario I, Duke A. The epidemiology of endometrial cancer in young women. Br J Cancer 47:749–756, 1983.
28. Centers for Disease Control, Cancer and Steroid Hormone Study. oral contraceptive use and the

risk of endometrial cancer. JAMA 249:1600–1604, 1983.
29. Escobedo LG, Lee NC, Peterson HB, Wingo PA. Infertility-associated endometrial cancer risk may be limited to specific subgroups of infertile women. Obstet Gynecol 77:124–128, 1991.
30. Brinton LA, Berman ML, Mortel R, Twiggs LB, Barrett RJ, Wilbanks GD, Lannom L, Hoover RN. Reproductive, menstrual, and medical risk factors for endometerial cancer: results from a case-control study. Am J Obstet Gynecol 167:1317–1325, 1992.
31. Venn A, Watson L, Lumley J, Giles G, King C, Healy D. Breast and ovarian cancer incidence after infertility and in vitro fertilization. Lancet 346:995–1000, 1995.
32. Modan B, Ron E, Lerner-Geva L, Blumstein T, Menczer J, Rabinovici J, Oelsner G, Freedman L, Mashiach S, Lunenfeld B. Cancer incidence in a cohort of infertile women. Am J Epidemiol 147:1038–1042, 1998.
33. Gusberg SB. Precursors of corpus carcinoma estrogens and adenomatous hyperplasia. Am J Obstet Gynecol 54:905–927, 1947.
34. The Writing Group for the PEPI Trial: Effects of estrogen or estrogen/progestin regimens on heart disease risk factors in postmenopausal women: the Postmenopausal Estrogen/Progestin Interventions (PEPI) Trial. JAMA 273:199–208, 1995.
35. Speroff L, Rowan J, Symons J, Genant H, Wilborn W. The comparative effect on bone density, endometrium, and lipids of continuous hormones as replacement therapy (CHART Study). JAMA 276:1397–1403, 1996.
36. Grady D., Gebretsadik T, Kerlikowske K., Ernster V, Petitti D. Hormone replacement therapy and endometrial cancer risk: a meta-analysis. Obstet Gynecol 85:304–313, 1995.
37. Herrington LJ, Weiss NS. Postmenopausal unopposed estrogens: characteristics of use in relation to the risk of endometrial carcinoma. Ann Epidemiol 3:308–318, 1993.
38. Weiss NS, Szekely DR, Austin DF. Increasing incidence of endometrial cancer in the United States. N Engl J Med 294:1259–1262, 1976.
39. Austin DF, Roe KM. The decreasing incidence of endometrial cancer: public health implications. Am J Public Health 72:65–68, 1982.
40. Kennedy DL, Baum C, Forbes MB. Noncontraceptive estrogens and progestins: use patterns over time. Obstet Gynecol 65:441–446, 1985.
41. Thom MH, White PJ, Williams RM, Sturdee DW, Paterson ME, Wade-Evans T, Studd JW. Prevention and treatment of endometrial disease in climacteric women receiving oestrogen therapy. Lancet 2:455–457, 1979.
42. Whitehead MI, McQueen J, Beard RJ, Minardi J, Campbell S. The effects of cyclical oestrogen therapy and sequential oestrogen/progestogen therapy on the endometrium of postmenopausal women. Acta Obstet Gynecol Scand Suppl 65:91–101, 1977.
43. Beresford SAA, Weiss NS, Voigt LF, McKnight B. Risk of endometrial cancer in relation to use of oestrogen combined with cyclic progestagen in postmenopausal women. Lancet 349:458–461, 1997.
44. Pike MC, Peters RK, Cozen W, Probst-Hensch NM, Felix JC, Wan PC, Mack TM. Estrogen–progestin replacement therapy and endometrial cancer. J Natl Cancer Inst 89:1110–1116, 1997.
45. Voigt LF, Weiss NS, Chu J, Daling JR, McKnight B, Van Belle G. Progestagen supplementation of exogenous oestrogens and risk of endometrial cancer. Lancet 338:274–277, 1991.
46. Weiderpass E, Baron JA, Adami H-O, Magnusson C, Lindgren A, Bergström R, Correia N, Persson I. Low-potency oestrogen and risk of endometrial cancer: a case-control study. Lancet 353:1824–1828. 1999.
47. Hill DA, Weiss NS, Beresford AA, Voigt LF, Daling JR, Stanford JL, Self S. Continuous combined hormone replacement therapy and risk of endometrial cancer. Am J Obstet Gynecol 183:1456–1461, 2000.
48. Weiderpass E, Adami H-O, Baron JA, Magnusson C, Bergström R, Lindgren A, Correia N, Persson I. Risk of endometrial cancer following estrogen replacement with and without progestins. J Natl Cancer Inst 91:1131–1137, 1999.
49. Hulka BS, Chambless LE, Kaufman DG, Fowler WC, Greenberg BG. Protection against endometrial carcinoma by combination-product oral contraceptives. JAMA 247:475–477, 1982.
50. Rosenblatt KA, Thomas DB, The WHO Collaborative Study of Neoplasia and Steroid Contraceptives. Hormonal content of combined oral contraceptives in relation to the reduced risk of endometrial cancer. Int J Cancer 49:870–874, 1991.
51. Voigt LF, Deng Q, Weiss NS. Recency, duration, and progestin content of oral contraceptives in relation to the incidence of endometrial cancer (Washington, USA). Cancer Causes Control 5:227–233, 1994.
52. Gorodeski GI, Beery R, Lunenfeld B, Geier A. Tamoxifen increases plasma estrogen-binding equivalents and has estradiol agonistic effect on histologically normal premenopausal and postmenopausal endometrium. Fertil Steril 57:320–327, 1992.
53. Satyaswaroop PG, Zaino RJ, Mortel R. Estrogen-like effects of tamoxifen on human endometrial carcinoma transplanted into nude mice. Cancer Res 44:4006–4010, 1984.
54. Fornander T, Rutqvist LE, Cedermark B, Glas U, Mattsson A, Silfversward C, Skoog L, Somell A, Theve T, Wilking N, Askergren J, Hjalmar M-L. Adjuvant tamoxifen in early breast cancer:

occurrence of new primary cancers. Lancet 1:117–120, 1989.
55. Andersson M, Storm HH, Mouridsen HT. Incidence of new primary cancers after adjuvant tamoxifen therapy and radiotherapy for early breast cancer. J Natl Cancer Inst 83:1013–1017, 1991.
56. Fisher B, Costantino JP, Wickerham DL, Redmond C, Redmond C, Kavanah M, Cronin WM, Vogel V, Dimitrov N, Atkins J, Daly M, Wieand S, Tan-Chiu E, Ford L, Wolmark N. Tamoxifen for prevention of breast cancer: report of the National Surgical Adjuvant Breast and Bowel Project P-1 Study. J Natl Cancer Inst 90:1371–1388, 1998.
57. Fisher B, Costantino JP, Redmond CK, Fisher ER, Wickerham DL, Cronin WM. Endometrial cancer in tamoxifen-treated breast cancer patients: findings from the National Surgical Adjuvant Breast and Bowel Project (NSABP) B-14. J Natl Cancer Inst 86:27–37, 1994.
58. van Leeuwen FE, Benraadt J, Coebergh JWW, Kiemeney LALM, Gimbrere CHF, Otter R, Schouten LJ, Damhui RAM, Bontenbal M, Diepenhorst FW, van den Belt-Dusebout AW, van Tinteren H. Risk of endometrial cancer after tamoxifen treatment of breast cancer. Lancet 343:448–452, 1994.
59. Sasco AJ, Chaplin G, Amoros E, Saez S. Endometrial cancer following breast cancer: effect of tamoxifen and castration by radiotherapy. Epidemiology 7:9–13, 1996.
60. Jordan VC. Tamoxifen: toxicities and drug resistance during the treatment and prevention of breast cancer. Annu Rev Pharmacol Toxicol 35:195–211, 1995.
61. Boss SM, Huster WJ, Neild JA, Glant MD, Eisenhut CC, Draper MW. Effects of raloxifene hydrochloride on the endometrium of postmenopausal women. Am J Obstet Gynecol 177:1458–1464, 1997.
62. Fugere P, Scheele WH, Shah A, Strack TR, Glant MD, Jones GS. Uterine effects of raloxifene in comparison with continuous-combined hormone replacement therapy in postmenopausal women. Am J Obstet Gynecol 182:568–574, 2000.
63. Goldstein SR, Scheele WH, Rajagopalan SK, Wilkie JL, Walsh BW, Parsons AK. A 12-month comparative study of raloxifene, estrogen, and placebo on the postmenopausal endometrium. Obstet Gynecol 95:95–103, 2000.
64. Cummings SR, Eckert S, Krueger KA, Grady D, Powles TJ, Cauley JA, Norton L, Nickelsen T, Bjarnason NH, Morrow M, Lippman ME, Black D, Glusman JE, Costa A, Jordan VC. The effect of raloxifene on risk of breast cancer in postmenopausal women. Results from the MORE randomized trial. JAMA 281:2189–2197, 1999.
65. Kelsey JL, LiVolsi VA, Holford TR, Fischer DB, Mostow ED, Schwartz PE, O'Connor T, White C. A case-control study of cancer of the endometrium. Am J Epidemiol 116:333–342, 1982.
66. Parazzini F, La Vecchia C, Moroni S, Chatenoud L, Ricci E. Family history and the risk of endometrial cancer. Int J Cancer 59:460–462, 1994.
67. Olson JE, Sellers TA, Anderson KE, Folsom AR. Does a family history of cancer increase the risk for postmenopausal endometrial carcinoma? A prospective cohort study and a nested case-control family study of older women. Cancer 85:2444–2449, 1999.
68. Parslov M, Lidegaard O, Klintorp S, Pedersen B, Jonsson L, Eriksen PS, Ottesen B. Risk factors among young women with endometrial cancer: a Danish case-control study. Am J Obstet Gynecol 182:23–239, 2000.
69. Gruber SB, Thompson WD, Cancer and Steroid Hormone Study Group. A population-based study of endometrial cancer and familial risk in younger women. Cancer Epidemiol Biomarkers Prev 5:411–417, 1996.
70. Hemminki K, Vaittinen P, Dong C. Endometrial cancer in the family-cancer database. Cancer Epidemiol Biomarkers Prev 8:1005–1010, 1999.
71. Marra G, Boland CR. Hereditary nonpolyposis colorectal cancer: the syndrome, the genes, and historical perspectives. J Natl Cancer Inst 87:1114–1125, 1995.
72. Aarnio M, Sankila R, Pukkala E, Salovaara R, Aaltonen LA, de la Chapelle A, Peltomaki P, Mecklin JP, Jarvinen HJ. Cancer risk in mutation carriers of DNA-mismatch-repair genes. Int J Cancer 81:214–218, 1999.
73. Boyd J. Estrogen as a carcinogen: the genetics and molecular biology of human endometrial carcinoma. Prog Clin Biol Res 394:151–173, 1996.
74. Sandles LG. Familial endometrial adenocarcinoma. Clin Obstet Gynecol 41:167–171, 1998.
75. Grodin JM, Siiteri PK, MacDonald PC. Source of estrogen production in postmenopausal women. J Clin Endocrinol Metab 36:207–214, 1973.
76. Nimrod A, Ryan KJ. Aromatization of androgens by human abdominal and breast fat tissue. J Clin Endocrinol Metab 40:367–372, 1975.
77. Dunning AM, Healey CS, Pharoah PD, Teare MD, Ponder BA, Easton DF. A systematic review of genetic polymorphisms and breast cancer risk. Cancer Epidemiol Biomarkers Prev 8:843–854, 1999.
78. Stone D, Hechter O. Studies on ACTH action in perfused bovine adrenals: aspects of progesterone as an intermediary in corticosteroidogenesis. Arch Biochem Biophys 54:121, 1955.
79. Halkerston IDK, Eichhorn J, Hechter O. A requirement for reduced triphosphopyridine nucleotide for cholesterol side-chain cleavage by

mitochondrial fractions of bovine adrenal cortex. J Biol Chem 236:374, 1961.
80. Sparkes RS, Klisak I, Miller WL. Regional mapping of genes encoding human steroidogenic enzymes: P450scc to 15q23–q24; adrenodoxin to 11q22; adrenodoxin reductase to 17q24–q25; and P450c17 to 10q24–q25. DNA Cell Biol 10:359–365, 1991.
81. Gharani N, Waterworth DM, Batty S, White D, Gilling-Smith C, Conway GS, McCarthy M, Franks S, Williamson R. Association of the steroid synthesis gene *CYP11a* with polycystic ovary syndrome and hyperandrogenism. Hum Mol Genet 6:397–402, 1997.
82. Diamanti-Kandarakis E, Bartzis MI, Bergiele AT, Tsianateli TC, Kouli CR. Microsatellite polymorphism (tttta)(n) at −528 base pairs of gene *CYP11alpha* influences hyperandrogenemia in patients with polycystic ovary syndrome. Fertil Steril 73:735–741, 2000.
83. San Millan JL, Sancho J, Calvo RM, Escobar-Morreale HF. Role of the pentanucleotide (tttta)(n) polymorphism in the promoter of the *CYP11a* gene in the pathogenesis of hirsutism. Fertil Steril 75:797–802, 2001.
84. Verreault H, Dufort I, Simard J, Labrie F, Luu-The V. Dinucleotide repeat polymorphisms in the *HSD3B2* gene. Hum Mol Genet 3:384, 1994.
85. Devgan SA, Henderson BE, Yu MC, Shi CY, Pike MC, Ross RK, Reichardt JK. Genetic variation of 3beta-hydroxysteroid dehydrogenase type II in three racial/ethnic groups: implications for prostate cancer risk. Prostate 33:9–12, 1997.
86. Feigelson HS, Coetzee GA, Kolonel LN, Ross RK, Henderson BE. A polymorphism in the *CYP17* gene increases the risk of breast cancer. Cancer Res 57:1063–1065, 1997.
87. Nedelcheva Kristensen V, Haraldsen EK, Anderson KB, Lonning PE, Erikstein B, Karesen R, Gabrielsen OS, Borresen-Dale AL. *CYP17* and breast cancer risk: the polymorphism in the 5′ flanking area of the gene does not influence binding to Sp-1. Cancer Res 59:2825–2828, 1999.
88. Feigelson HS, Shames LS, Pike MC, Coetzee GA, Stanczyk FZ, Henderson BE. Cytochrome P450c17alpha gene (*CYP17*) polymorphism is associated with serum estrogen and progesterone concentrations. Cancer Res 58:585–587, 1998.
89. Haiman CA, Hankinson SE, Spiegelman D, Colditz GA, Willett WC, Speizer FE, Kelsey KT, Hunter DJ. The relationship between a polymorphism in *CYP17* with plasma hormone levels and breast cancer. Cancer Res 59:1015–1020, 1999.
90. Haiman CA, Hankinson SE, Colditz GA, Hunter DJ, De Vivo I. A polymorphism in *CYP17* and endometrial cancer risk. Cancer Res 61:3955–3960, 2001.
91. Feigelson HS, McKean-Cowdin R, Pike MC, Coetzee GA, Kolonel LN, Nomura AM, Le Marchand L, Henderson BE. Cytochrome P450c17alpha gene (*CYP17*) polymorphism predicts use of hormone replacement therapy. Cancer Res 59:3908–3910, 1999.
92. Dunning AM, Healey CS, Pharoah PD, Foster NA, Lipscombe JM, Redman KL, Easton DF, Day NE, Ponder BA. No association between a polymorphism in the steroid metabolism gene *CYP17* and risk of breast cancer. Br J Cancer 77:2045–2047, 1998.
93. Helzlsouer KJ, Huang HY, Strickland PT, Hoffman S, Alberg AJ, Comstock GW, Bell DA. Association between *CYP17* polymorphisms and the development of breast cancer. Cancer Epidemiol Biomarkers Prev 7:945–949, 1998.
94. Weston A, Pan CF, Bleiweiss IJ, Ksieski HB, Roy N, Maloney N, Wolff MS. *CYP17* genotype and breast cancer risk. Cancer Epidemiol Biomarkers Prev 7:941–944, 1998.
95. Bergman-Jungestrom M, Gentile M, Lundin AC, Wingren S. Association between *CYP17* gene polymorphism and risk of breast cancer in young women. Int J Cancer 84:350–353, 1999.
96. Huang CS. Breast cancer risk associated with genotype polymorphism of the estrogen-metabolizing genes *CYP17*, *CYP1A1*, and *COMT*: a multigenic study on cancer susceptibility. Cancer Res 59:4870–4875, 1999.
97. Mitrunen K, Jourenkova N, Kataja V, Eskelinen M, Kosma V-M, Benhamou S, Vainio H, Uusitupa M, Hirvonen A. Steroid metabolism gene *CYP17* polymorphism and the development of breast cancer. Cancer Epidemiol Biomarkers Prev 9:1343–1348, 2000.
98. Miyoshi Y, Iwao K, Ikeda N, Egawa C, Noguchi S. Genetic polymorphism in *CYP17* and breast cancer risk in Japanese women. Eur J Cancer 36:2375–2379, 2000.
99. Hamajima N, Iwata H, Obata Y, Matsuo K, Mizutani M, Iwase T, Miura S, Okuma K, Ohashi K, Tajima K. No association of the 5′ promoter region polymorphism of *CYP17* with breast cancer risk in Japan. Jpn J Cancer Res 91:880–885, 2000.
100. Spurdle AB, Hopper JL, Dite GS, Chen X, Cui J, Mccredie MRE, Giles GG, Southey MC, Venter DJ, Easton DF, Chenevix-Trench G. *CYP17* promoter polymorphism and breast cancer in Australian women under age forty years. J Natl Cancer Inst 92:1674, 2000.
101. Feigelson HS, McKean-Cowdin R, Coetzee GA, Stram DO, Kolonel LN, Henderson BE. Building a multigenic model of breast cancer susceptibility: *CYP17* and *HSD17B1* are two important candidates. Cancer Res 61:785–789, 2001.
102. Olson SH, Elahi A, Roy P, Berwick M. *CYP17* and *COMT* genotypes in endometrial cancer [abstract]. Am J Epidemiol 149:S7, 1999. Abstract nr 26.

103. McKean-Cowdin R, Feigelson HS, Pike MC, Coetzee GA, Kolonel LN, Henderson BE. Risk of endometrial cancer and estrogen replacement therapy history by CYP17 genotype. Cancer Res 61:848–849, 2001.
104. Simpson ER, Zhao Y, Agarwal VR, Michael MD, Bulun SE, Hinshelwood MM, Graham-Lorence S, Sun T, Fisher CR, Qin K, Mendelson CR. Aromatase expression in health and disease. Recent Prog Horm Res 52:185–213, 1997.
105. Polymeropoulos MH, Xiao H, Rath DS, Merril CR. Tetranucleotide repeat polymorphism at the human aromatase cytochrome P-450 gene (CYP19). Nucleic Acids Res 19:195, 1991.
106. Siegelmann-Danieli N, Buetow KH. Constitutional genetic variation at the human aromatase gene (CYP19) and breast cancer risk. Br J Cancer 79:456–463, 1999.
107. Miyoshi Y, Iwao K, Ikeda N, Egawa C, Noguchi S. Breast cancer risk associated with polymorphisms in CYP19 in Japanese women. Int J Cancer 89:325–328, 2000.
108. Watanabe J, Harada N, Suemasu K, Higashi Y, Gotoh O, Kawajiri K. Arginine-cysteine polymorphism at codon 264 of the human CYP19 gene does not affect aromatase activity. Pharmacogenetics 7:419–424, 1997.
109. Probst-Hensch NM, Ingles SA, Diep AT, Haile RW, Stanczyk FZ, Kolonel LN, Henderson BE. Aromatase and breast cancer susceptibility. Endocr Relat Cancer 6:165–173, 1999.
110. Haiman CA, Hankinson SE, Spiegelman D, De Vivo I, Colditz GA, Willett WC, Speizer FE, Hunter DJ. A tetranucleotide repeat polymorphism in CYP19 and breast cancer risk. Int J Cancer 87:204–210, 2000.
111. Kristensen VN, Andersen TI, Lindblom A, Erikstein B, Magnus P, Borresen-Dale AL. A rare CYP19 (aromatase) variant may increase the risk of breast cancer. Pharmacogenetics 8:43–48, 1998.
112. Berends, M. J. W., Kleibeuker, J. H., de Vries, E. G. E., Mourits, J. J. E., Hollema, H., Pras, E., and van der Zee, A. G. J. The importance of family history in young patients with endometrial cancer. Eur J Obstet Gynecol Reprod Biol 82:139–141, 1999.
113. Healey CS, Dunning AM, Durocher F, Teare D, Pharoah PD, Luben RN, Easton DF, Ponder BA. Polymorphisms in the human aromatase cytochrome P450 gene (CYP19) and breast cancer risk. Carcinogenesis 21:189–193, 2000.
114. Berstein LM, Imyanitov EN, Suspitsin EN, Grigoriev MY, Sokolov EP, Togo A, Hanson KP, Poroshina TE, Vasiljev DA, Kovalevskij AY, Gamajunova VB. CYP19 gene polymorphism in endometrial cancer patients. J Cancer Res Clin Oncol 127:135–138, 2001.
115. Kristensen VN, Harada N, Yoshimura N, Haraldsen EK, Lonning PE, Erikstein B, Karesen R. Genetic variants of CYP19 (aromatase) and breast cancer risk. Oncogene 19:1329–1333, 2000.
116. Dumont M, Luu-The V, de Launoit Y, Labrie F. Expression of human 17beta-hydroxysteroid dehydrogenase in mammalian cells. J Steroid Biochem Mol Biol 41:605–608, 1992.
117. Miettinen MM, Mustonen MV, Poutanen MH, Isomaa VV, Vihko RK. Human 17beta-hydroxysteroid dehydrogenase type 1 and type 2 isoenzymes have opposite activities in cultured cells and characteristic cell- and tissue-specific expression. Biochem J 314:839–845, 1996.
118. Poutanen M, Miettinen M, Vihko R. Differential estrogen substrate specificities for transiently expressed human placental 17beta-hydroxysteroid dehydrogenase and an endogenous enzyme expressed in cultured COS-m6 cells. Endocrinology 133:2639–2644, 1993.
119. Wu L, Einstein M, Geissler WM, Chan HK, Elliston KO, Andersson S. Expression cloning and characterization of human 17beta-hydroxysteroid dehydrogenase type 2, a microsomal enzyme possessing 20alpha-hydroxysteroid dehydrogenase activity. J Biol Chem 268:12964–12969, 1993.
120. Adamski J, Normand T, Leenders F, Monte D, Begue A, Stehelin D, Jungblut PW, de Launoit Y. Molecular cloning of a novel widely expressed human 80 kDa 17beta-hydroxysteroid dehydrogenase IV. Biochem J 311:437–443, 1995.
121. Tseng L, Gurpide E. Induction of human endometrial estradiol dehydrogenase by progestins. Endocrinology 97:825–833, 1975.
122. Satyaswaroop PG, Wartell DJ, Mortel R. Distribution of progesterone receptor, estradiol dehydrogenase, and 20alpha-dihydroprogesterone dehydrogenase activities in human endometrial glands and stroma: progestin induction of steroid dehydrogenase activities in vitro is restricted to the glandular epithelium. Endocrinology 111:743–749, 1982.
123. Casey ML, MacDonald PC, Andersson S. 17beta-Hydroxysteroid dehydrogenase type 2: chromosomal assignment and progestin regulation of gene expression in human endometrium. J Clin Invest 94:2135–2141, 1994.
124. Luu-The V, Labrie C, Simard J, Lachance Y, Zhao HF, Couet J, Leblanc G, Labrie F. Structure of two in tandem human 17beta-hydroxysteroid dehydrogenase genes. Mol Endocrinol 4:268–275, 1990.
125. Mäentausta O., Sormunen R, Isomaa V, Lehto VP, Jouppila P, Vihko R. Immunohistochemical localization of 17beta-hydroxysteroid dehydrogenase in the human endometrium during the menstrual cycle. Lab Invest 65:582–587, 1991.
126. Labrie F, Luu-The V, Lin SX, Labrie C, Simard J, Breton R, Belanger A. The key role of 17 beta-hydroxysteroid dehydrogenases in sex steroid biology. Steroids 62:148–158, 1997.
127. Husen B, Psonka N, Jacob-Meisel M, Keil C,

Rune GM. Differential expression of 17beta-hydroxysteroid dehydrogenases types 2 and 4 in human endometrial epithelial cell lines. J Mol Endocrinol 24:135–144, 2000.
128. Martel C, Rheaume E, Takahashi M, Trudel C, Couet J, Luu-The V, Simard J, Labrie F. Distribution of 17beta-hydroxysteroid dehydrogenase gene expression and activity in rat and human tissues. J Steroid Biochem Mol Biol 41:597–603, 1992.
129. Corbould AM, Judd SJ, Rodgers RJ. Expression of types 1, 2, and 3 17beta-hydroxysteroid dehydrogenase in subcutaneous abdominal and intra-abdominal adipose tissue of women. J Clin Endocrinol Metab 83:187–194, 1998.
130. Peltoketo H, Piao Y, Mannermaa A, Ponder BAJ, Isomaa V, Poutanen M, Winqvist R, Vihko R. A point mutation in the putative TATA box, detected in nondiseased individuals and patients with hereditary breast cancer, decreases promoter activity of the 17β-hydroxysteroid dehydrogenase type 1 gene 2 (*EDH17B2*) in vitro. Genomics 23:250–252, 1994.
131. Normand T, Narod S, Labrie F, Simard J. Detection of polymorphisms in the estradiol 17beta-hydroxysteroid dehydrogenase II gene at the *EDH17B2* locus on 17q11–q21. Hum Mol Genet 2:479–483, 1993.
132. Mannermaa A, Peltoketo H, Winqvist R, Ponder BA, Kiviniemi H, Easton DF, Poutanen M, Isomaa V, Vihko R. Human familial and sporadic breast cancer: analysis of the coding regions of the 17beta-hydroxysteroid dehydrogenase 2 gene (*EDH17B2*) using a single-strand conformation polymorphism assay. Hum Genet 93:319–324, 1994.
133. Puranen TJ, Poutanen MH, Peltoketo HE, Vihko PT, Vihko RK. Site-directed mutagenesis of the putative active site of human 17beta-hydroxysteroid dehydrogenase type 1. Biochem J 304:289–293, 1994.
134. Yen SSC, Jaffe RB. Reproductive Endocrinology: Physiology, Pathophysiology, and Clinical Management, 3rd ed. Philadelphia: Saunders, 1991.
135. Zhu BT, Conney AH. Functional role of estrogen metabolism in target cells: review and perspectives. Carcinogenesis 19:1–27, 1998.
136. Ball P, Reu G, Schwab J, Knuppen R. Radioimmunoassay of 2-hydroxyesterone and 2-methoxyestrone in human urine. Steroids 33:563–576, 1979.
137. Ball P, Knuppen R. Catecholoestrogens (2- and 4-hydroxyoestrogens): chemistry, biogenesis, metabolism, occurrence and physiological significance. Acta Endocrinol (Copenh) 232(Suppl):1–127, 1980.
138. Lipsett M, Merriam G, Kono S, Brandon D, Pheiffer D, Loriaux D. Metabolic clearance of catechol estrogens. In: Merriam GR, Lipsett MB (eds). Catechol Estrogens. New York: Raven Press, 1983, pp 105–114.
139. Cavalieri EL, Stack DE, Devanesan PD, Todorovic R, Dwivedy I, Higginbotham S, Johansson SL, Patil KD, Gross ML, Gooden JK, Ramanathan R, Cerny RL, Rogan EG. Molecular origin of cancer: catechol estrogen-3,4-quinones as endogenous tumor initiators. Proc Natl Acad Sci USA 94:10937–10942, 1997.
140. Wang MY, Liehr JG. Induction by estrogens of lipid peroxidation and lipid peroxide-derived malonaldehyde-DNA adducts in male Syrian hamsters: role of lipid peroxidation in estrogen-induced kidney carcinogenesis. Carcinogenesis 16:1941–1945, 1995.
141. Yager JD, Liehr JG. Molecular mechanisms of estrogen carcinogenesis. Annu Rev Pharmacol Toxicol 36:203–232, 1996.
142. Yager JD. Endogenous estrogens as carcinogens through metabolic activation. J Natl Cancer Inst Monogr 27:67–73, 2000.
143. Cavalieri E, Frenkel K, Liehr JG, Rogan E, Roy D. Estrogens as endogenous genotoxic agents—DNA adducts and mutations. J Natl Cancer Inst Monogr 75–93, 2000.
144. Butterworth M, Lau SS, Monks TJ. Formation of catechol estrogen glutathione conjugates and gamma-glutamyl transpeptidase-dependent nephrotoxicity of 17beta-estradiol in the golden Syrian hamster. Carcinogenesis 18:561–567, 1997.
145. Lind C, Hochstein P, Ernster L. DT-diaphorase as a quinone reductase: a cellular control device against semiquinone and superoxide radical formation. Arch Biochem Biophys 216:178–185, 1982.
146. Schulz WA, Krummeck A, Rosinger I, Eickelmann P, Neuhaus C, Ebert T, Schmitz-Drager BJ, Sies H. Increased frequency of a null-allele for NAD(P)H:quinone oxidoreductase in patients with urological malignancies. Pharmacogenetics 7:235–239, 1997.
147. Kirkman H. Estrogen-induced tumors of the kidney in Syrian hamster. III. Growth characteristics in the Syrian hamster. NCI Monogr 1:1–57, 1959.
148. Li JJ, Li SA. Estrogen carcinogenesis in Syrian hamster tissues: role of metabolism. Fed Proc 46:1858–1863, 1987.
149. Bradlow HL, Hershcopf R, Martucci C, Fishman J. 16Alpha-hydroxylation of estradiol: a possible risk marker for breast cancer. Ann NY Acad Sci 464:138–151, 1986.
150. Fishman J, Martucci C. Biological properties of 16alpha-hydroxyestrone: implications in estrogen physiology and pathophysiology. J Clin Endocrinol Metab 51:611–615, 1980.
151. Fishman J, Schneider J, Hershcope RJ, Bradlow HL. Increased estrogen-16alpha-hydroxylase activity in women with breast and endometrial cancer. J Steroid Biochem 20:1077–1081, 1984.

152. Meilahn EN, De Stavola B, Allen DS, Fentiman I, Bradlow HL, Sepkovic DW, Kuller LH. Do urinary oestrogen metabolites predict breast cancer? Guernsey III cohort follow-up. Br J Cancer 78:1250–1255, 1998.
153. Muti P, Bradlow HL, Micheli A, Krogh V, Freudenheim JL, Schunemann HJ, Stanulla M, Yang J, Sepkovic DW, Trevisan M, Berrino F: Estrogen metabolism and risk of breast cancer: a prospective study of the 2:16alpha-hydroxyestrone ratio in premenopausal and postmenopausal women. Epidemiology 11:635–640, 2000.
154. Adlercreutz H, Gorbach SL, Goldin BR, Woods MN, Dwyer JT, Hamalainen E. Estrogen metabolism and excretion in Oriental and Caucasian women [published erratum appears in J Natl Cancer Inst 87:147, 1995]. J Natl Cancer Inst 86:1076–1082, 1994.
155. Liehr JG, Fang WF, Sirbasku DA, Ari-Ulubelen A. Carcinogenicity of catechol estrogens in Syrian hamsters. J Steroid Biochem 24:353–356, 1986.
156. van Aswegen CH, Purdy RH, Wittliff JL. Binding of 2-hydroxyestradiol and 4-hydroxyestradiol to estrogen receptor human breast cancers. J Steroid Biochem 32:485–492, 1989.
157. MacLusky NJ, Barnea ER, Clark CR, Naftolin F. Catechol estrogens and estrogen receptors. In: Merriam GR, Lipsett MB (eds). Catechol Estrogens. New York: Raven Press, 1983, pp 151–165.
158. Martucci C, Fishman J. Uterine estrogen receptor binding of catecholestrogens and of estetrol (1,3,5(10)-estratriene-3,15alpha,16alpha,17beta-tetrol). Steroids 27:325–333, 1976.
159. Roy D, Weisz J, Liehr JG. The O-methylation of 4-hydroxyestradiol is inhibited by 2-hydroxyestradiol: implications for estrogen-induced carcinogenesis. Carcinogenesis 11:459–462, 1990.
160. Liehr JG, Ricci MJ, Jefcoate CR, Hannigan EV, Hokanson JA, Zhu BT. 4-Hydroxylation of estradiol by human uterine myometrium and myoma microsomes: implications for the mechanism of uterine tumorigenesis. Proc Natl Acad Sci USA 92:9220–9224, 1995.
161. Weisz J, Clawson GA, Creveling CR. Biogenesis and inactivation of catecholestrogens. Adv Pharmacol 42:828–833, 1998.
162. Weisz J, Bui QD, Roy D, Liehr JG. Elevated 4-hydroxylation of estradiol by hamster kidney microsomes: a potential pathway of metabolic activation of estrogens. Endocrinology 131:655–661, 1992.
163. Stack DE, Byun J, Gross ML, Rogan EG, Cavalieri EL. Molecular characteristics of catechol estrogen quinones in reactions with deoxyribonucleosides. Chem Res Toxicol 9:851–859, 1996.
164. Spink DC, Eugster HP, Lincoln DW, Schuetz JD, Schuetz EG, Johnson JA, Kaminsky LS, Gierthy JF. 17Beta-estradiol hydroxylation catalyzed by human cytochrome P450 1A1: a comparison of the activities induced by 2,3,7,8-tetrachlorodibenzo-p-dioxin in MCF-7 cells with those from heterologous expression of the cDNA. Arch Biochem Biophys 293:342–348, 1992.
165. Vadlamuri SV, Glover DD, Turner T, Sarkar MA. Regiospecific expression of cytochrome P4501A1 and 1B1 in human uterine tissue. Cancer Lett 122:143–150, 1998.
166. Parkinson A. Biotransformation of xenobiotics. In: Klaassen MO (ed). Casarett and Doull's Toxicology. The Basic Science of Poisons. New York: McGraw-Hill, 2001, pp 192–196.
167. Whitlock JP. Induction of cytochrome P4501A1. Annu Rev Pharmacol Toxicol 39:103–125, 1999.
168. Kawajiri K, Nakachi K, Imai K, Yoshii A, Shinoda N, Watanabe J. Identification of genetically high risk individuals to lung cancer by DNA polymorphisms of the cytochrome P450IA1 gene. FEBS Lett 263:131–133, 1990.
169. Hayashi S, Watanabe J, Nakachi K, Kawajiri K. Genetic linkage of lung cancer–associated MspI polymorphisms with amino acid replacement in the heme binding region of the human cytochrome P450IA1 gene. J Biochem (Tokyo) 110:407–411, 1991.
170. Crofts F, Taioli E, Trachman J, Cosma GN, Currie D, Toniolo P, Garte SJ. Functional significance of different human *CYP1A1* genotypes. Carcinogenesis 15:2961–2963, 1994.
171. Cascorbi I, Brockmoller J, Roots I. A C4887A polymorphism in exon 7 of human *CYP1A1*: population frequency, mutation linkages, and impact on lung cancer susceptibility. Cancer Res 56:4965–4969, 1996.
172. Petersen DD, McKinney CE, Ikeya K, Smith HH, Bale AE, McBride OW, Nebert DW. Human *CYP1A1* gene: cosegregation of the enzyme inducibility phenotype and an RFLP. Am J Hum Genet 48:720–725, 1991.
173. Landi MT, Bertazzi PA, Shields PG, Clark G, Lucier GW, Garte SJ, Cosma G, Caporaso NE. Association between *CYP1A1* genotype, mRNA expression and enzymatic activity in humans. Pharmacogenetics 4:242–246, 1994.
174. Cosma G, Crofts F, Taioli E, Toniolo P, Garte S. Relationship between genotype and function of the human *CYP1A1* gene. J Toxicol Environ Health 40:309–316, 1993.
175. Kiyohara C, Hirohata T, Inutsuka S. The relationship between aryl hydrocarbon hydroxylase and polymorphisms of the *CYP1A1* gene. Jpn J Cancer Res 87:18–24, 1996.
176. Kelly EJ, Adman ET, Eaton DL. Functional significance of the human *CYP1A1* M4 polymorphism: relevance to endometrial cancer risk [abstract]. In: Program, 9th North American

International Society for the Study of Xenobiotics Meeting, Oct 24–28, 1999, Nashville, TN. Abstract nr 265.

177. Kawajiri K, Nakachi K, Imai K, Watanabe J, Hayashi S. The *CYP1A1* gene and cancer susceptibility. Crit Rev Oncol Hematol 14:77–87, 1993.

178. Rebbeck TR, Rosvold EA, Duggan DJ, Zhang J, Buetow KH. Genetics of *CYP1A1*: coamplification of specific alleles by polymerase chain reaction and association with breast cancer. Cancer Epidemiol Biomarkers Prev 3:511–514, 1994.

179. Ambrosone CB, Freudenheim JL, Graham S, Marshall JR, Vena JE, Brasure JR, Laughlin R, Nemoto T, Michalek AM, Harrington A. Cytochrome P4501A1 and glutathione S-transferase (M1) genetic polymorphisms and postmenopausal breast cancer risk. Cancer Res 55:3483–3485, 1995.

180. Ishibe N, Hankinson SE, Colditz GA, Spiegelman D, Willett WC, Speizer FE, Kelsey KT, Hunter DJ. Cigarette smoking, cytochrome P450 1A1 polymorphisms, and breast cancer risk in the Nurses' Health Study. Cancer Res 58:667–671, 1998.

181. Dialyna IA, Arvanitis DA, Spandidos DA. Genetic polymorphisms and transcriptional pattern analysis of *CYP1A1*, *AhR*, *GSTM1*, *GSTP1* and *GSTT1* genes in breast cancer. Int J Mol Med 8:79–87, 2001.

182. Taioli E, Trachman J, Chen X, Toniolo P, Garte SJ. A *CYP1A1* restriction fragment length polymorphism is associated with breast cancer in African-American women. Cancer Res 55:3757–3758, 1995.

183. Bailey LR, Roodi N, Verrier CS, Yee CJ, Dupont WD, Parl FF. Breast cancer and *CYP1A1*, *GSTM1*, and *GSTT1* polymorphisms: evidence of a lack of association in Caucasians and African Americans. Cancer Res 58:65–70, 1998.

184. Huang CS, Shen CY, Chang KJ, Hsu SM, Chern HD. Cytochrome P4501A1 polymorphism as a susceptibility factor for breast cancer in postmenopausal Chinese women in Taiwan. Br J Cancer 80:1838–1843, 1999.

185. Esteller M, Garcia A, Martinez-Palones JM, Xercavins J, Reventos J. Susceptibility to endometrial cancer: influence of allelism at p53, glutathione S-transferase (*GSTM1* and *GSTT1*) and cytochrome P-450 (*CYP1A1*) loci. Br J Cancer 75:1385–1388, 1997.

186. Esteller M, Garcia A, Martinez-Palones JM, Xercavins J, Reventos J. Germ line polymorphisms in cytochrome-P450 1A1 (C4887 *CYP1A1*) and methylenetetrahydrofolate reductase (*MTHFR*) genes and endometrial cancer susceptibility. Carcinogenesis 18:2307–2311, 1997.

187. Olson SH, Elahi A, Tang G, Roy P, Song Y, Ambrosone CB, Kadlubar FF, Thompson PA, Stone A, Berwick M. Polymorphisms in *CYP1A1* and *CYP1B1* in endometrial cancer [abstract]. Proc Am Assoc Cancer Res 41:23, 2000. Abstract nr 149.

188. Landi MT, Sinha R, Lang NP, Kadlubar FF. Human cytochrome P4501A2. In: Vineis P, Malats N, Lang M, d'Errico A, Caparaso N, Cuzick J, and Boffetta P (eds). Metabolic Polymorphisms and Susceptibility to Cancer. IARC Scientific Publication 148. Lyon: IARC, 148: 1999, pp 173–195.

189. Aoyama T, Korzekwa K, Nagata K, Gillette J, Gelboin HV, Gonzalez FJ. Estradiol metabolism by complementary deoxyribonucleic acid-expressed human cytochrome P450s. Endocrinology 126:3101–3106, 1990.

190. Shou M, Korzekwa KR, Brooks EN, Krausz KW, Gonzalez FJ, Gelboin HV. Role of human hepatic cytochrome P450 1A2 and 3A4 in the metabolic activation of estrone. Carcinogenesis 18:207–214, 1997.

191. Yamazaki H, Shaw PM, Guengerich FP, Shimada T. Roles of cytochromes P450 1A2 and 3A4 in the oxidation of estradiol and estrone in human liver microsomes. Chem Res Toxicol 11:659–665, 1998.

192. MacLeod S, Sinha R, Kadlubar FF, Lang NP. Polymorphisms of *CYP1A1* and *GSTM1* influence the in vivo function of *CYP1A2*. Mutat Res 376:135–142, 1997.

193. Nakajima M, Yokoi T, Mizutani M, Kinoshita M, Funayama M, Kamataki T. Genetic polymorphism in the 5'-flanking region of human *CYP1A2* gene: effect on the *CYP1A2* inducibility in humans. J Biochem (Tokyo) 125:803–808, 1999.

194. MacLeod SL, Tang YM, Yokoi T. The role of recently discovered genetic polymorphism in the regulation of the human *CYP1A2* gene. Proc Am Assoc Cancer Res 39:396, 1998.

195. Sachse C, Brockmoller J, Bauer S, Roots I. Functional significance of a C $\rightarrow$ A polymorphism in intron 1 of the cytochrome P450 *CYP1A2* gene tested with caffeine. Br J Clin Pharmacol 47:445–449, 1999.

196. Chida M, Yokoi T, Fukui T, Kinoshita M, Yokota J, Kamataki T. Detection of three genetic polymorphisms in the 5'-flanking region and intron 1 of human *CYP1A2* in the Japanese population. Jpn J Cancer Res 90:899–902, 1999.

197. Hayes CL, Spink DC, Spink BC, Cao JQ, Walker NJ, Sutter TR. 17Beta-estradiol hydroxylation catalyzed by human cytochrome P450 1B1. Proc Natl Acad Sci USA 93:9776–9781, 1996.

198. Hakkola J, Pasanen M, Pelkonen O, Hukkanen J, Evisalmi S, Anttila S, Rane A, Mantyla M, Purkunen R, Saarikoski S, Tooming M, Raunio H. Expression of *CYP1B1* in human adult and fetal tissues and differential inducibility of *CYP1B1* and *CYP1A1* by Ah receptor ligands in

human placenta and cultured cells. Carcinogenesis 18:391–397, 1997.
199. Shimada T, Hayes CL, Yamazaki H, Amin S, Hecht.S.S., Guengerich FP, Sutter TR. Activation of chemically diverse procarcinogens by human cytochrome P-450 1B1. Cancer Res 56:2979–2984, 1996.
200. Walker NJ, Crofts F, Li Y, Lax SF, Hayes CL, Strickland PT, Lucier GW, Sutter TR. Induction and localization of cytochrome P450 1B1 (CYP1B1) protein in the livers of TCDD-treated rats: detection using polyclonal antibodies raised to histidine-tagged fusion proteins produced and purified from bacteria. Carcinogenesis 19:395–402, 1998.
201. Kress S, Greenlee WF. Cell-specific regulation of human *CYP1A1* and *CYP1B1* genes. Cancer Res 57:1264–1269, 1997.
202. Stoilov I, Akarsu AN, Alozie I, Child A, Barsoum-Homsy M, Turacli ME, Or M, Lewis RA, Ozdemir N, Brice G, Aktan SG, Chevrette L, Coca-Prados M, Sarfarazi M. Sequence analysis and homology modeling suggest that primary congenital glaucoma on 2p21 results from mutations disrupting either the hinge region or the conserved core structures of cytochrome P4501B1. Am J Hum Genet 62:573–584, 1998.
203. Tang YM, Green BL, Chen GF, Thompson PA, Lang NP, Shinde A, Lin DX, Tan W, Lyn-Cook BD, Hammons GJ, Kadlubar FF. Human *CYP1B1* Leu^{432}Val gene polymorphism: ethnic distribution in African-Americans, Caucasians and Chinese; oestradiol hydroxylase activity; and distribution in prostate cancer cases and controls. Pharmacogenetics 10:761–766, 2000.
204. Shimada T, Watanabe J, Kawajiri K, Sutter TR, Guengerich FP, Gillam EM, Inoue K. Catalytic properties of polymorphic human cytochrome P450 1B1 variants. Carcinogenesis 20:1607–1613, 1999.
205. Hanna IH, Dawling S, Roodi N, Guengerich FP, Parl FF. Cytochrome P450 1B1 (*CYP1B1*) pharmacogenetics: association of polymorphisms with functional differences in estrogen hydroxylation activity. Cancer Res 60:3440–3444, 2000.
206. Spink DC, Spink BC, Zhuo X, Hussain MM, Gierthy JF, Ding X. NADPH- and hydroperoxide-supported 17beta-estradiol hydroxylation catalyzed by a variant form (432L, 453S) of human cytochrome P450 1B1. J Steroid Biochem Mol Biol 74:11–18, 2000.
207. Watanabe J, Shimada T, Gillam EM, Ikuta T, Suemasu K, Higashi Y, Gotoh O, Kawajiri K. Association of *CYP1B1* genetic polymorphism with incidence to breast and lung cancer. Pharmacogenetics 10:25–33, 2000.
208. Bailey LR, Roodi N, Dupont WD, Parl FF. Association of cytochrome p450 1b1 (*CYP1B1*) polymorphism with steroid receptor status in breast cancer. Cancer Res 58:5038–5041, 1998.
209. Zheng W, Xie D.W., Jin F., Cheng J.R., Dai Q., Wen W.Q., Shu X-O, Gao Y-T. Genetic polymorphism of cytochrome P450-1B1 and risk of breast cancer. Cancer Epidemiol Biomarkers Prev 9:147–150, 2000.
210. McLellan RA, Oscarson M, Hidestrand M, Leidvik B, Jonsson E, Otter C, Ingelman-Sundberg M. Characterization and functional analysis of two common human cytochrome P450 1B1 variants. Arch Biochem Biophys 378:175–181, 2000.
211. Niwa T, Yabusaki Y, Honma K, Matsuo N, Tatsuta K, Ishibashi F, Katagiri M. Contribution of human hepatic cytochrome P450 isoforms to regioselective hydroxylation of steroid hormones. Xenobiotica 28:539–547, 1998.
212. Huang Z, Guengerich FP, Kaminsky LS. 16Alpha-hydroxylation of estrone by human cytochrome P4503A4/5. Carcinogenesis 19:867–872, 1998.
213. Kerlan V, Dreano Y, Bercovici JP, Beaune PH, Floch HH, Berthou F. Nature of cytochromes P450 involved in the 2-/4-hydroxylations of estradiol in human liver microsomes. Biochem Pharmacol 44:1745–1756, 1992.
214. Shimada T, Yamazaki H, Mimura M, Inui Y, Guengerich FP. Interindividual variations in human liver cytochrome P-450 enzymes involved in the oxidation of drugs, carcinogens and toxic chemicals: studies with liver microsomes of 30 Japanese and 30 Caucasians. J Pharmacol Exp Ther 270:414–423, 1994.
215. Forrester LM, Henderson CJ, Glancey MJ, Back DJ, Park BK, Ball SE, Kitteringham NR, McLaren AW, Miles JS, Skett P. Relative expression of cytochrome P450 isoenzymes in human liver and association with the metabolism of drugs and xenobiotics. Biochem J 281:359–368, 1992.
216. Guengerich FP. Cytochrome P-450 3A4: regulation and role in drug metabolism. Annu Rev Pharmacol Toxicol 39:1–17, 1999.
217. Rebbeck TR, Jaffe JM, Walker AH, Wein AJ, Malkowicz SB. Modification of clinical presentation of prostate tumors by a novel genetic variant in *CYP3A4*. J Natl Cancer Inst 90:1225–1229, 1998.
218. Ball SE, Scatina J, Kao J, Ferron GM, Fruncillo R, Mayer P, Weinryb I, Guida M, Hopkins PJ, Warner N, Hall J. Population distribution and effects on drug metabolism of a genetic variant in the 5′ promoter region of *CYP3A4*. Clin Pharmacol Ther 66:288–294, 1999.
219. Wandel C, Witte JS, Hall JM, Stein CM, Wood AJ, Wilkinson GR. *CYP3A* activity in African American and European American men: population differences and functional effect of the *CYP3A4*°*1B* 5′-promoter region polymorphism. Clin Pharmacol Ther 68:82–91, 2000.
220. Ando Y, Tateishi T, Sekido Y, Yamamoto T, Satoh T, Hasegawa Y, Kobayashi S, Katsumata Y, Shimokata K, Saito H. Re: Modification of

clinical presentation of prostate tumors by a novel genetic variant in *CYP3A4*. J Natl Cancer Inst 91:1587–1588, 1999.
221. Rebbeck TR. More about: Modification of clinical presentation of prostate tumors by a novel genetic variant in *CYP3A4*. J Natl Cancer Inst 92:76, 2000.
222. Sata F, Sapone A, Elizondo G, Stocker P, Miller VP, Zheng W, Raunio H, Crespi CL, Gonzalez FJ. *CYP3A4* allelic variants with amino acid substitutions in exons 7 and 12: evidence for an allelic variant with altered catalytic activity. Clin Pharmacol Ther 67:48–56, 2000.
223. Inoue K, Tice LW, Creveling CR. Immunocytochemical localization of catechol-*O*-methyltransferase. In: Usdin E, Weiner N, Youdim MBH (eds). Structure and Function of Monoamine Enzymes. New York: Marcel Dekker, 1977, pp 835–859.
224. Zhu BT, Liehr JG. Quercetin increases the severity of estradiol-induced tumorigenesis in hamster kidney. Toxicol Appl Pharmacol 125:149–158, 1994.
225. Zhu BT, Liehr JG. Inhibition of catechol *O*-methyltransferase-catalyzed *O*-methylation of 2- and 4-hydroxyestradiol by quercetin. Possible role in estradiol-induced tumorigenesis. J Biol Chem 271:1357–1363, 1996.
226. Lotta T, Vidgren J, Tilgmann C, Ulmanen I, Melen K, Julkunen I, Taskinen J. Kinetics of human soluble and membrane-bound catechol *O*-methyltransferase: a revised mechanism and description of the thermolabile variant of the enzyme. Biochemistry 34:4202–4210, 1995.
227. Dawling S, Roodi N, Mernaugh RL, Wang X, Parl FF. Catechol-*O*-methyltransferase (COMT)-mediated metabolism of catechol estrogens: comparison of wild-type and variant COMT isoforms. Cancer Res 61:6716–6722, 2001.
228. Lavigne JA, Helzlsouer KJ, Huang HY, Strickland PT, Bell DA, Selmin O, Watson MA, Hoffman S, Comstock GW, Yager JD. An association between the allele coding for a low activity variant of catechol-*O*-methyltransferase and the risk for breast cancer. Cancer Res 57:5493–5497, 1997.
229. Thompson PA, Shields PG, Freudenheim JL, Stone A, Vena JE, Marshall JR, Graham S, Laughlin R, Nemoto T, Kadlubar FF, Ambrosone CB. Genetic polymorphisms in catechol-*O*-methyltransferase, menopausal status, and breast cancer risk. Cancer Res 58:2107–2110, 1998.
230. Millikan RC, Pittman GS, Tse CK, Duell E, Newman B, Savitz D, Moorman PG, Boissy RJ, Bell DA. Catechol-*O*-methyltransferase and breast cancer risk. Carcinogenesis 19:1943–1947, 1998.
231. Mitrunen K, Jourenkova N, Kataja V, Eskelinen M, Kosma VM, Benhamou S, Kang D, Vainio H, Uusitupa M, Hirvonen A. Polymorphic catechol-*O*-methyltransferase gene and breast cancer risk. Cancer Epidemiol Biomarkers Prev 10:635–640, 2001.
232. Mackenzie PI, Mojarrabi B, Meech R, Hansen A. Steroid UDP glucuronosyltransferases: characterization and regulation. J Endocrinol 150:79–86, 1996.
233. Hum DW, Belanger A, Levesque E, Barbier O, Beaulieu M, Albert C, Vallee M, Guillemette C, Tchernof A, Turgeon D, Dubois S. Characterization of UDP-glucuronosyltransferases active on steroid hormones. J Steroid Biochem Mol Biol 69:413–423, 1999.
234. Bélanger A, Hum DW, Beaulieu M, Levesque E, Guillemette C, Tchernof A, Belanger G, Turgeon D, Dubois S. Characterization and regulation of UDP-glucuronosyltransferases in steroid target tissues. J Steroid Biochem Mol Biol 65:301–310, 1998.
235. Raftogianis R, Creveling C, Weinshilboum R, Weisz J. Estrogen metabolism by conjugation. J Natl Cancer Inst Monogr 27:113–124, 2000.
236. Senafi SB, Clarke DJ, Burchell B. Investigation of the substrate specificity of a cloned expressed human bilirubin UDP-glucuronosyltransferase: UDP-sugar specificity and involvement in steroid and xenobiotic glucuronidation. Biochem J 303:233–240, 1994.
237. Cheng Z, Rios GR, King CD, Coffman BL, Green MD, Mojarrabi B, Mackenzie PI, Tephly TR. Glucuronidation of catechol estrogens by expressed human UDP-glucuronosyltransferases (UGTs) 1A1, 1A3, and 2B7. Toxicol Sci 45:52–57, 1998.
238. Aono S, Yamada Y, Keino H, Hanada N, Nakagawa T, Sasaoka Y, Yazawa T, Sato H, Koiwai O. Identification of defect in the genes for bilirubin UDP-glucuronosyl-transferase in a patient with Crigler-Najjar syndrome type II. Biochem Biophys Res Commun 197:1239–1244, 1993.
239. Koiwai O, Nishizawa M, Hasada K, Aono S, Adachi Y, Mamiya N, Sato H. Gilbert's syndrome is caused by a heterozygous missense mutation in the gene for bilirubin UDP-glucuronosyltransferase. Hum Mol Genet 4:1183–1186, 1995.
240. Bosma PJ, Chowdhury JR, Bakker C, Gantla S, de Boer A, Oostra BA, Lindhout D, Tytgat GN, Jansen PL, Oude Elferink RP. The genetic basis of the reduced expression of bilirubin UDP-glucuronosyltransferase 1 in Gilbert's syndrome. N Engl J Med 333:1171–1175, 1995.
241. Beutler E, Gelbart T, Demina A. Racial variability in the UDP-glucuronosyltransferase 1 (*UGT1A1*) promoter: a balanced polymorphism for regulation of bilirubin metabolism? Proc Natl Acad Sci USA 95:8170–8174, 1998.
242. Guillemette C, Millikan RC, Newman B, Housman DE. Genetic polymorphisms in uridine diphospho-glucuronosyltransferase 1A1 and as-

sociation with breast cancer among African Americans. Cancer Res 60:950–956, 2000.
243. Guillemette C, De Vivo I, Hankinson SE, Haiman CA, Spiegelman D, Housman DE, Hunter DJ. Association of genetic polymorphisms in UGT1A1 with breast cancer and plasma hormone levels. Cancer Epidemiol Biomarkers Prev 10:711–714, 2001.
244. Jin CJ, Miners JO, Lillywhite KJ, Mackenzie PI. cDNA cloning and expression of two new members of the human liver UDP-glucuronosyltransferase 2B subfamily. Biochem Biophys Res Commun 194:496–503, 1993.
245. Turgeon D, Carrier J, Levesque E, Hum DW, Belanger A. Relative enzymatic activity, protein stability, and tissue distribution of human steroid-metabolizing UGT2B subfamily members. Endocrinology 142:778–787, 2001.
246. Levesque E, Beaulieu M, Hum DW, Belanger A. Characterization and substrate specificity of UGT2B4 (E458): a UDP-glucuronosyltransferase encoded by a polymorphic gene. Pharmacogenetics 9:207–216, 1999.
247. Ritter JK, Sheen YY, Owens IS. Cloning and expression of human liver UDP-glucuronosyltransferase in COS-1 cells. 3,4-Catechol estrogens and estriol as primary substrates. J Biol Chem 265:7900–7906, 1990.
248. Coffman BL, King CD, Rios GR, Tephly TR. The glucuronidation of opioids, other xenobiotics, and androgens by human UGT2B7Y(268) and UGT2B7H(268). Drug Metab Dispos 26:73–77, 1998.
249. Carlstrom K, von Uexkull AK, Einhorn N, Fredricsson B, Lunell NO, Sundelin P. Metabolism of estrone sulfate in human endometrium. Acta Obstet Gynecol Scand 62:519–524, 1983.
250. Falany JL, Azziz R, Falany CN. Identification and characterization of cytosolic sulfotransferases in normal human endometrium. Chem Biol Interact 109:329–339, 1998.
251. Tseng L, Liu HC. Stimulation of arylsulfotransferase activity by progestins in human endometrium in vitro. J Clin Endocrinol Metab 53:418–421, 1981.
252. Clarke CL, Adams JB, Wren BG. Induction of estrogen sulfotransferase in the human endometrium by progesterone in organ culture. J Clin Endocrinol Metab 55:70–75, 1982.
253. Faucher F, Lacoste L, Dufort I, Luu-The V. High metabolization of catecholestrogens by type 1 estrogen sulfotransferase (hEST1). J Steroid Biochem Mol Biol 77:83–86, 2001.
254. Her C, Szumlanski C, Aksoy IA, Weinshilboum RM. Human jejunal estrogen sulfotransferase and dehydroepiandrosterone sulfotransferase: immunochemical characterization of individual variation. Drug Metab Dispos 24:1328–1335, 1996.
255. Raftogianis RB, Wood TC, Otterness DM, Van Loon JA, Weinshilboum RM. Phenol sulfotransferase pharmacogenetics in humans: association of common SULT1A1 alleles with TS PST phenotype. Biochem Biophys Res Commun 239:298–304, 1997.
256. Spink BC, Katz BH, Hussain MM, Pang S, Connor SP, Aldous KM, Gierthy JF, Spink DC. SULT1A1 catalyzes 2-methoxyestradiol sulfonation in MCF-7 breast cancer cells. Carcinogenesis 21:1947–1957, 2000.
257. Raftogianis RB, Wood TC, Weinshilboum RM. Human phenol sulfotransferases SULT1A2 and SULT1A1: genetic polymorphisms, allozyme properties, and human liver genotype-phenotype correlations. Biochem Pharmacol 58:605–616, 1999.
258. Zheng W, Xie D, Cerhan JR, Sellers TA, Wen W, Folsom AR. Sulfotransferase 1A1 polymorphism, endogenous estrogen exposure, well-done meat intake, and breast cancer risk. Cancer Epidemiol Biomarkers Prev 10:89–94, 2001.
259. Seth P, Lunetta KL, Bell DW, Gray H, Nasser SM, Rhei E, Kaelin CM, Iglehart DJ, Marks JR, Garber JE, Haber DA, Polyak K. Phenol sulfotransferases: hormonal regulation, polymorphism, and age of onset of breast cancer. Cancer Res 60:6859–6863, 2000.
260. Brix LA, Nicoll R., Zhu X., McMannus M.E. Structural and functional characterisation of human sulfotransferases. Chem Biol Interact 109:123–127, 1998.
261. Engelke CE, Meinl W, Boeing H, Glatt H. Association between functional genetic polymorphisms of human sulfotransferases 1A1 and 1A2. Pharmacogenetics 10:163–169, 2000.
262. Wood TC, Her C, Aksoy I, Otterness DM, Weinshilboum RM. Human dehydroepiandrosterone sulfotransferase pharmacogenetics: quantitative Western analysis and gene sequence polymorphisms. J Steroid Biochem Mol Biol 59:467–478, 1996.
263. Cadenas E. Antioxidant and prooxidant functions of DT-diaphorase in quinone metabolism. Biochem Pharmacol 49:127–140, 1995.
264. Venugopal R, Joseph P, Jaiswal AK. Gene expression of DT-diaphorase in cancer cells. In: Forman HJ, Cadenas E (eds). Oxidative Stress and Signal Transduction. New York: Chapman and Hall, 1997, pp 441–475.
265. Joseph P, Xie T, Xu Y, Jaiswal AK. NAD(P)H:quinone oxidoreductase 1 (DT-diaphorase): expression, regulation, and role in cancer. Oncol Res 6:525–532, 1994.
266. Montano MM, Katzenellenbogen BS. The quinone reductase gene: a unique estrogen receptor–regulated gene that is activated by antiestrogens. Proc Natl Acad Sci USA 94:2581–2586, 1997.
267. Montano MM, Jaiswal AK, Katzenellenbogen BS. Transcriptional regulation of the human quinone reductase gene by antiestrogen-liganded estrogen receptor-alpha and estrogen re-

ceptor-beta. J Biol Chem 273:25443–25449, 1998.
268. Traver RD, Horikoshi T, Danenberg KD, Stadlbauer THW, Danenberg PV, Ross D, Gibson NW: NAD(P)H:quinone oxidoreductase gene expression in human colon carcinoma cell: characterization of a mutation which modulates DT-diaphorase activity and mitomycin sensitivity. Cancer Res 52:797–802, 1992.
269. Traver RD, Siegel D, Beall HD, Phillips RM, Gibson NW, Franklin WA, Ross D. Characterization of a polymorphism in NAD(P)H:quinone oxidoreductase (DT-diaphorase). Br J Cancer 75:69–75, 1997.
270. Siegel D, McGuinness SM, Winski SL, Ross D. Genotype–phenotype relationships in studies of a polymorphism in NAD(P)H:quinone oxidoreductase 1. Pharmacogenetics 9:113–121, 1999.
271. Kuehl BL, Paterson JW, Peacock JW, Paterson MC, Rauth AM. Presence of a heterozygous substitution and its relationship to DT-diaphorase activity. Br J Cancer 72:555–561, 1995.
272. Gaedigk A, Tydale RF, Jurima-Romet M, Sellers EM, Grant DM, Leeder JS. NAD(P)H:quinone oxidoreductase: polymorphisms and allele frequencies in Caucasian, Chinese and Canadian Native Indian and Inuit populations. Pharmacogenetics 8:305–313, 1998.
273. Pan SS, Forrest GL, Akman SA, Hu LT: NAD(P)H:quinone oxidoreductase expression and mitomycin C resistance developed by human colon cancer HCT 116 cells. Cancer Res 55:330–335, 1995.
274. Ryan KD. Hormones in women. In: Ness RB, Kuller LH (eds). Health and Disease Among Women. Biological and Environmental Influences. New York: Oxford University Press, 1999, pp 133–154.
275. Furui K, Suganuma N, Tsukahara S, Asada Y, Kikkawa F, Tanaka M, Ozawa T, Tomoda Y. Identification of two point mutations in the gene coding luteinizing hormone (LH) beta-subunit, associated with immunologically anomalous LH variants. J Clin Endocrinol Metab 78:107–113, 1994.
276. Pettersson K, Ding YQ, Huhtaniemi I. An immunologically anomalous luteinizing hormone variant in a healthy woman. J Clin Endocrinol Metab 74:164–171, 1992.
277. Haavisto AM, Pettersson K, Bergendahl M, Virkamaki A, Huhtaniemi I. Occurrence and biological properties of a common genetic variant of luteinizing hormone. J Clin Endocrinol Metab 80:1257–1263, 1995.
278. Nilsson C, Jiang M, Pettersson K, Iitia A, Makela M, Simonsen H, Easteal S, Herrera RJ, Huhtaniemi I. Determination of a common genetic variant of luteinizing hormone using DNA hybridization and immunoassays. Clin Endocrinol (Oxf) 49:369–376, 1998.
279. Suganuma N, Furui K, Kikkawa F, Tomoda Y, Furuhashi M. Effects of the mutations (Trp[8] → Arg and Ile[15] → Thr) in human luteinizing hormone (LH) beta-subunit on LH bioactivity in vitro and in vivo. Endocrinology 137:831–838, 1996.
280. Rajkhowa M, Talbot JA, Jones PW, Pettersson K, Haavisto AM, Huhtaniemi I, Clayton RN. Prevalence of an immunological LH beta-subunit variant in a UK population of healthy women and women with polycystic ovary syndrome. Clin Endocrinol (Oxf) 43:297–303, 1995.
281. Elter K, Erel CT, Cine N, Ozbek U, Hacihanefioglu B, Ertungealp E. Role of the mutations Trp[8] → Arg and Ile[15] → Thr of the human luteinizing hormone beta-subunit in women with polycystic ovary syndrome. Fertil Steril 71:425–430, 1999.
282. Akhmedkhanov A, Toniolo P, Zeleniuch-Jacquotte A, Pettersson K, Huhtaniemi I. Genetic variant of luteinizing hormone and risk of breast cancer in older women. Cancer Epidemiol Biomarkers Prev 9:839–842, 2000.
283. Cramer DW, Petterson KS, Barbieri RL, Huhtaniemi IT. Reproductive hormones, cancers, and conditions in relation to a common genetic variant of luteinizing hormone. Hum Reprod 15:2103–2107, 2000.
284. Siiteri PK. Extraglandular oestrogen formation and serum binding of oestradiol: relationship to cancer. J Endocrinol 89:119P–129P, 1981.
285. Becchis M, Frairia R, Ferrera P, Fazzari A, Ondei S, Alfarano A, Coluccia C, Biglia N, Sismondi P, Fortunati N. The additionally glycosylated variant of human sex hormone–binding globulin (SHBG) is linked to estrogen-dependence of breast cancer. Breast Cancer Res Treat 54:101–107, 1999.
286. Misao R, Itoh N, Mori H, Fujimoto J, Tamaya T. Sex hormone–binding globulin mRNA levels in human uterine endometrium. Eur J Endocrinol 131:623–629, 1994.
287. Misao R, Nakanishi Y, Fujimoto J, Tamaya T. Expression of sex hormone–binding globulin mRNA in uterine leiomyoma, myometrium and endometrium of human subjects. Gynecol Endocrinol 9:317–323, 1995.
288. Power SG, Bocchinfuso WP, Pallesen M, Warmels-Rodenhiser S, Van Baelen H, Hammond GL. Molecular analyses of a human sex hormone–binding globulin variant: evidence for an additional carbohydrate chain. J Clin Endocrinol Metab 75:1066–1070, 1992.
289. Cousin P, Dechaud H, Grenot C, Lejeune H, Pugeat M. Human variant sex hormone–binding globulin (SHBG) with an additional carbohydrate chain has a reduced clearance rate in rabbit. J Clin Endocrinol Metab 83:235–240, 1998.
290. Matsuzaki S, Fukaya T, Suzuki T, Murakami T, Sasano H, Yajima A. Oestrogen receptor alpha and beta mRNA expression in human en-

290. dometrium throughout the menstrual cycle. Mol Hum Reprod 5:559–564, 1999.
291. Tsukamoto K, Inoue S, Hosoi T, Orimo H, Emi M. Isolation and radiation hybrid mapping of dinucleotide repeat polymorphism at the human estrogen receptor beta locus. J Hum Genet 43:73–74, 1998.
292. Westberg L, Baghaei F, Rosmond R, Hellstrand M, Landen M, Jansson M, Holm G, Bjorntorp P, Eriksson E. Polymorphisms of the androgen receptor gene and the estrogen receptor beta gene are associated with androgen levels in women. J Clin Endocrinol Metab 86:2562–2568, 2001.
293. Andersen TI, Heimdal KR, Skrede M, Tveit K, Berg K, Borresen AL. Oestrogen receptor (ESR) polymorphisms and breast cancer susceptibility. Hum Genet 94:665–670, 1994.
294. Roodi N, Bailey LR, Kao WY, Verrier CS, Yee CJ, Dupont WD, Parl FF. Estrogen receptor gene analysis in estrogen receptor-positive and receptor-negative primary breast cancer. J Natl Cancer Inst 87:446–451, 1995.
295. Iwase H, Greenman JM, Barnes DM, Hodgson S, Bobrow L, Mathew CG. Sequence variants of the estrogen receptor (ER) gene found in breast cancer patients with ER negative and progesterone receptor positive tumors. Cancer Lett 108:179–184, 1996.
296. Southey MC, Batten LE, McCredie MR, Giles GG, Dite G, Hopper JL, Venter DJ. Estrogen receptor polymorphism at codon 325 and risk of breast cancer in women before age forty. J Natl Cancer Inst 90:532–536, 1998.
297. Schubert EL, Lee MK, Newman B, King MC. Single nucleotide polymorphisms (SNPs) in the estrogen receptor gene and breast cancer susceptibility. J Steroid Biochem Mol Biol 71:21–27, 1999.
298. Andersen TI, Wooster R, Laake K, Collins N, Warren W, Skrede M, Elles R, Tveit KM, Johnston SR, Dowsett M, Olsen AO, Moller P, Stratton MR, Borresen-Dale AL. Screening for ESR mutations in breast and ovarian cancer patients. Hum Mutat 9:531–536, 1997.
299. Garcia T, Lehrer S, Bloomer WD, Schachter B. A variant estrogen receptor messenger ribonucleic acid is associated with reduced levels of estrogen binding in human mammary tumors. Mol Endocrinol 2:785–791, 1988.
300. Yaich L, Dupont WD, Cavener DR, Parl FF. Analysis of the PvuII restriction fragment-length polymorphism and exon structure of the estrogen receptor gene in breast cancer and peripheral blood. Cancer Res 52:77–83, 1992.
301. del Senno L, Aguiari GL, Piva R. Dinucleotide repeat polymorphism in the human estrogen receptor (ESR) gene. Hum Mol Genet 1:354, 1992.
302. Weiderpass E, Persson I, Melhus H, Wedren S, Kindmark A, Baron JA. Estrogen receptor alpha gene polymorphisms and endometrial cancer risk. Carcinogenesis 21:623–627, 2000.
303. Attia GR, Zeitoun K, Edwards D, Johns A, Carr BR, Bulun SE. Progesterone receptor isoform A but not B is expressed in endometriosis. J Clin Endocrinol Metab 85:2897–2902, 2000.
304. Rowe SM, Coughlan SJ, McKenna NJ, Garrett E, Kieback DG, Carney DN, Headon DR. Ovarian carcinoma–associated TaqI restriction fragment length polymorphism in intron G of the progesterone receptor gene is due to an Alu sequence insertion. Cancer Res 55:2743–2745, 1995.
305. Kieback DG, Tong X-WWNL, Agoulnik IU. A genetic mutation in the progesterone receptor (PROGINS) leads to an increased risk of nonfamilial breast and ovarian cancer causing inadequate control of estrogn receptor driven proliferation. J Soc Gynecol Investig 5:40a, 1998.
306. Agoulnik I, Weigel N, Tong XW, Bingman WE, Estella NM, Blankenberg K, Runnebaum IB, Korner W, Fishman A, Atkinson EN, Jones LA, Kieback DG. Functional analysis of mutated progesterone receptor that cosegregates with sporadic ovarian cancer. Proc Am Assoc Cancer Res 38:453, 1997.
307. Manolitsas TP, Englefield P, Eccles DM, Campbell IG. No association of a 306-bp insertion polymorphism in the progesterone receptor gene with ovarian and breast cancer. Br J Cancer 75:1398–1399, 1997.
308. Lancaster JM, Berchuck A, Carney ME, Wiseman R, Taylor JA. Progesterone receptor gene polymorphism and risk for breast and ovarian cancer. Br J Cancer 78:277, 1998.
309. Wang-Gohrke S, Chang-Claude J, Becher H, Kieback DG, Runnebaum IB. Progesterone receptor gene polymorphism is associated with decreased risk for breast cancer by age 50. Cancer Res 60:2348–2350, 2000.

21

Ovarian Cancer: Epidemiology and Molecular Endocrinology

ALICE S. WHITTEMORE
VALERIE McGUIRE

Ovarian cancer is the sixth most common cancer and the sixth most common cause of cancer death among women worldwide, accounting for 4.4% of incident cases and 5.5% of cancer deaths.[1] In 1999, an estimated 166,000 new cases of ovarian cancer and 101,000 deaths from this disease occurred throughout the world.[1] Incidence and mortality rates are highest in North America and western Europe and lowest in developing countries and Japan. The overall 5-year survival probability is less than 40%, largely because most cancers are diagnosed at an advanced stage. Thus, ovarian cancer is a major health problem, and there is a need to improve our understanding of the etiology of this disease.

In this chapter, we focus on epithelial ovarian cancers, which comprise about 90% of all ovarian cancers. The cell of origin of epithelial ovarian cancers is thought to lie either in the surface epithelium covering the ovaries, in the epithelium lining ovarian inclusion cysts, or possibly in residual mullerian tissue in or near the ovaries.[2] Our objectives are to review the current evidence on the roles of steroid hormones and gonadotropins in the etiology of epithelial ovarian cancers and to discuss some of the obstacles to improving our understanding of these roles. We begin with a brief review of the anatomy and hormonal physiology of the normal ovary before and after menopause. This is followed by a synopsis of current epidemiological evidence suggesting etiological roles for hormones in general and a summary of hypotheses for specific hormones in the pathogenesis of the disease. We conclude with a discussion of issues in need of research.

ANATOMY AND HORMONAL PHYSIOLOGY OF THE PREMENOPAUSAL OVARY

Anatomy

The ovaries are walnut-sized glands consisting largely of stromal tissue covered by a layer of epithelial cells. They lie near the posterior and lateral pelvic wall and are attached to the posterior surface of the broad ligament by the mesovarium, the tissue that encloses and holds the ovary in place (Fig. 21.1). Blood vessels, nerves, and lymphatics cross the mesovarium and enter the ovaries at the hilum, a depression of the ovary giving entrance and exit to vessels and nerves. The ovaries contain the follicles, the organelles in which oocytes develop and mature. A follicle

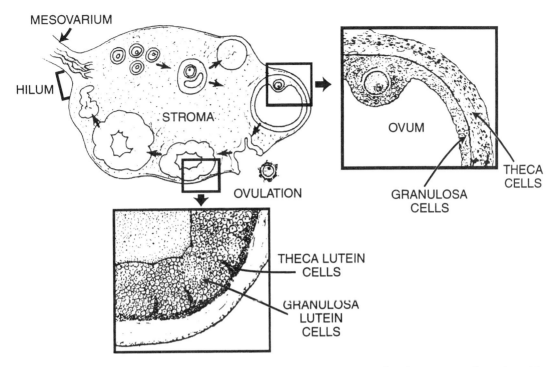

Figure 21.1 Follicle development during the menstrual cycle. (Reprinted with permission from Carr BR, Wilson JD. Disorders of the ovaries and female reproductive tract. In: Wilson JD, Braunwald E, Isselbach KJ (eds). Harrison's Principles of Internal Medicine. New York: McGraw-Hill, 1991, pp 1776–1795.)

typically contains a single immature oocyte covered by a layer of granulosa cells and a layer of theca cells. At birth, the ovaries contain some 1–2 million follicles, a number that diminishes through atresia to about 400,000 at puberty, 8000 at age 40 years, and essentially zero at menopause. At any age prior to menopause, about 10% of the existing follicles are in various stages of maturation. During each ovulatory menstrual cycle, one follicle matures, migrates to the ovarian surface, and ruptures the ovarian epithelium, releasing the unfertilized egg (Fig. 21.1). The residual follicle is then transformed into the corpus luteum by a series of biochemical and morphological changes occurring in the granulosa and theca cells called luteinization. The corpus luteum is the most active steroidogenic tissue in women. Further discussion of ovarian anatomy can be found in Clement[3] and Carr.[4]

Hormonal Physiology

The hypothalamus, the pituitary, and the ovaries interact to govern ovarian function (Fig. 21.2).

Gonadotropin-releasing hormone (GnRH), produced by the hypothalamus, induces pituitary release of follicle-stimulating hormone (FSH), which stimulates follicle growth, and of luteinizing hormone (LH), which regulates ovarian steroid biosynthesis. Pituitary release of FSH and LH also is influenced by estrogen and progesterone, produced by the ovaries. The nature of their effects (stimulation vs. inhibition) depends on the concentration and duration of pituitary exposure to these steroids. Sharply rising estrogen levels prior to ovulation trigger the LH surge required for ovulation. Progesterone is responsible for a surge in FSH just before ovulation. At least three ovarian protein hormones also modulate FSH release: activin appears to stimulate FSH, whereas inhibin and folliculostatin suppress it (Fig. 21.2).[5–7]

In response to FSH stimulation, the granulosa cells of growing follicles secrete estrogens. Luteinizing hormone stimulates the theca cells to produce androgens, which diffuse to the adjacent granulosa cells, where they are aromatized to estrogen. The amount of estrogen secreted by these cells is low during most of the

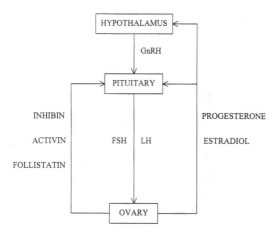

Figure 21.2 Schematic drawing of the feedback loop between the ovaries, the pituitary and the hypothalamus. GnRH, gonadotropin-releasing hormone; LH, luteinizing hormone; FSH, follicle-stimulating hormone.

follicular phase (Fig. 21.3). Toward ovulation, the rate of estrogen secretion rises rapidly, which stimulates secretion of FSH and LH by acting on the hypothalamus to increase its output of GnRH. Following ovulation, LH stimulates the transformation of the residual follicle into the corpus luteum. Under continued stimulation by LH during the luteal phase, cells in the corpus luteum secrete estrogen and progesterone. As estrogen and progesterone concentrations rise, LH concentrations decrease rapidly and the corpus luteum disintegrates. Consequently, estrogen and progesterone decline sharply at the end of the luteal phase. Thus, as seen in Figure 21.3, gonadotropins, estrogens, and progesterones undergo marked fluctuations during the menstrual cycle. In contrast, androgen concentrations undergo only minimal fluctuations.[4,8–14]

Steroidogenesis

All ovarian steroids are derived from low-density lipoprotein (LDL) cholesterol and are produced chiefly by theca cells, granulosa cells, corpus luteal cells, or stromal cells. Figure 21.4 shows the interrelationships among the progestagens produced by the cells of the corpus luteum (e.g., pregnenolone, progesterone, and 17-hydroxyprogesterone), the androgens produced by the theca cells [e.g., dehydroepiandrosterone (DHEA), androstenedione, and testosterone], and the estrogens produced by the granulosa cells (e.g., estradiol, estrone).

An ovarian cell's rate of steroid production is determined by its levels of LDL cholesterol and the LDL cholesterol receptor, of gonadotropins and their receptors, and of five major steroidogenic and/or steroid-metabolizing enzymes. In theca cells and corpus luteal cells, LH increases

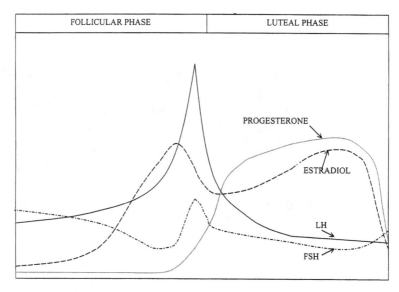

Figure 21.3 Hormonal fluctuations during the menstrual cycle. LH, luteinizing hormone; FSH, follicle-stimulating hormone.

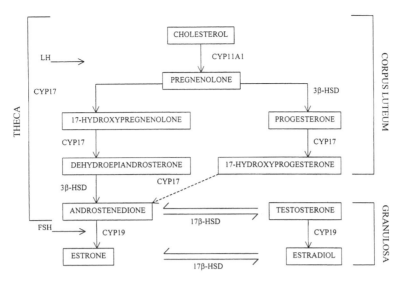

Figure 21.4 Steroids and steroidogenic enzymes produced by the theca cells, the granulosa cells and the corpus luteal cells of the ovary. LH, luteinizing hormone; FSH, follicle-stimulating hormone.

the binding and uptake of LDL cholesterol. It is thought that LH also induces theca cell production of androstenedione and small amounts of testosterone, while FSH induces granulosa cells to aromatize androstenedione to estrone. In the presence of estradiol, FSH also stimulates expression of LH receptors in granulosa cells (Fig. 21.4).

Figure 21.4 also shows the roles of five key enzymes involved in ovarian steroidogenesis: (1) cytochrome P-450 (CYP11A1), which catalyzes the conversion of LDL cholesterol to pregnenolone by the theca cells and corpus luteal cells; (2) 3β-hydroxysteroid dehydrogenase (3β-HSD), which catalyzes the conversion of pregnenolone to progesterone by corpus luteal cells; (3) 17α-hydroxylase (CYP17), which catalyzes the conversion of 17-hydroxypregnenolone to DHEA in the theca cells and of progesterone to 17-hydroxyprogesterone in the corpus luteal cells; (4) aromatase (CYP19), which promotes estradiol biosynthesis by the granulosa cells of mature follicles before ovulation; and (5) 17β-HSD, which facilitates the interconversion of androstenedione/testosterone and of estrone/estradiol.[4,15–19] For further details on the steroid pathways see Chapter 2.

In summary, the premenopausal ovary is a steroidogenic organ that is highly sensitive to the influences of pituitary gonadotropins.

HORMONES AND THE POSTMENOPAUSAL OVARY

By menopause the ovaries are nearly devoid of follicles, with their steroidogenic theca cells, granulosa cells, and corpus luteal cells. Instead, steroidogenesis in the postmenopausal ovary occurs in the ovarian corticostromal cells, which are derived from theca cells and produce steroidogenic enzymes, and in the hilar cells of the ovarian hilum. The latter cells are more apparent in the postmenopausal ovary than in the premenopausal ovary and appear to have steroidogenic properties.

These major anatomical and physiological changes are accompanied by striking changes in circulating levels of gonadotropins and steroid hormones. Serum levels of FSH increase 15-fold and reach a plateau within a year following cessation of menses, then decline slightly. However, FSH levels remain elevated compared to premenopausal levels, even in the very elderly. At menopause, LH levels rise threefold and remain elevated throughout life.

Ovarian secretion of estrogens declines dramatically at menopause. The ovaries of a small fraction (<10%) of postmenopausal women continue to secrete significant quantities of estradiol. However in most postmenopausal women, the major source of estradiol is aromatization of androgens by the adrenals. The effects of hysterectomy and oophorectomy on circulating estrogen levels in postmenopausal women were examined recently in a cross-sectional comparison of steroid hormones of 684 California women aged 50–89 years who were not using hormone-replacement therapy (HRT) at the time of blood donation.[20] Bilaterally oophorectomized women had slightly lower levels of total estradiol than did women with at least one ovary, after adjustment for age and body mass index. However, levels of bioavailable estradiol, estrone, and sex hormone–binding globulin (SHBG) did not differ by hysterectomy or oophorectomy status.

Menopause also is associated with decreased ovarian secretion of androstenedione. However, levels of testosterone secretion are similar to those in premenopausal women, so the ovaries remain the primary source of circulating testosterone. Laughlin et al.[20] found that among postmenopausal women with intact ovaries, total, but not bioavailable, testosterone levels increased with age, reaching premenopausal levels by age 70–79 years and stabilizing thereafter. Among oophorectomized women, by contrast, total and bioavailable testosterone levels did not vary with age throughout the 50–89 year age range, and they were more than 40% lower than those in women with intact ovaries. Similarly, androstenedione levels decreased and SHBG levels increased with age in women with intact ovaries but not in oophorectomized women.

Thus, circulating hormone levels in oophorectomized postmenopausal women differ from those in postmenopausal women with intact ovaries. These differences underscore the importance of the ovaries in steroidogenesis after menopause.[21–24]

EPIDEMIOLOGICAL EVIDENCE IMPLICATING HORMONES IN OVARIAN CANCER ETIOLOGY

The descriptive epidemiology of epithelial ovarian cancer provides indirect evidence for the involvement of hormones in the etiology of the disease. Figure 21.5 shows a semilog plot of ovarian cancer incidence rates vs. age. These rates are based on cross-sectional data obtained in the period 1968–1972, before the start of the strong cohort effects due to secular changes in oral contraceptive (OC) use, which distort the age dependence. It is evident from Figure 21.5 that the shape of the curve changes at about the median age of menopause: while rates continue to increase with age, they increase more slowly after menopause than before it. Since major hormonal changes occur at menopause, this deceleration suggests hormonal involvement in the disease. Pike[25] has hypothesized that these hormonal changes cause a decreased proliferation rate of epithelial cells and a consequent decrease in the rate of their transformation to malignancy.

However, it is difficult to study relationships between ovarian cancer and specific hormones in humans. The disease itself causes changes in hormone production, thus precluding any etiological inferences from case–control comparisons. Moreover, there are few data from prospective studies of serum or urinary hormone levels in relation to subsequent ovarian cancer risk. We shall describe these data in relation to hypotheses concerning specific hormones, discussed in the following section. Their interpretation is limited by small numbers of ovarian cancer cases and by uncertainty about the relation between hormonal concentrations in serum or urine and concentrations available to the ovarian epithelial cells at risk of malignancy. The ovarian surface epithelium is avascular,[3] suggesting that hormones exert local autocrine or paracrine, rather than endocrine, influences on the surface epithelial cells. The relevance of findings of differences in serum/urinary hormone concentrations between women who do and do not develop the disease is based on the assumption that these concentrations reflect ovarian production rates. However, this assumption, even if valid, ignores potential modifying autocrine and paracrine effects, particularly effects of hormone binding and metabolism at the cellular level. Thus, while positive associations observed prospectively between serum/urinary hormone levels and ovarian cancer risk are suggestive, negative findings are more difficult to interpret.

Because of these problems, inference of hormonal roles in ovarian cancer pathogenesis has been indirect, based either on in vitro assessment of normal or malignant ovarian tissue or

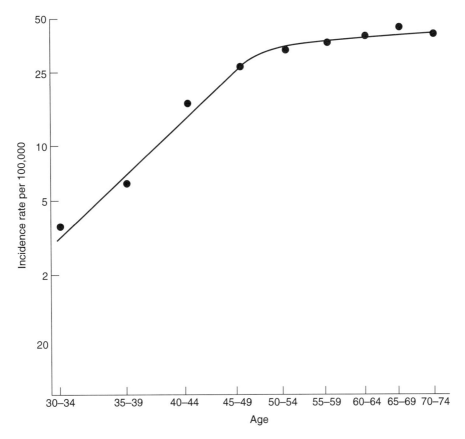

Figure 21.5 Age-specific incidence rates for ovarian cancer among women in Birmingham, England, during the period 1968–1972. (Reprinted with permission from Pike MC. Age-related factors in cancers of the breast, ovary, and endometrium. J Chronic Dis 40:59S–69S, 1987.)

on epidemiological associations between cancer risk and measurable reproductive factors. Attributes such as parity, lactation, OC use, and HRT alter endogenous hormone levels, so relationships between these attributes and ovarian cancer risk have implications for possible hormonal roles in the pathogenesis of the disease.

Parity

Epidemiological studies consistently have shown that pregnancies, even incomplete pregnancies, are associated with reduced risk of epithelial ovarian cancer.[26] The associations appear to reflect more than a correlation between low parity and some form of infertility that predisposes to the disease. Since pregnancies induce major changes in hormone production rates, evaluation of the nature of these changes may provide clues to hormonal bases for the protection afforded by pregnancy.

During pregnancy, the output of pituitary gonadotropins is suppressed.[27] Moreover, after fertilization, human chorionic gonadotropin stimulates progesterone production in the granulosa cells of the corpus luteum during pregnancy, with progesterone concentrations increasing 600%.[11] However, by the second trimester, the main source of progesterone secretion is the placenta.[3] Progesterone levels remain elevated during pregnancy.[28] Estrogen levels also increase, due to aromatization of circulating androgens by the placenta. Serum concentrations of DHEA decrease by 30%, serum testosterone levels increase approximately threefold, and serum androstenedione levels rise slightly.[27] In summary, the hormonal changes that occur during pregnancy include decreases in circulating gonadotropin levels and increases in levels of estrogen, progesterone, and testosterone. However, it is not known whether any of these transient changes are responsible for the reduced ovarian cancer risks associated with pregnancy.

Lactation

The epidemiological data also suggest that among women of similar parity duration of breast-feeding is associated negatively with ovarian cancer risk. Breast-feeding induces partial inhibition of ovulation, with the strength of the inhibition declining with time since childbirth. Breast-feeding also induces increased secretion of FSH but reduces secretion of LH,[29] estrogen, and progesterone. Lactation also induces increased secretion of prolactin, the principal hormone in milk biosynthesis, which inhibits ovulation by disrupting folliculogenesis.[30]

Oral Contraceptive Use

Nearly every study that has examined the association between OC use and ovarian cancer has found OC users, particularly long-term users, to be at reduced risk. A meta-analysis of 20 studies[31] estimated that each additional year of use is associated with an 11% reduction in ovarian cancer risk. The relatively low risk among OC users appears to be present at all intervals from first to last use and across all age groups and levels of parity[32] although the effect appears to be more pronounced in women aged 55 years and older.[33] It also is found consistently among both African-American and Caucasian women[34] and among both noncarriers and carriers of BRCA1 and BRCA2 mutations.[35] A recent analysis of data from a large US case-control study suggests that the OC formulations with high progestin potency are associated with greater reduction in ovarian cancer risk than those with low potency.[35a]

Estrogen-containing OCs suppress ovulation and reduce pituitary secretion of gonadotropins.[36] They also reduce circulating estradiol, progesterone, and androgen levels.[37–40]

Hormone-Replacement Therapy

Epidemiological data on the relationship between HRT and ovarian cancer risk are conflicting.[26] Studies have shown either no association between ovarian cancer and HRT[41–44] or a statistically significant increase in risk.[45,46] Two case-control studies have suggested that HRT may increase the risk of ovarian cancer with endometrioid histology.[43,47] In a recent meta-analysis of 20 case-control and cohort studies, the use of HRT for more than 10 years was associated with a statistically nonsignificant 27% increase in risk of developing invasive epithelial ovarian cancer.[48] These studies were based largely on women who had used unopposed estrogen. Research is needed on the associations between the disease and HRT that includes both estrogen and progestins. A recent cohort study of 44,241 postmenopausal US women found that those who had used estrogen-only replacement therapy, particularly for 10 or more years, were at significantly increased risk of ovarian cancer.[48a] Woman who used short-term estrogen-progestin-only replacement therapy were not at increased risk, but the authors noted that risk associated with such replacement therapy warrants further investigation.

Hormone-replacement therapy decreases the secretion of pituitary gonadotropin, although not to premenopausal levels.[49] Consequently, the hypothesis that circulating levels of gonadotropins increase ovarian cancer risk predicts that HRT decreases risk. Also, HRT increases circulating levels of estrone and estradiol.[50]

In conclusion, the protective effects of parity, lactation, and OC use and the lack of clear associations with HRT use provide provocative but inconclusive leads to possible associations between hormones and ovarian cancer. In the following section, we review hypotheses involving specific hormones and the evidence relevant to these hypotheses.

SPECIFIC HORMONES IN OVARIAN CARCINOGENESIS

Several investigators have speculated specific mechanisms for the pathogenesis of epithelial ovarian cancer, based on clues from the epidemiological findings. The most commonly involved mechanism implicates mutations due to increased cellular proliferation following rupture of the surface epithelium at ovulation. We shall not discuss this hypothesis further but, instead, shall focus on hypotheses concerning roles for gonadotropins, estrogens, androgens, and progesterone in the pathogenesis of epithelial ovarian cancer.

Gonadotropins

The hypothesis of a role for gonadotropins in human ovarian cancer dates back at least 40 years, when Gardner[51] and later Stadel[52] hypothesized that exposure of the ovarian epithelium to persistently high circulating levels of pituitary gonadotropins increased the likelihood of malignancy.

A number of animal experiments support this hypothesis. In domestic fowl, stimulating egg production induces ovarian adenomas,[53] and in rodents, ovarian transplantation that results in increased gonadotropin secretion enhances ovarian tumorigenesis.[54,55] Exposures and/or surgeries that destroy ovarian follicles and induce ovarian failure and consequent loss of feedback control of circulating gonadotropin levels also lead to ovarian tumors. However, because these experimentally induced tumors are nonepithelial, their relevance to epithelial ovarian cancer in humans is uncertain. Nevertheless, Cramer and Welch[56] argued that the presence of epithelial inclusion cysts in human ovarian stroma could induce a different type of tumor in humans than in rodents.[56]

Also in support of a deleterious role for circulating gonadotropins in ovarian cancer risk, investigators have found that cells in the ovarian surface epithelium express receptors for both LH[57] and FSH.[58] Moreover, tumor occurrence is reduced after treatment with GnRH agonists, which suppress the gonadotropins.[59] However, the limited prospective epidemiological data fail to support this hypothesis. A prospective study of some 20,000 female Maryland residents identified 31 ovarian cancer cases within 14 years of blood donation and matched them to 62 control women on age, menopausal status, and, for premenopausal women, time since start of last menstrual period. Cases and controls had similar baseline levels of LH, while cases had significantly lower serum FSH levels than controls.[60]

The indirect epidemiological evidence in support of this hypothesis is mixed. While parity and OC use suppress circulating levels of both LH and FSH, lactation, also associated with reduced ovarian cancer risk, reduces LH levels but increases FSH levels.[30]

Steroid Hormones

During ovulation, the surface epithelial cells surrounding the ovulatory wound are bathed in hormone-rich follicular fluid. This fact has suggested the hypothesis that estrogens and/or androgens in follicular fluid surrounding the epithelial cells at ovulation enhance proliferation and lead to mutations that increase the likelihood of malignancy. This hypothesis predicts that women with many ovulatory cycles are at increased risk because of greater ovulatory exposure of surface epithelial cells to specific hormones. A variant of this hypothesis[56] is that epithelial cells in ovarian inclusion cysts are exposed to these hormones because of their proximity to hormones in the ripening follicles. This hypothesis predicts that women who have ovulated more are at increased risk because of greater numbers of inclusion cysts exposing the epithelial cells to follicular hormones.

Estrogens

A role for estrogens in ovarian carcinogenesis gains plausibility from several observations. Follicular fluid surrounding surface epithelial cells at ovulation contains estradiol at concentrations some 10,000-fold greater than serum levels. Moreover, both α and β estrogen receptors have been found in ovarian epithelial cells.[61]

The prospective study by Helzlsouer et al.[60] described above in relation to gonadotropins found similar estrogen levels in stored sera from ovarian cancer cases and controls. However, because this null finding was based on only 31 cases, it cannot exclude a small or moderate association between estrogens and ovarian cancer risk. In addition, it does not exclude the possibility that paracrine effects of estrogen play a role in the disease.

The indirect epidemiological evidence concerning a deleterious role for estrogen is somewhat conflicting. Circulating levels of estradiol are increased during pregnancy. This fact appears to contraindicate a deleterious effect of estradiol on ovarian cancer risk since pregnancy is associated with reduced risk. However, pregnancy also suppresses ovulation, which may be necessary to expose the epithelial cells to estradiol at the site of follicular eruption. Use of OCs may protect by lowering circulating levels of estradiol. Lactation, which has been associated with reduced ovarian cancer risk, has been found to reduce circulating estrogen levels.[30] Serum estradiol and estrone levels are increased by HRT, but, as noted in the previous section, the relation between HRT and ovarian cancer is not clear.

Androgens

Risch[62] hypothesized that exposure of ovarian epithelial cells to androgens may increase their risk of transformation to malignancy. He cites several lines of evidence to support this hypothesis. First, cells in epithelial inclusion cysts may be exposed to the paracrine effects of androgens produced by the proximal theca cells. Second, androgens are found in follicular fluid, suggesting that surface epithelial cells may be exposed to androgens at ovulation. Third, androgen receptors have been identified in ovarian epithelial cells.[63] These cells express 17β-HSD and thus can convert androstenedione to testosterone,[4] which binds with higher affinity to the androgen receptor than does androstenedione. Edelson et al.[64] found that some epithelial ovarian cancers exhibit loss of a 1 centimorgan region on the X chromosome that contains the gene encoding the androgen receptor. However, the loss does not occur exclusively on the activated X chromosome, so the etiological significance of these findings is not clear.

The indirect epidemiological evidence concerning a role for androgens in ovarian cancer is mixed. In support of this hypothesis, polycystic ovary syndrome, which includes elevated androgen production, is associated with increased risk. In addition, central body adiposity, as measured by increased waist-to-hip ratios, is associated both with increased serum androgen levels and with increased risk.[65-68] However, results from a prospective study of urinary androgen metabolites and subsequent ovarian cancer risk among 1500 women from Guernsey conflict with this hypothesis. A total of 12 women who developed ovarian cancer within 5 years of donating urine samples had significantly lower urinary DHEA levels than 12 control women who were matched to cases on age and menopausal status.[69]

Progesterone

The preceding review indicates that little is known about the relationship between ovarian cancer risk and the production of estrogens and androgens. In contrast, several lines of molecular and epidemiological evidence suggest that endogenous and exogenous progesterone levels may reduce risk.

Two lines of molecular evidence support a protective role for progesterone. First, progesterone appears to play a role in controlling cellular proliferation. Increased apoptosis of surface ovarian epithelial cells was reported in monkeys treated with combined or progestin-only OCs.[70] This finding suggests that progesterone may increase cell death in premalignant cells. Moreover, progestin has been found to inhibit cellular proliferation, upregulate p53 expression, and induce apoptosis in ovarian carcinoma cell lines.[71] Progesterone also inhibits DNA synthesis and proliferation of cultured benign ovarian epithelial tumor cells.[72,73]

If progesterone regulates cellular proliferation, it may do so by interacting with *BRCA1* and *BRCA2* proteins. Treatment of ovariectomized mice with progesterone (alone or in combination with estrogen) induced *BRCA1* and *BRCA2* mRNA expression in mammary epithelium.[74,75] Since germline mutations inactivating *BRCA1* or *BRCA2* are associated with increased ovarian cancer risk, these data suggest that progesterone may prevent malignancy by upregulating wild-type *BRCA1* and *BRCA2*.

A second line of molecular evidence in support of a protective role for progesterone is the association noted between ovarian cancer and impaired function of the progesterone receptor. Ovarian cancer cells often exhibit loss of heterozygosity at chromosome 11q22–23, close to the progesterone receptor gene.[76-78] This loss of heterozygosity correlates with reduced receptor protein.[78] Progesterone receptor mRNA expression was downregulated in cultured ovarian cancer cells relative to cultured normal ovarian epithelial cells.[61] Thus, reduced expression of progesterone receptor is associated with the ovarian cancer phenotype. These findings suggest that inherited functional variants of the progesterone receptor gene may be associated with altered risk.

Indirect epidemiological evidence also supports a protective role for progesterone in the etiology of ovarian cancer. Progesterone levels are elevated during pregnancy, and pregnancy is associated with reduced risk for ovarian cancer. Moreover, serum progesterone levels are higher during twin pregnancies than singleton pregnancies,[79,80] and progesterone levels are elevated in mothers of dizygotic twins.[81] In two case-control studies, women with twins[42] and women with dizygotic twins[82] had lower risks for ovarian cancer than women with single births, after controlling for parity and age at diagnosis. Similarly, ovarian cancer risk is decreased among users of progestin-only OCs and of OCs containing both estrogen and progestin.[26]

DIRECTIONS FOR FUTURE RESEARCH

This chapter has indicated some of the rather formidable obstacles to advancing our understanding of the etiology of epithelial ovarian cancers and of the roles of specific hormones in its pathogenesis. In brief, inferences from epidemiological data require large prospective studies, which are limited by the infrequency of disease occurrence and by uncertainty about the relations between the hormone measurements obtained at baseline and etiologically relevant cellular levels. Despite these obstacles, there are opportunities and needs for research on several issues of public health importance. These include the effects of long-term use of recent formulations of exogenous hormones and chemopreventive agents on ovarian cancer risk, the effects of premenopausal physical activity on risk, and the relation between risk and germline variation in genes encoding key hormone receptors and metabolizing enzymes.

Effects of Exogenous Hormones and Chemopreventive Agents

Stimulation of the pituitary to secrete gonadotropins requires pulsatile, intermittent release of GnRH by the hypothalamus. Therefore, chronic administration of synthetic GnRH agonists suppresses pituitary gonadotropin production.[59] The resulting low levels of FSH and LH inhibit ovarian production of steroid hormones; indeed, sufficiently high doses of GnRH can suppress all ovarian steroid hormone production, producing biochemically the effects of oophorectomy. Pike and colleagues[83] have hypothesized that one can prevent cancers of the ovary, breast, and endometrium by administering to premenopausal women low doses of a GnRH agonist together with add-back estrogen (to prevent osteoporosis) and intermittent progestagen (to prevent endometrial cancer). The investigators reason that the doses of add-back steroid hormones are low enough to prevent cancers of the breast and endometrium. Pike[25] has speculated that the protection against ovarian cancer may be greater than that against breast cancer; he has estimated that ovarian cancer risk may be reduced by more than 90% if all women used the regimen starting at age 30 years.

In randomized pilot trials involving contraceptive use of this regimen, treated women experienced a beneficial rise in high-density lipoprotein cholesterol and a beneficial reduction in mammographic density[84] but a small annual loss of bone mineral density and, in some women, a loss of libido.[85] Clearly, there is need for information on the relation between chronic administration of this regimen and lifetime risk of ovarian cancer, as well as its effects on heart disease, breast cancer, endometrial cancer, and osteoporosis.

There also is need for data on ovarian cancer risk in relation to long-term use of antiestrogens, such as tamoxifen, prescribed to prevent breast cancer. One study indicated that tamoxifen use is associated with an increased occurrence of benign ovarian cysts.[86] The need for data on the effects of antiestrogens on ovarian cancer risk is particularly compelling in light of the fact that these agents are apt to be used extensively by carriers of mutations of *BRCA1* or *BRCA2*, who are at elevated risk of ovarian cancer.

Premenopausal Physical Activity Levels and Ovarian Cancer Risk

Since strenuous physical activity alters endogenous hormone levels,[87-89] an association between physical activity levels and risk of ovarian cancer could provide insight into specific hormonal mechanisms. However, few studies have assessed the strength of such an association.[90] In three cohort studies, risks of ovarian cancer among physically active women relative to risks among those with a sedentary lifestyle ranged from 0.3 to 1.1, with wide confidence intervals due to small numbers.[91–93] One cohort study of some 30,000 postmenopausal women in Iowa found physical activity to be associated with increased risk: risks among women with moderate and high activity levels were 1.4 and 2.1 times those of women with low levels.[68] There is need for more data on this issue.

Polymorphisms of Genes Encoding Hormonal Enzymes and Receptors

The obstacles to direct evaluation of the relationship between hormones and ovarian cancer risk indicate the need for alternative approaches. Useful new information may be gained by evaluating germline variation in ovarian cancer cases and controls of genes encoding hormone-metabolizing enzymes and hormone receptors. At present, however, this "candidate gene" approach has not been tested extensively, and the few findings published have been conflicting and inconclusive. Three major problems limit the approach; all three problems apply generally to chronic diseases and not just to ovarian cancer. The first problem is the dearth of polymorphisms whose variant alleles encode proteins that function differently from the wild type; most of the known polymorphisms occur within introns. There is need to identify new functional polymorphisms in genes encoding enzymes and receptors.

A second problem associated with the approach of identifying candidate polymorphisms associated with ovarian cancer risk is the need to stratify the analysis on ethnicity. Suppose e.g., that the locus of interest is not near a disease locus but that the population consists of two ethnic groups, with the first group having higher prevalence of both the disease and a specific allele than the second group. A random sample of cases would contain a higher fraction of the first ethnic group than either the general population or a sample of controls and, thus, a higher total count of the allele. In this case, the test statistic will lead to rejection of the null hypothesis more often than it should. Thus, failure to account for such *ethnic stratification*, either by a matched design or by a stratified analysis, could lead to invalid conclusions. Adjusting for population stratification requires collecting detailed race/ancestry information; however, this approach may leave residual confounding because many genetic traits vary in frequency within apparently homogenous ethnic groups. Race/ethnicity is also an incomplete surrogate for genetic makeup since there may be cultural and geographic influences on risk within subgroups of a given racial or ethnic group. Even within an apparently homogenous ethnic group, there may be different allele frequencies at many loci due to the effects of different geographical locations or migration patterns, and both of these factors may affect ethnic admixture within the apparently homogenous group. Thus, it is advisable to compare cases and controls within narrowly defined ethnic groups. The potential bias of concern here is a classic example of confounding (by ethnic origin), a problem that has been well studied by epidemiologists. In particular, the confounding factor (ethnic ancestry) can produce substantial bias only if it is strongly associated with both disease risk and genotype at the disease-associated locus. Moreover, large case-control differences in prevalences of suspect genotypes at the locus are less likely than small differences to be due to such bias.

Yet a third problem is the need to adjust the p values of statistical tests for the multiple testing arising when several polymorphisms in several genes are evaluated in one study. This need decreases the study's power to detect association of low to moderate magnitude. This potentially serious problem suggests that careful thought be given to plausible biological mechanisms that may motivate a priori hypotheses.

REFERENCES

1. Parkin DM, Pisani P, Ferlay J. Global cancer statistics. CA Cancer J Clin 49:33–64, 1999.
2. Dubeau L. The cell of origin of ovarian epithelial tumors and the ovarian surface epithelium dogma: does the emperor have no clothes? Gynecol Oncol 72:437–442, 1999.
3. Clement PB. Histology of the ovary. Am J Surg Pathol 11:277–303, 1987.
4. Carr BR. Disorders of the ovaries and female reproductive tract. In: Wilson JD, Foster DW, Kronenberg HM, Larser PR (eds). William's Textbook of Endocrinology. Philadelphia: WB Saunders, 1998, pp 751–817.

5. Fink G. Gonadotropin secretion and its control. In: Knobil E, Neill JD (eds). The Physiology of Reproduction. New York: Raven Press, 1988, pp 1349–1377.
6. Liu JH, Yen SS. Induction of midcycle gonadotropin surge by ovarian steroids in women: a critical evaluation. J Clin Endocrinol Metab 57:797–802, 1983.
7. Ying SY. Inhibins, activins, and follistatins: gonadal proteins modulating the secretion of follicle-stimulating hormone. Endocr Rev 9:267–293, 1988.
8. Erickson GF, Wang C, Hsueh AJ. FSH induction of functional LH receptors in granulosa cells cultured in a chemically defined medium. Nature 279:336–338, 1979.
9. Carr BR, Wilson JD. Disorders of the ovary and female reproductive tract. In: Braunwald E, Isselbacher KJ, Petersdorf RG, Wilson JD, Martin JB, Fauci AS (eds). Harrison's Principles of Internal Medicine. New York: McGraw-Hill, 1987, pp 1818–1837.
10. Apter D, Raisanen I, Ylostalo P, Vihko R. Follicular growth in relation to serum hormonal patterns in adolescent compared with adult menstrual cycles. Fertil Steril 47:82–88, 1987.
11. McNatty KP, Makris A, DeGrazia C, Osathanondh R, Ryan KJ. The production of progesterone, androgens, and estrogens by granulosa cells, thecal tissue, and stromal tissue from human ovaries in vitro. J Clin Endocrinol Metab 49:687–699, 1979.
12. Yeko TR, Khan-Dawood FS, Dawood MY. Human corpus luteum: luteinizing hormone and chorionic gonadotropin receptors during the menstrual cycle. J Clin Endocrinol Metab 68:529–534, 1989.
13. Yen SSC. The human menstrual cycle. In: Yen SSC, Jaffe RB (eds). Reproductive Endocrinology: Physiology, Pathophysiology and Clinical Management, Philadelphia: WB Saunders, 1986, pp 200–236.
14. Dorrington JH, Armstrong DT. Effects of FSH on gonadal functions. Recent Prog Horm Res 39:301–342, 1979.
15. Suzuki T, Sasano H, Kimura N, Tamura M, Fukaya T, Yajima A, Nagura H. Immunohistochemical distribution of progesterone, androgen and oestrogen receptors in the human ovary during the menstrual cycle: relationship to expression of steroidogenic enzymes. Hum Reprod 9:1589–1595, 1994.
16. Zhang Y, Word RA, Fesmire S, Carr BR, Rainey WE. Human ovarian expression of 17β-hydroxysteroid dehydrogenase types 1, 2, and 3. J Clin Endocrinol Metab 81:3594–3598, 1996.
17. McNatty KP, Smith DM, Makris A, Osathanondh R, Ryan KJ. The microenvironment of the human antral follicle: interrelationships among the steroid levels in antral fluid, the population of granulosa cells, and the status of the oocyte in vivo and in vitro. J Clin Endocrinol Metab 49:851–860, 1979.
18. Erickson GF. The ovary: basic principles and concepts. A: Physiology. In: Felig P, Baxter JD, Frohman LA (eds). Endocrinology and Metabolism. New York: McGraw-Hill, 1995, pp 973–1015.
19. Clark BJ, Soo SC, Caron KM. Hormonal and developmental regulation of the steroidogenic acute regulatory protein. Mol Cell Endocrinol 9:1346–1355, 1995.
20. Laughlin GA, Barrett-Connor E, Kritz-Silverstein D, von Muhlen D. Hysterectomy, oophorectomy, and endogenous sex hormone levels in older women: the Rancho Bernardo Study. J Clin Endocrinol Metab 85:645–651, 2000.
21. Judd HL, Judd GE, Lucas WE, Yen SS. Endocrine function of the postmenopausal ovary: concentration of androgens and estrogens in ovarian and peripheral vein blood. J Clin Endocrinol Metab 39:1020–1024, 1974.
22. Peluso JJ, Steger RW, Jaszczak S, Hafez ES. Gonadotropin binding sites in human postmenopausal ovaries. Fertil Steril 27:789–795, 1976.
23. Longscope C. The endocrinology of the menopause. In: Lobo RA (ed). Treatment of the Postmenopausal Woman. New York: Raven Press, 1994, pp 47–53.
24. Plouffe L Jr. Ovaries, androgens and the menopause: practical applications. Semin Reprod Endocrinol 16:117–120, 1998.
25. Pike MC. Age-related factors in cancers of the breast, ovary, and endometrium. J Chron Dis 40:59S–69S, 1987.
26. Riman T, Persson I, Nilsson S. Hormonal aspects of epithelial ovarian cancer: review of epidemiological evidence. Clin Endocrinol (Oxf) 49:695–707, 1998.
27. Lobo RA. The menstrual cycle. In: Mishell DR, Davagan V, Lobo RA (eds). Infertility, Contraception and Reproductive Endocrinology. Malden, MA: Blackwell Scientific, 1991, pp 104–124.
28. Yen SS. Endocrinology of pregnancy. In: Creasy RK, Resnik R (eds). Maternal–Fetal Medicine: Principles and Practice. Philadelphia: WB Saunders, 1994, pp 382–412.
29. Reyes FI, Winter JS, Faiman C. Pituitary–ovarian interrelationships during the puerperium. Am J Obstet Gynecol 114:589–594, 1972.
30. Shoupe D, Mishell DR. Endocrinology of lactation. In: Lobo RA, Mishell DR, Paulson RJ, Shoupe D (eds). Infertility, Contraception and Reproductive Endocrinology. Malden, MA: Blackwell Scientific, 1997, pp 207–223.
31. Hankinson SE, Colditz GA, Hunter DJ, Spencer TL, Rosner B, Stampfer MJ. A quantitative assessment of oral contraceptive use and risk of ovarian cancer. Obstet Gynecol 80:708–714, 1992.
32. Prentice RL, Thomas DB. On the epidemiology

of oral contraceptives and disease. Adv Cancer Res 49:285–401, 1987.
33. Whittemore AS. Personal characteristics relating to risk of invasive epithelial ovarian cancer in older women in the United States. Cancer 71:558–565, 1993.
34. John EM, Whittemore AS, Harris R, Itnyre J. Characteristics relating to ovarian cancer risk: collaborative analysis of seven U.S. case-control studies. Epithelial ovarian cancer in black women. Collaborative Ovarian Cancer Group. J Natl Cancer Inst 85:142–147, 1993.
35. Narod SA, Risch H, Moslehi R, Dorum A, Neuhausen S, Olsson H, Provencher D, Radice P, Evans G, Bishop S, Brunet JS, Ponder BA. Oral contraceptives and the risk of hereditary ovarian cancer. Hereditary Ovarian Cancer Clinical Study Group. N Engl J Med 339:424–428, 1998.
35a.Schildkraut JM, Calingaert B, Marchbanks PA, Moorman PG, Rodriguez GC. Impact of progestin and estrogen potency in oral contraceptives on ovarian cancer. J Natl Cancer Inst 94:32–38, 2002.
36. Lauritzen C. On endocrine effects of oral contraceptives. Acta Endocrinol (Copenh) 124 (Suppl):87–100, 1968.
37. Carr BR, Parker CR Jr, Madden JD, MacDonald PC, Porter JC. Plasma levels of adrenocorticotropin and cortisol in women receiving oral contraceptive steroid treatment. J Clin Endocrinol Metab 49:346–349, 1979.
38. Gaspard UJ, Romus MA, Gillain D, Duvivier J, Demey-Ponsart E, Franchimont P. Plasma hormone levels in women receiving new oral contraceptives containing ethinyl estradiol plus levonorgestrel or desogestrel. Contraception 27:577–590, 1983.
39. Murphy A, Cropp CS, Smith BS, Burkman RT, Zacur HA. Effect of low-dose oral contraceptive on gonadotropins, androgens, and sex hormone binding globulin in nonhirsute women. Fertil Steril 53:35–39, 1990.
40. van der Vange N, Blankenstein MA, Kloosterboer HJ, Haspels AA, Thijssen JH. Effects of seven low-dose combined oral contraceptives on sex hormone binding globulin, corticosteroid binding globulin, total and free testosterone. Contraception 41:345–352, 1990.
41. Whittemore AS, Harris R, Itnyre J. Characteristics relating to ovarian cancer risk: collaborative analysis of 12 US case-control studies. II. Invasive epithelial ovarian cancers in white women. Collaborative Ovarian Cancer Group. Am J Epidemiol 136:1184–1203, 1992.
42. Purdie D, Green A, Bain C, Siskind V, Ward B, Hacker N, Quinn M, Wright G, Russell P, Susil B. Reproductive and other factors and risk of epithelial ovarian cancer: an Australian case-control study. Survey of Women's Health Study Group. Int J Cancer 62:678–684, 1995.
43. Risch HA, Marrett LD, Jain M, Howe GR. Differences in risk factors for epithelial ovarian cancer by histologic type. Results of a case-control study. Am J Epidemiol 144:363–372, 1996.
44. Hempling RE, Wong C, Piver MS, Natarajan N, Mettlin CJ. Hormone replacement therapy as a risk factor for epithelial ovarian cancer. results of a case-control study. Obstet Gynecol 89:1012–1016, 1997.
45. Rodrigues C, Patel AV, Calle EE, Jacob EJ, Thun MJ. Estrogen replacement therapy and ovarian cancer mortality in a large prospective study of US women. JAMA 285:1460–1465, 2001.
46. Negri E, Tzonou A, Beral V, Lagiou P, Trichopoulos D, Parazzini F, Franceschi S, Booth M, La Vecchia C. Hormonal therapy for menopause and ovarian cancer in a collaborative re-analysis of European studies. Int J Cancer 80:848–851, 1999.
47. Weiss NS, Lyon JL, Krishnamurthy S, Dietert SE, Liff JM, Daling JR. Noncontraceptive estrogen use and the occurrence of ovarian cancer. J Natl Cancer Inst 68:95–98, 1982.
48. Garg PP, Kerlikowske K, Subak L, Grady D. Hormone replacement therapy and the risk of epithelial ovarian carcinoma: a meta-analysis. Obstet Gynecol 92:472–479, 1998.
48a.Lacey JV, Mink PJ, Lubin, JH, Sherman ME, Troisi R, Hartge P, Schatkin A, Schairer C. Menopausal hormone replacement therapy and risk of ovarian cancer. JAMA 288:334–341, 2002.
49. Larsson-Cohn U, Johansson EDB, Kagedal B, Vallentin L. Serum FSH, LH and oestrogen levels in postmenopausal patients on oestrogen therapy. BJOG 85:367–372, 1977.
50. Levrant SG, Barnes RB. Pharmacology of estrogens. In: Lobo R (ed). Treatment of the Postmenopausal Woman: Basic and Clinical Aspects. New York: Raven Press, 1994, pp 57–68.
51. Gardner WU. Tumorigenesis in transplanted irradiated and nonirradiated ovaries. J Natl Cancer Inst 26:829–854, 1961.
52. Stadel BV. The etiology and prevention of ovarian cancer. Am J Obstet Gynecol 123:772–774, 1975.
53. Wilson JE. Adenocarcinomas in hens kept in a constant environment. Poult Sci 37:1253–1266, 1958.
54. Biskind MS, Biskind GR. Development of tumors in the rat ovary after transplantation into the spleen. Proc Soc Exp Biol Med 55:176–179, 1944.
55. Biskind GR, Biskind MS. Atrophy of ovaries transplanted to the spleen in unilaterally castrated rats: proliferative changes following subsequent removal of intact ovary. Science 108:137–138, 1948.
56. Cramer DW, Welch WR. Determinants of ovarian cancer risk. II. Inferences regarding pathogenesis. J Natl Cancer Inst 71:717–721, 1983.
57. Lin J, Lei ZM, Lojun S, Rao CV, Satyaswaroop PG, Day TG. Increased expression of luteinizing hormone/human chorionic gonadotropin receptor gene in human endometrial carcinomas. J Clin Endocrinol Metab 79:1483–1491, 1994.

58. Zheng W, Magid MS, Kramer EE, Chen YT. Follicle-stimulating hormone receptor is expressed in human ovarian surface epithelium and fallopian tube. Am J Pathol 148:47–53, 1996.
59. Schally AV. Luteinizing hormone–releasing hormone analogs: their impact on the control of tumorigenesis. Peptides 20:1247–1262, 1999.
60. Helzlsouer KJ, Alberg AJ, Gordon GB, Longcope C, Bush TL, Hoffman SC, Comstock GW. Serum gonadotropins and steroid hormones and the development of ovarian cancer. JAMA 274:1926–1930, 1995.
61. Lau KM, Mok SC, Ho SM. Expression of human estrogen receptor-alpha and -beta, progesterone receptor, and androgen receptor mRNA in normal and malignant ovarian epithelial cells. Proc Natl Acad Sci USA 96:5722–5727, 1999.
62. Risch HA. Hormonal etiology of epithelial ovarian cancer, with a hypothesis concerning the role of androgens and progesterone. J Natl Cancer Inst 90:1774–1786, 1998.
63. al-Timimi A, Buckley CH, Fox H. An immunohistochemical study of the incidence and significance of sex steroid hormone binding sites in normal and neoplastic human ovarian tissue. Int J Gynecol Pathol 4:24–41, 1985.
64. Edelson MI, Lau CC, Colitti CV, Welch WR, Bell DA, Berkowitz RS, Mok SC. A one centimorgan deletion unit on chromosome Xq12 is commonly lost in borderline and invasive epithelial ovarian tumors. Oncogene 16:197–202, 1998.
65. Evans DJ, Hoffmann RG, Kalkhoff RK, Kissebah AH. Relationship of androgenic activity to body fat topography, fat cell morphology, and metabolic aberrations in premenopausal women. J Clin Endocrinol Metab 57:304–310, 1983.
66. Sonnichsen AC, Lindlacher U, Richter WO, Schwandt P. Obesity, body fat distribution and the incidence of breast, cervical, endometrial and ovarian carcinomas. Dtsch Med Wochenschr 115:1906–1910, 1990.
67. Kaye SA, Folsom AR, Soler JT, Prineas RJ, Potter JD. Associations of body mass and fat distribution with sex hormone concentrations in postmenopausal women. Int J Epidemiol 20:151–156, 1991.
68. Mink PJ, Folsom AR, Sellers TA, Kushi LH. Physical activity, waist-to-hip ratio, and other risk factors for ovarian cancer: a follow-up study of older women. Epidemiology 7:38–45, 1996.
69. Cuzick J, Bulstrode JC, Stratton I, Thomas BS, Bulbrook RD, Hayward JL. A prospective study of urinary androgen levels and ovarian cancer. Int J Cancer 32:723–726, 1983.
70. Rodriguez GC, Walmer DK, Cline M, Krigman H, Lessey BA, Whitaker RS, Dodge R, Hughes CL. Effect of progestin on the ovarian epithelium of macaques: cancer prevention through apoptosis? J Soc Gynecol Investig 5:271–276, 1998.
71. Bu SZ, Yin DL, Ren XH, Jiang LZ, Wu ZJ, Gao QR, Pei G. Progesterone induces apoptosis and up-regulation of p53 expression in human ovarian carcinoma cell lines. Cancer 79:1944–1950, 1997.
72. Luo MP, Granada E, Zheng W, Stallcup M, Dubeau L. Hormones that regulate the menstrual cycle modulate the growth of ovarian epithelium. Proc Am Assoc Cancer Res 39:481, 1998.
73. Luo MP, Stallcup M, Zheng W, Dubeau L. Hormones controlling the menstrual cycle influence cell cycle regulation and signal transduction in benign epithelial tumors. Proc Am Assoc Cancer Res 40:499, 1999.
74. Marquis ST, Rajan JV, Wynshaw-Boris A, Xu J, Yin GY, Abel KJ, Weber BL, Chodosh LA. The developmental pattern of *Brca1* expression implies a role in differentiation of the breast and other tissues. Nat Genet 11:17–26, 1995.
75. Rajan JV, Marquis ST, Gardner HP, Chodosh LA. Developmental expression of *Brca2* colocalizes with *Brca1* and is associated with proliferation and differentiation in multiple tissues. Dev Biol 184:385–401, 1997.
76. Foulkes WD, Campbell IG, Stamp GW, Trowsdale J. Loss of heterozygosity and amplification on chromosome 11q in human ovarian cancer. Br J Cancer 67:268–273, 1993.
77. Gabra H, Taylor L, Cohen BB, Lessels A, Eccles DM, Leonard RC, Smyth JF, Steel CM. Chromosome 11 allele imbalance and clinicopathological correlates in ovarian tumours. Br J Cancer 72:367–375, 1995.
78. Gabra H, Langdon SP, Watson JE, Hawkins RA, Cohen BB, Taylor L, Mackay J, Steel CM, Leonard RC, Smyth JF. Loss of heterozygosity at 11q22 correlates with low progesterone receptor content in epithelial ovarian cancer. Clin Cancer Res 1:945–953, 1995.
79. Batra S, Sjoberg NO, Aberg A. Human placental lactogen, estradiol-17β, and progesterone levels in the third trimester and their respective values for detecting twin pregnancy. Am J Obstet Gynecol 131:69–72, 1978.
80. Jawan B, Lee JH, Chong ZK, Chang CS. Spread of spinal anaesthesia for caesarean section in singleton and twin pregnancies. Br J Anaesth 70:639–641, 1993.
81. Gilfillan CP, Robertson DM, Burger HG, Leoni MA, Hurley VA, Martin NG. The control of ovulation in mothers of dizygotic twins. J Clin Endocrinol Metab 81:1557–1562, 1996.
82. Lambe M, Wuu J, Rossing MA, Hsieh CC. Twinning and maternal risk of ovarian cancer [letter]. Lancet 353:1941, 1999.
83. Spicer DV, Shoupe D, Pike MC. GnRH agonists as contraceptive agents: predicted significantly reduced risk of breast cancer. Contraception 44:289–310, 1991.
84. Spicer DV, Ursin G, Parisky YR, Pearce JG, Shoupe D, Pike A, Pike MC. Changes in mammographic densities induced by a hormonal con-

traceptive designed to reduce breast cancer risk. J Natl Cancer Inst 86:431–436, 1994.
85. Spicer DV, Pike MC, Pike A, Rude R, Shoupe D, Richardson J. Pilot trial of a gonadotropin hormone agonist with replacement hormones as a prototype contraceptive to prevent breast cancer. Contraception 47:427–444, 1993.
86. Cohen I, Figer A, Tepper R, Shapira J, Altaras MM, Yigael D, Beyth Y. Ovarian overstimulation and cystic formation in premenopausal tamoxifen exposure: comparison between tamoxifen-treated and nontreated breast cancer patients. Gynecol Oncol 72:202–207, 1999.
87. Pirke KM, Schweiger U, Broocks A, Tuschl RJ, Laessle RG. Luteinizing hormone and follicle stimulating hormone secretion patterns in female athletes with and without menstrual disturbances. Clin Endocrinol (Oxf) 33:345–353, 1990.
88. Beitins IZ, McArthur JW, Turnbull BA, Skrinar GS, Bullen BA. Exercise induces two types of human luteal dysfunction: confirmation by urinary free progesterone. J Clin Endocrinol Metab 72:1350–1358, 1991.
89. Greene JW. Exercise-induced menstrual irregularities. Compr Ther 19:116–120, 1993.
90. McTiernan A, Ulrich C, Slate S, Potter J. Physical activity and cancer etiology: associations and mechanisms. Cancer Causes Control 9:487–509, 1998.
91. Dosemeci M, Hayes RB, Vetter R, Hoover RN, Tucker M, Engin K, Unsal M, Blair A. Occupational physical activity, socioeconomic status, and risks of 15 cancer sites in Turkey. Cancer Causes Control 4:313–321, 1993.
92. Pukkala E, Poskiparta M, Apter D, Vihko V. Lifelong physical activity and cancer risk among Finnish female teachers. Eur J Cancer Prev 2:369–376, 1993.
93. Zheng W, Shu XO, McLaughlin JK, Chow WH, Gao YT, Blot WJ. Occupational physical activity and the incidence of cancer of the breast, corpus uteri, and ovary in Shanghai. Cancer 71:3620–3624, 1993.

22

Testicular Cancer: Epidemiology and Molecular Endocrinology

ANTHONY J. SWERDLOW

Testicular cancer typically accounts for 1%–2% of cancers occurring in men in white populations and a smaller percentage in most nonwhite populations. At young adult ages, however, it is the most common cancer in whites in many countries, constituting, e.g., a third of all cancers other than non-melanoma skin cancer in men aged 20–34 in England and Wales. The incidence in young white men has been increasing across the world for many decades, but its etiology remains largely unknown. Various hypotheses of hormonal influences in etiology have been put forward and considerable epidemiological research devoted to giving evidence about these, but none can yet be regarded as clearly confirmed. Complexity is added because the hypothesized relationships have been at three different periods of life (prenatally, childhood up to puberty, and adulthood) and in two different people (the case himself and his mother). This chapter first discusses the hypothesis that prenatal hormone exposures consequent on raised sex hormone concentrations in the mother may affect the risk of testicular cancer in the son, and the evidence relating to this hypothesis, and then discusses possible relations of testicular cancer risk to childhood growth and puberty and to adult hormone levels in men. The research published on genetic polymorphisms that may affect testicular cancer risks via endocrinological or metabolic mechanisms is then reviewed, but as this evidence is as yet very limited, the main emphasis of the chapter is on the potential hormonal relationships that might be the subject of future studies of polymorphisms.

The descriptive epidemiology of testicular cancer, which is not reviewed here (see Swerdlow[1]), suggests differences in etiology between tumors occurring in three different age groups: the small peak of incidence around age 2 years, the large peak in young adults, and cases occurring after about age 60 or 65 (Fig. 22.1). Cases in children and young adults are largely of germ cell histology, whereas those in elderly men are mainly non-germ cell, principally lymphomas. There is little epidemiological information about non-germ cell malignancies or about testicular cancer at older ages, and the epidemiology of the cancer in childhood has been hampered by small numbers. Unless otherwise specified, the material in this chapter relates to testicular cancer in young adults and, by presumption (although not all studies restricted their analyses solely to these), to germ cell malignancies at these ages.

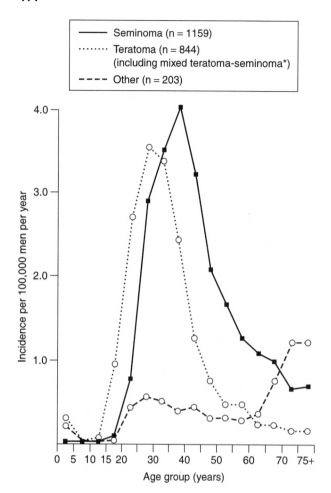

Figure 22.1 Age-specific incidence rates of testicular cancer, by histology, 1958–77, South Thames region, not including 42 testicular cancers of unknown histology and two seminomas of unknown age incident during the study period. *Teratoma and mixed teratoma–seminoma were coded as one in the registry files and could not therefore be analyzed separately. (From Swerdlow AJ, Skeet RG.[1a] with permission from the BMJ Publishing Group.)

THE PRENATAL ESTROGEN HYPOTHESIS

Several features of the epidemiology of testicular cancer strongly suggest that there is a prenatal etiological factor(s), although in the main they do not specifically indicate whether or not the factor is hormonal. Firstly, risk of testicular cancer is raised in men with cryptorchidism (the main known risk factor, accounting for about 10% of cases), childhood inguinal hernia, and probably other congenital genitourinary malformations. Aside from the obvious connection with a prenatal origin in the minority of cases that occur directly in these men, they might be a model for further cases occurring in men who have more subtle genitourinary abnormalities of prenatal origin that are not clinically diagnosed, e.g., histological abnormalities of the testis without overt clinical signs. It has been hypothesized that the raised risk of testicular cancer in relation to cryptorchidism (and other genitourinary abnormalities) might occur because the malformation and malignancy have prenatal etiological factors in common, rather than because the former causes the latter. Evidence on this hypothesis comes particularly from studies of the risk of testicular cancer in descended testes opposite maldescent and of risk of testicular malignancy after orchidopexy. This evidence is slightly supportive of the hypothesis but not clearly so. Relative risks in descended testes opposite maldescent compared with descended testes in noncryptorchid men have generally been between 1 and 1.5 (i.e., slightly raised but much less than in the undescended testis), and risks in relation to orchidopexy have been mixed and inconclusive, although with few data on risks after orchidopexy in early childhood, before degenerative histological changes to the testes have started.[1]

Secondly, the age distribution of testicular cancer in young adults (Fig. 22.1) is that which would be expected if there were a prenatal etiology with a 15- to 45-year induction period. Indeed, it is similar in shape (albeit at a slightly older age) to the age distribution of vaginal adenocarcinoma in young women caused by in utero exposure to diethylstilbestrol (DES) taken by the mother, almost always during the first trimester of pregnancy.

Thirdly, in Denmark, Norway, and Sweden, the rising incidence of testicular cancer leveled off for cohorts of men born during and just before the Second World War.[2] It is possible that this hiatus in the rising trend is a consequence of differences in in utero environment (e.g., because of changed maternal diet) for men born during the deprivations of the wartime years. The existence of this effect in men born just before the war as well as during it, however, might indicate effects in childhood rather than, or as well as, prenatally. Furthermore, no such leveling off of risk was observed for testicular cancer in men born in East Germany, Finland, Poland, Scotland, or England and Wales,[2-4] although there were considerable hardships and changes in diet during the war in several of these countries.

Based on the analogy with vaginal adenocarcinoma in young women, the relation to genitourinary malformations, and the age distribution of the tumor, Henderson et al.[5,6] developed a hormonal etiology hypothesis that might explain several features of testicular cancer epidemiology: they suggested that the underlying cause of both testicular cancer and cryptorchidism might be in utero exposure during the first trimester of pregnancy to raised levels of unbound maternal estrogens. Estrogen levels rise rapidly during the first trimester, and these authors hypothesized that exposure to this hormone might affect differentiating germ cells permanently, causing cryptorchidism in some instances, and causing, in these boys and in others in whom no overt genitourinary abnormality was manifest, abnormalities of germ cell development that led to increased risk of future malignancy. It was suggested that these germ cells then remained dormant until, under the stimulus of rising pituitary gonadotrophin levels at puberty, they divided and either became overtly malignant, reverted to normal, or died.

Several sources of raised maternal estrogen levels are possible. One is genetic variation, although this would not account for the rising secular trend of testicular cancer in white populations. Another is exogenous pharmaceutical estrogen intake. In the United States, DES was prescribed to almost 5 million women during pregnancy,[7] but its use was too uncommon in several other countries to explain the rising worldwide incidence of testicular cancer; e.g. in the United Kingdom, only about 12,000 pregnant women were treated with estrogens during 1940–1971.[8] Pharmacological exposures to estrogen in pregnancy have also arisen from the use of estrogen–progestin combinations as a test for pregnancy and inadvertent use of oral contraceptives after conception.[9] Other potential sources of estrogens that might explain the rising trend are intake of natural estrogens in foodstuffs such as soya beans, intake of dietary factors such as fiber, and fats, which may affect estrogen metabolism, excretion, and enterohepatic recycling, or intake of synthetic compounds with estrogenic properties, e.g., in foodstuffs contaminated with pesticides or from animals to which compounds such as DES were administered.[7,10] There are insufficient quantified data, however, to gauge the potential contribution (if any) of each; most, at least, are likely to be at a far lower level than endogenous estrogens.

In the 20 years since the prenatal estrogen hypothesis was put forward, a large number of epidemiological studies have been undertaken that provide relevant evidence, and these are reviewed in subsequent sections of this chapter. It will be noted, however, that with few exceptions they provide extremely indirect tests of the hypothesis, and there is the difficulty that there is uncertainty both as to whether the factors analyzed are related to risk of testicular cancer and often as to whether they relate to maternal estrogen concentrations. Thus, it is possible that even if these prenatal factors are shown to be etiologically related to subsequent risk of testicular cancer, it may not be via a hormonal mechanism.

Estrogen Concentrations in the Mother

The most direct test of the estrogen hypothesis would be to measure maternal estrogen concentrations during the index pregnancy and analyze the relation of these to subsequent risk of testicular cancer in the son. No such data have been published, however, and given the long lag period between exposure and malignancy and the uncommonness of testicular cancer, such data will be difficult to obtain. Evidence therefore rests at present on a range of more indirect sources.

The period between exposure and manifestation of disease is much shorter for cryptorchidism than for testicular malignancy, and as noted above, cryptorchidism is both a major risk factor for testicular cancer and a possible model for etiology in cases where it is not itself present. Three studies have reported risks of undescended testis in boys in relation to maternal estrogen concentrations in stored sera. All were small, ranging from inclusion of 18 to 24 cases with sera from the first 100 days or so of pregnancy. For total estradiol (E_2) concentrations, none found significant results, and it varied whether levels were greater in cases or controls.[11-13] In the one study that examined bioavailable E_2, there was significantly greater albumin-bound and nearly significantly greater free E_2 concentration in cases than controls.[12]

Exogenous Estrogen Intake by the Mother

In a strain of experimental mice with susceptibility to testicular teratoma, in utero exposure to ethinyl E_2 in early pregnancy gave a highly significant raised risk of cryptorchid testis and a nonsignificant doubled risk of testicular teratoma;[14] in mice exposed to DES in utero, testicular tumors were seen in 8% of exposed males but in no control males.[15] In humans, normal descent of the testis is under hormonal control,[16] and DES exposure in utero has been linked to raised prevalence of several structural abnormalities of the male reproductive tract, including cryptorchidism.[7]

Studies of the effect of prenatal exposure to exogenous estrogens on testicular cancer risk in humans, however, have been hampered by small numbers. A US cohort study of 1709 men with prenatal DES exposure found, tantalisingly, that they had a non-significantly raised risk of testicular cancer (relative risk of 2–3 depending on the comparison group).[16a] Limited data for a subset of the cohort did not show any clear association of risk with dose of DES or with first trimester exposure. A significant excess of testicular cancer was found in a subset of the cohort, after exclusion from the analysis of men exposed to progestins (which might reduce the effect of estrogen exposure) as well as DES in the first trimester. Several case-control studies, all from North America, have investigated the risk of testicular cancer in relation to history of maternal consumption of pharmacological estrogens.[5,17-22,22a] These studies were based on questionnaires to the men and/or their mothers, supplemented by obstetric record data in a small minority of instances in one study;[20] few subjects reported exposure, and there was no consistent evidence of raised risk. Exposures may have occurred at different times in pregnancy. A meta-analysis in 1996 of the available data showed relative risks of 2.1 [95% confidence interval (CI) 1.3–3.3] in relation to hormone exposure and 2.6 (1.1–6.1) in relation to DES exposure in utero.[7] A subsequent study found a significant relative risk of 4.9 (1.7–13.9) for hormone exposure.[22a] There is considerable potential for misclassification and bias in recall of drug intake 30 or so years earlier, however, so whether there is an etiological effect remains uncertain.

Other Obstetric Variables

A range of other obstetric variables have been investigated in relation to testicular cancer risk, several of which have potential associations with maternal estrogen concentrations. There is difficulty in taking these as evidence of prenatal hormonal etiology, however, for two reasons.

Firstly, the results on risk of the malignancy in relation to these variables have been inconsistent. In part, this is probably because of the difficulty in gaining reliable information about prenatal events that occurred decades before the malignancy. In particular, a difference in likely reliability of data should be noted between studies based on interviews with the cases or (better) their mothers after the testicular cancer had occurred, with considerable scope for recall misclassification and perhaps bias, and studies based on obstetric case notes from the time of the pregnancy with, hence, exposure data recorded before the testicular cancer had occurred. Table 22.1 shows for each published study of prenatal factors, the sources(s) of information used.

Secondly, the relation of many of the prenatal variables to maternal estrogen levels is uncertain; furthermore, these variables are often associated also with many other, nonhormonal obstetric factors and, indeed, in instances such as birth order, with postnatal factors too. Ideally, information is needed on the relation of the obstetric factors to biologically available plasma sex hormone concentrations [not bound to sex hormone–binding globulin (SHBG)] in the first trimester of pregnancy. Much of the evidence available, however, relates to total estrogen levels rather than bioavailable levels, to concentrations later in pregnancy or indeed in premenopausal women who are not currently pregnant rather than in the first trimester, and to urinary rather than plasma levels. Furthermore, the particular sex hormones assayed have varied between studies, and these hormones are of differing potential importance to testicular cancer risk. Estriol (E_3) is the predominant estrogen in the fetal circulation, whereas E_2 concentrations are greater than E_3 in maternal serum[36] and E_2 is the most biologically potent estrogen in hu-

Table 22.1 Epidemiological Studies of Prenatal Risk Factors for Testicular Cancer

Reference (country)	No. of Cases	No. of Controls	Sources of Data
Henderson et al., 1979 (US)[5]	131°	131°	Questions to subjects and mothers
Loughlin et al., 1980 (US)[18]	22	35°	Questions to subjects and mothers
Schottenfeld et al., 1980 (US)[17]	190	166 + 143†‡	Questions to subjects or parents
Coldman et al., 1982 (Canada)[23]	93°°	90°°	Questions to subjects
Depue et al., 1983 (US)[19]	108	108	Questions to subjects and mothers
Moss et al., 1986 (US)[20]	273°	273°	Questions to subjects and mothers (plus obstetric records where possible for hormone treatment)
Brown et al., 1986 (US)[21]	202	206	Questions to mothers (verification of birthweight against birth registrations where possible)[24]
Malone and Daling, 1986 (US)[25]	145	145	Birth certificates
Swerdlow et al., 1987 (England)[26]	259	238 + 251‡	Questions to subjects
Gershman and Stolley, 1988 (US)[22]	79	79	Questions to subjects or parents
Haughey et al., 1989 (US)[27]	250	250	Questions to subjects
Prener et al., 1992 (Denmark)§ [28]	183	366	School health records (questions to family at that time), plus birth certificates and midwives' records for subset
Akre et al., 1996 (Sweden)[29]	232	904	Hospital birth records
Møller and Skakkebaek, 1996, 1997 (Denmark)§ [30,31]	514°	720°	Questions to subjects and mothers
	296	287	Questions to mothers
Petridou et al., 1997 (Greece)[33]	97°	194°	Questions to subjects, mothers, and other family members, plus available documents including birth certificates and personal health booklets
Westergaard et al., 1998 (Denmark)§ [34]	626	—¶	Civil registers
Sabroe and Olsen, 1998 (Denmark)§ [32]	357	704	Midwives' records
Wanderås et al., 1998 (Norway)[35]	268	27,801°°°	National midwifery notifications
Weir et al., 2000 (Canada)[22a]	346	521	Questions to subjects and mothers
Dieckmann et al., 2001 (Germany)[35a]	418	636 + 120‡	Questions to subjects

°Number of respondents to questionnaires to subjects; fewer respondents to maternal questionnaires.
†In published table, 142.
‡Two different control sets.
°°Fewer respondents to second questionnaire asking more detail (unclear which questionnaire was used for prenatal factors).
§Probable overlap of subjects with other Danish studies.
¶N/A, cohort.
°°°Nested case-control study.

mans.[9] Thus, in interpreting the weight of evidence about hormone levels, which is described below in some detail because of its variety of measures and results, it is important to consider how much indication the data give, and how directly or indirectly they are a marker, of relevant bioavailable plasma estrogen concentrations in the first trimester.

Birth Order

In the one study that has examined it, free plasma E_2 concentrations in early pregnancy were significantly greater in women in their first than in the same women in their second pregnancy;[37] and in several studies, total plasma E_2 in early pregnancy[13,37] or late pregnancy[38] was greater in the first than the second pregnancy.

In nonpregnant premenopausal women, there has been inconsistent support for greater estrogen levels in nullipara. On day 11 of the cycle, total plasma E_2 and free E_2 levels were significantly greater in nulliparous than parous women in one study,[39] but in another, plasma estrone (E_1) and E_2 were only slightly and not significantly greater in nulliparous women.[40] In a further investigation, plasma E_2 was nonsignificantly greater in nulliparous than parous women in the luteal phase of the menstrual cycle but the opposite was found in the follicular phase.[41] In a prospective study in which early follicular-phase serum estrogen levels were measured in the same women before and after a first pregnancy, E_1 and E_2 levels were unchanged by pregnancy but E_3 levels were significantly increased.[42] When serum E_2 levels in relation to number of births were assessed at day 21 or 22, nulliparous women had greater concentrations compared with women of parity 1 or 2 but not with women of parity 3 or more.[43] Urinary E_1, E_2, and E_3 concentrations in nonpregnant premenopausal women in the follicular phase have been found greater in parous than nulliparous women at ages under 20 years and at ages 35 and above but the opposite at ages 25–34;[40,44] luteal-phase results showed similarities but were less consistent.[44] Cryptorchidism has generally been inversely associated with birth order.[31,45–47]

Thus, if the prenatal estrogen hypothesis is correct, one might expect greater risk of testicular cancer in firstborn than in second-born men and perhaps than in nonfirstborn men overall. In practice, it has generally been found, although not uniformly so, that risk of testicular cancer tends to diminish with birth order, but it has been less convincing that firstborn boys are at greater risk than second born boys. In four studies, a significant trend of lower testicular cancer risk with later birth order has been found.[26,28,34,35] Others have found nonsignificant results in the same direction[5,19,30,47a] or no consistent effect,[20,21,29,32,35a] and only in one small study was there an indication, and then not significant, of greater risk with greater maternal parity.[33] In general, despite a widespread perception that risk is raised in the firstborn, there has been no consistent or strong difference in risk between first- and second-born boys: relative risks for the former compared with the latter have mainly been in the range 0.9–1.1[5,20,22a,28,30,32,35,35a] and only slightly higher[26,34] or lower[33] in the remainder. (In studies reporting only relative risks for birth order 1 compared with 2+ or 2–3, risks were close to 1.0 in two instances[21,29] and raised in one.[19])

In contrast, there has been a considerable reduction in risk found for boys born fourth or later: the relative risk for fourth or later-born compared with firstborn has been 0.3 in two studies,[26,28] 0.6–0.8 in seven more[5,29,30,32,34,35,35a] (birth order 5+ in Sabroe and Olsen,[32] data not available for 4+ in this analysis; based on the larger and probably better of 2 control groups in Dieckmann et al.[35a]), and raised in only one study.[33] The effect appears to be greatest for seminoma: in all studies that have published data separately for risk of seminoma in boys born fourth or later,[26,28,30,32,34] the risk compared with firstborn has been substantially reduced and below that in the same analysis for testicular cancer overall. The risks found in relation to late birth order are of particular note because in a study of serum E_2 concentrations in nonpregnant premenopausal women, although levels decreased considerably from nulliparity to parity 2, they then increased with further parity to be well above that in nulliparous women for women of parity 4 or more.[43] Comparable data on bioavailable estrogen levels during early pregnancy would therefore be of great interest.

Maternal Age

Data are not available describing estrogen levels in early pregnancy in relation to maternal age. In one small study not publishing age-specific data, serum E_2 at 6–20 weeks was not significantly correlated with age.[13] At 17 weeks, serum E_3 has also been found unrelated to maternal age.[48] In late pregnancy, serum E_2 and total estrogens have been found lowest in women aged under 20 years, greatest in those aged 20–24, and intermediate in older women.[38]

In studies of premenopausal women not currently pregnant, there is inconsistent evidence that estrogen levels increase with age. On day 11 of the cycle, free E_2, total E_2, and total E_1 have been found to increase with age, in both parous and nulliparous women.[39,40] Another study found that plasma E_2 in nulliparous women increased with age in the follicular, midcycle, and luteal phases but in parous women increased only in the follicular phase, while SHBG averaged across the cycle increased with age; E_1 levels increased with age in the follicular but not the luteal phase in all women.[41] In another investigation, total plasma E_2 levels on day 21 or 22 increased inconsistently with age, but there were consistent decreases in the percentages of E_2 that were albumin-bound and free and an increase in SHBG concentration with age.[43] Urinary E_1, E_2, and E_3 have been found generally to increase with age in nulliparous women, especially in the follicular phase, but in parous women, this was so only at ages 35 and above.[40,44]

Studies of the relation of testicular cancer risk to maternal age have been inconsistent. Most have found no relation,[5,21,22a,23,26,27,29–31,34,35,35a] while one showed a strong trend of greater risk with greater maternal age[32] and another a nonsignificant trend in the same direction.[33] In analyses by histology, greater maternal age was a significant risk factor for seminoma in one study[31] and for teratoma in another;[32] in others, there was no consistent pattern.[5,29,34]

Men Born to Older Primiparous Mothers

Although the above evidence does not support an overall relation of maternal age to risk of testicular cancer, the evidence is more suggestive within sons of primiparous mothers. In a study, in England, a highly significant relation of greater maternal age to greater risk was found for nulliparous women,[26] and similar results were found in Denmark[31] and Norway.[35] In the first two studies, this effect was greater for seminoma than teratoma; in the third, the opposite applied. In a fourth study, no relation was found,[34] and in another the risk was greatest for first births to young mothers,[22a] but the oldest maternal age group examined in these latter two studies was 30 years and above, whereas the initial high risks had been for women aged 35 and above, a group contributing under half of the over-30s.[26]

The evidence on estrogen concentrations in older primiparous mothers has been inconclusive. There have been no data for early pregnancy. In late pregnancy, greatest blood levels of total E_2 have been found in women aged 20–24, in both parous and nulliparous women.[38] In nonpregnant premenopausal women, one study found that plasma E_2 and free E_2 concentrations on day 11 of the cycle were greatest in nulliparous older women;[39] a second found the same, although only slightly, for E_2 on day 11;[40] but a third found that, although this was probably so for E_2 at midcycle, it was not at other menstrual phases.[41] A study of urinary E_1, E_2, and E_3 found that these were clearly greatest in older nulliparous women in the follicular phase but less uniformly so in the luteal phase.[44]

Nausea/Hyperemesis in Index Pregnancy

Nausea in pregnancy occurs mainly in the first trimester, when there is a surge in hormone levels. Hyperemesis has been found associated with raised levels of total serum E_2 and non-protein-bound E_2 in the first trimester, an association supported by the finding that risk factors for hyperemesis are themselves associated with raised estrogen levels and by the clinical observation that administration of estrogens often leads to nausea.[49]

Risks of testicular cancer in relation to maternal nausea in the index pregnancy have been non-significantly raised in most studies,[5,19–21,22a,33] (significant in a multivariate but not a univariate analysis in Petridou et al.[33]) but the largest study on this found a borderline significantly reduced risk.[31] Restricting the analysis to first pregnancies showed greater (but still not significant) relative risks in relation to nausea in these than in subsequent pregnancies in some studies[19–21] but not all.[22a] There is some evidence for a stronger effect of nausea in relation to non-seminoma than seminoma.[20,33]

Preeclampsia

Decreased placental blood flow and other aspects of preeclampsia would lead one to expect decreased maternal estrogen levels in this condition, and urinary E_3 excretion is indeed reduced.[50] In some studies, but not all, decreased serum levels of total E_1 and E_3 have been found, but levels of nonconjugated E_1, E_2, and E_3 have not generally been decreased.[50]

Two studies have found a nonsignificantly reduced risk of testicular cancer in relation to maternal preeclampsia or eclampsia during the index pregnancy,[29,35] while one found a nonsignificantly increased risk[21] and another no association of risk with a history of preeclampsia.[31]

Maternal Diabetes

Risk of testicular cancer in men whose mothers had diabetes in pregnancy would be of particular interest because maternal diabetes leads to elevated cord serum testosterone and human chorionic gonadotropin (HCG) levels and to Leydig cell hyperplasia in the male infant of a degree that can result in macroscopically visible tumor-like nodules.[51] The placenta is large in pregnancies of diabetic mothers, which might lead one to expect raised estrogen levels; findings on maternal serum unconjugated E_1 levels have been inconsistent however.[52] Maternal diabetes has been found to be a risk factor for cryptorchidism.[45,46]

A case-control study based on obstetric records has found a significantly raised risk of testicular cancer overall and of seminoma in relation to maternal diabetes before the index pregnancy,[35] while one interview case-control study found a nonsignificantly increased risk in relation to maternal diabetes before the interview[26] and another reported 'no association', with data not presented.[31]

Maternal Weight

It is possible that estrogen levels in pregnancy are increased in obese mothers, although the one study to report directly on this found no indication of a correlation between serum E_2 at 6–20 weeks and Quetelet's index.[13] In postmenopausal women, conversion of androgens in adipose tissue is a major source of estrogens and estrogen levels are raised in heavier women; in premenopausal women, however, adipose tissue is a less important source and in nonpregnant premenopausal women total plasma E_2 levels have generally not been found to vary by weight or body mass index (BMI).[43,53] Plasma SHBG levels have been found reduced in premenopausal obese women, but the concentration of SHBG-bound E_2 was not related to weight or BMI.[53]

One study has found a significantly raised risk of testicular cancer in men whose mothers were obese before pregnancy;[19] however, two larger studies found no relation to prepregnant weight of the mother[22a,31] and a third found no relation to BMI of the mother at the time of the index birth.[29]

Ethnic Group

One of the most striking features of the descriptive epidemiology of testicular cancer is the much greater incidence rates in white than nonwhite groups, except Polynesians, when living in the same area. For instance, in Los Angeles County in 1988–1992, the annual incidence rate (age-standardized to the "World" standard population) of testicular cancer in non-Hispanic whites was 5.8 per 100,000 compared with 1.1 or less per 100,000 in blacks, Chinese, and other nonwhite groups.[54] Henderson et al.[55] therefore examined estrogen levels in early pregnancy in black compared with white women in their first pregnancies, to determine whether the differences in testicular cancer risk might be explained by greater estrogen levels in the white mothers. In fact, they found the opposite: greater total E_2 and free E_2 levels (although not significantly) were present in the black mothers, but there was a highly significantly greater testosterone level in the black mothers, which they speculated might counteract the estrogen levels and lead to reduced risks of testicular cancer. Similarly,

Hsieh et al[55a] compared serum E2 and unconjugated E3 levels at 16 weeks and at 27 weeks gestation between women with one or no previous births in Boston, US (white), and in Shanghai, China (Chinese), and found consistently and significantly greater levels in the Chinese women, despite a far greater incidence of testicular cancer in Boston than Shanghai.

Maternal Smoking

Smoking leads to decreased serum levels of total E_2 in early pregnancy,[13,56] decreasing with number of cigarettes smoked,[56] as well as reduced serum E_1 plus E_2 levels at 26 weeks[57] and reduced serum E_3 at 17 weeks and later, not depending on number of cigarettes.[48] However, SHBG binding capacity levels in early pregnancy have also been found reduced (nonsignificantly) in smokers,[56] and no data are available on bioavailable estrogen levels.

If bioavailable estrogen levels are reduced in smokers, then based on the prenatal estrogen hypothesis, risk of testicular cancer would be expected to be reduced in sons of women who smoked in pregnancy (provided that smoking does not affect risk by other mechanisms, such as direct carcinogenicity). Maternal smoking has not generally been found associated with testicular cancer risk,[5,21,26,31] although one study found a significant reduction in risk for sons of heavier smokers.[22a] A nearly significant *excess* of lung cancer has been reported in mothers of testicular cancer patients in a case-control study[26] and significant risks in the same direction in two cohort studies[58,59] but not a third.[60] (Data published in the latter only for respiratory cancers overall).

Maternal Alcohol Consumption

Maternal alcohol consumption at 26 weeks of pregnancy has been found significantly positively related to total serum estrogen concentration, with a nonsignificant relation to serum E_2.[61] In a trial in nonpregnant premenopausal women, alcohol consumption led to significantly raised plasma E_1 and E_2 (and raised amounts of bioavailable E_2) in midcycle but not other menstrual phases,[62] while an observational study found E_1, E_2, and bioavailable E_2 in all menstrual phases unrelated to alcohol consumption.[63] One study found a borderline significantly raised risk of testicular cancer in relation to maternal alcohol consumption,[21] but two found no relation.[22a,31]

Maternal Diet

A likely influence on maternal hormone levels, and one which has potential to explain geographical and secular trends in testicular cancer risks, is diet. (For a discussion of geographical associations and secular trends, see below, under (postnatal) Diet.) A Western-type, high-fat, high-protein, low-fiber diet is associated with high plasma sex hormone levels, low SHBG, and a high percentage of free E_2 in women in general[64] so it might have similar effects in pregnant women. There are no published studies, however, of the relation of maternal diet, even at the level of vegetarian/non-vegetarian, to risk of testicular cancer.

Maternal Exposure to Synthetic Nonpharmaceutical Compounds with Estrogenic Properties

Evidence on testicular cancer risk in relation to maternal exposure to estrogenic chemicals in pregnancy is very limited. Ekbom et al[65] noted that concentrations of the main isomer of dichlorodiphenyltrichloroethane (DDT) found in breast milk in different studies in the Nordic countries over time did not correspond with the time trends or geographical differences in testicular cancer incidence. In a record linkage study in Norway, Kristensen et al[66] found no association of testicular cancer risk with parental occupation in farms where pesticides with estrogenic effects might have been used.

Twins

A British study of testicular cancer in twins found a significantly raised risk for men born in dizygotic (DZ) rather than monozygotic (MZ) twinships;[67] the effect was consequent on a raised risk of seminoma but not teratoma. In Sweden, a borderline significantly raised risk of testicular cancer was found in DZ twins compared with expectations from national population incidence rates but no significant or similarly raised risk in MZ twins.[68] In other, much smaller studies, there were nonsignificant raised risks in DZ compared with MZ twins in two instances,[69,70] but not a third,[70a] and in a Norwegian study of twins, undifferentiated by zygosity, there was a non-significantly raised risk of testicular cancer in twins compared with the general population.[70b]

In conventional case-control studies of testicular cancer not focused on twins, where twins have been identified their zygosity has not been reported, and based on small numbers in individual studies, there has been no consistency as to whether risk of testicular cancer was raised in twins overall[19,21,26,31,34,35,35a,70c] (including one study of testicular cancer in children[70c]).

The findings of raised testicular cancer risk in DZ twins might accord with a prenatal hormonal etiology of testicular cancer as there is some evidence that gonadotrophin and sex hormone concentrations may be greater in DZ twin than MZ twin or singleton pregnancies. Gonadotrophin treatment for infertility often leads to multiple superovulated pregnancies. Maternal serum HCG,[71] E_2[71a] and testosterone[71a] levels in early pregnancy are much greater in twin than in singleton pregnancies. Plasma unconjugated E_2 has been shown to be greater in the third trimester[72] and serum unconjugated E_3 and HCG greater in the second trimester[73] in twin than in singleton pregnancies. In studies in Nigeria[74] and England,[71a] but not in Scotland,[74] follicle-stimulating hormone (FSH) levels in nonpregnant women who had had DZ twins were greater than in women who had had singletons.[74] In nonpregnant women who had borne one or more sets of DZ twins, FSH and luteinizing hormone (LH) concentrations were significantly greater and E_2 concentrations non-significantly greater than in parous women who had not borne DZ twins.[75]

A second way in which twins might illuminate the relation of testicular cancer to prenatal hormone exposure is the risk of the cancer in twin-born men in relation to the sex of their cotwin and, hence, to the fetal hormone output of the cotwin and the placenta. Levels of HCG in maternal serum at delivery and in cord blood have been found significantly twofold higher in male–female than in male–male twin pregnancies, possibly because of a specific "female effect."[76] In the one study of appreciable size that examined it, however, there was no relation of testicular cancer risk to sex of the cotwin,[77] although the confidence interval was compatible with moderate alterations.

Gestational Age

Most studies have found raised risk of testicular cancer in men born prematurely.[5,21,22,22a,29,31,32] One study that found no appreciable raised risk[35] was large and of a rigorous design, but numbers in the low-gestation group were small and the confidence interval would have been compatible with a substantially raised risk. The other negative studies reported only that there was 'no difference' in mean gestation[25] or not a significant finding for prematurity.[27] Limited data suggest that the relation of low gestation to risk may be for seminoma rather than nonseminoma.[22a,29,31]

Fewer studies have published data on the relation of testicular cancer risk to long gestation: three found no relation[22a,29,35] and two some evidence of a raised risk.[20,31]

Data on maternal estrogen levels in relation to subsequent gestational age of the baby at delivery are limited. One study found no relation between gestational age and serum E_3 levels in the mother at 17 weeks and at various later points in pregnancy.[48] Another found maternal salivary unconjugated E_3, which reflects free maternal serum levels, was significantly increased at 24–34 weeks of pregnancy in women who delivered preterm, compared with those with term delivery.[77a] There is inconclusive evidence for greater plasma E_2 levels at weeks 28–36 in mothers who subsequently went into preterm labour than those who did not.[77b,77c] Boys born at short gestations are exposed to in utero levels of estrogens for a shorter time than those born later, but this would be a reduction in third-trimester, not first-trimester, exposure.

Birthweight

Total plasma E_1 plus E_2 in week 26 of pregnancy[57] and E_3 in weeks 28–40[78] have been found significantly positively associated with birthweight, and there has been some evidence of an association for E_3 at week 17.[48] Concentrations of E_3 in the second and third trimesters[48,78,79] and free E_3 in the third trimester[78] tend to be reduced in pregnancies of growth-retarded fetuses, and E_3 tends to be raised in pregnancies giving rise to babies who are large for gestational age.[48,78] Risk of cryptorchidism is raised in boys of low birthweight (and in those of short gestation but probably not independently).[31,45–47]

The relation of birthweight to risk of testicular cancer has been described in six studies in which it was asked of patients and/or their mothers[19–21,22a,31,33] (Brown et al[21] verified against birth registrations where possible, with "the same estimates;"[24] possibly some recorded data also used in Petridou et al. but not clear from the publication[33]) and in four in which it was ascertained from records made at the time of birth.[25,29,32,35] A substantial raised risk for low birthweight was found in three of the studies[19,21,31] (i.e., the opposite relation to that which would accord with the prenatal estrogen etiology hypothesis), a U-shaped relation in one[29] (there was some indication of the same also in Brown et al.[21]), and no clear relation in six.[20,22a,25,32,33,35] It is notable that the results from studies with recorded data gave weak if any evidence of an association. Usually, an effect present in interview data but not in records would lead one to suspect that the former are biased (because of the fallibility of recall), but the reverse may be true to some extent here: the three studies with negative results from recorded data used a design in which controls were selected from birth records but not followed to ensure that the subjects were alive (and therefore eligible) at the ages when testicular cancer occurred in the cases.[25,32,35] Thus, the association with low birthweight may have been underestimated because low birthweight leads selectively to infant mortality. Analyses by histology have shown no consistent pattern.[20,29,31,32,35]

There has been little investigation of the joint effects of birthweight and gestation, but in the two studies that reported on this, risk was greatest for subjects who were both of low birthweight and short gestation.[21,32]

Placental Weight

The placenta is the principal source of hormone production during much of pregnancy, and there is some evidence that serum E_3 at week 17 and later increases with greater placental weight at delivery.[48] A Swedish case-control study found a nonsignificant positive trend in risk of testicular cancer overall with increasing placental weight and a borderline significant positive trend for seminoma.[29]

Handedness

It has been suggested that handedness may relate to sex hormone levels in utero.[80] One study found a nonsignificant excess of right-handed men among testicular cancer cases,[26] while two other studies found no indication of such a relation.[35a,81]

Hormone-Related Cancers in the Mothers and Sisters of Cases

If endogenous estrogen concentrations in the mother are etiological for testicular cancer, and if these raised levels persist beyond the index pregnancy, then one might expect the mothers themselves to show altered risks of hormone-related cancers, notably of the breast and endometrium. In three cohort studies in Scandinavia,[58–60] however, there was no consistent association of testicular cancer in the son with maternal endometrial or breast cancers, although there was a significantly raised risk in relation to maternal breast cancer in one study.[59] In case-control studies based on interviews with mothers[20] or with cases and controls,[35a] a raised risk of testicular cancer was found in relation to a maternal history of breast cancer (at the time of interview),[20,35a] significant in one study for non-seminoma.[20]

As altered maternal estrogen concentrations in pregnancy are also suspected as a prenatal cause of breast cancer in women,[82] if hormone concentrations were abnormal in the mother's pregnancies beyond the index one, then sisters of testicular cancer cases might be at altered risk of breast cancer. In one case-control study, a significant excess of breast cancer was reported in sisters of seminoma cases,[26] while a second found a borderline significant excess for sisters of testicular cancer cases[82a] and a third found no clear excess.[35a] In a cohort study, breast cancer risk was only slightly raised in sisters of testicular cancer patients.[58]

Other Hormone-Related Diseases in Mothers

Depue et al.[19] found a much lower incidence of surgically treated menorrhagia in mothers of testicular cancer cases than of controls. This menorrhagia was mainly perimenopausal, i.e., occurred years after the index pregnancy, and might have had a hormonal connection, although the authors did not have firm evidence of this.

POSTNATAL HORMONAL FACTORS

The age–incidence curve of testicular cancer in boys and young men (Fig. 22.1) shows parallels with testosterone and gonadotrophin concentrations by age. The peak of testicular cancer in young boys follows high levels of sex hormones and gonadotrophins in infancy (and indeed, almost adult levels of testosterone in male fetuses[83,84]), and the rising incidence of testicular cancer after puberty, reaching a peak in men aged around 30, is at ages when testosterone and free testosterone levels are increasing, and after which they decrease.[85,86] Levels of FSH and LH too rise rapidly after puberty, although unlike testicular cancer, they also increase progressively after age 40.[86] Thus, either raised gonadotrophin or testosterone levels might be etiologically associated with risk (e.g., by stimulating germ cell division), except that gonadotrophins would not in themselves explain the low risks in old age. (This might be explicable, however, if gonadotrophins had an etiological effect only in a limited pool of men whose germ cells had undergone an initial, prenatal, etiological stage.)

Hormonal abnormalities including raised serum HCG and E_2 have been found in a high proportion of men presenting with testicular cancer,[87] and partial Leydig cell insufficiency, with elevated LH and reduced testosterone levels, has been found in the contralateral testis after orchidectomy for unilateral testicular cancer.[88] In patients with unilateral testicular malignancy, a raised serum FSH level is predictive of the future development of malignancy or presence of cancer in situ in the contralateral testis.[89] These findings, however, are after incidence of a first malignancy and might be a consequence of it. Data on hormone levels before incidence of the tumor would be much more difficult to obtain, and no such data have been published. The available evidence is more indirect and, like the evidence for prenatal hormonal etiology, has the difficulty that there is often uncertainty about the relation of the factors investigated to both risk of testicular cancer and postnatal hormone levels. Data on a range of potentially relevant variables, such as age at puberty, baldness, and exercise, are discussed below. While the available information on many of these factors does not indicate the age at which any effects might occur, there are at least conceptually two different periods of postnatal life that might matter. Firstly, hormonal mechanisms might occur in relation to puberty and, because the occurrence of puberty is determined by events happening during the years before it, might also relate to factors during childhood. Secondly, hormonal factors might operate in adulthood.

Age at Puberty

As testicular cancer is rare in children before puberty and incidence rises rapidly in the late teens and as there is a potential analogy in the relation of the risk of breast cancer to age at menarche, it seems possible that age at puberty (and hence age at start of exposure to high adult testosterone and gonadotrophin concentrations) might relate to risk of subsequent testicular cancer. Several studies have investigated this possibility, with inconclusive results overall. The largest study[90] found a significant trend of increasing testicular cancer risk with younger age at occurrence of various markers of puberty, and another substantial study also found a raised risk with young age at puberty,[20] although this was confined to cases of malignancy incident at ages under 30. Two other large studies[30,91] found decreased risk in relation to both young and old ages at puberty, and another found decreased risk with old age at puberty and an inconsistently increased risk with young ages.[92] Several further investigations found no substantial relation,[19,27,93] and one in which all controls had leukemia[94] found a significantly older average age at puberty in cases than controls.

The timing of puberty in boys is difficult to

ascertain from retrospective questions to adult men, as there is no sharp defining event analogous to menarche in girls. The specific questions asked in the different studies varied, but all were potentially susceptible to substantial recall misclassification and perhaps bias. Thus, even if there is an association, it might plausibly not have been demonstrated consistently in studies to date. Age at female puberty has been declining over time, and the same may apply to men. (The age peak of the adolescent growth spurt in boys has become younger.[95]) Hence, decreasing age at puberty might be a factor contributing to the secular increase in testicular cancer incidence. Because attainment of puberty in boys, as in girls, is likely to be triggered by growth in childhood, childhood nutrition might well be a factor underlying any association with pubertal events.

Height

Although the literature is not entirely consistent, the balance of it is in favor of a raised risk of testicular cancer in taller men. Highly significant trends in this direction were seen in a large case-control study in Canada[93] and a Norwegian cohort,[96] and there was less strong evidence in three other case-control studies[91,97,98] and one cohort.[99] Other studies, some of which were large, found no clear relationship however.[33,81,100–102]

A study comparing men with testicular cancer who were twins with their unaffected male cotwins found a nonsignificantly raised risk in relation to being the taller twin (both in childhood and in adulthood) and significantly raised risks for men who had longer legs and arms than their co-twin.[103]

Adult height, like testicular cancer incidence, has been increasing throughout the twentieth century in Western countries.[95] This increase principally reflects prepubertal growth, probably of nutritional origin,[104] and in several countries has been almost entirely due to increased leg length.[105] Therefore, the association with height and the rising incidence of testicular cancer could be due to better nutrition in childhood leading to endocrinological effects on both height and risk of testicular cancer (height is also affected by intrauterine growth, but testicular cancer has been associated with *low* birthweight).

Diet

A nutritional etiology of testicular cancer is plausible from the general descriptive epidemiology of the cancer as well as the apparent association with height. Internationally, incidence rates of testicular cancer show a strong positive correlation with geographical variations in fat consumption,[106] and fat consumption and other aspects of the Western diet have increased over many decades, with, in Britain at least, a reversal during the Second World War.[4] (Fat consumption in Britain has fallen since 1969,[4] so a reversal of the increase in testicular cancer incidence might be expected in recent cohorts if there is indeed an etiological connection.)

Experimentally, change to a high-fat diet in Western men has been found to lead to reduced serum SHBG concentrations, and change to a low-fat diet to raised SHBG, reduced or unchanged total testosterone, and reduced free testosterone.[107–109] In Western men who are vegetarians or vegans compared with those who eat a usual Western diet, in which intake of fiber is low and 40% of calories are from fat, studies have found a significant reduction in total plasma testosterone[110] or no significant change[111,112] and a significant increase in SHBG,[111,112] so that free testosterone might be reduced, although estimates of this have been contradictory.[111,112] Vegetarians may differ from omnivores in other ways as well as diet, however.

Direct evidence on risk of testicular cancer in relation to diet has been limited, leaving it uncertain whether there is a relation to a Western-style diet. A hospital-based case-control study with a low response rate, potentially biased controls, and collection of data for different periods in cases than controls found large raised risks of testicular cancer, especially nonseminoma, in relation to greater consumption of total fat, saturated fat, and cholesterol, and lower consumption of fiber.[113] A further study found a significantly raised risk in relation to greater recalled milk consumption,[102] while another found no relation to intake of milk, other dairy products, or meat.[91] In a cohort study, there was no substantial relation to prior blood cholesterol and triglycerides.[99] In affected twins compared with their unaffected cotwins, there was no appreciable difference for reported dairy, red meat, or fruit consumption in adolescence, but a borderline significantly raised risk in relation to low vegetable consumption,[103] a variable related to risk in two other studies,[114,114a] in one significantly.[114]

Acne

Acne appears at an age when testosterone levels are rising. It is also often manifest in women who have hormonal abnormalities that include androgen overproduction, and there is evidence, although not conclusive, that acne in men may be a consequence of high testosterone levels.[115] Serum testosterone and dehydroepiandrosterone sulfate levels in males with acne have been found increased or normal and SHBG levels low.[115] It may be, however, that the key etiological factor for acne is the extent of testosterone metabolism in the skin, rather than circulating testosterone levels.[116]

An early study on the risk of testicular cancer in relation to acne found significant protection,[19] but others have not found significant results or any consistent indication of an association.[5,20,81,97,103]

Postnatal Sex Hormone Intake

If risk of testicular malignancy is affected by postnatal hormone concentrations, one might expect the risk of the cancer to be altered in individuals who have taken pharmaceutical sex hormones, at least if for a sustained period. In practice, however, few studies have reported on this, and as use is uncommon and recall imperfect, it is a difficult issue to investigate. In one study,[20] significantly decreased risks of testicular cancer were found in men who had used exogenous hormones (half of these exposures were cortisone or prednisone, and there was a deficit of cases who had used anabolic steroids); the authors considered, however, that the results might have been due to referral bias. In two other studies,[93,117] there was no association of hormonal drug use with risk of testicular cancer.

Baldness and Frequency of Shaving

Evidence from eunuchoid and prepubertally castrated men and from women with high androgen levels or who were administered androgens suggests that androgens are critical to the development of male-pattern baldness. In men with male-pattern baldness, measures of plasma androgen and free androgens have not been found raised but urinary steroid levels were increased.[118]

The only study to examine risk of testicular cancer in relation to baldness found a significantly reduced risk in bald men, especially for nonseminoma.[33] Three studies have investigated risks in relation to frequency of shaving as an adult.[19,27,91] None found a significant overall relationship, although in one there was a barely significantly reduced risk for men who needed to shave less than once a day compared with those who shaved once a day.[91]

Obesity

Plasma total non-SHBG-bound testosterone and free testosterone levels have been found to decrease with increasing BMI in men,[119] generally with normal gonadotrophin concentrations.[120,121] There has been no consistent relation of obesity to risk of testicular malignancy in epidemiological studies. One study found a significantly increased risk in obese men,[122] and another an effect in the same direction,[114a] whereas two others found borderline significantly decreased risks in such individuals,[33,96] and most found no relation.[81,91,93,99–101,113]

Ethnic Group

As noted earlier, rates of testicular cancer incidence are considerably greater in whites than in blacks. Young adult black men, however, have significantly higher concentrations of circulating free and total testosterone than whites,[123] the opposite of the pattern if androgens were to explain the ethnic difference in rates.

Alcohol Consumption

Consumption of alcohol leads to a decrease in plasma testosterone concentrations and a decrease in plasma testosterone binding capacity in men.[124] Three studies have investigated testicular cancer risk in relation to reported alcohol consumption;[81,91,98] they found no convincing evidence of an association.

Smoking

Serum total and free testosterone concentrations in men have generally been found positively associated with smoking.[125] A significantly raised risk of testicular cancer in relation to smoking was found in two case-control studies[97,114a] and near-significant raised risks in two more,[81,93] but others have not found any relation.[5,23,30,100,122]

Exercise

Vigorous regular exercise in men can significantly reduce resting levels of total serum testosterone, free testosterone, and non-SHBG-bound testosterone[126,127] and perhaps raise LH concentrations, although most studies have found LH to be normal.[127] Thus, an association of testicular cancer with lack of exercise would be in accord with a hypothesis of postnatal androgenic etiology. In two large case-control studies, significant relations of testicular cancer risk to lack of exercise were indeed shown.[90,93] In one of these studies, there was a significant downward trend in risk with numbers of hours of exercise, both at age 20 and at 1 year before diagnosis of malignancy.[90] An analysis of cancer registry data by job title showed a significant trend of increasing testicular cancer risk with decreasing estimated occupational exercise.[128] In two cohort studies, in which 47[99] and 67[100] testicular cancers occurred, there was no relation of risk to previous exercise. In a case-control study comparing twins with testicular cancer with their unaffected twin brothers, there was no relation of risk to reported childhood exercise, again based on modest numbers;[103] in another case-control study, cases and controls were reported to be similar with respect to "crude estimators of physical activity" at various ages,[113] and one case-control study found a significant relation of greater exercise in the mid-teens with *increased* risk of testicular cancer.[114a]

Testicular Atrophy

Raised risk of testicular cancer has been found in men with testicular atrophy without cryptorchidism,[27,81,117] with some evidence of raised risk in the testis contralateral to atrophy.[81,117] Whether the atrophy was of prenatal or postnatal origin is unclear. As atrophy can lead to raised gonadotrophin levels by feedback, and cryptorchidism is also associated with raised gonadotrophin levels,[129] this gives evidence compatible with an etiological role of gonadotrophins in testicular malignancy.

GENETIC CONDITIONS INVOLVING ABNORMAL HORMONE LEVELS

In principle, much could be learned about the effects of endogenous hormones on risk of testicular cancer from studying men with genetic conditions leading to abnormal sex hormone levels or action. Studies to date have been hampered by small numbers, however, because of the uncommonness of these conditions.

Men with Klinefelter's syndrome have on average reduced serum testosterone and greatly increased gonadotrophin levels, and characteristically have testicular atrophy.[130] Their risk of testicular cancer would therefore be of considerable interest in relation to theories of the effects of postnatal androgen and gonadotrophin levels. The only published data on risk of testicular cancer in patients with Klinefelter's syndrome, however, were too few to resolve the risk: one case occurred compared with 1.75 expected in a Danish cohort of 696 Klinefelter's syndrome patients,[131] and in testicular biopsies from 35 boys and men with Klinefelter's syndrome, no cases of carcinoma in situ were found.[132]

Although there are no epidemiological data to quantify the risk, it is clear from case reports that men with XY gonadal dysgenesis are at much raised risk of gonadal malignancy; case reports do not, however, obviously indicate a raised risk in gonadal dysgenesis without a Y chromosome, even though gonadotrophin levels are elevated.[133]

Cancer in situ has been found in testes of 4 of 12 (25%) patients with androgen insensitivity syndrome,[134] and several cases of testicular cancer have been reported in patients with complete androgen insensitivity, suggesting that risk is increased,[133] although this has not been quantified.

In men with Down's syndrome, there is a raised incidence of cryptorchidism, and often testicular atrophy and dysgenesis. Plasma concentrations of gonadotrophins are often elevated, and such elevations have also been shown in two Down's patients with testicular cancer.[135] Based on case reports and on two cohort studies in which a nonsignificant excess of testicular cancer occurred,[136,136a] there is modest evidence for a raised testicular cancer risk in men with Down's syndrome.

GENETIC POLYMORPHISMS

Studies of testicular cancer risk in first-degree relatives of men with the disease show a strong familial component, and indeed larger familial relative risks than for most other malignancies. Relative risks of testicular cancer in fathers of cases have been about 2–4,[137–139] in brothers about 10,[137–140] in sons 4,[140] in men who were the son and sib of a case 29,[140] in DZ twins up to 36,[67,140a] and in MZ twins 60–76.[67,140a] (the greater risk in DZ twins than nontwin brothers, although with wide confidence intervals, and for brothers than for fathers or sons, would be compatible with shared intrauterine environment as part of the reason for the raised sibling risk). There is evidence that the familial tendency may sometimes relate to undescended testis and other genitourinary abnormalities as well as testicular malignancy.[26,103,141]

These strong familial risks could be due to polymorphisms for genes controlling hormones or metabolic enzymes relevant to risk. The study of such polymorphisms in relation to testicular cancer risk has only just begun, however.

Glutathione S-transferase μ (GSTμ) is an enzyme important in the detoxification and deactivation of carcinogens. In a case-control study in Denmark,[98] there was no relation between GSTμ-negative phenotype or GSTM-1 null genotype and risk of testicular cancer, with confidence intervals ruling out (at the 95% level) relative risks of greater than 1.7. In analyses by histological type, there were no significant results.

In a large Norwegian case-control study, the allelic frequencies of three polymorphisms within the estrogen receptor gene were no different in testicular cancer patients than in controls,[142] nor was there any indication that allele frequencies differed in familial or cryptorchid subgroups of cases, although there was a nonsignificant increase in variant B allele in firstborn cases.

CONCLUSIONS

The hypothesis that prenatal exposure to elevated levels of maternal free estrogen in the first trimester of pregnancy is etiological for testicular cancer remains plausible but unproven. The epidemiological evidence to date has mainly, but not uniformly, been compatible with this hypothesis, although not strongly supportive of it. The most consistent evidence for prenatal relationships has been for a reduced risk with late birth order and for raised risks with DZ twinship and prematurity (especially prematurity with low birthweight). In each instance there is some evidence, varying in strength, that the risk is greater for seminoma than nonseminoma histologies, which would accord with the evidence that the risk after cryptorchidism is particularly high for seminoma.[1] There is some inconsistent evidence in favor of an association of testicular cancer risk with low birthweight, which would be contrary to the prenatal estrogen etiology hypothesis. The greater levels of maternal estrogens found in black and Chinese pregnant women than in whites, despite the much greater testicular cancer incidence in the latter, appears to argue against an etiology largely due to maternal estrogen levels, but interpretation of the estrogen data is not straightforward,[142a] and genetic differences complicate interpretation.

The nearest to convincing evidence of postnatal associations of testicular cancer, although not consistently found, have been trends of rising risk in relation to tallness and lack of exercise, and a diminished risk for men who had an older age at puberty. These could be related to hormonal factors, but do not at present give strong evidence for a postnatal hormonal etiology.

There is therefore reason to investigate the relation of testicular cancer risk to polymorphisms in genes in the mother concerning the synthesis, metabolism, and availability of estrogens, and in the son concerning the synthesis, metabolism, and action of testosterone and gonadotrophins.

Finally, there is another potential type of hormonal etiology of testicular cancer which does not appear to have been investigated, but may be worth consideration as it might explain several of the epidemiological features of testicular cancer discussed above. Insulin-like growth factor-I (IGF-I) is associated with height, in prepubertal children at least, and with pubertal development, and levels in children and adolescents are associated with testicular volume.[143] Serum concentrations reach a steep peak at puberty before declining throughout adulthood.[143] Strenuous exercise that produces a negative caloric balance diminishes IGF-I levels,[144] as does restriction of protein or energy intake.[145] Furthermore, IGF-I levels in the later stages of puberty are greater if puberty occurs at a younger age (although the reverse is true of levels in the early stages of puberty),[143] giving a potential reason for age at puberty to affect risk of testicular cancer. In the light of recent findings of an association between IGF-I concentrations and subsequent risks of premenopausal breast cancer[146] and prostate cancer,[147] two other height-related cancers, it would be worthwhile to explore whether testicular cancer risk relates to prior levels of IGF-I, and IGF binding proteins (especially IGFBP3, the main binding protein in the circulation), and to genes regulating IGF-I levels.

REFERENCES

1. Swerdlow AJ. Epidemiology of testicular cancer. In: Raghavan D, Scher HI, Leibel SA, Lange PH (eds). Principles and Practice of Genitourinary Oncology. Philadelphia: Lippincott-Raven, 1997, pp 643–652.
1a. Swerdlow AJ, Skeet RG. Occupational associations of testicular cancer in south east England. Br J Indust Med 45:225–230, 1988.
2. Bergström R, Adami H-O, Möhner M, et al. Increase in testicular cancer incidence in six European countries: a birth cohort phenomenon. J Natl Cancer Inst 88:727–733, 1996.
3. Swerdlow AJ, dos Santos Silva I, Reid A, Qiao Z, Brewster DH, Arrundale J. Trends in cancer incidence and mortality in Scotland: description and possible explanations. Br J Cancer 77(Suppl 3):1–54, 1998.
4. Swerdlow AJ, dos Santos Silva I, Doll R. Cancer Incidence and Mortality in England and Wales: Trends and Risk Factors. Oxford: Oxford University Press, 2001.
5. Henderson BE, Benton B, Jing J, Yu MC, Pike MC. Risk factors for cancer of the testis in young men. Int J Cancer 23:598–602, 1979.
6. Henderson BE, Ross RK, Pike MC, Depue RH. Epidemiology of testis cancer. In: Skinner DG (ed). Urological Cancer. New York: Grune and Stratton, 1983, pp 237–250.
7. Toppari J, Larsen JC, Christiansen P, et al. Male reproductive health and environmental xenoestrogens. Environ Health Perspect 104(Suppl 4):741–803, 1996.
8. Kinlen LJ, Badaracco MA, Moffett J, Vessey MP. A survey of the use of oestrogens during pregnancy in the United Kingdom and of the genito-urinary cancer mortality and incidence rates in young people in England and Wales. J Obstet Gynaecol Br Commonwealth 81:849–855, 1974.
9. Henderson BE, Ross R, Bernstein L. Estrogens as a cause of human cancer: the Richard and Hinda Rosenthal Foundation Award lecture. Cancer Res 48:246–253, 1988.
10. Sharpe RM, Skakkebaek NE. Are oestrogens involved in falling sperm counts and disorders of the male reproductive tract? Lancet 341:1392–1395, 1993.
11. Burton MH, Davies TW, Raggatt PR. Undescended testis and hormone levels in early pregnancy. J Epidemiol Community Health 41:127–129, 1987.
12. Bernstein L, Pike MC, Depue RH, Ross RK, Moore JW, Henderson BE. Maternal hormone levels in early gestation of cryptorchid males: a case-control study. Br J Cancer 58:379–381, 1988.
13. Key TJA, Bull D, Ansell P, et al. A case-control study of cryptorchidism and maternal hormone concentrations in early pregnancy. Br J Cancer 73:698–701, 1996.
14. Walker AH, Bernstein L, Warren DW, Warner NE, Zheng X, Henderson BE. The effect of in utero ethinyl oestradiol exposure on the risk of cryptorchid testis and testicular teratoma in mice. Br J Cancer 62:599–602, 1990.
15. Newbold RR, Bullock BC, McLachlan JA. Testicular tumors in mice exposed in utero to diethylstilbestrol. J Urol 138:1446–1450, 1987.
16. Hutson JM. A biphasic model for the hormonal

control of testicular descent. Lancet 2:419–421, 1985.
16a. Strohsnitter WC, Noller KL, Hoover RN, et al. Cancer risk in men exposed in utero to diethylstilbestrol. J Natl Cancer Inst 93:545–551, 2001.
17. Schottenfeld D, Warshauer ME, Sherlock S, Zauber AG, Leder M, Payne R. The epidemiology of testicular cancer in young adults. Am J Epidemiol 112:232–246, 1980.
18. Loughlin JE, Robboy SJ, Morrison AS. Risk factors for cancer of the testis. N Engl J Med 303:112–113, 1980.
19. Depue RH, Pike MC, Henderson BE. Estrogen exposure during gestation and risk of testicular cancer. J Natl Cancer Inst 71:1151–1155, 1983.
20. Moss AR, Osmond D, Bachetti P, Torti FM, Gurgin V. Hormonal risk factors in testicular cancer: a case-control study. Am J Epidemiol 124:39–52, 1986.
21. Brown LM, Pottern LM, Hoover RN. Prenatal and perinatal risk factors for testicular cancer. Cancer Res 46:4812–4816, 1986.
22. Gershman ST, Stolley PD. A case-control study of testicular cancer using Connecticut Tumour Registry data. Int J Epidemiol 17:738–742, 1988.
22a. Weir HK, Marrett LD, Kreiger N, Darlington GA, Sugar L. Pre-natal and peri-natal exposures and risk of testicular germ-cell cancer. Int J Cancer 87:438–443, 2000.
23. Coldman AJ, Elwood JM, Gallagher RP. Sports activities and risk of testicular cancer. Br J Cancer 46:749–756, 1982.
24. Depue RH, Pike MC, Henderson BE. Birth weight and the risk of testicular cancer. J Natl Cancer Inst 77:829–830, 1986.
25. Malone KE, Daling JR. Birth weight and the risk of testicular cancer. J Natl Cancer Inst 77:829, 1986.
26. Swerdlow AJ, Huttly SRA, Smith PG. Prenatal and familial associations of testicular cancer. Br J Cancer 55:571–577, 1987.
27. Haughey BP, Graham S, Brasure J, Zielezny M, Sufrin G, Burnett WS. The epidemiology of testicular cancer in upstate New York. Am J Epidemiol 130:25–36, 1989.
28. Prener A, Hsieh C-C, Engholm G, Trichopoulos D, Jensen OM. Birth order and risk of testicular cancer. Cancer Causes Control 3:265–272, 1992.
29. Akre O, Ekbom A, Hsieh C-C, Trichopoulos D, Adami H-O. Testicular nonseminoma and seminoma in relation to perinatal characteristics. J Natl Cancer Inst 88:883–889, 1996.
30. Møller H, Skakkebaek NE. Risks of testicular cancer and cryptorchidism in relation to socio-economic status and related factors: case-control studies in Denmark. Int J Cancer 66:287–293, 1996.
31. Møller H, Skakkebaek NE. Testicular cancer and cryptorchidism in relation to prenatal factors: case-control studies in Denmark. Cancer Causes Control 8:904–912, 1997.
32. Sabroe S, Olsen J. Perinatal correlates of specific histological types of testicular cancer in patients below 35 years of age: a case-cohort study based on midwives' records in Denmark. Int J Cancer 78:140–143, 1998.
33. Petridou E, Roukas KI, Dessypris N, et al. Baldness and other correlates of sex hormones in relation to testicular cancer. Int J Cancer 71:982–985, 1997.
34. Westergaard T, Andersen PK, Pedersen JB, Frisch M, Olsen JH, Melbye M. Testicular cancer risk and maternal parity: a population-based cohort study. Br J Cancer 77:1180–1185, 1998.
35. Wanderås EH, Grotmol T, Fosså SD, Tretli S. Maternal health and pre- and perinatal characteristics in the etiology of testicular cancer: a prospective population- and register-based study on Norwegian males born between 1967 and 1995. Cancer Causes Control 9:475–486, 1998.
35a. Dieckmann K-P, Endsin G, Pichlmeier U. How valid is the prenatal estrogen excess hypothesis of testicular germ cell cancer? A case control study on hormone-related factors. Eur Urol 40:677–683, 2001.
36. Kaplan SL, Grumbach MM. Pituitary and placental gonadotrophins and sex steroids in the human and sub-human primate fetus. Clin Endocrinol Metab 7:487–511, 1978.
37. Bernstein L, Depue RH, Ross RK, Judd HL, Pike MC, Henderson BE. Higher maternal levels of free estradiol in first compared to second pregnancy: early gestational differences. J Natl Cancer Inst 76:1035–1039, 1986.
38. Panagiotopoulou K, Katsouyanni K, Petridou E, Garas Y, Tzonou A, Trichopoulos D. Maternal age, parity, and pregnancy estrogens. Cancer Causes Control 1:119–124, 1990.
39. Bernstein L, Pike MC, Ross RK, Judd HL, Brown JB, Henderson BE. Estrogen and sex hormone-binding globulin levels in nulliparous and parous women. J Natl Cancer Inst 74:741–745, 1985.
40. Yu MC, Gerkins VR, Henderson BE, Brown JB, Pike MC. Elevated levels of prolactin in nulliparous women. Br J Cancer 43:826–831, 1981.
41. Dorgan JF, Reichman ME, Judd JT, et al. Relationships of age and reproductive characteristics with plasma estrogens and androgens in premenopausal women. Cancer Epidemiol Biomarkers Prev 4:381–386, 1995.
42. Musey VC, Collins DC, Brogan DR, et al. Long term effects of a first pregnancy on the hormonal environment: estrogens and androgens. J Clin Endocrinol Metab 64:111–118, 1987.
43. Ingram DM, Nottage EM, Willcox DL, Roberts A. Oestrogen binding and risk factors for breast cancer. Br J Cancer 61:303–307, 1990.

44. Trichopoulos D, Cole P, Brown JB, Goldman MB, MacMahon B. Estrogen profiles of primiparous and nulliparous women in Athens, Greece. J Natl Cancer Inst 65:43–46, 1980.
45. Hjertkvist M, Damber J-E, Bergh A. Cryptorchidism: a registry based study in Sweden on some factors of possible aetiological importance. J Epidemiol Community Health 43:324–329, 1989.
46. Jones ME, Swerdlow AJ, Griffith M, Goldacre MJ. Prenatal risk factors for cryptorchidism: a record linkage study. Paediatr Perinat Epidemiol 12:383–396, 1998.
47. Akre O, Lipworth L, Cnattingius S, Sparén P, Ekbom A. Risk factor patterns for cryptorchidism and hypospadias. Epidemiology 10:364–369, 1999.
47a. Hemminki K, Mutanen P. Birth order, family size, and the risk of cancer in young and middle-aged adults. Br J Cancer 84:1466–1471, 2001.
48. Kaijser M, Granath F, Jacobsen G, Cnattingius S, Ekbom A. Maternal pregnancy oestriol levels in relation to anamnestic and fetal anthropometric data. Epidemiology 11:315–319, 2000.
49. Depue RH, Bernstein L, Ross RK, Judd HL, Henderson BE. Hyperemesis gravidarum in relation to estradiol levels, pregnancy outcome, and other maternal factors: a seroepidemiologic study. Am J Obstet Gynecol 156:1137–1141, 1987.
50. Rosing U, Carlström K. Serum levels of unconjugated and total oestrogens and dehydroepiandrosterone, progesterone and urinary oestriol excretion in pre-eclampsia. Gynecol Obstet Invest 18:199–205, 1984.
51. Barbieri RL, Saltzman D, Phillippe M, et al. Elevated β-human chorionic gonadotropin and testosterone in cord serum of male infants of diabetic mothers. J Clin Endocrinol Metab 61:976–979, 1985.
52. Axelsson O, Lindberg BS, Nilsson BA, Johansson EDB. Plasma levels of non-conjugated oestrone in high risk pregnancies. Acta Obstet Gynecol Scand 57:113–119, 1978.
53. Dorgan JF, Reichman ME, Judd JT, et al. The relation of body size to plasma levels of estrogens and androgens in premenopausal women (Maryland, United States). Cancer Causes Control 6:3–8, 1995.
54. Bernstein L, Boone J, Deapen D, Ross R. USA, California, Los Angeles County. In: Parkin DM, Whelan SL, Ferlay J, Raymond L, Young J (eds). Cancer Incidence in Five Continents, vol VII. IARC Scientific Publication 143. Lyon: IARC, 1997, pp 198–225.
55. Henderson BE, Bernstein L, Ross RK, Depue RH, Judd HL. The early in utero oestrogen and testosterone environment of blacks and whites: potential effects on male offspring. Br J Cancer 57:216–218, 1988.
55a. Hsieh CC, Lambe M, Trichopoulos D, Ekbom A, Akre O, Adami H-O. Early life exposure to estrogen and testicular cancer risk: evidence against an aetiological hypothesis. Br J Cancer 86:1363–1364, 2002.
56. Bernstein L, Pike MC, Lobo RA, Depue RH, Ross RK, Henderson BE. Cigarette smoking in pregnancy results in marked decrease in maternal HCG and oestradiol levels. Br J Obstet Gynaecol 96:92–96, 1989.
57. Petridou E, Panagiotopoulou K, Katsouyanni K, Spanos E, Trichopoulos D. Tobacco smoking, pregnancy estrogens, and birth weight. Epidemiology 1:247–250, 1990.
58. Heimdal K, Olsson H, Tretli S, Flodgren P, Børresen A-L, Fosså SD. Risk of cancer in relatives of testicular cancer patients. Br J Cancer 73:970–973, 1996.
59. Vaittinen P, Hemminki K. Familial cancer risks in offspring from discordant parental cancers. Int J Cancer 81:12–19, 1999.
60. Kroman N, Frisch M, Olsen JH, Westergaard T, Melbye M. Oestrogen related cancer risk in mothers of testicular-cancer patients. Int J Cancer 66:438–440, 1996.
61. Petridou E, Katsouyanni K, Spanos E, Skalkidis Y, Panagiotopoulou K, Trichopoulos D. Pregnancy estrogens in relation to coffee and alcohol intake. Ann Epidemiol 2:241–247, 1992.
62. Reichman ME, Judd JT, Longcope C, et al. Effects of alcohol consumption on plasma and urinary hormone concentrations in premenopausal women. J Natl Cancer Inst 85:722–727, 1993.
63. Dorgan JF, Reichman ME, Judd JT, et al. The relation of reported alcohol ingestion to plasma levels of estrogens and androgens in premenopausal women (Maryland, United States). Cancer Causes Control 5:53–60, 1994.
64. Adlercreutz H. Diet, breast cancer, and sex hormone metabolism. Ann NY Acad Sci 595:281–290, 1990.
65. Ekbom A, Wicklund-Glynn A, Adami H-O. DDT and testicular cancer. Lancet 347:553–554, 1996.
66. Kristensen P, Andersen A, Irgens LM, Bye AS, Sundheim L. Cancer in offspring of parents engaged in agricultural activities in Norway: incidence and risk factors in the farm environment. Int J Cancer 65:39–50, 1996.
67. Swerdlow AJ, De Stavola BL, Swanwick MA, Maconochie NES. Risks of breast and testicular cancers in young adult twins in England and Wales: evidence on prenatal and genetic aetiology. Lancet 350:1723–1728, 1997.
68. Braun MM, Ahlbom A, Floderus B, Brinton LA, Hoover RN. Effect of twinship on incidence of cancer of the testis, breast, and other sites (Sweden). Cancer Causes Control 6:519–524, 1995.
69. Braun MM, Caporaso NE, Brinton L, Page WF. Re: Twin membership and breast cancer risk. Am J Epidemiol 140:575–576, 1994.

70. Braun MM, Caporaso NE, Page WF, Hoover RN. Prevalence of a history of testicular cancer in a cohort of elderly twins. Acta Genet Med Gemellol (Roma) 44:189–192, 1995.
70a. Verkasalo PK, Kaprio J, Koskenvuo M, Pukkala E. Genetic predisposition, environment and cancer incidence: a nationwide twin study in Finland, 1976–1995. Int J Cancer 83:743–749, 1999.
70b. Iversen T, Tretli S, Kringlen E. An epidemiological study of cancer in adult twins born in Norway 1905–1945. Br J Cancer 84:1463–1465, 2001.
70c. Li FP, Fraumeni JF Jr. Testicular cancers in children: epidemiologic characteristics. J Natl Cancer Inst 48:1575–1581, 1972.
71. Jovanic L, Landesman R, Saxena BB. Screening for twin pregnancy. Science 198:738, 1977.
71a. Thomas HV, Murphy MF, Key TJ, Fentiman IS, Allen DS, Kinlen LJ. Pregnancy and menstrual hormone levels in mothers of twins compared to mothers of singletons. Ann Hum Biol 25:69–75, 1998.
72. TambyRaja RL, Ratnam SS. Plasma steroid changes in twin pregnancies. In: Twin Research 3: Twin Biology and Multiple Pregnancy. New York: Alan R Liss, 1981, pp 189–195.
73. Wald N, Cuckle H, Wu T, George L. Maternal serum unconjugated oestriol and human chorionic gonadotrophin levels in twin pregnancies: implications for screening for Down's syndrome. Br J Obstet Gynaecol 98:905–908, 1991.
74. Nylander PPS. The factors that influence twinning rates. Acta Genet Med Gemellol (Roma) 30:189–202, 1981.
75. Martin NG, El Beaini JL, Olsen ME, Bhatnagar AS, Macourt D. Gonadotropin levels in mothers who have had two sets of DZ twins. Acta Genet Med Gemellol (Roma) 33:131–139, 1984.
76. Steier JA, Myking OL, Ulstein M. Human chorionic gonadotropin in cord blood and peripheral maternal blood in singleton and twin pregnancies at delivery. Acta Obstet Gynecol Scand 68:689–692, 1989.
77. Swerdlow AJ, De Stavola B, Maconochie N, Siskind V. A population-based study of cancer risk in twins: relationships to birth order and sexes of the twin pair. Int J Cancer 67:472–478, 1996.
77a. McGregor JA, Jackson GM, Lachelin GCL, Goodwin TM, Artal R, Hastings C, Dullien V. Salivary estriol as risk assessment for preterm labor: a prospective trial. Am J Obstet Gynecol 173:1337–1342, 1995.
77b. TambyRaja RL, Turnbull AC, Ratnam SS. Predictive oestradiol surge of premature labour and its suppression by glucocorticoids. Aust N Z J Obstet Gynaec 15:191–203, 1975.
77c. Block BSB, Liggins GC, Creasy RK. Preterm delivery is not predicted by serial plasma estradiol or progesterone concentration measurements. Am J Obstet Gynecol 150:716–722, 1984.
78. Gerhard I, Fitzer C, Klinga K, Rahman N, Runnebaum B. Estrogen screening in evaluation of fetal outcome and infant's development. J Perinat Med 14:279–291, 1986.
79. Klopper A, Jandial V, Wilson G. Plasma steroid assay in the assessment of fetoplacental function. J Steroid Biochem 6:651–656, 1975.
80. Geschwind N, Behan P. Left-handedness: association with immune disease, migraine, and developmental learning disorder. Proc Natl Acad Sci USA 79:5097–5100, 1982.
81. UK Testicular Cancer Study Group. Social, behavioural and medical factors in the aetiology of testicular cancer: results from the UK study. Br J Cancer 70:513–520, 1994.
82. Trichopoulos D. Hypothesis: does breast cancer originate in utero? Lancet 335:939–940, 1990.
82a. Bajdik CD, Phillips N, Huchcroft S, Hill GB, Gallagher RP. Cancer in the mothers and siblings of testicular cancer patients. Can J Urol 8:1229–1233, 2001.
83. Forest MG, Sizonenko PC, Cathiard AM, Bertrand J. Hypophyso-gonadal function in humans during the first year of life. I. Evidence for testicular activity in early infancy. J Clin Invest 53:819–828, 1974.
84. Vermeulen A. Plasma levels and secretion rates of steroids with anabolic activity in man. In: Lu FC, Rendel J (eds). Environmental Quality and Safety Supplement, vol V. Anabolic Agents in Animal Production. Stuttgart: Thieme, 1976, pp 171–180.
85. Pirke KM, Doerr P. Age related changes and interrelationships between plasma testosterone, oestradiol and testosterone-binding globulin in normal adult males. Acta Endocrinol (Copenh) 74:792–800, 1973.
86. Baker HWG, Burger HG, de Kretser DM, et al. Changes in the pituitary–testicular system with age. Clin Endocrinol (Oxf) 5:349–372, 1976.
87. Carroll PR, Whitmore WF Jr, Herr HW, et al. Endocrine and exocrine profiles of men with testicular tumors before orchiectomy. J Urol 137:420–423, 1987.
88. Willemse PHB, Sleijfer DT, Sluiter WJ, Schraffordt Koops H, Doorenbos H. Altered Leydig cell function in patients with testicular cancer: evidence for a bilateral testicular defect. Acta Endocrinol (Copenh) 102:616–624, 1983.
89. Wanderas EH, Fosså SD, Heilo A, Stenwig AE, Norman N. Serum follicle stimulating hormone—predictor of cancer in the remaining testis in patients with unilateral testicular cancer. Br J Urol 66:315–317, 1990.
90. United Kingdom Testicular Cancer Study Group. Aetiology of testicular cancer: association with congenital abnormalities, age at puberty, infertility, and exercise. BMJ 308:1393–1399, 1994.

91. Swerdlow AJ, Huttly SRA, Smith PG. Testis cancer: post-natal hormonal factors, sexual behaviour and fertility. Int J Cancer 43:549–553, 1989.
92. Weir HK, Kreiger N, Marrett LD. Age at puberty and risk of testicular germ cell cancer (Ontario, Canada). Cancer Causes Control 9:253–258, 1998.
93. Gallagher RP, Huchcroft S, Phillips N, et al. Physical activity, medical history, and risk of testicular cancer (Alberta and British Columbia, Canada). Cancer Causes Control 6:398–406, 1995.
94. Gorzynski JG, Lebovits A, Holland JC, Vugrin D. Significant antecedent psychosexual differences between testicular cancer patients and leukemic controls. Proc. 17th Annual Meeting of the American Society of Clinical Oncology. April 30–May 2, 1981. Washington DC. Vol. 22. p. 337. Abstract C-18. 1981.
95. Tanner JM. Foetus into Man. Physical Growth from Conception to Maturity, 2nd ed. Ware, UK: Castlemead, 1989.
96. Akre O, Ekbom A, Sparén P, Tretli S. Body size and testicular cancer. In: Akre O. Etiological Insights into the Testicular Cancer Epidemic. PhD Dissertation, Karolinska Institutet, Stockholm, 1999.
97. Brown LM, Pottern LM, Hoover RN. Testicular cancer in young men: the search for causes of the epidemic increase in the United States. J Epidemiol Community Health 41:349–354, 1987.
98. Vistisen K, Priemé H, Okkels H, Vallentin S, Loft S, Olsen JH, Poulsen HE. Genotype and phenotype of glutathione S-transferase μ in testicular cancer patients. Pharmacogenetics 7:21–25, 1997.
99. Thune I, Lund E. Physical activity and the risk of prostate and testicular cancer: a cohort study of 53,000 Norwegian men. Cancer Causes Control 5:549–556, 1994.
100. Whittemore AS, Paffenbager RS Jr, Anderson K, Lee JE. Early precursors of urogenital cancers in former college men. J Urol 132:1256–1261, 1984.
101. Davies TW, Prener A, Engholm G. Body size and cancer of the testis. Acta Oncol 29:287–290, 1990.
102. Davies TW, Palmer CR, Ruja E, Lipscombe JM. Adolescent milk, dairy product and fruit consumption and testicular cancer. Br J Cancer 74:657–660, 1996.
103. Swerdlow AJ, De Stavola BL, Swanwick MA, Mangtani P, Maconochie NES. Risk factors for testicular cancer: a case-control study in twins. Br J Cancer 80:1098–1102, 1999.
104. Proos LA. Anthropometry in adolescence—secular trends, adoption, ethnic and environmental differences. Horm Res 39(Suppl 3):18–24, 1993.
105. Tanner JM, Hayashi T, Preece MA, Cameron N. Increase in length of leg relative to trunk in Japanese children and adults from 1957 to 1977: comparison with British and with Japanese Americans. Ann Hum Biol 9:411–423, 1982.
106. Armstrong B, Doll R. Environmental factors and cancer incidence and mortality in different countries, with special reference to dietary practices. Int J Cancer 15:617–631, 1975.
107. Hill PB, Wynder EL. Effect of a vegetarian diet and dexamethasone on plasma prolactin, testosterone and dehydroepiandrosterone in men and women. Cancer Lett 7:273–282, 1979.
108. Hämäläinen E, Adlercreutz H, Puska P, Pietinen P. Diet and serum sex hormones in healthy men. J Steroid Biochem 10:459–464, 1984.
109. Reed MJ, Cheng RW, Simmonds M, Richmond W, James VHT. Dietary lipids: an additional regulator of plasma levels of sex hormone binding globulin. J Clin Endocrinol Metab 64:1083–1085, 1987
110. Howie BJ, Shultz TD. Dietary and hormonal interrelationships among vegetarian Seventh-Day Adventists and non-vegetarian men. Am J Clin Nutr 42:127–134, 1985.
111. Bélanger A, Locong A, Noel C, et al. Influence of diet on plasma steroid and sex plasma binding globulin levels in adult men. J Steroid Biochem 32:829–833, 1989.
112. Key TJA, Roe L, Thorogood M, Moore JW, Clark GMG, Wang DY. Testosterone, sex hormone-binding globulin, calculated free testosterone, and oestradiol in male vegans and omnivores. Br J Nutr 64:111–119, 1990.
113. Sigurdson AJ, Chang S, Annegers JF, et al. A case-control study of diet and testicular carcinoma. Nutr Cancer 34:20–26, 1999.
114. Gallagher R, Huchcroft S, Hill G, Phillips N. Physical activity and dietary factors in testicular cancer. In: Abstracts of the 28th Annual Meeting of the Society for Epidemiologic Research, Snowbird, Utah, June 21–24, 1995. Abstract nr 108.
114a.Srivastava A, Kreiger N. Relation of physical activity to risk of testicular cancer. Am J Epidemiol 151:78–87, 2000.
115. Pochi PE. Acne: endocrinologic aspects. Cutis 30:212–222, 1982.
116. Sansone G, Reisner RM. Differential rates of conversion of testosterone to dihydrotestosterone in acne and in normal human skin—a possible pathogenic factor in acne. J Invest Dermatol 56:366–372, 1971.
117. Swerdlow AJ, Huttly SRA, Smith PG. Testicular cancer and antecedent diseases. Br J Cancer 55:97–103, 1987.
118. Phillipou G, Kirk J. Significance of steroid measurements in male pattern alopecia. Clin Exp Dermatol 6:53–56, 1981.
119. Zumoff B, Strain GW, Miller LK, et al. Plasma free and non-sex-hormone-binding-globulin-bound testosterone are decreased in obese men

in proportion to their degree of obesity. J Clin Endocrinol Metab 71:929–931, 1990.
120. Glass AR, Swerdloff RS, Bray GA, Dahms WT, Atkinson RL. Low serum testosterone and sex-hormone-binding globulin in massively obese men. J Clin Endocrinol Metab 45:1211–1219, 1977.
121. Kley HK, Solbach HG, McKinnan JC, Krüskemper HL. Testosterone decrease and oestrogen increase in male patients with obesity. Acta Endocrinol (Copenh) 91:553–563, 1979.
122. Lin RS, Kessler II. Epidemiologic findings in testicular cancer. Am J Epidemiol 110:357, 1979.
123. Ross R, Bernstein L, Judd H, Hanisch R, Pike M, Henderson B. Serum testosterone levels in healthy young black and white men. J Natl Cancer Inst 76:45–48, 1986.
124. Gordon GG, Altman K, Southren AL, Rubin E, Lieber CS. Effect of alcohol (ethanol) administration on sex-hormone metabolism in normal men. N Engl J Med 295:793–797, 1976.
125. Dai WS, Gutai JP, Kuller LH, Cauley JA, for the MRFIT Research Group. Cigarette smoking and serum sex hormones in men. Am J Epidemiol 128:796–805, 1988.
126. Wheeler GD, Wall SR, Belcastro AN, Cumming DC. Reduced serum testosterone and prolactin levels in male distance runners. JAMA 252:514–516, 1984.
127. Hackney AC, Sinning WE, Bruot BC. Reproductive hormonal profiles of endurance-trained and untrained males. Med Sci Sports Exerc 20:60–65, 1988.
128. Brownson RC, Chang JC, Davis JR, Smith CA. Physical activity on the job and cancer in Missouri. Am J Public Health 81:639–642, 1991.
129. Werder EA, Illig R, Torresani T, et al. Gonadal function in young adults after surgical treatment of cryptorchidism. BMJ 2:1357–1359, 1976.
130. Paulsen CA, Plymate SR. Klinefelter's syndrome. In: King RA, Rotter JI, Motulsky AG. (eds). The Genetic Basis of Common Diseases. New York: Oxford University Press, 1992, pp 876–894.
131. Hasle H, Mellemgaard A, Nielsen J, Hansen J. Cancer incidence in men with Klinefelter syndrome. Br J Cancer 71:416–420, 1995.
132. Müller J, Skakkebaek NE. Gonadal malignancy in individuals with sex chromosome anomalies. In: Evans JA, Hamerton JL, Robinson A (eds). Children and Young Adults with Sex Chromosome Aneuploidy: Follow-Up, Clinical, and Molecular Studies. Proceedings of the 5th International Workshop on Sex Chromosome Anomalies, Minaki, Ontario, Canada, June 7–10, 1989. March of Dimes Birth Defects Foundation. Birth Defects: Original Article Series 26:247–255, 1991.
133. Verp MS, Simpson JL. Abnormal sexual differentiation and neoplasia. Cancer Genet Cytogenet 25:191–218, 1987.
134. Giwercman A, von der Maase H, Skakkebaek NE. Epidemiological and clinical aspects of carcinoma in situ of the testis. Eur Urol 23:104–110, 1993.
135. Sasagawa I, Nakada T, Hashimoto T, et al. Hormone profiles and contralateral testicular histology in Down's syndrome with unilateral testicular tumor. Arch Androl 30:93–98, 1993.
136. Hasle H, Clemmensen IH, Mikkelsen M. Risks of leukaemia and solid tumours in individuals with Down's syndrome. Lancet 355:165–169, 2000.
136a. Hermon C, Alberman E, Beral V, Swerdlow AJ for the Collaborative Study Group of Genetic Disorders. Mortality and cancer incidence in persons with Down's syndrome, their parents and siblings. Ann Hum Genet 65:167–176, 2001.
137. Forman D, Oliver RTD, Brett AR, et al. Familial testicular cancer: a report of the UK Family Register, estimation of risk and an HLA class 1 sib-pair analysis. Br J Cancer 65:255–262, 1992.
138. Heimdal K, Olsson H, Tretli S, Flodgren P, Børresen A-L, Fosså SD. Familial testicular cancer in Norway and southern Sweden. Br J Cancer 73:964–969, 1996.
139. Westergaard T, Olsen JH, Frisch M, Kroman N, Nielsen JW, Melbye M. Cancer risk in fathers and brothers of testicular cancer patients in Denmark. A population-based study. Int J Cancer 66:627–631, 1996.
140. Hemminki K, Vaittinen P, Dong C, Easton D. Sibling risks in cancer: clues to recessive or X-linked genes? Br J Cancer 84:388–391, 2001.
140a. Lichtenstein P, Holm NV, Verkasalo PK, et al. Environmental and heritable factors in the causation of cancer. Analyes of cohorts of twins from Sweden, Denmark, and Finland. N Eng J Med 343:78–85, 2000.
141. Tollerud DJ, Blattner WA, Fraser MC, et al. Familial testicular cancer and urogenital developmental anomalies. Cancer 55:1849–1854, 1985.
142. Heimdal K, Andersen TI, Skrede M, Fosså SD, Berg K, Børresen A-L. Association studies of estrogen receptor polymorphisms in a Norwegian testicular cancer population. Cancer Epidemiol Biomarkers Prev 4:123–126, 1995.
142a. Falk RT, Fears TR, Hoover RN, et al. Does place of birth influence endogenous hormone levels in Asian-American women? Br J Cancer 87:54–60, 2002.
143. Juul A, Bang P, Hertel NT, et al. Serum insulin-like growth factor-I in 1030 healthy children, adolescents, and adults: relation to age, sex, stage of puberty, testicular size, and body mass index. J Clin Endocrinol Metab 78:744–752, 1994.

144. Smith AT, Clemmons DR, Underwood LE, Ben-Ezra V, McMurray R. The effect of exercise on plasma somatomedin-C/insulinlike growth factor I concentrations. Metabolism 36:533–537, 1987.
145. Clemmons DR, Underwood LE. Nutritional regulation of IGF-I and IGF binding proteins. Annu Rev Nutr 11:393–412, 1991.
146. Hankinson SE, Willett WC, Colditz GA, et al. Circulating concentrations of insulin-like growth factor-I and risk of breast cancer. Lancet 351:1393–1396, 1998.
147. Chan JM, Stampfer MJ, Giovannucci E, et al. Plasma insulin-like growth factor-I and prostate cancer risk: a prospective study. Science 279:563–566, 1998.

Index

Aborelix, for prostate cancer, 346
Abortion, breast cancer risk and, 226–227
N-Acetyl-transferase-1 (NAT1), 202
Acne, testicular cancer risk and, 426
ACS (American Cancer Society), mammographic screenings and, 211
Acute promyelocytic leukemia (APL), retinoic acid receptors and, 71–72, 72f
Adipose tissue, aromatase expression, 172–173, 173f
  in breast cancer and, 173–175, 174f
  promoter II regulation and, 175, 176f
Adjuvant hormonal therapy, for prostate cancer, 351–352, 352f
Adrenal aromatase inhibitors, new, 355
Adrenalectomy, for prostate cancer, 345
Adrenal gland
  androgens, in prostate cancer progression, 343–344
  hormones biosynthesis in. See Hormones, biosynthesis
Adrenal steroid metabolism inhibitors, for prostate cancer, 354–355
African Americans
  breast cancer
    incidence, 6, 6t
    mortality rates, 6–7, 8t
  ovarian cancer
    mortality, 8t
    risk, 7
  prostate cancer
    incidence, 6, 6t, 7f
    mortality rates, 8t
    risk, 4
  testicular cancer
    mortality rates, 8t
    risk, 426–427
  uterine cancer
    incidence, 7, 8f, 9
    mortality rates, 8t
Age
  cancer risk and, 3
  at menarche, breast cancer and, 122
  at menopause, endometrial cancer and, 371
  ovarian cancer incidence and, 402–403, 403f
  prostate cancer risk and, 273
  at puberty, testicular cancer risk and, 424–425
AKR1C gene family, 21
Alcohol consumption
  breast cancer risk and, 124–125
  maternal, testicular cancer risk and, 421
  prostate cancer risk and, 277
  reduction, for BRCA1/2 gene carriers, 202
  testicular cancer risk and, 427
Allelic association (linkage disequilibrium), 109–111
American Cancer Society (ACS), mammographic screenings and, 211
Aminoglutethimide, for prostate cancer, 354–355
Anabolic steroids, testicular cancer risk and, 426
Androgen ablation therapy
  combined, 347–349, 348f, 349f
  failure, 288
    altered interaction with coregulatory molecules, 305–306
    androgen receptor gene mutations and, 302–305, 303f
    androgen receptor levels and, 302
  response to, 300–301, 301f, 331
Androgen inhibitor, with LHRH agonist, for prostate cancer, 347–349, 348f, 349f
Androgen insensitivity, testicular cancer risk, 428
Androgen receptor (AR)
  activation
    lipid-dependent, 294–295, 296f
    lipid-independent, 295
  amino-terminal transactivation domain, 288, 291, 332
    structure of, 289, 291, 290f
    variants, 305
  androgen binding, 343
  in androgen-unresponsive prostate cancer, 302
  CAG repeat, prostate cancer risk and, 295, 297, 299, 298t
  classification, 289
  cofactors
    altered interactions in androgen-ablation therapy failure, 305–306
    coactivators, 293–294
    corepressors, 294
    recruitment of, 292–294

437

Androgen receptor (AR) (*Continued*)
    conformational maturation, 291–292
    DNA-binding domain, 332
        structure, 289, 290f, 292
        variants, 304–305
    gene. *See AR* gene
    ligand-binding domain, 332
        structure, 289, 290f, 291
        variants, 304
    ligand-independent activation, androgen ablation failure and, 306–307
    localized prostate cancer and, 299–300
    overexpression, 333
    promiscuous, 346
    protein structure, 289, 290f
    repeat, prostate cancer risk and, 299
    signaling cascade, 332
        ligand-dependent, 333–334
        ligand-independent, 334–336
Androgen receptor-associated proteins, 294
Androgen response elements (AREs), 319
Androgens, 331–332. *See also specific androgens*
    in carcinogenesis, of prostate cancer, 319
    metabolism of, 22–23, 23f, 23t
    ovarian carcinogenesis and, 405–406
    prostate cancer and, 3–4
    prostate cancer risk and, 319
Androgen-signaling axis, 288
Androgen transactivation pathway, in prostate cancer, 278–282, 279t, 279f, 282f
3α-Androstanediol conjugates, 23t
3α,17β-Androstanediol glucuronide, 4
5-Androstene-3β,17β-diol, 17–18
Androstenedione
    formation, 16
    postmenopausal levels, 402
Androsterone conjugates, 23t
Angiogenesis inhibitors, for prostate cancer, 358
Angiotensin II, regulation of cholesterol to pregnenolone, 15
Anthropometric factors
    breast cancer risk and, 123
    prostate cancer risk and, 277
    testicular cancer risk and, 425
Antiandrogens
    for prostate cancer, 346, 353
    withdrawal response, 346
Antiestrogens. *See also* Tamoxifen
    development of, 218–219
    ovarian cancer risk and, 407
    properties of, 219–220
    structures of, 219f
Antioxidant vitamins, prostate cancer risk and, 275–276, 275t
APL (acute promyelocytic leukemia), retinoic acid receptors and, 71–72, 72f
Apoptosis
    modulating, 336–337
    progesterone and, 406
Apoptosis regulators, for prostate cancer, 357–358
AREs (androgen receptor elements), 292

AR gene
    androgen ablation therapy failure and, 302–305, 303f
    characteristics, 279t, 280
    control, 63–67, 64t, 65t, 66f
    in hereditary prostate cancer, 321–322
    mutations, 302–305, 303f, 333–334
    polymorphisms, prostate cancer risk and, 295, 297, 299, 298t
    structure, 289, 290f
Aromatase
    estrogen biosynthesis and, 169–171, 170f
    expression
        in adipose tissue, 172–173, 173f
        in breast cancer, 173–175, 174f
        in ovary, 171–172, 171f, 175
    inhibition, 175–177
Aromatization, 18
Arzoxifene (LY 353,381), 236, 236f
Association studies, 107–108
    breast cancer, 161–163, 162f, 163f
    candidate gene choice, 164–165
    case and control sets, 164, 164f
    conceptual diagram, 162f
    design of, 163–165, 164f
    direct and indirect markers in, 162–163, 163f
    genetic variant, assessment criteria for, 167
    inconsistent results, sources of, 111–115, 112t
        inadequate statistical power, 111–112
        low statistical thresholds, 113–114
        population stratification, 112–113
        true population differences, 114–115
    interpretation of, 165
    phenotype and, 163–164
    power issues, 164, 164f
AT gene, breast cancer risk and, 4

Baldness, testicular cancer risk and, 426
Bcl-2, 337
3β-HSD. *See* 3β-Hydroxysteroid dehydrogenase
Bioavailable steroids, 12
Birth order, testicular cancer risk and, 418
Birthweight, testicular cancer risk and, 423
BMD. *See* Bone mineral density
Body mass index (BMI)
    breast cancer risk and, 123
    prostate cancer risk and, 277
Bone mineral density (BMD)
    GnRH analogues and, 250
    GnRH analogues with HRT and, 251–252, 252t
    raloxifene and, 234, 234f
    tamoxifen and, 229
BRCA1 gene
    breast cancer risk and, 4, 128, 199–200
    carriers, 200
        with breast cancer, 204–206
        breast cancer risk and, 199–200
        Jewish women with, 205
        lifestyle modifications for, 202–204
        male, 201
        oophorectomy for, 181
        unaffected, management of, 200–204

INDEX

genetic testing for, 200
loss of tamoxifen responsiveness and, 265
mutations
    androgen receptor coactivation, 293
    breast cancer risk and, 226
    familial clustering of breast cancer and, 157–158, 158f
progesterone and, 406
screening for, 200–201

BRCA2 gene
breast cancer risk and, 4, 128, 199–200
carriers, 200
    with breast cancer, 205
    breast cancer risk for, 199–200
    lifestyle modifications for, 202–204
genetic testing for, 200
mutations
    breast cancer risk and, 226
    familial clustering of breast cancer and, 157–158
progesterone and, 406
screening for, 200–201

BRCA3 gene, 160

Breast cancer
age-incidence curve, 248–249, 249f
aromatase expression in, 173–175, 174f
in BRCA1/2 carriers, 204–205
    chemosensitivity of, 205
    prophylactic surgery for, 205–206
    radiosensitivity of, 205
    screening for secondary cancers, 205–206
candidate genes, 5, 5t, 129f
    COMT, 132
    CYP17, 129–132, 130f, 131f
    CYP1A1, 132
    HSD17B1, 132
carcinogenesis
    estrogens and, 2, 3, 4–5, 4f, 5t
    transmissible agents and, 2
ER-positive
    selective estrogen receptor modulators for, 222–223
    tamoxifen for, 220, 230, 231f
familial clustering, 157–160, 158f
genes
    high-penetrance, 165
    low-penetrance, 165–167
hereditary, 128
hormonal therapy. *See also specific hormonal therapies*
    historical aspects, 255
    in vivo human studies, 255–256
hormone-independent, 256–257, 266–267
incidence, 6t, 6, 8f
inherited predisposition
    applications for prevention, 165–167
    association studies, 161–163, 162f, 163f, 163–165, 164f
    BRCA1/2, 157–158, 158f
    evidence for, 157–160, 158f
    founder population, 161
    linkage analysis, 160
    model fitting and, 159–160
    non-BRCA1/2, 158–159, 159f
    sibling pair analysis, 160–161
mammographic screenings, benefits of, 210–211
mortality rates, 6–7
    geographic differences in, 120, 121f
    race and, 8t
peroxisome proliferator-activated receptors and, 74–75
population-attributable fraction, 166–167
prevention
    future of, 237–238
    genetic predisposition applications for, 165–167
    GnRH-releasing hormone analogues for, 248–249, 249f
    raloxifene for, biological basis of, 233–235, 234f
    tamoxifen for. *See* Tamoxifen, chemoprevention
risk factors, 122t, 132–133
    abortion, 226–227
    age at menarche, 122
    age at menopause, 122
    alcohol consumption, 124–125
    anthropometric factors, 123
    BRCA1 gene mutation, 226
    BRCA2 gene mutation, 226
    combined oral contraceptives, 140–143, 141f–143f
    CYP1A1 gene, 379–380
    diethylstilbestrol, 152–154
    endogenous hormones, 120, 122, 226
    environmental, 227
    estrogen-replacement therapy, 127–128
    exogenous hormones, 139–140, 154–155, 227. *See also specific exogenous hormones*
    family history, 128, 226
    fat, dietary, 125–126
    fertility-enhancing drugs, 154
    fiber, dietary, 126
    hormone-replacement therapy, 127–128
    16α-hydroxylated estrogens, 26–29
    insulin-like growth factors, 125
    interactions among, 227
    lactation, 124, 227
    menopause, 248
    obesity, 227
    oral contraceptives, 128
    parity, 122–123
    physical inactivity, 123–124, 227
    phytoestrogens, 126–127
    xenobiotic pesticides, 127
in sisters of testicular cancer patients, 423–424
susceptibility
    inherited, 128
    multigenic models of, 129–132, 129f–131f

Breast cells
epithelial
    17-HSDs in, 184
    proliferation, estrogen plus progesterone hypothesis and, 249
    proliferation, estrogen plus progesterone hypothesis and, 249

Breast-feeding
breast cancer risk and, 2, 124
ovarian cancer risk reduction, 404

CA125, 201
CAG repeats
    hereditary prostate cancer, 321–322
    prostate cancer and, 280–281
Cancer incidence rates. *See also under specific cancers*
    international variation in, 1–2
Candidate genes
    choice, for association studies, 164–165
    hereditary prostate cancer, 320–325
    for hormone-related cancers, 5, 5t
    identification, 103f
        in polygenic disorders, 101–102
        using genomic sequence information, 102–104
        using linkage analysis in families, 10–1054
    ovarian cancer, 408
    positional, 105
Carcinogenesis
    hormonal. *See Hormonal carcinogenesis*
    multistage concept, 1
CARM1 (coactivator-associated arginine methyl-transferase-1), 259, 293
CAR receptor, 69t, 70t, 76
Castration
    with androgen inhibitor, for prostate cancer, 347–349, 349f
    side effects, 352–353
    surgical, for prostate cancer, 300, 301f
Catechol-*O*-methyltransferase (COMT)
    endometrial cancer risk and, 381
    in estrogen metabolism, 21
    gene, breast cancer and, 132
CBG (corticosteroid-binding globulin), 20
CCAAT/enhancer binding protein (C/EBP), 175
Cervical cancers, clear cell adenocarcinomas, diethylstilbestrol and, 153
Cetrorelix, for prostate cancer, 346
CGH (comparative genomic hybridization), 105
Chemoprevention. *See also under specific cancers*
    breast cancer, candidates, identification of, 227–228
    of breast cancer, gonadotropin-hormone-releasing hormone analogues for, 249–250
    hormonal, for *BRCA1/2* gene carriers, 201–202
Chemotherapy
    with estramustine, 356
    with hormonal therapy, for prostate cancer, 355–356
Cholesterol
    conversion to pregnenolone, 12–15, 13f
    storage pool, 14
    tamoxifen and, 225–226
    transfer into mitochondrion, 14–15
Clear cell adenocarcinomas, diethylstilbestrol and, 153
Clomiphene
    breast cancer risk and, 154
    mechanism of action, 222–223
Coactivator-associated arginine methyl-transferase-1 (CARM1), 259, 293
Coactivators
    androgen-independent prostate cancer and, 336
    androgen receptor, 293–294
    estrogen receptor, 224, 225f, 259–260

nuclear receptor cofactors, 50–54, 51f, 53f
    p160, 293
Cofactor molecules, 38
Colon cancer, peroxisome proliferator-activated receptors and, 75
Combined oral contraceptives, breast cancer risk and, 140–143, 141f–143f
    dosage formulation and, 142–143, 142f
    duration of use, 140–141, 142f, 143f
Comparative genomic hybridization (CGH), 105
Comparative genomics, 104
Complex traits
    definition of, 99
    genes, identification of, 100–101
COMT. *See Catechol-O-methyltransferase*
Conjugation, 20, 22
Corepressors, androgen receptor, 294
Corticosteroid-binding globulin (CBG), 20
Corticosteroids, formation, 18–19
Corticosterone, 18
Corticotropin, regulation of cholesterol to pregnenolone, 15
CP336,156, 236
Cryptorchidism, testicular cancer risk and, 414, 415
Cyclin D1, 294
CYP1A1 gene
    breast cancer risk and, 132, 379–380
    endometrial cancer risk and, 379–380
    in estrogen metabolism, 25
CYP1A2 gene
    endometrial cancer risk and, 380
    in estrogen metabolism, 25
CYP3A4 gene, endometrial cancer risk and, 381
CYP11A1 gene
    endometrial cancer risk and, 374
    expression, regulation of cholesterol to pregnenolone, 15
    in steroidogenesis, 401, 401f
CYP1B1 gene, endometrial cancer risk and, 380–381
CYP17 gene
    breast cancer risk and, 129–132, 130f, 131f
    endometrial cancer risk and, 376, 377t
    in hereditary prostate cancer, 324–325
    prostate cancer and, 279t, 281
    in steroidogenesis, 16, 401, 401f
CYP19 gene
    endometrial cancer risk and, 376, 378, 377t
    in estrogen biosynthesis, 18, 169–171, 170f
    in steroidogenesis, 401, 401f
    structure, 170, 170f
Cyproterone acetate, side effects, 352–353
Cytochrome P-450. *See also specific cytochrome P-450 genes*
    in steroidogenesis, 24–25, 401, 401f

DDT (dichlorodiphenyltrichloroethane), 421
Deconjugation, of steroid hormones, 22
Deep vein thrombosis (DVT), tamoxifen and, 232
Dehydroepiandrosterone (DHEA), 16, 343
Dehydroepiandrosterone sulfate (DHEAS), 25
Dehydrogenases, in steroid hormones metabolism, 20–21
11-Deoxycorticosterone (DOC), 18

DES. *See* Diethylstilbestrol
Desmoplastic reaction, 173
DHEA (dehydroepiandrosterone), 16, 343
DHT. *See* Dihydrotestosterone
Dichlorodiphenyltrichloroethane (DDT), 421
Dietary factors
   breast cancer risk and, 227
   in cancer etiology, 2
   modifications, for *BRCA1/2* gene carriers, 202
   prostate cancer risk and, 275–276, 275t
   testicular cancer risk and, 425–426
Diethylstilbestrol (DES)
   breast cancer risk and, 152–154
   structure of, 219f
   testicular cancer risk and, 415, 416, 421
Dihydrotestosterone (DHT)
   in androgen transactivation pathways, 278
   circulating markers of, 23
   formation of, 22, 23f
   metabolism, 22–23, 23t
   prostate cancer and, 3–4
   in prostate carcinogenesis, 319
   prostate development and, 343
3α-Diol, 23
3α-Diol G, 23
DMT (DNA methyltransferase), 261–262
DNA, binding, to androgen receptor, 292
DNA methylation, loss of ER expression and, 261–262
DNA methyltransferase (DMT), 261–262
Down's syndrome, testicular cancer risk, 428
DVT (deep vein thrombosis), tamoxifen and, 232

EGF (epidermal growth factor), 334
ELAC2/HPC2 gene, hereditary prostate cancer and, 318
EM 652, 237
EM 800, 237
Endometrial cancer, 384–385
   candidate genes, 5, 5t
   genetic factors, 383–384
      of estrogen biosynthesis, 374, 376, 378, 375f, 377t, 374, 376, 378, 375f, 377t
      of estrogen catabolism, 378–383
      familial aggregation, 373–374
   risk factors, 384–385
      age at menopause, 371
      3β-hydroxysteroid dehydrogenases, 374, 376, 375f
      17β-hydroxysteroid dehydrogenases, 378
      CYP17, 376, 377t
      CYP19, 376, 378, 377t
      CYP11A1, 374
      endogenous hormones, 371–372
      estrogen receptor, 383–384
      estrogen-replacement therapy, 2
      exogenous hormones, 372–373
      hereditary nonpolyposis colorectal cancer, 373–374
      hormone-replacement therapy, 372–373
      LH variant alleles, 383
      medical conditions, 371
      NQO1, 382–383
      nulliparity, 371
      obesity, 371–372
      oral contraceptives, 9, 373
      progesterone receptor, 384
      selective estrogen receptor modulators, 373
      SHBG, 383
      smoking, 372
      sulfotransferases, 382
      uridine diphosphate-glucuronosyltransferases, 381–382
   tamoxifen and, 226
Environmental factors, breast cancer risk and, 227
Epidermal growth factor (EGF), 334
ER. *See* Estrogen receptor
ERA-923, 237
ERE (estrogen response element), 257, 263
ERT. *See* Estrogen-replacement therapy
Estradiol
   binding, 223–224, 224f
   biosynthesis, 17-hydroxysteroid dehydrogenases and, 181–182, 182f
   in breast carcinogenesis, 2–3, 4–5, 4f
   CYP17 genotype and, 130, 130f
   4-hydroxylation of, 29
   metabolism of, 23–26, 24t
   testicular cancer risk and, 417–418
   in twin gestation, 422
Estramustine, with chemotherapy, 356
Estrogen biosynthesis gene. *See CYP19* gene
Estrogen metabolism genes. *See HSD17B1* gene; *HSD17B2* gene
Estrogen plus progesterone hypothesis, breast cell proliferation and, 249
Estrogen receptor (ER)
   antiestrogenic activity at, 223–224, 224f
   assays, 256, 256t
   cancer and, 67
   cell biology of, 257
   characteristics, 255
   coactivators, 224, 225f, 259–260
   coregulator proteins, altered interaction, lack of tamoxifen responsiveness and, 263–264
   discovery of, 219
   endometrial cancer risk and, 383–384
   ERα
      assays, 256
      discovery of, 224
      molecular biology of, 257–259, 258f
   ERβ
      assays, 256
      discovery of, 224–225
      increased expression, tamoxifen responsiveness and, 266, 266t
      molecular biology of, 257–259, 258f
   gene control, 63–67, 64t, 65t, 66f
   gene expression, 260–261
   genetic polymorphisms, 383–384
   hormonal therapy response predictions and, 256, 256t
   ligand-independent activation, 264–266
   loss of expression, 260–262
   loss of tamoxifen responsiveness and, 260–266, 266t
   as nuclear transcription factor, 220
   sequence changes, 262–263
   status, tamoxifen response and, 220–221, 221t

Estrogen-replacement therapy (ERT)
  breast cancer risk, 127–128, 150–151, 150t
  endometrial cancer and, 2
  with GnRH analogues, 250–252, 251t, 252t, 252f
  historical background, 145–147
  mammographic density and, 212–213, 212t
  for ovarian cancer, 407
  with progestin. See Hormone-replacement therapy
  uterine cancer risk and, 9
Estrogen response element (ERE), 257, 263
Estrogens. See also specific estrogens
  biosynthesis, 18, 169, 170f
    aromatase and, 169–171, 170f
    CYP19 gene and, 169–171, 170f
      genetic factors in endometrial cancer and, 374, 376, 378, 375f, 377t
  breast cancer carcinogenesis and, 2, 3
  breast cancer risk, 120
  catabolism, genetic factors, endometrial cancer risk and, 378–383
  catechol, 24, 24f
  endogenous, mammographic density and, 213–214
  exogenous
    maternal, testicular cancer risk and, 415, 416
    postmenopausal. See Estrogen-replacement therapy
  maternal age and, 419
  maternal concentrations, testicular cancer risk and, 415, 416
  metabolism of, 23–26, 24t, 24f–26f
  non-steroidal, 218–219, 219f
  ovarian carcinogenesis and, 405
  postmenopausal levels, 401–402
  prenatal, testicular cancer risk and, 413, 414–415
  for prostate cancer, with chemotherapy, 355
  for prostate cancer therapy, 346
  stimulation of ovarian hormones, 399–400
Estrone
  4-hydroxylation of, 29
  metabolism of, 23–26, 24t
Ethnicity
  prostate cancer risk and, 273–274
  testicular cancer risk and, 426–427
Ethnic stratification, candidate gene polymorphisms and, 408
Ewing sarcoma, orphan nuclear receptors and, 77–79
Exercise
  for BRCA1/2 gene carriers, 202–203
  breast cancer risk and, 123–124, 227
  premenopausal, ovarian cancer risk and, 407
  prostate cancer risk reduction and, 277
  testicular cancer risk and, 427
Exogenous hormones
  breast cancer risk and, 139–140, 154–155
  diethylstilbestrol, breast cancer risk and, 152-154
  fertility-enhancing drugs, breast cancer risk and, 154
  oral contraceptives. See Oral contraceptives
  postmenopausal hormone therapy. See Estrogen-replacement therapy; Hormone-replacement therapy

Familial clustering
  of breast cancer, 157–160, 158f
    association studies, 161–163, 162f, 163f, 163–165, 164f
    BRCA1/2, 157–158, 158f
    founder population, 161
    linkage analysis, 160
    model fitting and, 159–160
    non-BRCA1/2, 158–159, 159f
    sibling pair analysis, 160–161
  of endometrial cancer, 373–374
  of hormone-responsive cancers, 99–100
  of prostate cancer, 316–317
Family history
  breast cancer risk and, 128, 199–200, 226
  prostate cancer risk and, 274–275
  testicular cancer risk and, 428
Farnesoid X receptor, 69t, 70t, 75
Fat, dietary
  breast cancer risk and, 125–126
  testicular cancer risk and, 425–426
Ferredoxin, 15
Ferredoxin reductase, 15
Ferrodoxin gene expression, regulation of cholesterol to pregnenolone, 15
Ferrodoxin reductase gene expression, regulation of cholesterol to pregnenolone, 15
Fertility-enhancing drugs, breast cancer risk and, 154
Fiber, dietary
  breast cancer risk and, 126
  prostate cancer risk and, 276, 275t
Flutamide withdrawal syndrome, 333–334
Flutamide with leuprolide, for prostate cancer, 347–349, 348f
Follicle-stimulating hormone (FSH)
  gonadotropin-hormone-releasing hormone analogues and, 249–250
  postmenopausal levels, 401
  regulation of cholesterol to pregnenolone, 15
  release, 399
  in steroidogenesis, 401, 401f
Follicular stimulating hormone, in mothers of twins, 422
Founder population, breast cancer, 161
FSH. See Follicle-stimulating hormone

GCG repeats, hereditary prostate cancer, 321–322
General transcription machinery, 38
Genes. See also specific genes
  association studies. See Association studies
  candidate. See Candidate genes
  copy number variation, 105
  expression, global analysis of, 105–106
  function, large-scale manipulation of, 106–107
  high-penetrance, prostate cancer, 274–275
  identification methods, for hormone-responsive cancers, 100–101
  mutations, identification of, 100–101
  polymorphisms
    ovarian cancer risk and, 408
    single nucleotide, 109–111
    testicular cancer risk and, 428

predisposing, identifying, for breast cancer, 160–165, 162f–164f
variation, common
association studies, 107–108
human genome sequence variation, 108–111
Genetic testing, for *BRCA1/2* gene mutations, 200
Genitourinary abnormalities, testicular cancer risk and, 414, 415
Genomic approaches
correlating genetic variation to disease, 101–102
familial clustering and, 99–100
gene identification methods, 100–101
for identifying candidate genes, 103f
expression analysis, 105–106
functional tests, 106–107
genome sequencing, 102–104
linkage analysis, 104–105
variation in gene copy number, 105
Genotyping, of single nucleotide polymorphisms, 111
Glucocorticoid pathway, of corticosteroid formation, 18
Glucocorticoid receptor (GR), 38, 67–68
Glucosteroids, for prostate cancer, 354
Glucuronidation, 22
Glutathione *S*-transferase (GSTμ), 428
GnRH (gonadotropin-releasing hormone), 154, 399
Gonadotropin-hormone-releasing hormone analogues
agonists
with estrogen replacement therapy, for ovarian cancer, 407
induced temporary menopause, mammographic density and, 213
for breast cancer prevention, 248–249, 249f
in combination, 250
FSH and, 249–250
with hormone replacement therapy, 250–252, 251t, 252t, 252f
LH and, 249–250
Gonadotropin-releasing hormone (GnRH), 154, 399
Gonadotropins, ovarian carcinogenesis and, 404–405
Gonads, hormones biosynthesis in. *See* Hormones, biosynthesis
GR (glucocorticoid receptor), 38, 67–68
Granulosa cells, 399
Growth factors. *See specific growth factors*
GSTμ (glutathione *S*-transferase), 428
GW5638, 236–237, 236f

Handedness, testicular cancer risk and, 423
Haplotype structure, human, 109–111
HATs (histone acetyltransferases), 55
HDACs (histone deacetylases), 55–56
HDL (high-density lipoprotein), in free cholesterol pool, 14
Heart and Estrogen/progestagen Replacement Study (HERS), 149
Height
breast cancer risk and, 123
prostate cancer risk and, 277
testicular cancer risk and, 425
Hepatocellular carcinoma, thyroid receptors and, 68–69

Herbal therapy, for prostate cancer, 346–347
Hereditary nonpolyposis colorectal cancer (HNPCC), 373–374
Hereditary prostate cancer
androgen pathway genes and, 320
candidate genes
AR, 321–322
CYP17, 324–325
HSD3B1, 322–324
HSD3B2, 322–324
HER2/*neu* oncogene
androgen ablation failure and, 306–307
androgen-independent prostate cancer and, 335
loss of tamoxifen responsiveness and, 265
HER2/*neu* receptor tyrosine kinase, androgen-independent prostate cancer and, 335
HERS (Heart and Estrogen/progestagen Replacement Study), 149
Heterozygosity, 108, 109
High-density array technology, 106
High-density lipoprotein (HDL), in free cholesterol pool, 14
Histone acetyltransferases (HATs), 55
Histone deacetylases (HDACs), 55–56
HNPCC (hereditary nonpolyposis colorectal cancer), 373–374
Hormonal carcinogenesis. *See under specific cancers*
evidence of, 2, 9–10
genetic basis of, 4–5, 4f, 5t
genetic factors, 4–5, 4f, 5t
incidence/log age model and, 3
model of, 2–3, 3f
Hormonal therapies, for metastatic prostate cancer, 300–301, 301f
Hormone-related cancers. *See specific hormone-related cancers*
candidate genes, 5, 5t
in sisters of testicular cancer patients, 423–424
Hormone-replacement therapy (HRT)
breast cancer risk, 127–128, 227
in *BRCA1/2* gene carriers and, 203–204
comparison of different preparations and, 150–151, 150t
vs. tamoxifen, 229
CYP17 gene and, 130–131
endometrial cancer risk, 372–373
with GnRH analogues, 250–252, 251t, 252t, 252f
historical background, 145–147
mammographic density and, 212–213, 212t
Hormone response elements (HREs), 38, 42, 44, 42f–43f
Hormone-responsive cancers. *See also specific hormone-responsive cancers*
familial clustering of, 99–100
gene identification in, 100–101
Hormones. *See also specific hormones*
bioavailable, 12
biosynthesis, 12
conversion of cholesterol to pregnenolone, 12–15, 13f
receptor gene control for, 66–67, 66f

Hormones (*Continued*)
　chemopreventive. *See* Selective estrogen receptor
　　modulators; Tamoxifen
　endogenous
　　breast cancer risk and, 120, 122, 226
　　endometrial cancer risk and, 371–372
　　mammographic density and, 213–214
　exogenous
　　with chemopreventive agents, 407
　　maternal, testicular cancer risk and, 415
　　postmenopausal. *See* Estrogen-replacement therapy;
　　　Hormone-replacement therapy
　metabolism of, 20
　　conjugation and, 22
　　deconjugation and, 22
　　dehydrogenases and, 20–21
　　hydroxylases and, 21
　　methyltransferases and, 21
　　reductases and, 20–21
　ovarian carcinogenesis and, 405–407
　steroid, biosynthesis of
　　AR gene control, 66–67, 66f
　　ovarian hormones and, 400–401, 401f
　transport, 19–20
HPC1 gene, hereditary prostate cancer and, 317–318
HREs (hormone response elements), 38, 42, 44, 42f–43f
HRT. *See* Hormone-replacement therapy
3α-HSD (3α-hydroxysteroid dehydrogenase), 21
3β-HSD. *See* 3β-Hydroxysteroid dehydrogenase
HSD17B1 enhancer, 188–189, 188f
HSD3B2 gene, 279t, 281–282, 282f
HSD17B1 gene, 185–186, 185t
　breast cancer and, 132
　expression
　　regulation of, 186–187
　　transcriptional regulation of, 187–189, 188f
HSD17B2 gene, 185–186, 185t
HSD17B3 gene, 279t, 281–282, 282f
17-HSDs. *See* 17-Hydroxysteroid dehydrogenases
Human genome
　sequence, 102–103, 103f
　sequence variations, 108–111
　size of, 113
Human menopausal gonadotropin, breast cancer risk
　and, 154
2-Hydroxyestradiol, 29
2-Hydroxyestrone, 29, 131–132
16α-Hydroxyestrone
　breast cancer risk and, 26–29, 131–132
　endometrial cancer risk and, 378–379
　metabolism, 25–26, 26f
11β-Hydroxylase, 18, 19
18-Hydroxylase, 19
21-Hydroxylase, 18, 19
Hydroxylases, steroid hormone metabolism and, 21
2/16α-Hydroxylated estrogen-breast cancer risk
　hypothesis, 27–29
2-Hydroxylated estrogens, endometrial cancer risk and,
　378
4-Hydroxylated estrogens, endometrial cancer risk and,
　378, 379

16α-Hydroxylated estrogens
　breast cancer risk and, 26–29
　metabolism of, 25–26, 26f
17α-Hydroxypregnenolone, 16
17α-Hydroxyprogesterone, 16
3α-Hydroxysteroid dehydrogenase (3α-HSD), 21
3β-Hydroxysteroid dehydrogenase (3β-HSD)
　endometrial cancer risk and, 17, 374, 376, 375f
　formation, 16, 17
17-Hydroxysteroid dehydrogenases (17-HSDs)
　in breast epithelial cells, 184
　in ovarian function, 183–184
　in steroid hormones metabolism, 181–182, 182f
　type 1
　　in breast epithelial cells, 184
　　characteristics of, 182, 183f
　　encoding genes, 185–186, 185t
　　as endocrine therapy target, 189–190, 192, 191f, 192f
　　in endometrium, 185
　　function of, 182, 183f
　　ligand-entry loop interactions, 190, 192, 192f
　　ligand recognition, molecular basis for, 190, 191f
　　in ovarian function, 183
　　regulation of expression, 186–189, 188f
　　staining methods for, 184–185
　　structural studies, 189–190
　type 2
　　in animals, 185
　　in breast epithelial cells, 184
　　characteristics of, 183, 183f
　　encoding genes, 185–186, 185t
　　in endometrium, 185
　　functions of, 183, 183f
　　in humans, 185
17β-Hydroxysteroid dehydrogenases, endometrial cancer
　risk and, 378
Hypophysectomy, for prostate cancer, 345

IGF. *See* Insulin-like growth factor
IGFBPs (insulin-like growth factor binding proteins), 125
Incessant ovulation hypothesis, 9
Inguinal hernia, testicular cancer risk and, 414
Initiators, properties of, 1
Insulin-like growth factor (IGF)
　androgen-independent prostate cancer and, 334
　breast cancer risk and, 125
　loss of tamoxifen responsiveness and, 264
　prostate cancer etiology and, 277–278, 278f
　receptors, breast cancer risk and, 125
　signaling pathways, in prostate cancer, 282–283
Insulin-like growth factor binding proteins (IGFBPs), 125
Insulin signaling, peroxisome proliferator-activated
　receptors and, 74
Interaction assays, orphan nuclear receptors and, 80

Japan, breast cancer mortality rates, 121f
Jewish women, with *BRCA1/2* mutation, 205
Jun amino-terminal kinase signaling pathway (JNK), 62

Kennedy's disease, 280
Keoxifene. *See* Raloxifene

Ketoconazole, for prostate cancer, 354
Klinefelter's syndrome, testicular cancer risk and, 427

Lactation
  breast cancer risk and, 124, 227
  ovarian cancer protective effect, 404
LDL. *See* Low-density lipoprotein
Leuprolide, with flutamide, for prostate cancer, 347–348, 348*f*
Leydig cell, testosterone secretion, 343
LH. *See* Luteinizing hormone
LHRH agonists
  with androgen inhibitor, for prostate cancer, 347–349, 348*f*, 349*f*
  with flutamide, for prostate cancer, 347
  for hormone-refractory prostate cancer, 353
  for prostate cancer, 300–301, 301*f*
LHRH antagonists, for prostate cancer, 300–301, 301*f*, 346
Liarazole, 355
Lifestyle modification, for *BRCA1/2* gene carriers, 202–204
Ligand-binding assays, orphan nuclear receptors and, 79
Ligand-binding domain, of androgen receptor, 289, 290*f*, 291
Linkage analysis
  hereditary prostate cancer
    AR gene and, 321–322
    CYP17 gene and, 324–325
    HSD3B1 gene and, 322–324
    HSD3B2 gene and, 322–324
    susceptibility loci, 317–319
  hereditary prostate cancer candidate genes, 320
  in identifying candidate genes, 104–105
  in multiple-breast cancer case families, 160
Linkage disequilibrium (allelic association), 109–111
Lipid receptors, 73–75
Lipids, raloxifene and, 234–235
Liver X receptor, 69*t*, 70*t*, 75
Loss of heterozygosity (LOH), 105
Low-density lipoprotein (LDL)
  in free cholesterol pool, 14
  raloxifene and, 234–235
  tamoxifen and, 225–226
Luteinizing hormone (LH)
  endometrial cancer risk and, 383
  gonadotropin-hormone-releasing hormone analogues and, 249–250
  physiology of, 399
  postmenopausal levels, 401
  regulation of cholesterol to pregnenolone, 15
  in steroidogenesis, 400–401, 401*f*
  variant, 383
Luteinizing hormone-releasing hormone analogs
  agonists, for prostate cancer, 345–356
  monotherapy, 331
LY156758. *See* Raloxifene
LY 353,381 (arzoxifene), 236, 236*f*

Major susceptibility genes, prostate cancer, 317
Mammography
  breast cancer incidence and, 6, 6*t*
  density
    endogenous hormone levels and, 213–214
    estrogen-replacement therapy and, 212–213, 212*t*
    GnRHA-induced temporary menopause, 213
    GnRH analogues with HRT and, 252, 252*f*
    hormone-replacement therapy and, 212–213, 212*t*
    hormones and, 212, 212*t*
    improving, hormonal strategies for, 214–215
    ovarian hormones and, 214
    role of, 212
    tamoxifen and, 213
  screenings
    benefits of, 210–211, 214
    cost factors, 211
    for older women, 211–212
    positive predictive value, 211
    sensitivity of, 211
    specificity of, 211
    for women with family history, 214
  sensitivity improvement, hormonal strategies for, 214–215
MAPK (mitogen-activated protein kinase), 335
Mastectomy, prophylactic, breast cancer risk in *BRCA1/2* gene carriers and, 204
Maternal age, testicular cancer risk and, 419
Maternal factors, in testicular cancer risk, diet, 421
Melanoma, risk, *BRCA2* gene carriers and, 201
Menarche, age at, 122
Menopause
  age at, breast cancer risk and, 122, 248
  breast cancer risk and, estrogen plus progesterone hypothesis and, 249
  cancer risk and, 3
  GnRHA-induced temporary, mammographic density and, 213
  ovarian hormones in, 401–402
Menorrhagia, in mothers of testicular cancer patients, 424
Menstrual cycle
  follicle development in, 398–399, 399*f*
  mammographic density and, 214
  ovarian hormone fluctuations in, 399–400, 400*f*
MER-25, 219, 219*f*
Metaloproteinase inhibitors (MMPs), 358
Methyltransferases, steroid hormone metabolism and, 21
Mineralocorticoid pathway, of corticosteroid formation, 18
Mineralocorticoid receptors, salt balance, 68
Mitogen-activated protein kinase (MAPK), 335
Mitomycin, and aminoglutethimide, for prostate cancer, 355
MMPs (metaloproteinase inhibitors), 358
Mothers
  alcohol consumption, testicular cancer risk in sons and, 421
  cigarette smoking, testicular cancer risk in sons and, 20
  estrogen levels, gestational age and, 422–423
  older primiparous, testicular cancer risk in sons and, 419
  of testicular cancer patients, menorrhagia in, 424
Mutagenesis, genome-wide, 107

NAD(P)H:quinone oxidoreductase 1 (NQO1), endometrial cancer risk and, 382–383
NAT1 (*N*-acetyl-transferase-1), 202
National Cancer Institute (NCI), mammographic screenings and, 211
National Surgical Adjuvant Breast and Bowel Project P-1 (NSABP-1), 201–202, 229–230, 232, 230*f*, 231*f*
Nausea/hyperemesis index, in pregnancy, testicular cancer risk and, 419–420
Neoadjuvant hormonal therapy, for prostate cancer, 350–351, 351*f*
NF-κB/Rel family of transcription factors, 62–63
Nitromifene, 222–223
Nonsteroidal antiinflammatory drugs (NSAIDs), prostate cancer risk reduction and, 277
North American Intergroup trial, 347
NQO1 (NAD(P)H:quinone oxidoreductase 1), endometrial cancer risk and, 382–383
NR corepressor, androgen-independent prostate cancer and, 336
NSABP-1 (National Surgical Adjuvant Breast and Bowel Project P-1), 201–202, 229–230, 232, 230*f*, 231*f*
NSAIDs (nonsteroidal antiinflammatory drugs), prostate cancer risk reduction and, 277
Nuclear receptor cofactors, 49–50
  atypical, 56
  chromatin link and, 55–56, 56*f*
  coactivators, 50–52, 51*f*
    antagonism, structural basis for, 52–53
    structure of coactivator:receptor complex, 52, 53*f*
    in transcriptional activation of AF-1 domain, 53–54
  corepressors, 54–55, 54*t*
Nuclear receptors
  AF-1 domain, 47–48, 47*t*, 53–54
  atypical, 46–47
  classification
    I, 44, 46, 44*f*–45*f*
    II, 46, 44*f*–45*f*, 56, 56*f*
    III, 46, 44*f*–45*f*
    IV, 46, 44*f*–45*f*
    by mode of action, 44–47, 44*f*–45*f*
  corticoid, 67–68
  crosstalk, 59, 61–63, 62*f*
  definition of, 38
  DNA-binding domain, 38, 39, 40, 42, 44, 42*f*–43*f*
  gene control
    corticoid, 67–68
    retinoid, 69, 71–73, 72*f*
    thyroid, 68–69, 69*t*, 70*t*
    vitamin D_3, 68, 69*t*, 70*t*
  hormone response elements and, 38
  ligand-binding domain, 39, 40, 42*f*
    ligand-dependent AF-2 function and, 48
    structure of, 48–49, 49*f*
  ligand-independent transactivation by, 47–48, 47*t*
  ligands
    nonsteroid receptor, 57, 60*f*–61*f*
    steroid and thyroid receptor, 57, 58*f*–59*f*
  modular functional domains, 40, 42*f*
  orphan. *See* Orphan nuclear receptors
  physiology of, 63–67, 64*t*, 65*t*, 66*f*
  sex steroid hormone. *See* Sex steroid hormone receptors
  species differences in, 40
  superfamily, evolution of, 39–40, 41*f*–42*f*
  transcriptional control
    allosterism and, 57, 59
    heterodimerization and, 59
    ligands and, 57, 58*f*–61*f*
Nulliparity, endometrial cancer and, 371
Nurses Health Study, 227

Obesity
  breast cancer risk and, 123, 227
  endometrial cancer risk and, 371–372
  peroxisome proliferator-activated receptors and, 74
  prostate cancer risk and, 277
  testicular cancer risk and, 426
Oophorectomy
  breast cancer risk and, 181, 248–249, 249*f*
  mammographic density and sensitivity, 214–215
  for metastatic breast cancer, 218, 219
  prophylactic
    for *BRCA1/2* gene carriers with breast cancer, 205–206
    breast cancer risk in *BRCA1/2* gene carriers and, 204
Oral contraceptives
  BRCA1/2 gene carriers and, 203
  breast cancer risk, 128
    formulation differences and, 140–144, 141*f*–146*f*
    implications of, 144–145, 147*f*
  combined estrogen-progestagen, 140–143, 141*f*–143*f*
  endometrial cancer risk, 9, 373
  ovarian cancer protective effect, 7, 404
  progestagen-only, 140, 143–144, 144*f*–146*f*
Orchiectomy, for prostate cancer, 344–345
  with androgen inhibitor, 347–349, 349*f*
  with chemotherapy, 355
Orphan nuclear receptors
  assays
    ligand-binding, 79
    proximity, 79–80
  cancer and, 77–79
  CAR, 69*t*, 70*t*, 76
  characteristics of, 39, 69*t*, 70*t*, 73, 76–77, 77*t*, 78*t*
  farnesoid X, 69*t*, 70*t*, 75
  functional assessment, through gene targeting in mice, 81
  lipid, 73–75
  liver X, 69*t*, 70*t*, 75
  peroxisome proliferator-activated receptors, 73–74
  pharmacology, 79–81
  PXR/SXR, 69*t*, 70*t*, 75–76
  reverse endocrinology, 79–81
  target gene identification, 80–81
Osteoporosis
  raloxifene and, 233–234, 234*f*
  tamoxifen and, 230, 232, 231*f*
Ovarian ablation, for breast cancer prevention, 250
Ovarian cancer
  candidate genes, 5, 5*t*

etiology, hormones in, 402–404, 403f
future research directions, 407–408
hormonal carcinogenesis
    gonadotropins and, 404–405
    steroid hormones and, 405–407
incidence, 6t, 7, 398
    age-specific, 402–403, 403f
    race and, 8f
mortality rates, 8t, 398
protective factors
    lactation, 404
    oral contraceptives, 404
    parity, 403–404
    progesterone, 406–407
risk factors
    antiestrogens, 407
    gonadotropins, 404–405
    hormone replacement therapy, 404
    oral contraceptive use by *BRCA1/2* gene carriers, 203
    premenopausal exercise, 407
screening, 201
Ovarian hormones. *See also specific ovarian hormones*
    biosynthesis, 5, 65
    mammographic density and, 214
    in menopause, 401–402
    menstrual cycle fluctuations, 399–400, 400f
    physiology of, 399–400, 400f
    postmenopausal, 401–402
    steroidogenesis, 400–401, 401f
Ovaries
    aromatase expression in, 171–172, 171f, 175
    estradiol and, 181
    follicle development in, menstrual cycle, 398–399, 399f
    function of, 7-HSDs and, 183–184
    postmenopausal, 401–402
    premenopausal, anatomy of, 398–399, 399f
18-Oxidase, in corticosteroid synthesis, 19

p63, 337
PAF (population-attributable fraction), 166–167
Parity
    breast cancer risk and, 122–123
    breast cancer risk in *BRCA1/2* gene carriers and, 203
    ovarian cancer risk and, 403–404
PBRs (peripheral-type benzodiazepine receptors), 14
PCAF, 50, 51f
p300/CBP, androgen receptor coactivation, 293
p160 coactivators, 293
PC-SPES, for prostate cancer, 346–347
Peripheral-type benzodiazepine receptors (PBRs), 14
Peroxisome proliferator-activated receptors (PPARs), 73–75, 176
Pesticides, xenobiotic, breast cancer risk and, 127
Phenotype, association studies and, 163–164
Phosphoinositide 3-kinase-AKT pathway, androgen-independent prostate cancer and, 335–336
Physical inactivity, breast cancer risk and, 123–124
Phytoestrogens
    breast cancer and, 126–127
    prostate cancer and, 276, 275t

PIAS, androgen-independent prostate cancer and, 336
PIN (prostate intraepithelial neoplasia), 276
Placental weight, testicular cancer risk and, 423
Polycystic ovary syndrome, 371
Polygenetic disorders, identification of genes in, 100–101
Polymorphisms, disease risk and, 102
Population-attributable fraction (PAF), 166–167
Population stratification, 112–113
Positional candidates, 105
Postmenopausal hormone therapy. *See also* Estrogen-replacement therapy; Hormone-replacement therapy
    breast cancer risk
        comparison of different preparations and, 150–151, 150t
        current *vs.* recent users, 147–148, 149f
        dosage and, 150t
        duration of, 147–148, 149f, 150f, 150t
        ever-users *vs.* never-users, 147, 148f, 152, 153f
        implications of, 151–152, 153f
        localized *vs.* spreading tumors, 149–150, 150f
    estrogen-only. *See* Estrogen-replacement therapy
    estrogen-progestin combinations. *See* Hormone-replacement therapy
    historical background, 145–147
PPARs (peroxisome proliferator-activated receptors), 73–75, 176
PR. *See* Progesterone receptor
Pregnancy
    breast cancer risk in *BRCA1/2* gene carriers and, 203
    diet during, testicular cancer risk and, 421
    estrogen levels, gestational age and, 422–423
    exogenous estrogen exposure during, testicular cancer risk and, 415
    exposure to synthetic estrogenic chemicals, testicular cancer risk and, 421
    nausea/hyperemesis index, testicular cancer risk and, 419–420
    ovarian cancer risk reduction and, 403–404
Pregnenolone
    biosynthesis of, 12–15, 13f
    conversion to progesterone, 16
Prevention. *See also under specific cancers*
    chemoprevention. *See* Chemoprevention, hormonal
    Lacassagne's principle of, 220
    secondary, 218
Progestagen-only oral contraceptives, breast cancer risk, 143–144, 144f–146f
Progesterone
    biosynthesis, 16
    CYP17 genotype and, 130, 131f
    metabolism of, 22
    ovarian carcinogenesis and, 406–407
Progesterone receptor (PR)
    cancer and, 67
    endometrial cancer risk and, 384
    gene control, 63–67, 64t, 65t, 66f
    genetic polymorphisms, endometrial cancer and, 384
PROGINS variant allele, 384
Promoter II, regulation of aromatase expression in adipose tissue, 175, 176f

Promoters, 1
Prophylactic surgery, breast cancer risk in *BRCA1/2* gene carriers and, 204
Prostate cancer
 adjuvant hormonal therapy, 351–352, 352f
 advanced metastatic, 344
 androgen independent, 331, 332
  AR-independent signaling pathways, 336–337
  ligand-dependent AR-mediated signaling, 333–334
  ligand-independent AR-mediated signaling pathways, 334–336
 androgens and, 3–4
 antigen-independent, 337–338
 candidate genes, 5, 5t
 carcinogenesis, androgens in, 319
 cellular biology, 343–344
 etiological pathways for, 277–278, 278f
 etiology
  androgen transactivation pathways, 278–282, 279t, 279f, 28f
  insulin-like growth factor signaling pathways, 282–283
  polygenic, 278
  vitamin D metabolism pathways, 283
 genetic factors
  familial aggregation and, 316–317
  linkage analysis, 317–319
  segregation analyses, 317
  twin studies, 317
 hereditary, 316
  candidate genes, 320
  familial aggregation and, 316–317
  linkage analysis, 317–319
  segregation analyses, 317
  twin studies, 317
 high-penetrance genes, 274–275
 hormonal therapies, 345t
  adrenalectomy, 345
  adrenal steroid metabolism inhibitors, 354–355
  antiandrogens, 346
  5α-reductase enzyme inhibitors, 347
  combined androgen ablation, 347–349, 348f, 349f
  differences in single-treatment modalities, 349–350
  estrogens, 346
  future strategies, 356–358
  hypophysectomy, 345
  intermittent *vs.* continuous, 353
  LHRH agonists, 345–346
  LHRH antagonists, 346
  neoadjuvant, 350–351, 351f
  new strategies, 355–356
  orchiectomy, 344–345
  PC-SPES, 346–347
  second-line, 353
  side effects of, 352–353
  timing of, 350
 hormone-refractory, LHRH agonist therapy for, 353
 incidence, 6, 6t, 7f
 localized, androgen receptor and, 299–300
 metastases, 344
 metastatic, hormonal therapies for, 300–301, 301f

 molecular epidemiology, future role of, 283–284
 mortality rates, 6, 8t
 natural history, 344
 risk factors
  age, 273
  androgen receptor polymorphisms, 295, 297, 299, 298t
  androgens, 319
  BRCA1/2 gene carriers, 201
  dietary, 275–276, 275t
  ethnicity, 273–274
  family history, 274–275
  nondietary, 276–277
  race, 273–274
 risk reduction, 277
 screening, 5–6
 susceptibility, family-based candidate gene study, 320–325
 treatment
  androgen ablation therapy. *See* Androgen ablation therapy
Prostate intraepithelial neoplasia (PIN), 276
Prostate specific antigen (PSA), 6, 344
Prostate stem cell antigen (PSCA), 337
Protein kinase A (PKA), loss of tamoxifen responsiveness and, 265
Proximity assays, orphan nuclear receptors and, 79–80
PSA (prostate specific antigen), 6, 344
PSCA (prostate stem cell antigen), 337
P450scc (side-chain cleavage cytochrome P-450), 12, 13, 13f, 15
PXR/SXR receptor, 69t, 70t, 75–76

Race. *See also* African Americans
 ovarian cancer and, 8f
 prostate cancer and, 8t, 273–274
 testicular cancer and, 8t
 uterine cancer and, 7, 8f
Radiation Therapy Oncology Group (RTOG) 85-31 trial, 351–352
Raloxifene, 222–223
 antitumor actions, 233–234, 234f
 binding, 223–224, 224f
 bone density and, 234, 234f
 for breast cancer prevention, biological basis of, 233–235, 234f
 endometrial cancer risk and, 373
 lipids and, 234–235
 osteoporosis and, 233–234, 234f
 STAR trial, 235
 structure of, 219f
 uterine effects, 235
Raloxifene Use for the Heart (RUTH), 235
Receptor-interacting protein 140 (RIPA140), 50
5α-Reductase enzyme inhibitors, for prostate cancer, 347
Reductases, in steroid hormones metabolism, 20–21
Reproductive factors, *BRCA1/2* gene carriers and, 203
Retinoic receptors
 RAR, 69, 69t, 70t, 71–72, 72f
  acute promyelocytic leukemia and, 71–72, 72f
  genetic expression, 70t, 71
 RXR, 68, 71, 69t, 70t, 72–73

INDEX

RIPA140 (receptor-interacting protein 140), 50
Risk. See under specific cancers
    candidate genes and, 101–102
    familial clustering and, 99–100
    genetic variation and, 102
RNA interference (RNAi), 107
Royal Marsden Pilot Study, 228–229
RUTH (Raloxifene Use for the Heart), 235

SAMs (selective aromatase modulators), 175–176
Segregation analyses, prostate cancer, 317
Selective aromatase modulators (SAMs), 175–176
Selective estrogen receptor modulators (SERMs)
    alternate estrogen receptor and, 224–225
    binding, 223–224, 224f
    endometrial cancer risk and, 373
    for ER-positive breast cancer, 222–223
    estrogen receptor coactivators and, 224, 225f
    new, 235–237, 236f
    raloxifene. See Raloxifene
    structure of, 236f
    tamoxifen. See Tamoxifen
Selenium, prostate cancer risk and, 276, 275t
SERMs. See Selective estrogen receptor modulators
Sex hormone-binding globulin (SHBG)
    breast cancer risk, 120, 122
    endometrial cancer risk and, 383
    gene mutations, endometrial cancer and, 383
    in menopause, 402
    physiologic role of, 19–20
    properties, 19
    prostate cancer risk and, 319
    regulation of, 20
Sex steroid hormone receptors. See also specific sex steroid hormone receptors
    cancer and, 67
    control of gene networks, 63–67, 64t, 65t, 66f
    gene control for hormone biosynthesis, 66–67, 66f
Sexual activity, prostate cancer risk and, 277
Shaving frequency, testicular cancer risk and, 426
SHBG. See Sex hormone-binding globulin
SHP (short heterodimer partner), 294
Sibling pair analysis, in breast cancer families, 160–161
Side-chain cleavage cytochrome P-450 (P450scc), 12, 13, 13f, 15
Silencing mediator of retinoid and thyroid receptors (SMRT), 54–55, 54t, 264, 294
Singapore, breast cancer mortality rates, 121f
Single nucleotide polymorphisms (SNPs), 109–111
Sisters of testicular cancer patients, hormone-related cancers in, 423–424
Small molecule screening, 107
Smoking
    endometrial cancer risk and, 372
    maternal, testicular cancer risk and, 20
    prostate cancer risk and, 277
    testicular cancer risk and, 427
Smoking cessation, for BRCA1/2 gene carriers, 202
SMRT (silencing mediator of retinoid and thyroid receptors), 54–55, 54t, 264, 294
SNPs (single nucleotide polymorphisms), 109–111

Spinal and bulbar muscular atrophy (SBMA), 291
SRD5A2 gene, 278–279, 279t, 279f, 280
StAR (steroidogenic acute regulatory protein), 13, 14
STAR trial, 235
State migration, 350
Stein-Leventhal syndrome, 371
Stem cells, 337
Steroid hormone-related diseases, 38–39
Steroid hormones. See Hormones
Steroidogenic acute regulatory protein (StAR), 13, 14
Sulfotransferases, endometrial cancer risk and, 382
Suramin, 355–356

Tamoxifen, 222–223
    bone density and, 229
    for BRCA1/2 gene carriers, 201–202
    breast cancer risk, vs. hormone replacement therapy, 229
    chemoprevention, 228, 232–233
        biologic basis of, 225–226
        candidates, identification of, 227–228
        duration of therapy, 221–222, 221f, 222t, 222f
        Italian Study, 232
        NSABP-1 Study, 229–230, 232, 230f, 231f, 229–230, 232, 230f, 231f
        response, estrogen receptor status and, 220–221, 221t
        Royal Marsden Pilot Study, 228–229
        testing, evidence for, 220–222, 221t, 221f, 222t, 222f
    deep vein thrombosis and, 232
    development of, 219
    endometrial cancer risk and, 226, 373
    for ER-positive breast cancer, 220
    loss of responsiveness, 260–266, 266t
        altered ER-coregulator interaction, 263–263
        crosstalk between growth factors and ER pathways and, 264–266
        ER sequence changes, 262–263
        estrogen response element changes and, 263
        increased ERβ, 266, 266t
        ligand-independent activation of ER, 264–266
        loss of ER expression and, 260–262
        reduced uptake and, 260
    mammographic density and, 213, 214–215
    mammographic sensitivity and, 214–215
    osteoporosis and, 230, 232, 231f
    STAR trial, 235
    structure of, 219f
Testicular atrophy, testicular cancer risk and, 427
Testicular cancer
    candidate genes, 5, 5t
    etiology, prenatal estrogen hypothesis, 413, 414–415
    genetic polymorphisms, 428
    hormonal abnormalities, genetic conditions of, 427–428
    incidence, 6t, 9
        age-specific, 413, 414f, 415, 424
        for WWII cohort, 415
    mortality rates, race and, 8t
    postnatal hormonal factors, 424–427

Testicular cancer (*Continued*)
  prevalence, 413
  risk factors, 9, 414, 428–429
    acne, 426
    age at puberty, 424–425
    alcohol consumption, 427
    alcohol consumption, maternal, 421
    androgen insensitivity, 428
    baldness, 426
    birth order, 418
    birthweight, 423
    cryptorchidism, 414, 415
    diet, maternal t, 421
    diet and, 425–426
    diethylstilbestrol, 415, 416
    Down's syndrome, 428
    ethnicity, 426–427
    exogenous estrogens, maternal, 415, 416
    family history, 428
    genitourinary abnormalities, 414, 415
    gestational age, 422–423
    handedness, 423
    height, 425
    hormone-related disease, maternal, 424
    Klinefelter's syndrome, 427
    maternal age, 419
    of men born to older primiparous mothers, 419
    nausea/hyperemesis index in pregnancy, 419–420
    obesity, 426
    obstetric variables, 416–418, 417t
    physical inactivity, 427
    placental weight, 423
    postnatal sex hormone intake, 426
    shaving frequency, 426
    smoking, 427
    smoking, maternal, 420
    testicular atrophy, 427
    in twins, 422
    in undescended *vs.* descended testes, 414–415
    XY gonadal dysgenesis, 427–428
  sisters, hormone-related cancers in, 423–424
Testosterone
  biosynthesis, 17–18, 181–182, 182f
  conjugates, 23t
  prostate cancer and, 3–4
  prostate development and, 343
TGFs (transforming growth factors), loss of tamoxifen responsiveness and, 264–266
Thyroid receptors
  cancer and, 68–69
  functions, 68, 69t, 70t
  target genes, 68, 69t, 70t
Tissue inhibitors of metalloproteinases (TIMPs), 358
TP53 gene, breast cancer risk and, 4
Transcription factors
  NF-?B/Rel family, 62–63
  nuclear receptors. *See* Nuclear receptors
Transforming growth factors (TGFs), loss of tamoxifen responsiveness and, 264–266
Triphenylethylene, 219f
TSE-424, 237
Twin studies
  of hormone-responsive cancers, 100
  prostate cancer, 317
  testicular cancer risk, 422

UGTs (uridine diphosphate-glucuronosyltransferases), 381–382
United Kingdom, breast cancer mortality rates, 121f
United States, breast cancer mortality rates, 121f
Uridine diphosphate-glucuronosyltransferases (UGTs), 381–382
Uterine cancer
  estrogen replacement therapy and, 9
  incidence, 6t, 7, 8f, 9
  mortality rates, 8t, 9
  race and, 7, 8f
Uterus, raloxifene effects on, 235

Vaginal clear cell adenocarcinomas, diethylstilbestrol and, 153
Vitamin D
  metabolism pathways, in prostate cancer, 283
  prostate cancer etiology and, 277–278, 278f
  prostate cancer risk and, 276, 275t
Vitamin D_3 receptors
  calcium deposition and, 68
  functions, 68, 69t, 70t
  target genes, 68, 69t, 70t
Vitamin E, prostate cancer risk and, 275–276, 275t

Weight, breast cancer risk and, 123
WEST (Women's Estrogen for Stroke Trial), 149
WHI (Women's Health Initiative), 149
Women's Estrogen for Stroke Trial (WEST), 149
Women's Health Initiative (WHI), 149

XY gonadal dysgenesis, testicular cancer risk, 427–428

Zolzdex in Premenopausal Patients (ZIPP), 250